Volume II

Probability Theory and Mathematical Statistics

Proceedings of the
Fifth Vilnius Conference
June 25 - July 1, 1989

Edited by

**B. Grigelionis, Yu. V. Prohorov,
V. V. Sazonov and V. Statulevičius**

MOKSLAS
Vilnius, Lithuania

///VSP///
Utrecht, The Netherlands

VSP BV
Post Box 346
3700 AH Zeist
The Netherlands

MOKSLAS
Žvaigždžiu 23
Vilnius
Lithuania

First published in 1990

CIP–DATA KONINKLIJKE BIBLIOTHEEK, DEN HAAG
Probability theory and mathematical statistics:
Proceedings of the Fifth Vilnius Conference:
Vilnius, Lithuania, June 25 – July 1, 1989
ed. by B. Grigelionis... [*et al.*]
– Utrecht: VSP BV/ Vilnius: Mokslas.
ISBN 90-6764-129-4 (vol.II)
ISBN 90-6764-130-8 (set)
SISO 517 UDC 519.2(063)
Subject headings: probability theory/ mathematical statistics.

Typeset in Lithuania by Baltic Amadeus / Publishing Service Group of IMI, Vilnius
Printed in Lithuania by Spindulys, Kaunas

CONTENTS

Prob. Theory and Math. Stat., Vol. 2, pp. 1–9
B.Grigelionis *et al.* (Eds.)
1990 VSP/Mokslas

WEAK CONVERGENCE OF SEMI–MARKOV RANDOM EVOLUTIONS (MARTINGALE APPROACH)

V.S. KOROLJUK and A.V. SWISHCHUK
Institute of Mathematics Ukr. Ac. Sci., Repina str. 3, 252601 Kiev, USSR

ABSTRACT

Two types of limit theorems for semi–Markov random evolutions (SMRE) in the series scheme are discussed: the averaging and the diffusion approximation. The weak convergence of SMRE is established by reducing it to a martingale problem for the infinitesimal operator of the limit evolution. Applications to storage theory, traffic processes and switching processes are considered.

1. SEMI–MARKOV RANDOM EVOLUTION IN THE SERIES SCHEME

SMRE in the series scheme (with a small parameter ε) is defined by the equation

$$V^\varepsilon(t) = I + \int_0^t V^\varepsilon(s)\Gamma\big(x(s/\varepsilon)\big)ds + \sum_{k=1}^{\nu(t/\varepsilon)} V^\varepsilon(\varepsilon\tau_k)[D^\varepsilon(x_k) - I]. \tag{1}$$

Here $\Gamma(x), x \in X$, is a family of generators of strongly continuous semigroups in a separable Banach space B with a common domain B_0 independent of x, B_0 is dense in B; $D^\varepsilon(x), x \in X$, is a family of linear bounded operators on $B : x(t) = x_{\nu(t)}$ is the semi–Markov processes (SMP) in the measurable phase space $(X, \mathcal{X})$ which is defined by a Markov renewal process $(x_n, \theta_n; n \geqslant 0)$ with semi–Markov kernel

$$\mathbf{Q}(x, A, t) = \mathbf{P}\{x_{n+1} \in A, \quad \theta_{n+1} \leqslant t \mid x_n = x\},$$

$\tau_n = \sum_{k=1}^{n} \theta_k$, $\nu(t) = \max\{n : \tau_n \leqslant t\}$ is a counting process, $\tau(t) = \tau_{\nu(t)}$ is a point process.

Example 1. The storage process with transfer speed $v(u,x)$ and jumps function $a(x)$ (the output and the input respectively) is defined by the integral equation (Koroljuk and Swishchuk, 1985)

$$U_t^\varepsilon = u_0 + \int_0^t v\big(U_s^\varepsilon, x(s/\varepsilon)\big)ds + \varepsilon \sum_{k=1}^{\nu(t/\varepsilon)} a(x_{k-1}). \tag{2}$$

Let $U_t(x, u_0)$ be the solution of the Cauchy problem

$$\frac{dU_t}{dt} = v(U_t, x), \quad U_0(x, u_0) = u_0, \quad x \in X. \tag{3}$$

Choose generators $\Gamma(x)$, $x \in X$, on a Banach space B of functions in the following way

$$\Gamma(x)\varphi(u) = v(u, x)\frac{d\varphi(u)}{du}. \tag{4}$$

Define also linear operators $D^\varepsilon(x)$, $x \in X$, as follows

$$D^\varepsilon(x)\varphi(u) = \varphi\big(u + \varepsilon a(x)\big). \tag{5}$$

Then SMRE (1) is defined by the expression

$$V^\varepsilon(t)\varphi(u) = \varphi(U_t^\varepsilon). \tag{6}$$

Example 2. Switching processes are defined by the equation

$$d\xi_t^\varepsilon = d\alpha_t\big(x(t/\varepsilon)\big) + \varepsilon a\big(x(t/\varepsilon)\big)d\nu(t/\varepsilon). \tag{7}$$

Here $\alpha_t(x)$, $x \in X$, is a family of homogeneous processes with independent increments which are mutually independent for every finite set of values $x \in X$ $\big(\alpha_0(x) \equiv 0, \forall x \in X\big)$.
The integral form of switching processes is expressed by

$$\xi_t^\varepsilon = \sum_{k=1}^{\nu(t/\varepsilon)} \big[\alpha_{\theta_k}(x_{k-1}) + \varepsilon a(x_k)\big] + \alpha_{t-\tau(t/\varepsilon)}\big(x(t/\varepsilon)\big). \tag{8}$$

Let $\Gamma(x)$, $x \in X$, be a family of infinitesimal operators of semigroups generated by Markov processes $\alpha_t(x)$.
Then the SMRE (1) can be given by the expression

$$V^\varepsilon(t)\varphi(u) = \mathbf{E}\big[\varphi(u + \xi_t^\varepsilon) \mid x(s),\, 0 \leqslant s \leqslant t/\varepsilon\big]. \tag{9}$$

In particular, if $\Gamma(x)$ are the cumulants of the processes $\alpha_t(x)$ and $D^\varepsilon(x) = \exp[i\varepsilon z a(x)]$, then the SMRE (1) defines the conditional characteristic function of the switching process (7)

$$V^\varepsilon(t) = \mathbf{E}\big[\exp(iz\xi_t^\varepsilon) \mid x(s),\, 0 \leqslant s \leqslant t/\varepsilon\big]. \tag{10}$$

Example 3. The *traffic process* is defined by the solution of the Cauchy problem

$$\frac{dU_t^\varepsilon}{dt} = v\big(U_t^\varepsilon, x(t/\varepsilon)\big), \quad U_0^\varepsilon = u_0, \tag{11}$$

and consequently is a particular case of the *storage process* (2), when $a(x) \equiv 0$, $\forall x \in X$.
Other models of stochastic systems with semi–Markov switchings, which describe the evolution of stochastic objects in semi–Markov random media, can also be represented by SMRE.

2. LIMIT AVERAGING THEOREM (SWISHCHUK, 1990).

Let the following conditions be satisfied:

C1. The embedded Markov chain $(x_n; n \geqslant 0)$ with transition probabilities $\mathbf{P}(x, A) = \mathbf{Q}(x, A, +\infty)$ is uniformly ergodic with stationary distribution $\rho(A)$, $A \in X$; the first two moments of sojourn times in states x are uniformly continuous and bounded on x :

$$m_1(x) = \int_0^\infty tG_x(dt), \quad m_2(x) = \int_0^\infty t^2 G_x(dt);$$

$$G_x(t) = Q(x, X, t).$$

C2. $D^\varepsilon(x) = I + \varepsilon D(x) + o(\varepsilon)$ as $\varepsilon \to 0$, where $D(x), x \in X$, are closed linear operators with domain B_0.

C3. Operator $\Gamma^* = \frac{1}{m} \int_X \rho(dx)[m_1(x)\Gamma(x) + D(x)]$ is closed,

$$m = \int_X \rho(dx) m_1(x).$$

C4. $\int_X \rho(dx)[m_1(x)\|\Gamma(x)f\| + \|D(x)f\|] \leqslant C_1 < +\infty, \forall f \in B_0 \subseteq \mathrm{Dom}(\Gamma^2(x))$.

Then SMRE $V^\varepsilon(t)$ converges weakly as $\varepsilon \to 0$ to the averaged evolution $V^*(t)$, which is defined as the solution of the equation

$$V^*(t) - I - \int_0^t V^*(s)\Gamma^* ds = 0. \tag{12}$$

Remark. If Γ^* is a bounded operator, then limit evolution $V^*(t) = \exp\{t\Gamma^*\}$. In the general case equation (12) is considered on functions from the domain of operators $\Gamma^2(x)$, $x \in X$.

3. DIFFUSION APPROXIMATION (KOROLJUK AND SWISHCHUK, 1989)

Let instead of conditions C2—C4 the following conditions be satisfied

C′2. $D^\varepsilon(x) = I + \varepsilon D(x) + \varepsilon^2 D_2(x)$ as $\varepsilon \to 0$; here $D(x), D_2(x)$, $x \in X$, are closed linear operators with domain B_0.

C′3. (the balance condition) $\Gamma^* f = 0, \forall f \in B_0$.

C′4. for all $f \in B_0 \subseteq \mathrm{Dom}(\Gamma^4(x))$,

$$\int_X \rho(dx)\Big[m_2(x)\|\Gamma^2(x)\| + m_1(x)\|\Gamma(x)PD(x)f\| + \|D_2(x)f\|\Big] \leqslant C_2 < +\infty.$$

If conditions C1, C'2—C'4 are satisfied, then SMRE $V^\varepsilon(t/\varepsilon)$ converges weakly as $\varepsilon \to 0$ to the solution $W^*(t)$ of the martingale problem

$$W^*(t) - I - \int_0^t W^*(s)L^*ds = M(t), \tag{13}$$

Here $M(t)$ is a F_t–martingale, $F_t = \sigma\{x(s) : 0 \leqslant s \leqslant t\}$,

$$L^* = \frac{1}{m}\int_X \rho(dx)L(x),$$

$$L(x) = c(x)(R_0 - I)c(x) + m_2(x)\Gamma^2(x)/2 + m_1(x)\Gamma(x)PD(x) + D_2(x),$$
$$c(x) = m_1(x)\Gamma(x) + PD(x), \tag{14}$$

where $R_0 = (I - P + \Pi)^{-1} - \Pi$ is the potential of the embedded Markov chain, P is the operator of transition probabilities, Π is the stacionary projector.

4. APPLICATIONS OF THE LIMIT AVERAGING THEOREM

Taking into account (5) for the *storage processes* (2) we obtain

$$D^\varepsilon(x)\varphi(u) = \varphi(u + \varepsilon a(x)) = \varphi(u) + \varepsilon a(x)d\varphi(u)/du + o(\varepsilon), \tag{15}$$

hence

$$D(x) = a(x)\frac{d}{du} \qquad \text{(condition C2)}. \tag{16}$$

Further,

$$\Gamma^* = [v^*(u) + a^*]\frac{d}{du},$$

$$v^*(u) = \int_X \rho(dx)m_1(x)v(u,x)/m, \qquad a^* = \frac{1}{m}\int_X \rho(dx)a(x). \tag{17}$$

Hence the limit averaging theorem of SMRE for the *storage processes* (2) is just the weak convergence of U_t^ε to the deterministic evolution U_t^* defined by the equation

$$U_t^* = u_0 + \int_0^t v^*(U_s^*)ds + a^*t. \tag{18}$$

The *traffic process* (11) in the averaging scheme converges weakly to the deterministic dynamical system

$$\frac{dU_t^*}{dt} = v(U_t^*), \quad U_0^* = u_0, \tag{19}$$

where $v^*(u) = \frac{1}{m}\int_X \rho(dx)m_1(x)v(u,x)$.

Switching process (7) in the averaging scheme converges weakly to the process with independent increments α_t^* with cumulant

$$\Gamma^* = \frac{1}{m}\int_X \rho(dx)m_1(x)\Gamma(x). \tag{20}$$

5. APPLICATIONS OF THE DIFFUSION APPROXIMATION

In the limit diffusion theorem for the *storage process* (2) the balance condition C′3 means

$$\int_X \rho(dx)m_1(x)v(u,x) = -\int_X \rho(dx)a(x), \quad \forall u \in U. \tag{21}$$

Taking into account (4) and condition C′2 and computing the infinitesimal operator L^* of the limit diffusion by formula (14) we obtain the following result. Under conditions C1, C′2—C′4 the *storage process* (2) in the time scale t/ε converges weakly as $\varepsilon \to 0$ to the diffusion process U_t with drift $b(u)$ and diffusion coefficient $\sigma^2(u)$, given by the formulae

$$b(u) = \frac{1}{m}\int_X \rho(dx)b(u,x), \quad \sigma^2(u) = \frac{1}{m}\int_X \rho(dx)\sigma^2(u,x), \tag{22}$$

where

$$b(u,x) = c(u,x)(R_0 - I)c'_u(u,x) + m_2(x)v(u,x)v'_u(u,x)/2,$$

$$\sigma^2(u,x) = 2[c(u,x)(R_0 - I)c(u,x) + a^2(x)/2 + m_2(x)v^2(x)/2+$$

$$+m_1(x)a(x)v(u,x)], \ c(u,x) = m_1(x)v(u,x) + Pa(x). \tag{23}$$

The diffusion approximation of the *traffic process* gives the following result. U_t^ε converges weakly as $\varepsilon \to 0$ in the time scale t/ε to the diffusion process U_t^0 with drift $a_0(u)$ and diffusion coefficient $\sigma_0^2(u)$:

$$a_0(u) = \frac{1}{m}\int_X \rho(dx)a(u,x), \quad \sigma_0^2(u) = \frac{1}{m}\int_X \rho(dx)m_1(x)\sigma^2(u,x),$$

$$a(u,x) = m_1(x)v(u,x)(R_0 - I)m_1(x)v'_u(u,x)+$$

$$+m_2(x)v(u,x)v'_u(u,x)/2, \tag{24}$$

$$\sigma^2(u,x) = 2[m_1(x)v(u,x)(R_0 - I)m_1(x)v(u,x) + m_2(x)v^2(u,x)/2].$$

The diffusion approximation of the *switching processes* (7) is constructed in the following way. The process (7) should be centered and normalized :

$$\eta_t^\varepsilon = \sqrt{\varepsilon}\,[\xi_t^\varepsilon - \Lambda t/\varepsilon]. \tag{25}$$

The parameter of the drift Λ is defined from the balance condition C′3 and from the fact that in this case

$$\Gamma^\varepsilon(x) = \gamma(\sqrt{\varepsilon}\, z, x) - i\sqrt{\varepsilon}\, z\Lambda. \tag{26}$$

Here $\gamma(z, x)$ are the cumulants of processes $\alpha_t(x)$. We suppose that there exist two semiinvariants of the processes $\alpha_t(x)$ so that

$$\gamma(z, x) = iz\gamma_1(x) - z^2\gamma_2(x)/2 + o(z^2), \quad z \to 0. \tag{27}$$

Then the asymptotic representation of the infinitesimal operator (25) can be written as follows

$$\Gamma^\varepsilon(x) = \Gamma_1(x) + \varepsilon\Gamma_2(x) + o(\varepsilon), \quad \text{as } \varepsilon \to 0,$$

where

$$\Gamma_1(x) = iz\big(\gamma_1(x) - \Lambda\big), \quad \Gamma_2(x) = -z^2\gamma_2(x)/2.$$

The balance condition C′3 enables us to calculate the drift

$$\Lambda = \int_X \rho(dx)[m_1(x)\gamma_1(x) + a(x)]. \tag{28}$$

From (14) and condition (27) it follows that the normalized process (25) converges weakly as $\varepsilon \to 0$ to the Wiener process W_t with variance

$$\sigma^2 = \frac{1}{m}\int_X \rho(dx)\sigma_1^2(x),$$

$$\begin{aligned}\sigma_1^2(x) = 2m_1(x)\big(\gamma_1(x) - \Lambda\big)(R_0 - I)m_1(x)\big(\gamma_1(x) - \Lambda\big) + \\ + \big(m_2(x) - m_1^2(x)\big)\big(\gamma_1(x) - \Lambda\big)^2 + m_1(x)\gamma_2(x) + a^2(x).\end{aligned} \tag{29}$$

6. HOW HAVE THE RESULTS BEEN OBTAINED

In order to obtain the limit averaging theorem (12) and the diffusion approximation (13), one has to solve two problems:

1) to establish the weak convergence of the SMRE;

2) to calculate the infinitesimal operators Γ^* and L^* of the limit evolutions.

It seems to us, that the most natural way to calculate the infinitesimal operators of the limit evolutions would be to apply the method of limit inversing of the operator which defines the Markov renewal equation for the initial evolution (2). In order to establish the compactness of the SMRE in the series scheme we apply the martingale approach proposed in a similar situation in (Watkins, 1984; 1985).

6.1. For the Laplace transform of the averaged SMRE

$$\Phi_\varepsilon(x) = \int_0^\infty e^{-\lambda t}\mathbf{E}_x\big[V^\varepsilon(t)\varphi(\cdot, x(t/\varepsilon))\big]dt, \tag{30}$$

we construct the Markov renewal equation

$$[I - Q_\varepsilon]\Phi_\varepsilon = m_\varepsilon. \tag{31}$$

Under the conditions of the limit averaging theorem, the following asymptotic representation holds

$$Q_\varepsilon = P + \varepsilon Q_1 + o(\varepsilon), \quad m_\varepsilon = \varepsilon m + o(\varepsilon). \tag{32}$$

Further, we apply the theorem of limit inversing operator perturbed on the spectrum (Koroljuk and Turbin, 1978)

$$\lim_{\varepsilon\to 0} \varepsilon[I - Q_\varepsilon]^{-1} m = \lim_{\varepsilon\to 0} \varepsilon[I - P - \varepsilon Q_1 + o(\varepsilon)]^{-1} m = \Gamma^{(-1)} m, \tag{33}$$

where

$$\Gamma = -\Pi Q_1 \Pi = \Gamma^* \Pi, \quad \Gamma^{(-1)} = (\Gamma^*)^{-1}\Pi.$$

The form of the infinitesimal operator Γ^* of the limit evolution in the averaging scheme is defined exactly in that way.
To calculate the infinitesimal operator L^* we can realize a similar (but more complicated) procedure under the balance condition. We construct a more complete asymptotical representation of the operator Q_ε :

$$Q_\varepsilon = P + \varepsilon Q_1 + \varepsilon^2 Q_2 + o(\varepsilon^2) \quad \text{as} \quad \varepsilon \to 0. \tag{34}$$

Under the balance condition $\Pi Q_1 \Pi = 0$, the corresponding theorem of limit inversing operator perturbed on the spectrum gives:

$$\lim_{\varepsilon\to 0} \varepsilon^2 [I - Q_\varepsilon]^{-1} = \lim_{\varepsilon\to 0} \varepsilon^2 \left[I - P - \varepsilon Q_1 - \varepsilon^2 Q_2 + o(\varepsilon^2)\right]^{-1} = L_\lambda^{(-1)}, \tag{35}$$

where

$$L_\lambda = \Pi Q_2 \Pi - \Pi Q_1 R_0 Q_1 \Pi = L_\lambda^* \Pi; \quad L_\lambda^{(-1)} = (L_\lambda)^{-1}\Pi; \quad L_\lambda^* = \lambda I - L^*. \tag{36}$$

6.2. In order to establish the compactness of random evolutions in the series scheme, we introduce the following martingales

$$M_\varepsilon(t) = V^\varepsilon(t)\varphi_\varepsilon(\cdot, x(t/\varepsilon)) - \varphi(\cdot, x) -$$

$$- \sum_{n=1}^{\nu(t/\varepsilon)} \left[V_n^\varepsilon \varphi_\varepsilon(\cdot, x_n) - E_\rho\big(V_n^\varepsilon \varphi_\varepsilon(\cdot, x_n)/F_{n-1}\big)\right], \tag{37}$$

where

$$V_n^\varepsilon = V^\varepsilon(\tau_n), \quad n \geqslant 0; \quad F_n = \sigma\{x(t) : 0 \leqslant t \leqslant \tau_n\}.$$

Now, taking into account conditions C1—C4 of the limit averaging theorem, the asymptotic representation (32) for the operator Q_ε, and choosing the functions φ_ε as the solution of the equation

$$[I - Q_\varepsilon]\varphi_\varepsilon = o(\varepsilon), \tag{38}$$

we obtain the following asymptotic representation for the martingales (37)

$$M_\varepsilon(t) = \left[V^\varepsilon(t) - I - \varepsilon \sum_{n=1}^{\nu(t/\varepsilon)} V_n^\varepsilon \Gamma^* m\right]\varphi_0(\cdot) + o(\varepsilon). \qquad (39)$$

Further, we establish the compactness of the martingales (37) and the convergence of their quadratic variations to zero, which leads to limit equation (12) for the averaging evolutions.
Under the conditions of the diffusion approximation C1, C′2–C′4 the martingales $M_\varepsilon(t/\varepsilon)$ in (37) are considered on functions φ_ε which are defined by the relation

$$[I - Q_\varepsilon]\varphi_\varepsilon = o(\varepsilon^2). \qquad (40)$$

Then the asymptotic representation of the martingales (37) takes the form

$$M_\varepsilon(t/\varepsilon) = \left[V^\varepsilon(t/\varepsilon) - I - \varepsilon^2 \sum_{n=1}^{\nu(t/\varepsilon^2)} V_n^\varepsilon L^* m\right]\varphi_0(\cdot) + o(\varepsilon^2), \qquad (41)$$

and the compactness of the martingales and their limit representation (13) follow from the representation (41).

7. SEMI–MARKOV RANDOM EVOLUTIONS IN THE SCHEME OF ASYMPTOTIC PHASE MERGING

These SMRE are defined by the equation

$$V_\varepsilon(t) = I + \int_0^t V_\varepsilon(s)\Gamma^\varepsilon(x^\varepsilon(s/\varepsilon))ds + \sum_{k=1}^{\nu^\varepsilon(t/\varepsilon)} V_\varepsilon(\varepsilon\tau_k^\varepsilon)\left[D^\varepsilon(x_k^\varepsilon) - I\right]. \qquad (42)$$

Here switching semi–Markov process $x^\varepsilon(t)$ depends on the series parameter ε and converges weakly to a Markov process $x^*(t)$ on the merged (factorized) phase space of states $(Y, \mathcal{Y})$ as $\varepsilon \to 0$ (Koroljuk and Turbin, 1978). Here we suppose that there exists

$$\lim_{\varepsilon\to 0} \varepsilon^{-1}\Pi(P_\varepsilon - I)\Pi = Q^*\Pi, \qquad (43)$$

where Q^* is the infinitesimal operator of the merged Markov process $x^*(t)$.
In the limit averaging theorem in the scheme of phase merging the limit SMRE is defined as the solution of the equation

$$V^*(t) - I - \int_0^t V^*(s)\Gamma^*(x^*(s))ds = 0, \qquad (44)$$

where the infinitesimal operators $\Gamma^*(y)$, $y \in Y$, are defined by analogy with the formula from condition C3.

The diffusion approximation of the SMRE in the scheme of phase merging leads to a martingale problem of the type

$$W^*(t) - I - \int_0^t W^*(s)L^*(x^*(s))ds = M(t), \tag{45}$$

where $M(t)$ is a martingale and infinitesimal operators $L^*(y)$, $y \in Y$, are calculated by analogy with (14).

The conditions of asymptotic phase merging extend essentially the class of limit random evolutions with Markov switching.

REFERENCES

Koroljuk, V.S. and Swishchuk, A.V. (1988). Semi- Markov random evolutions–phase agregations and applications In: *V Intern. Summer School on Prob. Th. Math. Stat., Varna, 1985.*, Bulg. Acad. Sci., Sofia, pp. 156–171.

Koroljuk, V.S. and Swishchuk, A.V. (1989). Limit representation of continuous semi–Markov random evolutions in the series scheme *Ukrain. Math. J.* (to appear).

Koroljuk, V.S. and Turbin, A.F. (1978). *Mathematical foundations of the phase merging of complex systems.* Naukova dumka, Kiev.

Swishchuk, A.V. (1990). Weak convergence of semi–Markov random evolutions in the averaging scheme (a martingale approach). *Ukrain. Math. J.* (to appear).

Watkins, J.A. (1984). Central limit theorems in random evolutions. *Ann. Prob.*, **12**, 480–513.

Watkins, J.A. (1985). Limit theorems for stationary random evolutions. *Stoch. Proc. Appl.*, **19**, 189–224.

Prob. Theory and Math. Stat., Vol. 2, pp. 10–21
B.Grigelionis *et al.* (Eds.)
1990 VSP/Mokslas

CONVERGENCE OF MOMENTS OF RANDOMLY INDEXED RANDOM SEQUENCES

V.M. KRUGLOV and V.Yu. KOROLEV
Computer Mathematics and Cybernetics Department, Moscow State University, 119899 Moscow, USSR

ABSTRACT

Several variants of necessary and sufficient conditions for the convergence of moments of randomly indexed sequences of random variables are presented in accordance with different types of asymptotic behaviour of centering and normalizing constants. The random indices are supposed to be independent of the original sequence. Some results concerning random sums are obtained as corollaries including the final form of H.Robbins' theorem.

Recently many papers appeared which deal with the convergence of moments of sums of a random number of random variables (r.v.'s), see references in (Korolev, 1985; Kruglov, 1988; Sakalauskas, 1988; Gut,1987). We generalize and sharpen these results. First, we consider arbitrary sequences of r.v.'s. Second, we present conditions which are necessary and sufficient. We especially point out the remarkable fact that it was the way of investigation of the asymptotic behaviour of the moment characteristics of randomly indexed random sequences that lead us to the exhaustive solution of some problems connected with the asymptotic behaviour of the distributions of randomly indexed random sequences, when the random index is supposed to be independent of the original random sequence.

Let $\{Z_n\}_{n=1}^{\infty}$ be a sequence of r.v.'s and $\{N = N_t\}_{t>0}$ be a family of positive integer r.v.'s. Without loss of generality t may be treated as a location parameter, say, mean or median of N_t. Suppose that for each t, the r.v. N_t is independent of the sequence $\{Z_n\}_{n=1}$. Let $\{a_n\}_{n=1}^{\infty}$ and $\{b_n\}_{n=1}^{\infty}$ be two sequences of real numbers such that $b_n > 0$ $(n \geqslant 1)$, $b_n \to \infty$ $(n \to \infty)$ and

$$b_n^{-1}(Z_n - a_n) \Rightarrow Z \quad \text{as} \quad n \to \infty, \tag{1}$$

where Z is a r.v. Here and in what follows symbol $\Rightarrow$ will stand for convergence in distribution. Let $d(t)$ be an increasing to infinity positive function.

At first we shall assume that $a_n = 0$ $(n \geqslant 1)$ and state the following result.

THEOREM 1. *Suppose that*
(A) *Condition* (1) *holds;*

(B) *There exists a r.v. Y such that*

$$b_N/d(t) \Rightarrow Y \text{ as } t \to \infty$$

and either
(C1) $P(Y=0)=0$
or
(C2) *The sequence $\{b_n\}_{n=1}^{\infty}$ is strictly monotonic and $N_t \to \infty$ in probability as $t \to \infty$.*
Then

$$Z_N/d(t) \Rightarrow Z^*Y^* \quad \text{as} \quad t \to \infty,$$

where r.v.'s Z^* and Y^* are independent and equidistributed with Z and Y, respectively.

The proof of this theorem can be found in (Kruglov and Korolev, 1990)*. Theorem 1 generalizes one result from (Dobrushin, 1955). The next statement describes the behaviour of moments in Theorem 1.

THEOREM 2. *Suppose that $q \in \mathbf{R}^1$ and*

$$\mathbf{E} \mid Z_n/b_n \mid^q \longrightarrow \mathbf{E} \mid Z \mid^q \quad \text{as} \quad n \to \infty, \tag{2}$$

$0 < \mathbf{E} \mid Z \mid^q < \infty$. *Then*

$$\mathbf{E} \mid Z_N/d(t) \mid^q \longrightarrow c\mathbf{E} \mid Z \mid^q \quad \text{as} \quad t \to \infty, \tag{3}$$

iff

$$\mathbf{E}(b_N/d(t))^q \longrightarrow c \quad \text{as} \quad t \to \infty. \tag{4}$$

Proof. Fix an arbitrary $\varepsilon > 0$. Condition (2) implies the existence of $n_0 = n_0(\varepsilon)$ such that for all $n \geqslant n_0$

$$\Big|\mathbf{E} \mid Z_n/b_n \mid^q - \mathbf{E} \mid Z \mid^q \Big| < \varepsilon.$$

Without loss of generality, suppose that

$$M_q = \sup_n \mathbf{E} \mid Z_n/b_n \mid^q < \infty.$$

Then we have

$$\mathbf{E} \mid Z_N/d(t) \mid^q = \sum_{n=1}^{\infty} \mathbf{P}(N=n)\mathbf{E} \mid Z_n/b_n \mid^q (b_n/d(t))^q =$$

$$= \sum_{n=1}^{n_0} \mathbf{P}(N=n)\mathbf{E} \mid Z_n/b_n \mid^q (b_n/d(t))^q +$$

* Recently we obtained the proof of Theorem 1 free from conditions C1 and C2. However this proof has not been published yet.

$$+\sum_{n=n_0+1}^{\infty} \mathbf{P}(N=n)\mathbf{E} \mid Z_n/b_n \mid^q (b_n/d(t))^q \leqslant$$

$$\leqslant M_q(\sup\{b_n : 1 \leqslant n \leqslant n_0\}/d(t))^q +$$

$$+(\mathbf{E} \mid Z \mid^q + \varepsilon) \sum_{n=n_0+1}^{\infty} \mathbf{P}(N=n)(b_n/d(t))^q. \tag{5}$$

Since $d(t)$ infinitely increases as $t \to \infty$, there exists $t_1 = t_1(\varepsilon)$ such that for all $t \geqslant t_1$

$$M_q \sup\{b_n^q : 1 \leqslant n \leqslant n_0\} \leqslant \varepsilon(d(t))^q.$$

Therefore (5) implies

$$\sum_{n=n_0+1}^{\infty} \mathbf{P}(N=n)(b_n/d(t))^q \geqslant (\mathbf{E} \mid Z_N/d(t) \mid^q - \varepsilon)/(\mathbf{E} \mid Z \mid^q + \varepsilon) \tag{6}$$

for all $t \geqslant t_1$. Further we have

$$\delta_q \equiv \mathbf{E}(b_N/d(t))^q - \sum_{n=n_0+1}^{\infty} \mathbf{P}(N=n)(b_n/d(t))^q =$$

$$= \sum_{n=1}^{n_0} \mathbf{P}(N=n)(b_n/d(t))^q \leqslant \sup\{b_n^q : 1 \leqslant n \leqslant n_0\}/(d(t))^q.$$

Moreover, according to the definition of $d(t)$ there exists $t_2 = t_2(\varepsilon)$ such that $\delta_q < \varepsilon$ for all $t \geqslant t_2$. Thus from (6) we get the inequality

$$\mathbf{E}(b_N/d(t))^q \geqslant (\mathbf{E} \mid Z_N/d(t) \mid^q - \varepsilon)/(\mathbf{E} \mid Z \mid^q + \varepsilon) - \varepsilon, \tag{7}$$

which is valid for all $t \geqslant \max\{t_1, t_2\}$. On the other hand, for $t \geqslant t_2$ we have

$$\mathbf{E} \mid Z_N/d(t) \mid^q \geqslant \sum_{n=n_0+1}^{\infty} \mathbf{P}(N=n)\mathbf{E} \mid Z_n/b_n \mid^q (b_n/d(t))^q \geqslant$$

$$\geqslant (\mathbf{E} \mid Z \mid^q - \varepsilon) \sum_{n=n_0+1}^{\infty} \mathbf{P}(N=n)(b_n/d(t))^q \geqslant (\mathbf{E} \mid Z \mid^q - \varepsilon)(\mathbf{E}(b_N/d(t))^q - \varepsilon)$$

and hence,

$$\mathbf{E}(b_N/d(t))^q \leqslant \varepsilon + \mathbf{E} \mid Z_N/d(t) \mid^q /(\mathbf{E} \mid Z \mid^q - \varepsilon) \tag{8}$$

if $t \geqslant t_2$. Thus (3) and (4) are equivalent because in (7) and (8) ε is arbitrary. Notice that the statement of the theorem remains valid if $c = \infty$. The proof is completed.

COROLLARY 1. *Let*

$$Z_n = S_n = X_1 + \ldots + X_n$$

where r.v.'s $\{X_j\}_{j=1}$ *are independent and identically distributed with* $\mathbf{E}X_1 = 0$, $\mathbf{E}X_1^2 = \sigma^2$ *and* $\mathbf{E} \mid X_1 \mid^q < \infty$, $q \geqslant 2$. *Then*

$$\mathbf{E} \mid S_N/d(t) \mid^q \longrightarrow C \quad \text{as} \quad t \to \infty,$$

iff

$$\mathbf{E}N^{q/2}/(d(t))^q \longrightarrow C\pi^{1/2}\sigma^{-q}2^{-q/2}\Gamma((q+1)/2) \quad \text{as} \quad t \to \infty.$$

Proof. According to (Bernstein, 1964, p. 358), condition $\mathbf{E} \mid X_1 \mid^q < \infty$ implies (2) with $Z_n = S_n$, $b_n = \sigma n^{1/2}$ and Z being a r.v. with the standard normal distribution.
Corollary 1 now follows from Theorem 2.

THEOREM 3. *Suppose that conditions* (1) *and* (2) *hold for some* $q > 0$ *and* $0 < \mathbf{E} \mid Z \mid^q < \infty$. *Then*

$$Z_N/d(t) \Rightarrow Z \quad \text{as} \quad t \to \infty, \tag{9}$$

and simultaneously

$$\mathbf{E} \mid Z_N/d(t) \mid^q \longrightarrow \mathbf{E} \mid Z \mid^q \quad \text{as} \quad t \to \infty, \tag{10}$$

iff

$$b_N/d(t) \Rightarrow 1 \quad \text{as} \quad t \to \infty, \tag{11}$$

and simultaneously

$$\mathbf{E}(b_N/d(t))^q \longrightarrow 1 \quad \text{as} \quad t \to \infty. \tag{12}$$

Proof. Necessity. According to the moments convergence theorem (Loève, 1963, p. 184) conditions (9) and (10) imply

$$\mathbf{E} \mid Z_N/d(t) \mid^p \longrightarrow \mathbf{E} \mid Z \mid^p \quad \text{as} \quad t \to \infty,$$

for all $p \in [0, q]$. Therefore Theorem 2 gives us

$$\mathbf{E}(b_N/d(t))^p \longrightarrow 1 \quad \text{as} \quad t \to \infty. \tag{13}$$

Thus using Markov's inequality, we obtain

$$\lim_{x\to\infty} \sup_t \mathbf{P}(b_N > xd(t)) \leqslant \lim_{x\to\infty} \sup_t x^{-p}\mathbf{E}(b_N/d(t))^p = 0,$$

i.e., the family of distribution functions (d.f.'s) $\{G_t(x) = \mathbf{P}(b_N < xd(t))\}_{t>0}$ is weakly relatively compact. Let $\{G_{t_k}\}_{k\geqslant 1}$ be a sequence of d.f.'s weakly converging to a d.f., say, G of a r.v., say, U. Since (13) holds, we have

$$\mathbf{E}(b_N/d(t_k))^p \longrightarrow 1 \quad \text{as} \quad k \to \infty,$$

and therefore $\mathbf{E}U^p = 1$ for each $p \in [0, q]$, which is possible only if $\mathbf{P}(U = 1) = 1$. Thus the necessity of (11) and (12) is proved.

Sufficiency. According to Theorem 1 conditions (1) and (11) imply (9). Now note that

$$\mathbf{E} \mid Z_N/d(t) \mid^q = \sum_{n=1}^{\infty} \mathbf{P}(N=n)(b_n/d(t))^q \mathbf{E} \mid Z_n/b_n \mid^q . \tag{14}$$

Since (2) and (12) hold, relation (14) by Toeplitz's lemma implies (10). The proof is completed.

COROLLARY 2. *Let* r.v.'s $\{X_j\}_{j=1}$ *be the same as in Corollary 1. Suppose that* $\mathbf{E}N_t \to \infty$ *as* $t \to \infty$. *Let* Z_0 *be a* r.v. *with the standard normal distribution. Then*

$$(\mathbf{D}S_N)^{-1/2} S_N \Rightarrow Z_0 \quad \text{as} \quad t \to \infty,$$

and simultaneously

$$\mathbf{E} \mid (\mathbf{D}S_N)^{-1/2} S_N \mid^q \longrightarrow \mathbf{E} \mid Z_0 \mid^q \quad \text{as} \quad t \to \infty,$$

iff

$$N_t/\mathbf{E}N_t \Rightarrow 1 \quad \text{as} \quad t \to \infty,$$

and simultaneously

$$\mathbf{E}N_t^q/(\mathbf{E}N_t)^q \longrightarrow 1 \quad \text{as} \quad t \to \infty.$$

Proof. Note that here $\mathbf{D}S_N = \sigma^2 \mathbf{E}N_t$. Application of Bernstein's theorem mentioned above and Theorem 3 concludes the proof.
Now we turn to the case when $a_n \neq 0$ for infinitely many $n \geqslant 1$. Consider two auxiliary functions: a positive infinitely increasing function $b(t)$ and an arbitrary real function $a(t)$.

THEOREM 4. *Suppose that* (1) *holds and*

$$\limsup_{t\to\infty} b(t)/d(t) < \infty.$$

If

$$(b_N - b(t))/d(t) \Rightarrow 0 \qquad \text{as} \qquad t \to \infty, \tag{15}$$

then for each $s \in \mathbf{R}^1$

$$\lim_{t\to\infty} \Big| \mathbf{E}\exp\{is(Z_N - a(t))/d(t)\} -$$

$$- \mathbf{E}\exp\{is(a_N - a(t))/d(t)\}\, \mathbf{E}\exp\{isZb(t)/d(t)\} \Big| = 0.$$

This theorem is a generalization of a result due to Dobrushin.
We also mention that a particular variant of condition (15) was used by Z.Rychlik in the random-sum version of the central limit theorem (Rychlik, 1976). The proof of Theorem 4 can be found in (Kruglov and Korolev, 1990).

COROLLARY 3. *Suppose that* (1) *and* (15) *hold and*
(D) $b(t)/d(t) \to k$ as $t \to \infty$;
(E) $(a_N - a(t))/d(t) \Rightarrow W$ as $t \to \infty$.
Then

$$(Z_N - a(t))/d(t) \Rightarrow W^* + kZ^* \quad \text{as} \quad t \to \infty,$$

where r.v.'s W^* *and* Z^* *are independent and equidistributed with* W *and* Z, *respectively.*

We shall consider convergence of moments in Theorem 4 in a less general situation which however turns out to be general enough to obtain some important corollaries. We shall introduce some additional notation. Put

$$A_n = \mathbf{E}Z_n,\ A = \mathbf{E}A_N\ (= \mathbf{E}Z_N),\ B_n^2 = \mathbf{D}Z_n,\ B^2 = \mathbf{E}B_N^2,\ \Sigma^2 = \mathbf{D}Z_N.$$

It is quite easy to verify that $\Sigma^2 = B^2 + \mathbf{D}A_N$. Note that A, B^2 and Σ^2 are functions of t. Suppose that
(F) for some $q = 2p \geqslant 2$

$$\mathbf{E} \mid (Z_n - A_n)/B_n \mid^q \to \mathbf{E} \mid Z \mid^q \quad \text{as} \quad n \to \infty; \tag{16}$$

(G) The following coordination condition holds:

$$0 < \inf_n \mid (B_n^2 - B^2)/(A_n - A) \mid \leqslant \sup_n \mid (B_n^2 - B^2)/(A_n - A) \mid < \infty; \tag{17}$$

(H) $\Sigma^2 \to \infty$ as $t \to \infty$;
(I) $N_t \to \infty$ in probability, as $t \to \infty$;
(J) $\Sigma^{-2}\mathbf{D}A_N \to m$ as $t \to \infty$.

THEOREM 5. *Suppose that conditions* (1) *with* $a_n = A_n$, $b_n = B_n$ *and* (F), (G), (H), (I), (J) *hold. Consider two cases:*
(i) $m = 0$. *Then*

$$\Sigma^{-1}(Z_N - A) \Rightarrow Z \quad \text{as} \quad t \to \infty, \tag{18}$$

and simultaneously

$$\mathbf{E} \mid \Sigma^{-1}(Z_N - A) \mid^{2p} \longrightarrow \mathbf{E} \mid Z \mid^{2p} \quad \text{as} \quad t \to \infty, \tag{19}$$

iff

$$\mathbf{E} \mid \Sigma^{-1}(A_N - A) \mid^{2p} \to 0 \quad as \quad t \to \infty. \tag{20}$$

(ii) $0 < m \leqslant 1$. *Then there exists a r.v.* Y_1 *such that*

$$\Sigma^{-1}(Z_N - A) \Rightarrow Y_1 \quad \text{as} \quad t \to \infty, \tag{21}$$

and simultaneously

$$\mathbf{E} \mid \Sigma^{-1}(Z_N - A) \mid^{2p} \longrightarrow \mathbf{E} \mid Y_1 \mid^{2p} \quad \text{as} \quad t \to \infty, \tag{22}$$

iff there exists a r.v. Y_2 *such that*

$$(A_N - A)(\mathbf{D}A_N)^{-1/2} \Rightarrow Y_2 \quad \text{as} \quad t \to \infty, \tag{23}$$

and simultaneously

$$\mathbf{E} \mid (A_N - A)(\mathbf{D}A_N)^{-1/2} \mid^{2p} \longrightarrow \mathbf{E} \mid Y_2 \mid^{2p} \quad \text{as} \quad t \to \infty. \tag{24}$$

Moreover, $Y_1 = (1-m)^{1/2} Z^* + m^{1/2} Y_2^*$, *where r.v.'s* Z^* *and* Y_2^* *are independent and equidistributed with* Z *and* Y_2, *respectively.*

Proof. Necessity. By Jensen's inequality we have

$$\mathbf{E} \mid Z_n - A \mid^{2p} \geqslant \mid A_n - A \mid^{2p} .$$

Therefore

$$\mathbf{E} \mid \Sigma^{-1}(Z_N - A) \mid^{2p} = \sum_{n=1}^{\infty} \mathbf{P}(N = n) \mathbf{E} \mid \Sigma^{-1}(Z_N - A) \mid^{2p} \geqslant$$

$$\geqslant \sum_{n=1}^{\infty} \mathbf{P}(N = n) \mid \Sigma^{-1}(A_N - A) \mid^{2p} = \mathbf{E} \mid \Sigma^{-1}(A_N - A) \mid^{2p} . \tag{25}$$

From (19) and (25) we get

$$\sup_t \mathbf{E} \mid \Sigma^{-1}(A_N - A) \mid^{2p} < \infty. \tag{26}$$

Thus

$$\lim_{t \to \infty} \mathbf{E} \mid \Sigma^{-2}(A_N - A) \mid^{2p} = 0. \tag{27}$$

Since

$$\mathbf{E} \mid \Sigma^{-1}(Z_N - A_N) \mid^{2p} = \mathbf{E} \mid \Sigma^{-1}(Z_N - A) + \Sigma^{-1}(A - A_N) \mid^{2p} \leqslant$$

$$\leqslant 2^{2p-1}(\mathbf{E} \mid \Sigma^{-1}(Z_N - A) \mid^{2p} + \mathbf{E} \mid \Sigma^{-1}(A_N - A) \mid^{2p}),$$

relations (19) and (26) imply

$$\sup_t \mathbf{E} \mid \Sigma^{-1}(Z_N - A_N) \mid^{2p} < \infty. \tag{28}$$

Using the same reasoning as in Theorem 2 we obtain

$$\lim_{t \to \infty} \left| \mathbf{E} \mid \Sigma^{-1}(Z_N - A_N) \mid^{2p} - \mathbf{E} \mid Z \mid^{2p} \mathbf{E}(\Sigma^{-1} B_N)^{2p} \right| = 0. \tag{29}$$

The following reasoning depends on the value of m. Let at first $m = 0$. In this case by Chebyshev's inequality we have

$$\Sigma^{-1}(A_N - A) \Rightarrow 0 \quad \text{as} \quad t \to \infty,$$

and consequently (since (18) holds)

$$\Sigma^{-1}(Z_N - A_N) = \Sigma^{-1}(Z_N - A) - \Sigma^{-1}(A_N - A) \Rightarrow Z \quad \text{as} \quad t \to \infty. \tag{30}$$

Denote

$$K(t) = \sup_n \left|(B_n^2 - B^2)/(A_n - A)\right|.$$

Using (17) and (27) we get

$$\lim_{t\to\infty} \mathbf{E} \mid B^{-2}B_N - 1 \mid^{2p} = \lim_{t\to\infty} (B^{-2}\Sigma^2)^{2p} \lim_{t\to\infty} \mathbf{E} \mid \Sigma^{-2}(B_N^2 - B^2) \mid^{2p} =$$

$$= (1-m)^{-2p} \lim_{t\to\infty} \mathbf{E} \mid \Sigma^{-2}(B_N^2 - B^2) \mid^{2p} =$$

$$= (1-m)^{-2p} \lim_{t\to\infty} \mathbf{E} \mid (B_N^2 - B^2)/(A_N - A) \mid^{2p} (\Sigma^{-2} \mid A_N - A \mid)^{2p} \leqslant$$

$$\leqslant (1-m)^{-2p} \sup_t K^{2p}(t) \lim_{t\to\infty} \mathbf{E} \mid \Sigma^{-2}(A_N - A) \mid^{2p} = 0, \tag{31}$$

i.e.,

$$B^{-2}B_N^2 \Rightarrow 1 \quad \text{as} \quad t \to \infty, \tag{32}$$

and

$$\mathbf{E}(B^{-1}B_N)^{2p} \to 1 \quad \text{as} \quad t \to \infty. \tag{33}$$

According to Theorem 3 conditions (32) and (33) imply

$$\Sigma^{-1}(Z_N - A_N) = (\Sigma^{-1}B)B^{-1}(Z_N - A_N) \Rightarrow (1-m)^{1/2}Z \tag{34}$$

and

$$\mathbf{E} \mid \Sigma^{-1}(Z_N - A_N) \mid^{2p} = (\Sigma^{-1}B)^{2p}\mathbf{E} \mid B^{-1}(Z_N - A_N) \mid^{2p} \to (1-m)^p \mathbf{E} \mid Z \mid^{2p} \tag{35}$$

as $t \to \infty$. By the uniform integrability theorem (UIT) (Loève, 1963, p.183), conditions (34) and (35) guarantee that

$$\lim_{x\to\infty} \sup_t \mathbf{E} \mid \Sigma^{-1}(Z_N - A_N) \mid^{2p} \chi(\mid \Sigma^{-1}(Z_N - A_N) \mid > x) = 0. \tag{36}$$

Here and further $\chi(S)$ is the indicator function of a set S. According to the UIT, conditions (18) and (19) imply

$$\lim_{x\to\infty} \sup_t \mathbf{E} \mid \Sigma^{-1}(Z_N - A) \mid^{2p} \chi(\mid \Sigma^{-1}(Z_N - A) \mid > x) = 0. \tag{37}$$

Since

$$\Sigma^{-1}(A_N - A) = \Sigma^{-1}(Z_N - A) - \Sigma^{-1}(Z_N - A_N),$$

from (36) and (37) we get

$$\lim_{x\to\infty} \sup_t \mathbf{E} \mid \Sigma^{-1}(A_N - A) \mid^{2p} \chi(\mid \Sigma^{-1}(A_N - A) \mid > x) = 0. \tag{38}$$

By Chebyshev's inequality, we have

$$\mathbf{P}(|\,\Sigma^{-1}(A_N - A)\,| > c) \leqslant c^{-2}\Sigma^{-2}\mathbf{D}A_N$$

for an arbitrary $c > 0$, and thus with the account of $m = 0$, we get

$$\Sigma^{-1}(A_N - A) \Rightarrow 0 \quad \text{as} \quad t \to \infty. \tag{39}$$

By the UIT mentioned above (38) and (39) imply (20).
Let now $0 < m \leqslant 1$. Since (17) and (27) hold, we have

$$\lim_{t\to\infty} \mathbf{E}\,|\,\Sigma^{-2}(B_N^2 - B^2)\,|^{2p} = \lim_{t\to\infty} \mathbf{E}\,|\,(B_N^2 - B^2)/(A_N - A)\,|^{2p}|\,\Sigma^{-2}(A_N - A)\,|^{2p} \leqslant$$

$$\leqslant \sup_t K^{2p}(t) \lim_{t\to\infty} \mathbf{E}\,|\,\Sigma^{-2}(A_N - A)\,|^{2p} = 0$$

and, consequently,

$$\Sigma^{-2}(B_N^2 - B^2) \Rightarrow 0 \quad \text{as} \quad t \to \infty.$$

This condition, by Theorem 4, implies

$$\lim_{t\to\infty} |\,\mathbf{E}\exp(is(Z_N - A)/\Sigma) - h(s(1 - \Sigma^{-2}\mathbf{D}A_N)^{1/2})w_t(s(\Sigma^{-2}\mathbf{D}A_N)^{1/2})\,| = 0$$

for each $s \in \mathbf{R}^1$ where

$$h(s) = \mathbf{E}\exp(isZ), \quad w_t(s) = \mathbf{E}\exp(is(A_N - A)(\mathbf{D}A_N)^{-1/2}).$$

Therefore with the account of (J), we have

$$\lim_{t\to\infty} |\,\mathbf{E}\exp(is(Z_N - A)/\Sigma) - h(s(1 - m)^{1/2})w_t(sm^{1/2})\,| = 0$$

for each $s \in \mathbf{R}^1$, and hence, by the continuity theorem for characteristic functions, condition (21) implies (23).
If $0 < m < 1$, then in the same way as before, using (31), we deduce that the conditions of Theorem 3 hold, and hence we have (34) and (35), and consequently (36) and (37) hold. Since

$$(A_N - A)(\mathbf{D}A_N)^{-1/2} = (\Sigma^2/\mathbf{D}A_N)^{1/2}((Z_N - A)/\Sigma - (Z_N - A_N)/\Sigma),$$

condition $0 < m < 1$, together with (36) and (37), imply

$$\lim_{x\to\infty} \sup_t \mathbf{E}\,|\,(\mathbf{D}A_N)^{-1/2}(A_N - A)\,|^{2p}\,\chi(|\,(\mathbf{D}A_N)^{-1/2}(A_N - A)\,| > x) = 0$$

which together with (23), by the UIT, gives (24).
If $m = 1$ then $B^2\Sigma^{-2} \to 0$ as $t \to \infty$, and in accordance with (29) we have

$$\lim_{t\to\infty} \mathbf{E}\,|\,\Sigma^{-1}(Z_N - A_N)\,|^{2p} = 0, \tag{40}$$

and hence, by Markov's inequality

$$\Sigma^{-1}(Z_N - A_N) \Rightarrow 0 \quad \text{as} \quad t \to \infty,$$

which together with (38) and (21) gives

$$(A_N - A)/(\mathbf{D}A_N) \Rightarrow Y_1 \quad \text{as} \quad t \to \infty,$$

which, by (40), implies (39) and consequently, by the UIT, implies (24).

Sufficiency. Consider the case $m = 0$. Conditions $\Sigma^2 \to \infty$ as $t \to \infty$, and (20) imply (27), which in its turn implies (32) and (33), which by Theorem 3 give (34) and (35). These conditions, by the UIT, imply (36). Condition (20) implies (39). Therefore (18) holds, since

$$(Z_N - A)/\Sigma = (Z_N - A_N)/\Sigma + (A_N - A)/\Sigma.$$

Now (19) is a consequence of (18) and (36), by the UIT.
Let finally $0 < m \leqslant 1$. Conditions $\Sigma^2 \to \infty$ as $t \to \infty$, and (25) imply (27). By Theorem 4 conditions $\Sigma^2 \to \infty$ as $t \to \infty$, (17) and (23) are sufficient for (21). Conditions (24) and (F) imply (38), which in its turn with the account of (27) implies (22). The proof is completed.
Now we shall state one important corollary of Theorem 5. In our opinion it is one of the main results of this paper. That's why we state it as a separate theorem. Let $\{X_j\}_{j\geqslant 1}$ be a set of independent identically distributed r.v.'s with $\mathbf{E}X_j = a$ and $\mathbf{D}X_j = \sigma^2 < \infty$. Put $S_n = X_1 + \ldots + X_n$.

THEOREM 6. *Suppose that* $\mathbf{D}N < \infty$ *and condition* (J) *holds with* $0 < m < 1$. *Then the following statements are equivalent:*
(K) *There exists a r.v.* Y *such that*

$$(N - \mathbf{E}N)(\mathbf{D}N)^{-1/2} \Rightarrow Y \quad as \quad t \to \infty;$$

(L) *For some* $c \in (0,1)$

$$\mathbf{E}\min(c, | N/\mathbf{E}N - 1 |) \longrightarrow 0 \quad as \quad t \to \infty;$$

(M) *There exists a r.v.* U *such that*

$$(S_N - \mathbf{E}S_N)(\mathbf{D}S_N)^{-1/2} \Rightarrow U \quad as \quad t \to \infty.$$

Moreover, $U = (1-m)^{1/2}Z + m^{1/2}Y^*$, *where r.v.* Y^* *is equidistributed with* Y *and independent of r.v.* Z, *which has the standard normal distribution.*

Proof. Implications (K) $\subset$ (M) and (M) $\subset$ (K) are proved in Theorem 5, because $\mathbf{D}U = \mathbf{D}((S_N - \mathbf{E}S_N)(\mathbf{D}S_N)^{-1/2}) = 1$.
Implication (K) $\subset$ (L) holds, because

$$\mathbf{P}(| N/\mathbf{E}N - 1 | > x) = \mathbf{P}(| N - \mathbf{E}N | /(\mathbf{D}N)^{1/2} > x\mathbf{E}N(\mathbf{D}N)^{-1/2}) =$$

$$= \mathbf{P}\Big(\mid Y \mid > x\big((m^{-1} - 1)\sigma^2 \mathbf{E}N(1 + o(1))\big)^{1/2}\Big) + o(1) = o(1) \quad \text{as} \quad t \to \infty,$$

for each $x > 0$.
Implication (L) $\subset$ (M) follows from the estimate

$$\sup_x \left|\mathbf{P}\big((S_N - \mathbf{E}S_N)(\mathbf{D}S_N)^{-1/2} < x\big) - \mathbf{P}\big((1-m)^{1/2}Z + m^{1/2}Y^* < x\big)\right| \leqslant$$

$$\leqslant \sup(\Delta_n : n \geqslant (1-c)\mathbf{E}N) + Q(c)\mathbf{E}\min(c, \mid N/\mathbf{E}N - 1 \mid), \tag{41}$$

which is valid for each $c \in (0,1)$, where

$$\Delta_n = \sup_x \left|\mathbf{P}\big((S_n - \mathbf{E}S_n)(\mathbf{D}S_n)^{-1/2} < x\big) - \mathbf{P}(Z < x)\right|,$$

$$Q(c) = \max\big((1-c)^{-1}, (2\pi e)^{-1/2}(c + c^{1/2})^{-1}\big).$$

The estimate (41) was proved in (Korolev, 1989). This completes the proof.
Theorem 6 is a final form of the famous theorem from (Robbins, 1948), which was later generalized in dozens of papers but never since 1948 necessary conditions were presented. By the way, it follows from Theorem 6 that the estimate (41) appears to be unimprovable in the sense that the LHS and the RHS of (41) must tend or not tend to zero, simultaneously.
In all above stated theorems concerning the convergence of moments, we supposed that we had the convergence of moments of non–randomly indexed random sequence. If we want to avoid such assumptions, it is quite evident that some versions of the uniform integrability conditions should be inserted into the formulations of the theorems. We present one example of such statements concerning random sums.
Let r.v.'s $\{X_j\}_{j\geqslant 1}$ be independent. Denote by $l_t(q)$ the q–quantile of the r.v. N_t. Put $S_n \doteq X_1 + \ldots + X_n$.

THEOREM 7. *Suppose that there exist a r.v.* X *and an infinitely increasing function* $d(t)$ *such that* $S_N/d(t) \Rightarrow X$ *as* $t \to \infty$, $p \geqslant 2$, $\mathbf{E}X_j = 0$, $\mathbf{E} \mid X_j \mid^p < \infty$ *for all* $j \geqslant 1$, $\mathbf{E} \mid S_N \mid^p < \infty$. *Then*

$$\mathbf{E} \mid S_N/d(t) \mid^p \longrightarrow \mathbf{E} \mid X \mid^p \quad \textit{as} \quad t \to \infty,$$

iff

$$\lim_{q \uparrow 1} \sup_t \sum_{k > l_t(q)} \mathbf{P}(N = k)\mathbf{E} \mid S_k/d(t) \mid^p = 0,$$

and for each $q \in (0,1)$,

$$\lim_{r \to \infty} \sup_t (d(t))^{-p} \sum_{k \leqslant l_t(q)} \mathbf{E} \mid X_k \mid^p \chi\,(\mid X_k \mid > r)d(t) = 0.$$

The proof of this theorem as well as its different generalizations and modifications can be found in (Kruglov and Korolev, 1990).

REFERENCES

Bernstein, S.N. (1964). *Selected Works.* **4**, Nauka, Moscow.

Dobrushin, R.L. (1955). A lemma about the limit of the superposition of random functions. *Uspekhi Mat. Nauk.* **10**, 157–159.

Gut, A. (1987). *Stopped random walks.Limit theorems and applications.* Springer, Berlin–Heidelberg etc.

Korolev, V.Yu. (1985). Convergence of moments of random sums of independent random variables. *Theory Probab. Appl.* **30**, 386–390.

Korolev, V.Yu. (1989). On the accuracy of the approximation of the distribution of sums of a random number of random variables by mixtures of normal laws. *Theory Probab. Appl.* **34**, 581–588.

Kruglov, V.M. (1988). Convergence of moments of random sums. *Theory Probab. Appl.* **33**, 360–363.

Kruglov, V.M. and Korolev, V.Yu. (1990). *Limit Theorems for Random Sums.* Moscow State University, Moscow.

Loève, M. (1960). *Probability Theory.* Van Nostrand, Princeton.

Robbins, H. (1948). The asymptotic distribution of the sum of a random number of random variables. *Bull. Amer. Math. Soc.* **54**, 1151–1161.

Rychlik, Z. (1976). A central limit theorem for sums of a random number of independent random variables. *Colloq. Math.* **35**, 147–158.

Sakalauskas, V. (1988). On the asymptotics of the absolute moments of sums of a random number of independent random variables. *Liet. Mat. Rink.* **28**, 352–359.

Prob. Theory and Math. Stat., Vol. 2, pp. 22–34
B.Grigelionis *et al.* (Eds.)
1990 VSP/Mokslas

ON THE RATE OF CONVERGENCE OF THE DISTRIBUTIONS OF SEMIMARTINGALES TO THE DISTRIBUTION OF STABLE PROCESS

K. KUBILIUS
Institute of Mathematics and Informatics, Lithuanian Ac. Sci., Akademijos 4, 232600 Vilnius, Lithuania.

ABSTRACT

An estimate of the Lévy—Prokhorov distance between the distributions of a semimartingale and a stable process in terms of their predictable characteristics is given.

INTRODUCTION

In the previous papers (Kubilius and Mikulevičius, 1987; Kubilius, 1990; Jacod and Mano, 1988) estimates of the Lévy—Prokhorov distance between distributions of semimartingales and diffusion processes were obtained. In this paper the limiting process is δ–stable with $\delta \in (0,2)$, i.e., we consider a particular case of this problem. We don't actually discuss the case $\delta = 2$, because the corresponding result can be found in (Kubilius and Mikulevičius, 1987; Kubilius, 1990).
Here we'll give only a sketch of the proof of the main results. Complete proofs will be published in Lithuanian Mathematical Journal.

1. THE MAIN RESULTS

Let $(\Omega, \mathcal{F}, \mathbf{P})$ be a complete probability space with filtration $\mathsf{F} = (\mathcal{F}_t)_{t\geqslant 0}$ satisfying usual conditions. Let Y be a d–dimensional $(\mathbf{P}, \mathsf{F})$–semimartingale. It is well–known that it can be represented uniquely in the form

$$Y_t = Y_0 + \alpha_t + Y_t^c + \int_0^t \int_{|x|\leqslant 1} x(p - \Pi)(ds, dx) + \int_0^t \int_{|x|>1} xp(ds, dx), \quad t \geqslant 0,$$

where α is a predictable process with locally integrable variation, Y^c is a continuous local martingale, $p(ds, dx)$ is the measure of jumps of the process Y, $\Pi(ds, dx)$ is the dual predictable projection of the measure p. Denote

$$C = (C^{(i,j)})_{1\leqslant i,j\leqslant d}\ ,\ C^{(i,j)} = \langle Y^{(i)c}, Y^{(j)c}\rangle.$$

We call (α, C, Π) the triplet of predictable characteristics (t.p.ch.) of Y.
Let $\tilde{\mathbf{F}} = (\tilde{\mathcal{F}}_t)_{t\geqslant 0}$ be another filtration on $(\Omega, \mathcal{F}, \mathbf{P})$ satisfying usual conditions. Let X be a d-dimensional $(\mathbf{P}, \tilde{\mathbf{F}})$-semimartingale with t.p.ch. $(\tilde{\alpha}, \tilde{C}, \tilde{\Pi})$ such that

$$\tilde{\alpha}_t = \mathbf{1}_{\delta\in(0,1)} \int_0^t \int_{|x|\leqslant 1} x\tilde{\Pi}(ds, dx) + \gamma t \mathbf{1}_{\delta=1} - \mathbf{1}_{\delta\in(1,2)} \int_0^t \int_{|x|>1} x\tilde{\Pi}(ds, dx),$$

$$\tilde{C}_t^{(i,j)} = a^{(i,j)} t \mathbf{1}_{\delta=2}, \quad \tilde{\Pi}(ds, dx) = \frac{dx}{|x|^{d+\delta}} dt,$$

where $\mathbf{1}_A$ is the indicator of the set A, $a = (a^{(i,j)})$ is a symmetric non-negative definite $d \times d$ real matrix, $\gamma \in \mathbf{R}^d$. Hence X is a homogeneous process with independent increments, in fact, X is δ-stable with $\delta \in (0, 2]$.
Let $\exp\{t\psi^{(\delta)}(\xi)\}$ be the characteristic function of X. The function $\psi^{(\delta)}(\xi)$ equals

$$\psi^{(\delta)}(\xi) = (\xi, \gamma)\mathbf{1}_{\delta=1} + (a\xi, \xi)\mathbf{1}_{\delta=2} +$$

$$+\mathbf{1}_{\delta\in(0,2)} \int \left[e^{i(\xi,x)} - 1 - i(\xi, x)\mathbf{1}_{\{|x|\leqslant 1\}}\mathbf{1}_{\delta=1} - i(\xi, x)\mathbf{1}_{\delta\in(1,2)}\right] \frac{dx}{|x|^{d+\delta}}.$$

We fix a constant $L > 0$ and suppose that for each $(t, x) \in \mathbf{R}_+ \times \mathbf{R}^d$ a d-dimensional vector $b(t, x)$ is given. Suppose that for each $\delta \in (0, 2]$, $t \in \mathbf{R}_+$, $x, y \in \mathbf{R}^d$ a real-valued function $m^{(\delta)}(t, x, y) \geqslant 0$ is given. Assume that for each $\delta \in (0, 2]$, $t \in \mathbf{R}_+$, $x \in \mathbf{R}^d$, the function $m^{(\delta)}(t, x, \cdot)$ is 0-homogeneous, and has partial derivatives $\partial^\gamma m^{(\delta)}$ ($|\gamma| \leqslant [d/2] + 1$). Moreover, for each $T > 0$, $R > 0$

$$\inf_{(t,x)\in[0,T]\times\{|x|\leqslant R\}} \sup_{|y|=1} m^{(\delta)}(t, x, y) > 0.$$

We assume that the functions b, $m^{(\delta)}$ are measurable and for each $\delta \in (0, 2]$, $t \in \mathbf{R}_+$, $x, y \in \mathbf{R}^d$,

$$|b(t, x)| + m^{(\delta)}(t, x, y) \leqslant K.$$

Definition. Let Y be a semimartingale with t.p.ch. (α, C, Π). We say that Y belongs to the class S^δ, $\delta \in (0, 2]$, if

$$\alpha_t = A_t \mathbf{1}_{\delta\in[1,2]} + \mathbf{1}_{\delta\in(0,1)} \int_0^t \int_{|y|\leqslant 1} y m^{(\delta)}(s, Y_s, y) \frac{dy}{|y|^{d+\delta}} ds +$$

$$+\mathbf{1}_{\delta\in[1,2]} \int_0^t b(s, Y_s) ds - \mathbf{1}_{\delta\in(1,2]} \int_0^t \int_{|y|>1} y m^{(\delta)}(s, Y_s, y) \frac{dy}{|y|^{d+\delta}} ds,$$

$$C_t^{(i,j)} = B_t^{(i,j)} \mathbf{1}_{\delta=2}, \quad \Pi(dt, dy) = m^{(\delta)}(t, Y_t, y) \frac{dy}{|y|^{d+\delta}} dt + \pi(dt, dy),$$

where A is a process with locally integrable variation, B is a symmetric non-negative definite $d \times d$ real matrix,

$$\int_0^t \int_{\mathsf{E}} |y|^\delta \wedge 1 \; \pi(ds, dy) < \infty, \; \mathsf{E} = \mathbf{R}^d \backslash \{0\}.$$

To formulate the results we need the following notation:

$$\alpha_t^\varepsilon = \alpha_t - \int_0^t \int_{\varepsilon < |x| \leqslant 1} x \Pi(ds, dx), \quad \tilde{\alpha}_t^\varepsilon = \tilde{\alpha}_t - \int_0^t \int_{\varepsilon < |x| \leqslant 1} x \tilde{\Pi}(ds, dx),$$

$$R_t^\varepsilon = \tilde{\alpha}_t^\varepsilon - \alpha_t^\varepsilon, \; \delta_t^l(\Pi) = \int_0^t \int_{|x| > l} \Pi(ds, dx),$$

$$\varkappa_t^{\varepsilon,\delta,\beta}(\Pi) = \int_0^t \int_{\mathsf{E}} |x_\varepsilon - \Delta\alpha_s^\varepsilon|^{\delta+\beta} \Pi(ds, dx) +$$

$$+ \sum_{s \leqslant t} \Big(1 - \Pi\big(\{s\}, \mathsf{E}\big)\Big) |\Delta\alpha_s^\varepsilon|^{\delta+\beta},$$

$$\rho_{(s,t)}(\Pi^\varepsilon, \tilde{\Pi}^\varepsilon) = \sup_{g \in \mathbf{C}_1^1} \left| \int_s^t \int_{\mathsf{E}} g(x) \Pi^\varepsilon(du, dx) - \int_s^t \int_{\mathsf{E}} g(x) \tilde{\Pi}^\varepsilon(du, dx) \right|,$$

$$\rho_t(\Pi^\varepsilon, \tilde{\Pi}^\varepsilon) = \rho_{0,t}(\Pi^\varepsilon, \tilde{\Pi}^\varepsilon), \quad \Pi^\varepsilon(ds, dx) = \mathbf{1}_{\{|x| > \varepsilon\}} \Pi(ds, dx),$$

where $\varepsilon \in (0,1]$, $\delta \in (0,2]$, $\beta \in (0,1)$, $x_\varepsilon = x \mathbf{1}_{\{|x| \leqslant \varepsilon\}}$

$$\mathbf{C}_1^1 = \big\{ g \in \mathbf{C}^1(\mathbf{R}^d) : \; |g|_\infty \leqslant 1, \; |g(x) - g(y)| \leqslant |x - y| \big\}.$$

THEOREM 1. *For any $\varepsilon \in (0,1]$, $l > 1$, there exist constants $C_1 = C_1(d)$ and $C_2 = C_2(d, \varepsilon, \delta, l)$ such that for any $t_0 < t_1 < \ldots < t_n$, $n \geqslant 1$,*

a) *for $\delta \in (0,1)$, $\beta \in (0, 1-\delta)$,*

$$\pi\Big((Y_{t_0}, \ldots, Y_{t_n}), (X_{t_0}, \ldots, X_{t_n})\Big) \leqslant C_1 \Big\{ \mathbf{E}^{1/(1+\delta)} \Big[|Y_{t_0} - X_{t_0}|^\delta \wedge 1 \Big] +$$

$$+ \mathbf{E}\Big[\delta_{t_n}^l(\Pi) - \delta_{t_0}^l(\Pi)\Big] + \Big[\delta_{t_n}^l(\tilde{\Pi}) - \delta_{t_0}^l(\tilde{\Pi})\Big] +$$

$$+ \mathbf{E}^{1/(1+\delta+\beta)} \left[\int_{t_0}^{t_n} \int_{|y| \leqslant \varepsilon} |y|^{\delta+\beta} \Pi(du, dy) \right] + \left[\int_{t_0}^{t_n} \int_{|y| \leqslant \varepsilon} |y|^{\delta+\beta} \tilde{\Pi}(du, dy) \right]^{1/(1+\delta+\beta)} \Bigg\} +$$

$$+C_2\Big\{\sum_{k=0}^{n-1}\mathbf{E}^{\delta/((d+1+\delta)(1+\delta))}\Big[\Big(\sup_{t_k<u\leqslant t_{k+1}}\rho_{t_k,u}(\Pi^\varepsilon,\tilde\Pi^\varepsilon)\Big)\wedge 1\Big]+$$

$$+\mathbf{E}^{1/(1+\delta)}\Big[\sup_{t_k<u\leqslant t_{k+1}}\Pi^\varepsilon(\{u\},\mathsf{E})\Big]^\delta\Big\},$$

b) *for* $\delta\in[1,2),\beta\in(0,2-\delta)$

$$\pi\Big((Y_{t_0},\dots,Y_{t_n}),(X_{t_0},\dots,X_{t_n})\Big)\leqslant$$

$$\leqslant C_1\Big\{\pi^{1/2}(Y_{t_0},X_{t_0})+\mathbf{E}\Big[\delta^l_{t_n}(\Pi)-\delta^l_{t_0}(\Pi)\Big]+\Big[\delta^l_{t_n}(\tilde\Pi)-\delta^l_{t_0}(\tilde\Pi)\Big]+$$

$$+\sum_{k=0}^{n-1}\mathbf{E}^{1/2}\Big[|R^\varepsilon_{t_{k+1}}-R^\varepsilon_{t_k}|\wedge 1\Big]+\mathbf{E}^{1/2}\Big[\varkappa^{\varepsilon,2,0}_{t_n}(\Pi)-\varkappa^{\varepsilon,2,0}_{t_0}(\Pi)\Big]\Big\}+$$

$$+C_2\Big\{\sum_{k=0}^{n-1}\Big(\mathbf{E}^{1/(2(d+2))}\Big[\Big(\sup_{t_k<u\leqslant t_{k+1}}\rho_{t_k,u}(\Pi^\varepsilon,\tilde\Pi^\varepsilon)\Big)\wedge 1\Big]+$$

$$+\mathbf{E}^{1/2}\Big[\sup_{t_k<u\leqslant t_{k+1}}|\Delta\alpha^\varepsilon_u|\Big]+\mathbf{E}^{1/2}\Big[\sup_{t_k<u\leqslant t_{k+1}}\Pi^\varepsilon\Big(\{u\},\mathsf{E}\Big)\Big]\Big)+$$

$$+\Big[\varkappa^{\varepsilon,\delta,\beta}_{t_n}(\tilde\Pi)-\varkappa^{\varepsilon,\delta,\beta}_{t_0}(\tilde\Pi)\Big]^{1/(1+\delta+\beta)}+\mathbf{E}^{1/(1+\delta+\beta)}\Big[\varkappa^{\varepsilon,\delta,\beta}_{t_n}(\Pi)-\varkappa^{\varepsilon,\delta,\beta}_{t_0}(\Pi)\Big]\Big\},$$

where π *is the Lévy—Prokhorov distance.*

THEOREM 2. *For any* $\varepsilon\in(0,1)$, $l>1$, $T>0$ *there exist constants* $C_1=C_1(d,T)$ *and* $C_2=C_2(d,\varepsilon,\delta,l,T)$ *such that*

a) *for* $\delta\in(0,1)$, $\beta\in(0,1-\delta)$

$$\pi_T(Y,X)\leqslant C_1\Big\{\mathbf{E}^{1/(1+\delta)}\Big[|Y_0-X_0|^\delta\wedge 1\Big]+\mathbf{E}\Big[\delta^l_T(\Pi)\Big]+\delta^l_T(\tilde\Pi)+$$

$$+\mathbf{E}^{1/2}\Big[\Big|\int_0^T\int_{|y|\leqslant\varepsilon}|y|\Pi(dt,dy)\Big|\wedge 1\Big]+\Big[\int_0^T\int_{|y|\leqslant\varepsilon}|y|^{\delta+\beta}\tilde\Pi(dt,dy)\Big]^{1/(1+\delta+\beta)}+$$

$$+\mathbf{E}^{1/(1+\delta+\beta)}\Big[\int_0^T\int_{|y|\leqslant\varepsilon}|y|^{\delta+\beta}\Pi(dt,dy)\Big]\Big\}+C_2\Big\{\mathbf{E}^{1/(7(1+\delta))}\Big[\sup_{t\leqslant T}\Pi^\varepsilon(\{t\},\mathsf{E})\Big]^\delta+$$

$$+\mathbf{E}^{\delta/(7(1+\delta)(d+1+\delta))}\Big[\Big(\sup_{t\leqslant T}\rho_t(\Pi^\varepsilon,\tilde\Pi^\varepsilon)\Big)\wedge 1\Big]\Big\},$$

b) *for* $\delta\in[1,2)$, $\beta\in(0,2-\delta)$

$$\pi_T(Y,X)\leqslant C_1\Big\{\pi^{1/2}(Y_0,X_0)+\mathbf{E}\Big[\delta^l_T(\Pi)\Big]+\delta^l_T(\tilde\Pi)+$$

$$+\mathbf{E}^{1/6}\left[\left|\varkappa_T^{\varepsilon,2,0}(\Pi)\right| \wedge 1\right] + \left[\varkappa_T^{\varepsilon,\delta,\beta}(\tilde{\Pi})\right]^{1/(1+\delta+\beta)} + \mathbf{E}^{1/(1+\delta+\beta)}\left[\varkappa_T^{\varepsilon,\delta,\beta}(\Pi)\right]\Big\}+$$

$$+C_2\Big\{\mathbf{E}^{1/26}\left[\left(\sup_{t\leqslant T}|R_t^\varepsilon|\right)\wedge 1\right] + \mathbf{E}^{1/(26(d+2))}\left[\left(\sup_{t\leqslant T}\rho_t(\Pi^\varepsilon,\tilde{\Pi}^\varepsilon)\right)\wedge 1\right]+$$

$$+\mathbf{E}^{1/26}\left[\sup_{t\leqslant T}|\Delta\alpha_t^\varepsilon|\right]+$$

$$+\mathbf{E}^{1/26}\left[\sup_{t\leqslant T}\Pi^\varepsilon\Big(\{t\},\mathsf{E}\Big)\right]\Big\},$$

where π_T is the Lévy—Prokhorov distance between the distributions of the processes $(Y_t)_{t\leqslant T}$ and $(X_t)_{t\leqslant T}$.

2. MAIN NOTATIONS AND AUXILIARY RESULTS

2.1. *Fourier transform, classes of functions*

The Fourier transform $\mathcal{Z}$ and the inverse Fourier transform $\mathcal{Z}^{-1}$ are defined by

$$\mathcal{Z}f(x) = \int e^{-i(\xi,x)}f(x)dx,$$

$$\mathcal{Z}^{-1}\phi(x) = (2\pi)^{-d}\int e^{i(\xi,x)}\phi(\xi)d\xi.$$

Let $\mathsf{C}_0^\infty(\mathbf{R}^d)$ denote the space of functions $f : \mathbf{R}^d \to \mathbf{R}^d$ with compact support having continuous derivatives of all orders. It is well-known that for $f \in \mathsf{C}_0^\infty(\mathbf{R}^d)$ the generator $A^{(\delta)}$ of the stable process X is given by

$$A^{(\delta)}f(x) = \mathcal{Z}^{-1}\left[\psi^{(\delta)}\mathcal{Z}f\right](x),$$

where

$$A^{(\delta)}f(x) = \mathbf{1}_{\delta=1}\sum_{i=1}^d \gamma^{(i)}\partial_{x_i}f(x) + \mathbf{1}_{\delta=2}\sum_{i,j=1}^d a^{(i,j)}\partial^2_{x_ix_j}f(x)+$$

$$+\mathbf{1}_{\delta\in(0,2)}\int\left[f(x+y)-f(x)-\sum_{i=1}^d\partial_{x_i}f(x)\Big(y^{(i)}\mathbf{1}_{\delta\in(1,2)}+y^{(i)}\mathbf{1}_{\{|y|\leqslant 1\}}\mathbf{1}_{\delta=1}\Big)\right]\frac{dy}{|y|^{d+\delta}}.$$

Fix $T > 0$. For $\delta \in (0,2]$, $u \in \mathsf{C}_0^\infty(\mathsf{H}_T)$, $\mathsf{H}_T = [0,T]\times\mathbf{R}^d$ denote

$$\partial^\delta u(t,x) = \mathcal{Z}^{-1}\left[|\xi|^\delta \mathcal{Z}u(t,\xi)\right](x).$$

As can be easily seen,

$$\partial^\delta u(f,x) = c(\delta)\int\Big[u(t,x+y) - u(t,x)-$$

$$-\sum_{i=1}^{d}\partial_{x_i}u(t,x)\big(y^{(i)}\mathbf{1}_{\{|y|\leqslant 1\}}\mathbf{1}_{\delta=1}+y^{(i)}\mathbf{1}_{\delta\in(1,2)}\big)\Big]\frac{dy}{|y|^{d+\delta}},$$

if $\delta \in (0,2)$, and

$$\partial^2 u(t,x)=\sum_{i,j=1}^{d}\partial_{x_ix_j}u(t,x),\qquad \partial^0 u(t,x)=u(t,x).$$

For $\delta \in (0,2]$, $\beta \in (0,1)$ let us introduce the following norms:

$$|u|_\infty=\sup_{(t,x)\in\mathbf{H}_T}|u(t,x)|,$$

$$|u|_\beta=|u|_\infty+\langle u\rangle_\beta=|u|_\infty+\sup_{(t,x),(t,z)\in\mathbf{H}_T,\ x\neq z}\frac{|u(t,x)-u(t,z)|}{|x-z|^\beta},$$

$$|u|^{\delta,0}=|u|_\infty+|\partial^\delta u|_\infty,\qquad |u|^{\delta,\beta}=|u|_\infty+|\partial^\delta u|_\beta.$$

$\mathbf{C}^\beta(\mathbf{H}_T)$ (resp., $\mathbf{C}^{\delta,0}(\mathbf{H}_T)$, $\mathbf{C}^{\delta,\beta}(\mathbf{H}_T)$) denotes the completion of the space $\mathbf{C}_0^\infty(\mathbf{H}_T)$ with respect to the norm $|\cdot|^\beta$ (resp., $|\cdot|^{\delta,0}$, $|\cdot|^{\delta,\beta}$). Let $\mathbf{C}^\beta(\mathbf{R}^d)$, $\mathbf{C}^{\delta,0}(\mathbf{R}^d)$, $\mathbf{C}^{\delta,\beta}(\mathbf{R}^d)$ denote the spaces of functions on $\mathbf{R}^d$ obtained from the previous definitions by replacing $\mathbf{H}_T$ by $\mathbf{R}^d$.
It is not difficult to prove the following lemmas.

LEMMA 1. *The following norms are equivalent*

$$|u|^{1,\beta}\sim|u|_\infty+|\partial_x u|^\beta,$$

$$|u|^{\delta,\beta}\sim|u|_\infty+|\partial_x u|_\infty+|\partial^{\delta-1}\partial_x u|^\beta\quad for\ \delta\in(1,2).$$

LEMMA 2 (see (Komatsu, 1984)). *Let $\delta \in (0,1)$. There are constants c_1 and c_2 such that*

$$\int\big||x+y|^{\delta-d}-|y|^{\delta-d}\big|dy=c_1|x|^\delta,$$

$$f(x+z)-f(x)=c_2\int\Big(|y+z|^{\delta-d}-|y|^{\delta-d}\Big)\partial^\delta f(x-y)dy$$

for each $f\in\mathbf{C}^{\delta,\beta}(\mathbf{R}^d)$.

LEMMA 3. *Let $\delta\in(0,1)$, $\gamma\in(0,1-\delta)$. There is a constant c such that*

$$|f(x+z)-f(x)|\leqslant c\langle\partial^\delta f\rangle_\gamma\,|z|^{\delta+\gamma}$$

for each $f\in\mathbf{C}^{\delta,\gamma}(\mathbf{R}^d)$.

2.2. *The Lévy—Prokhorov metric*

Let $\mathfrak{X}$ be a Polish space with a metric d and the Borel σ-algebra $\mathcal{B}(\mathfrak{X})$. The Lévy—Prokhorov metric $\pi(\lambda,\mu)$ for any non-negative bounded measures on $(\mathfrak{X},\mathcal{B}(\mathfrak{X}))$ is defined by the following relation:

$$\pi(\lambda,\mu)=\inf\{\varepsilon:\lambda(B)\leqslant\mu(B^\varepsilon)+\varepsilon\quad\text{for all closed } B\in\mathcal{B}(\mathfrak{X})\},$$

where $B^\varepsilon=\{x:d(x,B)<\varepsilon\}$.
Now we examine the case $\mathfrak{X}=D$, where $D=D_{[0,T]}(\mathbf{R}^d)$ is a space of cádlàg functions $x:[0,T]\to\mathbf{R}^d$ endowed with Skorokhod's J_1 topology. Let $n\in\mathbf{N}$. We define a mapping $f^n:D\to D$ by $f_t^n(x)=x_{t_i}$, if $t_i\leqslant t<t_{i+1}$, $t_i=iT/n$ $(i=0,\dots,n)$, $x\in D$.
Denote

$$I_{t_0,\dots,t_n}:x\to(x_{t_0},\dots,x_{t_n}),$$

$$w'_T(x,\delta)=\inf\{\max_{1\leqslant i\leqslant r}w(x,[s_{i-1},s_i)):0=s_0<\dots<s_r=T,\inf_{i\leqslant r-1}(s_i-s_{i-1})\geqslant\delta\},$$

$$w(x,[s_{i-1},s_i))=\sup\{|x(s)-x(t)|:s,t\in[s_{i-1},s_i)\}.$$

LEMMA 4 (see (Kubilius, 1990)). *If λ,μ are probability measures on D, then for any $\varepsilon>0$, $n\in\mathbf{N}$*

$$\pi_T(\lambda,f^n(\lambda))\leqslant\varepsilon\vee\lambda(4w'_T(x,2T/n)\geqslant\varepsilon-T/n),$$

$$\pi_T(f^n(\lambda),f^n(\mu))\leqslant\pi\big(I_{t_0,\dots,t_n}(\lambda),I_{t_0,\dots,t_n}(\mu)\big).$$

LEMMA 5. *Let ν_1,ν_2 be probability measures on $\mathbf{R}^d$, c_i $(i=0,\dots,3)$ be non-negative numbers, $\beta\in(0,1]$, and let for every $f\in C_0^\infty(\mathbf{R}^d)$, $f\geqslant0$, the following relations be true:*

$$|\nu_1(f)-\nu_2(f)|\leqslant c_0|f|_\infty+c_1|\partial^\delta f|_\infty+c_2\langle\partial^\delta f\rangle_\beta\quad\text{for }\delta\in(0,1),$$

$$|\nu_1(f)-\nu_2(f)|\leqslant c_0|f|_\infty+c_1|\partial_x f|_\infty+c_2\langle\partial_x f\rangle_\beta,$$

$$|\nu_1(f)-\nu_2(f)|\leqslant c_0|f|_\infty+c_1|\partial_x f|_\infty+c_2|\partial^{\delta-1}\partial_x f|_\infty+c_3\langle\partial^{\delta-1}\partial_x f\rangle_\beta\quad\text{for }\delta\in(1,2).$$

Then for some constants K_i, $1\leqslant i\leqslant2$, which do not depend on ν_1,ν_2,c_i $(i=0,\dots,3)$, β, the inequalities

$$\pi(\nu_1,\nu_2,)\leqslant K_1(c_0+c_1^{1/(1+\delta)}+c_2^{1/(1+\delta+\beta)})\quad\text{for }\delta\in(0,1],$$

$$\pi(\nu_1,\nu_2,)\leqslant K_2(c_0+c_1^{1/2}+c_2^{1/(1+\delta)}+c_3^{1/(1+\delta+\beta)})\quad\text{for }\delta\in(1,2)$$

hold.

3. ESTIMATION OF THE DISTANCE BETWEEN FINITE DIMENSIONAL DISTRIBUTIONS

Let $f \in \mathbf{C}^{\delta,\beta}(\mathbf{R}^d)$, and
$$u(s,x) = \mathbf{E}f(x + X_t - X_s),$$
for $s \leqslant t$, where $0 \leqslant t \leqslant T$ are fixed. Then it is not difficult to show that

$$(\partial_s + A^{(\delta)})u(s,x) = 0, \ \ u(t,x) = f(x).$$

Moreover, $u \in \mathbf{C}^{\delta,\beta}(\mathbf{H}_t)$ and there exists a constant c such that

$$|\partial_s u|_\infty \leqslant c.$$

Let
$$Q_{s,x}(A) = \mathbf{P}(X_t - X_s + x \in A),$$
where $A \in \sigma\{u \geqslant s : x_u\}$. Then

$$u(s,x) = Q_{s,x} f(\hat{X}_t)$$

for the canonical process $\hat{X} = (\hat{X}_t)_{0 \leqslant t \leqslant T}$, i.e., $\hat{X}_t = \hat{X}_t(w) = w_t, \ w \in D$.

LEMMA 6. *Suppose that $f(x) = f(\omega, x)$ is $\mathcal{F}_s \otimes \mathcal{B}(\mathbf{R}^d)$ measurable function on $\Omega \times \mathbf{R}^d$ such that $\sup_\omega |f(\omega,\cdot)|^{\delta,\beta} < \infty$ for $\delta \in (0,1)$, $\beta \in (0, 1-\delta)$, or $\delta \in [1,2)$, $\beta \in (0, 2-\delta)$. Then for any $\varepsilon \in (0,1]$, $l > 1$ there exist constants $C_1 = C_1(d)$ and $C_2 = C_2(d,\varepsilon,l)$ such that for any $s < t$ and $Y \in S^\delta$*

a) *for $\delta \in (0,1)$*

$$|\mathbf{E}f(Y_t) - \mathbf{E}Q_{s,Y_s} f(\hat{X}_t)| \leqslant C_1 \Big\{ \sup_\omega |f(\omega,\cdot)|_\infty \Big(\mathbf{E}\Big[|\delta_t^l(\Pi) - \delta_s^l(\Pi)| \wedge 1\Big] +$$
$$+ \Big[|\delta_t^l(\tilde{\Pi}) - \delta_s^l(\tilde{\Pi})| \wedge 1\Big] \Big) + \sup_\omega \langle \partial^\delta f(\omega,\cdot)\rangle_\beta \Big(\mathbf{E}\Big[\Big| \int_s^t \int_{|y| \leqslant \varepsilon} |y|^{\delta+\beta} \Pi(du,dy) \Big| \wedge 1\Big] +$$
$$+ \Big[\Big| \int_s^t \int_{|y| \leqslant \varepsilon} |y|^{\delta+\beta} \tilde{\Pi}(du,dy) \Big| \wedge 1\Big] \Big) \Big\} + C_2 \sup_\omega |f(\omega,\cdot)|^{\delta,0} \Big\{ \mathbf{E}\Big[\sup_{s<u\leqslant t} \Pi^\varepsilon\Big(\{u\}, \mathrm{E}\Big)\Big]^\delta +$$
$$+ \mathbf{E}^{\delta/(d+1+\delta)} \Big[\Big(\sup_{s<u\leqslant t} \rho_{s,u}(\Pi^\varepsilon, \tilde{\Pi}^\varepsilon) \Big) \wedge 1 \Big] \Big\},$$

b) *for $\delta \in [1,2)$*

$$|\mathbf{E}f(Y_t) - \mathbf{E}Q_{s,Y_s} f(\hat{X}_t)| \leqslant C_1 \Big\{ \sup_\omega |f(\omega,\cdot)|_\infty \Big(\mathbf{E}\Big[|\delta_t^l(\Pi) - \delta_s^l(\Pi)| \wedge 1\Big] +$$

$$+\Big[|\delta_t^l(\tilde\Pi)-\delta_s^l(\tilde\Pi)|\wedge 1\Big]\Big)+\sup_\omega|\partial_x f(\omega,\cdot)|_\infty\Big(\mathbf{E}\Big[|R_t^\varepsilon-R_s^\varepsilon|\wedge 1\Big]+$$

$$+\mathbf{E}\Big[|\varkappa_t^{\varepsilon,2,0}(\Pi)-\varkappa_s^{\varepsilon,2,0}(\Pi)|\wedge 1\Big]\Big)\Big\}+C_2\Big\{\sup_\omega|f(\omega,\cdot)|^{1,0}\Big(\mathbf{E}\Big[\sup_{s<u\leqslant t}|\Delta\alpha_u^\varepsilon|\Big]+$$

$$+\mathbf{E}\Big[\sup_{s<u\leqslant t}\Pi^\varepsilon(\{u\},\mathrm{E})\Big]+\mathbf{E}^{1/(d+2)}\Big[\Big(\sup_{s<u\leqslant t}\rho_{s,u}(\Pi^\varepsilon,\tilde\Pi^\varepsilon)\Big)\wedge 1\Big]\Big)+$$

$$+\sup_\omega\langle\partial^{\delta-1}\partial_x f(\omega,\cdot)\rangle_\beta\Big(\Big[|\varkappa_t^{\varepsilon,\delta,\beta}(\tilde\Pi)-\varkappa_s^{\varepsilon,\delta,\beta}(\tilde\Pi)|\wedge 1\Big]+$$

$$+\mathbf{E}\Big[|\varkappa_t^{\varepsilon,\delta,\beta}(\Pi)-\varkappa_s^{\varepsilon,\delta,\beta}(\Pi)|\wedge 1\Big]\Big)\Big\}.$$

Proof. For any $s<t$ denote

$$\bar Y_t=Y_s+(Y_t-Y_s)-(\alpha_t^\varepsilon-\alpha_s^\varepsilon)\mathbf{1}_{\delta\in(0,1)}+(\tilde\alpha_t^\varepsilon-\tilde\alpha_s^\varepsilon)\mathbf{1}_{\delta\in(0,1)}.$$

Let

$$G(u,x)=G(\omega,u,x)=Q_{u,x}f(\hat X_t).$$

Then for any $\omega\in\Omega$

$$(\partial_u+A^{(\delta)})G(\omega,u,x)=0,\ \ G(\omega,t,x)=f(\omega,x).$$

It is clear that

$$\mathbf{E}f(Y_t)-\mathbf{E}Q_{s,Y_s}f(\hat X_t)=\mathbf{E}[f(Y_t)-f(\bar Y_t)]+\mathbf{E}[G(t,\bar Y_t)-G(s,\bar Y_s)]=I_1+I_2.$$

Note that $I_1=0$ for $\delta\in(0,1)$. For $\delta\in[1,2)$ the inequality

$$|I_1|\leqslant 2\sup_\omega|f(\omega,\cdot)|^{1,0}\mathbf{E}[|Y_t-\bar Y_t|\wedge 1]=2\sup_\omega|f(\omega,\cdot)|^{1,0}\mathbf{E}[|R_t^\varepsilon-R_s^\varepsilon|\wedge 1]$$

holds.
For the estimation of I_2 we'll introduce some new notation.
Let N be a Poisson process with compensator t. Let $(N^i)_{1\leqslant i\leqslant p}$ be independent copies of N. Let

$$g\in C_0^\infty(\mathbf{R}^d),\ \operatorname{supp} g\subset\{|x|\leqslant 1\},\ g\geqslant 0,\ \int g(x)dx=1.$$

Fix $0<\sigma<(\varepsilon/3)\wedge\Lambda$, where

$$\Lambda=\max\Big\{1,\int_{|x|\leqslant 1}|\partial_x g(x)|dx\Big\}.$$

Divide the set

$$\Big\{x:-\Big([l/\sigma]+2\Big)\sigma\leqslant x^{(i)}\leqslant\Big([l/\sigma]+2\Big)\sigma,\ 1\leqslant i\leqslant d\Big\}$$

into d-dimensional cubes with the edge's length σ. Let (E_k), $1 \leqslant k \leqslant p$, be the smallest family of such cubes such that

$$\{x : \varepsilon - \sigma < |x| \leqslant l + \sigma\} = E^\sigma \subset \bigcup_{k=1}^{p} E_k.$$

It is clear that

$$p < 2^d([l/\sigma] + 2)^d.$$

Denote by λ_k an arbitrary point in E_k, $1 \leqslant k \leqslant p$. Let $z = (z^{(1)}, \dots, z^{(p)}) \in \mathbf{R}^p_+$, $V(z) = (V_z^{(1)}, \dots, V_z^{(d)})$, $V_z^{(i)} = V^{(i)}(z^{(1)}, \dots, z^{(p)})$,

$$V_z^{(i)} = \sum_{k=1}^{p} \lambda_k^{(i)} N_{z_k}^k.$$

Define the function

$$F(u, x, z) = F(\omega, u, x, z) = \mathbf{E}(G(u, x + V(z))|\mathcal{F}_s).$$

Let

$$\varphi_k(x) = \mathbf{1}_{\{E_k\}} * g_\sigma(x),\ 1 \leqslant k \leqslant p,$$

where $g_\sigma(x) = \sigma^{-d} g(x/\sigma)$. Then

$$\sum_{k=1}^{p} \varphi_k(x) = \mathbf{1}_{\{\bigcup_{k=1}^{p} E_k\}} * g_\sigma(x)$$

and

$$\mathbf{1}_{\{\bigcup_{k=1}^{p} E_k\}} * g_\sigma(x) = 1 \quad \text{for } \varepsilon < |x| \leqslant l$$

since $\bigcup_{k=1}^{p} E_k \supset E^\sigma$. So

$$\sum_{k=1}^{p} \varphi_k(x) \leqslant 1 \quad \text{for } x \in \mathbf{R}^d \quad \text{and} \ \sum_{k=1}^{p} \varphi_k(x) = 1 \quad \text{for } \varepsilon < |x| \leqslant l.$$

Denote

$$\hat{B}_u^{(i)s} = (\tilde{\Pi}_u^\varepsilon(\varphi_i) - \Pi_u^\varepsilon(\varphi_i)) - (\tilde{\Pi}_s^\varepsilon(\varphi_i) - \Pi_s^\varepsilon(\varphi_i)),\ 1 \leqslant i \leqslant p,$$

$$B_u^{(i)s} = \sup_{s \leqslant z \leqslant u} |\hat{B}_z^{(i)s}| + \hat{B}_u^{(i)s},\ B^s = (B^{(i)s})_{1 \leqslant i \leqslant p},$$

where the functions (φ_i) are defined above.
Now we estimate I_2. It is clear that

$$|I_2| = |\mathbf{E}F(t, \bar{Y}_t, 0) - \mathbf{E}F(s, \bar{Y}_s, B_s^s)| \leqslant |\mathbf{E}F(t, \bar{Y}_t, 0) - \mathbf{E}F(t, \bar{Y}_t, B_t^s)| +$$

$$+|\mathbf{E}F(t,\bar{Y}_t,B_t^s)-\mathbf{E}F(s,\bar{Y}_s,B_s^s)|=I_{21}+I_{22}.$$

So

$$I_{21}\leqslant\sum_{k=1}^{p}(2\sup_{\omega}|F(\omega,\cdot,\cdot,\cdot)|_\infty+\sup_{\omega}|\partial_{z_k}F(\omega,\cdot,\cdot,\cdot)|_\infty)\mathbf{E}[|B_t^{(k)s}|\wedge 1]\leqslant$$

$$\leqslant 8p\sup_{\omega}|f(\omega,\cdot)|_\infty\max(1,\Lambda\sigma^{-1})\mathbf{E}\Big[\Big(\sup_{s<u\leqslant t}\rho_{s,u}(\Pi^\varepsilon,\tilde{\Pi}^\varepsilon)\Big)\wedge 1\Big].$$

By Ito's formula,

$$F(t,Y_t,B_t^s)-F(s,Y_s,B_s^s)=\Bigg\{\int_s^t\partial_uF(u,Y_{u-},B_{u-}^s)du+$$

$$+\sum_{k=1}^{p}\int_s^t\partial_{z_k}F(u,Y_{u-},B_{u-}^s)d\tilde{\Pi}_u^\varepsilon(\varphi_k)\Bigg\}+\sum_{k=1}^{p}\int_s^t\partial_{z_k}F(u,Y_{u-},B_{u-}^s)d(B_u^{(k)s}-\hat{B}_u^{(k)s})+$$

$$+\sum_{s<u\leqslant t}\Big[F(u,Y_u,B_u^s)-F(u,Y_{u-},B_{u-}^s)-\sum_{k=1}^{p}\int_s^t\partial_{z_k}F(u,Y_{u-},B_{u-}^s)\Delta B_u^{(k)s}\Big].$$

Now it remains to estimate the summands. This can be done in the same manner as in (Kubilius, 1990).
The proof of the case $\delta\in[1,2)$ is similar to the case $\delta\in(0,1)$.

Proof of Theorem 1. Let $\delta\in(0,1)$. Let f be a function on $\mathbf{R}^{d(n+1)}$ such that $|f|^{\delta,\beta}<\infty$ for $\beta\in(0,1-\delta)$. Then there exist constants $C_1=C_1(d)$ and $C_2=C_2(d,\varepsilon,l)$ such that

$$|\mathbf{E}f(Y_{t_0},\dots,Y_{t_n})-\mathbf{E}f(X_{t_0},\dots,X_{t_n})|\leqslant C_1\Bigg\{|f|_\infty\Big(\mathbf{E}[\delta_{t_n}^l(\Pi)-\delta_{t_0}^l(\Pi)]+$$

$$+[\delta_{t_n}^l(\tilde{\Pi})-\delta_{t_0}^l(\tilde{\Pi})]\Big)+\langle\partial^\delta f\rangle_\beta\Big(\mathbf{E}\Big[\int_{t_0}^{t_n}\int_{|y|\leqslant\varepsilon}|y|^{\delta+\beta}\Pi(du,dy)\Big]+$$

$$+\int_{t_0}^{t_n}\int_{|y|\leqslant\varepsilon}|y|^{\delta+\beta}\tilde{\Pi}(du,dy)\Big)\Bigg\}+C_2|f|^{\delta,0}\sum_{k=1}^{n-1}\Bigg\{\mathbf{E}\Big[\sup_{t_k<u\leqslant t_{k+1}}\Pi^\varepsilon\Big(\{u\},\mathrm{E}\Big)\Big]^\delta+$$

$$+\mathbf{E}^{\delta/(d+1+\delta)}\Big[\Big(\sup_{t_k<u\leqslant t_{k+1}}\rho_{t_k,u}(\Pi^\varepsilon,\tilde{\Pi}^\varepsilon)\Big)\wedge 1\Big]\Bigg\}.$$

This estimate can be proved in a similar way as in (Kubilius, 1990). From it and Lemma 5 one can get the estimate of Theorem 1 in the case $\delta\in(0,1)$.
The proof in the case $\delta\in[1,2)$ is similar to the case $\delta\in(0,1)$.

4. PROOF OF THEOREM 2

For the proof of Theorem 2 we need some estimates of $\pi(Y, f^n(Y))$ and $\pi(X, f^n(X))$, which are derived below.
Let $\hat{S}_0(x) = 0$,

$$\hat{S}_{k+1}(x) = \begin{cases} \inf\{t > \hat{S}_k(x) : |x_t - x_{\hat{S}_k(x)}| \geqslant \eta\}, \\ T, \text{ if } \{\cdot\} = \varnothing, \end{cases}$$

where $x \in D$. Now for any càdlàg process Z and any $q \in \mathbb{N}$, $\theta > 0$, $\eta > 0$, $\varkappa > 0$ one can get

$$\mathbf{P}(\omega'_T(Z,\theta) > 2\eta) \leqslant (\varkappa/(\varkappa q - T)) \sum_{k=1}^{q} \mathbf{P}(S_k \leqslant T,\ S_k - S_{k-1} \leqslant \varkappa) +$$

$$+ \sum_{k=1}^{q} \mathbf{P}(S_k \leqslant T,\ S_k - S_{k-1} \leqslant \theta), \tag{1}$$

where $S_k = \hat{S}_k(Z)$. One can obtain this inequality from the proof of the Aldous criterion for tightness (see (Jacod, 1985)).

LEMMA 7. *For any $\varepsilon \in (0,1]$, there exist constants $C_1 = C_1(T)$ and $C_2 = C_2(\varepsilon, \delta, T)$ such that*

a) *for $\delta \in (0,1)$*

$$\pi_T(Y, f^n(Y)) \leqslant 4\mathbf{E}^{1/2}\Big[\Big|\int_0^T \int_{|y|\leqslant\varepsilon} |y|\Pi(ds,dy)\Big| \wedge 1\Big] +$$

$$+ C_1\Big\{\mathbf{E}^{1/4}\Big[\Big(\sup_{t\leqslant T} \rho_t(\Pi^\varepsilon,\ \tilde{\Pi}^\varepsilon)\Big) \wedge 1\Big] + n^{-1/2}\Big\},$$

$$\pi_T(X, f^n(X)) \leqslant C_1 n^{-1/6},$$

b) *for $\delta \in [1,2)$*

$$\pi_T(Y, f^n(Y)) \leqslant 4\mathbf{E}^{1/3}\Big[|\varkappa_T^{\varepsilon,2,0}(\Pi)| \wedge 1\Big] +$$

$$+ C_1\Big\{\mathbf{E}^{1/4}\Big[\Big(\sup_{t\leqslant T}|R_t^\varepsilon|\Big) \wedge 1\Big] + \mathbf{E}^{1/4}\Big[\Big(\sup_{t\leqslant T} \rho_t(\Pi^\varepsilon,\ \tilde{\Pi}^\varepsilon)\Big) \wedge 1\Big] + n^{-1/2}\Big\},$$

$$\pi_T(X, f^n(X)) \leqslant C_1 n^{-1/12}.$$

Proof of the Lemma follows from Lemma 4 and the inequality (1).
Proof of Theorem 2 follows immediately from Theorem 1 and Lemma 7.

REFERENCES

Jacod, J. (1985). Theoremes limites pour les processus. *Lect. Notes in Math.* **1117.** Springer, Berlin.

Jacod, J. and Mano, P. (1988). Une evaluation de la distance entre les lois d'une semimartingale et d'un processus à accroissements indépendants. *Stochastics* **25**, 87–124.

Komatsu, T. (1984). On the martingale problem for generators of stable processes with perturbations. *Osaka J. Math.* **21**, 113–132.

Kubilius, K. (1990). On the rate of convergence of the distributions of semimartingales to the distribution of diffusion process with the jumps. *Liet. Matem. Rink.* **30.** (to appear).

Kubilius, K. and Mikulevičius, R. (1987). On the rate of convergence of distributions of semimartingales. In: *Proc. Fourth Vilnius Conference,* **2**, VNU Sci. Press, Utrecht. pp. 139–159.

Prob. Theory and Math. Stat., Vol. 2, pp. 35–46
B.Grigelionis *et al.* (Eds.)
1990 VSP/Mokslas

ANTICIPATING STOCHASTIC DIFFERENTIAL EQUATIONS

H.-H. KUO and J. POTTHOFF
Department of Mathematics, Louisiana State University,
Baton Rouge, Louisiana 70803, USA

ABSTRACT

An anticipating stochastic integral equation (1) is considered. Using the operator ∂_s^* in white noise calculus it is rewritten in the form (2). The existence and uniqueness of a solution of (2) under certain conditions is showed utilizing the U–functional method in white noise calculus and the iteration method (the contraction principle cannot be applied). Several examples show that the U–functional provides a direct way to solve (2). Some possible extensions are also discussed.

1. INTRODUCTION AND APPROACH

Since 1972 several extensions of Itô's integral and Itô's lemma to anticipating integrands have been obtained with various methods by many authors. However, only quite recently, some results about anticipating stochastic differential equations of special types have appeared (Buckdahn, 1987; Ocone and Pardoux, 1988; Pardoux and Protter, 1988; Shiota, 1986). The general case of such equations seems to be very hard to handle. In this paper, we will introduce an approach to study the general case. This approach comes originally from Hida's idea of using the U–functional to solve a linear stochastic partial differential equation (Hida, 1983). For the clarity of exposition, we will consider in this paper a very simple form of anticipating stochastic integral equations, i.e.,

$$X_t = x + \int_0^t f(s, X_s, \cdot)\delta B(s), \tag{1}$$

where f may be anticipating and the integral is the stochastic integral introduced in (Hitsuda, 1973; King (a student of K. Itô), 1975; Skorokhod, 1975). As it turns out, this integral can be expressed in terms of the operator ∂_s^* in white noise calculus (Hida, 1975; 1985; Kubo and Takenaka, 1980; Kuo, 1983; 1988a; 1988b) as follows

$$\int_0^t f(s, X_s, \cdot)\delta B(s) = \int_0^t \partial_s^* f(s, X_s, \cdot)ds.$$

Thus, equation (1) can be rewritten in the following form

$$X_t = x + \int_0^t \partial_s^* f(s, X_s, \cdot) ds. \tag{2}$$

If one tries the classical L^2-theory to solve equation (2), then terms involving the unbounded white noise differentiation ∂_s appear (Kuo, 1975, Theorem 3) so that neither the contraction principle nor the iteration method will work. Our approach to show the existence and uniqueness of a solution of (2) under certain conditions is to utilize the U-functional method in white noise calculus. It should be pointed out that in this approach the iteration method works quite well, while the contraction principle can not be applied. Moreover, the U-functional provides a direct way to solve (2), as to be seen from the examples in Section 4. We remark also that our approach can be used to solve more general anticipating stochastic integral equations than (2). Some possible extensions will be mentioned in Section 5.

2. WHITE NOISE CALCULUS

In this section, we give a very brief description of white noise calculus which will be sufficient for this paper. We will use the space $(S)^*$ of generalized Brownian functionals which is a special choice in the class of spaces introduced in (Kubo and Takenaka, 1980) (cf. also (Hida et al., 1988)). Let S be the Schwartz space of rapidly decreasing functions on $\mathbf{R}$. The dual space of S is the space S^* of tempered distributions on $\mathbf{R}$. Let μ be the standard Gaussian measure on S^*, i.e., its characteristic function is given by

$$\int_{S^*} e^{i\langle x, \xi \rangle} d\mu(x) = \exp[-|\xi|^2/2], \quad \xi \in S,$$

where $|\cdot|$ is the $L^2(\mathbf{R})$ norm. Then $B(t,x) = \langle x, \mathbf{1}_{(0,t]} \rangle$, $t \geqslant 0$, $x \in S^*$ a.s., is a Brownian motion. Thus (S^*, μ) can be regarded as a sample space of white noise. By the well-known Wiener–Itô theorem (Itô, 1951), the space (L^2) of μ-square integrable random variables on S^* has the following orthogonal decomposition

$$(L^2) = \bigoplus_{n=0}^{\infty} K_n,$$

where $K_0 = \mathbf{R}$ and K_n is the space of n-fold Wiener integrals $I_n(f)$, $f \in \widehat{L^2}(\mathbf{R}^n)$, with respect to the Brownian motion $B(t,x)$.

Let $A = -d^2/dx^2 + x^2 + 1$. For each real number p, let S_p be the completion of S with respect to the following norm

$$|f|_p = |A^p f|. \tag{3}$$

Note that for $p > 0$, $S_p \subset L^2(\mathbf{R}) \subset S_{-p}$ and $S_{-p} = S_p^*$. Let $K_{n,p}$ be the space of n-fold Wiener integrals $I_n(f)$, $f \in \widehat{\otimes}^n S_p$, with respect to the Brownian motion $B(t,x)$. Let

$$(S)_p = \bigoplus_{n=0}^{\infty} K_{n,p}$$

be the orthogonal direct sum of $K_{n,p}$, $n = 0, 1, 2, \ldots$, with the following norm

$$\left\| \sum_{n=0}^{\infty} I_n(f_n) \right\|_p = \left\{ \sum_{n=0}^{\infty} n! |f_n|^2_{\hat{\otimes}^n S_p} \right\}^{1/2}. \tag{4}$$

Note that for $p > 0$, $K_{n,p} \subset K_n \subset K_{n,-p}$ and $K_{n,-p} = K_{n,p}^*$ is the space of n-fold generalized multiple Wiener integrals (Hida, 1978). Hence for $p > 0$, $(S)_{-p} = (S)_p^*$ and, by (3) and (4), we have

$$(S)_p \subset (L^2) \subset (S)_p^*.$$

Let (S) be the projective limit of $\{(S)_p;\ p \in \mathbb{N}\}$ and $(S)^*$ the inductive limit of $\{(S)_p^*;\ p \in \mathbb{N}\}$. Then we have the following Gel'fand triple

$$(S) \subset (L^2) \subset (S)^*. \tag{5}$$

Both (S) and $(S)^*$ are nuclear spaces. We will call (S) the space of test functionals and $(S)^*$ the space of generalized Brownian functionals.

For $\varphi \in (L^2)$, define its S-transform by

$$(S\varphi)(\xi) = \int_{S^*} \varphi(x + \xi) d\mu(x), \quad \xi \in S.$$

This S-transform was introduced in (Kubo and Takenaka, 1980) for white noise analysis. It was studied earlier in several different contexts, e.g. (Gross, 1967; Hida and Ikeda, 1965; Krée, 1974; Krée and Raczka, 1978; Kuo, 1975; Wiener, 1930). By the translation formula for the Gaussian measure μ, $S\varphi$ can be rewritten as follows:

$$(S\varphi)(\xi) = e^{-|\xi|^2/2} \int_{S^*} e^{\langle x, \xi \rangle} \varphi(x) d\mu(x), \quad \xi \in S.$$

Note that for each $\xi \in S$, the function $e^{\langle \cdot, \xi \rangle}$ is in (S). Thus the S-transform can be extended to $(S)^*$ by using the pairing $\ll \cdot, \cdot \gg$ of $(S)^*$ and (S). This extension is called the U-functional, i.e., we make the following

Definition 1. For Φ in $(S)^*$, its U-functional is defined to be the function

$$U[\Phi](\xi) = e^{-|\xi|^2/2} \ll \Phi, e^{\langle \cdot, \xi \rangle} \gg, \quad \xi \in S.$$

It can be checked that the function $G(\xi) = \xi(t) U[\Phi](\xi)$, $\xi \in S$, is also a U-functional, i.e., there exists an element in $(S)^*$ whose U-functional is given by G. Therefore, the following definition for the operator ∂_t^* acting on $(S)^*$ makes sense.

Definition 2. For $\Phi \in (S)^*$, $\partial_t^* \Phi$ is defined to be the generalized Brownian functional in $(S)^*$ with U-functional given by

$$U[\partial_t^* \Phi](\xi) = \xi(t) U[\Phi](\xi), \quad \xi \in S.$$

It turns out that ∂_t^* is the adjoint operator of the white noise differentiation ∂_t (Hida, 1983; 1985; Kubo and Takenaka, 1980; Kuo, 1983; 1988b). Moreover, if $f(t,x)$, $t \in [0,1]$, $x \in S^*$, is nonanticipating with $\mathbf{E}\int_0^1 |f(t,\cdot)|^2 dt < \infty$, then

$$\int_0^1 f(t)dB(t) = \int_0^1 \partial_t^* f(t)dt$$

(Kubo and Takenaka, 1980).

For some of the examples in Section 4, we need the renormalized multiplication (Hida and Ikeda, 1965) in $(S)^*$. It can be checked that the product of two U–functionals is also a U–functional (Kubo and Takenaka, 1980). Thus we can make the following

Definition 3. For Φ and Ψ in $(S)^*$, the renormalized (or Wick–) product $\Phi \diamond \Psi$ of Φ and Ψ is defined to be the generalized Brownian functional in $(S)^*$ with U–functional given by

$$U[\Phi \diamond \Phi](\xi) = U[\Phi](\xi)U[\Psi](\xi), \quad \xi \in S.$$

In fact, it is this renormalized multiplication that suggests the conditions for the existence and uniqueness of a solution of equation (2).

3. MAIN THEOREM

In this section, we state a very simple version of theorems concerning equation (2) and sketch its proof. The details of the proof and some extensions to more general equations will appear elsewhere.

THEOREM. *Suppose $f : [0,1] \times (S)^* \longrightarrow (S)^*$ satisfies the following conditions:*

1) *For any $t \in [0,1]$, $x \in \mathbf{R}$, and $\xi \in S$,*

$$\int_0^1 \Big| U[f(t,x)](\xi) \Big| dt < \infty.$$

2) *For each $\Phi \in (S)^*$, there exists $p \geqslant 1$ such that $f(t,\Phi) \in (S)_p^*$ for all $t \in [0,1]$.*

3) *There exists a constant $c > 0$ such that for any Φ, $\Psi \in (S)^*$, $\xi \in S$, and $t \in [0,1]$,*

$$\big|U[f(t,\Phi)](\xi) - U[f(t,\Psi)](\xi)\big| \leqslant c\big|U[\Phi](\xi) - U[\Psi](\xi)\big|.$$

Then the stochastic integral equation

$$X_t = x + \int_0^t \partial_s^* f(s, X_s)ds$$

has a unique solution X_t in $(S)^*$.

Remark. It is possible to state a Lipschitz condition without referring to the U–functional. For instance, in view of Definition 1, the following condition implies condition 3) above: There exists a constant $c > 0$ such that for any $\Phi, \Psi \in (S)^*$, $\varphi \in (S)$, and $t \in [0,1]$,

$$| \ll f(t,\Phi) - f(t,\Psi), \varphi \gg | \leqslant c \, | \ll \Phi \dot{-} \Psi, \varphi \gg |.$$

Proof. The uniqueness of a solution is easy to show. Suppose Y_t is another solution. Let F_t and G_t be the U–functionals of X_t and Y_t, respectively. Then, by Definition 2 and the Lipschitz condition 3), we have

$$|F_t(\xi) - G_t(\xi)| \leqslant c|\xi|_\infty \int_0^t |F_s(\xi) - G_s(\xi)| ds,$$

where $|\xi|_\infty = \sup\{|\xi(t)|;\ 0 \leqslant t \leqslant 1\}$. It follows from the Gronwall inequality that $F_t(\xi) - G_t(\xi) = 0$ for all $t \in [0,1]$ and $\xi \in S$. Thus $X_t = Y_t$.
Now, we sketch a proof for the existence of a solution. We use Picard's iteration method, i.e., define

$$X_t^{(0)} = x,$$
$$X_t^{(n+1)} = x + \int_0^t \partial_s^* f(s, X_s^{(n)}) ds, \quad n \geqslant 0.$$

Let $F_t^{(n)}$ be the U–functional of $X_t^{(n)}$. Then

$$F_t^{(0)} = x,$$
$$F_t^{(n+1)}(\xi) = x + \int_0^t \xi(s) U[f(s, X_s^{(n)})](\xi) ds. \tag{6}$$

By using the Lipschitz condition 3), we get

$$\left|F_t^{(n+1)}(\xi) - F_t^{(n)}(\xi)\right| \leqslant c|\xi|_\infty \int_0^t \left|F_s^{(n)}(\xi) - F_s^{(n-1)}(\xi)\right| ds.$$

Therefore, for $t \in [0,1]$ and $\xi \in S$,

$$\left|F_t^{(n+1)}(\xi) - F_t^{(n)}(\xi)\right| \leqslant (n!)^{-1} c^n t^n |\xi|_\infty^{n+1} \int_0^1 \left|U[f(s,x)](\xi)\right| ds. \tag{7}$$

Hence

$$\left|F^{(n+1)}(\xi) - F^{(n)}(\xi)\right|_{\infty} \leqslant (n!)^{-1} c^{n} |\xi|_{\infty}^{n+1} \int_{0}^{1} \left|U[f(s,x)](\xi)\right| ds,$$

where $|\cdot|_{\infty}$ is the sup-norm in the t-variable. Thus, for each $\xi \in S$, $\{F^{n}(\xi); n \geqslant 0\}$ converges uniformly on $[0,1]$. Let

$$F_t(\xi) = \lim_{n\to\infty} F_t^{(n)}(\xi).$$

Note that

$$F_t^{(n)}(\xi) = e^{-|\xi|^2/2} \ll X_t^{(n)}, e^{\langle \cdot,\xi\rangle} \gg .$$

Therefore,

$$\lim_{n\to\infty} \ll X_t^{(n)}, e^{\langle \cdot,\xi\rangle} \gg = F_t(\xi) e^{|\xi|^2/2}. \tag{8}$$

Note that the collection $\{e^{\langle \cdot,\xi\rangle};\ \xi \in S\}$ spans a dense subspace of (S). Hence, the sequence $\{X_t^{(n)}; n \geqslant 0\}$ converges on a dense subset of (S). On the other hand, it follows from (7) and condition 2) that

$$|F_t^{(n)}(\xi)| \leqslant c_1 \exp(c_2 |\xi|_p), \tag{9}$$

where $|\cdot|_p$ is the norm given in (3). The inequality (9) implies that there exist positive constants q and L such that for all t and n,

$$\|X_t^{(n)}\|_{-q} \leqslant L.$$

Therefore, $\{X_t^{(n)};\ n \geqslant 0\}$ converges weakly in $(S)^*$ to a limit, say X_t, i.e.,

$$\lim_{n\to\infty} \ll X_t^{(n)}, \varphi \gg = \ll X_t, \varphi \gg \quad \text{for all} \quad \varphi \in (S).$$

Thus, by (8), we have

$$\ll X_t, e^{\langle \cdot,\xi\rangle} \gg = F_t(\xi) e^{|\xi|^2/2}$$

or

$$F_t(\xi) = e^{-|\xi|^2/2} \ll X_t, e^{\langle \cdot,\xi\rangle} \gg .$$

This says that F_t is the U-functional of X_t. Finally, letting $n \to \infty$ in (6), we obtain

$$F_t(\xi) = x + \int_0^t \xi(s) U[f(s,X_s)](\xi) ds.$$

Therefore, X_t satisfies the following equation in weak sense

$$X_t = x + \int_0^t \partial_s^* f(s,X_s) ds.$$

4. EXAMPLES

In this section, we will use the U-functional method to solve some stochastic differential equations.

Example 1. $dX_t = \partial_t^* X_t dt$ (or $dX_t = X_t dB(t)$ in Itô's sense), $X_0 = x$. It is well-known that the solution is $X_t = x\exp[B(t) - t/2]$, which can be checked by using Itô's lemma. However, we can derive this solution directly by the U-functional method. Let F_t be the U-functional of X_t. Then F_t satisfies the following equation for each ξ in S:

$$\frac{d}{dt}F_t(\xi) = \xi(t)F_t(\xi), \quad F_0(\xi) = x.$$

Thus $F_t(\xi)$ is given by

$$F_t(\xi) = x\exp\left[\int_0^t \xi(s)ds\right] = x\sum_{n=0}^{\infty}\frac{1}{n!}\left(\int_0^t \xi(s)ds\right)^n.$$

Note that

$$U\left[H_n(B(t)/\sqrt{2t})\right](\xi) = (2/t)^{n/2}\left(\int_0^t \xi(s)ds\right)^n,$$

(see e.g., (Kuo, 1983)). Here H_n is the Hermite polynomial of degree n, i.e., $H_n(x) = (-1)^n\exp(x^2) \times (d/dx)^n \exp(-x^2)$. Therefore, X_t is given by

$$X_t = x\sum_{n=0}^{\infty}\frac{1}{n!}\left(\frac{t}{2}\right)^{n/2} H_n\left(B(t)/\sqrt{2t}\right).$$

But, by using the generating function of the Hermite polynomials, we see easily that

$$X_t = x\exp\left[B(t) - \frac{1}{2}t\right].$$

Example 2. $dX_t = \partial_t^*(B(1)\diamond X_t)dt$, $X_0 = x$. Let F_t be the U-functional of X_t. Then F_t satisfies the following equation for each ξ in S:

$$\frac{d}{dt}F_t(\xi) = \xi(t)\left(\int_0^1 \xi(s)ds\right)F_t(\xi), \quad F_0(\xi) = x.$$

The solution is given by

$$F_t(\xi) = x\exp\left[\int_0^1 \xi(s)ds\int_0^t \xi(u)du\right] = x\exp\left[-\frac{1}{2}(\xi,\ L\xi)\right],$$

where $(\cdot\,,\,\cdot)$ is the $L^2(\mathbf{R})$ inner product and L is the following operator

$$Lf(s) = -2\cdot \mathbf{1}_{(0,1]}(s)\int_0^t f(u)du, \quad f\in L^2(\mathbf{R}).$$

Now, if K is a trace class operator on $L^2(\mathbf{R})$, then the function $\exp[-(z,Kz)/2]$, $z\in S^*$, is an ordinary Brownian function with expectation $\{\det(I+K)\}^{-1/2}$ ((Streit and Hida, 1983) or (Kuo, 1975, p. 106, a misprint of minus sign in the exponent)). Thus we can consider the following functional in (L^2)

$$\Phi(z) = \{\det(I+K)\}^{1/2}\exp\left[-\frac{1}{2}(z,Kz)\right], \quad z\in S^*. \tag{10}$$

The U-functional of Φ is given by

$$U[\Phi](\xi) = \exp\left[-\frac{1}{2}(\xi,\ K(I+K)^{-1}\xi)\right], \quad \xi\in S.$$

By applying this result to the above operator $L = K(I+K)^{-1}$, i.e., $K = L(I-L)^{-1}$, we obtain that

$$X_t(z) = x\Big\{\det[I+L(I-L)^{-1}]\Big\}^{1/2}\exp\left[-\frac{1}{2}(z,\ L(I-L)^{-1}z)\right], \quad z\in S^*.$$

A direct calculation shows that

$$\det\left[I+L(I-L)^{-1}\right] = (1+2t)^{-1}$$

and

$$\exp\left[-\frac{1}{2}(z,\ L(I-L)^{-1}z)\right] = \exp\left[(1+2t)^{-1}\langle z,\ \mathbf{1}_{(0,1]}\rangle\langle z,\ \mathbf{1}_{(0,t]}\rangle\right] =$$
$$= \exp\left[(1+2t)^{-1}B(1)B(t)\right].$$

Here we have substituted z by the white noise B' so that

$$\langle z,\ \mathbf{1}_{(0,1]}\rangle = \int_0^1 B'(u)du = B(1) \quad \text{and} \quad \langle z,\ \mathbf{1}_{(0,t]}\rangle = B(t).$$

Therefore, the solution X_t of the equation $dX_t = \partial_t^*(B(1)\diamond X_t)dt$, $X_0 = x$, is given by

$$X_t = x(1+2t)^{-1/2}\exp\left[(1+2t)^{-1}B(1)B(t)\right].$$

Example 3. $dX_t = \partial_t^*\big(B(t)\diamond X_t\big)dt$, $X_0 = x$. The U-functional F_t of X_t satisfies the following equation for each $\xi\in S$:

$$\frac{d}{dt}F_t(\xi) = \xi(t)\left(\int_0^t \xi(s)ds\right)F_t(\xi), \quad F_0(\xi) = x.$$

Thus $F_t(\xi)$ is given by

$$F_t(\xi) = x \exp\left[\frac{1}{2}\left(\int_0^t \xi(s)ds\right)^2\right] = x \exp\left[-\frac{1}{2}(\xi,\ J\xi)\right],$$

where J is the following operator

$$Jf(s) = -\mathbf{1}_{(0,t]}(s) \int_0^t f(u)du, \quad f \in L^2(\mathbf{R}).$$

We can apply the result about Φ in (10) to $J = K(I+K)^{-1}$, i.e., $K = J(I-J)^{-1}$. In this case, we have

$$\det\left[I + J(I-J)^{-1}\right] = (1+t)^{-1}$$

and

$$\exp\left[-\frac{1}{2}(z,\ J(I-J)^{-1}z)\right] = \exp\left[(1+t)^{-1}B(t)^2\right].$$

Therefore, the solution X_t is given by

$$X_t = (1+t)^{-1/2} \exp\left[(1+t)^{-1}B(t)^2\right].$$

Example 4. $dX_t = \partial_t^*(B'(t) \diamond X_t)dt$, $X_0 = x$. Since $U[B'(t)](\xi) = \xi(t)$, the U–functional F_t of X_t satisfies the equation for each $\xi \in S$:

$$\frac{d}{dt}F_t(\xi) = \xi(t)^2 F_t(\xi), \quad F_0(\xi) = x.$$

Hence $F_t(\xi)$ is given by

$$F_t(\xi) = x \exp\left[\int_0^t \xi(s)^2 ds\right] = x \exp\left[-\frac{1}{2}(\xi, Q\xi)\right],$$

where Q is the integral operator given by the kernel $-2 \cdot \mathbf{1}_{(0,t]}(u)\mathbf{1}_{(0,t]}(v)\delta(u-v)$. Just as in the last two examples, let $Q = K(I+K)^{-1}$. Then K is the integral operator with kernel $-\frac{2}{3}\mathbf{1}_{(0,t]}(u)\mathbf{1}_{(0,t]}(v)\delta(u-v)$. Since K is no longer a trace class operator, we cannot expect to obtain an ordinary Brownian functional like Φ in (10). Rather, we need a multiplicative renormalization of the expression in (10), which cancels the divergent determinant. We denote this renormalization by a "factor" $\mathcal{N}$ and get

$$\mathcal{N} \exp\left[-\frac{1}{2}(z,\ Kz)\right] = \mathcal{N} \exp\left[\frac{1}{3}\int_0^{\bullet t} B'(s)^2 ds\right].$$

Still, the last expression is slightly informal, but it is the visualization of a well-defined generalized Brownian functional (Kuo, 1983; 1988a; Streit and Hida, 1983) with U-functional given by

$$U\left[\mathcal{N}\exp\left(\frac{1}{3}\int_0^t B'(s)^2 ds\right)\right](\xi) = \exp\left[\int_0^t \xi(s)^2 ds\right], \quad \xi \in S.$$

Therefore, the solution X_t of $dX_t = \partial_t^*(B'(t) \diamond X_t)\,dt$, $X_0 = x$, is given by

$$X_t = x\mathcal{N}\exp\left[\frac{1}{3}\int_0^t B'(s)^2 ds\right].$$

5. SOME EXTENSIONS

In this section, we make some remarks about possible extensions of our theorem. The details will appear elsewhere.

1) A trivial extension of our theorem is to add a drift term, i.e.,

$$X_t = x + \int_0^t \partial_s^* f(s, X_s)ds + \int_0^t g(s, X_s)ds. \tag{11}$$

If $g : [0,1] \times (S)^* \to (S)^*$ satisfies the same conditions as f in the theorem, then equation (11) has a unique solution in $(S)^*$.

2) If we observe that $\partial_s^* f(s, X_s) = B'(s) \diamond f(s, X_s)$, then it becomes clear that we can generalize our theorem to the following equation

$$X_t = x + \int_0^t \theta(s) \diamond f(s, X_s)ds + \int_0^t \varrho(s) \diamond g(s, X_s)ds, \tag{12}$$

where f and g satisfy the same conditions as in the theorem, θ and $\varrho : [0,1] \to (S)^*$ satisfy the following conditions ($\lambda = \theta, \varrho$):

a) $\sup_{0 \leqslant s \leqslant 1} |U[\lambda(s)](\xi)| < \infty, \quad \xi \in S.$

b) There exists $q \geqslant 1$ such that $\lambda(s) \in (S)_q^*$ for all $s \in [0,1]$.

A special case of (12) is the famous Oosawa equation (see Hida's paper in these proceedings):

$$X_t = x + \int_0^t \left[- k_+ E_s(-\dot{f})X_s + k_- E_s(\dot{f})(1 - X_s)\right]ds,$$

where $k_{\pm} > 0$ and $E_t(f) = \exp\left[\int_0^t f(t-u)\partial_u^* du\right]$.

3) The U–functional method can be applied to equations involving not only ∂_t^*, but also the one–sided white–noise differentiation ∂_{t+} and ∂_{t-} (Kuo and Russek, 1988), e.g.,

$$X_t = x + \int_0^t \partial_s^* f(s, X_s)ds + \int_0^t \partial_{s+} g(s, X_s)ds + \int_0^t \partial_{s-} h(s, X_s)ds,$$

where f satisfies the same conditions as in the theorem, while g and h satisfy certain conditions.

4) The use of the renormalized product in the examples in Section 4 makes the equations easy to solve. On the other hand, instead of the equation in Example 2, we can consider the equations $dX_t = \partial_t^*(B(1)X_t)dt$ (Buckdahn, 1987) and $dX_t = B(1)X_t \circ dB(t)$, where $\circ$ denotes the Stratonovich integral. These two equations cannot be handled by our theorem. Nevertheless, they can be solved in terms of series by the U–functional method.

5) The U–functional method transfers a problem about generalized Brownian functionals in $(S)^*$ into another problem about functions defined on S. The analysis of functions on S was initiated by Lévy (Lévy, 1951). For further development, see the book (Polishchuk, 1988) and references therein. On the other hand, for the analysis of generalized Brownian functionals, there is an important question as when the solution X_t is in the space (L^2) of μ–square integrable functions on S^*, or when X_t is an ordinary random variable on S^*. This type of regularity problem will be addressed elsewhere.

REFERENCES

Asch, J. and Potthoff, J. (1987). A generalization of Itô's lemma. *Proc. Japan Acad.* **63A**, 289–291.

Berger, M. and Mizel, V. (1982). An extension of the stochastic integral. *Ann. Probab.* **10**, 435–450.

Buckdahn, R. (1987). Skorohod's integral and linear stochastic differential equations. *Preprint.*

Gross, L. (1967). Potential theory on Hilbert space. *J. Func. Anal.* **1**, 123–181.

Hida, T. (1975). *Analysis of Brownian Functionals.* (Carleton Math. Lect. Notes, **13**). Carleton Univ., Ottawa.

Hida, T. (1978). Generalized multiple Wiener integrals. *Proc. Japan Acad.* **54A**, 55–58.

Hida, T. (1983). Causal analysis and an application to prediction theory. *Pesi Masani Volume*, 123–130.

Hida, T. (1985). Brownian motion and its functionals. *Ricerche di Matematics* **34**, 183–222.

Hida, T. and Ikeda, N. (1965). Analysis on Hilbert space with reproducing kernel arising from multiple Wiener integral. In: *Proc. Fifth Berkeley Symp. Math. Stat. Probab.* **2**, pp. 117–143.

Hida, T., Potthoff, J. and Streit, L. (1988). Dirichlet forms and white noise analysis. *Commun. Math. Phys.* **116**, 235–245.

Hitsuda, M. (1973). Formula for Brownian partial derivatives. *Lect. Notes in Math.* **550**, 111–114.

Itô, K. (1951). Multiple Wiener integral. *J. Math. Soc. Japan* **3**, 157–169.

Itô, K. (1976). Extension of stochastic integrals. In: *Proc. Intern. Symp. Stochastic Differential Equations*, Kyoto, pp. 95–109.

King, R. (1975). Stochastic integrals and metadistributions. Applications to stochastic partial differential equations and quantum field theory. *Ph. D. Dissertation.* Cornell Univ.

Krée, M. (1974). Propriété de trace en dimension infinie d'éspaces du type Sobolev. *C. R. Acad. Sc. Paris, Sér.* A, **279**, 157–160.

Krée, P. and Raczka, R. (1978). Kernels and symbols of operators in quantum field theory. *Ann. Inst. Henri Poincaré, Sec.* A, **28**, 41–73.

Kubo, I. and Takenaka, S. (1980). Calculus on Gaussian white noise I. *Proc. Japan Acad.* **56A**, 376–380.

Kuo, H.-H. (1975). Gaussian measures in Banach spaces. *Lect. Notes in Math.* **463** .

Kuo, H.-H. (1983). Brownian functionals and applications. *Acta Applicandae Mathematicae* **1**, 175–188.

Kuo, H.-H. (1988a). Brownian motion, diffusions and infinite dimensional calculus. *Lect. Notes in Math.* **1316**, 130–169.

Kuo, H.-H. (1988b). White noise calculus. In: *Algebra, Analysis and Geometry.* World Scientific, Singapoure, pp. 77–119.

Kuo, H.-H. and Russek, A. (1988). White noise approach to stochastic integration. *J. Multivariate Analysis* **24**, 218–236.

Lévy, P. (1951). *Problémes Concrets d'Analyse Fonctionnelle.* Gauthier–Villars, Paris.

Nualart, D. and Pardoux, E. (1988). Stochastic calculus with anticipating integrands. *Probab. Th. Rel. Fields* **78**, 535–581.

Nualart, D. and Zakai, M. (1986). Generalized stochastic integrals and the Malliavin calculus. *Probab. Th. Rel. Fields* **73**, 255–280.

Ocone, D. and Pardoux, E. (1988). Non–adapted solutions of stochastic differential equations. *Preprint.*

Pardoux, E. and Protter, P. (1988). Stochastic Volterra equations with anticipating coefficients. *Preprint.*

Polishchuk, E. M. (1988). *Continual Means and Boundary Value Problems in Function Spaces.* Birkhäuser Verlag, Basel etc.

Shiota, Y. (1986). A linear stochastic integral equation containing the extended Itô integral. *Math. Rep. Toyama Univ.* **9**, 43–65.

Skorokhod, A. V. (1975). On a generalization of a stochastic integral. *Theory Probab. Appl.* **20**, 219–233.

Streit, L. and Hida, T. (1983). Generalized Brownian functionals and the Feynman integral. *Stoch. Proc. Appl.* **16**, 55–69.

Wiener, N. (1930). Generalized harmonic analysis. *Acta Math.* **55**, 117–258.

Prob. Theory and Math. Stat., Vol. 2, pp. 47–58
B.Grigelionis *et al.* (Eds.)
1990 VSP/Mokslas

SECOND ORDER MINIMAX ESTIMATION: THE NUISANCE PARAMETERS AND HYPOELLIPTIC OPERATORS

Z.M. LANDSMAN and B.Ya. LEVIT
IEIMAC, Chair of Applied Mathematics, Tashkent, Uzbek SSR, USSR
Glavmosoblstroj, 117000 Moscow, USSR

ABSTRACT

The second order evaluation of the minimax risk in the presence of nuisance parameters is considered and second order optimal estimates are constructed. The second order asymptotics of minimax risk is expressed through the solution of the first eigen-value problem for the semielliptic differential operator. The operator becomes elliptic when nuisance parameters are absent. That is the essential distinction between these two cases.

Let $x_1, \dots, x_n$ be independent observation with values in a measurable space $(\mathfrak{X}, \mathfrak{A})$ having common density function $f(x, \theta)$ with respect to a σ-finite measure ν and, depending on s parameters $\theta = \|\theta_1, \dots, \theta_s\|^T \in \Theta \subset \mathbf{R}^s$. Here $\theta_{p1} = \|\theta_1, \dots, \theta_p\|^T$ are parameters of interest (or structural parameters) and $\theta_{q1} = \|\theta_{p+1}, \dots, \theta_s\|^T$ are nuisance parameters.
Let

$$\varepsilon^2 = \frac{1}{n}, \quad x_\varepsilon = (x_1, \dots, x_n), \quad f_\varepsilon(x_\varepsilon, \theta) = \prod_{i=1}^{n} f(x_i, \theta),$$

$$\mathfrak{X}_\varepsilon = \underbrace{\mathfrak{X} \times \dots \times \mathfrak{X}}_{(n \text{ times})}, \quad \nu_\varepsilon = \underbrace{\nu \times \dots \times \nu}_{(n \text{ times})}.$$

We denote by $\theta_\varepsilon^{p1} = \theta_\varepsilon^{p1}(x_1, \dots, x_n) = \|\theta_\varepsilon^1 \dots \theta_\varepsilon^p\|^T$ the p-dimensional estimator of the p-dimensional structural parameters $\theta_{p1} = \|\theta_1, \dots, \theta_p\|^T$.
Denote by

$$R_\varepsilon(\theta_\varepsilon^{p1}, \theta) = \mathbf{E}_\theta W\big(\varepsilon^{-1}(\theta_\varepsilon^{p1} - \theta_{p1})\big) =$$

$$= \int_{\mathfrak{X}_\varepsilon} W\big(\varepsilon^{-1}(\theta_\varepsilon^{p1} - \theta)\big) f_\varepsilon(x_\varepsilon, \theta)\, d\nu_\varepsilon$$

the risk function of the estimator θ_ε^{p1}, where $W(y_{p1})$ is a non-negative, symmetric and quasi-convex loss function, depending only on the first p arguments $y_{p1} = (y_1, \dots, y_p)^T$. This is the main peculiarity of the presence of nuisance parameters.

1. FIRST ORDER ASYMPTOTIC OPTIMALITY IN THE PRESENCE OF THE NUISANCE PARAMETERS

Let

$$l(x,\theta) = \ln f(x,\theta), \quad \frac{\partial}{\partial\theta} = \left\| \frac{\partial}{\partial\theta_1}, \dots, \frac{\partial}{\partial\theta_s} \right\|^T$$

and

$$I(\theta) = \mathbf{E}_\theta \left(\frac{\partial}{\partial\theta} l, \frac{\partial^T}{\partial\theta} l \right)$$

be the Fisher information matrix. Denote by

$$J(\theta) = I(\theta)^{-1}, \quad \varphi_\theta(y) = \left|I(\theta)\right|^{1/2} (2\pi)^{-s/2} \exp\left\{ -\frac{1}{2} y^T I y \right\}.$$

If the nuisance parameters are absent, i.e., $p = s$, it is known that the value

$$R(\theta) = \int_{\mathbf{R}^s} W(y)\varphi_\theta(y)\, dy$$

plays an important role in the first order asymptotic estimation since for the maximal likelihood estimators $\widehat{\theta}_\varepsilon = (\widehat{\theta}^1_\varepsilon \dots \widehat{\theta}^s_\varepsilon)$ the following asymptotic expansion

$$R_\varepsilon(\widehat{\theta}_\varepsilon, \theta) = R(\theta) + o(1), \quad \varepsilon \to 0,$$

holds uniformly in $\theta \in \Theta$ if Θ is compact. So, according to the Hajek—Le Cam—Ibragimov—Hasminskii theory we have

$$\inf_{\theta_\varepsilon} \sup_{\theta\in\Theta} \left(R_\varepsilon(\theta_\varepsilon, \theta) - R(\theta) \right) = o(1), \quad \varepsilon \to 0,$$

for compact Θ.

Now suppose that the nuisance parameters are present, i.e., $p < s$. We shall split the $(s \times s)$–matrices I, J into blocks

$$I = \left\| \begin{matrix} I_{pp} & I_{pq} \\ I_{qp} & I_{qq} \end{matrix} \right\|, \quad J = \left\| \begin{matrix} J_{pp} & J_{pq} \\ J_{qp} & J_{qq} \end{matrix} \right\|,$$

where p means that we take only the first p lines or columns, while q means that we take the other $q = s - p$ lines or columns.

Then the $(p \times p)$–matrix $\widetilde{I}(\theta) = J_{pp}^{-1}$ plays the role of the Fisher information matrix with structural parameters $\theta_{p1} = \|\theta_1, \dots, \theta_p\|^T$ in the presence of nuisance parameters $\theta_{q1} = \|\theta_{p+1}, \dots, \theta_s\|^T$ since $R(\theta)$ is represented as

$$\widetilde{R}(\theta) = \int_{\mathbf{R}^p} W(y_{p1})\, \widetilde{\varphi}_\theta(y_{p1})\, dy_{p1},$$

where

$$\varphi_\theta(y_{p1}) = \left|I(\theta)\right|^{1/2} (2\pi)^{-p/2} \exp\left\{ -\frac{1}{2} y_{p1}^T\, \widetilde{I}(\theta)\, y_{p1} \right\}.$$

So, from the general theory it follows that

$$\inf_{\theta_\varepsilon^{p1}} \sup_{\theta\in\Theta} \left(R_\varepsilon(\theta_\varepsilon^{p1},\theta) - \widetilde{R}(\theta)\right) = o(1), \quad \varepsilon \to 0. \tag{1.1}$$

This relation we call the first order minimaxity in the presence of nuisance parameters, and the first p MLE estimators $\widehat{\theta}_\varepsilon^{p1} = (\widehat{\theta}_\varepsilon^1, \ldots, \widehat{\theta}_\varepsilon^p)^T$ are the first order minimax estimators of the structural parameters.
Note, that another method of definition I and $\widetilde{I}(\theta)$ exists, as $I(\theta)$ may be represented in the form

$$I(\theta) = \left\| \mathbf{E}_\theta \left[\left(\frac{\partial l}{\partial\theta_i} - \widehat{\mathbf{E}}_\theta \left(\frac{\partial l}{\partial\theta_i} \Big| \frac{\partial l}{\partial\theta_{p+1}}, \ldots, \frac{\partial l}{\partial\theta_s} \right) \right) \times \right.\right.$$

$$\left.\left. \times \left(\frac{\partial l}{\partial\theta_j} - \widehat{\mathbf{E}}_\theta \left(\frac{\partial l}{\partial\theta_j} \Big| \frac{\partial l}{\partial\theta_{p+1}}, \ldots, \frac{\partial l}{\partial\theta_s} \right) \right) \right] \right\|_{i,j=1}^{p},$$

where $\widehat{\mathbf{E}}$ is a projector (see (Rao, 1965; Klebanov and Melamed, 1978)).
For our goal it is more convenient to define $I(\theta)$ with the help of p vector fields

$$X_{p1} = \|X_1, \ldots, X_p\|^T = \frac{\partial}{\partial\theta_{p1}} + c_{pq}\frac{\partial}{\partial\theta_{q1}}, \tag{1.2}$$

$$X_{1p} = X_{p1}^T,$$

where

$$\frac{\partial}{\partial\theta_{p1}} = \left\| \frac{\partial}{\partial\theta_1}, \ldots, \frac{\partial}{\partial\theta_p} \right\|^T, \quad \frac{\partial}{\partial\theta_{q1}} = \left\| \frac{\partial}{\partial\theta_{p+1}}, \ldots, \frac{\partial}{\partial\theta_s} \right\|^T,$$

$$c_{pq} = J_{pp}^{-1} \cdot J_{pq} = -I_{pq} \cdot I_{qq}^{-1}.$$

The matrix I is represented as

$$\widetilde{I}(\theta) = \mathbf{E}_\theta\left(X_{p1} l(x,\theta) \cdot X_{1p} l(x,\theta)\right).$$

The vector fields $X_1, \ldots, X_p$ play even more distinguished role in the second order minimaxity problem.

2. THE SECOND ORDER ASYMPTOTIC MINIMAXITY IN THE PRESENCE OF NUISANCE PARAMETERS. STATEMENT OF A PROBLEM

As it has been shown in Section 1, in the theory of first order optimality in the presence of nuisance parameters the essential problems haven't been observed, as the general theory doesn't exclude the case, when the loss function W depends on only $p < s$ arguments. Absolutely other picture we may observe in the second order minimax problem, which is to find the value

$$\widetilde{r}_\varepsilon(\Theta) = \inf_{\theta_\varepsilon^{p1}} \sup_{\theta\in\Theta} \varepsilon^{-2}\left(R_\varepsilon(\theta_\varepsilon^{p1},\theta) - \widetilde{R}(\theta)\right)$$

for $\varepsilon \to 0$.
Now we introduce $(s \times s)$–matrix playing the distinguished role in the second order minimax theory

$$A(\theta) = \int_{\mathbf{R}^s} (yy^T - J(\theta))W(y_{p1})dy$$

and we shall split this matrix into blocks

$$A(\theta) = \begin{Vmatrix} A_{pp} & A_{pq} \\ A_{qp} & A_{qq} \end{Vmatrix}.$$

Among the first order optimal estimation we distinguish the special estimators $\theta^{p1}_{\varepsilon,\omega} = \|\theta^1_{\varepsilon,\omega}, \ldots, \theta^p_{\varepsilon,\omega}\|^T$ of structural parameters

$$\theta^{p1}_{\varepsilon,\omega} = \widehat{\theta}^{p1}_{\varepsilon,\omega} + \varepsilon^2 V^{p1}_{\varepsilon}(\widehat{\theta}^{p1}_{\varepsilon}, \widehat{\theta}^{q1}_{\varepsilon}), \tag{2.1}$$

where $\omega(\theta)$ is positive function, $\lambda(\theta) = \omega(\theta)^2$ is integrable on Θ

$$V^{(p1)}_{\varepsilon}(\theta) = \begin{cases} V^{(p1)}(\theta) = J(2X_{p1}\ln\omega + \widetilde{q}A^{-1}_{pp}(\theta)), & \text{if } \left|V^{(p1)}\right| < \varepsilon^{-1}, \\ O_{p1}, & \text{if } \left|V^{(p1)}(\theta)\right| \geqslant \varepsilon^{-1}. \end{cases} \tag{2.2}$$

Here O_{p1} is a vector–column of p zeros, $\widetilde{J} = \widetilde{I}^{-1}$

$$\widetilde{g} = (Ig)_{pp}\widetilde{J}, \quad g = \|g_1, \ldots, g_s\|^T$$

(here and further we mean the summation for the same index),

$$g_k = \int_{\mathbf{R}^s} M_{ijm} y_i y_j (y_i y_k - 3J_{ik}) W(y_{p1}) \varphi_\theta(y)\, dy, \quad k = 1, \ldots, s,$$

$$M_{ijm} = \mathbf{E}_\theta\, l_{ijm}, \quad l_{ijm} = \frac{\partial}{\partial\theta_i}\frac{\partial}{\partial\theta_j}\frac{\partial}{\partial\theta_m}\, l(x,\theta).$$

We shall call the estimators $\theta^{p1}_{\varepsilon,\omega}$ the second order asymptotical Bayes estimators of structural parameters $\theta_{p1} = \|\theta_1, \ldots, \theta_p\|^T$ with respect to non–negative integrable weight function $\lambda(\theta) = \omega(\theta)^2$. Let $\omega(\theta)$ satisfies the following conditions

$$\omega(\theta) \in C^{2+\beta}(\Theta) \ \text{ for some } \ 0 < \beta \leqslant 1,$$

$$C_1\rho(\theta) \leqslant \omega(\theta) \leqslant C_2\rho(\theta),$$

where $\rho(\theta) = d(\theta, \partial\Theta)$ is the distance from θ to bound $\partial\Theta$ of subset Θ. Denote by

$$R_\varepsilon(\lambda) = \inf_{\theta^{p1}_\varepsilon} \int_\Theta R_\varepsilon(\theta^{p1}_\varepsilon, \theta)\lambda(\theta)\, d\theta.$$

THEOREM 1. *Let Θ be connected open subset of $\mathbf{R}^s$ with bound $\partial\Theta \in C^{2+\beta}$, $\bar{\Theta} \subset \Theta'$, where Θ' is open, and some regularity conditions are fulfilled. Then*

$$\int_{\Theta} R_\varepsilon(\theta^{p1}_{\varepsilon,\omega}, \theta)\lambda(\theta)\, d\theta = R_\varepsilon(\lambda) + o(\varepsilon^2), \quad \varepsilon \to 0.$$

Denote by $\Theta_{\delta(\varepsilon)} = \{\theta \in \Theta \mid \rho(\theta) > \varepsilon\}$.

THEOREM 2. *Let the conditions of the Theorem 1 and condition* (C) *of* (Levit, 1982) *be fulfilled.*
Then for the risk function of estimators $\theta^{p1}_{\varepsilon,\omega}$ the asymptotic expansion

$$R_\varepsilon(\theta^{p1}_{\varepsilon,\omega}, \theta) = \widetilde{R}(\theta) + \varepsilon^2 \frac{L\omega}{\omega} + o(\omega^{-2}\varepsilon^{2+\gamma}), \quad \varepsilon \to 0,$$

is valid uniformly with respect to $\theta \in \Theta_{\delta(\varepsilon)}$. Here

$$L\omega = 2\frac{\partial}{\partial\theta_i} a_{ij}(\theta) \frac{\partial}{\partial\theta_j} \omega(\theta) + a(\theta)$$

is the second order linear symmetric differential operator with coefficient matrix $A = \|a_{ij}(\theta)\|^s_{i,j=1}$.

If $p = s$, i.e., there are no nuisance parameters, the matrix A is non-singular and positive definite (Levit, 1982), that is why the operator L is elliptic in all domain Θ. Then the Dirichlet problem

$$L\omega + \alpha\omega = 0,$$
$$\omega \mid_{\partial\Theta} = 0$$

has eigen-value $0 < \alpha_1 < \alpha_2 \leqslant \ldots$ and using the Dirichlet variational principle it may be obtained (Levit, 1982)

$$\widetilde{r}_\varepsilon(\Theta) = -\alpha_1 + o(1), \quad \varepsilon \to 0.$$

If $p < s$, i.e., there are the nuisance parameters, as we shall see from following Lemma 2.1, the matrix A is not positive definite but only non-negative definite, because A is singular. That is why the operator L is not elliptic and only semi-elliptic and it is degenerate in all Θ. The Dirichlet problem formulation on eigen-value is not clear.
Thus there is an essential distinction with the case of the absence of nuisance parameters.

LEMMA 2.1. *Let $I(\theta)$ and $I_{qq}(\theta)$ be nonsingular, $A(\theta) \in C^2(\Theta)$. Then*

$$A(\theta) = \left\| \begin{matrix} A_{pp} & A_{pp}C_{pq}(\theta) \\ C_{qp}A_{pp} & C_{qp}A_{pp}C_{pq} \end{matrix} \right\|,$$

where $C_{qp} = C^T_{pq}$, and

$$A_{pp} = \int\limits_{\mathbf{R}^p} (y_{p1}y^T_{p1} - J_{pp})\widetilde{\varphi}_\theta(y_{p1}).dy_{p1}$$

is positive definite matrix and the operator L is represented by

$$L = \sum_{i,j=1}^{p} a_{ij}(\theta)X_iX_j + \sum_{j=1}^{p} b_jX_j + a(\theta). \tag{2.3}$$

From Lemma 2.1 we can see that the matrix $A(\theta)$ as a matter of fact is singular and range $A(\theta) = p < s$ and that operator L may be reduced to Hormander's form (see (Hormander, 1967))

$$L = Z_1^2 + \ldots + Z_p^2 + f_j(\theta)Z_j + a(\theta), \tag{2.4}$$

where vector–fields $Z_1, \ldots, Z_p$ are the linear combination of the vector–fields $X_1, \ldots, X_p$.
As it has been known, the main properties of operator L depend on range value of Lie algebra $\mathcal{L}(X_1, \ldots, X_p)$ generated by $X_1, \ldots, X_p$ at each point $\theta \in \bar{\Theta}$.
Now we consider Θ as the s–dimensional C^∞–manifold.
We remind that here the vector fields X is understood as the mapping $X : \Theta \ni \theta \to X_\theta = \xi_j(\theta)\frac{\partial}{\partial\theta_j} \in T_\theta$ is tangent space to manifold Θ at point $\theta \in \Theta$.
On the other hand X is linear differential operator, operated on $f(\theta) \in C^\infty(\Theta)$.
The Lie brackets (or commutator) of two vector fields

$$X_i = \sum_{j=1}^{s} \xi^i_j(\theta)\frac{\partial}{\partial\theta_j}, \quad i = 1, 2,$$

is the vector field

$$[X_1, X_2] = (X_1\xi^2_j - X_2\xi^1_j)\frac{\partial}{\partial\theta_j}.$$

The Lie algebra, generated by vector fields $X_1, \ldots, X_p$ at point $\theta \in \Theta$ is linear space generated by $X_1, \ldots, X_p$ Lie brackets $[X_i, X_j]$, $[X_i, [X_j, X_k]], \ldots$ at point $\theta \in \Theta$ with coefficients depending on θ and closed with respect to Lie brackets. The maximal number of linear independent vector fields at point $\theta \in \Theta$ is range of Lie algebra at point $\theta \in \Theta$.
Now we suppose that the range of Lie algebra generating by $X_1, \ldots, X_p$ is equal r at each point $\theta \in \Theta$.
It's obvious that $p \leqslant r \leqslant s$. In the following paragraph we shall consider in details of the very important case $r = s$. In this case the vector fields $X_1, \ldots, X_p$ together with their commutators span $\mathbf{R}^s$ at each point of Θ. Then the Hormander's condition (Hormander, 1967) is fulfilled and the operator L is hypoelliptic operator on $\bar{\Theta}$.

3. HYPOELLIPTIC OPERATOR L. METRICS. SECOND ORDER MINIMAXITY

Let the coefficients of operator L be $C^\infty(\Theta)$-smooth functions. Introduce the following condition.

(R) The range of Lie algebra, generated by $X_1, \dots, X_p$ is equal to s at each point $\theta \in \bar{\Theta}$.

As L is represented as (2.3) and consequently (2.4), L is hypoelliptic operator on Θ (see (Hormander, 1967)).

Now consider the Dirichlet problem

$$\begin{aligned} &L\omega + \alpha q(\theta)\omega = 0, \\ &\omega\big|_{\partial\Theta} = 0 \end{aligned} \tag{3.1}$$

with some weight function $q(\theta) > 0$.

Let the α_1 exists such that

$$\alpha(\theta) + \alpha_1 q(\theta) \leqslant a_0 < 0. \tag{3.2}$$

As $a(\theta)$ in L may be replaced by $a(\theta) + \alpha_1 q(\theta)$ then the eigen-value of problem (3.1) shifts from α to $\alpha - \alpha_1$ that is why we can assume $a(\theta) \leqslant a_0 < 0$ without loss of generality.

Let $\vec{n}(n_1, \dots, n_s)$ is normal to bound $\partial\Theta$ and

$$\sum_{i,j=1}^{s} a_{ij} n_i n_j > 0 \tag{3.3}$$

at each point of $\partial\Theta$.

Then (see (Bony, 1969; Negriny and Scornazzani, 1959)) for any $f \in C(\Theta)$ there exists a unique solution of the Dirichlet problem

$$Lu = -f \quad \text{on } \Theta, \tag{3.4}$$

$$u\big|_{\partial\Theta} = 0. \tag{3.5}$$

This solution can be represented in the form

$$u(x) = Gf = \int_\Theta g(x,y) f(y)\, dy, \tag{3.6}$$

where $g(x,y)$ is Green function, $g(x,y) = g(y,x)$, g is C^∞-smoothed with respect to x and y out the diagonal of $\Theta \times \Theta$ and positive.

If the condition (R) is fulfilled, to any operator (2.4) we can attach a distance d which is as natural for L as Euclidean distance is natural for the Laplace operator. We shall define the distance in the following way (Negriny and Scornazzani, 1959).

A continuous curve $\gamma \subseteq \Theta$ is *Z-admissible* if:

a) γ is piecewise C^1,

b) each piece C^1 of γ is an integral curve of one of the vector fields $\pm Z_1, \dots, \pm Z_p$.

If $\gamma : [0,T] \to \Theta$ is a parametrization of γ, we put $l(\gamma) = T$. The condition (R) guarantees that for any $x, y \in \Theta$ there exists an Z-admissible curve γ joining x and y (see (Rashevskii, 1938)). Finally, we define

$$d(x,y) = \inf\{l(\gamma) \mid \gamma \text{ is } Z-\text{admissible and joins } x \text{ and } y\}.$$

In Theorem 3 (Negriny and Scornazzani, 1959) for Green function g the following estimators

$$C_1 \frac{d^2(x,y)}{\left(S\left(x, d(x,y)\right)\right)} \leqslant g(x,y) \leqslant C_2 \frac{d(x,y)^2}{\left(S\left(x, d(x,y)\right)\right)} \tag{3.7}$$

is given for any (x,y) is a neighbourhood of the diagonal of $\Theta \times \Theta$, where $S(x,r)$ is the sphere in metric d with center x and radius r, μ is the Lebesgue measure on $\mathbf{R}$.

LEMMA 3.1. *Operator G (3.6) operates from $C^0(\Theta)$ to $C'(\Theta)$ and is bounded and completely continuous. Operator G is bounded and completely continuous from $L_2(\Theta)$.*

Proof. As

$$u(x') - u(x) \leqslant$$

$$\leqslant \|f\|_{C^0(\Theta)} \left[\int\limits_{S(x,\delta)} |g(x',y) - g(x,y)|\, dy + \int\limits_{\Theta \setminus S(x,\delta)} |g(x',y) - g(x,y)|\, dy \right]. \tag{3.8}$$

If $d(x',x) < \delta$ the second integral is proper in (3.8) so it is less than $\varepsilon\mu(\bar{\Theta})$ if $|x' - x| < \delta_1$ and δ_1 doesn't depend on x.
Denote by $C_k = S(x, \delta\lambda^k) \setminus S(x, \delta\lambda^{k+1})$, then from (3.7)

$$\int\limits_{S(x,\delta)} g(x,y)dy = \sum_{k=0}^{\infty} \int\limits_{C_k} g(x,y)dy \leqslant$$

$$\leqslant C\delta^2 \sum_{k=0}^{\infty} \lambda^{2k} \frac{\mu\left(S(x, \delta\lambda^k)\right)}{\mu\left(S(x, \delta\lambda^{k+1})\right)}. \tag{3.9}$$

Put $\lambda = 1/2$ and using the doubling property of d (Negriny and Scornazzani, 1959):

$$\mu\left(S(x, 2\delta)\right) \leqslant C\mu\left(S(x,\delta)\right)$$

from (3.9) we obtain

$$\int\limits_{S(x,\delta)} g(x,y)dy \leqslant C\delta^2 \sum_{k=0}^{\infty} \lambda^{2k} = C\delta^2 < \varepsilon. \tag{3.10}$$

With the help of (3.10) we can see that the first integral in (3.8) is less that ε. Hence we can conclude that $u(x)$ is continuous and even more the family of $u(x)$ corresponded to $f(\theta) \in C^0(\Theta)$ is equicontinuous.

As $\Theta \subset S(x,H)$, from (3.6) and (3.10) we get

$$\|u\|_{C^0(\Theta)} \leqslant C\|f\|_{C^0(\Theta)}\cdot H^2.$$

The first part of Lemma 3.1 is proved. The second part of Lemma 2.1 can be proved in just the same way as for the integral operators having the weak singular kernel (see (Michlin, 1968, Chapter 7, §3)) to take into account the estimators (3.7) for metrics and estimator (3.10). Lemma 3.1 is proved.
As the operator G is bounded, completely continuous and G has the symmetric kernel, its eigen–values are real and they are ranked to decrease of absolute values

$$|\mu_1| \geqslant |\mu_2| \geqslant |\mu_3| \geqslant \dots$$

Therefore the spectral range of operator G is $\rho(G) = |\mu_1|$.
As the kernel of G is positive, G is positive operator with cone K of non–negative functions, and from (3.7) we get

$$\int\limits_{S(x,\delta)} g(x,y)\,dy \geqslant C_1 \sum_{k=0}^{\infty} \int\limits_{C_k} \frac{d^2(x,y)}{\mu\Big(S(x,d(x,y))\Big)}\,dy \geqslant$$

$$\geqslant \delta^2\lambda^2 \sum_{k=0}^{\infty} \lambda^{2k} \geqslant \delta^2\lambda^2,$$

used the doubling property of d.
Then the condition

$$Gu \geqslant Cu$$

takes place for some $u \in K$ (for example $u = 1$) and there are the eigen–value $\mu_1 = \rho(G) > 0$ and its eigen–function $u_1(x) \in K$, i.e., $u_1(x) \geqslant 0$ according to the Theorem 5.8 in (Zabreiko, 1968, p. 150). Obviously $\lambda_1 = 1/\mu_1$ is eigen–value of operator $-L$ and $u(x)$ is eigen–function of $-L$.

THEOREM 3. *Let $\Theta \subset \Theta'$ and the conditions of Theorem 2 be fulfilled for Θ'. Let the conditions* (R), (3.2) *for $q = 1$ and* (3.3) *be fulfilled. Then*

$$\widetilde{r}(\Theta) = \inf_{\theta_\varepsilon^{p1}} \sup_{\theta\in\Theta} \varepsilon^{-2}\big(R_\varepsilon(\theta_\varepsilon^{p1},\theta) - \widetilde{R}(\theta)\big) = -\lambda_1 + o(1).$$

Proof. Let $\lambda(\theta) = \omega(\theta)^2$ is Bayes density, i.e., $\int\limits_\Theta \omega(\theta)^2\,d\theta = 1$ then from Theorems 1 and 2 it follows that

$$\widetilde{r}_\varepsilon(\Theta) \geqslant \inf_{\theta_\varepsilon^{p1}} \varepsilon^{-2} \int\limits_\Theta \big(R_\varepsilon(\theta_\varepsilon^{p1},\theta) - \widetilde{R}(\theta)\big)\omega(\theta)^2\,d\theta =$$

$$= \varepsilon^{-2} \int\limits_\Theta \big(R_\varepsilon(\theta_{\varepsilon,\omega}^{p1},\theta) - \widetilde{R}(\theta)\big)\omega(\theta)^2\,d\theta + o(1) =$$

$$= \int_{\Theta} (L\omega)\omega \, d\theta + o(1), \quad \varepsilon \to 0. \tag{3.11}$$

If we take $\omega(\theta) = u_1(\theta)$, where $u_1(\theta)$ is eigen–function of operator $-L$ with eigen–value λ_1 and $\int_{\Theta} u_1(\theta)^2 \, d\theta = 1$, then from (3.11) we get

$$\tilde{r}_\varepsilon(\Theta) \geqslant -\lambda_1 + o(1), \quad \varepsilon \to 0. \tag{3.12}$$

Now we show that the positive $C^\infty(\Theta)$–solution of equation

$$L\omega + (\lambda_1 - \delta)\omega = 0 \tag{3.13}$$

exists. In the matter of fact, suppose $\tilde{\omega}(\theta) = \omega(\theta) - C$, where $C > 0$, then (3.13) is equal to

$$L\tilde{\omega} + (\lambda_1 - \delta)\tilde{\omega} = -(\lambda_1 - \delta)C \tag{3.14}$$

(we assume $a(\theta) = 0$ for simplicity).
We introduce the boundary condition

$$\tilde{\omega}\Big|_{\partial\Theta} = 0 \tag{3.15}$$

and from (3.14) pass to integral equation

$$\frac{1}{\lambda_1 - \delta}\omega = G\tilde{\omega} + f, \tag{3.16}$$

where $f = G\big((\lambda_1 - \delta)C\big) > 0$.
So as $\mu = 1/(\lambda_1 - \delta) > \mu_1 = 1/\lambda_1 = \rho(G)$ and $f = G\big((\lambda_1 - \delta)C\big) > 0$ from Theorem 5.7 (Zabreiko, 1968) the solution of (3.16) $\tilde{\omega}_\delta \geqslant 0$, $\tilde{\omega}_\delta \in C^\infty(\Theta)$ exists. Hence $\omega_\delta = \tilde{\omega}_\delta + C > 0$, $\omega_\delta \in C^\infty(\Theta)$ and ω_δ satisfies to (3.16). Then assumed $\Theta \subset \Theta'_{\delta(\varepsilon)}$ from Theorem 2 get

$$\tilde{r}_\varepsilon(\Theta) \leqslant \sup_{\theta\in\Theta} \varepsilon^{-2}\big(R_\varepsilon(\theta^{p1}_{\varepsilon,\omega_\delta}, \theta) - \tilde{R}(\theta)\big) =$$

$$= \frac{L\omega_\delta}{\omega_\delta} + o(1) = -(\lambda_1 - \delta) + o(1), \quad \varepsilon \to 0.$$

Theorem 3 is proved.

4. THE OPERATOR L AND FOLIATION OF Θ

Now we shall consider the case, when the range of Lie algebra generated by $X_1, \ldots, X_p$ is equal to r at each point $\theta \in \bar{\Theta}$. We are going to tell about this case in details in the individual report and here we shall consider it in a general way.
Now we remind some information of differential geometry. The linear space generated by $X_1, \ldots, X_p$ with coefficients–functions from $C^\infty(\Theta)$ is called the distribution D. The distribution D is called the involutive if

$$[X_i, X_j] = \sum_{l=1}^{p} \gamma_l(\theta) X_l, \quad i, j = 1, \ldots, p. \tag{4.1}$$

If the (4.1) is not hold we include these Lie brackets and so on. Finally, we constructed the involutive distribution $D' \supset D$ and D' has the dimension r. Then it's known that the r–dimensional integral submanifold exists at each point $\theta \in \Theta$. This submanifold is called the leaf of D'.
Denote by $\pi(\theta)$ the projection which carries $\theta \in \Theta$ onto leaf of D' containing θ. Introduce the condition

(M_1) For each leaf the global coordinate system exists, i.e., if $\theta' \in \pi^{-1}(\pi(\theta))$, then $\theta' = f(\theta, t_1, \dots, t_p)$, where $f \in C^\infty$ with respect to θ and $t = (t_1, \dots, t_p)$. Now we introduce the equivalence relation (which we denote by D' too): $\theta' \sim \theta''$ if both θ' and θ'' belong to the same leaf. Then $\Theta \mid D'$ is quotient space with respect to D' and it is called the foliation defined by D'. As it was shown in the foliation $\Theta \mid D'$ is $(s-r)$–dimensional submanifold if the distribution D' is "regularity". We require, however, the more

(M_2) There is such $(s-r)$–dimensional C^∞–smooth surface

$$\{\theta \in \Theta \mid \theta = \psi(v), \quad v \in \mathbf{R}^{s-r}\}$$

in Θ that the each leaf intersects this surface in only one point.
The following condition requires to have a non–equal to zero angle between the tangent space to leaf and tangent space to surface $\{\theta = \psi(v)\}$.

(S) The determinant

$$\det \left\| \frac{d\psi(v)}{dv} \xi(\psi(v)) \right\| \neq 0,$$

where $\xi(\theta)$ is the matrix of coefficients of vector fields $X_1, \dots, X_p, X_{p+1}, \dots, X_r$ which generated D'.

THEOREM 4. *Let $\Theta \subset \Theta'$ where Θ' is compact, and conditions* (M_1), (M_2) *and* (S) *be fulfilled. Then*
1) *one–to–one map $\eta : \Theta \to \eta(\Theta) = \Omega$ exists such that*

$$\det \|\eta'(\theta)\| \neq 0,$$

$$L = \widetilde{a}_{ij} \widetilde{X}_i \widetilde{X}_j + \widetilde{b}_j \widetilde{X}_j + \widetilde{a}, \quad i, j = 1, \dots, p,$$

where

$$\widetilde{X}_i = \sum_{j=1}^{r} \beta_{ij}(\eta_1, \dots, \eta_r, h) \frac{\partial}{\partial \eta_j}, \quad h \in \mathbf{R}^{s-r}, \ i = 1, \dots, p,$$

2) *range of Lie algebra, generated by $\widetilde{X}_1, \dots, \widetilde{X}_p$ is equal to r at each point of section Ω_h*

$$\underline{\eta}_1 = (\eta_1, \dots, \eta_r) \in \Omega_h, \quad h \in H = \left\{ h \in \mathbf{R}^{s-r} \middle| \Omega_h \neq \varnothing \right\},$$

3) *L is hypoelliptic operator with respect to $\eta_1, \ldots, \eta_r$ for each fixed $h \in H$.*

From Theorem 4 and the results of Section 3 it follows that for $h \in H$ there exists the first eigen-value $\lambda_1(h)$ of Dirichlet problem

$$L\omega(\eta_1, h) + \lambda(h)\omega = 0,$$
$$\omega(\underline{\eta}_1, h)\Big|_{\underline{\eta}_1 \in \partial\Omega_h} = 0,$$

Let $\lambda_1 = \inf\limits_{h \in H} \lambda_1(h)$.

THEOREM 5. *Let the conditions of Theorems 1, 2 and 4 be fulfilled, and the conditions of Theorem 3 in connection to $\mathbf{R}^r$ be fulfilled for each $h \in H$ and some condition on subset Ω holds (see (*Landsman and Levit, 1989*)).*
Then

$$\widetilde{r}_\varepsilon(\Theta) = -\lambda_1 + o(1), \quad \varepsilon \to 0.$$

REFERENCES

Bony, J.M. (1969). Principe du maximum inegalite de Harnack et unicite du problem de Cauchy pour les operateurs elliptiques degeneres. *Ann. Inst. Fourier, Grenoble* **19**, 277–304.

Hormander, L. (1967). Hypoelliptic second order differential equations. *Acta Math.* **119**, 147–171.

Klebanov, L.B. and Melamed, I.A. (1978). Several notes on Fisher information in presence of nuisance parameters. *Math. Operations-forsch. Statist. Ser. Statist.* **9**, 85–90.

Landsman, Z.M. and Levit, B.Ya. (1989). Second order minimax estimation. The nuisance parameters and hypoelliptic operators. In: *Abstr. Fifth Intern. Vilnius Conf. Probab. Theory and Math. Statist.*, **1**, pp. 300–301.

Levit, B.Ya. (1982). Minimax estimation and positive solution of elliptic equations. *Theory Probab. Appl.* **27**, 525–546.

Michlin, S.G. (1968). *Mathematical Physics.* Nauka, Moscow.

Negriny, P. and Scornazzani, V. (1959). Viener criterion for class of degenerate elliptic operators. *J. Diff. Equations* **66**, 151–164.

Palais, R.S. (1957). A global formulation of the Lie theory of transformation of groups. *Mem. Amer. Math. Soc.* **22**, 1–123.

Rao, C.R. (1965). *Linear Statistical Inference and its Applications.* Wiley, New York.

Rashevskii, P.K. (1938). About join of any two points of completely non-holonomic space of admissible curve. *Uchen. Zapiski Moskovsk. Pedagog. Inst. Ser. Phys.-Math.* **2.**

Zabreiko, P.P. (1968). *Integral Equations.* Ref. Math. Library, Moscow.

Prob. Theory and Math. Stat., Vol. 2, pp. 59–69
B.Grigelionis *et al.* (Eds.)
1990 VSP/Mokslas

LIMIT THEOREMS FOR THE RIEMANN ZETA–FUNCTION IN THE COMPLEX SPACE

A. LAURINČIKAS
Faculty of Mathematics, Vilnius University,
Naugarduko str., 24, 232006 Vilnius, Lithuania

ABSTRACT

The asymptotic behaviour of the Riemann zeta–function in the half–plane $\sigma \geqslant \frac{1}{2}$ is considered. It is proved that the Riemann zeta–function in this half–plane has a limiting distribution in a sense of weak convergence of probability measures.

Let $s = \sigma + it$ be a complex variable and let $\zeta(s)$ denote the Riemann zeta–function. The idea of using probabilistic methods for studying the distribution of the values of the zeta–function belongs to H. Bohr. In (Bohr, 1922) he has noted, that asymptotically the behaviour of the zeta–function in the half–plane $\sigma > 1/2$ is governed by some probabilistic laws. This was confirmed in (Bohr and Jessen, 1930). Let K be any closed rectangle in the complex plane $\mathbf{C}$ with the edges parallel to the axes, and $L(T, K)$ be the Jordan measure of the set $\{t \in [0, T], \ln\zeta(\sigma + it) \in K\}$. Suppose, $\sigma > 1$. It is proved in (Bohr and Jessen, 1930) that there exists the limit $\lim\limits_{T\to\infty} L(T, K)/T$, which depends only on σ and K. In (Bohr and Jessen, 1932) an analoguous result is obtained for $\sigma > 1/2$. The studies of (Bohr and Jessen, 1930, 1932) were continued in (Jessen and Wintner, 1935) by applying another method.
Let meas$\{A\}$ be the Lebesgue measure of a set A and $\nu_T(\ldots) = (1/T)$ meas$\{t \in [0, T], \ldots\}$, where by dots we understand some condition satisfied by t. By $\mathcal{B}(\mathbf{S})$ we shall denote the class of Borel sets of the space $\mathbf{S}$. On $(\mathbf{C}, \mathcal{B}(\mathbf{C}))$ we define the probability measure

$$\mathbf{P}_{\sigma,T}(A) = \nu_T\left(\zeta(\sigma + it) \in A\right).$$

THEOREM 1. *Let $\sigma > 1/2$. Then there exists a probability measure $\mathbf{P}_\sigma$ on $(\mathbf{C}, \mathcal{B}(\mathbf{C}))$ such that the measure $\mathbf{P}_{\sigma,T}$, as $T \to \infty$, converges weakly to $\mathbf{P}_\sigma$.*

The behaviour of the function $\zeta(s)$ is very interesting when $\sigma = 1/2$ or $\sigma \to 1/2+0$. In (Laurinčikas, 1987a; 1987b; 1989) the normed modulus of the function $\zeta(s)$ was considered and it was proved that the distribution function

$$\nu_T\left(\left|\zeta\left(\frac{1}{2} + it\right)\right|^{(2^{-1}\ln\ln T)^{-1/2}} < x\right),$$

as $T \to \infty$, converges to the function $G(x)$, where

$$G(x) = \begin{cases} \Phi(\ln x), & x > 0, \\ 0, & x \leqslant 0, \end{cases}$$

$$\Phi(x) = \frac{1}{\sqrt{2\pi}} \int_{-\infty}^{x} e^{-u^2/2} du.$$

A similar result remains valid near the critical line $\sigma = 1/2$ (Laurinčikas, 1988; 1989b). Let $0 < l_T \leqslant \ln T$, $l_T \to \infty$, $\psi_T > 0$, $\psi_T \to \infty$ and $\ln \psi_T = o(\ln l_T)$, as $T \to \infty$. We put $\sigma_T = 1/2 + \psi_T \sqrt{\ln l_T}/l_T$.

THEOREM A (Laurinčikas, 1988; 1989b). *The distribution function*

$$\nu_T \left(|\zeta(\sigma_T + it)|^{(2^{-1} \ln l_T)^{-1/2}} < x \right),$$

as $T \to \infty$, converges to the function $G(x)$.

Let $1/2 \leqslant \sigma \leqslant \sigma_T$, $l_T = \ln T$.

THEOREM B (Laurinčikas, 1990). *The distribution function*

$$\nu_T \left(|\zeta(\sigma + it)|^{(2^{-1} \ln \ln T)^{-1/2}} < x \right),$$

as $T \to \infty$, converges to the function $G(x)$.

Since the function $\zeta(s)$ is complex-valued the distribution of its values is more completely reflected by limit theorems in the complex space $\mathbf{C}$.
If $1/2 \leqslant \sigma < 1$ and $\zeta(s) \neq 0$, $a \in \mathbf{R}$, we shall understand $\zeta^a(s)$ as $\exp\{a \ln \zeta(s)\}$, where $\arg \zeta(s)$ is defined by a continuous displacement from the point $s = 2$ along a path, joining the points 2, $2 + it$ and $\sigma + it$. Taking into account the estimate $\nu_T(\zeta(\sigma + it) = 0) = o(1)$, which is valid for $\sigma = \sigma_T$ and for $1/2 \leqslant \sigma \leqslant \sigma_T$ when $l_T = \ln T$, we agree that $\zeta^a(s) = 0$ if $\zeta(s) = 0$.

For sets $A \in \mathcal{B}(\mathbf{C})$ we define the probability measure

$$\mathbf{P}_T(A) = \nu_T \left(\zeta^{(2^{-1} \ln l_T)^{-1/2}}(\sigma_T + it) \in A \right).$$

THEOREM 2. *The measure $\mathbf{P}_T$, as $T \to \infty$, weakly converges to the measure defined by the characteristic transform* $\exp\left\{ -(\tau^2/2) - (k^2/2) \right\}$.

Now let $1/2 \leqslant \sigma \leqslant \sigma_T$, $l_T = \ln T$. For sets $A \in \mathcal{B}(\mathbf{C})$ we define the probability measure

$$Q_T(A) = \nu_T \left(\zeta^{(2^{-1} \ln \ln T)^{-1/2}}(\sigma_T + it) \in A \right).$$

THEOREM 3. *The measure Q_T, as $T \to \infty$, weakly converges to the measure defined by the characteristic transform* $\exp\{-(\tau^2/2)-(k^2/2)\}$.

We remind that the characteristic transform of a probability measure $\mathbf{P}$ on $(\mathbf{C}, \mathcal{B}(\mathbf{C}))$ is

$$\int\limits_{\mathbf{C}\setminus\{0\}} |s|^{i\tau} e^{ik\arg s}\, d\mathbf{P},$$

where $\tau \in \mathbf{R}$, $k \in \mathbf{Z}$. From the convergence of the characteristic transforms to a function continuous at $\tau = 0$ the weak convergence of probability measures follows.

Since the function h: $\mathbf{R}^2 \to \mathbf{C}$ defined by the formula $h(x,y) = x + iy$, $x, y \in \mathbf{R}$, is continuous we can study the measure

$$\widetilde{\mathbf{P}}_{\sigma,T}(A) = \nu_T((\mathrm{Re}\,\zeta(\sigma+it), \mathrm{Im}\,\zeta(\sigma+it)) \in A), \quad A \in \mathcal{B}(\mathbf{R}^2),$$

instead of the measure $\mathbf{P}_{\sigma,T}$. For this we shall need a two–dimensional limit lemma for trigonometric polynomials. Let

$$p_n(t) = \sum_{m=1}^{n} a_m\, e^{i\lambda_m t},$$

$\varphi_m = \arg a_m$ and let $J_k(\cdot)$ denote the Bessel function. We define the probability measure

$$\mathbf{P}_{p_n,T}(A) = \nu_T((\mathrm{Re}\, p_n(t), \mathrm{Im}\, p_n(t)) \in A), \quad A \in \mathcal{B}(\mathbf{R}^2).$$

LEMMA 1. *The measure $\mathbf{P}_{p_n,T}$, as $T \to \infty$, weakly converges to the measure defined by the characteristic function*

$$\varphi_{p_n}(\tau_1,\tau_2) = \sum_{\substack{k_1,\dots,k_n=-\infty}}^{\infty} \sum_{\substack{l_1,\dots,l_n=-\infty \\ (k_1+l_1)\lambda_1+\dots+(k_n+l_n)\lambda_n=0}}^{\infty} J_{k_1}(|a_1|\tau_1)\dots J_{k_n}(|a_n|\tau_1)\times$$

$$\times J_{l_1}(|a_1|\tau_2)\dots J_{l_n}(|a_n|\tau_2)\exp\left\{i\left(\sum_{m=1}^{n} k_m\left(\varphi_m+\frac{\pi}{2}\right)+\sum_{m=1}^{n} l_m\varphi_m\right)\right\}.$$

Proof. The characteristic function $\varphi_{p_n,T}(\tau_1,\tau_2)$ of the measure $\mathbf{P}_{p_n,T}$ is

$$\frac{1}{T}\int_0^T \exp\{i\tau_1\,\mathrm{Re}\, p_n(t) + i\tau_2\,\mathrm{Im}\, p_n(t)\}\, dt. \tag{1}$$

Since

$$\mathrm{Re}\, p_n(t) = \sum_{m=1}^{n} |a_m| \cos(\lambda_m t + \varphi_m),$$

$$\mathrm{Im}\, p_n(t) = \sum_{m=1}^{n} |a_m| \sin(\lambda_m t + \varphi_m)$$

and

$$e^{ix\cos\theta} = \sum_{l=-\infty}^{\infty} i^l J_l(x)\, e^{il\theta}, \quad e^{ix\sin\theta} = \sum_{l=-\infty}^{\infty} J_l(x)\, e^{il\theta},$$

we obtain from (1) that

$$\varphi_{p_n,T}(\tau_1,\tau_2) = \sum_{\substack{k_1,\dots,k_n=-\infty \\ (k_1+l_1)\lambda_1+\dots+(k_n+l_n)\lambda_n=0}}^{\infty} \sum_{l_1,\dots,l_n=-\infty}^{\infty} J_{k_1}(|a_1|\tau_1)\dots J_{k_n}(|a_n|\tau_1)\times$$

$$\times J_{l_1}(|a_1|\tau_2)\dots J_{l_n}(|a_n|\tau_2)\exp\left\{i\left(\frac{\pi}{2}\sum_{m=1}^{n}k_m + \sum_{m=1}^{n}(k_m+l_m)\varphi_m\right)\right\}+$$

$$+\sum_{\substack{k_1,\dots,k_n=-\infty \\ (k_1+l_1)\lambda_1+\dots+(k_n+l_n)\lambda_n\neq 0}}^{\infty} \sum_{l_1,\dots,l_n=-\infty}^{\infty} J_{k_1}(|a_1|\tau_1)\dots J_{k_n}(|a_n|\tau_1)\times$$

$$\times J_{l_1}(|a_1|\tau_2)\dots J_{l_n}(|a_n|\tau_2)\exp\left\{i\left(\frac{\pi}{2}\sum_{m=1}^{n}k_m + \sum_{m=1}^{n}(k_m+l_m)\varphi_m\right)\right\}\times$$

$$\times\left(\exp\left\{iT\sum_{m=1}^{n}(k_m+l_m)\lambda_m\right\}-1\right)\Big/\left(T\sum_{m=1}^{n}(k_m+l_m)\lambda_m\right). \tag{2}$$

It is well-known that uniformly in $|x| \leqslant c_1$ the estimate $J_l(x) = B/l!$ is valid. Here and further on, $c_1, c_2, \dots$ are positive constants, B denotes a number (not always the same) which is bounded by a constant. Let ε be an arbitrary positive number. Then there exists a number $M = M(\varepsilon)$ such that for all $|\tau_1| \leqslant c_2$, $|\tau_2| \leqslant c_3$ the inequality

$$\sum_{\substack{k_1,\dots,k_n=-\infty \\ (|k_1|+|l_1|)+\dots+(|k_n|+|l_n|)>M}}^{\infty} \sum_{l_1,\dots,l_n=-\infty}^{\infty} |J_{k_1}(|a_1|\tau_1)\dots$$

$$\dots J_{k_n}(|a_n|\tau_1)J_{l_1}(|a_1|\tau_2)\dots J_{l_n}(|a_n|\tau_2)| < \frac{\varepsilon}{2} \tag{3}$$

is satisfied. Divide the second sum in (2) into two parts $\sum_1$ and $\sum_2$ where in $\sum_1$ the summation is over k_m and l_m such that $(|k_1|+|l_1|)+\dots+(|k_m|+|l_m|) > M$ and in $\sum_2$ it is over the remaining k_m and l_m. Then from (3) it follows that for all $|\tau_1| \leqslant c_2$, $|\tau_2| \leqslant c_3$

$$\left|\sum\nolimits_1\right| < \frac{\varepsilon}{2}. \tag{4}$$

In $\sum_2$ we choose $T_0 = T_0(\varepsilon)$ such that at $T \geqslant T_0$ for all $|\tau_1| \leqslant c_2$, $|\tau_2| \leqslant c_3$ the inequality

$$\left|\sum\nolimits_2\right| < \frac{\varepsilon}{2} \tag{5}$$

is satisfied. Consequently, by (2), (4) and (5) we have that $\varphi_{p_n,T}(\tau_1,\tau_2)$, as $T \to \infty$, converges to $\varphi_{p_n}(\tau_1,\tau_2)$ uniformly in τ_1 and τ_2 in every finite interval. This proves the lemma.

LEMMA 2. *For* $0 < \sigma_0 < \sigma \leqslant 2$, $x \geqslant |t|/\pi$,

$$\zeta(s) = \sum_{m \leqslant x} \frac{1}{m^s} + \frac{x^{1-s}}{s-1} + Bx^{-\sigma},$$

where the constant bounding B depends only on σ_0.

The proof of the lemma can be found, for example, in (Karatsuba, 1979).

LEMMA 3 (Theorem of Montgomery—Vaughan). *Let* $a_1, \dots, a_n$ *be arbitrary complex numbers. Then*

$$\int_0^T \left| \sum_{m \leqslant n} \frac{a_m}{m^{it}} \right|^2 dt = \sum_{m \leqslant n} |a_m|^2 + B \sum_{m \leqslant n} m |a_m|^2.$$

For a proof see (Ivič, 1985, p. 130—134).
Let

$$\zeta_n(\sigma + it) = \sum_{m \leqslant n} \frac{1}{m^{\sigma + it}}.$$

LEMMA 4. *Let* $\sigma > 1/2$. *Then for an arbitrary* $\varepsilon > 0$ *there exists a number* $n = n(\varepsilon, \sigma)$ *such that*

$$\overline{\lim_{T \to \infty}} \frac{1}{T} \int_0^T |\zeta(\sigma + it) - \zeta_n(\sigma + it)|^2 \, dt < \varepsilon, \quad \sigma \neq 1,$$

$$\overline{\lim_{T \to \infty}} \frac{1}{T} \int_1^T |\zeta(\sigma + it) - \zeta_n(\sigma + it)|^2 \, dt < \varepsilon.$$

Proof. Since for $\sigma \geqslant 2$ the Dirichlet series for the function $\zeta(s)$ converges uniformly, the proof is obvious. Consequently, we can limit ourselves only by the case $1/2 < \sigma < 2$. Let $\sigma \neq 1$. We put $x = T$ in Lemma 2. Then applying Lemmas 2 and 3 we find that

$$\int_0^T |\zeta(\sigma + it) - \zeta_n(\sigma + it)|^2 \, dt =$$

$$= B \int_0^T \left| \sum_{n < m \leqslant T} \frac{1}{m^{\sigma + it}} \right|^2 dt + BT^{2-2\sigma} =$$

$$= BT \sum_{n < m \leqslant T} \frac{1}{m^{2\sigma}} + B \sum_{n < m \leqslant T} \frac{1}{m^{2\sigma - 1}} + BT^{2-2\sigma} =$$

$$= BT\sum_{m>n}\frac{1}{m^{2\sigma}} + BT^{2-2\sigma}.$$

Hence, since $\sigma > 1/2$, the first assertion of Lemma 4 follows. The case $\sigma = 1$ is treated similarly.

Proof of Theorem 1. As it was remarked above it is sufficient to consider the weak convergence of the measure $\widetilde{\mathbf{P}}_{\sigma,T}$.
Let $\varphi_{\zeta_n,T}(\tau_1,\tau_2)$ denote the characteristic function of the measure $\mathbf{P}_{\zeta_n,T}$ and let $\varphi_T(\tau_1,\tau_2)$ is the characteristic function of the measure $\widetilde{\mathbf{P}}_{\sigma,T}$. By the Cauchy inequality we have

$$\left|\varphi_T(\tau_1,\tau_2) - \varphi_{\zeta_n,T}(\tau_1,\tau_2)\right| \leqslant$$

$$\leqslant \frac{|\tau_1|}{T}\int_0^T |\mathrm{Re}\,(\zeta(\sigma+it) - \zeta_n(\sigma+it))|\,dt +$$

$$+\frac{|\tau_2|}{T}\int_0^T |\mathrm{Im}\,(\zeta(\sigma+it) - \zeta_n(\sigma+it))|\,dt \leqslant$$

$$\leqslant 2(|\tau_1| + |\tau_2|)\left(\frac{1}{T}\int_0^T |\zeta(\sigma+it) - \zeta_n(\sigma+it)|^2\,dt\right)^{1/2} \qquad (6)$$

Let $\sigma \neq 1$ and τ_0 be an arbitrary positive number. Then from (6) and Lemma 4 it follows that for every $\varepsilon > 0$ there exist numbers $n = n(\varepsilon,\sigma,\tau_0)$ and $T_0 = T_0(\varepsilon,\sigma,\tau_0)$ such that when $T > T_0$ the inequality

$$|\varphi_T(\tau_1,\tau_2) - \varphi_{\zeta_n,T}(\tau_1,\tau_2)| < \varepsilon \qquad (7)$$

is satisfied for all $|\tau_1| \leqslant \tau_0$ and $|\tau_2| \leqslant \tau_0$. In view of Lemma 1, $\varphi_{\zeta_n,T}(\tau_1,\tau_2)$, as $T \to \infty$, converges to the characteristic function $\varphi_{\zeta_n}(\tau_1,\tau_2)$, and this convergence is uniform in every finite interval. Consequently, there exists $T_0' = T_0'(\varepsilon,\sigma,\tau_0)$ such that for $T > T_0'$ the inequality

$$|\varphi_{\zeta_n,T}(\tau_1,\tau_2) - \varphi_{\zeta_n}(\tau_1,\tau_2)| < \varepsilon \qquad (8)$$

is valid for all $|\tau_1| \leqslant \tau_0$ and $|\tau_2| \leqslant \tau_0$. Let us choose $T_1, T_2 > \max(T_0, T_0')$. Then taking into account the inequalities (7) and (8) we find that for all $|\tau_1| \leqslant \tau_0$ and $|\tau_2| \leqslant \tau_0$

$$|\varphi_{T_1}(\tau_1,\tau_2) - \varphi_{T_2}(\tau_1,\tau_2)| < 4\varepsilon.$$

The above inequality proves that the characteristic function $\varphi_T(\tau_1,\tau_2)$, as $T \to \infty$, converges to some function $\varphi(\tau_1,\tau_2)$ which is continuous at $(0,0)$. Hence, the weak convergence of the measure $\widetilde{\mathbf{P}}_{\sigma,T}$ follows. This proves the theorem.
If $\sigma = 1$, then uniformly for $A \in \mathcal{B}(\mathbf{C})$

$$\nu_T(\zeta(1+it) \in A) = \frac{1}{T-1}\mathrm{meas}\,\{t \in [1,T],\ \zeta(1+it) \in A\} + o(1).$$

Thus the theorem is a consequence of a slight modification of Lemmas 1 and 4. For the proof of Theorems 2 and 3 we shall use Theorems A and B.

LEMMA 5. *Let $T \to \infty$. Then uniformly in $|u| \leqslant c_4$*

$$\int_0^T \exp\left\{u\frac{\arg\zeta(\sigma_T+it)}{\sqrt{\ln l_T}}\right\}dt = BT.$$

Let $1/2 \leqslant \sigma \leqslant \sigma_T$, $l_T = \ln T$. Then uniformly in $|u| \leqslant c_5$

$$\int_0^T \exp\left\{u\frac{\arg\zeta(\sigma+it)}{\sqrt{\ln\ln T}}\right\}dt = BT.$$

Proof. If $\sigma = 1/2$ the assertion of the lemma follows from the estimation

$$\frac{1}{T}\int_0^T \frac{|\arg\zeta(1/2+it)|^m}{(\sqrt{\ln\ln T})^m}\,dt = B\Gamma\left(\frac{m+1}{2}\right), \quad T \to \infty, \tag{9}$$

obtained in (Ghosh, 1981, 1983). Here $\Gamma(x)$ denotes the Euler gamma-function. The proof of the estimation of the type (9) for other values of σ repeats the considerations of (Ghosh, 1981, 1983).

Let $\eta_T > 0$, η_T tends, as $T \to \infty$, to infinity sufficiently slowly (more precisely η_T will be defined below) and let

$$A_T = \left\{t \in [0,T],\ |\zeta(\sigma_T+it)| \geqslant \exp\left\{-\eta_T\sqrt{2^{-1}\ln l_T}\right\}\right\},$$

and

$$\tilde{A}_T = \left\{t \in [0,T],\ |\zeta(\sigma+it)| \geqslant \exp\left\{-\eta_T\sqrt{2^{-1}\ln\ln T}\right\}\right\}$$

for $1/2 \leqslant \sigma \leqslant \sigma_T$, $l_T = \ln T$.

Theorems A and B give us

$$\begin{aligned} \text{meas}\{A_T\} &= T + o(T), \\ \text{meas}\{\tilde{A}_T\} &= T + o(T), \quad T \to \infty. \end{aligned} \tag{10}$$

LEMMA 6. *Let $T \to \infty$. Then uniformly in $|u| \leqslant c_6$*

$$\int_{A_T} |\zeta(\sigma_T+it)|^{u(\ln l_T)^{-1/2}}\,dt = BT.$$

Let $1/2 \leqslant \sigma \leqslant \sigma_T$, $l_T = \ln T$. Then uniformly in $|u| \leqslant c_7$

$$\int_{\tilde{A}_T} |\zeta(\sigma+it)|^{u(\ln\ln T)^{-1/2}}\,dt = BT.$$

Proof. If $u \geqslant 0$, the assertions of the lemma are corollary of the proofs of Theorems A and B. Let $\varkappa = (\ln l_T)^{-1/2}$.
Assume that $u < 0$. Note, that we can suppose $|u| > 1$. It is clear that for $t \in A_T$ the inequality

$$|\zeta(\sigma_T + it)|^{u\varkappa} < 2e^{|u|\eta_T}$$

is satisfied. Let us put

$$\omega_T = \sup_{-\infty<x<\infty} \left|\nu_T\left(|\zeta(\sigma_T + it)|^{\varkappa/\sqrt{2}} < x\right) - G(x)\right|.$$

Since the function $G(x)$ is continuous, we have from Theorem A that $\omega_T \to 0$ as $T \to \infty$. Now we choose η_T, putting $\eta_T = \sqrt{\ln(1/\omega_T)}$.
Let

$$\Omega_j = \left\{t \in A_T,\ |\zeta(\sigma_T + it)|^{u\varkappa} \geqslant 2e^{|u|\eta_T}e^{-j}\right\},\quad j = 0, 1, \ldots, j_0.$$

It is easily seen that $\varnothing = \Omega_0 \subset \Omega_1 \subset \ldots \subset \Omega_{j_0}$. Then

$$\frac{1}{\text{meas}\{A_T\}} \int\limits_{A_T} |\zeta(\sigma_T + it)|^{u\varkappa}\, dt \leqslant$$

$$\leqslant 2e^{|u|\eta_T}e^{-j_0} + \frac{2}{\text{meas}\{A_T\}} \sum_{j=1}^{j_0} e^{|u|\eta_T}e^{1-j}\text{meas}\{\Omega_j\}. \tag{11}$$

Let us estimate meas$\{\Omega_j\}$. From the definition of the set Ω_j it follows in virtue of Theorem A that

$$\text{meas}\{\Omega_j\} \leqslant \text{meas}\left\{t \in [0, T], |\zeta(\sigma_T + it)|^{|u|\varkappa} < \frac{e^j}{2e^{|u|\eta_T}}\right\} \leqslant$$

$$\leqslant \text{meas}\left\{t \in [0, T], |\zeta(\sigma_T + it)|^{|u|\varkappa} < \frac{e^{j/|u|}}{e^{\eta_T}}\right\} =$$

$$= TG\left(\frac{e^{j/|u|}}{e^{\eta_T}}\right) + BT\omega_T.$$

Taking into account the form of the function $G(x)$ and (10), putting $j_0 = [|u|\eta_T]$, and observing (11) we find that

$$\frac{1}{\text{meas}\{A_T\}} \int\limits_{A_T} |\zeta(\sigma_T + it)|^{u\varkappa}\, dt = B.$$

This proves the first part of the lemma. The second assertion of the lemma is obtained in a similar way using Theorem B.

Proof of Theorem 2. Let

$$F_T(x) = \frac{1}{\text{meas}\{A_T\}}\text{meas}\left\{t \in A_T, |\zeta(\sigma_T + it)|^{\varkappa/\sqrt{2}} < x\right\}.$$

From Theorem A and (10) it follows that, as $T \to \infty$, $x \neq 0$

$$v_T(\tau) = \int_{-\infty}^{\infty} |x|^{i\tau} \, dF_T(x) =$$

$$= \frac{1}{\text{meas}\{A_T\}} \int_{A_T} |\zeta(\sigma_T + it)|^{i\tau\varkappa/\sqrt{2}} \, dt = e^{-\tau^2/2} + o(1)$$

uniformly in $|\tau| \leqslant c_8$ for any c_8. From the definition of the modulus we have

$$v_T(\tau) = \frac{1}{\text{meas}\{A_T\}} \int_{A_T} \zeta(\sigma_T + it)^{i\tau\varkappa/(2\sqrt{2})} \bar{\zeta}(\sigma_T + it)^{i\tau\varkappa/(2\sqrt{2})} \, dt.$$

Let z be a complex variable and

$$v_T(z, i\tau) = \frac{1}{\text{meas}\{A_T\}} \int_{A_T} \zeta(\sigma_T + it)^{z\varkappa/(2\sqrt{2})} \bar{\zeta}(\sigma_T + it)^{i\tau\varkappa/(2\sqrt{2})} \, dt.$$

For $t \in A_T$ we have

$$\zeta(\sigma_T + it)^{z\varkappa/(2\sqrt{2})} = \sum_{m=0}^{\infty} \frac{z^m}{m!} \left(\frac{\varkappa}{2\sqrt{2}}\right)^m \left(\ln|\zeta(\sigma_T + it)| + i \arg \zeta(\sigma_T + it)\right)^m.$$

Thus

$$v_T(z, i\tau) = \sum_{m=0}^{\infty} a_T(m) z^m,$$

where

$$a_T(m) = \frac{1}{m!} \left(\frac{\varkappa}{2\sqrt{2}}\right)^m \frac{1}{\text{meas}\{A_T\}} \int_{A_T} \bar{\zeta}(\sigma_T + it)^{i\tau\varkappa/(2\sqrt{2})} \times$$

$$\times \left(\ln|\zeta(\sigma_T + it)| + i \arg \zeta(\sigma_T + it)\right)^m \, dt.$$

Since, for $t \in A_T$,

$$\ln|\zeta(\sigma_T + it)| = B \ln T, \quad \arg \zeta(\sigma_T + it) = B \ln T,$$

for every $T \geqslant c_9$

$$\overline{\lim_{m \to \infty}} |a_T(m)|^{1/m} = 0.$$

Consequently, the function $v(z, i\tau)$ is regular in every disk $|z| \leqslant c_{10}$. By the Cauchy inequality

$$v_T(z, i\tau) = \frac{B}{\text{meas}\{A_T\}} \int_{A_T} |\zeta(\sigma_T + it)|^{\varkappa \text{Re} z/(2\sqrt{2})} \times$$

$$\times \exp\left\{\frac{\varkappa(-\operatorname{Im} z+\tau)}{2\sqrt{2}}\arg\zeta(\sigma_T+it)\right\}dt=$$

$$=\frac{B}{\text{meas}\{A_T\}}\left(\int\limits_{A_T}|\zeta(\sigma_T+it)|^{\varkappa\operatorname{Re} z/\sqrt{2}}\,dt\right)^{1/2}\times$$

$$\times\left(\int\limits_{A_T}\exp\left\{\frac{\varkappa(-\operatorname{Im} z+\tau)}{\sqrt{2}}\arg\zeta(\sigma_T+it)\right\}dt\right)^{1/2}$$

This and Lemmas 5 and 6 imply that there exists a constant c_{13} such that for all $|z|\leqslant c_{11}$, $|\tau|\leqslant c_{12}$ and $T\geqslant T_0(\sigma_T)$

$$|v_T(z,i\tau)|\leqslant c_{13}.$$

Thus to the function $v_T(z,i\tau)$ in every disk $|z|\leqslant c_{14}$ the theorem of Vitali (Titchmarsh, 1939, p. 178) can be applied. Hence, taking into account the uniqueness of the analytic function and (11), we find that for $|z|\leqslant c_{14}$

$$v_T(z,i\tau)=e^{i\tau z/2}+o(1)\quad(T\to\infty). \tag{12}$$

Applying similar considerations for the function

$$v_T(z_1,z_2)=\frac{1}{\text{meas}\{A_T\}}\int\limits_{A_T}\zeta(\sigma_T+it)^{z_1\varkappa/(2\sqrt{2})}\bar{\zeta}(\sigma_T+it)^{z_2\varkappa/(2\sqrt{2})}\,dt$$

by (12) we obtain uniformly in $|z_1|\leqslant c_{15}$, $|z_2|\leqslant c_{16}$

$$v_T(z_1,z_2)=e^{z_1z_2/2}+o(1)\quad(T\to\infty). \tag{13}$$

The characteristic transform $w_T(\tau,k)$ of the probability measure

$$\mathbf{P}_T^*(A)=\frac{1}{\text{meas}\{A_T\}}\text{meas}\left\{t\in A_T,\zeta^{\varkappa/\sqrt{2}}(\sigma_T+it)\in A\right\},\quad A\in\mathcal{B}(\mathbf{C}),$$

is equal to

$$\int\limits_{\mathbf{C}\setminus\{0\}}|s|^{i\tau}e^{ik\arg s}\,d\mathbf{P}_T^*=$$

$$=\frac{1}{\text{meas}\{A_T\}}\int\limits_{A_T}\zeta(\sigma_T+it)^{(i\tau+k)\varkappa/(2\sqrt{2})}\bar{\zeta}(\sigma_T+it)^{(i\tau-k)\varkappa/(2\sqrt{2})}\,dt.$$

Hence by (13) for all $\tau\in\mathbf{R}$ and $k\in\mathbf{Z}$

$$w_T(\tau,k)=\exp\{-(\tau^2/2)-(k^2/2)\}+o(1)\quad(T\to\infty).$$

Consequently, the measure $\mathbf{P}_T^*$, as $T\to\infty$, weakly converges to the measure defined by the characteristic transform $\exp\{-(\tau^2/2)-(k^2/2)\}$. Applying (10) again, we deduce the assertion of the theorem.

The proof of Theorem 3 is similar to that of Theorem 2.

REFERENCES

Bohr, H. (1922). Über diophantische Approximationen und ihre Anwendungen auf Dirichletsche Reihen, besonders auf die Riemannsche Zetafunction. In: *Proc. 5th Congress of Scand. math.* Helsingfors. 131–154.

Bohr, H. and Jessen, B. (1930). Über die Wertverteilung der Riemannschen Zetafunction. Erste Mitteilung. *Acta Math.* **54**, 1–35.

Bohr, H. and Jessen, B. (1932). Über die Wertverteilung der Riemannschen Zetafunction. Zweite Mitteilung. *Acta Math.* **58**, 1–55.

Jessen, B. and Wintner, A. (1935). Distribution functions and the Riemann zeta-function. *Trans. Amer. Math. Soc.* **38**, 48–88.

Laurinčikas, A. (1975). The distribution of values of complex function. *Liet. Matem. Rink.* **29**, 123–134.

Laurinčikas, A. (1987a). A limit theorem for the Riemann zeta-function on the critical line. I. *Liet. Matem. Rink.* **27**, 113–132.

Laurinčikas, A. (1987b). A limit theorem for the Riemann zeta-function on the critical line. II. *Liet. Matem. Rink.* **27**, 485–500.

Laurinčikas, A. (1988). A limit theorem for the Riemann zeta-function near the critical line. *Math. USSR Sb.* **177**, 3–11.

Laurinčikas, A. (1989a). A limit theorem for the Riemann zeta-function on the critical line. III. *Liet. Matem. Rink.* **29**, 83–89.

Laurinčikas, A. (1989b). A limit theorem for the Riemann zeta-function near the critical line. II. *Math. USSR Sb.* **180**, 733–749.

Laurinčikas, A. (1990). A limit theorem for the Riemann zeta-function near the critical line in the complex space. *Acta Aritm.* (to appear).

Karatsuba, A.A. (1979). *Principles of Analytic Number Theory.* Nauka, Moscow.

Ivič, A. (1985). *The Riemann Zeta-Function.* Wiley, New York.

Ghosh, A. (1981). On the Riemann's zeta-function – sign-changes of $S(t)$. In: *Recent Progress in Analytic number theory.* **1**, Academic Press, New York, pp. 25–46.

Ghosh, A. (1983). On the Riemann zeta-function – mean value theorems and the distribution of $S(t)$. *J. Number Theory.* **17**, 93–102.

Titchmarsh, E.C. (1939). *The Theory of Functions.* Oxford Univ. Press, Oxford.

Prob. Theory and Math. Stat., Vol. 2, pp. 70–78
B.Grigelionis *et al.* (Eds.)
1990 VSP/Mokslas

STOCHASTIC INTEGRALS WITH RESPECT TO SEMIMARTINGALE MEASURES AND CHANGE OF THE FILTRATION

V.A. LEBEDEV
Moscow State University, 119899 Moscow, USSR

ABSTRACT

The main results of this paper (Theorem 1 and 2), characterise behaviour of stochastic integrals with respect to μ, ν and $\mu - \nu$ as semimartingale random measures, that is as σ–finite L^p–valued random measures in the sense of (Bichteler and Jacod, 1983), under restriction of the filtration keeping the proper measurability of the integrated function.

Let $(\Omega, \mathcal{F}, \mathbf{P})$ be a complete probability space with two filtrations $\underline{\mathcal{F}} = (\mathcal{F}_t)_{t\in\mathbf{R}_+}$ and $\underline{\mathcal{G}} = (\mathcal{G}_t)_{t\in\mathbf{R}_+}$, satisfying the usual conditions and such that $\mathcal{G}_t \subset \mathcal{F}_t$ for every $t \in \mathbf{R}_+$. Let $\mathcal{P}(\underline{\mathcal{F}})$ and $\mathcal{P}(\underline{\mathcal{G}})$ be the predictable σ–algebras on $\Omega \times \mathbf{R}_+$ with respect to $\underline{\mathcal{F}}$ and $\underline{\mathcal{G}}$, respectively, and let $\mathcal{B}(M)$ be the Borel σ–algebra of a metric space M. If $(N, \mathcal{N})$ is a measurable space, then by $\mathcal{N}$ we shall also mean the class of $\mathcal{N}$–measurable functions on N.
Let $(E, \mathcal{E})$ be a measurable space. As a general enough concept of a driving system for stochastic integration of $\mathcal{P}(\underline{\mathcal{G}})\otimes\mathcal{E}$–measurable functions on $\Omega\times\mathbf{R}_+\times E$, Bichteler and Jacod consider a *σ-finite L^p-valued random measure* on $(\Omega\times\mathbf{R}_+\times E,\ \mathcal{P}(\underline{\mathcal{G}})\otimes\mathcal{E})$ for some $p \geqslant 0$, namely, a family $\theta = (\theta_t)_{t\in\mathbf{R}_+}$ satisfying the following conditions:

(i) for each $t \in \mathbf{R}_+$, θ_t is a σ–finite measure on $(\Omega\times\mathbf{R}_+\times E,\ \mathcal{P}(\underline{\mathcal{G}})\otimes\mathcal{E})$ with values in $L^p(\Omega, \mathcal{G}_t, \mathbf{P})$ (this means that there is a strictly positive $\mathcal{P}(\underline{\mathcal{G}}) \otimes \mathcal{E}$–measurable function V such that if $\widetilde{\mathcal{P}}_V(\underline{\mathcal{G}}) = \{\varphi : \mathcal{P}(\underline{\mathcal{G}}) \otimes \mathcal{E}\text{–measurable, } \varphi/V \text{ bounded}\}$ and then we have: (i1) θ_t is a linear mapping from $\widetilde{\mathcal{P}}_V(\underline{\mathcal{G}})$ into $L^p(\Omega, \mathcal{G}_t, \mathbf{P})$, (i2) if (φ_n) is a sequence in $\widetilde{\mathcal{P}}_V(\underline{\mathcal{G}})$ with $|\varphi_n| \leqslant V$ converging pointwise to 0, then $\theta_t(\varphi_n) \to 0$ in L^p);

(ii) $\theta_s(\varphi) = \theta_t(\varphi \mathbf{1}_{\Omega\times[0,s]\times E})$ for all $\varphi \in \widetilde{\mathcal{P}}_V(\underline{\mathcal{G}})$, $s \leqslant t$;

(iii) $\theta_t(\varphi\mathbf{1}_{A\times I\times E}) = \mathbf{1}_A\theta_t(\varphi\mathbf{1}_{\Omega\times I\times E})$ for all $\varphi \in \widetilde{\mathcal{P}}_V(\underline{\mathcal{G}})$, $t \in \mathbf{R}_+$, if $A \in \mathcal{G}_0$ and $I = \mathbf{R}_+$, or if $A \in \mathcal{G}_s$ and $I =]s, s']$ with $s < s'$.
In particular, when $V \equiv 1$, θ is called a *finite L^p-valued random measure* on $(\Omega \times \times\mathbf{R}_+ \times E,\ \mathcal{P}(\underline{\mathcal{G}}) \otimes \mathcal{E})$. Let us denote by $\tilde{S}^p_\sigma(\underline{\mathcal{G}})$ (respectively, by $\tilde{S}^p(\underline{\mathcal{G}})$) the space of all σ–finite (respectively, finite) L^p–valued random measures on $(\Omega \times \mathbf{R}_+ \times \times E,\ \mathcal{P}(\underline{\mathcal{G}}) \otimes \mathcal{E})$.
For $\theta \in \tilde{S}^p_\sigma(\underline{\mathcal{G}})$ we put

$$\|\varphi\|_{L^{1,p}(\theta)}=\sum_{n=1}^{\infty}2^{-n}(1\wedge\sup_{\psi\in\tilde{\mathcal{P}}_V(\mathcal{G}),\,|\psi|\leqslant|\varphi|}\|\theta_n(\psi)\|_p),$$

and by $L^{1,p}(\theta)$ denote the set of $\mathcal{P}(\mathcal{G})\otimes\mathcal{E}$–measurable functions φ on $\Omega\times\mathbf{R}_+\times E$ for each of which there exists a sequence $(\varphi_n)\in\widetilde{\mathcal{P}}_V(\mathcal{G})$ with $\|\varphi_n-\varphi\|_{L^{1,p}(\theta)}\to 0$. Then, for each $t\in\mathbf{R}_+$, $\theta_t(\varphi_n)$ tends to a limit in L^p which does not depend on the chosen (φ_n) for the given φ and is denoted by $\theta_t(\varphi)$. Besides that, $\theta\in\tilde{S}^p_\sigma(\mathcal{G})$ and $\widehat{L}^{1,q}(\theta)\subset L^{1,p}(\theta)$ if $p\leqslant q$ and $\theta\in\tilde{S}^q_\sigma(\mathcal{G})$.
Let us note that for every $\varphi\in L^{1,p}(\theta)$, the random variables $\theta_t(\varphi)$ at various $t\in\mathbf{R}_+$ constitute a semimartingale denoted by $\varphi*\theta$ and called the ***stochastic integral process*** of φ with respect to θ. By $\widehat{L}^p(\theta)$ we denote the set of $\mathcal{P}(\mathcal{G})\otimes\mathcal{E}$–measurable functions φ on $\Omega\times\mathbf{R}_+\times E$ for each of which there is a strictly positive $\mathcal{G}$–predictable process K such that $K\varphi\in L^{1,p}(\theta)$ and $1/K\in L^{1,p}(K\varphi*\theta)$ with $E=\{1\}$ and the semimartingale $K\varphi*\theta$ which is regarded as a finite L^p–valued random measure on $(\Omega\times\mathbf{R}_+,\mathcal{P}(\mathcal{G}))$ (for $p=0$ this coincides with the usual stochastic integrability with respect to a semimartingale). Then for $\varphi\in\widehat{L}^p(\theta)$ we define $\varphi*\theta=1/K\cdot(K\varphi*\theta)$, and this obviously does not depend on the chosen K. We have of course, that $L^{1,p}(\theta)\subset\widehat{L}^p(\theta)$ and $\widehat{L}^q(\theta)\subset\widehat{L}^p(\theta)$ for $p\leqslant q$.
According to (Lebedev, 1986), $\tilde{\theta}\in\tilde{S}^p_\sigma(\mathcal{F})$ is called an *extension* of some $\theta\in\tilde{S}^p_\sigma(\mathcal{G})$ to $(\Omega\times\mathbf{R}_+\times E,\ \mathcal{P}(\mathcal{F})\otimes\mathcal{E})$, if there exists a strictly positive $\mathcal{P}(\mathcal{G})\otimes\mathcal{E}$–measurable function W on $\Omega\times\mathbf{R}_+\times E$, belonging simultaneously to $L^{1,p}(\theta)$ and to $L^{1,p}(\tilde{\theta})$ and such that for any $\varphi\in\widetilde{\mathcal{P}}_w(\mathcal{G})$ we have $\varphi*\theta=\varphi*\tilde{\theta}$ up to $\mathbf{P}$–indistinguishability. If, for the measure $\theta\in\tilde{S}^p_\sigma(\mathcal{G})$ and for the filtration $\mathcal{F}$, there exists such a measure $\tilde{\theta}\in\tilde{S}^p_\sigma(\mathcal{F})$, then we say that θ admits an extension to $(\Omega\times\mathbf{R}_+\times E,\ \mathcal{P}(\mathcal{F})\otimes\mathcal{E})$ as a σ–finite L^p–valued random measure. According to the assertion a) of the theorem from (Lebedev, 1986), if $\theta\in\tilde{S}^p_\sigma(\mathcal{G})$ admits an extension $\tilde{\theta}$ to $(\Omega\times\mathbf{R}_+\times E,\ \mathcal{P}(\mathcal{F})\otimes\mathcal{E})$ then $\tilde{\theta}$ is unique, $(\mathcal{P}(\mathcal{G})\otimes\mathcal{E})\bigcap L^{1,p}(\tilde{\theta})\subset L^{1,p}(\theta)$, and for $\varphi\in(\mathcal{P}(\mathcal{G})\otimes\mathcal{E})\bigcap L^{1,p}(\tilde{\theta}))$

$$\varphi*\theta=\varphi*\tilde{\theta}$$

up to $\mathbf{P}$–indistinguishability.
Now let $(E,\mathcal{E})$ be a separable metric space, which is universally measurable in its completion, with $\mathcal{E}=\mathcal{B}(E)$, μ be a $\mathcal{G}$–optional $(\mathcal{P}(\mathcal{G})\otimes\mathcal{B}(E))$ and σ–finite integer–valued random measure on $(\mathbf{R}_+\times E,\ \mathcal{B}(\mathbf{R}_+)\otimes\mathcal{B}(E))$, ν and $\tilde{\nu}$ be its dual predictable projections on $\mathcal{G}$ and $\mathcal{F}$, respectively. Then μ, ν and $\mu-\nu$ can be regarded as σ–finite L^0–valued random measures on $(\Omega\times\mathbf{R}_+\times E,\ \mathcal{P}(\mathcal{G})\otimes\mathcal{B}(E))$ by

$$\mu_t(h)=\int_{[0,t]}h_{s,u}\,\mu(ds\times du),\tag{1}$$

and so on, for a $\mathcal{P}(\mathcal{G})\otimes\mathcal{B}(E)$–measurable h which is $\mathbf{P}$–a.s. absolutely integrable up to every $t\in\mathbf{R}_+$ with respect to μ, ν and $|\mu-\nu|$ (the variation of $\mu-\nu$), respectively. For a $\mathcal{P}(\mathcal{F})\otimes\mathcal{B}(E)$–measurable h, (1) gives the extension of μ, ν and $\mu-\nu$ to $(\Omega\times\mathbf{R}_+\times E,\ \mathcal{P}(\mathcal{F})\otimes\mathcal{B}(E))$ and defines $\tilde{\nu}$ and $\mu-\tilde{\nu}$ as σ–finite L^0–valued random measures on $(\Omega\times\mathbf{R}_+\times E,\ \mathcal{P}(\mathcal{F})\otimes\mathcal{B}(E))$. Let ψ be a strictly positive $\mathcal{P}(\mathcal{G}) \otimes \mathcal{B}(E)$–measurable function on $\Omega \times \mathbf{R}_+ \times E$, belonging to $L^{1,0}(\nu)$. By Theorem 4.10 from (Bichteler and Jacod, 1983), it means that the integral

$$\nu_t(\psi) = \int_{[0,t]} \int_E \psi_{s,u} \nu(ds \times du)$$

is finite $\mathbf{P}$–a.s. for every $t \in \mathbf{R}_+$.
Let us define the set $J = \{(\omega, t) \in \Omega \times \mathbf{R}_+ : \nu(\omega, \{t\} \times E) > 0\}$. Without loss of generality (by extending E if necessary), we shall assume that

$$\nu(\omega, \{t\} \times E) = \mathbf{1}_J(\omega, t), \tag{2}$$

and point out it below. For $A = \psi * \nu$ introduce the factorization

$$\nu(\omega, ds \times du) = \varrho_s(\omega, du) dA_s(\omega) \tag{3}$$

with a regular transition measure ϱ from $(\Omega \times \mathbf{R}_+, \mathcal{P}(\mathcal{G}))$ to $(E, \mathcal{B}(E))$. Notice that instead of $\psi * \nu$, as A we can take any $\mathcal{G}$–predictable increasing process with respect to which $\psi * \nu$ is absolutely continuous $\mathbf{P}$–a.s.
Let λ be the dual $\mathcal{G}$–predictable projection of $|\tilde{\nu} - \nu|$. Obviously $0 \leqslant \lambda \leqslant 2\nu$, and there exists a $\mathcal{P}(\mathcal{G}) \otimes \mathcal{B}(E)$–measurable function g on $\Omega \times \mathbf{R}_+ \times E$ with $0 \leqslant g \leqslant 2$ such that $\lambda = g \cdot \nu$.
According to (Lebedev, 1990), if μ, ν and $\mu - \nu$ are regarded as L^0–valued σ–finite random measures on $(\Omega \times \mathbf{R}_+ \times E,\ \mathcal{P}(\mathcal{G}) \otimes \mathcal{B}(E))$ then the spaces $\widehat{L}^0(\nu, \mathcal{G})$, $\widehat{L}^0(\mu, \mathcal{G})$ and $\widehat{L}^0(\mu - \nu, \mathcal{G})$ coincide with the sets of $\mathcal{P}(\mathcal{G}) \otimes \mathcal{B}(E)$–measurable functions on $\Omega \times \mathbf{R}_+ \times E$, for which integrals in right–hand sides are well–defined:

$$h * \nu = \int_E h_{\cdot,u} \varrho(du) \cdot A, \tag{4}$$

$$h * \mu = h\mathbf{1}_{\{|h|>1\}} * \mu + h\mathbf{1}_{\{|h|\leqslant 1\}} * (\mu - \nu) + h\mathbf{1}_{\{|h|\leqslant 1\}} * \nu \tag{5}$$

and (under(2))

$$h * (\mu - \nu) = (h - \bar{h})\mathbf{1}_{\{|h-\bar{h}|>1\}} * \mu - (h - \bar{h})\mathbf{1}_{\{|h-\bar{h}|>1\}} * \nu +$$

$$+ (h - \bar{h})\mathbf{1}_{\{|h-\bar{h}|\leqslant 1\}} * (\mu - \nu) \tag{6}$$

with

$$\bar{h}_t(\omega) = \int_E h_{t,u}(\omega) \nu(\omega, \{t\} \times du).$$

In (5) and (6) integrals being defined with respect to ν in the sense of (4) and with respect to $\mu - \nu$ as a martingale measure. Then so defined stochastic integrals with respect to ν, μ and $\mu - \nu$ in the sense of $\widehat{L}^0$, that is with respect to semimartingale measures, are expressed by (4), (5) and (6), respectively.
According to (Jacod, 1980, Theorem 9), for $\tilde{\nu} = \nu$ it is necessary and sufficient that $\tilde{\nu}$ would be $\mathcal{G}$–predictable and in this case $h \in \mathcal{P}(\mathcal{G}) \otimes \mathcal{B}(E)$ admits the stochastic

integral with respect to $\mu - \nu$ as a martingale measure on $\underline{\mathcal{F}}$, if and only if h is such on $\underline{\mathcal{G}}$, and those stochastic integrals coincide. The following more general result is valid.

THEOREM 1. a) *We have the equality*

$$(\mathcal{P}(\underline{\mathcal{G}}) \otimes \mathcal{B}(E)) \bigcap \widehat{\mathcal{L}}^0(\nu, \underline{\mathcal{F}}) = \widehat{\mathcal{L}}^0(\nu, \underline{\mathcal{G}}),$$

and the stochastic integrals with respect to ν *on* $\underline{\mathcal{G}}$ *and* $\underline{\mathcal{F}}$ *coincide.*

b) *On* $\widehat{\mathcal{L}}^0(\mu, \underline{\mathcal{G}}) \bigcap \widehat{\mathcal{L}}^0(\mu, \underline{\mathcal{F}})$ *the stochastic integrals with respect to* μ *on* $\underline{\mathcal{G}}$ *and* $\underline{\mathcal{F}}$ *coincide, and for* $h \in (\mathcal{P}(\underline{\mathcal{G}}) \otimes \mathcal{B}(E)) \bigcap \widehat{\mathcal{L}}^0(\mu, \underline{\mathcal{F}})$ *we have* $h \in \widehat{\mathcal{L}}^0(\mu, \underline{\mathcal{G}})$ *iff*

$$\int_E |h| \mathbf{1}_{\{|h| \leqslant 1\}} d\varrho < \infty \qquad \mathbf{P} \times dA - a.e.$$

c) *We have the inclusion*

$$(\mathcal{P}(\underline{\mathcal{G}}) \otimes \mathcal{B}(E)) \bigcap \widehat{\mathcal{L}}^0(\mu - \nu, \underline{\mathcal{F}}) \subset \widehat{\mathcal{L}}^0(\mu - \nu, \underline{\mathcal{G}})$$

and, if

$$\int_E |h - \bar{h}| \mathbf{1}_{\{|h - \bar{h}| \leqslant 1\}} \mathbf{1} d\varrho < \infty \qquad \mathbf{P} \times dA - a.e.,$$

then on the left-hand side the stochastic integrals $h * (\mu - \nu)$ *on* $\underline{\mathcal{G}}$ *and* $\underline{\mathcal{F}}$ *coincide.*

d) *For* $\tilde{\nu} = \nu$ *it is necessary and sufficient that* $\tilde{\nu}$ *would be* $\underline{\mathcal{G}}$*-predictable, in this case we have the equalities*

$$(\mathcal{P}(\underline{\mathcal{G}}) \otimes \mathcal{B}(E)) \bigcap \widehat{\mathcal{L}}^0(\mu, \underline{\mathcal{F}}) = \widehat{\mathcal{L}}^0(\mu, \underline{\mathcal{G}})$$

and

$$(\mathcal{P}(\underline{\mathcal{G}}) \otimes \mathcal{B}(E)) \bigcap \widehat{\mathcal{L}}^0(\mu - \nu, \underline{\mathcal{F}}) = \widehat{\mathcal{L}}^0(\mu - \nu, \underline{\mathcal{G}}),$$

and the stochastic integrals with respect to μ *and* $\mu - \nu$*, respectively, defined on* $\underline{\mathcal{G}}$ *and* $\underline{\mathcal{F}}$ *coincide.*

Proof. a) The factorization (3) on $\underline{\mathcal{G}}$ remains the same on $\underline{\mathcal{F}}$ for the measure ν, hence and from (4) the result follows for ν.

b) Let $h \in (\mathcal{P}(\underline{\mathcal{G}}) \otimes \mathcal{B}(E)) \bigcap \widehat{\mathcal{L}}^0(\mu, \underline{\mathcal{F}})$. If $h \in \widehat{\mathcal{L}}^0(\mu, \underline{\mathcal{G}})$ then from (5) $h\mathbf{1}_{\{|h| \leqslant 1\}} \in \widehat{\mathcal{L}}^0(\nu, \underline{\mathcal{F}})$, hence and from (4)

$$\int_E |h| \mathbf{1}_{\{|h| \leqslant 1\}} d\varrho < \infty \qquad \mathbf{P} \times dA - \text{a.e.}$$

Conversely, let

$$\int_E |h| \mathbf{1}_{\{|h| \leqslant 1\}} d\varrho < \infty \qquad \mathbf{P} \times dA - \text{a.e.}$$

Since if

$$K = 1\Big/\left(\int_E |h|\mathbf{1}_{\{|h|\leqslant 1\}}d\varrho\right)\bigvee 1,$$

then $K|h|\mathbf{1}_{\{|h|\leqslant 1\}} * \nu \leqslant A$, so that

$$Kh\mathbf{1}_{\{|h|\leqslant 1\}} \in L^{1,0}(\nu,\underline{\mathcal{G}}),\quad Kh\mathbf{1}_{\{|h|\leqslant 1\}} \in L^{1,0}(\mu,\underline{\mathcal{G}}) \subset L^{1,0}(\mu,\underline{\mathcal{F}})$$

and

$$h\mathbf{1}_{\{|h|\leqslant 1\}} * \mu = 1/K \cdot (Kh\mathbf{1}_{\{|h|\leqslant 1\}} * \mu) \tag{7}$$

on $\underline{\mathcal{F}}$. Since, by (Jacod, 1980, Theorem 7), (7) holds on $\underline{\mathcal{G}}$ also and the integral in the right–hand side is the same for $\underline{\mathcal{G}}$ and $\underline{\mathcal{F}}$, so that $h\mathbf{1}_{\{|h|\leqslant 1\}} \in \widehat{L}^0(\mu,\underline{\mathcal{G}})$. At last

$$h\mathbf{1}_{\{|h|>1\}} \in (\mathcal{P}(\underline{\mathcal{G}})\otimes\mathcal{B}(E))\bigcap L^{1,0}(\mu,\underline{\mathcal{F}}) = L^{1,0}(\mu,\underline{\mathcal{G}})$$

and

$$h * \mu = h\mathbf{1}_{\{|h|>1\}} * \mu + h\mathbf{1}_{\{|h|\leqslant 1\}} * \mu \tag{8}$$

on $\underline{\mathcal{G}}$ and $\underline{\mathcal{F}}$ simultaneously, so that $h \in \widehat{L}^0(\mu,\underline{\mathcal{G}})$, and by the preceding proof the stochastic integrals $h * \mu$ on $\underline{\mathcal{G}}$ and $\underline{\mathcal{F}}$ coincide.

c) Let $h\mathbf{1}_{\{|h|\leqslant 1\}} \in (\mathcal{P}(\underline{\mathcal{G}})\otimes\mathcal{B}(E))\bigcap\widehat{L}^0(\mu-\nu,\underline{\mathcal{F}})$. Analogously to (6), for the integration with respect to $\underline{\mathcal{F}}$ of $h \in \widehat{L}^0(\mu-\nu,\underline{\mathcal{F}})$, we have the representation

$$h*(\mu-\nu) = (h-\bar{h})\mathbf{1}_{\{|h-\bar{h}|>1\}} * \mu - (h-\bar{h})\mathbf{1}_{\{|h-\bar{h}|>1\}} * \nu +$$

$$+(h-\bar{h})\mathbf{1}_{\{|h-\bar{h}|\leqslant 1\}} * (\mu-\tilde{\nu}) + (h-\bar{h})\mathbf{1}_{\{|h-\bar{h}|\leqslant 1\}} * (\tilde{\nu}-\nu), \tag{9}$$

with respect to ν and $\tilde{\nu}-\nu$ in the sense of (4) on $\underline{\mathcal{F}}$. Thus, similarly to the preceding proof,

$$(h-\bar{h})\mathbf{1}_{\{|h-\bar{h}|>1\}} \in (\mathcal{P}(\underline{\mathcal{G}})\otimes\mathcal{B}(E))\bigcap L^{1,0}(\mu,\underline{\mathcal{F}}) = L^{1,0}(\mu,\underline{\mathcal{G}}),$$

$$(h-\bar{h})\mathbf{1}_{\{|h-\bar{h}|>1\}} \in (\mathcal{P}(\underline{\mathcal{G}})\otimes\mathcal{B}(E))\bigcap\widehat{L}^0(\nu,\underline{\mathcal{F}}) = \widehat{L}^0(\nu,\underline{\mathcal{G}})$$

and

$$(h-\bar{h})^2\mathbf{1}_{\{|h-\bar{h}|\leqslant 1\}} \in (\mathcal{P}(\underline{\mathcal{G}})\otimes\mathcal{B}(E))\bigcap L^{1,0}(\mu,\underline{\mathcal{F}}) = L^{1,0}(\mu,\underline{\mathcal{G}}).$$

Hence $(h-\bar{h})^2\mathbf{1}_{\{|h-\bar{h}|\leqslant 1\}} \in L^{1,0}(\nu,\underline{\mathcal{G}})$ and $(h-\bar{h})\mathbf{1}_{\{|h-\bar{h}|\leqslant 1\}} \in L^{1,0}(\mu-\nu,\underline{\mathcal{G}})$, so that by (6) $h \in \widehat{L}^0(\mu-\nu,\underline{\mathcal{G}})$. Now let

$$\int_E |h-\bar{h}|\mathbf{1}_{\{|h-\bar{h}|\leqslant 1\}}g\,d\varrho < \infty.$$

Since if

$$K = 1\Big/\left(\int_E |h-\bar{h}|\mathbf{1}_{\{h-\bar{h}|\leqslant 1\}}g\,d\varrho\right)\bigvee 1,$$

then $K|h-\bar{h}|\mathbf{1}_{\{|h-\bar{h}|\leqslant 1\}} * \lambda \leqslant A$ so that $K(h-\bar{h})\mathbf{1}_{\{|h-\bar{h}|\leqslant 1\}} \in L^{1,0}(\tilde{\nu}-\nu,\underline{\mathcal{F}})$. Hence $K(h-\bar{h})\mathbf{1}_{\{|h-\bar{h}|\leqslant 1\}} \in L^{1,0}(\mu-\nu,\underline{\mathcal{F}})$, and so

$$(h-\bar{h})\mathbf{1}_{\{|h-\bar{h}|\leqslant 1\}} * (\mu-\nu) = 1/K \cdot \left[K(h-\bar{h})\mathbf{1}_{\{|h-\bar{h}|\leqslant 1\}} * (\mu-\nu)\right] \tag{10}$$

on $\underline{\mathcal{F}}$, but by (Jacod, 1980, Theorem 7) (10) holds also on $\underline{\mathcal{G}}$ and the integral in the right-hand side is the same for $\underline{\mathcal{G}}$ and $\underline{\mathcal{F}}$. At last for $h \in \widehat{L}^0(\mu-\nu,\underline{\mathcal{F}})$ (6) is also valid with the integral $h\mathbf{1}_{\{|h|\leqslant 1\}} * (\mu-\nu)$ with respect to $\mu-\nu$ as a semimartingale measure on $\underline{\mathcal{F}}$, and by the preceding proof the stochastic integrals $h * (\mu - \nu)$ on $\underline{\mathcal{G}}$ and $\underline{\mathcal{F}}$ also coincide.

d) At last if the measure $\tilde{\nu}$ is $\underline{\mathcal{G}}$–predictable then, by (Jacod, 1980, Theorem 9), it is the dual $\underline{\mathcal{G}}$–predictable projection of ν and hence it coincides with ν. In this case instead of inclusions, by (5) and (6) we have equalities and the stochastic integrals also coincide by the above proof together with $\tilde{\nu} = \nu$.

COROLLARY. *Let $\mu^d = \mathbf{1}_{J\times E}\cdot\mu$ and $\nu^d = \mathbf{1}_{J\times E}\cdot\nu$. Then:*

a) *the inclusion $(\mathcal{P}(\underline{\mathcal{G}})\otimes\mathcal{B}(E))\bigcap\widehat{L}^0(\mu^d,\underline{\mathcal{F}}) \subset \widehat{L}^0(\mu^d,\underline{\mathcal{G}})$ holds;*

b) *on $(\mathcal{P}(\underline{\mathcal{G}})\otimes\mathcal{B}(E))\bigcap\widehat{L}^0(\mu^d-\nu^d,\underline{\mathcal{F}})$ the stochastic integrals with respect to $\mu^d-\nu^d$ on $\underline{\mathcal{G}}$ and $\underline{\mathcal{F}}$ coincide.*

Proof. a) Since the set J is $\mathcal{P}(\underline{\mathcal{G}})$–measurable, ν^d is the dual $\underline{\mathcal{G}}$–predictable projection of μ^d. Let $\varrho^d = \mathbf{1}_{J\times E}\cdot\varrho$; then we have the factorization

$$\nu^d(\omega, ds\times du) = \varrho^d_s(\omega,du)dA_s(\omega) = \varrho^d_s(\omega,du)dA^d_s(\omega),$$

where $A^d = \mathbf{1}_J\cdot A = \psi * \nu^d$. Now we have

$$\int_E |h_{s,u}|\mathbf{1}_{\{|h_{s,u}|\leqslant 1\}}\varrho^d_s(du) \leqslant \varrho^d_s(E) = \frac{\nu^d(\{s\},E)\mathbf{1}_J(s)}{\Delta A_s} = \frac{\mathbf{1}_J(s)}{\Delta A_s}$$

(the last equality by (2)). So if

$$h \in (\mathcal{P}(\underline{\mathcal{G}})\otimes\mathcal{B}(E))\bigcap\widehat{L}^0(\mu^d,\underline{\mathcal{F}}),$$

then $h \in \widehat{L}^0(\mu^d,\underline{\mathcal{G}})$.

b) Let $\tilde{\nu}^d = \mathbf{1}_{J\times E}\cdot\tilde{\nu}$. Since the set J is $\mathcal{P}(\underline{\mathcal{G}})$–measurable, then $\tilde{\nu}^d$ is the dual $\underline{\mathcal{F}}$–predictable projection of μ^d. Therefore $|\tilde{\nu}^d-\nu^d| = \mathbf{1}_{J\times E}\cdot|\tilde{\nu}-\nu|$ and $|\tilde{\nu}^d-\nu^d|\mathcal{P}(\underline{\mathcal{G}}) = \mathbf{1}_{J\times E}\cdot\lambda = g\mathbf{1}_{J\times E}\cdot\nu = g\cdot\nu^d$. Now we have

$$\int_E |h_{s,u}-\bar{h}_s|\mathbf{1}_{\{|h_{s,u}-\bar{h}_s|\leqslant 1\}}g_{s,u}\varrho^d_s(du) \leqslant 2\varrho^d_s(E) = \frac{2\mathbf{1}_J(s)}{\Delta A_s},$$

so that for $h \in (\mathcal{P}(\underline{\mathcal{G}})\otimes\mathcal{B}(E))\bigcap\widehat{L}^0(\mu^d-\nu^d,\underline{\mathcal{F}})$ its stochastic integrals with respect to $\mu^d-\nu^d$ on $\underline{\mathcal{G}}$ and $\underline{\mathcal{F}}$ coincide.

Theorem 1 is a more precise version of the main result of (Lebedev, 1989), where primary proof there was a flaw. Below for the case when $\tilde{\nu}$ is an abstract $\underline{\mathcal{F}}$-predictable $\mathcal{P}(\underline{\mathcal{G}}) \otimes \mathcal{B}(E)$ — σ-finite random measure on $(\mathbf{R}_+ \times E, \mathcal{B}(\mathbf{R}_+) \otimes \mathcal{B}(E))$ for which ν is its dual $\underline{\mathcal{G}}$-predictable projection, we shall adduce counter-examples where $h \in (\mathcal{P}(\underline{\mathcal{G}}) \otimes \mathcal{B}(E)) \bigcap \widehat{\mathcal{L}}^0(\tilde{\nu}, \underline{\mathcal{F}})$ (and even $h \in L^{1,0}(\tilde{\nu}, \underline{\mathcal{F}})$) but $h \notin \widehat{\mathcal{L}}^0(\nu, \underline{\mathcal{G}})$, and for $h \in (\mathcal{P}(\underline{\mathcal{G}}) \otimes \mathcal{B}(E)) \bigcap \widehat{\mathcal{L}}^0(\tilde{\nu} - \nu, \underline{\mathcal{F}})$ we have

$$[h * (\tilde{\nu} - \nu)]^{\mathcal{P}(\underline{\mathcal{G}})} \neq 0$$

(by (6) and (9) the condition $[(h - \bar{h})\mathbf{1}_{\{|h-\bar{h}| \leqslant 1\}} * (\tilde{\nu} - \nu)]^{\mathcal{P}(\underline{\mathcal{G}})} = 0$ is obviously necessary for $h \overset{\underline{\mathcal{F}}}{*} (\mu - \nu) = h \overset{\underline{\mathcal{G}}}{*} (\mu - \nu)$). However in our original case when $\tilde{\nu}$ is the dual $\underline{\mathcal{F}}$-predictable projection of some $\underline{\mathcal{G}}$-optional integer-valued random measure μ, the situation is more complicated and the question of existence of such counter-examples remains open. Let us give the following result which cannot be extended to the abstract case.

THEOREM 2. a) *For*

$$h \in (\mathcal{P}(\underline{\mathcal{G}}) \otimes \mathcal{B}(E)) \bigcap \widehat{\mathcal{L}}^0(\mu, \underline{\mathcal{F}}),$$

we have $h \in \widehat{\mathcal{L}}^0(\mu, \underline{\mathcal{G}})$ if and only if there exists a strictly positive $\mathcal{P}(\underline{\mathcal{G}})$-measurable process K such that

$$K|h|\mathbf{1}_{\{|h| \leqslant 1\}} \in L^{1,0}(\tilde{\nu}, \underline{\mathcal{F}}).$$

b) *If for $h \in (\mathcal{P}(\underline{\mathcal{G}}) \otimes \mathcal{B}(E)) \bigcap \widehat{\mathcal{L}}^0(\mu - \nu, \underline{\mathcal{F}})$ there exists a strictly positive $\mathcal{P}(\underline{\mathcal{G}})$-measurable process K such that*

$$K|h - \bar{h}|\mathbf{1}_{\{|h-\bar{h}| \leqslant 1\}} \in L^{1,0}(|\tilde{\nu} - \nu|, \underline{\mathcal{F}}),$$

then the stochastic integrals $h * (\mu - \nu)$ on $\underline{\mathcal{G}}$ and $\underline{\mathcal{F}}$ coincide.

Proof. a) If $h \in \widehat{\mathcal{L}}^0(\mu, \underline{\mathcal{G}})$ then by (5) $h\mathbf{1}_{\{|h| \leqslant 1\}} \in \widehat{\mathcal{L}}^0(\nu, \underline{\mathcal{G}})$, so there exists a strictly positive $\mathcal{P}(\underline{\mathcal{G}})$-measurable process K such that $Kh\mathbf{1}_{\{|h| \leqslant 1\}} \in L^{1,0}(\nu, \underline{\mathcal{G}})$, that is $K|h|\mathbf{1}_{\{|h| \leqslant 1\}} \in L^{1,0}(\nu, \underline{\mathcal{G}})$, and hence $K|h|\mathbf{1}_{\{|h| \leqslant 1\}} \in L^{1,0}(\tilde{\nu}, \underline{\mathcal{F}})$.
Conversely, if such K exists then without loss of generality we may assume that $K \leqslant 1$ taking otherwise $K \bigwedge 1$ instead of K. We have

$$h\mathbf{1}_{\{|h|>1\}} \in L^{1,0}(\mu, \underline{\mathcal{G}}), \quad h^2\mathbf{1}_{\{|h| \leqslant 1\}} \in L^{1,0}(\mu, \underline{\mathcal{G}}), \quad h^2\mathbf{1}_{\{|h| \leqslant 1\}} \in L^{1,0}(\nu, \underline{\mathcal{G}}),$$

hence $h\mathbf{1}_{\{|h| \leqslant 1\}} \in L^{1,0}(\mu - \nu, \underline{\mathcal{G}})$. Then $K|h|\mathbf{1}_{\{|h| \leqslant 1\}} \in L^{1,0}(\tilde{\nu}, \underline{\mathcal{F}})$ implies

$$K|h|\mathbf{1}_{\{|h| \leqslant 1\}} \in (\mathcal{P}(\underline{\mathcal{G}}) \otimes \mathcal{B}(E)) \bigcap L^{1,0}(\mu, \underline{\mathcal{F}}) = L^{1,0}(\mu, \underline{\mathcal{G}})$$

and hence, by $|K| \leqslant 1$,

$$K|h|\mathbf{1}_{\{|h|\leqslant 1\}} \in L^{1,0}(\nu,\underline{\mathcal{G}}), \quad Kh\mathbf{1}_{\{|h|\leqslant 1\}} \in L^{1,0}(\nu,\underline{\mathcal{G}}),$$

$$h\mathbf{1}_{\{|h|\leqslant 1\}} \in \widehat{L}^0(\nu,\underline{\mathcal{G}}),$$

that is $h \in \widehat{L}^0(\mu,\underline{\mathcal{G}})$.

b) We may assume again that $K \leqslant 1$. We have

$$K(h-\bar{h})\mathbf{1}_{\{|h-\bar{h}|\leqslant 1\}} \in L^{1,0}(\tilde{\nu}-\nu,\underline{\mathcal{F}}), \quad K(h-\bar{h})\mathbf{1}_{\{|h-\bar{h}|\leqslant 1\}} \in L^{1,0}(\mu-\tilde{\nu},\underline{\mathcal{F}}),$$

so

$$K(h-\bar{h})\mathbf{1}_{\{|h-\bar{h}|\leqslant 1\}} \in (\mathcal{P}(\underline{\mathcal{G}})\otimes\mathcal{B}(E))\bigcap L^{1,0}(\mu-\nu,\underline{\mathcal{F}}) \subset L^{1,0}(\mu-\nu,\underline{\mathcal{G}})$$

and the stochastic integrals $K(h-\bar{h})\mathbf{1}_{\{|h-\bar{h}|\leqslant 1\}} * (\mu-\nu)$ on $\underline{\mathcal{G}}$ and on $\underline{\mathcal{F}}$ coincide. Now as in Theorem 1 we have (10) on $\underline{\mathcal{F}}$ and, by (Jacod, 1980, Theorem 7), also on $\underline{\mathcal{G}}$. Therefore and the integral in the right–hand side is the same for $\underline{\mathcal{G}}$ and $\underline{\mathcal{F}}$ and as in Theorem 1 we obtain the required statement.

Counter examples. Let us show, as we have promised above, that for an abstract measure $\tilde{\nu}$, Theorem 2 a) in the sense of $h \in \widehat{L}^0(\tilde{\nu},\underline{\mathcal{F}})$ instead of $h \in \widehat{L}^0(\mu,\underline{\mathcal{F}})$ and b) in the sense of $[|h-\bar{h}|\mathbf{1}_{\{|h-\bar{h}|\leqslant 1\}} * (\tilde{\nu}-\nu)]^{\mathcal{P}(\underline{\mathcal{G}})} = 0$ instead of $h \overset{\underline{\mathcal{G}}}{*} (\mu-\nu) = h \overset{\underline{\mathcal{F}}}{*} (\mu-\nu)$) is not valid.

a) Let $(\Omega,\mathcal{F},\mathbf{P}) = ([0,1],\mathcal{B}^{\Lambda}([0,1]),\Lambda)$, where Λ is the Lebesgue measure and $\mathcal{B}^{\Lambda}([0,1])$ is the completion of $\mathcal{B}([0,1])$ with respect to Λ. For every $t \in \mathbf{R}_+$, let $\mathcal{G}_t = \{\varnothing,\Omega\}$ and $\mathcal{F}_t = \mathcal{F}$. Let

$$E =]0,1], \quad \tilde{\nu}(\omega, dt\times du) = \frac{\mathbf{1}_{\{u\geqslant\omega^2\}}}{u^{s/2}}dtdu$$

and $h_{t,u}(\omega) = u$. Then $h \in L^{1,0}(\tilde{\nu},\underline{\mathcal{F}})$ and $h * \tilde{\nu}_t = 2(1/\omega - 1)t$, so that the condition of a) in Theorem 2 holds. But $\nu(dt\times du) = dtdu/u^2$ and $h \notin L^{1,0}(\nu,\underline{\mathcal{G}})$, so that since $h \geqslant 0$

$$h \notin \widehat{L}^0(\nu,\underline{\mathcal{G}}).$$

b) As before, let

$$(\Omega,\mathcal{F},\mathbf{P}) = ([0,1],\mathcal{B}^{\Lambda}([0,1]),\Lambda), \ \mathcal{G}_t = \{\varnothing,\Omega\}, \ \underline{\mathcal{F}}_t = \mathcal{F}, \ E =]0,1]$$

and $h_{t,u}(\omega) = u$, but now

$$\tilde{\nu}(\omega, dt\times du) = \left[\frac{1}{u^{5/2}} + \frac{u-\omega^2}{u^{3/2}(u+\omega^2)^2}\right]dtdu$$

and hence

$$\nu(dt \times du) = \left[\frac{1}{u^{5/2}} + \frac{2}{u^{3/2}(u+1)}\right] dtdu.$$

Then $h \in L^{1,0}(|\tilde{\nu}-\nu|, \underline{\mathcal{F}})$ and $h * |\tilde{\nu}-\nu|_t = 2(\frac{1}{\omega} - \frac{1}{\omega^2+1})t$, so that the condition of b) in Theorem 2 holds. But $h*(\tilde{\nu}-\nu)_t = -2t/(\omega^2+1)$, so that $[h*(\tilde{\nu}-\nu)]_t^{\mathcal{P}(\underline{\mathcal{G}})} =$ $= -\pi t/2$.

REFERENCES

Bichteler, K. and Jacod, J. (1983). Random measures and stochastic integration. *Lect. Notes in Control and Inf. Sci.* **49**, 1–18.

Jacod, J. (1980). Intégrales stochastiques par rapport a une semimartingale vectorielle et changements de filtration. *Lect. Notes in Math.* **784**, 161–172.

Lebedev, V.A. (1986). Extension of L^p-valued random measures. In: *Abstr. First World Congr. Bernoulli Soc. Math. Statist. and Probab. Theory,* **2**, Nauka, Moscow, pp. 734.

Lebedev, V.A. (1989). Stochastic integrals with respect to semimartingale measuren and change of the filtration. In: *Abstr. Fifth Intern. Vilnius Conf. Probab. Theory and Math. Statist.*, **3**, IMC, Vilnius, pp. 351–352.

Lebedev, V.A. (1990). Stochastic integration with respect to semimartingale measures. *Theory Probab. Appl.* (to appear).

Prob. Theory and Math. Stat., Vol. 2, pp. 79–89
B.Grigelionis *et al.* (Eds.)
1990 VSP/Mokslas

LIMIT DISTRIBUTIONS OF MEASURES OF SOJOURNS OF VECTOR GAUSSIAN FIELDS WITH STRONG DEPENDENCE

N. N. LEONENKO
Department of Mathematics, Kiev State University,
252017 Kiev, Vladimirskaya 64, Ukrainian SSR, USSR

ABSTRACT

Asymptotic distributions are considered for the functionals

$$\left|\left\{x \in v_n(r) : \xi(x) \notin v_p\left(f^{1/2}(r)\right)\right\}\right|,$$

where $v_k(r) = \{x \in \mathbf{R}^k : |x| < r\}$ is a ball in $\mathbf{R}^k, |\cdot|$ is the Lebesgue measure, $f : (0,\infty) \to (0,\infty)$ is a non-random continuous function such that $f(r) \uparrow \infty$ as $r \to \infty$, and $\xi(x) = [\xi_1(x),\dots,\xi_p(x)]'$ is a homogeneous isotropic vector Gaussian random field with strongly dependent components.

1. INTRODUCTION

We consider limit distributions of the sojourns of multidimensional Gaussian random fields with dependent components such that the integral of the correlation function diverges. Limit theorems for non-linear functionals of Gaussian fields were obtained in (Dobrushin and Major, 1979). Problems involving the sojourns of vector stationary Gaussian processes with a long range dependence have been studied in (Berman, 1984; Maejima, 1985, 1986). Limit theorems for functionals of geometric type of homogeneous isotropic Gaussian fields with strong dependence were obtained in (Leonenko, 1984, 1987; Ivanov and Leonenko, 1989).

2. THE MAIN THEOREM

Let $\mathbf{R}^n$ be the n–dimensional Euclidean space, $\mathcal{B}^n$ be the σ–algebra of Borel subsets of $\mathbf{R}^n$ and define

$$v_n(r) = \left\{x \in \mathbf{R}^n : |x| < r\right\},\ c_1(n) = 2\pi^{n/2}\Big/\left[n\Gamma(n/2)\right].$$

We introduce the special functions

$$\varphi(u) = (2\pi)^{-1/2}\exp\left\{-u^2/2\right\},\quad u \in \mathbf{R}^1;\quad \Phi(t) = \int_{-\infty}^{t} \varphi(u)du;$$

$$I_\mu(p,q) = \int_0^\mu t^{p-1}(1-t)^{q-1}dt/B(p,q), \quad p>0, \quad q>0, \quad \mu \in [0,1];$$

$$J_\nu(z) = \sum_{k=0}^{\infty} (-1)^k z^{2k+\nu} \Big/ \Big[2^{2k+\nu} k!\Gamma(k+\nu+1)\Big], \ \nu > -1/2. \tag{1}$$

Let $\{H_k(u)\}_{k=0}^{\infty}$ be the Hermite polynomials with leading coefficient 1, which form a complete orthogonal system in

$$L_2\Big(\mathbf{R}^1, \varphi(u)du\Big) \quad \Big(H_0(u) = 1, H_1(u) = u, H_2(u) = u^2 - 1, \dots\Big).$$

Let $\mathcal{L}$ be a class of functions $L(t)$, $t \in (0,\infty)$, slowly varying at infinity and bounded on each finite interval.
(A) Let $\xi(x) = [\xi_1(x), \dots, \xi_p(x)]', x \in \mathbf{R}^n$, be a measurable mean–square continuous homogeneous and isotropic Gaussian random field with

$$\mathbf{E}\xi(x) = 0, \mathbf{R}\Big(|x|\Big) = \mathbf{E}\xi(0)\xi(x)' = \Big(\mathbf{R}_{ij}(|x|)\Big)_{1\leqslant i,\ j\leqslant p},$$

$$\mathbf{R}_{ii}\big(|x|\big) = a\big(|x|\big), \quad i=1,\dots,p; \ \mathbf{R}_{ij}(|x|) = b\big(|x|\big), \quad i \neq j, \quad i,j = 1,\dots,p;$$

$$a(0) = 1, \quad b(0) = \rho_0 \in \big[0,1\big), \quad a\big(|x|\big) \sim L\big(|x|\big)/|x|^\alpha, \quad |x| \to \infty;$$

$$b\big(|x|\big) \sim \rho_\infty L\big(|x|\big)/|x|^\alpha, \ |x| \to \infty; \ \rho_\infty \in [0,1), \quad L \in \mathcal{L}, \quad \alpha > 0.$$

We denote by $\mathbf{1}\{\cdot\}$ the indicator function. Consider the functional

$$G(r) = \Big|\Big\{x \in v_n(r) : \xi(x) \notin v_p\Big(f^{1/2}(r)\Big)\Big\}\Big| =$$

$$= \int_{v_n(r)} \mathbf{1}\Big\{\xi(x) \in \mathbf{R}^p \backslash v_p\Big(f^{1/2}(r)\Big)\Big\}dx,$$

where $f(r)$, $r > 0$, is a real non–random continuous function such that $f(r) \uparrow \infty$, $r \to \infty$, $n \geqslant 1$, $p \geqslant 1$.
Let

$$d_1 = \Big[1 + (p-1)\rho_\infty\Big] \Big/ \Big[1 + (p-1)\rho_0\Big], \quad d_2 = \dots = d_p = (1-\rho_\infty)/(1-\rho_0), \tag{2}$$

$$\mu_1 = \Big[1 + (p-1)\rho_0\Big]^{-1/2}, \quad \mu_2 = \dots = \mu_p = (1-\rho_0)^{-1/2} \tag{3}$$

and

$$k_* = \frac{1}{2}\Big(\mu_1^2 - \mu_2^2\Big) \geqslant 0. \tag{4}$$

If $k_* = 0$, $\alpha \in (0, n/2)$, then

$$\mathrm{Var}\ G(r) \sim A_1^2(r) = c_2(n,2,\alpha)c_1^2(p)(2\pi)^{-p}(2p)e^{-f(r)}\Big[f(r)\Big]^p r^{2(n-\alpha)}L^2(r), \tag{5}$$

as $r \to \infty$, where

$$c_2(n,m,\alpha) = m!\ 2^{n-m\alpha+1}\pi^{n-1/2}\Gamma\Big(\frac{n-m\alpha+1}{2}\Big)\times$$

$$\times\Big[\Big(n-m\alpha\Big)\Gamma\Big(\frac{n}{2}\Big)\Gamma\Big(\frac{2n-m\alpha+2}{2}\Big)\Big]^{-1},\ n \geqslant 1,\ m \geqslant 1,\ 0<\alpha<n/m. \quad (6)$$

If $k_* > 0$, $\alpha \in (0, n/2)$, $p \geqslant 2$, then

$$\mathrm{Var}\ G(r) \sim A_2^2(r) = c_2(n,2,\alpha)(p-1)^2\mu_1^2\mu_2^{2(p-1)}\times$$
$$\times k_*^{1-p} d_2^2 \pi 2^{p-1} f(r) r^{2(n-\alpha)} L^2(r) \quad (7)$$

as $r \to \infty$.

Denote $V = \mathrm{diag}(d_1,\dots,d_p)$ the diagonal matrix with $d_j, j=1,\dots,p$ defined by (2), and let $T = (t_{ij})_{1\leqslant i,j\leqslant p}$ be the $(p\times p)$–matrix with

$$t_{1j} = \Big[p\big(1+(p-1)\rho_0\big)\Big]^{-1/2},\quad 1\leqslant j\leqslant p;$$

$$t_{ij} = \Big[i(i-1)(1-\rho_0)\Big]^{-1/2}, 2\leqslant i\leqslant p,\ 1\leqslant j\leqslant i;$$

$$t_{ii} = -(i-1)\Big[i(i-1)(1-\rho_0)\Big]^{-1/2},\ 2\leqslant i<p;\ t_{ij}=0,\ 2\leqslant i\leqslant p-1,\ i<j\leqslant p.$$

Note that $T^{-1} = (s_{ij})_{1\leqslant i,j\leqslant p}$, where $s_{i1} = \sqrt{(1+(p-1)\rho_0)/p}$, $1\leqslant i<p$; $s_{ij} = \sqrt{(1-\rho_0)/i(i-1)}$, $2\leqslant i<p$, $1\leqslant i\leqslant j$; $s_{ii} = (1-i)\sqrt{(1-\rho_0)/i(i-1)}$, $2\leqslant i\leqslant p$; $s_{ij}=0$, $2\leqslant i\leqslant p-1$, $j<i\leqslant p$.

If condition (A) holds, then the random field $\eta(x) = [\eta_1(x),\dots,\eta_p(x)]' = T\xi(x)$ has independent components, i.e.,

$$\mathbf{E}\eta(x) = 0,\ \widetilde{\mathbf{R}}(|x|) = (\widetilde{\mathbf{R}}_{ij}(|x|))_{1\leqslant i,j\leqslant p} = \mathbf{E}\eta(0)\eta(x)' =$$
$$= V\big(L(|x|)/|x|^\alpha\big) \text{ as } |x|\to\infty,\ L\in\mathcal{L},\ \alpha>0,\ \widetilde{\mathbf{R}}_{ii}(0)=1.$$

The proof it based on the procedure of orthogonalization.

THEOREM. *Suppose that assumption* (A) *holds for* $\alpha \in (0, n/2)$ *and assume that* $f(r) = o(\log r)$ *as* $r\to\infty$.
If $k_* = 0$, *then the limit distribution of the random variables* $[G(r)-\mathbf{E}G(r)]/A_1(r)$ *is the same as the limit distribution of the variables*

$$\sum_{j=1}^{p} d_j^{-1} \int\limits_{v_n(r)} H_2\Big(\eta_j(x)\Big)dx\Big/r^{n-\alpha}L(r)\sqrt{c_1(n,2,\alpha)2p} \quad (8)$$

(if one of them exists).
If $k_* > 0$, $p\geqslant 2$ *then the limit distribution of the random variables* $[G(r) - \mathbf{E}G(r)]/A_2(r)$ *is the same as the limit distribution of the variables*

$$\sum_{j=2}^{p} \int\limits_{v_n(r)} H_2\Big(\eta_j(x)\Big)dx\Big/r^{n-\alpha}L(r)\sqrt{c_2(n,2,\alpha)(p-1)d_2} \quad (9)$$

(if one of them exists).

From (Ivanov and Leonenko, 1989, Theorem 2.10) it follows that if $\alpha \in (0, n/2)$, $p \geqslant 2$ and $r \to \infty$ the limit distribution of the random variables (8) and (9) exists and has the form

$$\sum_{j=1}^{p} d_j^{-1} \bar{Z}_j \Big/ \sqrt{2pc_2(n,2,\alpha)}$$

and

$$\sum_{j=2}^{p} \bar{Z}_j \Big/ \sqrt{2c_2(n,2,\alpha)(p-1)d_2}$$

respectively, where $\bar{Z}_j, j = 1, \ldots, p$, are independent copies of the multiple stochastic integrals

$$\bar{Z} = (2\pi)^{n/2} \int_{\mathbf{R}^{2n} \setminus \{\lambda_i = \pm\lambda_j, i,j=1,2\}} J_{n/2} \times$$

$$\times \Big(|\lambda_1 + \lambda_2| \Big) |\lambda_1 + \lambda_2|^{-n/2} Z_{F_0}\Big(d\lambda_1\Big) Z_{F_0}\Big(d\lambda_2\Big), \tag{10}$$

here $J_\nu(z)$ is the ν–th order Bessel functions of the first kind (see (1)), and $Z_{F_0}(\cdot)$ is a Gaussian random measure subordinate to the measure $F_0(\cdot)$ described in (Ivanov and Leonenko, 1989, p.125) ($\mathbf{E}\bar{Z}^2 < \infty$).

3. PROOF OF THE THEOREM

LEMMA 1. *Let $(\xi_1, \ldots, \xi_{2p})$ be a 2p–dimensional Gaussian vector with $\mathbf{E}\xi_j = 0$, $\mathbf{E}\xi_j^2 = 1$, $1 \leqslant j \leqslant 2p$,*

$$\mathbf{E}\xi_j \xi_{j+p} = r_j, \ 1 \leqslant j \leqslant p;$$

$$\mathbf{E}\xi_j \xi_k = 0, \ (j,k) \notin \{(i,j) : (j+p, j+p), (j, j+p), 1 \leqslant i \leqslant p\}.$$

Then

$$\mathbf{E} \prod_{j=1}^{p} H_{k_j}(\xi_j) H_{m_j}(\xi_{j+p}) = \prod_{j=1}^{p} \delta_{k_j}^{m_j} k_j! \, r_j^{k_j},$$

where δ_m^k is the Kronecker symbol.

Proof. See (Taqqu, 1977).

LEMMA 2. *Let $\bar{\alpha}$ and $\bar{\beta}$ be random points chosen in the ball $v_n(r)$ independently according to the uniform law. The the density of the distribution of the distance $\rho = |\bar{\alpha} - \bar{\beta}|$ between $\bar{\alpha}$ and $\bar{\beta}$ can be written as*

$$p_\rho(z) = nr^{-n} z^{n-1} I_{\mu(z,r)}\left(\frac{n+1}{2}, \frac{1}{2}\right), \quad 0 \leqslant z \leqslant 2r,$$

$$\mu(z,r) = 1 - \left(\frac{z}{2r}\right)^2.$$

Proof. See (Santalo, 1976).

If $f(\rho), \rho \geqslant 0$, is a Borel function, then by Lemma 2

$$\int\limits_{v_n(r)} \int\limits_{v_n(r)} f\Big(|x-y|\Big) dx dy = \Big|v_n(r)\Big|^2 \mathbf{E} f\Big(|\bar{\alpha} - \bar{\beta}|\Big) =$$

$$= \frac{4\pi^n}{n\Gamma^2\left(\frac{n}{2}\right)} r^n \int\limits_0^{2r} z^{n-1} f(z) I_{\mu(z,r)}\Big(\frac{n+1}{2}, \frac{1}{2}\Big) dz. \tag{11}$$

LEMMA 3. *Let $\alpha > 0$, $\beta > 0$, $0 < a \leqslant \infty$ and $f \in C([0, d], \mathbf{R})$. If $f(x) = f(0) + O(x)$, $x \to 0$, then*

$$\int\limits_0^a x^{\beta-1} f(x) e^{-\lambda x^\alpha} dx = \frac{1}{\alpha} f(0) \Gamma\Big(\frac{\beta}{\alpha}\Big) \lambda^{-\beta/\alpha} + O\Big(\lambda^{-(\beta+1)/\alpha}\Big)$$

as $\lambda \to \infty$.

Proof. See (Zorich, 1984).

Let $\nu = (k_1, \dots, k_p)$ be the multi–index, $u = (u_1, \dots, u_p) \in \mathbf{R}^p$,

$$\mathbf{E}_\nu(u) = \prod_{j=1}^p H_{k_j}(u_j).$$

Then the polynomials $\{E_\nu(u)\}_\nu$ form a complete orthogonal system in the Hilbert space

$$L_2(\mathbf{R}^p, \Phi^p) = \Big\{ g(u), u \in \mathbf{R}^p : \int\limits_{\mathbf{R}^p} g^2(u) \Big\{ \prod_{j=1}^p \varphi(u_j) \Big\} du < \infty \Big\}.$$

Let $k \geqslant 0$ be an integer $S_k = \{\nu = (k_1, \dots, k_p) : k_j \geqslant 0,\ 1 \leqslant j \leqslant p,\ k_1 + \ldots + k_p = k\}$. The function $g_r(u) \in L_2(\mathbf{R}^p, \Phi^p)$ can be expanded in the Fourier series:

$$g_r(u) = \sum_{k \geqslant 0} \sum_{S_k} C_r(\nu) E_\nu(u) =$$

$$= \sum_{k \geqslant 0} \sum_{(k_1, \dots, k_p) \in S_k} C_r(k_1, \dots, k_p) E_\nu(u), \tag{12}$$

where

$$C_r(\nu) = C_r(k_1, \dots, k_p) = \Big(\prod_{j=1}^p k_j!\Big)^{-1} \int\limits_{\mathbf{R}^p} g_r(u) E_\nu(u) \Big\{ \prod_{j=1}^p \varphi(u_j) \Big\} du.$$

(B) Let $g_r(u) \in L_2(\mathbf{R}^p, \Phi^p)$ and assume that there exists an integer $m \geqslant 1$ such that $C_r(k_1, \ldots, k_p) = C_r(\nu) = 0$ if $\nu = (k_1, \ldots, k_p) \in S_k$, $1 \leqslant k \leqslant m-1$, but $C_r(m_1, \ldots, m_p) \neq 0$ for some $(m_1, \ldots, m_p) \in S_m$, $m_1 + \ldots + m_p = m$.
Let $\eta(x) \in \mathbf{R}^p, x \in \mathbf{R}^n$, be the random field with independent components constructed earlier and $g_r(u) \in L_2(\mathbf{R}^p, \Phi^p)$. We consider the random processes

$$K_r(t) = \int\limits_{v_n(rt^{1/n})} g_r\Big(\eta(x)\Big)dx, \ t \in [0,1];$$

$$L_r^{(m)}(t) = \sum_{S_m} C_r(\nu) \int\limits_{v_n(rt^{1/n})} E_\nu\Big(\eta(x)\Big)dx, \ t \in [0,1],$$

where $g_r(\eta(x)) = g_r\,(\eta_1(x), \ldots, \eta_p(x)), x \in \mathbf{R}^n$.

LEMMA 4. *Let assumptions* (A), (B) *hold for* $\alpha \in (0, n/m)$ *and*

$$\zeta_r = \Big\{\sum_{S_m} C_r^2\Big(m_1, \ldots, m_p\Big)\Big\}^{-1} \int\limits_{\mathbf{R}^p} g_r^2(u)\Big\{\prod_{j=1}^{p} \varphi(u_j)\Big\}du = o\Big(\frac{r^\alpha}{L(r)}\Big)$$

as $r \to \infty$. *Then the finite–dimensional distributions of the random process*

$$X_r(t) = \big[K_r(t) - \mathbf{E}K_r(t)\big]/\sqrt{\mathrm{Var}K_r(1)}, \ t \in [0,1]$$

are the same as the limit finite–dimensional distributions of the random process

$$X_r^{(m)}(t) = L_r^{(m)}(t)/\sqrt{\mathrm{Var}L_r^{(m)}(1)}, \ t \in [0,1]$$

(if one of them exists).

Proof. Obviously $\mathbf{E}K_r(t) = tr^n|v_n(1)|C_r(0, \ldots, 0)$. Let $U_r(t_j) = X_r(t) - -X_r^{(m)}(t_j)$, $a_j \in \mathbf{R}^1, t_j \in [0,1]$, $j = 1, \ldots, q$.
We will show that

$$\lim_{r\to\infty} \mathbf{E}\Big[\sum_{j=1}^{q} a_j U_r(t_j)\Big]^2 = 0,$$

or $\lim\limits_{r\to\infty} \mathbf{E}U_r^2(t) = 0, \quad t \in [0,1]$. From (B) we have

$$K_r(t) - \mathbf{E}K_r(t) = L_r^{(m)}(t) + A_r(t)$$

where, in the Hilbert space $L_2(\Omega) = \{\xi : \mathbf{E}\xi^2 < \infty\}$,

$$A_r(t) = \sum_{\nu \geqslant m+1} \sum_{S_\nu} C_r(\nu) \int\limits_{v_n(rt^{1/n})} E_\nu\Big(\eta(x)\Big)dx.$$

By Lemma 1

$$\mathbf{E}L_r^{(m)}(t)A_r(t) = \sum_{\nu,\nu'\geqslant m+1}\sum_{S_\nu}\sum_{S_{\nu'}} C_r(\nu)C_r(\nu')\times$$

$$\times \int\limits_{v_n(rt^{1/n})}\int\limits_{v_n(rt^{1/n})} \mathbf{E}\Big\{E_\nu\big(\eta(x)\big)E_\nu\big(\eta(y)\big)\Big\}\, dxdy = 0.$$

Furthermore Var $K_r(t) =$ Var $L_r^{(m)}(t) +$ Var $A_r(t)$.
Using (A), (11) and Lemma 1 for $\alpha \in (0, n/m)$, $\quad r \to \infty$ we have

$$\text{Var } L_r^{(m)}(t) = \sum_{\nu\in S_m}\sum_{\nu'\in S_m} C_r(\nu)C_r(\nu')\times$$

$$\times \int\limits_{v_n(rt^{1/n})}\int\limits_{v_n(rt^{1/n})} \mathbf{E}\Big\{E_\nu\big(\eta(x)\big)E_{\nu'}\big(\eta(y)\big)\Big\}dxdy =$$

$$= \sum_{\nu\in S_m} C_r^2(\nu) \int\limits_{v_n(rt^{1/n})}\int\limits_{v_n(rt^{1/n})} \left(\Big[a(|x-y|) - (p-1)b(|x-y|)\Big]\Big/\mu_1^2\right)^{m_1}\times$$

$$\times m_1!\prod_{2\leqslant j\leqslant p} m_j!\left(\Big[a\big(|x-y|\big) - b\big(|x-y|\big)\Big]\Big/\mu_j^2\right)^{m_j} dxdy =$$

$$= c_3(n)tr^n \sum_{\nu\in S_m} C_r^2(\nu)\int\limits_0^{2rt^{1/n}} z^{n-1}I_{\mu(z,r)}\Big(\frac{n+1}{2},\frac{1}{2}\Big)m_1!\Big(\Big[z^{-\alpha}L(z)+$$

$$+(p-1)\rho_\infty z^{-\alpha}L(z)\Big]\Big/\mu_1^2\Big)^{m_1},$$

$$\prod_{2\leqslant j\leqslant p} m_j!\Big(\Big[z^{-\alpha}L(z) - \rho_\infty z^{-\alpha}L(z)\Big]\Big/\mu_j^2\Big)^{m_j} dz \sim$$

$$\sim c_2(n,m,\alpha)\Big[\sum_{\nu\in S_m} C_r^2(\nu)\prod_{1\leqslant j\leqslant m} m_j!d_j^{m_j}\Big]\times$$

$$\times t^{2-(m\alpha)/n}r^{2n-m\alpha}L^m(rt^{1/n}),\ r\to\infty,\ t\in[0,1],$$

$$c_3(n) = 4\pi^n\Big/\Big[n\Gamma^2\Big(\frac{n}{2}\Big)\Big], \tag{13}$$

and

$$\text{Var } A_r(t) = c_3(n)tr^n\sum_{l\geqslant m+1}\sum_{\nu\in S_l} C_r^2(\nu)\int\limits_0^{2rt^{1/n}} z^{n-1}\times$$

$$\times I_{\mu(z,r)}\Big(\frac{n+1}{2},\frac{1}{2}\Big)k_1!\Big(\Big[a(z)+(p-1)b(z)\Big]\Big/\mu_1^2\Big)^{k_1}\times$$

$$\times \prod_{2\leqslant j\leqslant p} k_j!\Big(\big[a(z)-b(z)\big]/\mu_j^2\Big)^{k_j} dz. \tag{14}$$

For each multi–index $\nu = (k_1,\ldots,k_p)$ such that $k_1+\ldots+k_p \geqslant m+1$ there exists multi–index $q=(q_1,\ldots,q_p)$ depending on ν such that $q_j \leqslant k_j, q_1+\ldots+q_p \geqslant m+1$. We have from (14)

$$\operatorname{Var} A_r(t) \leqslant c_3(n)tr^n \sum_{l\geqslant m+1} \sum_{\nu\in S_l} C_r^2(\nu) \int_0^{2rt^{1/n}} z^{n-1} I_{\mu(z,r)}\Big(\frac{n+1}{2},\frac{1}{2}\Big)k_1!\times$$

$$\times\Big(\big[a(z)-(p-1)b(z)\big]/\mu_1^2\Big)^{q_1} \prod_{2\leqslant j\leqslant p} k_j!\Big(\big[a(z)-b(z)\big]/\mu_j^2\Big)^{q_j} dz \leqslant$$

$$\leqslant c_3(n)tr^n \sum_{l\geqslant m+1}\sum_{\nu\in S_l} C_r^2(\nu)\Big(\prod_{1\leqslant j\leqslant p} k_j! d_j^{q_j}\Big) \int_0^{2rt^{1/n}} z^{n-1}L(z)\times$$

$$\times z^{-(m+1)\alpha} I_{\mu(z,r)}\Big(\frac{n+1}{2},\frac{1}{2}\Big)dz. \tag{15}$$

By Parseval equality

$$\sum_{l\geqslant 0}\sum_{\nu\in S_l} C_r^2(\nu)\prod_{j=1}^p k_j! = \int_{\mathbf{R}^p} q_r^2(u)\Big\{\prod_{j=1}^p \varphi(u_j)\Big\}du. \tag{16}$$

Using (13), (16) we obtain

$$\operatorname{Var} A_r(t)/\operatorname{Var} L_r^{(m)}(t) \leqslant c_4 L(r) r^{-\alpha}\zeta_r, \quad c_4>0. \tag{17}$$

Therefore $\operatorname{Var} A_r(t) = o(\operatorname{Var} L_r^{(m)}(t))$, $t \in [0,1]$ and $\lim_{r\to\infty} \mathbf{E}U_r^2(t) = 0$, $\operatorname{Var} K_r(1) \sim \operatorname{Var} L_r^{(m)}(1)$ as $r\to\infty$.

Proof of the Theorem. Let $\tilde{\Delta}_r\{y\in\mathbf{R}^p : T^{-1}y\in\Delta_r\}$, then $(\Delta_r\in\mathcal{B}^p)$

$$G(r) = \int_{v_n(r)} \mathbf{1}\Big\{\xi(x)\in\Delta_r\Big\}dx = \int_{v_n(r)} \mathbf{1}\Big\{\eta(x)\in\tilde{\Delta}_r\Big\}dx.$$

Let $\Delta_r = \mathbf{R}^p\backslash v_p(f^{1/2}(r))$, then

$$\tilde{\Delta}_r = \Big\{(y_1,\ldots,y_p)' : \sum_{j=1}^p \Big(\frac{y_j}{\mu_j}\Big)^2 < f(r)\Big\}$$

where $\mu_j, j=1,\ldots,p$, are given by (3).
We consider the case $k_*=0$, i.e., $\rho_0=0$ (see condition (A)).

If $q(u) = \mathbf{1}\{|u|^2 > f(r)\}$, then $C_r(\nu) = 0$, $\nu = (k_1, \dots, k_p) \in S_1$, i.e., $k_1 + \dots + k_p = 1$ or $(k_1, \dots, k_p) \in S_2$ and $k_i = k_j = 1$ for some $i \neq j$.
If $k_* = 0$, then

$$C_r(2,0,\dots,0) = c_1(p)(2\pi)^{-p/2} e^{-f(r)/2} [f(r)]^{p/2},$$

$$C_r(0,2,0,\dots,0) = \dots = C_r(0,\dots,0,2) = C_r(2,0,\dots,0).$$

The condition (B) holds for $m = 2$,

$$\int_{\mathbf{R}^p} g_r^2(u) \Big\{ \prod_{j=1}^{p} \varphi(u_j) \Big\} du \sim c_5 \Big[f(r) \Big]^{(p-2)/2} e^{-f(r)/2}, \ c_5 > 0,$$

as $r \to \infty$
Using (12) we have ($c_6 > 0, c_7 > 0$)

$$\zeta_r \sim c_6 \Big[f(r) \Big]^{(p-2)/2} e^{-f(r)/2} \Big/ \Big[f(r) \Big]^p e^{-f(r)} =$$

$$= c_7 e^{f(r)/2} \Big/ \Big(f(r) \Big)^{p/2+1} = o\Big(\frac{r^\alpha}{L(r)} \Big),$$

as $r \to \infty$.
Thus by Lemma 4 the limit distribution of the random variables $[G(r) - \mathbf{E}G(r)]/A_1(r)$ as $r \to \infty$ is the same as that of the random variables

$$C_r\Big(2,0,\dots,0\Big) \sum_{j=1}^{p} d_j^{-1} \int_{v_n(r)} H_2\Big(\eta_j(x)\Big)\, dx \Big/ \Big[c_2(n,2,\alpha) 2p C_r(2,0,\dots,0) \times$$

$$\times r^{2n-2\alpha} L^2(2) \Big]^{1/2} = \sum_{j=1}^{p} d_j^{-1} \int_{v_n(r)} H_2\Big(\eta_j(x)\Big)\, dx \Big/ \ r^{n-\alpha} L(r) \sqrt{c_2(n,2,\alpha)2p}.$$

If $k_* > 0$ then $C_r(k_1, \dots, k_p) = 0$, if $(k_1, \dots, k_p) \in S_1$ or $(k_1, \dots, k_p) \in S_2$ and $k_i = k_j = 1$ for some $i \neq j$. We have

$$C_r(2,0,\dots,0) = \frac{1}{2} \int_{\tilde{\Delta}_r} H_2(u_1) \Big\{ \prod_{j=1}^{p} \varphi(u_j) \Big\} du =$$

$$= -\mu_1 \mu_2^{p-1} \int_{y_2^2+\dots+y_p^2<f(r)} \prod_{j=2}^{p} \varphi(\mu_2 y_j) dy_j \int_0^{(f(r)-y_2-\dots-y_p)^{1/2}} H_2(\mu_1 y_1) \times$$

$$\times \varphi(\mu_2 y_1) dy_1 = (2\pi)^{-p/2} \mu_1 \mu_2^{p-1} \int_{y_2^2+\dots+y_p^2<f(r)} e^{-\mu_2^2(y_2^2+\dots+y_p^2)/2} \times$$

$$\times \exp\left\{-\mu_1^2\Big(f(r)-(y_2^2+\ldots+y_p^2)/2\Big)\right\}\sqrt{f(r)-y_2^2-\ldots-y_p^2}\,dy_2\ldots dy_p =$$

$$= (2\pi)^{-p/2}\mu_1\mu_2^{p-1}e^{-\mu_1^2 f(r)/2}2\pi^{(p-1)/2}\Gamma^{-1}\Big(\frac{p-1}{2}\Big)\int\limits_0^{\sqrt{f(r)}}\rho^{p-2}e^{-k_*\rho^2}\times$$

$$\times\sqrt{f(r)-\rho^2}d\rho = (2\pi)^{-p/2}\mu_1\mu_2^{p-1}e^{-\mu_1^2 f(r)/2}c_1(p)\Big[f(r)\Big]^{p/2}\times$$

$$\times\int\limits_0^1 x^{p-2}e^{-x^2k_* f(r)}\sqrt{1-x^2}dx.$$

By Lemma 3

$$\int\limits_0^1 x^{p-2}e^{-x^2\lambda}\sqrt{1-x^2}dx\sim\frac{1}{2}\Gamma\Big(\frac{p-1}{2}\Big)\lambda^{-(p-1)/2}$$

as $\lambda\to\infty$. Hence,

$$C_r(2,0,\ldots,0)\sim\mu_1\mu_2^{p-1}(2\pi)^{-p/2}\pi^{(p-1)/2}k_*^{-(p-1)/2}\sqrt{f(r)}\;e^{-\mu_1^2 f(r)/2}$$

as $r\to\infty$.
Analogously

$$C_r(0,2,0,\ldots,0)=C_r(0,0,2,0,\ldots,0)=\ldots=C_r(0,0,\ldots,0,2)\sim$$

$$\sim\mu_1\mu_2^{p-1}(2\pi)^{-p/2}k_*^{-(p-1)/2}\sqrt{f(r)}e^{-\mu_2^2 f(r)/2},\ \text{as } r\to\infty.$$

Therefore

$$\text{Var } G(r)\sim c_2(n,2,\alpha)r^{2n-2\alpha}L^2(r)2[C_r^2(2,0,\ldots,0)d_1^2+(p-1)d_2^2\times$$

$$\times C_r^2(0,2,0,\ldots,0)]\sim c_2(n,2,\alpha)r^{2n-2\alpha}L^2(r)(p-1)d_2^2C_r^2(0,2,0,\ldots,0),\ \text{as } r\to\infty,$$

since

$$C_r(2,0,\ldots,0)/C_r(0,2,0,\ldots,0)=e^{-k_* f(r)}\to 0 \text{ as } r\to\infty. \tag{17}$$

The condition (13) holds because

$$\zeta_r\sim c_8\Big[\Big(f(r)\Big)^{(p-2)/2}e^{-f(r)/2}\Big]\Big/\Big[f(r)e^{-\mu_2^2 f(r)}\Big]=$$

$$=c_8\exp\left\{-\Big[\Big(\mu_2^2-\frac{1}{2}\Big)f(r)-\Big(\frac{p}{2}-2\Big)\log\ f(r)\Big]\right\}=o\Big(r^\alpha/L(r)\Big)\ \text{as } r\to\infty.$$

By Lemma 4 the limit distribution of the random variables $\left[G(r)-\mathbf{E}G(r)\right]/A_2(r)$ is the same as that of the random variables

$$\left[C_r(2,0,\ldots,0)\int\limits_{v_n(r)} H_2\left(\eta_1(x)\right)dx + C_r(0,2,0,\ldots,0)\sum_{j=2}^{p}\int\limits_{v_n(r)} H_2\left(\eta_j(x)\right)dx\right]\times$$

$$\times\left[c_2(n,2,\alpha)r^{2n-2\alpha}L^2(r)(p-1)d_2^2C_r(0,2,0,\ldots,0)\right]^{-1/2}.$$

Using (17) we have

$$\operatorname{Var}\left[C_r(2,0,\ldots,0)\int\limits_{v_n(r)} H_2\left(\eta_1(x)\right)dx\times\right.$$

$$\left.\times C_r(0,2,0,\ldots,0)\sqrt{c_2(n,2,\alpha)(p-1)}\right]\to 0$$

as $r\to\infty$. Thus the limit distribution of $\left[G(r)-\mathbf{E}G(r)\right]\Big/A_2(r)$ is the same as that of the random variables

$$\sum_{j=2}^{p}\int\limits_{v_n(r)} H_2\left(\eta_j(x)\right)dx\Big/\ r^{n-\alpha}L(r)\sqrt{c_2(n,2,\alpha)(p-1)d_2}.$$

The Theorem is proved.

REFERENCES

Berman, S.M. (1984). Sojourns of vector Gaussian process inside and outside sphere. *Z. Wahrsch. verw. Geb.* **6**, 529–542.

Dobrushin, R.L. and Major, P. (1979). Non–central limit theorem for non–linear functionals of Gaussian fields. *Z. Wahrsch. verw. Geb.* **50**, 27–52.

Ivanov, A.V. and Leonenko, N.N. (1989). *Statistical Analysis of Random Fields.* Kluwer, Dordrecht.

Leonenko, N.N. (1984). On measure of exceses of an isotropic Gaussian random fields over a level. *Theory Probab. and Math. Statist.* **31**, 64–82.

Leonenko, N.N. (1987). Limit theorems for functional of geometric type of homogeneous isotropic random fields. In: *Proc. Fourth Intern. Vilnius Conf. Probab. Theory Math. Statist.,* **2**, VNU Science Press, Utrecht, pp. 173–202.

Maejima, M. (1985). Sojourns of multidimensional Gaussian processes with dependente components. *Yokohama Math. J.* **33**, 121–130.

Maejima, M. (1986). Sojourns of multidimensional Gaussian processes. In: *Dependence in Probability and Statistics,* Birkhauser, Basel, pp. 91–108.

Santalo, C. (1976). *Integral Geometry and Geometric Probability.* Addison–Wesley.,

Taqqu, M.S. (1977). Law of iterated logarithm for sums of non–linear functions of Gaussian variables. *Z. Wahrsch. verw. Geb.* **40**, 203–238.

Zorich, W.A. (1984). *Mathematical Analysis.* **II**, Nauka, Moscow.

Prob. Theory and Math. Stat., Vol. 2, pp. 90–96
B.Grigelionis *et al.* (Eds.)
1990 VSP/Mokslas

LIMIT THEOREMS FOR ONE–DIMENSIONAL RANDOM WALKS IN RANDOM ENVIRONMENTS

A.V. LETCHIKOV
Department of Mathematics, Udmurt University, 426057 Izhevsk, USSR

ABSTRACT

The limit theorems for one–dimensional random walks in random environments are formulated, where the jumps of the moving particle are bounded (not exceeding R for jumps to the right and not exceeding L for jumps to the left).

We consider the following model of one–dimensional random walks in random environments.
Let R and L be natural numbers. Suppose G is the set of vectors $p = (p_{-L}, p_{-L+1}, \dots, p_R) \in \mathbf{R}^{R+L+1}$ such that

$$\sum_{i=-L}^{R} p_i = 1, \quad p_i \geqslant 0, \quad i = -L, \dots, R.$$

The set $\Pi = G^{\mathbf{Z}}$ is called the *space of environments,* and each element of this set is called an environment. Thus an environment is a sequence of G–valued vectors $p = \{\mathbf{p}(x), x \in \mathbf{Z}\}$.
For a fixed arbitrary environment p a random walk in the environment p is the homogeneous Markov chain $\{x(t), t = 0, 1, \dots\}$ with the state space $\mathbf{Z}$, the initial state 0 and the transition matrix $A = \{a_{ij}\}$, $i, j \in \mathbf{Z}$, with elements defined by

$$a_{ij} = \begin{cases} p_{j-i}(i), & \text{if } j - i = -L, -L+1, \dots, R, \\ 0, & \text{otherwise.} \end{cases}$$

The vector $\mathbf{p}(x)$ gives the distribution of the jumps from x for the random walk in the environment p.
Suppose $\mathbf{X}$ is the space of paths of one–dimensional random walks, $F_{\mathbf{X}}$ is the σ–algebra of its subsets generated by the cylinder subsets of $\mathbf{X}$. We denote by $\Pr_p$ the probability measure corresponding to the random walk in the environment p. The randomness of environments expresses itself in such a way that p is a realization of a sequence of independent identically distributed (i.i.d.) random vectors, defined on the probability space $(\Pi, F_\Pi, \mathbf{P})$. We introduce the notation $\Omega = \Pi \times \mathbf{X}$, $F = F_\Pi \times F_{\mathbf{X}}$. On the measurable space (Ω, F) we define the probability measure

P in the following way:

$$P(B_1 \times B_2) = \int_{B_1} \mathrm{Pr}_p(B_2)\mathbf{P}(dp)$$

for any $B_1 \in F_\Pi$, $B_2 \in F_\mathbf{X}$.

Definition. A random walk in the random environment (RWIRE) is the stochastic process defined on the probability space (Ω, F, P) in the following way:

$$\text{for} \quad \omega = (p, x(*)) \in \Omega, \quad x(t,\omega) = x(t), \quad t = 0, 1, 2, \ldots .$$

Let us remark that the conditional distribution of RWIRE on the space of paths $\mathbf{X}$ relative to a fixed environment p is the distribution Pr_p of the random walk in the environment p.
This definition of RWIRE was given in (Key, 1984). In the general case random walks in the random environment were studied in (Kalikow, 1981).
For $x \in \mathbf{Z}$ let

$$q_i(x) = \begin{cases} \sum\limits_{j=i+1}^{R} p_i(x), & \text{if } 0 \leqslant i \leqslant R-1, \\ \sum\limits_{j=-L}^{i} p_i(x), & \text{if } -L \leqslant i \leqslant -1. \end{cases}$$

Assume that a sequence $\{r(x), x \in \mathbf{Z}\}$ solves the system of equations:

$$\sum_{i=0}^{R-1} q_i(x) r(x+i) = \sum_{i=-L}^{-1} q_i(x) r(x+i). \tag{1}$$

Introduce the vector

$$\mathbf{r}(x) = \begin{pmatrix} r(x+R-1) \\ \vdots \\ r(x-L+1) \end{pmatrix} \in \mathbf{R}^{R+L+1}.$$

Then $\mathbf{r}(x-1) = V(x)\mathbf{r}(x)$, where $V(x)$ is a $(R+L-1) \times (R+L-1)$ matrix whose elements are defined by

$$v_{ij}(x) = \begin{cases} q_{R-j}(x)/q_{-L}(x), & \text{if } 1 \leqslant j \leqslant R, \ i = R+L, \\ -q_{R-j}(x)/q_{-L}(x), & \text{if } R+1 \leqslant j \leqslant R+L-1, \ i = R+L-1, \\ 1, & \text{if } i+1 = j, \ 1 \leqslant i \leqslant R+L-2, \\ 0, & \text{otherwise.} \end{cases} \tag{2}$$

We see from this formula that the matrix $V(x)$ depends only on the vector $\mathbf{p}(x)$. Therefore $\{V(x), x \in \mathbf{Z}\}$ is a sequence of i.i.d. random matrices. Using formula (2), it is easy to show that the conditions

$$\mathbf{E_P} \ln p_R(x) > -\infty, \quad \mathbf{E_P} \ln p_{-L}(x) > -\infty, \tag{3}$$

imply that $\mathbf{E_p}\big|\ln\|V(x)\|\big| < +\infty$, $\mathbf{E_p}\big|\ln\|V^{-1}(x)\|\big| < +\infty$ (here $\mathbf{E_p}$ means the mathematical expectation with respect to the measure $\mathbf{P}$). So Oseledets's multiplicative ergodic theorem (Oseledets, 1968) can be applied to the sequence $\{V(x)\}$. This theorem says that there exist non-random numbers $\lambda_1 \leqslant \lambda_2 \leqslant \ldots \leqslant \lambda_{R+L-1}$, called Lyapunov exponents, and a strictly increasing non-stochastic sequence of natural numbers $1 = i_1 < i_2 < \ldots < i_{k+1} = R+L$ such that $\lambda_{i_k} < \lambda_{i_{k+1}}$, $\lambda_i = \lambda_j$, if $i_k \leqslant i \leqslant j \leqslant i_{k+1}$, $k = 1,2,\ldots,n$. Moreover, there exists a unique ($\mathbf{P}$-a.s.) random partition of the space $\mathbf{R}^{R+L-1}$ into subspaces

$$W_1 \otimes W_2 \otimes \ldots \otimes W_n = \mathbf{R}^{R+L-1} \tag{4}$$

such that $\dim W_k = i_{k+1} - i_k$, and if $v \in W_k$, then

$$\lim_{x\to-\infty} \frac{1}{|x|} \ln\|V(1+x)V(2+x)\ldots V(0)v\| = \lambda_{i_k} \qquad (\mathbf{P}\text{-a.s.}),$$

$$\lim_{x\to+\infty} \frac{1}{|x|} \ln\|V^{-1}(x)V^{-1}(x-1)\ldots V^{-1}(1)v\| = -\lambda_{i_k} \qquad (\mathbf{P}\text{-a.s.})$$

Theorem 1 (A RWIRE recurrence criterion). *Assume that conditions* (3) *hold. Let $\lambda_1 \leqslant \lambda_2 \leqslant \ldots \leqslant \lambda_{R+L-1}$ be the Lyapunov exponents of the sequence of random matrices $\{V(x)\}$. Then:*

1) $\lambda_R > 0 \Leftrightarrow \mathbf{P}\{\lim_{t\to+\infty} x(t) = +\infty\} = 1.$

2) $\lambda_R = 0 \Leftrightarrow \mathbf{P}\{\varliminf_{t\to+\infty} x(t) = -\infty,\quad \varlimsup_{t\to+\infty} x(t) = +\infty\} = 1.$

3) $\lambda_R < 0 \Leftrightarrow \mathbf{P}\{\lim_{t\to+\infty} x(t) = -\infty\} = 1.$

In the particular case $R = L = 1$, $\{V(x)\}$ is a sequence of random variables and this criterion reduces to the well-known result (Solomon, 1975). For the general case we have Key's recurrence criterion for RWIRE (Key, 1984). Theorem 1 is proved as a corollary of Key's criterion.

Part 2 of this theorem corresponds to a recurrent RWIRE. This means that for almost all p the initial state 0 belongs to the class of recurrent states of the random walk in the environment p. In cases 1 and 3 the RWIRE is obviously transient.

Let $\varepsilon_0 > 0$ be a sufficiently small real number. We assume the following three conditions:

(C1) $\mathbf{P}\{p_{-L}(0) \geqslant \varepsilon_0,\quad p_R(0) \geqslant \varepsilon_0,\quad p_1(0) \geqslant \varepsilon_0,\quad p_{-1}(0) \geqslant \varepsilon_0\} = 1,$

(C2) for almost each environment p there exist a constant $\Gamma > 1$ and a positive solution $r = \{r(x),\ x \in \mathbf{Z}\}$ of the system of equations (1) such that $r(0) = 1$ and for all $x \in \mathbf{Z}$

$$\Gamma^{-1} \leqslant \frac{r(x+1)}{r(x)} \leqslant \Gamma,$$

(C3) the random vector $\mathbf{p}(0)$ has an absolutely continuous distribution.

Let us remark that under the condition (C1) RWIRE is irreducible and the communicating set is the whole lattice $\mathbb{Z}$.
We use the following Lemma to formulate the main results.

LEMMA. *Let conditions* (C1—C3) *hold. Then for almost each (w.r.t.* $\mathbf{P}$*) environment* p *there exists a unique positive solution* $r = \{r(x)\}$ *of the system of equations* (1) *such that* $r(0) = 1$,

$$\lim_{x\to\pm\infty} \frac{1}{x} \ln r(x) = -\lambda_R, \tag{5}$$

$$\lim_{x\to\pm\infty} \frac{1}{|x|}\Big(\mathbf{E_P} \ln^2 r(x) - \big(\mathbf{E_P} \ln r(x)\big)^2\Big) = \sigma^2 > 0. \tag{6}$$

The scheme of the proof is the following. At first, we show that for a random walk in a nonrandom environment p the positive solution of the system (1) is in effect the potential in which the random walk occurs. This means that the main probability characteristics, such as the probability of leaving an interval, the mean time to leave an interval and the σ–finite stationary measure on $\mathbb{Z}$, may be estimated from the sequence $\{r(x)\}$. Using this fact we prove that the limit (5) takes place for the positive solution $r = \{r(x)\}$. But an arbitrary solution of the system (1) is uniquely reconstructed from the initial vector $\mathbf{r}(0)$ and the sequence of matrices $\{V(x)\}$. Therefore the initial vector $\mathbf{r}(0)$ belongs to the subspace of the partition (3) corresponding to the R–th Lyapunov exponent λ_R of the sequence $\{V(x)\}$. Condition (C3) implies that all the Lyapunov exponents $\lambda_1, \lambda_2, \dots, \lambda_{R+L-1}$ are distinct. This follows from the theory of limit theorems for products of random matrices, which has been developed in (Tutubalin, 1977). Consequently, all subspaces of random decomposition (3) are one–dimensional. This implies the uniqueness of the positive solution r in the Lemma. Methods due to V.N. Tutubalin permit us to prove the positivity of the limit (6).
We introduce the following random variables:

$$r_+ = \sum_{x=0}^{+\infty} \frac{1}{r(x)}, \quad r_- = \sum_{x=-\infty}^{0} \frac{1}{r(x)}.$$

THEOREM 2 (A criterion of RWIRE's linear drift). *Assume that conditions* (C1—C3) *hold. Then*

1) *if* $\mathbf{E_P} r_- < +\infty$, *then* $\lambda_R < 0$ *and there exists a non–random positive number* α *such that*

$$\mathbf{P}\Big\{\lim_{t\to+\infty} x(t)/t = \alpha\Big\} = 1,$$

2) *if* $\mathbf{E_P} r_- = \mathbf{E_P} r_+ = +\infty$, *then* $\mathbf{P}\Big\{\lim_{t\to+\infty} x(t)/t = 0\Big\} = 1,$

3) *if* $\mathbf{E_p} r_+ < +\infty$, *then* $\lambda_R > 0$ *and there exists a non-random negative number* α *such that*

$$\mathbf{P}\Big\{\lim_{t\to+\infty} x(t)/t = \alpha\Big\} = 1.$$

This theorem extends the well-known criterion obtained in (Solomon, 1975) for RWIRE in the case $R = L = 1$, and solves the problem formulated in (Kesten, 1984). The proof of Theorem 2 applies the main ideas of (Solomon, 1975) and the tools used to prove the Lemma.

Let us remark that the inequalities $\mathbf{E_p} r_+ < +\infty$ and $\mathbf{E_p} r_- < +\infty$ do not hold simultaneously.

In the last two theorems we study the asymptotic behaviour of the recurrent RWIRE.

Let $x_0(t) = x(t)\ln^{-2} t$, $\quad y_0(t) = \ln^{-2} t \max\{x(s), 0 \leqslant s \leqslant t\}$.

Theorem 3. *Assume that conditions* (C1—C3) *hold and* $\lambda_R = 0$. *Then for any integer* $t > 1$ *there exist random variables* $m(t)$ *and* $M(t)$ *dependent only on* t *and the environment* p *such that*

$$x_0(t) - m(t) \xrightarrow{\mathbf{P}} 0, \quad y_0(t) - M(t) \xrightarrow{\mathbf{P}} 0 \quad \text{as } t \to\to +\infty$$

If σ^2 *is the constant defined by* (6), *then*

$$\lim_{t\to+\infty} \mathbf{P}\{\sigma^2 m(t) \leqslant u\} = \begin{cases} \frac{1}{2} + \frac{1}{2}\int\limits_0^u h_1(v)dv, & \text{if } u \geqslant 0, \\ \frac{1}{2} - \frac{1}{2}\int\limits_0^{-u} h_1(v)dv, & \text{if } u < 0, \end{cases}$$

and

$$\lim_{t\to+\infty} \mathbf{P}\{\sigma^2 M(t) \leqslant u\} = \int\limits_0^u h_2(v)dv, \quad u \geqslant 0.$$

The densities of limit distributions $h_1(v)$, $h_2(v)$ and their Laplace transforms $\psi_1(\lambda)$, $\psi_2(\lambda)$ are given by the following formulae:

$$h_1(v) = \frac{2}{\pi}\sum_{k=0}^{+\infty}\frac{(-1)^k}{2k+1}\exp\left\{-\frac{(2k+1)^2}{8}\pi^2 v\right\}, \quad \psi_1(\lambda) = (\mathrm{ch}\sqrt{2\lambda} - 1)/(\lambda\,\mathrm{ch}\sqrt{2\lambda}),$$

$$h_2(v) = \frac{2}{\pi}\sum_{k=0}^{+\infty}\exp\left\{-\frac{(2k+1)^2}{8}\pi^2 v\right\}, \quad \psi_2(\lambda) = (\mathrm{th}\sqrt{2\lambda})/(\sqrt{2\lambda}).$$

For the case $R = L = 1$, the existence of the limit distributions of the random variables $x_0(t)$ and $y_0(t)$ as $t \to +\infty$ was established in (Sinai, 1982) and (Ritter, 1976), respectively. Explicit expressions for the limit distributions in this case were obtained in (Golosov, 1983; Kesten, 1986). The case $R = 1, L = 2$ was considered in (Letchikov, 1989). Theorem 3 is a generalization of these results. Theorems 1 and 3 were published in (Letchikov, 1989).

We use the Lemma and the methods of (Sinai, 1982) and (Golosov, 1983) to prove Theorem 3.
Introduce now random variables: $\rho_0 = 0$, $\rho_{n+1} = \min\{t > \rho_n,\ x(t) = 0\}$, $S(n) = \max\{|x(t)|,\ 0 \leqslant t \leqslant \rho_n\}$, $\varkappa(t) = \{k \geqslant 0,\ x(t) = 0,\ k \leqslant t\}$.
We shall say that a condition holds for all sufficiently large (a.s.l.) t, if there exists a number t_0 such that this condition holds for all $t \geqslant t_0$. We shall say that a condition holds infinitely often (i.o.) as $t \to +\infty$, if there exists a sequence $\{t_k\}_{k=1}^{+\infty}$, for which this condition holds. Let $\ln_{(k)} n$ be the k–th iteration of the logarithm of $n (k \geqslant 2)$.

THEOREM 4. *Assume that conditions* (C1—C3) *hold and* $\lambda_R = 0$. *Then for any* $\varepsilon > 0$ *with* **P**–*probability* 1:

1) $S(n) \geqslant \dfrac{1-\varepsilon}{2\sigma^2} \dfrac{(\ln n)^2}{\ln_{(3)} n}$ for a.s.l. n,

2) $S(n) \leqslant \dfrac{1+\varepsilon}{2\sigma^2} \dfrac{(\ln n)^2}{\ln_{(3)} n}$ i.o. as $n \to +\infty$,

3) $S(n) \leqslant (\ln n)^2 (\ln_{(2)} n)^{2+\varepsilon}$ for a.s.l. n,

4) $S(n) \geqslant (\ln n)^2 (\ln_{(2)} n)^2$ i.o. as $n \to +\infty$,

5) $\lim\limits_{t \to +\infty} \varkappa(t)/t = 0$,

6) $\varkappa(t) \geqslant \exp\{(\ln t)(\ln_{(2)} t)^{-1-\varepsilon}\}$ for a.s.l. t.

Theorem 4 extends the results obtained in (Deheuvels and Revesz, 1986) for RWIRE, when $R = L = 1$.

REFERENCES

Deheuvels, P. and Revesz, P. (1986). Simple random walk on the line in random environment. *Probab. Th. Rel. Fields.* **72**, 215–230.

Golosov, A.O. (1983). The limit distribution for random walks in random environments. *Sov. Math. Dokl.* **28**, 19–22.

Golosov, A.O. (1986). About limit distributions for the random walk in the critical one–dimensional random environment. *Uspekhi Mat. Nauk.* **41**, 189–190.

Kalikow, S.A. (1981). Generalized random walk in a random environment. *Ann. Probab.* **9**, 753–768.

Kesten, H. (1984). Open problems. Random walk in random environment. *Contemp. Math.* **50**, 332.

Kesten, H. (1986). The limit distribution of Sinai's random walk in a random environment. *Phys. A.* **138**, 299–309.

Key, E. (1984). Recurrence and transience criteria for random walk in a random environment. *Ann. Probab.* **12**, 529–560.

Letchikov, A.V. (1988). One limit theorem for a random walk in a random environment. *Theory Probab. Appl.* **33**, 246–256.

Letchikov, A.V. (1989). A limit theorem for a recurrent random walk in a random environment. *Soviet Math. Dokl.* **304**, 25–28.

Oseledec, V.I. (1968). A multiplicative ergodic theorem. Lyapunov characteristic numbers for dynamical systems. *Trans. Moscow Math. Soc.* **19**, 197–231.

Ritter, G.A. (1976). Random walk in a random environment, critical case. *Ph. D. thesis.* Cornell University, New York.

Sinai, Ya.G. (1982). The limiting behaviour of one–dimensional random walk in a random medium. *Theory Probab. Appl.* **27**, 256–268.

Solomon, F. (1975). Random walks in a random environment. *Ann. Probab.* **3**, 1–31.

Tutubalin, V.N. (1977). A central limit theorem for products of random matrices. *Symp. Math.* **21**, 101–116.

Prob. Theory and Math. Stat., Vol. 2, pp. 97–104
B.Grigelionis *et al.* (Eds.)
1990 VSP/Mokslas

ON THE NORM DISTRIBUTION OF GAUSSIAN AND OTHER STABLE VECTORS

M.A. LIFSHITS
Department of Computer Science, Leningrad Institute of Finance and Economy,
190000 Leningrad, Griboedov channel 30, USSR

ABSTRACT

We obtain the lower and upper bounds for the density of the norm distribution for the Gaussian vector in the separable Banach space.
The generalization of Zak—Lewandowsky formula is given for the density of the norm distribution for the α–stable not necessary symmetric vector, $\alpha < 1$.

Let X be a random vector taking values in the separable Banach space $(B, \|\cdot\|)$; $F(r) = \mathbf{P}\{\|X\| \leqslant r\}$ be the norm distribution function; $p(r) = F'(r)$ be the density of the norm distribution; $\mathcal{P}$ be the distribution law of X ($\mathcal{P}$ is a Radon measure on B); d be the dimension of the topological support N of $\mathcal{P}$ (we consider $d > 0$).
In the *first part* we deal with a *Gaussian* vector. Let $b \in B$ and $K : B^* \to B$ be the barycentre and the covariance operator of $\mathcal{P}$. Tsirelson (1975) proved that the density p exists and is continuous on $(0, \infty)$ with the exception of a countable set of real numbers; p is uniformly bounded on each segment $[\varepsilon, \infty)$. Some examples (Paulauskas, 1987; Paulauskas and Rachkauskas, 1982; Rhee, 1984; Davydov and Lifshits, 1984) show that p may be not bounded at zero if the space B is not uniformly convex and $\mathcal{P}$ has bad entropic properties. Lifshits in (Lifshits, 1986) proved the boundedness of p for $\mathcal{P}$ with "moderate" entropy. Rhee and Talagrand in (Rhee and Talagrand, 1986) proved the same for the uniformly convex space B of power type (earlier Davydov obtained boundedness for $B = L^p$ (Davydov and Lifshits, 1984)). We recall that $(B, \|\cdot\|)$ is uniformly convex of the power type $\beta > 1$, if for some $\Delta > 0$ and all $\rho \in (0, 2]$

$$\delta(\rho) = 1 - \sup\{\|x+y\|/2\,,\quad \|x\| = \|y\| = 1,\ \|x-y\| \geqslant \rho\} \geqslant \Delta\rho^{\beta}. \tag{1}$$

The next problem, arising when boundedness is proved, is the investigation of the dependence of the upper bound on the barycentre and covariance. The first theorem solves this problem for the case of fixed covariance. This result generalizes one estimate (Paulauskas and Rachkauskas, 1982, Chapter 4) for the case when $B = l^p$

and independent components. One can see from (Paulauskas and Rachkauskas, 1982) also, that the exponent $\beta - 1$ in the following result can not be improved.

THEOREM 1. *If $(B, \|\cdot\|)$ is uniformly convex of the power type β and $d \geqslant \beta$, then for some finite $C(K)$ depending on covariance*

$$\operatorname*{ess\,sup}_{r\in(0,\infty)} p(r) \leqslant C(K)(\|b\|+1)^{\beta-1}. \tag{2}$$

Proof. Let $\mathcal{P}_0$ be the centred measure: $\mathcal{P}_0\{A\} = \mathcal{P}\{A+b\}$, and Φ be the distribution function of the standard normal law $\mathcal{N}(0,1)$. The function

$$\psi(r) = \Phi^{-1}(F(r)) \tag{3}$$

is very useful for our investigation. Ehrhard's theorem (Ehrhard, 1983) shows that $\psi(\cdot)$ is a concave increasing function. Hence it's derivative

$$\psi'(r) = p(r)/\Phi'(\Phi^{-1}(F(r))) = (2\pi)^{1/2} p(r)\exp\{\psi^2(r)/2\}$$

is not increasing. Let $r \geqslant \|b\|+1$. Then the inequality $\psi'(r) \leqslant \psi'(\|b\|+1)$ may be rewriten in the form

$$p(r) \leqslant p(\|b\|+1)\exp\left\{[\psi^2(\|b\|+1) - \psi^2(r)]/2\right\}. \tag{4}$$

If $F(\|b\|+1) \geqslant 1/2$ then $0 \leqslant \psi(\|b\|+1) \leqslant \psi(r)$, and we obtain

$$p(r) \leqslant p(\|b\|+1). \tag{5}$$

If $F(\|b\|+1) \leqslant 1/2$ then we use trivial inequalities

$$F(\|b\|+1) = \mathbf{P}\{\|X\| \leqslant \|b\|+1\} \geqslant \mathcal{P}_0\{x \mid \|x\| \leqslant 1\},$$

$$0 \geqslant \psi(\|b\|+1) \geqslant \Phi^{-1}(\mathcal{P}_0\{x \mid \|x\| \leqslant 1\}).$$

In this case, (4) means that

$$p(r) \leqslant p(\|b\|+1)\exp\left\{[\Phi^{-1}\mathcal{P}_0\{x \mid \|x\| \leqslant 1\}]^2/2\right\}. \tag{6}$$

Measure $\mathcal{P}_0$ does not depend on b, so (5) and (6) show: we need to prove only the uniform estimate for $p(r)$, $0 \leqslant r \leqslant \|b\|+1$.
Let $m \geqslant \beta$; $l_1, \ldots, l_m$ be orthonormal vectors in the kernel $H \subset B$ of the Gaussian measure $\mathcal{P}_0$ (it is well–known, that H is a Hilbert space, including all admissible shifts of $\mathcal{P}_0$). Let's construct the partition $\mathcal{T}$ of the space B by m–dimensional parallel subspaces $\gamma = x_\gamma + L$, $L = \operatorname{Span}\{l_1, \ldots, l_m\}$. We choose the "origin" x_γ so, that

$$\|x_\gamma\| = \inf_{x\in\gamma} \|x\|.$$

Conditional measures $\mathcal{P}_\gamma$ generated by the partition $\mathcal{T}$ will be shifted standard normal $\mathbf{R}^m$–distributions. Fix γ and construct "internal" partition Θ of the subspace γ. Elements $\theta \in \Theta$ are the radial lines, passing through the point x_γ. Let $\theta = \{x_\gamma + cl, c \in \mathbf{R}^1\}$ be an element of Θ. We normalize the vector l so, that $|l|_H = 1$. Then the conditional measure on θ is

$$\mathcal{P}_\theta(dc) = q_\theta(c)dc = A_\theta|c|^{m-1} \exp\{-(c-\alpha_\theta)^2/2\}dc$$

for some A_θ, α_θ.

We have to estimate the conditional density of the norm distribution under the condition $X \in \theta$. The function $f_\theta(c) = \|x_\gamma + cl\|$ is increasing on $(0,\infty)$ and decreasing on $(-\infty, 0)$, that's why the conditional norm distribution $F_\theta = \mathcal{P}_\theta f^{-1}$ has a density

$$F'_\theta(r) = \sum_{\{c:f_\theta(c)=r\}} q_\theta(c)|f'_\theta(c)|^{-1}, \quad r > \|x_\gamma\| = f_\theta(0). \tag{7}$$

There are exactly two summands in (7) for each $r > \|x_\gamma\|$. The condition (1) implies inequalities

$$|c| \leqslant r\|l\|^{-1}\Delta^{-1/\beta}(1-\|x_\gamma\|/r)^{1/\beta}, \quad r = f_\theta(c); \tag{8}$$

$$|f'_\theta(c)| \geqslant \|l\|(1-\|x_\gamma\|/r)^{1-1/\beta}\Delta^{1/\beta}. \tag{9}$$

The upper bounds for q_θ we get from the following elementary proposition.

LEMMA 1. *Let $q(c)$ be the density of the probability measure on $\mathbf{R}^1$:*

$$q(c) = A|c|^{m-1} \exp\{-(c-\alpha)^2/(2\sigma^2)\}$$

for some A, α, m, σ. Then

$$q(c) \leqslant \min\left\{A_m|c|^{m-1}\sigma^{-m};\ 2\pi^{-1/2}\sigma^{-1}\right\}, \tag{10}$$

where $A_m = 2^{1-m/2}/\Gamma(m/2)$.

We use (10) for q_θ. After the substitution $\sigma = 1$ and (8) we get

$$q_\theta(c) \leqslant \min\left\{A_m\left[r\|l\|^{-1}\Delta^{-1/\beta}(1-\|x_\gamma\|/r)^{1/\beta}\right]^{m-1},\ 2\pi^{-1/2}\right\}. \tag{11}$$

The next step is the substitution of (11) and (9) into (7):

$$F'_\theta(r) \leqslant 2\min\left\{A_m r^{m-1}\|l\|^{-m}\Delta^{-m/\beta}\left(1-\|x_\gamma\|/r\right)^{m/\beta-1},\right.$$

$$\left. 2\pi^{-1/2}\|l\|^{-1}\left(1-\|x_\gamma\|/r\right)^{1/\beta-1}\Delta^{-1/\beta}\right\}.$$

We can eliminate $(1-\|x_\gamma\|/r)$ and get more compact estimate:

$$F'_\theta(r) \leqslant A(m,\beta)\Delta^{-1}\|l\|^{-\beta}r^{\beta-1}, \tag{12}$$

where

$$A(m,\beta) = 4\pi^{-1/2}(A_m\pi^{1/2}/2)^{(\beta-1)/(m-1)}.$$

The integration of (12) over Θ and then over T implies

$$p(r) = F'(r) \leqslant A(m,\beta)\Delta^{-1}\Big\{\inf_{l\in L, |l|_H=1} \|l\|\Big\}^{-\beta} r^{\beta-1}. \tag{13}$$

This inequality gives the estimate for the density, which does not depend on barycentre. The substitution of (13) with $r = \|b\| + 1$ into (5) and (6) gives (2).

Remarks. 1. Together with mentioned above Tsirelson's result, (13) implies the uniform boundedness of $p(\cdot)$ on $\mathbf{R}^1$. So we obtain a proof of Rhee—Talagrand theorem.

2, The Theorem 1 holds in spaces $L^p(1 < p < \infty; \beta = \max\{p,2\})$ and in Hilbert space ($\beta = 2$), where it is well-known (Paulauskas and Rachkauskas, 1987).

3. The dependence on covariance is hidden in the factor $\inf\limits_{l\in L,|l|_H=1} \|l\|$, so one can study it, optimizing the choise of the space L.

The next theorem illustrates the dependence between the density and the tail probability of the norm distribution. I hope, it will be useful in frequent situations, when tails of F have been deeply investigated and it's necessary to obtain some information about the density.

We use notations

$$\sigma^2 = \|K\| = \sup_{f\in B^*} \|Kf\|\|f\|^{-1} = \sup_{f\in B^*} \mathbf{E}f^2(X-b)\|f\|^{-2}$$

for the norm of covariance operator K and $m = F^{-1}(1/2)$ for the median of F. We also use the function

$$S(u) = \int_0^\infty \exp\{-v - v^2u^{-2}/2\}dv, \quad 0 < u < \infty,$$

and the function ψ, defined in (3). To explain the role of S, let's note, that

$$S(u) = u[1-\Phi(u)]/\Phi'(u), \; u > 0. \tag{14}$$

We see, that S is a ratio of the tail probability to the density of $\mathcal{N}(0,1)$. The idea of the following inequality is to show that the similar situation takes place in general situation.

THEOREM 2. *For any $r > m$*

$$\frac{\psi(r)[1-F(r)]}{\sigma S(\psi(r))} \leqslant p(r) \leqslant \frac{r^2[1-F(r)]}{\sigma^2(r-m)S(\psi(r))}. \tag{15}$$

Proof. We may consider F as a distribution function of the image of the measure $\mathcal{N}(0,1)$ by the mapping $Q = \psi^{-1} = F^{-1}\Phi$: $\mathbf{R}^1 \to \mathbf{R}^+$. It was mentioned yet that ψ is concave, so Q is convex. Hence,

$$(r-m)/\psi(r) = (Q(\psi(r)) - Q(0))/\psi(r) \leqslant Q'(\psi(r)) \leqslant \lim_{u\to\infty} Q'(u). \tag{16}$$

It is obvious, that $p(r) = F'(r) = \Phi'(\psi(r))/Q'(\psi(r))$. We substitute here (14) with $u = \psi(r)$ and (16), and obtain:

$$\frac{\psi(r)[1-\Phi(\psi(r))]}{[\lim_{u\to\infty} Q'(u)]S(\psi(r))} \leqslant p(r) \leqslant \frac{\psi(r)[1-\Phi(\psi(r))]\psi(r)}{S(\psi(r))(r-m)}. \tag{17}$$

The last part of the proof will be based on the lemma which is well-known for the specialists. This lemma is an evident consequence of the results on tail probabilities of the norm distribution (see, for example (Talagrand, 1984)).

LEMMA 2.

$$\lim_{r\to\infty} (\sigma\psi(r) - r) = 0. \tag{18}$$

This statement means, that concave function ψ is asymptotically linear at infinity: $\psi(r) \approx r/\sigma$, $F(r) \approx \Phi(r/\sigma)$. From concavity and (18) we obtain $\psi(r) \leqslant \sigma^{-1}r$,

$$\lim_{u\to\infty} Q'(u) = \sigma.$$

Now (17) implies (15).

COROLLARY.

$$\lim_{r\to\infty} p(r)/(\sigma^{-2}r(1-F(r))) = 1. \tag{19}$$

Obviously, $\lim_{r\to\infty} S(\psi(r)) = \lim_{u\to\infty} S(u) = 1$. Hence (19) is the direct consequence of (15) and (18).

Remarks. 1. Generalization. The result close to the Theorem 2 holds for distributions of the wide class of convex functionals. Let $G \subset B^*$ be a set of linear functionals, $f = \sup_{g\in G}\{g(X)\}$ be a.s. finite random variable; F, p, m be the distribution function, density and median for f, respectively; $\psi = \Phi^{-1}F$. We must change the definition of σ only. Now

$$\sigma^2 = \sup_{g\in G} \mathbf{E}(g(X-b))^2.$$

We introduce one new parameter

$$\tau = \lim_{r\to\infty} (\sigma\psi(r) - r) \in (-\infty, +\infty).$$

In this situation the lower bound (15) is valid again; the upper bound remains valid after the change r^2 by $(r+\tau)^2$. The proof works without any changes.

2. There is no hope to prove something like (15) for the domain $\{r < m\}$, because p may be very irregular (oscillating with infinite amplitude) near the starting point zero even in the case $b=0$.

We begin the *second part* of the paper. Now X will be a stable vector with $\alpha \in (0,1)$. In (Ryznar, 1990; Lifshits, 1987; Smorodina and Lifshits, 1989) one can find information on the boundedness, smoothness and other properties of the distribution of $\|X\|$.

Let $t = \inf\{r | \mathbf{P}\{\|X\| \leqslant r\} > 0\}$ be the starting point of F, ν be the Lévy measure for the stable law $\mathcal{P}$. Recently Lewandowsky and Zak discovered a nice equation, connecting $\mathcal{P}, p$ and ν (Lewandowsky and Zak, 1987): if the law $\mathcal{P}$ is *symmetric*, then for $r > 0$

$$p(r) = \alpha r^{-1} \int_B \Big[\mathbf{P}\{\|X\| \leqslant r\} - \mathbf{P}\{\|X+y\| \leqslant r\}\Big]\nu(dy). \tag{20}$$

This infinite–dimensional result corresponds with the one–dimensional equation (Kanter, 1975). We generalize (20) to the entire class of α–stable laws, $\alpha < 1$. For this purpose we need a representation $X = X_0 + b$, where X_0 is the strictly stable vector. In our case $\alpha < 1$ this representation exists (and is unique). Vector $b \in B$ is a kind of barycentre for the law $\mathcal{P}$.

THEOREM 3. *If $p(r) = F'(r)$ exists for some $r > t$ then*

$$p(r) = \alpha r^{-1} \int_B \Big[\mathbf{P}\{\|X\| \leqslant r\} - \mathbf{P}\{\|X+y\| \leqslant r\}\Big]\nu(dy) -$$

$$-r^{-1}\frac{\partial}{\partial b}\mathbf{P}\{\|X\| \leqslant r\}. \tag{21}$$

Remarks. 1. If X is *strictly* stable, then $b=0$. Thus the second summand disappears and we obtain (20).

2. The integral in (21) exists for any $r > 0$. The derivatives $p(r) = F'(r)$ and b–derivative in (21) exist simultaneously (for almost all r). I hope, they exist for *any* r (as F' in symmetric case).

Proof. We consider the convolution semigroup of measures $\mathcal{P}_c$: for $c > 0$, $\mathcal{P}_c$ is the distribution law of r.v. $X_c = c^{1/\alpha}X + \Delta_c$; $\Delta_c = (c - c^{1/\alpha})b$. Particularly, $\mathcal{P}_1 = \mathcal{P}$. Then

$$\mathbf{P}\{\|X_c - \Delta_c\| \leqslant r\} = \mathbf{P}\{\|X\| \leqslant rc^{-1/\alpha}\} = F(rc^{-1/\alpha}),$$

$$F'(r) = \lim_{c\to 1} \left[F(r) - F(rc^{-1/\alpha})\right] / (r(1-c^{-1/\alpha})) =$$

$$= \lim_{c\to 1} \alpha\Big[\mathbf{P}\{\|X\| \leqslant r\} - \mathbf{P}\{\|X_c - \Delta_c\| \leqslant r\}\Big] / (r(c-1)).$$

We split this expression by intermediate member $\mathbf{P}\{\|X-\Delta_c+\Delta_{c-1}\|\leqslant r\}$. Let $s=c-1$. Then

1)

$$\alpha r^{-1}\lim_{c\to 1}\Big[\mathbf{P}\{\|X\|\leqslant r\}-\mathbf{P}\{\|X-\Delta_c+\Delta_s\|\leqslant r\}\Big]\Big/(c-1)=$$

$$=r^{-1}\lim_{u\to 0}\Big[\mathbf{P}\{\|X\|\leqslant r\}-\mathbf{P}\{\|X+ub\|\leqslant r\}\Big]/u=-r^{-1}\frac{\partial}{\partial b}\mathbf{P}\{\|X\|\leqslant r\}$$

is the second summand in (21).

2)

$$\alpha r^{-1}\lim_{c\to 1}\Big[\mathbf{P}\{\|X-\Delta_c+\Delta_s\|\leqslant r\}-\mathbf{P}\{\|X_c-\Delta_c\|\leqslant r\}\Big]\Big/(c-1)=$$

$$=\alpha r^{-1}\lim_{c\to 1}\int_B\Big[\mathbf{P}\{\|X-\Delta_c+\Delta_s\|\leqslant r\}-\mathbf{P}\{\|X-\Delta_c+z\|\leqslant r\}\Big]\frac{\mathcal{P}_s(dz)}{s}.$$

We change the variable $z=y+\Delta_s$; denote by $\hat{\mathcal{P}}_s$ the law of $s^{1/\alpha}X$. Then $\mathcal{P}_s(dz)=\hat{\mathcal{P}}_s(dy)$, and we obtain

$$\alpha r^{-1}\lim_{c\to 1}\int_B\Big[\mathbf{P}\{\|X-\Delta_c+\Delta_s\|\leqslant r\}-\mathbf{P}\{\|X-\Delta_c+\Delta_s+y\|\leqslant r\}\Big]\frac{\hat{\mathcal{P}}_s(dy)}{s}. \quad (22)$$

Let's interchange the limit and the integral. Note that

$$\lim_{s\to 0}h_s(y)=\lim_{s\to 0}\Big[\mathbf{P}\{\|X-\Delta_c+\Delta_s\|\leqslant r\}-\mathbf{P}\{\|X-\Delta_c+\Delta_s+y\|\leqslant r\}\Big]=$$

$$=h(y)=\mathbf{P}\{\|X\|\leqslant r\}-\mathbf{P}\{\|X+y\|\leqslant r\},$$

and measures $\nu_s(dy)=\hat{\mathcal{P}}_s(dy)/s$ weakly converge to $\nu(dy)$ on each set $\{x\in B,\ \|x\|>\varepsilon\}$, where ν is finite.
This weak convergence is a consequence of well–known results on the large deviations of stable measures.
Omitting the rigorous elimination of the small set $\{x\in B/\|x\|\leqslant\varepsilon\}$ we may rewrite the limit (22) as

$$\alpha r^{-1}\int_B h\,d\nu=\alpha r^{-1}\int_B\Big[\mathbf{P}\{\|X\|\leqslant r\}-\mathbf{P}\{\|X\|\leqslant r\}\Big]\nu(dy).$$

It's exactly the first summand of (21). Now we must only explain, why $\int h_s d\nu_s\to$ $\to\int h\,d\nu$. This is a standard problem of the theory of weak convergence. It is enough (Billingsley, 1968, § 5) to establish the "joint continuity":

$$h_s(y_s)\to h(y)\ \text{ if }\ y_s\to y.$$

In fact, for $\beta_s=\|\Delta_c-\Delta_s\|+\|y_s-y\|$,

$$|h_s(y_s) - h(y)| \leqslant \mathbf{P}\{r - \beta_s \leqslant \|X\| \leqslant r + \beta_s\} + \mathbf{P}\{r - \beta_s \leqslant \|X + y\| \leqslant r + \beta_s\}.$$

Hence as $\beta_s \to 0$, so

$$\limsup_{s \to 0} |h_s(y_s) - h(y)| \leqslant \mathbf{P}\{\|X\| = r\} + \mathbf{P}\{\|X + y\| = r\}$$

and the joint continuity is the consequence of the absolute continuity of the norm distribution.

Remark. What occurs if $1 \leqslant \alpha < 2$? Only symmetric variant of (20) is proved for Hilbert space by Zak and for some uniformly convex spaces by Ryznar and Zak. In general case formula (21) needs some modification (usual in the theory of stable measures) to make integral in (21) finite.

REFERENCES

Billingsley, P. (1968). *Convergence of probability measures.* Wiley, New York.

Davydov, Yu.A. and Lifshits, M.A. (1984). The fibre method in some probability problems. *Itogi Nauki i Techniki. Theory Probab. Math. Statist. Theor.Cybern.* **22**, 61–157.

Ehrhard, A. (1983). Symmetrisation dans l'espace de Gauss. *Math. Scand.* **53**, 281–301.

Kanter, M. (1975). Stable densities under change of scale and total variation inequalities. *Ann. Probab.* **3**, 697–707.

Lewandowsky, M. and Zak, T. (1987). On the density of the distribution of p-stable seminorms. *Proc. Amer. Math. Soc.* **100**, 345–351.

Lifshits, M.A. (1986). On the distribution of the maximum of a Gaussian process. *Theory Probab. Appl.* **31**, 132–142.

Lifshits, M.A. (1987). On the density of the distribution of the norm of a stable vector. *Notes Sci. Seminars Leningrad Branch of Math. Inst.* **158**, 105–114.

Paulauskas, V. (1982). On the density of the norm of Gaussian vector in Banach spaces. *Lect. Notes in Math.* **990**, 179–197.

Paulauskas, V. and Rachkauskas, A. (1987). *Accuracy of Approximation in the Central Limit Theorem in Banach Spaces.* Mokslas, Vilnius.

Rhee, W.S. (1984). On the distribution of the norm for a Gaussian measure. *Ann. Inst. H.Poincare. Probab. et Statist.* **10**, 277–286.

Rhee, W.S. and Talagrand, M. (1986). Uniform convexity and the distribution of the norm for a Gaussian measure. *Probab. Theory Rel. Fields* **71**, 59–67.

Ryznar, M. (1990). Geometrical properties of Banach spaces and the distribution of the norm for a stable measure. *Studia Math.* (to appear).

Smorodina, N.V. and Lifshits, M.A. (1989). On the distribution of the norm of a stable vector. *Theory Probab. Appl.* **34**, 304–313.

Talagrand, M. (1984). Sur l'integrability des vecteurs Gaussiens. *Z. Wahrsch. verw. Geb.* **68**, 1–8.

Tsirelson, B.S. (1975). The density of the distribution of the maximum of a Gaussian process. *Theory Probab. Appl.* **20**, 865–873.

Prob. Theory and Math. Stat., Vol. 2, pp. 105–123
B.Grigelionis *et al.* (Eds.)
1990 VSP/Mokslas

ON SET CONVOLUTIONS AND INTEGRAL–SUM KERNEL OPERATORS

J.M. LINDSAY
Mathematics Department, King's College, Strand,
London WC2R 2LS, England

ABSTRACT

The purpose of this paper is to collect together the basic properties of a class of convolution–like products and integral–kernel–like operators on Fock space. The main new results here are injectivity of the map from three–argument kernel to integral–sum operator, and the non–representability of finite rank operators by means of three–argument kernels (in the non–atomic case).

INTRODUCTION

The purpose of this paper is to collect together the basic properties of a class of convolution–like products and integral–kernel–like operators on Fock space. The former arose from the desire to *algebrise* the well–known isomorphism between Wiener space and Fock space, which is given by orthogonal expansion in terms of multiple Wiener—Itô integrals (Lindsay and Maassen, 1990). The latter arose from an approach to quantum stochastic differential equations. Both have their origin in a paper of H. Maassen (Maassen, 1985), and have been expounded upon by P.–A. Meyer (Meyer, 1986). Quantum stochastic calculus from this point of view is developed in (Lindsay and Maassen, 1988).
Among the applications of these operations currently in mind are non–causal stochastic analysis, both classical (e.g., (Nualart and Zakai, 1986)) and quantum (Lindsay, 1990), and the Malliavin calculus; quantum stochastic differential equations (Lindsay and Maassen, 1990; Lindsay and Parthasarathy, 1989); Bose—Fermi relations for quantum fields (Lindsay and Parthasarathy, 1989); operator cocycles on Wiener—Fock space (Hudson and Lindsay, 1987; Journé, 1987; Accardi, Journé and Lindsay, 1989; Arveson, 1987); hypercontractivity (Lindsay, 1990); classical martingales determined by structural equations and chaos completeness (Azema, 1985; Parthasarathy, 1988; Meyer, 1988).
The hybrid Fock space considered here is constructed in Paragraph 1 and an elementary, but fundamental, combinatorial identity is highlighted in Paragraph 2 which also contains the germ of the duality between the Skorokhod integral (Skorokhod, 1975) and the Malliavin gradient/stochastic derivative. The basic estimate for the set convolutions is established in Paragraph 2, and integral–sum kernel operators are treated in Paragraphs 3 and 4. The main new results here

are injectivity of the map from three–argument kernel to integral–sum operator, and the non–representability of finite rank operators by means of three–argument kernels (in the non–atomic case). The final section is designed to share some of the intuition behind the form of these operators, and also to serve as an appetizer for (Lindsay 1991).

Conventions. When a map $f : X \to Y$ is introduced, and X, Y are measurable spaces, f will be understood to be measurable. An unspecified norm on a function space will always be the L^2–norm. The domain of a (possibly unbounded) ***Hilbert space*** operator T is denoted Dom(T), and inner products $<,>$ are linear in their ***second*** arguments.

1. THE SYMMETRIC FUNCTOR

For a set S let $\Gamma(S)$ be the ***finite power set of*** S : $\{\sigma \subset S : \#\sigma < \infty\}$, where $\#\sigma$ denotes the cardinality of σ, and let $\underline{S} = \bigcup_{n \geqslant 0} S^{(n)}$, where $S^{(n)} = \{\underline{s} \in S^n : s_i \neq s_j$ if $i \neq j\}$ with $S^{(0)} = S^0$ understood as the singleton set with element 0. Then to any function f on $\Gamma(S)$ corresponds a unique *symmetric* function $f^\vee$ on $\underline{S}$, through the relation $f^\vee(\underline{s}) = f(\{s_1, \ldots, s_n\})$ for $\underline{s} \in S^{(n)}, f^\vee(0) = f(\emptyset)$, and conversely. Similarly, the class of subsets of $\Gamma(S)$ is in bijective correspondence with the family of symmetric subsets of $\underline{S}$.

We shall need to deal with products of Γ's, in particular in relation to the union operation, and therefore define, for $d \geqslant 1, \Gamma^{(d)}(S) = \{\underline{\sigma} \in \Gamma(S)^d : \sigma_i \bigcap \sigma_j = \emptyset$ if $i \neq j\}$ — there should be no danger of conflict with the notation above. For $\underline{\sigma} \in \Gamma^{(d)}(S)$ write $|\underline{\sigma}|$ for $\sigma_1 \bigcup \sigma_2 \bigcup \cdots \bigcup \sigma_d$. For $\varphi : S \to \mathbf{C}$ the ***multiplicative*** function $\sigma \in \Gamma_S \mapsto \prod_{s \in S} \varphi(s)$ will be denoted π_φ.

To any σ–algebra $\mathcal{M}$ on S there corresponds a natural σ–algebra $\Gamma(\mathcal{M})$ on $\Gamma(S)$ defined as follows. Let $\sum_n = \mathcal{M}^n_{\text{sym}} \bigcap S^{(n)}$, where $\mathcal{M}^n_{\text{sym}}$ is the collection of symmetric sets in the product σ–algebra $\mathcal{M}^n$, then $\Gamma(\mathcal{M}) = \bigvee_{n \geqslant 0} \widehat{\Sigma}^n$. There is an alternative description of $\Gamma(\mathcal{M})$ which is sometimes useful. For a subset A of S let $\eta_A : \Gamma(S) \to \mathbf{N}$ denote the function $\sigma \mapsto \#(\sigma \bigcap A)$, then $\Gamma(\mathcal{M})$ is also the σ–algebra generated by $\{\eta_A : A \in \mathcal{M}\}$. This may be seen from the following two identities:

$$\{\eta_A = k\} = \left\{ \bigcup_{n \geqslant 0} (\overbrace{A \times \cdots \times A}^{k} \times \overbrace{A^c \times \cdots \times A^c}^{n})_{\text{sym}} \cap S^{(k+n)} \right\}^\wedge,$$

$$(A_1 \times \cdots \times A_k)_{\text{sym}} \cap S^{(k)} = \left\{ \bigcap_{i=1}^{k} \{\eta_{A_i} \geqslant 1\} \cap \Gamma_k(S) \right\}^\vee,$$

where $\Gamma_k(S) := \{\sigma \in \Gamma : \#\sigma = k\} = \{\eta_S = k\}$ and, for $B \subset S^m, B_{\text{sym}}$ means the symmetrised set $\{\underline{s} \in S^m : (s_{\pi(1)}, \ldots, s_{\pi(m)}) \in B$ for some permutation $\pi\}$.

Before showing how a measure m on $(S, \mathcal{M})$ naturally determines a measure on $\Gamma(S, \mathcal{M})$, certain technical questions must be faced. These arise from the possible non–measurability of $S^{(2)}$. All measure spaces will be assumed σ–finite and, to avoid labouring the measurability questions, the following *assumptions* (automatically satisfied when m is non–atomic) will be made:

(i) $S^{(2)}$ belongs to the m^2–completion of $\mathcal{M}^2$;

(ii) $S^{(2)}\backslash S^{\langle 2\rangle}$ is m^2–null,

where $S^{\langle 2\rangle} := \cup_{A\in\mathcal{M}} A \times A^c$ is the collection of all points in S^2 which lie in a measurable rectangle having disjoint sides. An immediate *consequence* of this is the existence of an $\mathcal{M}^2$–measurable subset $S^{[2]}$ of $S^{\langle 2\rangle}$ with the following properties. Put $S^{[k]} = \{\underline{s} \in S^k : (s_i, s_j) \in S^{[2]}$ for each distinct $i,j\}$ for $k \geqslant 2, S^{[k]} = S^k$ for $k < 2$ and $\Gamma^{[k]} = \{\underline{\sigma} \in \Gamma^k : |\underline{\sigma}| \in \widehat{S^{[n]}}$ for some $n\}$ for $k \geqslant 1$. Then for all $k \geqslant 0$,

$$\text{(a) } S^{(k)}\backslash S^{[k]} \text{ is } m^k\text{–null;} \qquad \text{(b) } \Gamma^{(k)}\backslash\Gamma^{[k]} \text{ is } \mu^k\text{–null.}$$

Clearly these sets depend on the choice $S^{[2]}$ only up to null sets — in particular the following definition makes sense. The *symmetric measure* of m is the measure $\Gamma(m)$ on $\Gamma(\mathcal{M})$ given by the prescription

$$U \mapsto \iota_\emptyset(U) + \sum_{n\geqslant 1}(n!)^{-1} m^n\Big(U^\vee \cap S^{[n]}\Big)$$

where $\iota_\emptyset(U)$ equals 1 if $\emptyset \in U$ and 0 otherwise. $\Gamma(X) = (\Gamma(S), \Gamma(\mathcal{M}), \Gamma(m))$ is the *symmetric measure space of* $X = (S, \mathcal{M}, m)$ discussed by Guichardet in the non–atomic case (Guichardet, 1972).

If the measure m is non–atomic then $L^2(\Gamma_X)$ is naturally isomorphic to Bose (or, symmetric) Fock space over $L^2(X)$. In general one needs the extra structure of a (measurable) total order on X — then $L^2(\Gamma_X)$ is naturally isomorphic to Fermi (or, anti–symmetric) Fock space over $L^2(X)$. In fact what one requires more precisely is an "almost everywhere defined total order", thus for a.a. $(x,y) \in X^2$, either $x < y$ or $y < x$. This extra leeway is in fact useful — for example $\underline{x} < \underline{y}$ if $x_1 < y_1$ gives an a.e., defined total order on $\mathbf{R}^d$.

When a measure space $X = (S, \mathcal{M}, m)$ is fixed, the following abbreviations are used: Γ^d or Γ^d_X for $(\Gamma(X))^d$, μ for $\Gamma(m)$, $d\sigma$ for $d\mu(\sigma)$, $d\underline{\sigma}$ for $d\mu^n(\underline{\sigma})$ and $\Gamma^{(d)}$ for $\Gamma^{(d)}(X)$. The indicator function of $\Gamma^{[k]}$ will be denoted $\chi_{[k]}$.

Remarks. When the measure space is *separated* ($\{s\} = \cap\{U \in \mathcal{M} : U \text{contains } s\}$ for each s), and each $S^{(k)}$ is measurable — for example, when X is the Borel space of a Hausdorff topological space — all square (and pointed) brackets may be softened to round brackets. If, furthermore, each singleton of S is measurable then the map $(\alpha, \beta) \in \Gamma^2 \mapsto \alpha \cup \beta$ is measurable without restriction to $\Gamma^{(2)}$. However we shall be interested in operations which exclude the 'diagonals' $\Gamma^d\backslash\Gamma^{(d)}$ and so this would be of no great advantage here.

The *number operator* N is the multiplication operator given by $Nf(\sigma) = \#\sigma f(\sigma)$. For functions $\varphi : \mathbf{N} \to \mathbf{C}$, $\varphi(N)$ is the operator on $L^2(\Gamma_X)$ defined through the functional calculus. Thus $\text{Dom}(\varphi(N)) = \{f \in L^2(\Gamma_X) : \int |\varphi(\#\sigma) f(\sigma)|^2 d\sigma < \infty\}$. The following linear subspaces of $L^2(\Gamma_X)$ have particular importance:

$$\mathcal{F}_n(X) = \{f \in L^2(\Gamma_X) : \text{ supp } f \subset \Gamma_n\} \qquad n = 0, 1, \ldots;$$

$$\mathcal{F}(X) = \{f \in L^2(\Gamma_X) : \text{ supp } f \subset \Gamma_{\leqslant n}, \text{ for some } n\};$$

$$\Psi(X) = \text{linear span of } \{\pi_\varphi : \varphi \in L^2(X)\};$$

$$\mathcal{K}(X) = \bigcap_{c>0} \text{Dom } (c^N);$$

elements of $\mathcal{F}$ are called ***finite particle functions***, and elements of $\mathcal{F}_n$ ***n–particle functions***.
From now on X will be a fixed σ–finite measure space satisfying conditions (i) and (ii) above, and reference to X in the notation will frequently be dropped.

2. UNION–PARTITION IDENTITIES

The following basic combinatorial identity is very useful.

LEMMA 2.1. *Let function $g : \Gamma_X^d \to \mathbb{C}$ be non–negative, or integrable, then $G : \sigma \in \Gamma \mapsto \sum_{|\underline{\alpha}|=\sigma} g(\underline{\alpha})$ is measurable. Suppose that g vanishes almost everywhere on $\Gamma^d \backslash \Gamma^{(d)}$, then*

$$\int g d\mu^d = \int G d\mu.$$

Proof. The measurability of G may be verified by considering the corresponding function on S — a ***complete symmetrisation*** of g. The proof now proceeds exactly as for a non–atomic measure space (Lindsay and Parthasarathy, 1989).
Next consider the union operation $\Phi : \underline{\sigma} \in \Gamma^{(d)} \mapsto |\underline{\sigma}| = \sigma_1 \cup \cdots \cup \sigma_d \in \Gamma$. In view of the identity

$$\Phi^{-1}(\{\eta_A = n\}) = \Gamma^{(d)} \bigcap \bigcup \{\eta_A = j_1\} \times \cdots \times \{\eta_A = j_d\},$$

where the union is over those d–tuples $(j_1, \ldots, j_d)$ from $\mathbb{N}$ whose sum is n, Φ is measurable for the σ–algebra $\Gamma(\mathcal{M})^d \cap \Gamma^{(d)}$. For $h : \Gamma \to \mathbb{C}$ and $d \geqslant 1$, let $h^{[d]} : \Gamma^d \to \mathbb{C}$ be defined by

$$h^{[d]}(\underline{\sigma}) = \chi_{[d]}(\underline{\sigma}) h(|\underline{\sigma}|), \tag{2.1}$$

so that, for example, $h^{[2]}(\alpha, \beta) = \chi_{\Gamma^{[2]}}(\alpha, \beta) h(\alpha \cup \beta)$. By the above, if h is measurable then each $h^{[d]}$ is $\Gamma(\mathcal{M})^d$–measurable (and of course $h^{[d]}$ only depends on the choice of $S^{[2]}$ up to a $\dot{\mu}^d$–null set). The Lemma 2.1 has the following immediate consequences.

PROPOSITION 2.2. *Let $h : \Gamma_X \to \mathbb{C}$ be measurable, then for $d \in \mathbb{N}$,*

$$\int h^{[d]} d\mu^d = \int d^{\#\sigma} h(\sigma) d\sigma \tag{2.2}$$

if either h is positive, or $d^N h$ is integrable.

Another version of (2.2) is sometimes useful.

PROPOSITION 2.3. *Let $f : \Gamma \to \mathbb{C}$ be measurable and $a_1, \ldots, a_d$ be non–negative numbers, then*

$$\int \cdots \int a_1^{\#\omega_1} \cdots a_d^{\#\omega_d} \left| f^{[d]}(\underline{\omega}) \right|^2 d\underline{\omega} = \int (a_1 + \cdots + a_d)^{\#\sigma} \left| f(\sigma) \right|^2 d\sigma,$$

in other words,

$$\left\| \left(\prod_i c_i^{N_i} \right) f^{[d]} \right\| = \left\| \sqrt{\sum_i |c_i|^2}^{N} f \right\|$$

for $\underline{c} \in \mathbf{C}^d$, where N_i is the operator of multiplication by the function $\underline{\sigma} \mapsto \#\sigma_i$.

The next two propositions give the structure and relations of the union and partition operations.

DEFINITION 2.4. *For $d \in \mathbf{N}$, the partition operator $\mathcal{P} = \mathcal{P}_{(d)} : L^2(\Gamma^d) \to L^2(\Gamma)$ and union operator $\mathcal{U} = \mathcal{U}_{(d)} : L^2(\Gamma) \to L^2(\Gamma^d)$ are defined by*

$$\mathcal{P}x : \sigma \mapsto \sum_{|\underline{\alpha}| = \sigma} x(\underline{\alpha}) : \quad \mathcal{U}f = f^{[d]} : \underline{\sigma} \mapsto \chi_{[d]}(\underline{\sigma}) f(|\underline{\sigma}|)$$

with maximal domains.

PROPOSITION 2.5. *Fix $d \in \mathbf{N}$. Then $\mathcal{U}$ and $\mathcal{P}$ are closed, mutually adjoint operators with polar decompositions*

$$\mathcal{U} = V\sqrt{d}^{N}; \quad \mathcal{P} = V^*(\sqrt{d}^{\Sigma N_i} P_{\mathrm{sym}})$$

in which V is the isometric embedding given by $Vf(\underline{\sigma}) = \sqrt{d}^{-\#|\underline{\sigma}|} f^{[d]}(\underline{\sigma})$, and P_{sym} is the orthogonal projection given by

$$(P_{\mathrm{sym}}x)(\underline{\sigma}) = \chi_{[d]} d^{-\#|\underline{\sigma}|} \sum_{|\underline{\alpha}| = |\underline{\sigma}|} x(\underline{\alpha}).$$

PROPOSITION 2.6. *In the above notation,*

$$\mathcal{U} = \sqrt{d}^{\Sigma N_i} V; \quad \mathcal{P} = \sqrt{d}^{N} V^*; \quad VV^* = P_{\mathrm{sym}},$$

$$\mathcal{P}\mathcal{U} = d^N; \quad \mathcal{U}\mathcal{P} = d^{\Sigma N_i} P_{\mathrm{sym}}; \quad L^2(\Gamma^d) = H_S \oplus H_A,$$

where $H_S = P_{\mathrm{sym}}, L^2(\Gamma^d) = \{x : \text{for a.a. } \underline{\sigma},\ x(\underline{\sigma}) = \chi_{[d]}(\underline{\tau}) x(\underline{\tau}),\ \text{if } |\underline{\sigma}| = |\underline{\tau}|\}$ and $H_A = \{x : \sum_{|\underline{\alpha}| = \sigma} x(\underline{\alpha}) = 0 \text{ for a.a. } \sigma\}$.

The basic facts about Hilbert space operators which are used in the proof are:

(i) TB is closed and $(BT)^* = T^*B^*$ when T is closed and B is bounded,

(ii) if $T = VA$, where T is closed, A is non-negative self-adjoint and V is partially isometric with initial space equal to the closure of the range of A, then VA is the polar decomposition of T.

Proof of Proposition 2.5. From the definition of V we have the relation

$$\mathcal{U} = \sqrt{d}^{\Sigma N_i} V.$$

Now for $a > 0$,

$$\int a^{\#|\underline{\sigma}|} \left| f^{[d]}(\underline{\sigma}) \right|^2 d\underline{\sigma} = \int (ad)^{\#\sigma} \left| f(\sigma) \right|^2 d\sigma,$$

by the Lemma 2.1. Putting $a = d^{-1}$ we see that V is isometric, and therefore $\mathcal{U}$ is closed. Putting $a = 1$ gives Dom$\mathcal{U}$ = Dom$\sqrt{d}^N$, so that $V\sqrt{d}^N$ is the polar decomposition of $\mathcal{U}$. Another application of the Lemma 2.1 gives

$$\int \sqrt{d}^{-\#|\underline{\sigma}|} f^{[d]}(\underline{\sigma}) x(\underline{\sigma}) d\underline{\sigma} = \int \left\{ \sum_{|\underline{\alpha}|=\sigma} f(\sigma) \sqrt{d}^{-\#\sigma} x(\underline{\alpha}) \right\} d\sigma,$$

so that V^* is given by

$$V^* x : \sigma \mapsto \sqrt{d}^{-\#\sigma} \sum_{|\underline{\alpha}|=\sigma} x(\underline{\alpha}).$$

In particular, $\mathcal{P} = \sqrt{d}^N V^*$. Thus $\mathcal{P} = \mathcal{U}^*$ and, since $\mathcal{U}$ is closed, $\mathcal{P}^* = \mathcal{U}$. Finally, the identity

$$\int \left| \sum_{|\underline{\alpha}|=\sigma} x(\underline{\alpha}) \right|^2 d\sigma = \int d^{\#\sigma} \sum_{|\underline{\alpha}|=\sigma} \left| x_{\text{sym}}(\underline{\alpha}) \right|^2 d\sigma = \int d^{\#|\underline{\sigma}|} \left| x_{\text{sym}}(\underline{\sigma}) \right|^2 d\underline{\sigma}$$

implies that $V^*(\sqrt{d}^{\Sigma N_i} P_{\text{sym}})$ is the polar decomposition of $\mathcal{P}$ and completes the proof.

In the course of the above proof we have also established Proposition 2.6.

Very close relatives of these operators are the Skorokhod integral for non–adapted processes (Skorokhod, 1975), and the Malliavin gradient. Their respective closure and mutual adjointness (Gaveau and Trauber, 1982) may be easily seen from the above approach. This approach will be elaborated upon in a forthcoming article (Lindsay, 1991). The integration by parts formula of Malliavin calculus is one consequence of the mutual adjointness of the above operators.

Clearly the product function $\pi_\varphi : \Gamma \to \mathbf{C}$ is $\Gamma(\mathcal{M})$–measurable if φ is $\mathcal{M}$–measurable, moreover its dependence on φ is only up to null sets. We next list its basic L^2–properties.

PROPOSITION 2.7. *Let $\varphi, \varphi' \in L^2(X)$.*

$$\text{(i) } \|\pi_\varphi\|^2 \leqslant \exp \|\varphi\|^2; \ \|\varphi - \varphi'\| \leqslant \|\pi_\varphi - \pi_{\varphi'}\| \leqslant \|\varphi - \varphi'\| \|\sqrt{N+1}\, \pi_{|\varphi| \vee |\varphi'|}\| \quad (2.3)$$

in particular, $\varphi \mapsto \pi_\varphi$ is bicontinuous.

(ii) *If X is non–atomic, then*

$$\langle \pi_\varphi, \pi_{\varphi'} \rangle = \exp\langle \varphi, \varphi' \rangle \tag{2.4}$$

and $\{\pi_\varphi : \varphi \in L^2(X)\}$ is linearly independent and total in $L^2(\Gamma_X)$. If Y is another non–atomic measure space then $\pi_{(\varphi,\psi)} \mapsto \pi_\varphi \otimes \pi_\psi$ extends to an isometric isomorphism from $L^2(\Gamma_{X\otimes Y})$ to $L^2(\Gamma_X) \otimes L^2(\Gamma_Y)$.

(iii) *If X is discrete then $\{\pi_{\chi_\sigma} : \sigma \in \Gamma_S\}$ is linearly independent and total in $L^2(\Gamma_X)$.*

Proof. The estimate (2.3) follows simply from the Lemma 2.1. For linear independence, in the non–atomic case, suppose that $\{\pi_{\varphi_i}\}$ is a non–empty minimal linearly dependent set, choose $\psi \in L^2(X)$ so that $\{\langle \varphi_i, \psi \rangle\}$ are distinct, and use (2.4) and the fact that the C^∞–functions $s \mapsto \langle \pi_{\varphi_i}, \pi_{s\psi} \rangle$ are eigenvectors of the differential operator $f \mapsto f'$ with distinct eigenvalues, and therefore cannot be linearly dependent, to obtain a contradiction. The rest of the proof of the proposition is straightforward.

3. TWISTED SET CONVOLUTIONS

Before defining convolution–type products, for functions on a finite power set, one more element of structure will be introduced. It has proved useful to incorporate a measurable involution j on $S(j^2 = \text{id.})$. Involution j induces an involution $\widetilde{\cdot}$ on Γ, Γ^d and $\mathcal{F}(\Gamma^d; \mathbb{C})$ in turn, as follows:

$$\widetilde{f}(\underline{\sigma}) = \overline{f(\underline{\widetilde{\sigma}})}; \quad (\sigma_1, \ldots, \sigma_d)\widetilde{\ } = (\widetilde{\sigma}_d, \ldots, \widetilde{\sigma}_1); \quad \widetilde{\sigma} = \{j(s) : s \in \sigma\}.$$

Write $\Gamma(Z) = (\Gamma(X), \widetilde{\cdot})$ for the *symmetric twisted measure space* of $Z = (S, \mathcal{M}, m, j)$, and let $\mu^\dagger$ denote the symmetric measure of the measure $(m + \widetilde{m})$.

DEFINITION 3.1. *The twisted set convolution $f *_\sim g$ of $f, g : \Gamma_X \to \mathbb{C}$ is defined almost everywhere by*

$$f *_\sim g : \sigma \mapsto \int_\Gamma \Big[\sum_{\alpha \subset \sigma} f^{[2]}(\alpha, \widetilde{\omega}) g^{[2]}(\omega, \bar{\alpha}) \Big] d\omega,$$

where $\bar{\alpha}$ denotes the complement of α in σ, for pairs (f, g) for which the bracketed expression is integrable for a.a. σ.

For $X = I_1 \coprod I_2$, the disjoint union of subintervals $I_1 = I_2$ of the real line, $m = a\lambda_1 + b\lambda_2$, where $a \geqslant b > 0$, and λ_i is Lebesque measure on I_i, $*_\sim$ is the *bose product* representing the CCR algebra. Quantum stochastic calculus based on this product is extensively developed in (Lindsay and Maassen, 1988). Deformations of the twisted convolution are investigated in (Lindsay and Parthasarathy, 1989), where homotopies between CCR and CAR algebra representations are established. In view of Proposition 2.3 the convolution is defined when, for example, $\widetilde{f}, g \in \text{Dom}(\sqrt{2}^N)$. Note that the summation is a finite one so that the order of summing and integrating may be interchanged at will. When there is no twist, i.e., the involution is the identity map, the set convolution has been called the ***Wiener product*** (Meyer, 1986) and will be denoted simply $*$. The basic estimate for twisted set convolutions is given next.

PROPOSITION 3.2. *Let $\widetilde{f}, g \in \mathrm{Dom}(\sqrt{2}^N)$ and $a \geqslant 0$, then*

$$\left\|a^N(f *_\sim g)\right\| \leqslant \left\|(a\sqrt{2})^{N_1} f'\right\| \left\|\sqrt{1+2a^2}^N g\right\|,$$

where $f' : (\alpha, \beta) \in \Gamma^2 \mapsto f^{[2]}(\alpha, \widetilde{\beta})$. In particular, if $g \in \mathrm{Dom}(\sqrt{3}^N)$ and $f' = h^{[2]}$, for some $h \in \mathrm{Dom}(\sqrt{3}^N)$, then $f *_\sim g$ is square integrable.

Proof. Two applications of Cauchy—Schwarz and one of the Lemma 2.1 gives

$$\int \left| a^{\#\sigma} \sum_{\alpha \subset \sigma} \int f^{[2]}(\alpha, \widetilde{\omega}) g^{[2]}(\omega, \bar{\alpha}) d\omega \right|^2 d\sigma \leqslant$$

$$\leqslant \int (2a^2)^{\#\sigma} \sum_{\alpha \subset \sigma} \left\{ \int \left|f^{[2]}(\alpha, \widetilde{\omega})\right|^2 d\omega \int \left|g^{[2]}(\omega, \bar{\alpha})\right|^2 d\nu \right\} d\sigma =$$

$$= \int\int (2a^2)^{\#\alpha} \left|f^{[2]}(\alpha, \widetilde{\omega})\right|^2 d\omega d\alpha \int\int (2a^2)^{\#\beta} \left|g^{[2]}(\nu, \beta)\right|^2 d\nu d\beta.$$

The result therefore follows from Proposition 2.3.

DEFINITION 3.3. *For a twisted measure space Z, let*

$$\mathcal{K}(Z) = \left\{ f : \Gamma_X \to \mathbf{C} \text{ s.t. } \int |a^N f|^2 d\mu^\dagger < \infty, \quad \forall a > 0 \right\},$$

also viewed as a subspace of $L^2(\Gamma_X)$.

A similar argument to the above proof shows that $\mathcal{K}(Z)$ is closed under $*_\sim$. The principal identities enjoyed by $*_\sim$ are given next.

PROPOSITION 3.4. *The following identities hold (almost everywhere) whenever each expression is defined:*

(i) $f *_\sim (g *_\sim h) = (f *_\sim g) *_\sim h$;

(ii) $\widetilde{f *_\sim g} = \widetilde{g} *_\sim \widetilde{f}$;

(iii) $\langle \widetilde{g} *_\sim f,\ h \rangle = \langle f,\ g\ *_\sim h \rangle = \widetilde{f} *_\sim g *_\sim h(\emptyset)$.

Proof. (ii) may be seen by inspection. Repeated application of the Lemma 2.1 yields the common expressions

$$\int \sum_{|\underline{\alpha}|=\sigma} \chi_{[6]}(\underline{\alpha}, \underline{\omega}) f(\tilde{\omega}_1 \cup \tilde{\omega}_2 \cup \alpha_3) g(\omega_1 \cup \alpha_2 \cup \tilde{\omega}_3) h(\alpha_1 \cup \omega_2 \cup \omega_3) d\underline{\omega}$$

and

$$\int \chi_{[3]}(\underline{\omega}) f(\omega_1 \cup \omega_2) g(\omega_1 \cup \omega_3) h(\omega_2 \cup \omega_3) d\underline{\omega},$$

respectively for (i) at σ and (iii).

4. TWO ARGUMENT INTEGRAL–SUM KERNEL OPERATORS

In this section the class of operators introduced by Maassen will be discussed (Maassen, 1985). Their elementary properties will be given and how they relate to the convolutions of the previous section. In the finite dimensional case this class exhausts the full algebra of operators, and in the non–atomic case it gives a strongly dense family. Some deficiencies of this class will then be mentioned as motivation for the discussion of three argument integral–sum kernel operators in the next section.
For $f : \Gamma_X^2 \to \mathbf{C}$ let $f *_1 k : \Gamma_X \to \mathbf{C}$ be defined almost everywhere by

$$\sigma \mapsto \int_\Gamma \left(\sum_{\alpha \subset \sigma} f(\alpha, \omega) k^{[2]}(\omega, \bar{\alpha}) \right) d\omega$$

for those $k : \Gamma_X \to \mathbf{C}$ for which the bracketed expression is integrable in ω (for a.a. σ). In view of the Fubini theorem and Lemma 2.1, $f *_1 k$ depends on f and k only up to null sets. Since $h^{[2]} *_1 k = h * k$ (trivial twist), when each is defined, we have extended $*$. The lopsidedness of this extension will be rectified soon but the immediate task here is to define *operators*.
Given $f : \Gamma_X^2 \to \mathbf{C}$, $F_2 : k \mapsto f *_1 k$ will denote the operator on $L^2(\Gamma)$ whose domain consists of those k for which $f *_1 k$ is both defined and square integrable. The Lemma 2.1 provides the basic estimate for these operators:

$$\|a^N F_2 k\| \leqslant \|(a\sqrt{2})^{N_1} f\| \left\| \sqrt{1+2a^2}^N k \right\| \quad (a \geqslant 0).$$

In particular, a sufficient condition for F_2 to be densely defined is

$$f \in \mathrm{Dom}(\sqrt{2}^{N_1}) \subset L^2(\Gamma^2).$$

The next proposition demonstrates the injectivity of the map $f \mapsto F_2$.

PROPOSITION 4.1. *Suppose that F_2 vanishes on a dense set $\mathcal{D}$ within its domain. If f has the local square integrability property: $f_{\chi\{\mathrm{supp} g'\} \times \{\mathrm{supp} k\}}$ is square* *integrable for all $k \in \mathcal{D}$ and $g' \in \mathcal{D}'$, where $\mathcal{D}'$ is any dense set in $L^2(\Gamma)$, then $f = 0$ almost everywhere.*

Proof. By the Lemma 2.1

$$\int\int\int g^{[2]}(\alpha,\beta)f(\alpha,\omega)k^{[2]}(\omega,\beta)d\alpha d\omega d\beta \tag{4.1}$$

is identically zero for $g \in L^2(\Gamma)$ and $k \in \mathcal{D}$. Putting $g = \delta_\emptyset$ establishes the following inductive hypothesis for $n = 0$:"f vanishes a.e. on $\Gamma_{\leqslant n} \otimes \Gamma$" . Suppose this holds for $n = j$ then putting $g = \chi\Gamma_{j+1}g'$ in (4.1), where $g' \in \mathcal{D}'$, gives $\int (g'\chi\Gamma_{j+1} \otimes k) f d\mu^2 = 0$ from which the inductive step follows.
In plainer language.

COROLLARY 4.2. *If $f \in \mathrm{Dom}(\sqrt{2}^{N_1})$ and $F_2|_{\mathcal{F}} = 0$, where $\mathcal{F}$ is the collection of finite particle functions, then $f = 0$ a.e.*

Under appropriate conditions on the pair (f, g) (see below), the product F_2G_2 is given by an integral–sum kernel:

$$f *_M g : (\sigma,\tau) \mapsto \sum_{\alpha\subset\sigma,\beta\subset\tau} \int \chi_{[2]}(\bar\beta,\omega)\chi_{[2]}(\omega,\bar\alpha)f(\alpha,\bar\beta\cup\omega)g(\omega\cup\bar\alpha,\beta)d\omega. \tag{4.2}$$

The next proposition lists the basic properties of these operators — each one easily proved using the Lemma 2.1.

PROPOSITION 4.3.
(i) $\left\|a^{N_1}b^{N_2}(f \;*_M g)\right\| \leqslant \left\|(a\sqrt{2})^{N_1}\sqrt{1+2b^2}^{N_2} f\right\|\left\|\sqrt{1+2a^2}^{N_1}(b\sqrt{2})^{N_2}g\right\|$ *for* $a, b \geqslant 0$;

(ii) *If $f \in \mathrm{Dom}(2^{N_1}\sqrt{3}^{N_2}), g \in \mathrm{Dom}(\sqrt{6}^{N_1}\sqrt{2}^{N_2})$ and $k \in \mathrm{Dom}(\sqrt{7}^{N})$ then $F_2G_2k = H_2k$, where $h = f *_M g$;*

(iii) $F_2^* \supset E_2$, *where* $e = f : (\sigma,\tau) \mapsto \overline{f(\tau,\sigma)}$.
Thus, putting $\mathcal{K}^2 = \bigcap_{c>0} \mathrm{Dom}(c^{N_1+N_2}) \subset L^2(\Gamma_X^2), \{F_2 : f \in \mathcal{K}^2\}$ forms a symmetric algebra of unbounded operators on $L^2(\Gamma_X)$ leaving the common domain $\mathcal{K}$ invariant.

Note that (ii) relieves us of the obligation of verifying the associativity of $*_M$ in full. If $g = k^{[2]}$ then $f *_M g(\sigma,\emptyset) = f *_1 k(\sigma)$, and if $f = h^{[2]}$, then $f *_M g = (h * k)^{[2]}$ — the Maassen convolution $*_M$ therefore extends the Wiener convolution $*$ and rectifies the lopsidedness of $*_1$. However, the aim now is to show how the Maassen convolution is a particular twisted convolution. This follows from the simple observation that the map $\Phi : \rho \mapsto (\rho \cap S, \rho \cap S')$ defines a bijection from $\Gamma(S \coprod S')$ to $\Gamma(S) \times \Gamma(S')$ which determines an isomorphism of the measurable space $\Gamma((S,\mathcal{M}) \oplus (S',\mathcal{M}'))$ and $\Gamma(S,\mathcal{M}) \times \Gamma(S',\mathcal{M}')$.

Let $Y = X \oplus X'$, where $X = (S, \mathcal{M}, m)$ and $X' = (S, \mathcal{M}, 0)$ — 0 being the zero measure (thus $\Gamma(0) = \iota_\emptyset$) — and let j be the involutive map on $S \coprod S$ given by interchange of copies. Then, for $f, g \in \mathrm{Dom}(\sqrt{3}^N) \subset L^2(\Gamma_Y)$,

$$f *_\sim g(\rho) = \sum_{\alpha \subset \rho} \int_{\Gamma_Y} \chi_{[2]}(\alpha, \tilde{\omega}) \chi_{[2]}(\omega, \bar{\alpha}) f(\alpha \cup \tilde{\omega}) g(\omega \cup \bar{\alpha}) d\mu(\omega) =$$

$$= \sum_{\alpha_1 \subset \rho_1, \alpha_2 \subset \rho_2} \int_{\Gamma_X} \chi_{[2]}(\alpha_{[2]}, j(\omega_1)) \chi_{[2]}(\omega_1, \bar{\alpha}_1) f(\alpha_1 \cup \alpha_2 \cup j(\omega_1)) g(\omega_1 \cup \bar{\alpha}_1 \cup \bar{\alpha}_2) d\omega_1$$

or, writing h' for $h \circ \Phi^{-1}$,

$$(f *_\sim g)'(\sigma, \tau) = \sum_{\alpha \subset \sigma, \beta \subset \tau} \int_{\Gamma_X} \chi_{[2]}(\beta, \omega) \chi_{[2]}(\omega, \bar{\alpha}) f(\alpha, \beta \cup \omega) g(\omega \cup \bar{\alpha}, \beta) d\omega =$$

$$= f' *_M g'(\sigma, \tau).$$

In other words the Maassen convolution for $X = (S, \mathcal{M}, m)$ is the twisted convolution for $Z = ((S \coprod S, \mathcal{M} \oplus \mathcal{M}, m \oplus 0)$, interchange) under the identification $\Phi : \Gamma_S \times \Gamma_S \to \Gamma_{S \coprod S}$.

If the set S is finite then $L^2(\Gamma)$ may be identified with $\otimes^S \mathbf{C}^2$ (with appropriate inner product) and the operator F_2 with kernel $\delta_\sigma \otimes \delta_\tau$ is then $a_\sigma^* a_\tau$ where $a_\tau = \otimes_{s \in S} x_s$ and $x_s = a$ if $s \in \tau$ and I otherwise. Here

$$a = \begin{pmatrix} 0 & 1 \\ 0 & 0 \end{pmatrix} \text{ and } I = \begin{pmatrix} 1 & 0 \\ 0 & 1 \end{pmatrix}.$$

Since these operators linearly generate $\mathcal{L}(\otimes^S \mathbf{C}^2)$ every operator on $L^2(\Gamma)$ is given by an integral–sum kernel operator. This is what Meyer and J.–L. Journé call *toy Fock space* (Meyer, 1986), thinking of the above as an approximation to the Fock CCR representation.

On the other hand, when X is non–atomic, the collection of bounded kernel operators is strongly dense in the full algebra $\mathcal{L}(L^2(\Gamma_X))$. This is so since, for $\varphi \in L^2(X)$, $[\exp(-\|\varphi\|^2/2)\pi_\varphi \otimes \pi_{-\varphi}]$ is the kernel of the Weyl operator in the Fock representation of the canonical commutation relations. Maassen made the important observation that the solutions of linear quantum stochastic differential equations (of which the Weyl operators are a very special case) are also expressible via integral–sum kernels (Maassen, 1985).

We next indicate various operators that fail to be representable via two– argument kernels. First note that the identity operator is always representable: its kernel is $\delta_\emptyset \otimes \delta_\emptyset$.

PROPOSITION 4.4. *Let $f : \Gamma_X^2 \to C$ be locally square integrable in the sense that $f\chi_{\Gamma_i \times \Gamma_j}$ is square integrable for each $i, j \in \mathbb{N}$. Suppose that $F_2 = \varphi(\mathbb{N})$ on $\mathcal{F}$, for some function $\varphi : \mathbb{N} \to \mathbf{C}$, then either φ is constant, so that F_2 is a multiple of the identity, or $L^2(X)$ is finite dimensional.*

Proof. Suppose that $L^2(X)$ is infinite dimensional and let $h \in \mathcal{F}$. Then for almost all $\sigma \in \Gamma$,

$$F_2 h(\sigma) = \sum_{\alpha \subset \sigma} \int f(\alpha, \omega) h^{[2]}(\omega, \bar{\alpha}) d\omega = \varphi(\#\sigma) h(\sigma). \tag{4.3}$$

Putting $h = \delta_\emptyset$ we see that the following inductive hypothesis holds for $n = 0$:

$$'f - f(\emptyset, \emptyset)\delta_\emptyset \otimes \delta_\emptyset \text{ vanishes a.e. on } \Gamma \times \Gamma_{\leqslant n}'.$$

Suppose that it holds for $n = j$, then, for h with support in Γ_{j+1}, (4.3) reads:

$$f(\emptyset, \emptyset) h(\sigma) + \int f(\sigma, \omega) h(\omega) d\omega = \varphi(j+1) h(\sigma)$$

for a.a. $\sigma \in \Gamma_{j+1}$. In other words the Hilbert—Schmidt operator on $L^2(\Gamma_{j+1})$ with integral kernel f (restricted to $\Gamma_{j+1} \times \Gamma_{j+1}$) is a multiple of the identity, and therefore zero since $L^2(X)$, and therefore $L^2(\Gamma_{j+1})$, is infinite dimensional. This clinches the inductive step and (4.3) now reads

$$F_2 h(\sigma) = \varphi(\#\sigma) h(\sigma) = f(\emptyset, \emptyset) h(\sigma),$$

so we are done.

PROPOSITION 4.5. *Let X be non–atomic and let f be locally square integrable in the sense of the previous proposition. Suppose $F_2 = M_g$ on $\mathcal{F}$ where M_g is the operator of multiplication by $g \in L^\infty(\Gamma_X)$. Then F_2 is a multiple of the identity and g is constant.*

Proof. There are no non–zero multiplication operators on $L^2(\Gamma_{j+i})$ which are Hilbert—Schmidt when m (and hence $\mu|_{\Gamma_{j+i}}$) is non–atomic (and non–trivial). The inductive argument of the above proof therefore applies with obvious modifications.

In particular, stochastic integrals with respect to the number process (of sure integrands) (Hudson and Parthasarathy, 1984) are not representable since these have the form

$$\left[\int f d\Lambda k\right](\sigma) = \left(\sum_{s \in \sigma} f(s)\right) k(\sigma). \tag{4.4}$$

If X is non–atomic then no finite rank operators are representable. The next result shows a little more.

PROPOSITION 4.6. *Let X be non–atomic. If f is square integrable and the domain of F_2 contains, say $\Psi(X)$ (the linear span of the (L^2–) product functions), then F_2 cannot be the restriction of a non–zero Hilbert—Schmidt operator.*

Proof. Suppose that F_2 is the restriction of a Hilbert—Schmidt operator with integral kernel H, then for $k \in \Psi$ and $g \in L^2(\Gamma)$

$$\int\int\int g^{[2]}(\alpha,\beta)f(\beta,\omega)k^{[2]}(\omega,\alpha)d\alpha d\beta d\omega = \int\int g(\sigma)H(\sigma,\tau)k(\tau)d\sigma d\tau. \quad (4.5)$$

Choosing g and k with supports in $\Gamma(A)$ and $\Gamma(A^c)$, respectively, (4.5) may be written

$$\int g \otimes k[f - H]d\mu^2 = 0.$$

Since X is non-atomic the collection of such $g \otimes k$, for varying $A \in \mathcal{M}$, is total, and so $f = H\mu^2$-a.e. Now choosing $g = \pi_\varphi$ and $k = \pi_\psi$, (4.5) becomes

$$\left(\exp\left\{\textstyle\int \varphi\psi dm\right\} - 1\right)\int(\pi_\varphi \otimes \pi_\psi)d\mu^2 = 0.$$

But since $\left\{(\varphi,\psi) \in L^2(X) \oplus L^2(X) : \int \varphi\psi dm \neq 0\right\}$ is clearly dense in $L^2(X) \oplus L^2(X)$, function f must vanish almost everywhere by Proposition 2.7.

5. THREE–ARGUMENT INTEGRAL–SUM KERNEL OPERATORS

Motivated by the non–representability of stochastic integrals with respect to the number process by two–argument integral–sum kernel operators, and also the form of the Poisson convolution, Meyer was led to consider three–argument kernels (Meyer, 1986). Let $f : \Gamma_X^3 \to \mathbf{C}$, then an operator F_3 on $L^2(\Gamma_X)$ is defined (almost everywhere) by

$$F_3 k : \sigma \mapsto \int \sum_{|\underline{\alpha}|=\sigma} f(\alpha_1,\alpha_2,\omega)k^{[3]}(\omega,\alpha_2,\alpha_3)d\omega$$

with domain consisting of those $h \in L^2(\Gamma_X)$ for which, firstly, the bracketed expression is integrable in ω, and, secondly, $F_3 h$ is square–integrable. Notice that if f is supported by $\Gamma \times \{\emptyset\} \times \Gamma$ then $F_3 = G_2$, where $g(\sigma,\tau) = f(\sigma,\emptyset,\tau)$, so that this class of operator includes those of the previous section. The basic estimate here is:

PROPOSITION 5.1. *Let $f : \Gamma_X^3 \to \mathbf{C}$ satisfy $|f(\alpha,\beta,\gamma)| \leqslant K^{\#\beta}g(\alpha,\gamma)$ for some $g : \Gamma_X^2 \to \mathbf{C}$ and $K > 0$. Then*

$$\|a^N F_3 k\| \leqslant \left\|(a\sqrt{3})^{N_1} g\right\| \left\|\sqrt{3a^2K^2 + 3a^2 + 1}^{\,N} k\right\|.$$

If $f : \Gamma_X^3 \to \mathbf{C}$ is such that F_3 is densely defined then F_3^* is an extension of the three–argument integral–sum kernel operator $F_3^\dagger$ with kernel $\widetilde{f} : (\alpha,\beta,\gamma) \mapsto \overline{f(\gamma,\beta,\alpha)}$. If $f = h^{[3]}$ for some $h : \Gamma_X \to \mathbf{C}$, then $F_3 k = h *_P k$, where $*_P$

is the Poisson convolution (Meyer, 1986; Lindsay and Parthasarathy, 1990). The question of injectivity for the map $f \mapsto F_3$ will be tackled next.
From now on the transform of $f : \Gamma^3 \mapsto \mathbf{C}$ given by

$$(\alpha, \beta, \gamma) \mapsto \sum_{\sigma \subset \beta} f(\alpha, \sigma, \gamma) \tag{5.1}$$

will be written f'. Note that f is recovered from f' by a Möbius–type inversion:

$$f(\alpha, \beta, \gamma) = \sum_{\sigma \subset \beta} (-1)^{\#\bar{\sigma}} f'(\alpha, \sigma, \gamma), \tag{5.2}$$

and that the operator F_3 has the alternative representation (Meyer, 1986).

$$F_3 k(\sigma) = \int_{\Gamma} \sum_{\alpha \subset \sigma} f'(\alpha, \bar{\alpha}, \omega) k(\omega \cup \bar{\alpha}) d\omega. \tag{5.3}$$

PROPOSITION 5.2. *Let X be a non–atomic, separable measure space. Suppose that $f : \Gamma_X^3 \to \mathbf{C}$ is such that*

(i) *for* a.a. (α, β), $f(\alpha, \beta, \cdot)$ *and* $f(\cdot, \beta, \alpha)$ *are square–integrable on each* Γ_n;

(ii) $\mathrm{Dom}(F_3) \cap \mathrm{Dom}(F_3^\dagger) \supset \mathcal{F}$.

If F_3 annihilates $\mathcal{F}$ then $f = 0$ a.e.

Proof. Suppose that $F_3 k = 0, \forall k \in \mathcal{F}$. Let $\mathcal{D}$ be the complex rational span of the indicator functions of sets from a countable generating ring. By non–atomicity (and σ–finiteness), standard measure theoretic arguments show that there is a null set $\mathcal{N}_{\mathcal{D}}$ in Γ_X such that for $\tau \notin \mathcal{N}_{\mathcal{D}}$, $\mathcal{D}_{(\tau, u)}$ is dense in $L^2(X)$ for each $u \in [\mathbf{Q}+i\mathbf{Q}]^\tau$, where

$$\mathcal{D}_{\tau, u} = \{\varphi \in \mathcal{D} : \varphi(t) = u(t), \quad \forall t \in \tau\}. \tag{5.4}$$

Now for $g \in L^2(\Gamma_X), k \in \mathcal{F}$ we have, from (5.3),

$$0 = \langle \bar{g}, F_3 k \rangle = \int g(\tau) \sum_{\alpha \subset \tau} \int f'(\alpha, \bar{\alpha}, \omega) k^{[2]}(\omega, \bar{\alpha}) d\omega d\tau. \tag{5.5}$$

Consider the induction hypothesis:

$$f'_{j,k,l} = 0 \quad \text{a.e., if} \quad \min\{j + k, k + l\} \leqslant K,$$

where $f'_{j,k,l}$ is the restriction of f' (defined in (5.1)) to $\Gamma_j \times \Gamma_k \times \Gamma_l$. By putting in turn g and k equal to $\delta_\emptyset$ in (5.5), we see that the hypothesis is satisfied for $K = 0$. Suppose then, that it is satisfied for $K = n - 1$. If g is supported by Γ_n and $k = \pi_\varphi \chi_{\Gamma_m}$, where $m \geqslant n$, then (5.5) reads

$$\int_{\Gamma_n} g(\tau) \int \pi_\varphi(\omega) \sum_{\alpha \subset \tau} \sum_{j=0}^{n} f'_{n-j,j,m-j}(\alpha, \bar{\alpha}, \omega) \pi_\varphi(\bar{\alpha}) d\omega d\tau.$$

PROPOSITION 5.2'. *Let X be as above and suppose that the transform f' of $f : \Gamma_X \to \mathbf{C}$ is square integrable. If F_3 vanishes on Ψ then $f = 0$ a.e.*

This is proved by establishing the totality of the set

$$\{\pi_{(\dot{\varphi},\varphi\psi,\psi)} : \varphi, \psi \in (L^2 \cap L^4)(X)\}$$

in $L^2(\Gamma_X^3)$.
Again, under appropriate conditions on $f, g : \Gamma_X^3 \to \mathbf{C}$, F_3G_3 is given by the kernel $f \star g$:

$$(\alpha, \beta, \gamma) \mapsto$$

$$\mapsto \sum \int \chi_E(\underline{\alpha}, \underline{\beta}, \underline{\gamma}, \omega) f(\alpha_1, \beta_1 \cup \beta_2 \cup \alpha_3, \gamma_1 \cup \gamma_2 \cup \omega) g(\omega \cup \alpha_2 \cup \alpha_3, \gamma_1 \cup \beta_2 \cup \beta_3, \gamma_3) d\omega,$$

where the sum is over partitions $\{\alpha_i\}, \{\beta_i\}, \{\gamma_i\}$ of α, β, γ and E serves to separate α_3 from $\beta_1 \cup \beta_2$, ω from $\gamma_1 \cup \gamma_2$ etc.
Note,

(i) if $g = k \otimes \delta_\emptyset \otimes \delta_\emptyset$, then $f \star g(\alpha, \emptyset, \emptyset) = F_3 k$;
(ii) if $f = h^{[3]}, g = k^{[3]}$, then $f \star g = h *_P k$;
(iii) if $|f_i(\alpha, \beta, \gamma)| \leqslant K_i^{\#\beta} g_i(\alpha, \gamma)$ with $g_i : \Gamma_X^2 \to \mathbf{C}$ $(i = 1, 2)$, then

$$|f_1 \star f_2(\alpha, \beta, \gamma)| \leqslant K^{\#\beta} g(\alpha, \gamma),$$

where $K = (1 + K_1)(1 + K_2)$ and g satisfies

$$\|a^{N_1} b^{N_2} g\| \leqslant \|(a\sqrt{3})^{N_1} (bK_2\sqrt{3})^{N_2} g_1\| \|(aK_1\sqrt{3})^{N_1} (b\sqrt{3})^{N_2} g_2\|.$$

This motivates the definition

$$\mathcal{K}_b^3 := \{f : \Gamma_X^3 \to \mathbf{C} : |f(\alpha, \beta, \gamma)| \leqslant K^{\#\beta} g(\alpha, \gamma)\}$$

for some $K \geqslant 0$ and $g \in \mathcal{K}^2$, where

$$\mathcal{K}^2 = \bigcap_{b \geqslant 0} \mathrm{Dom}(b^{N_1 + N_2}).$$

Thus $\mathcal{K}_b^3$ is closed under $\star$ and $\tilde{\ }$, and therefore determines a further symmetric algebra of unbounded operators on $L^2(\Gamma_X)$ which leaves $\mathcal{K}$ invariant.

Example. The creation and annihilation operators of the Fock CAR representation over $L^2(\mathbf{R})$ are given by

$$F_\varphi f : \sigma \mapsto \int_{\mathbf{R}} (-1)^{n(t,\sigma)} \bar{\varphi}(t) f(\sigma \cup t) dt,$$

$$F_\varphi^* f : \sigma \mapsto \sum_{s \in \sigma} (-1)^{n(s,\sigma)} \varphi(s) f(\sigma \backslash s),$$

where $n(t, \sigma) = \#\{(t, s) : s \in \sigma, t > s\}$ (c.f. (Hudson and Parthasarathy, 1986)). These are representable by three-argument kernels: the kernel of F_φ, for example, is

$$(\sigma, \tau, \omega) \mapsto \chi_{\{\emptyset\}}(\sigma) \chi_{\Gamma_1}(\omega) \pi_{\bar{\varphi}}(\omega) \sum_{\alpha \subset \tau} (-1)^{\#\bar{\alpha} + n(\omega, \alpha)}.$$

With the increased freedom afforded by three arguments the negative conclusion of Proposition 4.6 may be rectified. The following is in keeping with the sufficiency of (A^*, Λ, A) for representation of Hilbert—Schmidt operators on $L^2(\Gamma)$ in terms of quantum stochastic integrals (Hudson, Lindsay and Parthsarathy, 1986).

Example. Let

$$f : (\alpha, \beta, \gamma) \mapsto (-1)^{\#\beta} h(\alpha, \gamma),$$

then

$$F_3 k(\sigma) = \int h(\sigma, \omega) k(\omega) d\omega.$$

Thus the linear space of three–argument integral–sum operators includes the Hilbert—Schmidt class. In particular the rank one operator $|g\rangle\langle k|$ has three–argument integral–sum kernel

$$(\alpha, \beta, \gamma) \mapsto g(\alpha)(-1)^{\#\beta} \overline{k(\gamma)}.$$

PROPOSITION 5.3. *Any multiplication operator on $L^2(\Gamma_X)$ is representable by a three–argument integral–sum kernel operator.*

Proof. Given $g : \Gamma_X \to \mathbf{C}$, define $f : \Gamma_X^3 \to \mathbf{C}$ by

$$(\rho, \sigma, \tau) \mapsto \delta_{(\emptyset,\emptyset)}(\sigma, \tau) \times \sum_{\alpha \subset \sigma} (-1)^{\#\bar{\alpha}} g(\alpha),$$

then

$$F_3 k(\sigma) = \sum_{\beta \subset \sigma} f(\emptyset, \beta, \emptyset) k(\sigma) = g(\sigma) k(\sigma)$$

for $k : \Gamma_X \to \mathbf{C}$ such that both k and gk are square integrable. In other words $F_3 = M_g$.

6. POST–SCRIPT

As noted by R. L. Hudson, the introduction of *four–argument* kernel operators would rectify the asymmetry between "time" and chance (with $X = \mathbf{R}$ as our model). To see this, let $f : \Gamma_X^4 \to \mathbf{C}$, and define an operator F_4 on $L^2(\Gamma_X)$ by

$$F_4 k : \sigma \mapsto \sum_{|\underline{\alpha}| = \sigma} \int f(\alpha_1, \alpha_2, \omega_1, \omega_2) k^{[3]}(\omega_1, \alpha_2, \alpha_3) d\underline{\omega}$$

for $\varphi \in \mathcal{D}_{(\sigma,u)}, u \in [\mathbf{Q}+i\mathbf{Q}]^{\sigma}$. By Proposition 2.7, f' vanishes on $\Gamma \times \Gamma_{\geqslant 1} \times \Gamma$, and hence f does also — in other words $F_3 = H_2$. The result therefore follows from Proposition 4.6.
This section ends on a more positive note.

PROPOSITION 5.4. *Any multiplication operator on $L^2(\Gamma_X)$ is representable by a three–argument integral–sum kernel operator.*

Proof. Given $g : \Gamma_X \to \mathbf{C}$, define $f : \Gamma_X^3 \to \mathbf{C}$ by

$$(\rho, \sigma, \tau) \mapsto \delta_{(\emptyset,\emptyset)}(\sigma, \tau) \times \times \sum_{\alpha \subset \sigma} (-1)^{\#\bar{\alpha}} g(\alpha),$$

then

$$F_3 k(\sigma) = \sum_{\beta \subset \sigma} f(\emptyset, \beta, \emptyset) k(\sigma) = g(\sigma) k(\sigma)$$

for $k : \Gamma_X \to \mathbf{C}$ such that both k and gk are square integrable. In other words $F_3 = M_g$.

6. POST–SCRIPT

As noted by R. L. Hudson, the introduction of *four–argument* kernel operators would rectify the asymmetry between "time" and chance (with $X = \mathbf{R}$ as our model). To see this, let $f : \Gamma_X^4 \to \mathbf{C}$, and define an operator F_4 on $L^2(\Gamma_X)$ by

$$F_4 k : \sigma \mapsto \sum_{|\underline{\alpha}| = \sigma} \int f(\alpha_1, \alpha_2, \omega_1, \omega_2) k^{[3]}(\omega_1, \alpha_2, \alpha_3) d\underline{\omega}$$

with the by–now–obvious domain. Such operators combine according to the rule $F_4 G_4 = H_4$, where

$$h = fg : (\alpha, \beta, \gamma, \tau) \mapsto \chi_{[4]}(\alpha, \beta, \gamma, \tau) \times$$

$$\times \sum f(\alpha_1, \beta_1 \cup \beta_2 \cup \alpha_3, \gamma_1 \cup \gamma_3 \cup \tau_1, \tau_2) g(\tau_1 \cup \alpha_2 \cup \alpha_3, \gamma_1 \cup \beta_2 \cup \beta_3, \gamma_2, \tau_3)$$

— the summation being over partitions $\alpha_1 \cup \alpha_2 \cup \alpha_3$ of α, etc. At the cost of a large summation ($3^{\#\alpha + \#\beta + \#\gamma + \#\tau}$ terms) all integration has been removed from the set convolution. Formally, when $X = \mathbf{R}$,

$$F_4 = \int\int\int\int f(\alpha, \beta, \gamma, \tau) dA_\alpha^* d\Lambda_\beta dA_\gamma d\tau,$$

where, for example, $d\Lambda_\beta = d\Lambda_{\beta_1} \ldots \Lambda_{\beta_k} (k = \#\beta)$ and these integrators are the *creation, preservation and annihilation processes* of (Hudson and Parthasarathy, 1984). The obvious problem here is that of making the (highly non–unique) choice of kernel f, to represent a given operator, in some consistent fashion.

For this class of operator all quantum stochastic integration — including ordinary (repeated) integration with respect to time — is realized by a simple operation on the corresponding kernel. Let $\{F_\tau : \tau \in \Gamma\}$ be a (suitably measurable) family of operators with corresponding kernels $\{f(\cdot\,;\tau) : \tau \in \Gamma\}$, then (c.f. (Maassen, 1985; Meyer, 1986))

$$I_k f(\underline{\alpha}) = \sum_{\tau \subset \alpha_k} f(\underline{\alpha} \backslash (\tau)_k; \bar{\tau}),$$

where $(\tau)_k$ is the array with τ in the k–th place and $\emptyset$ elsewhere, gives the kernel of the integrated processes: $A^*(F)$; $\Lambda(F)$; $A(F)$ and $T(F) = \int F_\tau d\tau$, respectively, for $k = 0, 1, 2, 3$ (Lindsay, 1991). It is not possible even to give the full definition of these generalized quantum stochastic integral operations in this space. They generalize both the quantum stochastic integrals of Hudson and Parthasarathy and the classical, but non–adapted, integral of Skorokhod.

Acknowledgement
Financial support from the Science and Engineering Research Council, England is gratefully acknowledged.

REFERENCES

Accardi, L., Journé, J.–L. and Lindsay, J.M. (1989). On multidimensional markovian cocycles. In: *Proceedings of the Rome Symposium on Quantum Probability and Applications*, 1986–87, Springer LNM, **1396**, pp. 59–67.

Arveson, W. (1987). Continuous analogues of Fock space. *Preprint.*

Azema, J. (1985). Sur les fermés aléatoires. *Sém. Prob. XIX, Lect. Notes in Math.* **1123**, 397–495.

Gaveau, B. and Trauber, P. (1982). L'integrale stochastique comme operateur de divergence dans l'éspace fonctionel. *J. Funct. Anal.* **46**, 230–238.

Guichardet, A. (1972). Symmetric Hilbert spaces and related topics. *Lect. Notes in Math.* **261**,

Hudson, R.L. and Lindsay, J.M. (1987). On characterising quantum stochastic evolutions. *Math. Proc. Camb. Phil. Soc.* **102**, 363–369.

Hudson, R.L., Lindsay, J.M. and Parthasarathy, K.R. (1986). Stochastic integral representation of some quantum martingales in Fock space. In: *From local times to global geometry, control and physics. Proceedings of Warwick Symposium,* 1984/85, Pitman RNM, London, pp. 121–131.

Hudson, R.L. and Parthasarathy, K.R. (1984). Quantum Itô's formula and stochastic evolutions. *Commun. Math. Phys.* **93**, 301–323.

Hudson, R.L. and Parthasarathy, K.R. (1986). Unification of fermion and boson stochastic calculus. *Commun. Math. Phys.* **104**, 457–470.

Journé, J.–L. (1987). Structure des cocycles markoviens sur l'éspace de Fock. *Prob. Th. Rel. Fields* **75**, 291–316.

Lindsay, J.M. (1990). Gaussian hypercontractivity revisited. *J. Funct. Anal.* (to appear)

Lindsay, J.M. (1991). On generalised quantum stochastic integrals (in preparation).

Lindsay, J. M. and Maassen H. (1988). The stochastic calculus of Bose noise. *Preprint.*

Lindsay, J.M. and Maassen, H. (1990). Duality transform as ∗–algebraic isomorphism. In: *Quantum Probability and Applications,* **V**, Springer LNM (to appear).,

Lindsay, J.M. and Parthasarathy, K.R. (1989). Cohomology of power sets with applications in quantum probability. *Commun. Math. Phys.* **124**, 337–364.

Lindsay, J.M. and Parthasarathy, K.R. (1990). Rigidity of the Poisson convolution. In: *Quantum Probability and Applications,* **V**, Springer LNM, (to appear).

Maassen, H. (1985). Quantum Markov processes on Fock space described by integral kernels. In: *Quantum Probability and Applications,* **II**, Springer LNM, **1136**, pp. 361–374.

Meyer, P.-A. (1986). Eléments de probabilités quantiques I, II. *Lect. Notes in Math.* **1204**, **1247**, 186– 312, 38–80.

Meyer, P.-A. (1988). New examples of chaotic représentation. *Sém. de Prob.*

Nualart, D. and Zakai, M. (1986). Generalised stochastic integrals and the Malliavin calculus. *Prob. Th. Rel. Fields* **73**, 255–280.

Parthasarathy, K.R. (1988). Remarks of the quantum stochastic differential equation $dX=(c-1)Xd\Lambda+dQ$. *Sém.de Prob.*

Parthasarathy, K.R. and Sinha, K.B. (1986). Stochastic integral representation of bounded quantum martingales in Fock space. *J. Funct. Anal.* **67**, 126–151.

Skorokhod, A.V (1975). On a generalisation of a stochastic integral. *Theory Prob. Appl.* **20**, 219–233.

Prob. Theory and Math. Stat., Vol. 2, pp. 124–133
B.Grigelionis *et al.* (Eds.)
1990 VSP/Mokslas

LARGE DEVIATION FOR MARTINGALES WITH INDEPENDENT AND HOMOGENEOUS INCREMENTS

R.S. LIPTSER and A.N. SHIRYAEV
Institute for Control Sciences, Profsoyuznaya str., 65, 117806 Moscow, USSR
Steklov Mathematical Institute, Vavilov str., 42, 117966 Moscow, USSR

ABSTRACT

The paper contains the upper bounds and the logarithmic asymptotic for the probabilities $\mathbf{P}(t^{-1}M_t \geqslant a)$, $\mathbf{P}(\sup_{s\geqslant t} s^{-1}M_s \geqslant a)$, $\mathbf{P}(\sup_{s\leqslant t} M_s \geqslant ta)$ for martingales with independent and homogeneous increments.

1. INTRODUCTION

Let $M = (M_t)_{t\geqslant 0}$ be a martingale with $M_0 = 0$ defined on some stochastic basis $(\Omega, \mathcal{F}, F, P)$ (without loss of generality it is possible to assume that the filtration $\mathbf{F}$ is generated by M). In the case of independent and homogeneous increments of M we give the upper bounds and the logarithmic asymptotic as $t \to \infty$ for the following probabilities:

$$\mathbf{P}(t^{-1}M_t \geqslant a),\ \mathbf{P}(\sup_{s\geqslant t} s^{-1}M_s \geqslant a),\ \mathbf{P}(\sup_{s\leqslant t} M_s \geqslant ta), \qquad a > 0.$$

In the discrete time case ($t = 0, 1, 2, \ldots$) the independent and homogeneous increments assumption is equivalent to the following representation for M:

$$M_t = \sum_{k=1}^{t} m_k,$$

where $(m_k)_{k\geqslant 1}$ is a sequence of i.i.d. random variables with $\mathbf{E}m_1 = 0$.
In addition, under the Cramér condition

$$\mathbf{E}e^{zm_1} < \infty, \quad z \in \mathbf{R}, \tag{1.1}$$

for the indicated above probabilities we have the following upper bounds (see for example (Shiryaev, 1989)):

$$\left.\begin{array}{c} \mathbf{P}(t^{-1}M_t \geqslant a) \\ \mathbf{P}(\sup\limits_{s\geqslant t} s^{-1}M_s \geqslant a) \\ \mathbf{P}(\sup\limits_{s\leqslant t} M_s \geqslant a) \end{array}\right\} \leqslant e^{-tL(a)}, \quad t \geqslant 0, \tag{1.2}$$

and it is well-known (see Bahadur and Ranga Rao, 1960; Chernoff, 1952) that

$$\lim_{t\to\infty} t^{-1} \ln \mathbf{P}(t^{-1}M_t \geqslant a) = -L(a), \tag{1.3}$$

where $L = L(y)$ is the Cramér—Legendre—Fenchel transform of the convex function $H(z) = \ln \mathbf{E}e^{zm_1}$, defined by

$$L(y) = \sup_z (zy - H(z)).$$

In the continuous time case we establish the upper bounds of the type (1.2) and the logarithmic asymptotic which is the same for all three probabilities.

2. THE MAIN RESULT

1. Now we describe some properties of M which are needed for the formulation of the main result.
Since increments of M are independent and homogeneous the triplet $(B,\ C,\ \nu)$ of the predictable characteristics of M corresponding to the canonical semimartingale representation is deterministic (Theorem 4.4.1 and Corollary to it in (Liptser and Shiryaev, 1989)) and (see (Jacod and Shiryaev, 1987, Theorem II.4.19))

$$\nu(dt,\ dx) = dtK(dx),$$

$$B_t = -t \int\limits_{|x|>1} xK(dx), \tag{2.1}$$

$$C_t = tc, \tag{2.1}$$

where $c \geqslant 0$ and $K(dx)$ is a measure on $(\mathbf{R},\ \mathcal{B}(\mathbf{R}))$ such that $K(\{0\}) = 0$ and

$$\int\limits_{\mathbf{R}} (|x| \wedge x^2)K(dx) < \infty$$

(see Problems 4.1.6 and 4.1.10 in (Liptser and Shiryaev, 1989)).
The following condition we shall call the Cramér condition for the measure $K(dx)$ (cf. (1.1)):

$$\int\limits_{\mathbf{R}} \left(e^{zx} - 1 - zx\right)K(dx) < \infty, \quad z \in \mathbf{R}. \tag{2.2}$$

Under the condition (2.2) define the cumulant

$$H(z) = \frac{1}{2}z^2c + \int\limits_{\mathbf{R}} \left(e^{zx} - 1 - zx\right)K(dx). \tag{2.3}$$

The function $H = H(z)$ is continuous by differentiable and its derivative $H'(z)$ may be written as

$$H'(z) = \int\limits_{\mathbf{R}} x\left(e^{zx} - 1\right)K(dx). \tag{2.4}$$

The functions $H(z)$ and $H'(z)$ have the properties $H(0) = 0$ and $\operatorname{sign} H'(z) = = \operatorname{sign} z$. That means $H = H(z)$ is a non-negative strictly convex function.
Let L be the Cramér—Legendre—Fenchel transform of the cumulant H:

$$L(y) = \sup_z \left(zy - H(z)\right).$$

2. Now we are ready to state the main result.

THEOREM. *Let $M = (M_t)_{t\geqslant 0}$ be a martingale with $M_0 = 0$ and with independent and homogeneous increments. Under the Cramér condition (2.2), if $\lim\limits_{z\to\infty} H(z) = \infty$ then we have, for $a > 0$,*

$$1)\ \mathbf{P}(t^{-1}M_t \geqslant a) \leqslant \left\{ \begin{matrix} \mathbf{P}(\sup\limits_{s\geqslant t} s^{-1}M_s \geqslant a) \\ \mathbf{P}(\sup\limits_{s\leqslant t} M_s \geqslant ta) \end{matrix} \right\} \leqslant e^{-tL(a)}, \quad t > 0,$$

$$2)\ \lim_{t\to\infty} t^{-1}\ln \mathbf{P}(t^{-1}M_t \geqslant a) = \lim_{t\to\infty} t^{-1}\ln \mathbf{P}(\sup_{s\geqslant t} s^{-1}M_s \geqslant a) =$$

$$= \lim_{t\to\infty} t^{-1}\ln \mathbf{P}(\sup_{s\leqslant t} M_s \geqslant ta) = -L(a).$$

3. THE AUXILIARY RESULTS

1. The process $Z(M,z) = \left(Z_t(M,z)\right)_{t\geqslant 0}$ with

$$Z_t(M,z) = e^{zM_t - tH(z)} \tag{3.1}$$

($H(z)$ has been defined in (2.3)) plays an important role in the proof of the Theorem.

LEMMA 1. *Under the Cramér condition (2.2), for each $z \in \mathbf{R}$ the process $Z(M,z)$ is a square integrable martingale on the stochastic basis $(\Omega,\ \mathcal{F},\ \mathbf{F},\ P)$.*

Proof. By the Ito formula (see, e.g., Chapters 2, 3 in (Liptser and Shiryaev, 1989)) we have

$$Z_t(M,z) = 1 + z\int_0^t Z_{s-}(M,z)dM_s + \Big(\frac{1}{2}z^2c - H(z)\Big)\int_0^t Z_{s-}(M,z)ds +$$

$$+\sum_{s\leqslant t} Z_{s-}(M,z)\Big(e^{z\Delta M_s} - 1 - z\Delta M_s\Big) = 1 + z\int_0^t Z_{s-}(M,z)dM_s +$$

$$+\Big[\sum_{s\leqslant t} Z_{s-}(M,z)\big(e^{z\Delta M_s} - 1 - z\Delta M_s\big) -$$

$$-\int_0^t\int_{\mathbf{R}} Z_{s-}(M,z)\big(e^{zx} - 1 - zx\big)K(dx)ds\Big]. \tag{3.2}$$

The process in square brackets in the right side of (3.2) is a local martingale. Therefore $Z(M,z)$ is also a local martingale as well, with

$$\mathbf{E}Z_t(M,z) \leqslant 1, \quad t > 0. \tag{3.3}$$

Let us show that $\mathbf{E}Z_t^2(M,z) < \infty$, $t \geqslant 0$, i.e., $Z(M,z)$ is a square integrable martingale. By (3.3) we have

$$\mathbf{E}Z_t^2(M,z) = \mathbf{E}Z_t(M,2z)e^{t(H(2z)-2H(z))} \leqslant e^{t(H(2z)-2H(z))}.$$

2. Denote Q the distribution of M. The Q is a probability measure on the measurable space $(D,\ \mathcal{D})$ $(D = D[0,\infty)$ is the Skorokhod space of right continuous and with finite left limits functions $X = (X_t)_{t\geqslant 0}$, $\mathcal{D}$ is the Borel σ–algebra of D). The process X is a semimartingale defined on the stochastic basis $(D,\ \mathcal{D}^Q,\ \mathbf{P}_+^Q,\ Q)$, where D^Q is the completion of $\mathcal{D}$ with respect to the measure Q,

$$\mathbf{D}_+^Q = (\mathcal{D}_{t+}^Q)_{t>0}, \quad \mathcal{D}_{t+}^Q = \bigcap_{\varepsilon>0}\sigma(X_s,\ st+\varepsilon)\vee N,$$

N is the union of Q–zero sets from $\mathcal{D}^Q$. The semimartingale X has the triplet $(B,\ C,\ \nu)$ defined in (2.1). Define on the measurable space $(D,\ \mathcal{D})$ the probability measure Q_1^z by

$$dQ_1^z = Z_1(X,z)dQ,$$

where $Z_1(X,z) = e^{zX_1 - H(z)}$.

LEMMA 2. *Under the Cramer condition (2.2) the process X defined on the stochastic basis $(D,\ \mathcal{D}^Q,\ \mathbf{P}_+^Q,\ Q_1^z)$ is a semimartingale with the following triplet of predictable characteristics:*

$$B_t^{z,1} = B_{t\wedge 1} + (t\wedge 1)H'(z),$$

$$C_t^{z,1} = C_{t\wedge 1},$$

$$\nu^{z,1}(dt,dx) = I(t \leqslant 1)e^{zx}\nu(dt,dx).$$

The proof follows from Theorem 4.5.4 in (Liptser and Shiryaev, 1989) and from the definition of $H'(z)$ (see (2.4)).

3. From (2.1) and Lemma 2 it follows that the process $X^1 = (X_t)_{0\leqslant t\leqslant 1}$ is a semimartingale with deterministic and homogeneous triplet of predictable characteristics. Thus, by Theorem II.4.19 in (Jacod and Shiryaev, 1987) X^1 is a process with independent and homogeneous increments. In just the same way it is possible to define the probability measure Q_T^z for any $T > 1$. For this measure Q_T^z the triplet $(B^{z,T},\ C^{z,T},\ \nu^{z,T})$ is defined by the following formulas:

$$B_t^{z,T} = B_{t\wedge T} + (t\wedge T)H'(z),$$

$$C_t^{z,T} = C_{t\wedge T},$$

$$\nu^{z,T}(dt,dx) = I(t \leqslant T)e^{zx}\nu(dt,dx).$$

This triplet corresponds to the semimartingale $X^T = (X_t)_{0\leqslant t\leqslant T}$ which is a process with independent and homogeneous increments.
It is clear that the family (Q_T^z), $T > 0$, can be extended to a probability measure Q^z which is the distribution of the semimartingale $X = (X_t)_{t\geqslant 0}$ with the following triplet of predictable characteristics:

$$B_t^z = B_t + tH'(z),$$

$$C_t^z = C_t,$$

$$\nu^z(dt,dx) = e^{zx}\nu(dt,dx).$$

It should be noted that the measure Q^z is locally absolutely continuous with respect to Q ($Q^z \ll Q$) and the process $Z(X,z) = (Z_t(X,z))_{t\geqslant 0}$, where

$$Z_t(X,z) = e^{zX_t - tH(z)}$$

is the density process of Q^z with respect to Q. From the definition of $Z(X,z)$ it follows that $Z_t(X,z) > 0$ Q–a.s. for any $t > 0$. Therefore $Q \ll Q^z$ and so the measures Q and Q^z are locally equivalent ($Q \sim Q^z$).

LEMMA 3. *Under the Cramér condition* (2.2)

$$\lim_{t\to\infty} t^{-1}X_t = H'(z) \quad Q^z\text{–a.s.}, \quad z \in \mathbf{R}.$$

Proof. By Lemma 1, $Z(M,z)$ is a square integrable martingale. Thus $Z(X,z)$ is a square integrable martingale with respect to $(\mathbf{D}_+^Q,\ Q)$, and from the representation (3.2) it follows that

$$Z_t(X,z) = 1 + z\int_0^t Z_{s-}(X,z)dX_s +$$

$$+\Big[\sum_{s\leqslant t} Z_{s-}(X,z)(e^{z\Delta X_s} - 1 - z\Delta X_s) -$$

$$-\int_0^t\int_{\mathbf{R}} Z_{s-}(X,z)(e^{zx} - 1 - zx)K(dx)ds\Big].$$

(Note that X is a martingale with respect to $(\mathbf{D}_+^Q, Q)$.)
This representation implies that the mutual quadratic variation $[X, Z(X,z)]$ has the form

$$[X, Z(X,z)]_t = z\int_0^t Z_{s-}(X,z)d[X,X]_s +$$

$$+\sum_{s\leqslant t} Z_{s-}(X,z)(e^{z\Delta X_s} - 1 - z\Delta X_s)\Delta X_s.$$

The compensator of $[X, Z(X,z)]$ is the mutual quadratic characteristic

$$\langle X, Z(X,z)\rangle_t = z\int_0^t Z_{s-}(X,z)d\langle X^c\rangle_s +$$

$$+\int_0^t\int_{\mathbf{R}} Z_{s-}(X,z)x(e^{zx} - 1)\nu(ds,dx). \tag{3.4}$$

Since the quadratic characteristic $\langle X^c\rangle_t = C_t = tc$ and $\nu(ds,\ dx) = dsK(dx)$, it follows from (2.1) and (2.4) that

$$\langle X, Z(X,z)\rangle_t = H'(z)\int_0^t Z_{s-}(X,z)ds. \tag{3.5}$$

Since $Q^z \ll Q$, the process $X^z = (X^z)_{t\geqslant 0}$, where

$$X_t^z = X_t - \int_0^t Z_{s-}^{-1}(X,z)d\langle X, Z(X,z)\rangle_s$$

is a local martingale by Theorem 4.5.2 in (Liptser and Shiryaev, 1989). From this and (3.5) we have $X_t^z = X_t - tH'(z)$ and so

$$t^{-1}X_t = H'(z) + t^{-1}X_t^z.$$

The required statement follows now if

$$\lim_{t\to\infty} t^{-1}X_t^z = 0 \quad Q^{-z}\text{– a.s.} \tag{3.6}$$

To established (3.6) we observe that X^z has the triplet of predictable characteristics $(B^z,\ C^z,\ \nu^z)$ defined in (3.4).
The Cramér condition implies that

$$\int_{\mathbf{R}} x^2 e^{zx} K(dx) < \infty.$$

This means that

$$\int_0^t \int_{\mathbf{R}} x^2 \nu^z(ds, dx) = t \int_{\mathbf{R}} x^2 e^{zx} K(dx)$$

and hence X^z is a square integrable martingale with the quadratic characteristic

$$\langle X^z\rangle_t = t(c + \int_{\mathbf{R}} x^2 e^{zx} K(dx)) = b(z)t.$$

By Theorem 2.6.10 in (Liptser and Shiryaev, 1989) (3.6) is true if

$$\int_0^\infty (1+s)^{-1} d\langle X^z\rangle_s < \infty \qquad Q^z\text{– a.s.}$$

In our case

$$\int_0^\infty (1+s)^{-1} d\langle X^z\rangle_s = b(z).$$

4. PROOF OF THE THEOREM

1. The first part of statement 1) is evident. Let us show that

$$\mathbf{P}(\sup_{s\geqslant t} s^{-1}M_s \geqslant a) \leqslant e^{-tL(a)},\ t>0,\ a>0. \tag{4.1}$$

Define the Markov time

$$\tau = \inf(s \geqslant t\colon s^{-1}M_s \geqslant a)$$

with $\tau = \infty$ on $(s^{-1}M_s < a$ for all $s \geqslant t)$. By Lemma 1, $Z(M,z)$ is a martingale. Moreover $Z(M,z)$ is a non–negative process. Thus $Z(M,z)$ is a non–negative supermartingale and therefore there exists

$$Z_\infty(M,z) = \lim_{t\to\infty} Z_t(M,z) \qquad P\text{– a.s.}$$

and

$$\mathbf{E}Z_\tau(M,z) \leqslant \mathbf{E}Z_0(M,z) = 1, \tag{4.2}$$

where τ is the defined above Markov time and $Z_\tau(M,z) = Z_\infty(M,z)$ on the set $(\tau = \infty)$.
Define z^a by the equation $H'(z^a) = a$. It is evident that

$$(z^a a - H(z^a)) = L(a) > 0. \tag{4.3}$$

Observe that $\tau \geqslant t$ and

$$\mathbf{P}\Big(\sup_{s\geqslant t} s^{-1}M_s \geqslant a\Big) = \mathbf{P}\left(\tau^{-1}M_\tau \geqslant a\right) I \bigcap (\tau < \infty).$$

Using this equality, (4.2), and (4.3) we have

$$\begin{aligned}1 &\geqslant \mathbf{E}Z_\tau(M,z^a) \geqslant \mathbf{E}Z_\tau(M,z^a)I\left(\tau^{-1}M_\tau \geqslant a\right) I(\tau < \infty) \geqslant \\ &\geqslant \mathbf{E}\exp\left\{\tau(z^a a - H(z^a)\right\}I(\tau^{-1}M_\tau \geqslant a)I(\tau < \infty) = \\ &= \mathbf{E}\exp\left\{\tau L(a)\right\}I(\tau^{-1}M_\tau \geqslant a)I(\tau < \infty) \geqslant \\ &\geqslant \exp\left\{tL(a)\right\}\mathbf{P}(\sup_{s\geqslant t} s^{-1}M_s \geqslant a),\end{aligned}$$

i.e., (4.1) is true.
The inequality

$$\mathbf{P}(\sup_{s\leqslant t} M_s \geqslant ta) \leqslant e^{-tL(a)} \tag{4.4}$$

is proved in a similar way. For this purpose we define the stopping time $\sigma = \inf(s \leqslant t \colon M_s \geqslant ta) \wedge t$ and use the equality $\mathbf{P}(\sup_{s\leqslant t} M_s \geqslant ta) = \mathbf{P}(M_\sigma \geqslant ta)$. Evidently $\mathbf{E}Z_\sigma(M,z^a) \leqslant 1$, and therefore

$$\begin{aligned}1 &\geqslant \mathbf{E}Z_\sigma(M,z^a)I(M_\sigma \geqslant ta) \geqslant \mathbf{E}e^{t\left(z^a a - H(z^a)\right)}I(M_\sigma \geqslant ta) = \\ &= e^{tL(a)}\mathbf{P}(\sup_{s\leqslant t} M_s \geqslant ta),\end{aligned}$$

i.e., (4.4) holds true.

2. By virtue of the first statement of the Theorem we have

$$\left.\begin{array}{c}\overline{\lim\limits_{t\to\infty}}\, t^{-1}\ln \mathbf{P}(t^{-1}M_t \geqslant a) \\ \overline{\lim\limits_{t\to\infty}}\, t^{-1}\ln \mathbf{P}(\sup\limits_{s\geqslant t} s^{-1}M_s \geqslant a) \\ \overline{\lim\limits_{t\to\infty}}\, t^{-1}\ln \mathbf{P}(\sup\limits_{s\leqslant t} M_s \geqslant ta)\end{array}\right\} \leqslant -L(a).$$

Therefore it is sufficient to show that

$$\left.\begin{array}{c}\varliminf_{t\to\infty} t^{-1}\ln \mathbf{P}(t^{-1}M_t \geqslant a)\\ \varliminf_{t\to\infty} t^{-1}\ln \mathbf{P}(\sup_{s\geqslant t} s^{-1}M_s \geqslant a)\\ \varliminf_{t\to\infty} t^{-1}\ln \mathbf{P}(\sup_{s\leqslant t} M_s \geqslant ta)\end{array}\right\} \geqslant -L(a). \tag{4.5}$$

Since

$$\mathbf{P}(t^{-1}M_t \geqslant a) \leqslant \begin{cases}\mathbf{P}(\sup_{s\geqslant t} s^{-1}M_s \geqslant a)\\ \mathbf{P}(\sup_{s\leqslant t} M_s \geqslant ta)\end{cases}$$

the inequality (4.5) holds if

$$\varliminf_{t\to\infty} t^{-1}\ln \mathbf{P}(t^{-1}M_t \geqslant a) \geqslant -L(a). \tag{4.6}$$

Let us prove (4.6). Choose z^r such that $H'(z^r) = a + r/2$, $r > 0$. Let Q be the distribution of M and Q^{z^r} be the measure defined in Lemma 2 with $z = z^r$. By Lemma 2, $Q_t^{z^r} \sim Q_t$ for each $t \geqslant 0$, where $Q_t^{z^r}$ and Q_t are the restrictions of Q^{z^r} and Q on the σ–algebra $\mathcal{D}_{t+}^{Q}$. Recall that

$$\frac{dQ_t}{dQ_t^{z^r}} = Z_t^{-1}(X, z^r).$$

We have

$$\begin{aligned}\mathbf{P}(t^{-1}M_t \geqslant a) &= Q\big(t^{-1}X_t \geqslant a\big) = Q_t\big(t^{-1}X_t \geqslant a\big) =\\ &= \int_D I\big(t^{-1}X_t \geqslant a\big)Z_t^{-1}(X, z^r)dQ^{z^r} \geqslant\\ &\geqslant \int_D I\big(a + r \geqslant t^{-1}X_t \geqslant a\big)Z_t^{-1}(X, z^r)dQ^{z^r} =\\ &= \int_D I\big(a + r \geqslant t^{-1}X_t \geqslant a\big)\exp\Big\{-z^r X_t + tH\big(z^r\big)\Big\}dQ^{z^r} \geqslant\\ &\geqslant \int_D I\big(a + r \geqslant t^{-1}X_t \geqslant a\big)dQ^{z^r}\exp\Big\{t\Big(-z^r(a + r) + H\big(z^r\big)\Big)\Big\}.\end{aligned}$$

From this and from $z^r(a + r) - H\big(z^r\big) = L(a + r)$ it follows that

$$\mathbf{P}(t^{-1}M_t \geqslant a) \geqslant e^{-tL(a+r)}\int_D I\big(a + r \geqslant t^{-1}X_t \geqslant a\big)dQ^{z^r}. \tag{4.7}$$

By Lemma 3

$$\lim_{t\to\infty} t^{-1}X_t = H'(z^r) = a + r/2 \qquad Q^{z^r} - \text{a.s.},$$

and therefore

$$\lim_{t\to\infty} \int_D I(a+r \geqslant t^{-1}X_t \geqslant a)\,dQ^{z^r} = 1.$$

Therefore from (4.7) it follows that

$$\varliminf_{t\to\infty} \mathbf{P}(t^{-1}X_t \geqslant a) \geqslant -L(a+r/2).$$

Now (4.6) follows since $r > 0$ is arbitrary.

REFERENCES

Bahadur, R. and Ranga Rao, R. (1960). On deviation of the sample mean. *Ann. Math. Statist.* **31**, 1015—1027.

Chernoff, H. (1952). A measure of asymptotic efficiency for test a hypothesis based on the summ of observation. *Ann. Math. Statist.* **23**, 493—507.

Jacod, J. and Shiryaev, A.N. (1987). *Limit Theorems for Stochastic Processes.* Springer, Berlin—Heidelberg etc

Liptser, R.S. and Shiryaev, A.N. (1989). *Theory of Martingales.* Kluwer, Dordrecht.

Shiryaev, A.N. (1989). *Probability.* Nauka, Moscow.

Prob. Theory and Math. Stat., Vol. 2, pp. 134–138
B.Grigelionis *et al.* (Eds.)
1990 VSP/Mokslas

LAWS OF THE ITERATED LOGARITHM FOR I.I.D. CASE

A.I. MARTIKAINEN
Leningrad State University, Leningrad, USSR

ABSTRACT

If $\{X_n\}$ is a sequence of i.i.d. random variables, $S_n = \sum_{j=1}^{n} X_j$, when do there exist scaling constants b_n such that $\limsup(S_n - \text{med}(S_n))/b_n = 1$ almost certainly and how to find this scaling sequence if it exists? The answer on the first question is well-known but the second is open. There are many results in this field which permit to find scaling constants under some additional moment restrictions. In this paper we will give a criterion which can examine a wide class of scaling sequences. We will state the result without any additional restrictions on $\{X_n\}$. As a partial case we will discuss one-sided strong law of large numbers with $b_n = n^{1/d}$, $1 \leqslant d < 2$. Also, we will touch "the universal law of the iterated logarithm".

TOWARDS A UNIVERSAL LAW OF THE ITERATED LOGARITHM

Let $X, X_1, X_2, \ldots$ be a sequence of independent, identically distributed nondegenerate random variables, $S_n = X_1 + \ldots + X_n$, $n \geqslant 1$. The first object of this lecture is to study conditions for

$$0 < \limsup(S_n - \text{med}(S_n))/b_n < \infty \quad \text{a.s.} \tag{1}$$

with various scaling sequences b_n. The choice of medians of S_n as a centering constants is natural but not sole. For example, truncated means are good and simple centering constants. We will see that in various situations $\text{med}(S_n)$ can be changed in (1) by a sum of truncated means without altering of $\limsup$. Therefore we will use most universal centering — by medians.

There are some results in which (1) is established under various conditions on distribution function $F(x) = \mathbf{P}(X < x)$. A classical theorem due to Hartman and Wintner establishes (1) with $b_n = (n \log\log\, n)^{1/2}$ under condition $\mathbf{E}X^2 < \infty$. For these b_n in (Rosalsky, 1980) and independently in (Pruitt, 1981; Martikainen, 1980) it was proved that (1) implies finitness of $\mathbf{E}X^2$ (as for centering by zeros see (Martikainen, 1983), as for centering by medians (Martikainen, 1980)). In (Klass, 1976; 1977) scaling constants with (1) for mean zero distributions was constructed. M. Klass and H. Teicher studied the case when $\mathbf{E}|X| = \infty$ (but under some additional smoothness restrictions). D. Mason investigated X bounded above.

The purpose of "universal law of the iterated logarithm" (see (Klass, 1976)) is to establish (1) without any preliminary conditions on F.
Now we state necessary and sufficient conditions for (1) for a wide class of scaling sequences. Such conditions were found by author for $b_n = n^{1/d}$, $d < 2$ (see (Martikainen, 1983) for $d < 1$ and (Martikainen, 1988) for $d \geqslant 1$), i.e., for one-sided strong law of large numbers.
Suppose for a moment that $\mathbf{E}|X| < \infty$ and denote for $x > 0$

$$D(x) = \frac{1}{x}\int_{-x}^{0} y^2 dF(y) - \int_{-\infty}^{-x} y dF(y).$$

It is easy to see that $D(x)$ is continuous and non-increasing function ($D(x)$ decreases for sufficiently large x), $D(\infty) = 0$, $xD(x)$ nondecreases. So, the equation

$$D(L) = x^{-(d-1)/d}, \quad d > 1,$$

has the unique solution $L = L(x)$ for sufficiently large x.
It is not difficult to prove that $L(x)$ is a converse function for

$$G(x) = \left(\frac{1}{x}\int_{0}^{x} \mathbf{E}(|X| I\{-X > u\}) du\right)^{-d/(d-1)}$$

($G(x)$ is non-decreasing absolutely continuous function for $x > 0$).

THEOREM 1. *Let* $1 < d < 2$, $c > 1$. *Then* $\limsup(S_n - \operatorname{med}(S_n))n^{-1/d} < \infty$ *a.s. if and only if*

$$\mathbf{E}X^d I\{X > 0\} < \infty, \tag{2}$$

$$\limsup n^d \mathbf{P}(X < -n) < \infty, \tag{3}$$

$$\sum_{n=1}^{\infty} \exp\left(-\frac{ac^{n/d}}{L(c^n)}\right) < \infty \quad \textit{for some} \quad a > 0. \tag{4}$$

This result is a slight reformulation of the main result in (Martikainen, 1988):

THEOREM 1′. *Let* $1 < d < 2$, $c > 1$. *Then*

$$\limsup(S_n - \operatorname{med}(S_n))n^{-1/d} = 0 \quad \text{a.s.} \tag{5}$$

if and only if (2) *holds, a series in* (4) *converges for any* $a > 0$ *and*

$$\lim n^d \mathbf{P}(X < -n) = 0.$$

Note that lim sup in (5) is non-negative necessarily. Conditions (2) and (3) imply $\mathbf{E}X < \infty$, so $L(c^n)$ is well-defined in Theorems 1 and 1′ for large n.

Criterion for (1) is an immediate corollary of these theorems if $b_n = n^{1/d}$ and $1 < d < 2$. Cases $d < 1$ and $d = 1$ differ properly from above one. For example, if $d < 1$ then (5) holds if and only if (2) holds and $\lim n^d \mathbf{P}(X < -n) = 0$ (Martikainen, 1983). But a gap between one-sided strong law of large numbers ($b_n = n^{1/d}$) and the law of the iterated logarithm ($b_n = (n \log\log n)^{1/2}$) is still open. It would be very interesting to remove this gap, especially because qualitative peculiarities of distribution which lead to one-sided l.i.l. considerably differ from peculiarities which lead to one-sided s.l.l.n.

Let us now formulate the main result. We introduce a basic function defined for $u > 0, \quad 0 < x < u$

$$D_u(x) = \frac{1}{x} \int_{-x}^{x} y^2 dF(y) + \int_{x<|y|<u} |y| dF(y).$$

It is clear that D_u is in fact defined above D but for truncated random variable $X' = -|X| I\{|X| < u\}$. So, the properties of D pointed out above remain true for D_u. Note that $D_u(x)$ is finite for any distribution function F.

For $a > 0$ and $b > 0$ we define L_n as a least solution of

$$D_{ab_n}(L_n) = b b_n / n \tag{6}$$

if such solution exists in $(0, ab_n]$. Really, the solution of (6) is unique for large n if it exists in $(0, ab_n]$. For others n put $L_n = 0$, so that $C/L_n = +\infty$ if $C > 0$.

THEOREM 2. *Suppose that*

$$b_n / n^d \uparrow \quad \textit{for some} \quad d > 1/3, \tag{7}$$

$$\sup_{n \geqslant 1} b_{2n} / b_n < \infty. \tag{8}$$

Then

$$\limsup (S_n - \operatorname{med}(S_n)) / b_n < \infty \quad \text{a.s.} \tag{9}$$

if and only if

$$\sum_{n=1}^{\infty} \mathbf{P}(X > b_n) < \infty, \tag{10}$$

$$\left\{ (S_n - \operatorname{med}(S_n)) / b_n \right\} \quad \textit{is stochastically bounded,} \tag{11}$$

$$\sum_{n=1}^{\infty} \frac{1}{n} \exp\left(-\frac{c b_n}{L_n} \right) < \infty \quad \textit{for some} \quad a > 0,\ b > 0,\ c > 0. \tag{12}$$

Simple criterion for (11) is well-known.

Slight reformulation of the theorem yields necessary and sufficient conditions for

$$\limsup (S_n - \operatorname{med}(S_n)) / b_n = 0 \quad \text{a.s.} \tag{13}$$

THEOREM 2$'$. *Let b_n satisfy the conditions* (7) *and* (8), $a > 0$, $b > 0$. *Then* (13) *holds if and only if* (10) *holds, a series in* (11) *converges for all* $c > 0$ *and* $(S_n - \text{med}(S_n))/b_n \to 0$ *in probability.*

Combination of Theorems 2 and 2$'$ gives necessary and sufficient conditions for (1) under the assumptions (7) and (8). If (8) fails (as in a case of $b_n = 2^n$) then Theorems 2 and 2$'$ become true after some complication of condition (12) and its analogue in Theorem 2$'$.
Condition (12) has a complicated structure. Our next problem is to find sufficient conditions for (9) and (13) of more simple structure. It is easy to see that (12) holds if $L_n = O(b_n/\log\log n)$. A simple calculation leads to the following results.

COROLLARY 3. *Suppose that* (7) *and* (8) *hold. If*

$$\limsup n \log\log n \mathbf{P}(|X| > b_n \log\log n) < \infty, \tag{14}$$

$$\sum_{n=1}^{\infty} \mathbf{P}(X > b_{n/\log\log n} \log\log n) < \infty, \tag{15}$$

$$\int_{-b_n}^{b_n} y^2 dF(y) + b_n \int_{b_n \leqslant |y| < b_n \log\log n} |y| dF(y) = O(b_n^2/n) \tag{16}$$

then

$$\limsup \left(S_n - \text{med}(S_n)\right)/(b_{n/\log\log n} \log\log n) < \infty \quad a.s.$$

and this relation remains true if med(S_n) *is changed by*

$$n\mathbf{E}XI\{|X| \leqslant b_{n/\log\log n} \log\log n\}.$$

Here b_x denote $b_{[x]}$, where $[\cdot]$ is a largest integer function.

COROLLARY 3$'$. *Suppose that* (7),(8) *and* (15) *holds. If* lim sup *in* (14) *is equal to* 0 *and* (16) *holds with* $o(\cdot)$ *instead of* $O(\cdot)$, *then*

$$\limsup \left(S_n - \text{med}(S_n)\right)/(b_{n/\log\log n} \log\log n) = 0 \quad \text{a.s.}$$

and this relation remains true if med(S_n) *is changed by n truncated means as in Corollary 3.*

We can conclude now that the universal law of the iterated logarithm

$$0 < \limsup \left(S_n - \text{med}(S_n)\right)/(b_{n/\log\log n} \log\log n) < \infty \quad \text{a.s.}$$

holds if conditions (7), (8), (14), (15) and (16) are satisfied but at least one of conditions of Corollary 3$'$ fails. Note that we have two possible infringements of conditions of Corollary 3$'$, so we have two possible reasons for rising of the universal law. The first we can realize for some distributions if define b_n in Corollary 3 by

$$n \log\log n \mathbf{P}(X < -b_n \log\log n) = 1.$$

The second is more complicated. We will discuss these possibilities in one of future papers. The equation

$$\int_{-b_n}^{b_n} y^2 dF(y) + b_n \int_{b_n \leqslant |y| < b_n \log\log n} |y| dF(y) = b_n^2/n \tag{17}$$

has for large n at least one solution b_n. It is satisfied if F is continuous on the tails. Otherwise, b_n can be defined as a largest solution of inequality which forms if symbol = in (17) is changed by $\geqslant$ (this way is available in the first case of the universal law). So, (17) leads to the situation which was investigated by Klass under condition $\mathbf{E}|X| < \infty$. If $\mathbf{E}|X| < \infty$ then

$$\int_{b_n \leqslant |y| < b_n \log\log n} |y| dF(y)$$

can be changed in (17) by $\int_{b_n \leqslant |y|} |y| dF(y)$. After this replacement, b_n has a quite regular behaviour: $b_n/n \searrow 0$, $b_n^2/n \nearrow \mathbf{E}X^2$ (which may be $+\infty$). It simplifies proofs if $\mathbf{E}|X| < \infty$.
We will discuss these possibilities in detail in one of future papers.
At the end we note that the statements of Theorems 2 and 2′ remain true if we suppose that $\mathbf{E}|X| < \infty$ and change D_u by D_∞.

REFERENCES

Klass, M. (1976). Toward a universal law of the iterated logarithm. 1 *Z. Wahrsch. verw. Geb.* **36**, 165–178.

Klass, M. (1977). Toward a universal law of the iterated logarithm. 2. *Z. Wahrsch. verw. Geb.* **39**, 151–165.

Martikainen, A. (1980). A converse to the law of the iterated logarithm for random walk. *Theory Probab. Appl.* **25**, 364–366.

Martikainen, A. (1983). One–sided versions of strong limit theorems. *Theory Probab Appl.* **28**, 48–64.

Martikainen, A. (1988). One–sided strong law of large numbers for random walk: modern state. In: *Proc. Second All–Union Conf. Probab. Methods in Discrete Math.*, Petrozavodsk, pp 59–60.

Pruitt, W. (1981). General one–sided laws of the iterated logarithm. *Ann. Probab.* **9**, 1–48.

Rosalsky, A. (1980) On the converse to the iterated logarithm law *Sankhya.* **A 42**, 103–108.

Prob. Theory and Math. Stat., Vol. 2, pp. 139–149
B.Grigelionis *et al.* (Eds.)
1990 VSP/Mokslas

DISTRIBUTION OF VALUES OF ADDITIVE ARITHMETIC FUNCTIONS

E. MANSTAVIČIUS
Faculty of Mathematics, Vilnius University
Naugarduko str., 24, 232006 Vilnius, Lithuania

ABSTRACT

Necessary and sufficient conditions for the weak convergence of the distribution function

$$\Big(\sum_{m\leqslant x} m^{-1}\Big)^{-1} \sum_{m\leqslant x,\ h(m)<A(x)+vB(x)} m^{-1}$$

($A(x)$ and $B(x)$ are standard normalizing functions) to a limiting distribution with mean zero and variance one are obtained.

1. INTRODUCTION

The following two sequences of probability measures

$$\mathbf{P}_x(A) = [x]^{-1}\#\{m \leqslant x, \quad m \in A\}$$

and

$$\tilde{\mathbf{P}}_x(A) = \Big(\sum_{m\leqslant x} m^{-1}\Big)^{-1} \sum_{m\leqslant x,\ m\in A} m^{-1},$$

where $x \geqslant 1$, $A \subset \mathbf{N}$, are widely used in probabilistic number theory. The weak convergence problem of the corresponding distribution functions

$$F_x(v) = F_x(v, \alpha(x), \beta(x)) = \mathbf{P}_x\Big(h(m) < \alpha(x) + v\beta(x)\Big)$$

and

$$\Phi_x(v) = \Phi_x(v, \alpha(x), \beta(x)) = \tilde{\mathbf{P}}_x\Big(h(m) < \alpha(x) + v\beta(x)\Big)$$

as $x \to \infty$, constitutes one of the main problems in the field. Here $h : \mathbf{N} \to \mathbf{R}$ is an additive function, $\alpha(x)$ and $\beta(x) > 0$ are arbitrary normalizing functions. We prefer to consider x as a continuous parameter but $\alpha(x) = \alpha([x])$, $\beta(x) = \beta([x])$. General results on the convergence of $F_x(v)$ one can find in monographs (Kubilius, 1964; Elliott, 1979, 1980) and in more recent papers (Timofeev, 1982, 1983, 1985a, 1985b). We will deal with the second problem, and we will present now the known results concerning it. The problem of convergence of $\Phi_x(v, \alpha(x), 1)$ was solved in 1979 by V. Keinys (unpublished). His result is contained in Theorem 1 (Timofeev and Tuljaganov, 1982) which completely describes the case when $\beta(x) = O(1)$. The case $\beta(x) \to \infty$ as $x \to \infty$ remains unsolved up to now. Necessary and sufficient conditions for the existence of a proper limit distribution for $\Phi_x(v, \alpha(x), \beta(x))$ are known only under the extra restriction $\beta(x^u)/\beta(x) \to 1$ for each $u \in (0, 1]$ as $x \to \infty$ (see (Timofeev and Tuljaganov, 1982), Theorem 2). Note that all mentioned results including the analogue of the law of large numbers (Levin and Timofeev, 1977, Lemma 1) can be obtained by an elementary method without the use of the Dirichlet generating series and contour integration.

The investigation of the joint convergence of $F_x(v)$ and $\Phi_x(v)$ carried out in (Levin and Timofeev, 1977) is of great importance. Their paper had stimulated our approach.

Define

$$A(x) = \sum_{p \leqslant x} \frac{h(p)}{p}, \quad B^2(x) = \sum_{p \leqslant x} \frac{h^2(p)}{p}.$$

Here and in the following p denotes a prime number. In many cases we use limiting relations with respect to $x \to \infty$ omitting this notation. Now we formulate our results.

THEOREM 1. *In order that the sequence $\Phi_x(v, A(x), B(x))$ with $B(x) \to \infty$ converges to a limiting distribution with mean zero and variance one it is both necessary and sufficient that the following two conditions are satisfied:*

(I) *there exists a distribution function $K(v)$ such that*

$$\frac{1}{B^2(x)} \sum_{p \leqslant x;\ h(p) < vB(x)} \frac{h^2(p)}{p} \to K(v)$$

for each continuity point of $K(v)$,

(II) $\varphi_x(u) = B(x^u)/B(x) \to 1$ *for each $u \in (0, 1]$.*

If these conditions are satisfied, the logarithm of the characteristic function of the limit distribution can be expressed by the Kolmogorov formula

$$\log \tau(t) = \int_{-\infty}^{\infty} \frac{e^{itv} - 1 - itv}{v^2} \, dK(v).$$

Sometimes the condition (I) implies (II). As an example we present the following.

THEOREM 2. *Let $B(x) \to \infty$. The sequence $\Phi_x(v, A(x), B(x))$ converges to the standard normal law if and only if for each $\varepsilon > 0$*

$$\frac{1}{B^2(x)} \sum_{p \leqslant x;\ |h(p)| \geqslant \varepsilon B(x)} \frac{h^2(p)}{p} \to 0.$$

It is well-known that conditions (I) and (II) are only sufficient for the convergence of $F_x(v, A(x), B(x))$ (see (Kubilius, 1964, Theorem 4.1, and Timofeev, 1985)). In view of Theorems 2 (Timofeev and Tuljaganov, 1982) and 4.1 of (Kubilius, 1964) we concentrate our efforts on the necessity part of Theorem 1. Our method is a simple one. At first we note that the convergence of $\Phi_x(v)$ yields the relative compactness of the sequence of distribution functions $F_x(v)$. This enables us to use relations between limiting distributions. After that an investigation of the first two moments gives functional equations for the partial limits of the sequences $\varphi_x(u)$ and $\psi_x(u) = (A(x^u) - A(x))/B(x)$ uniquely determining the limits. Having the asymptotics $\varphi_x(u) = 1 + o(1)$ and $\psi_x(u) = o(1)$ for $u \in [M^{-1}, M]$, where $M > 1$ is arbitrary, we come to the aforementioned theorems.
The relative compactness of the sequences of distribution functions is considered in the second part of the paper. Lemma 1 gives a new elementary approach to the investigation of the normalizing constants but we use it partly.

2. COMPACTNESS OF DISTRIBUTIONS

In the distribution function space $\mathcal{F}$ we use the Lévy metric

$$L(F,G) = \inf \left\{ \varepsilon; F(v-\varepsilon) - \varepsilon \leqslant G(v) \leqslant F(v+\varepsilon) + \varepsilon,\ v \in \mathbf{R} \right\},$$

where $F, G \in \mathcal{F}$.
Let $h_x : \mathbf{N} \to \mathbf{R}$ be a sequence of additive functions and $\alpha_x \in \mathbf{R}$ be an arbitrary centralizing sequence. Denote

$$\nu_y(v) = \nu_{yx}(v) = \mathbf{P}_y(h_x(m) - \alpha_x < v). \tag{1}$$

Put $a \wedge b = \min\{a, b\}$, $a^* = (1 \wedge |a|) \operatorname{sign} a$ and

$$Q(y) = \sum_{p \leqslant y} \frac{h_x^{*2}(p)}{p}.$$

LEMMA 1. *Let* $x' = x^{1+\delta}$, $|\delta| \leqslant 1/5$, $x \geqslant 3$. *Then*

$$L(\nu_x, \nu_{x'}) = B(Q(x)+1)^{1/3}(|\log|\delta||^{-1/3} + (\log\log x)^{-1/3}).$$

(Here and in the following B stands for the quantity bounded by an absolute constant.)

Proof is based on A. Hildebrand's ideas (Hildebrand, 1986). We omit it.

LEMMA 2. *Let* $\nu_y(v) = \nu_{yx}(v)$ *be defined by* (1) *and* $y = y(x) \to \infty$. *The sequence* ν_y *is relatively compact if and only if there exists a bounded function* λ_x *such that*

$$\lim_{U\to\infty} \limsup_{x\to\infty} \frac{1}{U^2} \sum_{p^r \leqslant y} \frac{\hat{h}_x^2(p^2) \wedge U^2}{p^r} = 0 \tag{2}$$

and

$$\alpha_x = \alpha_{xy} = \lambda_x \log y + \sum_{p \leqslant y,\ |h_x(p)|<1} \frac{\hat{h}_x(p)}{p} + O(1), \tag{3}$$

where $\hat{h}_x(m) = h_x(m) - \lambda_x \log m$.

Proof (see (Šiaulys, 1987)). The necessity of (3) follows from (13) in (Šiaulys, 1987). The double indices x and y we used for convenience only.
Note that under the extra condition $h_x(p^r) = o(1)$ for each prime p and $r \geqslant 1$ the parameter λ_x in Lemma 2 has the property $\lambda_x = o(1)$ (Šiaulys,1987). If the sequence ν_x is relatively compact, so is

$$\nu_x(u,v) = \mathbf{P}_{x^u}(h_x(m) - \alpha_{x,x^u} < v)$$

for each fixed u, $0 < u \leqslant 1$, where α_{xy} is defined by (3). The following lemma yields much more.

LEMMA 3. *Suppose the condition $h_x(p^r) \to 0$ is satisfied and ν_x is relatively compact. Then for each sequence $z \to \infty$ there exist a subsequence $y \to \infty$ and a family of distribution functions $F(u,\cdot)$ such that for each $M > 1$ uniformly in $u \in [M^{-1}, M]$*

$$L(\nu_y(u,\cdot),\ \ F(u,\cdot)) \to 0, \quad y \to \infty, \tag{4}$$

and

$$L(F(u+\delta,\cdot),\ F(u,\cdot)) \to 0 \tag{5}$$

as $\delta \to 0$.

Proof. We can consider the distribution function of $\hat{h}_x(m) - \alpha_x(u) - \lambda_x u \log x$. By the routine choosing of subsequences we obtain (4) for a countable and dense subset of points $u \in [M^{-1}, M]$. Finally, Lemma 1 yields (4) for all u in the interval and the continuity in the form (5).
Put $\tilde{\Phi}_x(v) = \tilde{\mathbf{P}}_x(h_x(m) - \alpha_x < v)$.

LEMMA 4. *The sequence $\tilde{\Phi}_x$ is relatively compact if and only if the following conditions are satisfied:*

(I) $\displaystyle \lim_{U\to\infty} \limsup_{x\to\infty} U^{-2} \sum_{p^k \leqslant x} \frac{h_x^2(p^{(k)} \wedge U^2)}{p^k} = 0,$

(II) $\displaystyle \alpha_x = \sum_{p \leqslant x,\ |h_x(p)|<1} \frac{h_x(p)}{p} + O(1).$

Proof is even simpler then that of Theorem 1 in (Šiaulys, 1987).

COROLLARY. *The relative compactness of the sequence $\tilde{\Phi}_x$ yields the relative compactness of ν_x with the same centralizing function α_x.*

3. PROOF OF THEOREM 1

Sufficiency. Note that an additive function belongs to the class H defined in (Kubilius, 1964) and due to Theorem 4.1 (Kubilius, 1964) the sequence $F_x(v, A(x), B(x))$ weakly converges to the limit distribution with the characteristic function $\tau(t)$ expressed by the Kolmogorov formula. Using summation by parts, condition (II) of Theorem 1 and the estimate

$$\psi_x(u) = \Big(A(x^u) - A(x)\Big)/B(x) = o(1)$$

which is uniform with respect to $u \in [M^{-1}, M]$. For each $M > 1$ we find for the characteristic function

$$\tau_x(t) = \exp\left\{ - it\frac{A(x)}{B(x)}\right\}\Big(\sum_{m\leqslant x}\frac{1}{m}\Big)^{-1}\sum_{m\leqslant x}\frac{1}{m}\exp\left\{it\frac{h(m)}{B(x)}\right\},$$

the same limit expression $\tau(t)$ as $x \to \infty$.

Necessity. Let $L(\Phi_x, \Phi) \to 0$ for a distribution function Φ with mean zero and variance one. From Lemma 3 (Levin and Timofeev, 1977) we have that $\varphi_x(u)$ and $\psi_x(u)$ are bounded uniformly in $u \in [M^{-1}, M]$ for each $M > 1$. Here and in the following using Lemma 4 (Levin and Timofeev, 1977) we extend their propositions for the interval $[1, M]$.

Let $M > 1$ be fixed and $x_1 \to \infty$ be an arbitrary sequence. By our Lemma 3 and Lemma 3 from (Levin and Timofeev, 1977) we can choose a sequence $y \to \infty$ such that the following propositions are true:

(E_1) $\varphi_y(u) \to \varphi(u)$, $\psi_y(u) \to \psi(u)$ uniformly in $u \in [M^{-1}, M]$, where $\varphi(u)$ and $\psi(u)$ are continuous for $u > 0$ functions and $\varphi(u) > 0$;

(E_2) $L(F_{y^u},\ F(u,\cdot)) \to 0$ uniformly in $u \in [M^{-1}, M]$.

Here and in the following where there is no ambiguity we omit the sign $y \to \infty$. The functions $\varphi(u)$, $\psi(u)$ and $F(u, v)$ may depend on the sequence x_1. Let $\tau(t)$ and $f(u, t)$ be the characteristic functions of the distributions Φ and $F(u, \cdot)$, respectively. Note that $f(u, t)$ is continuous in u.

Summation by parts in the expression for $\tau_y(t)$ yields the limiting identity (see (Levin and Timofeev, 1977, p. 340))

$$v\tau(t\varphi(v))\exp\{it\psi(v)\} = \int_0^v f(u, t\varphi(u))\exp\{it\psi(u)\}du \tag{6}$$

for $v > 0$. In our case $\tau(\xi)$ is a twice differentiable function and $\varphi(u)$ is nondecreasing, therefore, much easier than in (Levin and Timofeev, 1977) we can conclude that $\varphi(u)$ is differentiable. The relation (see (Levin and Timofeev, 1977))

$$\Big(\mid\tau(\xi)\mid\Big)'_\xi\,\xi\frac{\varphi'(u)}{\varphi(u)} = \frac{b(\xi)}{u},$$

where $b(\xi) = (\mathrm{Re}\overline{\tau(\xi)}f(u,\xi) - \mid\tau(\xi)\mid^2),\mid\tau(\xi)\mid^{-1}$ and $\mid\xi\mid$ is sufficiently small but positive, yields $\varphi'(u)/\varphi(u) \leqslant c/u$ with some constant $c \geqslant 0$ depending on $\tau(t)$.

When $c = 0$, since $\varphi(1) = 1$ we have $\varphi(u) \equiv 1$. The last proposition will be proved in a more difficult case when $c > 0$. In the following we will use the estimate

$$\varphi(u) = O(u^c) \tag{7}$$

obtained by the integration of the above differential equation. Returning to (6) we note that $\psi(u)$ is differentiable too.
Let from now on $h_y(m) = h(m)/B(y)$, $s = 1 + (1 + i\tau)/\log y = 1 + z/\log y$, $|\tau| \leqslant K$, $K \geqslant 1$ and

$$Z(s) = \sum_{m=1}^{\infty} \frac{\exp\{ith_y(m)\}}{m^s}.$$

From the definition of $\tau_y(t)$, the convergence $\tau_y(t) = \tau(t) + o(1)$ and (E_1) we have

$$\exp\left\{ -it\frac{A(y)}{B(y)} \right\} (\log y)^{-1} Z\left(1 + \frac{z}{\log y}\right) =$$

$$= z \int_0^{\infty} u\tau_y u(t\varphi_y(u)) \exp\left\{it\psi_y(u)\right\} e^{-uz}\, du + o(1) =$$

$$= z \int_{M^{-1}}^{M} u\tau(t\varphi(u)) \exp\left\{it\psi(u)\right\} e^{-uz}\, du + O\left(\frac{1}{M}\right) + o(1). \tag{8}$$

The Euler identity and the well-known properties of the Riemann ζ-function applied to the left-hand side of (8) yield

$$P_y(z,t) = \sum_p \frac{\exp\{ith_y(p)\} - 1}{p} \exp\left\{ -z\frac{\log p}{\log y} \right\} - it\frac{A(y)}{B(y)} =$$

$$= \log\left(z^2 \int_{M^{-1}}^{M} u\tau(t\varphi(u)) \exp\left\{it\psi(u)\right\} e^{-uz}\, du \right) + O\left(\frac{1}{M}\right) + o(1) \tag{9}$$

uniformly in $|t| \leqslant t_0$, $|\tau| \leqslant K$, provided t_0 is sufficiently small, $M \geqslant M_0(K)$ and $y \geqslant y_0(K, M)$. Due to the expansion

$$\tau(t\varphi(u)) \exp\{it\psi(u)\} = 1 + it\psi(u) + Bt^2(\varphi^2(u) + \psi^2(u))$$

we have

$$P_y(z,t) = \log\left(1 + itz^2 \int_{M^{-1}}^{M} u\psi(u) e^{-uz}\, du + \right.$$

$$\left. + O\left(t^2 |z|^2 \int_{M^{-1}}^{M} u(\varphi^2(u) + \psi^2(u)) e^{-u}\, du\right) + O\left(\frac{1}{M}\right) + o(1)\right).$$

The estimate (7) and

$$\psi(u) = O\Big(\sqrt{|\log u|}\Big) \tag{10}$$

imply that the integrals in the parenthesis converge as $M \to \infty$. The first of them converges uniformly in $|\tau| \leqslant K$. The parameter M can be chosen arbitrarily therefore there exists $\varrho(y) = \varrho_K(y) \to 0$ such that for some sufficiently small $\varepsilon > 0$, $y \geqslant y_1(K)$ and

$$|t| K^2 \leqslant t_0 K^2 < \varepsilon \tag{11}$$

we obtain

$$P_y(z,t) = itz^2 \int_0^\infty u\psi(u) e^{-uz}\, du + O(t^2 |z|^4) + \varrho(u). \tag{12}$$

Now using the formula

$$\frac{1}{2\pi i} \int_{1-iK}^{1+iK} \frac{e^{\alpha z}}{z^k}\, dr = \frac{Be^\alpha}{K^{k-1}} + \begin{cases} \frac{\alpha^{k-1}}{(k-1)!} & \text{if } \alpha > 0, \\ 0, & \text{if } \alpha \leqslant 0, \end{cases}$$

where $k \geqslant 6$, we integrate the equality (12) multiplied by $(k-1)!\; e^z(2\pi i z^k)^{-1}$. We have

$$P_y(t) = \sum_{p \leqslant y} \frac{\exp\{ith_y(p)\} - 1 - ith_y(p)}{p} \left(\frac{\log y/p}{\log y}\right)^{k-1} +$$

$$+it \sum_{p \leqslant y} \frac{h_y(p)}{p} \frac{(k-1)!}{2\pi i} \int_{1-iK}^{1+iK} z^{-k} \left(\exp\left\{ z \frac{\log y/p}{\log y} \right\} - \exp\{z\} \right) dz =$$

$$= it(k-1)(k-2) \int_0^1 u\psi(u)(1-u)^{k-3}\, du +$$

$$+ O\Big(\frac{1}{K^{k-3}}\Big) + O(t^2) + \varrho(y). \tag{13}$$

If J denotes the second term in the left–hand side of the last estimate, then

$$J = it \sum_{p \leqslant y} \frac{h_y(p)}{p} \int_1^{\log(y/p)/\log y} \left(\frac{(k-1)!}{2\pi i} \int_{1-iK}^{1+iK} \frac{e^{vz}}{z^{k-1}}\, dz \right) dv =$$

$$= it \sum_{p \leqslant y} \frac{h_y(p)}{p} \left(\left(\frac{\log(y/p)}{\log y} \right)^{k-1} - 1 \right) + O\left(\frac{|t|}{K^{k-2}} \sum_{p \leqslant y} \frac{|h_y(p)| \log p}{p \log y} \right) =$$

$$= itS + O\Big(|t| K^{-k+2} \Big). \tag{14}$$

For the sum S we have

$$S = -\frac{A(y)}{B(y)} + \int_1^y \sum_{p \leqslant u} \frac{h_y(p)}{p}(k-1)\left(\frac{\log(y/u)}{\log y}\right)^{k-2} d\left(\frac{\log u}{\log y}\right) =$$

$$= (k-1)\int_0^1 \psi(u)(1-u)^{k-2}\,du + o(1). \tag{15}$$

To verify the last formula one can estimate the integral upon the interval $[0, M^{-1}]$ at first, as we have done earlier. Substituting (15) in (14) and the later in (13), letting $y \to \infty$, we obtain

$$\int_0^1 \psi(u)(1-u)^{k-2}\,du = \int_0^1 (u\psi(u))'(1-u)^{k-2}\,du + O(|\,t\,|) + O\Big(\frac{1}{|\,t\,|\,K^{k-3}}\Big).$$

Using the restriction (11) and choosing $t = K^{-3} \to 0$ for $k \geqslant 7$ we have

$$\int_0^1 u\psi'(u)(1-u)^{k-2}\,du = 0.$$

The completeness of the system of functions $1, v, v^2, \ldots,$ yields the differential equation $\psi'(u) = 0$ for $0 < u < 1$. Only the solution $\psi(u) = 0,\ 0 < u < 1$, satisfies the boundary condition $\psi(1) = 0$. Thus, from an arbitrary sequence $x_1 \to \infty$ we have chosen a subsequence $y \to \infty$ such that $\psi_y(u) = o(1)$, therefore

$$\lim_{x \to \infty} \psi_x(u) = 0, \quad 0 < u < 1. \tag{16}$$

If in the definition of $\psi_x(u)$ we change x^u by z, $0 < u < 1$, then we get the relation $\psi_x(u) = -\psi_z(1/u)\varphi_z(1/u)$ which due to (7) shows us that the limit (16) remains true for all $u > 0$.

In order to determine $\varphi(u)$ when $c > 0$ we return to (9) replacing $\psi(u)$ by zero. Now we use the estimate $|\,\tau(\xi) - 1 + \xi^2/2\,| \leqslant \delta\xi^2$ which is valid for each $\delta > 0$ when $|\,\xi\,| \leqslant \xi_0 = \xi_0(\delta)$. By the estimate (7) for sufficiently small $\delta_1 > 0$ and for $u < \delta_1\,|\,t\,|^{-1/c} = u_t$ we have $|\,t\,|\,\varphi(u) \leqslant \xi_0$, therefore

$$P_y(z,t) = \log\left(z^2\int_0^\infty u\tau(t\varphi(u))e^{-uz}\,du + \varrho(y)\right) =$$

$$= \log\left(1 - \frac{t^2z^2}{2}\int_0^{u_t} u\varphi^2(u)e^{-uz}\,du + O\Big(\delta t^2\,|\,z\,|^2 + |\,z\,|^2\exp\{-u_t/2\}\Big) + \varrho(y)\right).$$

Here and in the following $\varrho(y)$ denotes some remainder term which tends to zero as $y \to \infty$. Now for sufficiently large y provided $t_0 K < \varepsilon$ we have

$$P_y(z,t) = -\frac{t^2 z^2}{2} \int_0^\infty u\varphi^2(u) e^{-uz}\, du + o(t^2 \mid z \mid^4) + \varrho(y).$$

Repeating the integration with respect to z, $\mid \tau \mid \leqslant K$, we come to the asymptotics

$$P_y(t) = -\frac{t^2}{2}(k-1)(k-2) \int_0^1 u\varphi^2(u)(1-u)^{k-3}\, du + O(K^{5-k}) + o(t^2) + \varrho(y) \quad (17)$$

where $k \geqslant 6$ and $P_y(t)$ is defined in (13). This function is of greater interest now. Set

$$K^k_{xy}(v) = \sum_{p \leqslant x;\ h_y(p) < v} \frac{h_y^2(p)}{p} \left(\frac{\log(y/p)}{\log y} \right)^{k-1}, \quad K^1_{xy} = K_{xy}, \quad K_{xx} = K_x.$$

The sequence of distribution functions $K_x(v)$ is weakly relatively compact (partial limits need not be distribution functions). But the approach proposed in the second part of the paper is applicable.
In the space of non-negative, bounded, nondecreasing and left continuous functions we will consider the weak convergence of sequences denoted by $\Longrightarrow$. By the Helly theorem and the diagonalization procedure from the sequence y we choose a subsequence $y_1 \to \infty$ such that

$$K_{y_1^u}(v) \Longrightarrow K(u,v) \quad (18)$$

for all $u \in [M^{-1}, 1] \bigcap Q$. Here $M > 1$ is an arbitrary constant and $K(u,\cdot)$ is a family of functions in the aforementioned space. Note that

$$k(u) = K(u, +\infty) - K(u, -\infty) \leqslant 1.$$

We will prove that (18) holds for irrational u from the interval $[M^{-1}, 1]$ too. It is easy to check that for $u,\ u + \varepsilon \in [M^{-1}, 1]$

$$K_{y^{u+\varepsilon}}(v) = K_{y^u, y^{u+\varepsilon}}(v) + O(\eta(\varepsilon)) + o(1) =$$

$$= \sum_{p \leqslant y^u;\ h(p) < vB(y^{u+\varepsilon})} \frac{h^2(p)}{pB^2(y^u)} + O(\eta(\varepsilon)) + o(1)$$

uniformly in $v \in \mathbf{R}$ as $y \to \infty$. Here

$$\eta(\varepsilon) = \sup \Big\{ \mid \varphi(u+\varepsilon) - \varphi(u) \mid;\ u \in [M^{-1}, 1] \Big\} \to 0 \ \text{ as } \ \varepsilon \to 0.$$

Changing now the multiplier of v we have

$$K_{y^{u+\varepsilon}}(v) = K_{y^u}\Big(v(1 + O(\eta(\varepsilon)) + o(1)) \Big) + O(\eta(\varepsilon)) + o(1).$$

This yields our proposition about the convergence (18) for all $u \in [M^{-1}, 1]$ as well as the continuity of $K(u,v)$ with respect to u when v is a continuity point.
When $y = y_1 \to \infty$ (the index will be omitted later) for continuity points v of $K(u,v)$ we have

$$K_y^k(v) = \int\limits_{y^{1/M}}^{y} \left(\frac{\log(u/w)}{\log y} \right)^{k-1} d_w K_{w,y}(v) + O(M^{-2c}) =$$

$$= \int\limits_{M^{-1}}^{1} (1-u)^{k-1} d_u \Big(\varphi_y^2(u) K_{y^u}(v/\varphi_y(u)) \Big) + O(M^{-2c}) =$$

$$= (k-1) \int\limits_0^1 \varphi^2(u) K\Big(u, v/\varphi(u)\Big)(1-u)^{k-2}\, du + O(M^{-2c}) + o(1) =$$

$$= (k-1) K^k(v) + O(M^{-2c}) + o(1).$$

Therefore

$$P_y(t) = \int\limits_{-\infty}^{\infty} \frac{\exp\{itv\} - 1 - itv}{v^2}\, dK_y^k(v) =$$

$$= (k-1) \int\limits_{-\infty}^{\infty} \frac{\exp\{itv\} - 1 - itv}{v^2}\, dK^k(v) + \varrho(y) =$$

$$= (k-1) \varkappa_k(t) + \varrho(y). \tag{19}$$

Now $\varrho(y) \to 0$ as $y = y_1 \to 0$. Note that

$$\lim_{t\to 0} \varkappa_k(t) t^{-2} = -\Big(K^k(+\infty) - K^k(-\infty)\Big)/2 =$$

$$= -\frac{1}{2} \int\limits_0^1 \varphi^2(u) k(u) (1-u)^{k-2}\, du. \tag{20}$$

Using (19) in the equality (17) we let $y \to \infty$. Dividing then by t^2, choosing $t = K^{-3} \to 0$ and using (20) for $k \geqslant 12$ we obtain

$$\int\limits_0^1 \Big(\varphi^2(u) k(u) - (u\varphi^2(u))'(1-u)^{k-2} \Big)\, du = 0.$$

Whence for $0 < u < 1$

$$\varphi^2(u)(k(u) - 1) = 2u\varphi(u)\varphi'(u) \geqslant 0.$$

Therefore $k(u) = 1$ and $\varphi'(u) = 0$ which yields $\varphi(u) = 1$ for $0 < u < 1$.

Again from an arbitrary sequence $x_1 \to \infty$ we have chosen a subsequence $y_1 \to \infty$ such that $\varphi_{y_1}(u) = 1 + o(1)$. This implies the existence of the limit

$$\lim_{x \to \infty} \varphi_x(u) = 1 \quad \text{for} \quad 0 < u \leqslant 1 \tag{21}$$

as it was desired.
Using (16), (21) and (6) for an arbitrary sequence $x_1 \to \infty$ and $u > 0$ we find $f(u,t) = \tau(t)$. This gives the convergence $L(F_x, \Phi) \to 0$ as $x \to \infty$. Now the necessity of the conditions follows from (21) and Theorem 4.1 in (Kubilius, 1964). Theorem 1 is proved.

Acknowledgment

The author is indebted to prof. J. Kubilius and to the participants of his seminar for many valuable suggestions.

REFERENCES

Elliott, P.D.T.A. (1979). *Probabilistic Number Theory.* (R.Grundlehren math.Wiss., **239**), Springer, Berlin—Heidelberg etc.

Elliott, P.D.T.A. (1980). *Probabilistic Number Theory.* (R.Grundlehren math.Wiss., **240**), Springer, Berlin—Heidelberg etc.

Hildebrand, A. (1986). On Wirsing's mean value theorem for multiplicative functions. *Bull. London Math. Soc.* **18**, 147–152.

Kubilius, J. (1964). *Probabilistic Methods in the Theory of Numbers.* Amer. Math. Soc. Trans. of Math. Monographs, **11**, AMS.

Levin, B.V. and Timofeev, N.M. (1977). Distribution of values of additive functions. *Acta Arithm.* **33**, 327–351.

Šiaulys, J. (1987). Compactness of distributions of a sequence of additive functions. *Liet. Matem. Rink.* **27**, 369–382.

Timofeev, N.M. (1982). On convergence to discontinuous laws. *Liet. Matem. Rink.* **22**, 159–171.

Timofeev, N.M. (1983). The normal limit distribution for additive functions. In: *Abstr. All–Union Conf. Theory of Transcendent Functions,* Moscow, pp. 138–139.

Timofeev, N.M. (1985a). Stable limit laws for additive functions. *Math. Notes* **37**.

Timofeev, N.M. (1985b). Integral limit theorems for additive functions with standard normalizing and centering sequences. In: *Abstr. Fourth Intern. Vilnius Conf. Probab. Theory and Math. Statist.,* **3**, pp. 181–182.

Timofeev, N.M. and Tuljaganov, S.T. (1982). Distribution of values of additive functions with weight $1/n$. *Izvestia Acad. of Sci. Uzbek. SSR, Ser. phys.-math.* **4**, 31–37.

Prob. Theory and Math. Stat., Vol. 2, pp. 150–167
B.Grigelionis *et al.* (Eds.)
1990 VSP/Mokslas

STATISTICAL INFERENCES FOR SEMIMARTINGALE REGRESSION MODELS

A.V. MELNIKOV and A.A. NOVIKOV
Steklov Mathematical Institute, Vavilova str. 42, Moscow, USSR

ABSTRACT

We consider a regression model in which the observed process is a semimartingale depending on an unknown vector parameter. Exploiting the theory of martingales we study the following problems concerning least–squares (LS) estimates: strong consistency, asymptotic behavior (the iterated logarithm law), asymptotic normality, sequential estimators. This survey paper (which is in fact a record of the lecture presented at the Fifth Vilnius conference) contains both known results as well as some new ones.

INTRODUCTION

Let $(\Omega, \mathcal{F}, F = (\mathcal{F}_t)_{t\geqslant 0}, \mathbf{P})$ be a stochastic basis on which the following classes of cadlag processes are defined: $A(\mathbf{R}^k, \mathbf{R}^d)$ is a class of matrix–valued processes $A_t : \mathbf{R}^k \to \mathbf{R}^d$ which have bounded variation on any finite set; $A^+(\mathbf{R}^d)$ is a class of processes $A \in A(\mathbf{R}^d, \mathbf{R}^d)$ such that any difference $A_t - A_s$ is a symmetric positive matrix, $t > s$; $\mathcal{M}_{\mathrm{loc}}(\mathbf{R}^d)$ (respectively, $\mathcal{M}^2_{\mathrm{loc}}(\mathbf{R}^d)$, $\mathcal{M}^c_{\mathrm{loc}}(\mathbf{R}^d)$, $\mathcal{M}^d_{\mathrm{loc}}(\mathbf{R}^d)$) is the class of local (respectively, local square integrable, continuous, pure discontinuous) martingales whose values are in $\mathbf{R}^d$ with norm $\| x \| = (x,x)^{1/2}$, $x \in \mathbf{R}^d$; $\mathcal{P}$ is the class of predictable process.
In the case $M \in \mathcal{M}_{\mathrm{loc}}(\mathbf{R}^d)$ we shall use the canonical decomposition

$$M_t = M_t^c + M_t^d = M_t^c + \int_0^t \int_{\mathbf{R}^d\setminus\{0\}} x d(\mu - \nu),$$

where $M^d \in \mathcal{M}^d_{\mathrm{loc}}(\mathbf{R}^d)$, $M^c \in \mathcal{M}^c_{\mathrm{loc}}(\mathbf{R}^d)$, $\mu = \mu(dxds, \omega)$ is the jump measure generated by M^d, $\nu = \nu(x, s, \omega)$ is the compensator measure of μ.
We use in fact all the notations and definitions from (Liptser and Shiryaev, 1986).

The model of observation

We suppose that the observed process has the following representation:

$$X_t = A_t \Theta + M_t, \tag{0.1}$$

where $\Theta \in \mathbf{R}^k$ is an unknown parameter, $A \in A(\mathbf{R}^k, \mathbf{R}^d) \cap \mathcal{P}$, $X_t \in \mathbf{R}^d$, $M \in \mathcal{M}_{\mathrm{loc}}(\mathbf{R}^d)$.

Often we shall use the following assumption: there exists a process $V \in A^+(\mathbf{R}^1) \cap \mathcal{P}$ such that

$$A \ll V; \tag{0.2}$$

denote, under this assumption,

$$\frac{dA_t}{dV_t} = f_t^*, \quad \int_0^t \operatorname{tr} f_s f_s^* \, dV_s < \infty.$$

Using (0.2) we may define an estimate for the parameter Θ which may be considered as a LS–estimate:

$$\Theta_t = F_t^{-1} \int_0^t f_s \, dX_s, \quad \text{where} \quad F_t = \int_0^t f_s f_s^* dV_s \in A^+(\mathbf{R}^k) \cap \mathcal{P}. \tag{0.3}$$

We denote by $\lambda_2(F)$ and $\lambda_1(F)$ correspondingly the maximal and the minimal eigenvalues of the matrix F. Asymptotical properties of this estimate depends very essentially on a relation between its.

The following examples show that the model (0.1)—(0.2) covers many cases known before.

1) *Diffusion model* (see e.g., (Liptser and Shiryaev, 1974))

$$X_t = \int_0^t f_s^* \, ds\Theta + W_t, \tag{0.4}$$

where W is a Wiener process.

2) *Discrete model* (see e.g., (Lai and Wei, 1982))

$$x_t = f_t^* \Theta + e_t, \quad t = 0, 1, \ldots, \tag{0.5}$$

where e is a martingale–difference with respect to F. In order to see that this model can be treated as a partial case of (0.1)—(0.2) one has to consider the right–continuous modifications of the processes

$$X_t = \sum_{i=0}^t x_i, \quad M_t = \sum_{i=0}^t e_i$$

and take $V_t = [t]$, the integer part of t. Then

$$F_t = \sum_{i=0}^t f_i f_i^* \quad \text{and} \quad \Theta_t = F_t^{-1} \sum_{i=0}^t f_i x_i. \tag{0.6}$$

3) *Point processes* (see e.g., (Spreij, 1986)). Let X be a multivariate point process. Then by the Doob—Meyer decomposition

$$X_t = B_t + M_t, \quad (B \in \mathcal{P}, M \in \mathcal{M}_{\mathrm{loc}}(\mathbf{R}^1)).$$

If $B_t = \int_0^t f_s dV_s \Theta$ we have the model (0.1)—(0.2).

1. STRONG CONSISTENCY

There exist two approaches to study strong consistency of the LS–estimate (0.3) which are both based on the strong law of large numbers (SLLN) for martingales. The first one is based on an one–dimensional version of SLLN for martingales and was used in (Drygas, 1976; Lai and Wei, 1982, 1983; Liptser, 1980; Novikov, 1985; Christopeit, 1986). The second one is based on a multidimensional version and was used in (Anderson and Moore, 1976; Melnikov, 1986; Le Breton and Musiela, 1989). Here we formulate two results which illustrate both approaches. Let $M \in \mathcal{M}^2_{\mathrm{loc}}(\mathbf{R}^d)$, and $\langle M \rangle$ be the trace of its matrix of quadratic characteristics C^M, i.e.,

$$\|M\|^2 - \langle M \rangle \in \mathcal{M}_{\mathrm{loc}} \cap \mathcal{P}.$$

Denote G the class of increasing functions $g(x) > 1$, such that

$$\int_0^\infty g^{-1}(x)\, dx < \infty.$$

THEOREM 1 (Novikov, 1985). *Let in* (0.1) $M \in \mathcal{M}^2_{\mathrm{loc}}(\mathbf{R}^d)$, (0.2) *be fulfilled and* $\lambda_1(F_t) \to \infty$ a.s. *as* $t \to \infty$. *If there exists* $g \in G$ *such that*

$$g(\log \lambda_2(F_t))/\lambda_1(F_t) \to 0 \quad \text{a.s. } (t \uparrow \infty) \tag{1.1}$$

and

$$\sup_t \frac{d\langle M \rangle_t}{dV_t} < \infty \quad \text{a.s.}, \tag{1.2}$$

then $\theta_t \to \theta$ a.s.

The proof of this theorem is based on the following one–dimensional form of SLLN for martingales.

THEOREM 1.2 (Liptser, 1980). *Let* $N \in \mathcal{M}^2_{\mathrm{loc}}(\mathbf{R}^1)$, $A \in A^+(\mathbf{R}^1) \cap \mathcal{P}$ *be such that* $\lim_{t\to\infty} A_t = \infty$ a.s. *and*

$$\int_0^\infty A_s^{-2}\, d\langle N \rangle_s < \infty \quad \text{a.s.}$$

Then $A_t^{-1} N_t \to 0$ a.s. $(t \uparrow \infty)$.

To prove Theorem 1.1 we note that

$$\Theta_t - \Theta = F_t^{-1} N_t$$

and

$$\| \Theta_t - \Theta \|^2 \leqslant \lambda_1^{-1}(F_t) N_t^F,$$

where

$$N_t = \int_0^t f_s \, dM_s, \quad N_t^F = N_t^* F_t^{-1} N_t.$$

Using stochastic calculus we can show that

$$\text{trace} \left(\int_0^t F_s^{-1} f_s f_s^* dV_s \right) \leqslant \text{const} \cdot \log \lambda_2(F_t).$$

Now a simple application of Theorem 1.2 gives the result of Theorem 1.1.
In particular Theorem 1.1 implies the following result for the discrete time model (0.5)—(0.6) which was proved in somewhat less general form in (Lai and Wei, 1982) (but in fact their paper contains the method which was described above).

COROLLARY. *Let in* (0.5)—(0.6) $\sup_t \mathbf{E}(\| e_t \|^2 | \mathcal{F}_{t-1}) < \infty$ *and* $\lambda_1(F_t) \to \infty$ a.s. *as* $t \to \infty$. *If there exists* $g \in G$ *such that*

$$g(\log \lambda_2(F_t))/\lambda_1(F_t) \to 0 \quad \text{a.s.} \quad (t \uparrow \infty)$$

then $\Theta_t \to \Theta$ a.s. $(t \uparrow \infty)$.

Remark. In the recent paper (Le Breton and Musiela, 1989), the result of Theorem 1.1 was formulated under the condition

$$g \in G, \quad \sup_t g(\log \lambda_2(F_t)/\lambda_1(F_t)) < \infty \tag{1.3}$$

instead of (1.1). But it appears to be equivalent to (1.1) because of an obvious implication

$$(1.3) \text{ holds} \Rightarrow \exists \tilde{g} \in G \text{ such that } \tilde{g}(x)/g(x) \to 0 \ \ (x \uparrow \infty)$$

and so (1.1) is also fulfilled.
The second approach to study strong consistency is based on a multidimensional form of SLLN for martingales which is given by the following theorem.

THEOREM 1.3 (Melnikov, 1986). *Let* $N \in \mathcal{M}^2_{\text{loc}}(\mathbf{R}^d)$, $\Gamma \in A^+(\mathbf{R}^d) \cap \mathcal{P}$ *and* a.s. $\lambda_1(F_t) \to \infty$ *as* $t \to \infty$,

$$\limsup_t \lambda_1^{-1}(\Gamma_t) \lambda_2(\Gamma_t) < \infty, \tag{1.4}$$

$$\int_0^\infty \text{trace}\,(\Gamma_t^{-1} dC_t^N \Gamma_t^{-1}) < \infty. \tag{1.5}$$

Then a.s. $\| \Gamma_t^{-1} N_t \| \to 0 \;\; (t \uparrow \infty)$.

To prove this result we consider the process

$$n_t = \int_0^t \Gamma_s^{-1} dN_s.$$

Condition (1.5) implies the convergence of n_t to a finite limit n_∞ a.s. $(t \to \infty)$. Integrating by parts and using some algebra we have

$$\Gamma_t^{-1} N_t = \Gamma_t^{-1} \int_0^t d\Gamma_s (n_t - n_{s-}),$$

which implies for $u < t$

$$\|\Gamma_t^{-1} N_t\| \leqslant \|\Gamma_t^{-1}\| \cdot \left\| \int_0^t d\Gamma_s (n_t - n_{s-}) \right\| \leqslant \lambda_1^{-1}(\Gamma_t) \left\| \int_0^t d\Gamma_s (n_t - n_{s-}) \right\| \leqslant$$

$$\leqslant \lambda_1^{-1}(\Gamma_t) \|n_t - n_\infty\| \cdot d \cdot \lambda_2(\Gamma_t) + \lambda_1^{-1}(\Gamma_t) \left\| \int_0^u d\Gamma_s (n_\infty - n_{s-}) \right\| +$$

$$+ \lambda_1^{-1}(\Gamma_t) \left\| \int_u^t d\Gamma_s (n_\infty - n_{s-}) \right\|.$$

By choosing sufficiently large u and then sufficiently large t the three last terms here can be made arbitrarily small.

Let us consider the model (0.1) with $M \in \mathcal{M}^2_{\text{loc}}(\mathbf{R}^d)$ and under condition (0.2). Taking in Theorem 1.3

$$N_t = \int_0^t f_s \, dM_s, \quad \Gamma_t = F_t$$

we get as a direct consequence the following result:

THEOREM 1.4 (Melnikov, 1986). *Let* (a.s.) $\lambda_1(F_t) \to \infty \;\; (t \uparrow \infty)$,

$$\limsup_t \left(\lambda_1^{-1}(F_t) \lambda_2(F_t) \right) < \infty, \tag{1.6}$$

$$\int_0^\infty \text{trace}\,(F_t^{-1} \, dC_t^N F_t^{-1}) < \infty. \tag{1.7}$$

Then a.s. $\|\Theta_t - \Theta\| \to 0 \quad (t \uparrow \infty)$.

Remark. As it may be seen from the above exposition the second approach allows to relax the "variance condition" (1.2) but at the same time it needs more restrictive conditions on the relation between eigenvalues of the matrix F like (1.1). In (Le Breton and Musiela, 1989) some other results in this direction are given.
Consider now the LS–estimate in the case of a non–square integrable martingale noise process $M \in \mathcal{M}^d_{\text{loc}}(\mathbf{R}^d)$.

THEOREM 1.5 (Melnikov, 1988). *Let* a.s.

$$\lambda_1(F_t) \to \infty \; (t \uparrow \infty),$$

$$\limsup_t \lambda_1^{-1}(F_t)\lambda_2(F_t) < \infty,$$

$$\sup_t \frac{d}{dV_t} \int_0^t \int_{\mathbf{R}^d \setminus \{0\}} \|x\|^r \, d\nu < \infty, \quad \int_0^\infty \lambda_1^{-1}(F_s)\|f_s\|^r \, dV_s < \infty, \quad r \in (1,2].$$

Then (a.s.)

$$\|\Theta_t - \Theta\| \to 0 \; (t \uparrow \infty).$$

To prove this result a version of SLLN for non–square integrable martingales should be used (see (Melnikov, 1988)).

2. ASYMPTOTIC BEHAVIOR OF THE TRAJECTORIES OF WEIGHTED LS–ESTIMATES

We suppose here that in model (0.1) $M \in \mathcal{M}^2_{\text{loc}}(\mathbf{R}^d)$, the matrix $Q_t = dC^M_t / d\langle M\rangle_t$ is positive and

$$A_t = \int_0^t Q_s^{1/2} f_s^* \, d\langle M\rangle_s.$$

In this section we consider the asymptotic behavior of a weighted LS–estimate which is defined as follows

$$\Theta_t = F_s^{-1} \int_0^t f_s Q_s^{-1/2} \, dX_s, \quad F_t = \int_0^t f_s f_s^* \, d\langle M\rangle_s.$$

Obviously

$$Y_t = \Theta_t - \Theta = F_t^{-1} \int_0^t f_s Q_s^{-1/2} \, dM_s = F_t^{-1} N_t = (C_t^N)^{-1} N_t.$$

Denote by $v_i^{-1}(t)$ a diagonal element of the matrix $(C_t^N)^{-1}$, N_t^i is a component of the vector N_t, $i = 1, \dots, k$.
The following theorem is a slight modification of the results of (Darwich, 1988).

THEOREM 2.1. *Let the following conditions be fulfilled:*

$$\mathbf{E} \sup_t \|\Delta N_t\| < \infty, \tag{2.1}$$

$$\lim_t v_i(t) = \infty \quad \text{a.s.}, \tag{2.2}$$

$$\lim_t \langle N^i, N^i \rangle_t / v_i(t) = 1 \quad \text{a.s.}, \tag{2.3}$$

$$\lambda_2(F_t) = O(\lambda_1^\alpha(F_t)) \ (t \uparrow \infty) \quad \text{a.s.}, \quad \alpha \geqslant 1. \tag{2.4}$$

Then for each $i = 1, \dots, k$

$$\limsup_t |Y_t^i| v_i^{1/2}(t) (2 \log\log v_i(t))^{-1/2} = 1$$

a.s.

Proof. We consider only the case $k = 2$ (the general case can be treated with the help of Lemma 2 in (Le Breton and Musiela, 1987)). We have

$$Y_t^1 = v_1^{-1}(t) \left[N_t^1 - \frac{\langle N^1, N^2 \rangle_t}{\langle N^2 \rangle_t} N_t^2 \right],$$

$$Y_t^2 = v_2^{-1}(t) \left[N_t^2 - \frac{\langle N^2, N^1 \rangle_t}{\langle N^1 \rangle_t} N_t^1 \right]. \tag{2.5}$$

By simple calculations we have for Y_t^1 from (2.5)

$$\frac{Y_t^1 v_1^{1/2}(t)}{(2 \log\log v_1(t))^{1/2}} =$$

$$= \frac{N_t^1}{(2 v_1(t) \log\log v_1(t))^{1/2}} - \frac{\langle N^1, N^2 \rangle_t \cdot N_t^2}{\langle N^2 \rangle_t (2 v_1(t) \log\log v_1(t))^{1/2}}. \tag{2.6}$$

First let us show that the second term in (2.6) vanishes as $t \to \infty$.
Note that for $i = 1, 2$

$$\lambda_1(F_t) \leqslant \langle N^i \rangle_t, \quad v_i(t) \leqslant \lambda_2(F_t). \tag{2.7}$$

Furthermore,

$$\varphi_1 = |N_t^2| \Big((2 \langle N^2 \rangle_t \log\log v_1(t) \Big)^{-1/2} =$$

$$= \frac{|N_t^2|}{(2 \langle N^2 \rangle_t \log\log \langle N^2 \rangle_t)^{1/2}} \, \frac{(\log\log \langle N^2 \rangle_t)^{1/2}}{(\log\log v_1(t))^{1/2}},$$

$$\varphi_2 = \frac{|\langle N^1, N^2\rangle_t|}{(v_1(t)\langle N^2\rangle_t)^{1/2}} = \frac{(\langle N^1\rangle_t - v_1(t))^{1/2}}{v_1^{1/2}(t)}.$$

Now conditions (2.7), (2.2) and (2.3) imply that $\varphi_2(t) \to 0$ as $t \to \infty$. Applying the iterated logarithm law (ILL) for martingales (see (Lepingle, 1977)) we see that under the conditions (2.1), (2.2) and (2.4) the process $\varphi_1(t)$ is bounded. Therefore the second term in (2.6) vanishes as $t \to \infty$.
Now applying again ILL to the first term in (2.6) we get the result of Theorem 2.1.

Remark. Conditions (2.3) and (2.4) are deliberately fulfilled if

$$\lim_t \lambda_1^{-1}(F_t)\lambda_2(F_t) = 1 \quad \text{a.s.} \tag{2.8}$$

Note that

$$y_t = \Theta_t - \Theta = (C_t^m)^{-1} m_t,$$

where $m \in M^2_{\text{loc}}(\mathbf{R}^k)$. The following example shows that the conditions (2.4) and (2.8) are not necessary.

Example. Let

$$m_t = \left(W_t, \int_0^t (1+s)^{-1/2}\, dW_s\right)^*$$

be a two-dimensional martingale, where W is the standard Wiener process. It is easy to check that

$$C_t^m = \begin{pmatrix} t & 2[(1+t)^{1/2} - 1] \\ 2[(1+t)^{1/2} - 1] & \log(1+t) \end{pmatrix}.$$

So for the components of the weighted LS-estimate

$$\lim_t t^{-1} v_1(t) = \lim_t \left(\log(1+t)\right)^{-1} v_2(t) = 1$$

and the asymptotics of y_t^1 and y_t^2 are as in Theorem 2.1. On the other hand

$$\lambda_1(C_t^m)\Big/\frac{1}{2}\log(1+t) \to 1, \quad \lambda_2(C_t^m)/t \to 1 \quad \text{a.s.}, \quad t \uparrow \infty.$$

Therefore

$$\lambda_1^{-1}(C_t^m)\log\lambda_2(C_t^m) \to 1/2 > 0 \quad \text{a.s.}, \quad t \uparrow \infty,$$

and so conditions (2.4) and (2.8) are not satisfied.

3. ASYMPTOTIC NORMALITY

We consider here conditions for the asymptotic normality of a LS-estimate in the case $M \in \mathcal{M}^2_{\text{loc}}(\mathbf{R}^d)$. In fact we follow (Christopeit, 1986) but in his paper only the case of one-dimensional M and X was considered.

Suppose there exist non-random positive matrices P_T and Q such that

$$P_T^{-1} F_T^{1/2} \xrightarrow{\mathbf{P}} Q \quad (T \uparrow \infty), \tag{3.1}$$

$$P_T^{-1} \Big(\int_0^T f_s dC_s^M f_s^* \Big)^{1/2} \xrightarrow{\mathbf{P}} I \quad (T \uparrow \infty), \tag{3.2}$$

where I is the identity matrix and $\xrightarrow{\mathbf{P}}$ means the convergence in probability. Denote μ_T a the jump measure of the martingale

$$Y_t(T) = P_T^{-1} \int_0^{tT} f_s dM_s, \quad t \in [0,1],$$

(with respect to the family of σ–algebras $\mathcal{F}_{tT}$) and suppose that

$$\int_{(0,1]} \int_{\|x\|>\varepsilon} \|x\|^2 \, d\nu_T \xrightarrow{\mathbf{P}} 0 \quad (T \uparrow \infty), \tag{3.3}$$

where ν_T is the compensator of μ_T, $\varepsilon > 0$.
Below we use the notation $\xrightarrow{D}$ for the convergence in distribution.

THEOREM 3.1. *Let conditions* (3.1)—(3.3) *be fulfilled. Then*

$$F_T^{1/2}(\Theta_T - \Theta) \underset{T\to\infty}{\xrightarrow{D}} N(0, Q^{-1}),$$

where $N(0,Q)$ *is the normal distribution with covariation matrix* Q.

The proof consists in checking the conditions in the central limit theorem for martingales (Liptser and Shiryaev, 1986; Lavrentev, 1982). At first we observe that

$$F_T^{1/2}(\Theta_T - \Theta) = P_T P_T^{-1} F_T^{-1/2} \int_0^T f_s \, dM_s = P_T F_T^{-1/2} Y_1(T).$$

Condition (3.1) implies

$$P_T F_T^{-1/2} \underset{T\to\infty}{\xrightarrow{\mathbf{P}}} Q^{-1},$$

and so we need only to show that

$$Y_1(T) \underset{T\to\infty}{\xrightarrow{D}} N(0, I). \tag{3.4}$$

The matrix of quadratic characteristics of $Y_1(T)$ is

$$C_1^{Y_1(T)} = P_T^{-1} \int_0^T f_s \, dC_s^M f_s^* P_T^{-1}.$$

By (3.2)

$$C_1^{Y(T)} \xrightarrow[T\to\infty]{\mathbf{P}} I.$$

Taking into account (3.3) we get (3.4) and Theorem 3.1 follows.

4. MODELS WITH NONOBSERVABLE REGRESSORS

We consider here the case when condition (0.2) in the model (0.1) is fulfilled but the regressor process f is not observable and so we can not use the LS–estimate Θ_t. Sometimes it is natural to suppose that we can observe a matrix function g which is close in some sense to f. We may try to use the following LS–like estimate

$$\tilde{\Theta}_t = G_t^{-1} \int_0^t g_s \, dX_s,$$

where

$$G_t = \int_0^t g_s g_s^* \, dV_s, \quad \int_0^t \operatorname{tr} g_s g_s^* \, dV_s < \infty.$$

THEOREM 4.1. *Let in the model* (0.1) $M \in \mathcal{M}^2_{\mathrm{loc}}(\mathbf{R}^d)$, $\|\Theta\| \leqslant R$ *and the following assumptions be fulfilled* (a.s.):

$$\lambda_1(G_t) \to \infty \quad (t \uparrow \infty), \tag{4.1}$$

$$\limsup_t \lambda_1^{-1}(G_t)\lambda_2(G_t) < \infty, \tag{4.2}$$

$$\int_0^\infty \operatorname{tr} g_s \, dC_s^M g_s^* < \infty, \tag{4.3}$$

$$\lim_t \lambda_1^{-1}(G_t) \int_0^t \|f_s - g_s\|^2 \, dV_s = 0. \tag{4.4}$$

Then a.s. $\|\tilde{\Theta}_t - \Theta\| \to 0 \quad (t \uparrow \infty)$.

Proof. It is easy to check that

$$\|\tilde{\Theta}_t - \Theta\| \leqslant \left\| \left(G_t^{-1} \int_0^t g f^* \, dV - I \right) \Theta \right\| + \left\| G_t^{-1} \int_0^t g \, dM \right\| = I_1(t) + I_2(t).$$

By SLLN for martingales $I_2 \to 0$ a.s. Now taking into account (4.1), (4.2), (4.4), by the Hölder inequality we have

$$\|I_1(t)\| \leqslant \left\| G_t^{-1} \int_0^t (g f^* - g g^*) \, dV \right\| \cdot \|\Theta\| \leqslant$$

$$\leqslant R \cdot \lambda_1^{-1}(G_t^{-1}) \cdot \int_0^t \|g\| \cdot \|f^* - g^*\| \, dV \leqslant$$

$$\leqslant R \cdot \lambda_1^{-1}(G_t) \Big(\int_0^t \|g\|^2 dV \Big)^{1/2} \Big(\int_0^t \|f^* - g^*\|^2 \, dV \Big)^{1/2} \leqslant$$

$$\leqslant R \cdot \lambda_1^{-1/2}(G_t) \cdot \lambda_2^{1/2}(G_t) \cdot \lambda_1^{-1/2}(G_t) \Big(\int_0^t \|f_s - g_s\|^2 \, dV_s \Big)^{1/2},$$

$$\|I_1(t)\| \to 0 \quad (t \uparrow \infty).$$

The proof is completed.

5. MODELS WITH DETERMINISTIC REGRESSORS

We consider here the case when the observed process X and the noise processes M are scalar, $\Theta \in \mathbf{R}^k$, and the regressor function f is a k–dimensional deterministic vector; V is also a deterministic scalar function. Under these assumptions we can prove strong consistency of the LS–estimate Θ_t in the case when M is a semimartingale satisfying the following "convergence" condition: for any locally bounded deterministic function c

$$\int_0^\infty c_s^2 \, dV_s < \infty \Longrightarrow \int_0^t c_s \, dM_s \ \text{ converges to a finite limit (a.s.).} \tag{5.1}$$

This condition is satisfied obviously for $M \in M^2$ (with $V = \langle M \rangle$) and for $M_t = = \sum_{i=0}^{t} e_i$, $V = [t]$, where e is a stochastic sequence with uncorrelated increments.

The following theorem (proved in (Le Breton and Musiela, 1987)) generalizes some ideas used in considering the discrete time model in (Chen et al., 1981).

THEOREM 5.1. *Let condition* (5.1) *be fulfilled and*

$$\lambda_1(F_t) \to \infty \quad (t \uparrow \infty). \tag{5.2}$$

Then

$$\|\Theta_t - \Theta\| \to 0 \quad (t \uparrow \infty) \quad \text{a.s.}$$

A similar result was obtained independently by A.V. Melnikov in the paper presented at the IV-th European Young Statisticians Conference (Varna, 1985).
It is interesting to note that in the case when $V = \langle M \rangle$ is a deterministic process the LS–estimate Θ_t is strong consistent only under condition (5.2).
Now consider the case of infinite–dimensional parameter Θ (belonging to some Hilbert space). A direct generalization of Theorem 5.1 is impossible because the

operator F is nuclear and hence condition (5.2) can not be satisfied. But we may consider a sequence of regression models

$$X_t^n = \int_0^t f_n^*(s)\, dV_s^n + M_t^n \tag{5.3}$$

and try to prove strong consistency of the LS–estimate as n and t tend to infinity (see (Gadjiev, 1989) concerning the discrete time case). For simplicity we shall consider only the case of one–dimensional X^n, V^n, M^n, $\dim(f_n) = \dim(\Theta^n) = n$, $\langle M^n \rangle$ is a deterministic function such that $d\langle M^n\rangle_t / dV_t^n \leqslant \text{const}$.

Theorem 5.2. *Let for model (5.3) the following conditions be fulfilled:*

$$\lambda_1(F_t^n) \to \infty \quad ((n,t) \to \infty), \tag{5.4}$$

$$\sup_{n,t} \lambda_1^{-1}(F_t^n)\, \lambda_2(F_t^n) < \infty, \tag{5.5}$$

$$\lambda_1^{-1}(F_t^n) n \to 0 \quad ((n,t) \to \infty). \tag{5.6}$$

Then

$$\mathbf{P}\{\|\Theta_t^n - \Theta^n\| > \varepsilon\} \to 0 \quad ((n,t) \to \infty),\ \varepsilon > 0.$$

The proof of this theorem is based on the Tchebyshev inequality: for any $\varepsilon > 0$

$$\mathbf{P}\{\|\Theta_t^n - \Theta^n\| > \varepsilon\} \leqslant \varepsilon^{-2}\mathbf{E}\|\Theta_t^n - \Theta^n\|^2 =$$

$$= \varepsilon^{-2}\mathbf{E}\left\|(F_t^n)^{-1} \int_0^t f_s^n dM_s^n\right\|^2 \leqslant \text{const} \cdot \varepsilon^{-2}\lambda_1^{-1}(F_t^n)n.$$

6. SEQUENTIAL LS–ESTIMATES

At first consider the one–dimensional diffusion model (0.4). In this case strong consistency of Θ_t as $t \to \infty$ follows from the condition:

$$\int_0^\infty f_s^2\, ds = \infty \qquad \text{(a.s.)}$$

(see e.g., (Liptser and Shiryaev, 1974)). Novikov in (Novikov, 1972) considered the following sequential modification of this LS–estimate:

$$\bar\Theta_H = H^{-1} \int_0^{\tau_H} f_s\, dX_s, \quad \text{where } \tau_H = \inf\left\{t > 0 : \int_0^t f_s^2\, ds \geqslant H\right\}, \quad H > 0.$$

It can be easily shown that in this case

$$\mathbf{E}\bar{\Theta}_H = \Theta, \quad \mathrm{Var}\,(\bar{\Theta}_H) = H^{-1}$$

since

$$\int_0^{\tau_H} f_s^2\, ds = H \qquad \text{a.s.} \tag{6.1}$$

In fact in this case the sequential estimate $\bar{\Theta}_H$ even has normal distribution (see (Novikov, 1972)).
For the general model (0.1)—(0.2) equality (6.1) is not true and so it is necessary to use some correction terms for constructing a modification of $\bar{\Theta}_H$. Such modification was given for discrete time case in (Borisov and Konev, 1977; Konev, 1985).
For the general model (0.1) an analogous sequential estimate was considered in (Novikov and Melnikov, 1988).
We define for the one-dimensional model (0.1)—(0.2)

$$\hat{\Theta}_H = H^{-1} \int_0^{\tau_H -} f_s \gamma_s^{-1}\, dX_s + H^{-1} \beta_H \gamma_{\tau_H}^{-1} f_{\tau_H} \Delta x_{\tau_H},$$

where γ is a predictable positive function,

$$\tau_H = \inf\{t > 0 : \int_0^t f_s^2 \gamma_s^{-1} dV_s \geqslant H\}$$

and β_H is a $\mathcal{F}_{\tau_H -}$-measurable random variable such that $\beta_H \in (0,1]$,

$$\int_0^{\tau_H} f_s^2 \gamma_s^{-1}\, dV_s + \beta_H f_{\tau_H}^2 \gamma_{\tau_H}^{-1} \Delta V_{\tau_H} = H. \tag{6.2}$$

THEOREM 6.1 (Melnikov and Novikov, 1988). *Let the following conditions be fulfilled:* $M \in \mathcal{M}^2_{\mathrm{loc}}(\mathbf{R}^1)$

$$\frac{d\langle M\rangle_t}{dV_t} \leqslant \gamma_t \cdot \mathrm{const} \quad (\text{a.s.}), \tag{6.3}$$

$$\int_0^{\infty} f_s^2 \gamma_s^{-1}\, dV_s = \infty \qquad (\text{a.s.}). \tag{6.4}$$

Then $\mathbf{E}\hat{\Theta}_H = \Theta$ *and* $\mathrm{Var}\,\hat{\Theta}_H \leqslant \mathrm{const} \cdot H^{-1}$.

(The last property we name as "*fixed accuracy*".)

Proof. Note that $\hat{\Theta}_H = \Theta + H^{-1}N_{\tau_H}$, where

$$N_t = \int_0^t I_{\{s<\tau_H\}} f_s \gamma_s^{-1} dM_s + I_{\{t=\tau_H\}} \beta_H f_{\tau_H} \gamma_{\tau_H}^{-1} \Delta M_{\tau_H}.$$

The process $N \in \mathcal{M}^2_{\text{loc}}(\mathbf{R}^1)$ is a stochastic integral with respect to $M \in \mathcal{M}^2_{\text{loc}}(\mathbf{R}^1)$. Its quadratic characteristic is

$$\langle N \rangle_t = \int_{(0,t)} I_{\{s<\tau_H\}} f_s^2 \gamma_s^{-2} d\langle M \rangle_s + I_{\{t=\tau_H\}} \beta_H^2 f_{\tau_H}^2 \gamma_{\tau_H}^{-2} \Delta \langle M \rangle_{\tau_H}.$$

Now using conditions (6.2)—(6.4) we have

$$\langle N \rangle_{\tau_H} \leqslant \int_{(0,\tau_H)} f_s^2 \gamma_s^{-1} dV_s + \beta_H f_{\tau_H}^2 \gamma_{\tau_H}^{-1} \Delta V_{\tau_H} = \text{const} \cdot H$$

and so the theorem is proved.

Consider now the multidimensional model (0.1) with $M \in \mathcal{M}^2_{\text{loc}}(\mathbf{R}^d)$ and $\Theta \in \mathbf{R}^k$. Denote

$$a_t = (f_t f_t^*)^{-1} f_t, \quad \frac{dC_t^M}{dV_t} \leqslant Q_t, \quad Q_t^a = a_t Q_t a_t^*, \quad Q_t^{-a} = a_t Q_t^{-1} a_t^*$$

and define a sequential estimate as follows:

$$\hat{Q}_H = H^{-1} \Big[\int_{(0,\tau_H)} \lambda_1(Q_t^{-a}) a_t dX_t + \beta_H \lambda_1(Q_{\tau_H}^{-a}) a_{\tau_H} \Delta X_{\tau_H} \Big],$$

where H is a positive number,

$$\tau_H = \inf \Big\{ t : \int_0^t \lambda_1(Q_s^{-a}) dV_s \geqslant H \Big\},$$

β_H is a $\mathcal{F}_{\tau_H}$-measurable random variable such that $\beta_H \in (0,1]$,

$$\int_{(0,\tau_H)} \lambda_1(Q_t^{-a}) dV_t + \beta_H \lambda_1(Q_{\tau_H}^{-a}) \Delta V_{\tau_H} = H. \tag{6.5}$$

THEOREM 6.2 (see (Melnikov and Novikov, 1988); cf. (Konev, 1985), for the discrete time case). *Assume that*

$$\int_0^t \lambda_1(Q_s^{-a}) dV_s \in A^+(\mathbf{R}^1),$$

$$\int_0^\infty \lambda_1(Q_t^{-a})dV_t = \infty \quad \text{a.s.} \tag{6.6}$$

$$\lambda_1(Q_t^{-a})\lambda_2(Q_t^a) \leqslant \text{ const }. \tag{6.7}$$

Then

$$\mathbf{E}\hat{\Theta}_H = \Theta, \quad \text{Cov } (\hat{\Theta}_H, \hat{\Theta}_H) \leqslant \text{ const } \cdot H^{-1} \cdot I.$$

In particular,

$$\|\hat{\Theta}_H - \Theta\|^2 \leqslant \text{ const } (k) \cdot H^{-1}.$$

Proof. The first statement of the theorem follows from (6.5):

$$\mathbf{E}\hat{\Theta}_H = H^{-1}\mathbf{E}\Big\{ \int\limits_{(0,\tau_H)} \lambda_1(Q_s^{-a})a_s f_s^* dV_s \Theta + \beta_H \lambda_1(Q_{\tau_H}^{-a}) a_{\tau_H} f_{\tau_H}^* \Delta V_{\tau_H} \Theta +$$

$$+ \int\limits_{(0,\tau_H)} \lambda_1(Q_s^{-a})a_s dM_s + \beta_H \lambda_1(Q_{\tau_H}^{-a}) a_{\tau_H} \Delta M_{\tau_H} \Big\} =$$

$$= H^{-1}\mathbf{E}\Big\{ \int\limits_{(0,\tau_H)} \lambda_1(Q_s^{-a}) dV_s \Theta + \beta_H \lambda_1(Q_{\tau_H}^{-a}) \Delta V_{\tau_H} \Big\} = \Theta.$$

Let us use this statement to estimate Cov $(\hat{\Theta}_H, \hat{\Theta}_H)$. Using (6.5), (6.6), (6.7) we have

$$\text{Cov } (\hat{\Theta}_H, \hat{\Theta}_H) = H^{-2}\mathbf{E}\Big\{ \Big[\int\limits_{(0,\tau_H)} \lambda_1(Q_s^{-a})a_s dM_s + \beta_H \lambda_1(Q_{\tau_H}^{-a}) a_{\tau_H} \Delta M_{\tau_H} \Big] \times$$

$$\times \Big[\int\limits_{(0,\tau_H)} \lambda_1(Q_s^{-a})a_s dM_s + \beta_H \lambda_1(Q_{\tau_H}^{-a}) a_{\tau_H} \Delta M_{\tau_H} \Big]^* \Big\} =$$

$$= H^{-2}\mathbf{E}\Big\{ \int\limits_{(0,\tau_H)} \lambda_1^2 a_s Q_s a_s^* dV_s + \beta_H^2 \lambda_1^2 a_{\tau_H} Q_{\tau_H} a_{\tau_H}^* \Delta V_{\tau_H} \Big\} \leqslant$$

$$\leqslant H^{-2}\mathbf{E}\Big\{ \int\limits_{(0,\tau_H)} \lambda_1^2(Q_s^{-a})\lambda_2(Q_s^a) dV_s +$$

$$+ \lambda_1^2(Q_{\tau_H}^{-a})\lambda_2(Q_{\tau_H}^a)\beta_H \Delta V_{\tau_H} \Big\} \times I \leqslant \text{ const } \cdot H^{-1} \cdot I.$$

Example 1. ("Efficiency" of the sequential estimate.) Consider the two-dimensional model

$$X_t = \Theta \int_0^t s^r B X_s ds + W_t,$$

where W is a two-dimensional Wiener process, $B = \begin{pmatrix} 0 & -1 \\ 1 & 0 \end{pmatrix}$, $r > -1$.
Let us investigate the ratio

$$\rho = \text{Var } \Theta_t / \text{Var } \hat{\Theta}_H$$

under the condition $\mathbf{E}\hat{\Theta}_H = t$. We have

$$\rho = \mathbf{E}\eta^{-1} \Big/ \left[\mathbf{E}\eta^{-(2+2r)^{-1}}\right]^{2(1+r)}, \quad \eta = \int_0^1 s^{2r} |X_s|^2 \, ds,$$

$$r \geqslant -1/2 \Longrightarrow \text{Var } \hat{\Theta}_H \leqslant \text{Var } \Theta_t,$$
$$-1/2 \geqslant r \geqslant -1 \Longrightarrow \text{Var } \hat{\Theta}_H \geqslant \text{Var } \Theta_t$$

(see more details in (Melnikov and Novikov, 1988)).

Example 2. Let X be a point process with compensator

$$A_t = \Theta \int_0^t f_s ds, \quad \Theta \in [0, b] \quad \text{and} \quad \int_0^\infty f_s \, ds = \infty \quad \text{a.s.}$$

If we take $\gamma = f$, const $= b$ then we get

$$\widehat{\Theta}_H = H^{-1} X_{\tau_H}, \quad \text{Var } \widehat{\Theta}_H \leqslant bH^{-1}, \quad \tau_H = \inf \left\{ t : \int_0^t f_s \, ds \geqslant H \right\}.$$

Example 3. Let X be the Galton—Watson branching process,

$$X_t = \sum_{k=1}^{X_{t-1}} \eta_k(t), \quad X_0 = 1,$$

$$\{\eta_k(t)\} \text{ are i.d.d.}, \quad \eta_k(t) = 0, 1, \ldots, \quad \mathbf{E}\eta_k(t) = \Theta, \quad \text{Var } (\eta_k(t)) = \sigma^2.$$

We can use an another representation for X:

$$X_t = \Theta X_{t-1} + \varepsilon_t, \quad \varepsilon_t = \sum_{k=1}^{X_{t-1}} (\eta_k(t) - \Theta), \quad \mathbf{E}(\varepsilon_t^2 | \mathcal{F}_{t-1}) = \sigma^2 X_{t-1},$$

$$\mathcal{F}_t = \sigma\{\varepsilon_1, \ldots, \varepsilon_t\}.$$

Take

$$V_t = [t], \ \gamma_t = X_{t-1}, \quad \tau_H = \inf \left\{ n : \sum_{k=1}^{n} X_{k-1} \geqslant H \right\},$$

$$\widehat{\Theta}_H = H^{-1} \left[\sum_{k=1}^{\tau_H - 1} X_k + \beta_H X_{\tau_H} \right], \quad \beta_H = X_{\tau_H}^{-1} \left[H - \sum_{k=1}^{\tau_H - 1} X_k \right].$$

Then we get $\mathbf{E}\widehat{\Theta}_H = \Theta$ and Var $\widehat{\Theta}_H \leqslant \sigma^2 H^{-1}$.

Remark. The sequential approach described above can be applied to the construction of sequential tests for checking statistical hypotheses in semimartingale regression models (see (Melnikov and Novikov, 1988)).

REFERENCES

Anderson, B.D.O. and Moore, J.B. (1976). A matrix Kronecker Lemma. *Linear Algebra and its Appl.* **15**, 227–234.

Borisov, V.Z. and Konev, V.V. (1977). Sequential estimation of parameters of discrete processes. *Automatics and Remote Control* **10**, 58–64.

Chen Gui-Jing, Lai, T.L. and Wei, C.Z. (1981). Convergence systems and strong consistency of LS-estimates in regression models. *J.Multiv.Anal.* **11**, 319–333.

Christopeit, N. (1986). Quasi-least-squares estimation in semimartingale regression models. *Stochastics* **16**, 255–278.

Darwich, A.R. (1988). Une Loi du logarithme itere pour les martingales multidimensionnelles et son application en regressi linneare stochastiques. *Publications de l'Institute de Recherche de Mathematiques de Rennes* **1**, 89–96.

Drygas, H. (1976). Weak and strong consistency of LS-estimates in regressions models. *Z.Wahrsch. verw. Geb.* **34**, 119–127.

Gadjiev, A.G. (1989). On estimation in linear regression models. *Vestnik Moscow State Univ.* **1**, 8–13.

Konev, V.V. (1985). *Sequential Estimates of Parameters of Stochastic Dynamical Systems.* Tomsk State Univ., Tomsk.

Lai, T.L., Robbins, H. and Wei, C.Z. (1979). Strong consistency of LS-estimates in multiple regression. *J.Multiv.Anal.* **9**, 343–360.

Lai, T.L. and Wei, C.Z. (1982). Least-squares estimates in stochastic regression with applications to identification and control of dynamic systems. *Ann.Statist.* **10**, 154–166.

Lai, T.L. and Wei, C.Z. (1983). Asymptotic properties of general autoregressive models and strong consistency of LS-estimates of their parameters. *J.Multiv.Anal.* **13**, 1–23.

Lavrentev, V.V. (1982). Functional central limit theorem in Hilbert space. *Uspekhi Math.Sci.* **37**, 165–166.

Le Breton, A. and Musiela, M. (1987). Strong consistency of LS-estimates in linear regression models driven by semimartingales. *J.Multiv. Anal.* **23**, 77–92.

Le Breton, A. and Musiela, M. (1989). Laws of large numbers for semimartingales with application in stochastic regression. *Th. Prob. Rel. Fields* **81**, 275–290.

Lepingle, D. (1977). Sur le comportements asymptotiques des martingales locales. *Lect. Notes in Math.* **649**, 148–161.

Liptser, R.S. (1980). A strong law of large numbers for local martingales. *Stochastics* **3**, 217–228.

Liptser, R.S. and Shiryaev, A.N. (1974). *Statistics of Random Processes.* Nauka, Moscow.

Liptser, R.S. and Shiryaev, A.N. (1986). *Theory of Martingales.* Nauka, Moscow.

Melnikov, A.V. (1986). Strong law of large numbers for multidimensional martingales. *Soviet Math. Dokl.* **289**, 546–550.

Melnikov, A.V. (1988). On regression models with non-square integrable martingale-like errors. *Publications de l'Institute de Recherche de Mathematiques de Rennes* **1**, 97–107.

Melnikov, A.V. and Novikov, A.A. (1988). Sequential fixed accuracy inferences for semimartingales. *Theory Probab. Appl.* **3**, 480–494.

Novikov, A.A. (1972). Sequential estimations of parameters of diffusion processes. *Mathem. Notes* **12**, 627–638.

Novikov, A.A. (1985). *Consistency of LS-estimates in Regression Models with Martingale Errors, Statistics and Control of Stochastic Processes.* Springer, Berlin—Heidelberg etc.

Spreij, P. (1986). Recursive parameter estimation for counting processes with linear intensity. *Stochastics* **18**, 277–312.

Prob. Theory and Math. Stat., Vol. 2, pp. 168–175
B.Grigelionis *et al.* (Eds.)
1990 VSP/Mokslas

ON THE MARTINGALE PROBLEM ASSOCIATED WITH INTEGRO–DIFFERENTIAL OPERATORS

R. MIKULEVIČIUS and H. PRAGARAUSKAS
Institute of Mathematics and Cybernetics, Akademijos 4,
232600 Vilnius, Lithuania

ABSTRACT

The existence and uniqueness of solution to the martingale problem associated with two types of Lévy's generators is considered. The first type corresponds to non–degenerate operators, whose principal part has continuous (in spatial variable) coefficients. The second type corresponds to degenerate operators with sufficiently smooth coefficients.

Let $H = [0,1] \times \mathbf{R}^d, \mathbf{R}_0^d = \mathbf{R}^d \backslash \{0\}$. Suppose that for each $(t,x) \in H$ we are given a symmetric non–negative definite $d \times d$–matrix $a(t,x)$, d–dimensional vector $b(t,x)$ and non–negative measure $\pi(t,x,dy)$ on $\mathbf{R}_0^d$. Throughout the paper we shall assume that the functions a, b, π are measurable and there exists a constant K such that for each $(t,x) \in H$

$$\|a(t,x)\| + |b(t,x)| + \int |y|^2 \wedge 1 \quad \pi(t,x,dy) \leqslant K.$$

For $u \in C_0^\infty(\mathbf{R}^d)$, $(t,z) \in H$ let us introduce the operator

$$L_{t,z}u(x) = \sum_{i,j=1}^{d} a_{ij}(t,z)u_{x_i x_j}(x) + \sum_{i=1}^{d} b_i(t,z)u_{x_i}(x) +$$

$$+ \int \left[u(x+y) - u(x) - \sum_{i=1}^{d} u_{x_i}(x) y_i \mathbf{1}_{\{|y| \leqslant 1\}} \right] \pi(t,z,dy),$$

and for brevity denote

$$Lu = L_t u(x) = L_{t,x}u(x).$$

Let $D = D\left([0,1], \mathbf{R}^d\right)$ be the Skorokhod space, with the canonical process $X_t(\omega) = \omega_t \, (\omega \in D, t \in [0,1])$ and the σ–algebras $\mathcal{D}_t = \sigma\{X_r : 0 \leqslant r \leqslant t\}$, $\mathcal{D} = \bigvee_{t \in [0,1]} \mathcal{D}_t$.

We shall say that a probability measure $\mathbf{P}$ on $(D, \mathcal{D})$ is a solution to the (s, x, L)–martingale problem if

$$\mathbf{P}\{X_r = x, 0 \leqslant r \leqslant s\} = 1$$

and for all $u \in C_0^\infty(\mathbf{R}^d)$ the process

$$u(X_t) - \int_s^t L_r u(X_r) dr$$

is a $((\mathcal{D}_{t+}), \mathbf{P})$–martingale.

We shall consider two cases of the (s, x, L)–martingale problem. The first case deals with a non–degenerate operator which means that either the matrix–valued function a is strictly positive definite, or $a \equiv 0$ and the principal part of the Levy's measure has a non-degenerate density with respect to the Levy's measure of the sferically symmetric stable process. The main result in this case (Theorem 1) generalizes the corresponding results given in (Stroock, 1975) and (Komatsu, 1984). The second case deals with a possibly degenerate operator L with sufficiently smooth coefficients. The main result in this case (Theorem 2) generalizes the corresponding results given in (Negoro and Tsuchiya, 1989).

Let $C_R = [0,1] \times \{x \in \mathbf{R}^d : |x| < R\}, d_0 = [d/2] + 1$, where $[r]$ denotes the integer part of a number r. Let S^{d-1} be the unit sphere in $\mathbf{R}^d$ with the area element $\mu(d\omega)$ on it. If $(U, \mathcal{U})$ is a measurable space, then $\mathcal{R}(U)$ (resp., $\mathcal{R}^+(U)$) denotes the space of σ–finite (resp., non–negative σ–finite) measures on $\mathcal{U}$. Let $|\nu|$ denote the total variation of an element $\nu \in \mathcal{R}(U)$.

The first group of assumptions concerns with a non–degenerate operator L.

Suppose that for each $\alpha \in (0,2]$ we are given a real–valued measurable function $m^{(\alpha)}(t, x, y) \geqslant 0$ on H and a measurable function $\nu^{(\alpha)}(t, x, dy)$ on H with values in $\mathcal{R}(\mathbf{R}_0^d)$. We shall assume that there exists a constant K such that for any $\alpha \in (0,2]$, $(t, x) \in H$, $y \in \mathbf{R}^d$

$$m^{(\alpha)}(t, x, y) + \int |y|^\alpha \wedge 1 \left|\nu^{(\alpha)}\right|(t, x, dy) \leqslant K.$$

ASSUMPTION 1. *For each $\alpha \in (0,2]$, $(t, x) \in H$ the function $m^{(\alpha)}(t, x, y)$ is homogeneous with respect to y with index zero, and*

$$\int_{S^{d-1}} \omega m^{(1)}(t, x, \omega) \mu(d\omega) = 0, \quad m^{(2)} \equiv 0.$$

For each $\alpha \in (0,2]$, the function $m^{(\alpha)}$ and its partial derivatives $\partial_y^\gamma m^{(\alpha)}, |\gamma| \leqslant d_0$, with respect to y are continuous in (x, y) uniformly in $t \in [0,1]$.

ASSUMPTION 2. *For each $\alpha \in (0,2]$, $R > 0$, $\varepsilon > 0$, there exist measurable functions $\nu_i^{(\alpha)}(t, x, dy)$, $i = 1,2$, with values in $\mathcal{R}^+(\mathbf{R}_0^d)$ such that for all $(t, x) \in C_R$*

$$\int |y|^\alpha \nu_1^{(\alpha)}(t, x, dy) \leqslant \varepsilon, \quad \sup_{C_R} \nu_2^{(\alpha)}(t, x, \mathbf{R}_0^d) < \infty,$$

$$|\nu^{(\alpha)}|(t,x,dy) \leqslant \left(\nu_1^{(\alpha)} + \nu_2^{(\alpha)}\right)(t,x,dy).$$

ASSUMPTION 3. *Let one of the two following conditions* a) *and* b) *be satisfied:*

a) *The function a is continuous in x uniformly with respect to $t \in [0,1]$. Moreover, for each $R > 0$*

$$\inf_{C_R} \inf_{|\xi|=1} (a\xi, \xi) > 0 \quad \text{and} \quad \pi \equiv \nu^{(2)};$$

b) *$a \equiv 0$ and there exists $\alpha \in (0,2)$ such that for each $(t,x) \in H$, $R > 0$*

$$\pi(t,x,dy) = m^{(\alpha)}(t,x,y)\frac{dy}{|y|^{d+\alpha}} + \nu^{(\alpha)}(t,x,dy),$$

$$\inf_{C_R} \sup_{|y|=1} m^{(\alpha)}(t,x,y) > 0.$$

Moreover $b \equiv \int_{|y|\leqslant 1} y\pi(\cdot,dy)$ in the case $\alpha \in (0,1)$, $b \equiv \tilde{b} + \int_{|y|\leqslant 1} y\nu^{(1)}(\cdot,dy)$ in the case $\alpha = 1$, and $\tilde{b}$ is continuous in x uniformly with respect to $t \in [0,1]$.

ASSUMPTION 4.

$$\lim_{R\to\infty} \int_0^1 \sup_{x\in\mathbf{R}^d} \pi\left(t,x,\{y : |y| > R\}\right) dt = 0.$$

THEOREM 1. *Let Assumptions 1–3 (resp., 1–4) be satisfied. Then for each $(s,x) \in H$ there exists a unique solution $\mathbf{P}_{s,x}$ to the (s,x,L)–martingale problem, the function $(s,x) \to \mathbf{P}_{s,x}$ is measurable (resp., continuous) and $(X_t, (\mathcal{D}_{t+}), \mathbf{P}_{s,x})$ is strongly Markovian.*

The assumption 2 is satisfied if, for example,

$$\lim_{\delta\downarrow 0} \sup_{(t,x)\in H} \int_{|y|\leqslant\delta} |y|^\alpha \left|\nu^{(\alpha)}\right|(t,x,dy) = 0.$$

In (Komatsu, 1984) the case was considered when the functions a, b, π are independent of t,

$$\left|\nu^{(\alpha)}\right|(x,dy) \leqslant N^{(\alpha)}(dy)$$

on $\mathbf{R}_0^d$ for any $x \in \mathbf{R}^d, \alpha \in (0,2]$ and

$$\int |y|^\alpha \wedge 1 N^{(\alpha)}(dy) < \infty.$$

The uniqueness of the solution to the martingale problem was proved under the following additional assumption: $m^{(\alpha)}$ is strictly positive for any $\alpha \in (0,2)$, $\nu^{(\alpha)}(x,dy) = n^{(\alpha)}(x,y)dy$ in the case $\alpha \in (0,1)$ and for any $\mu \geqslant 1$ the measure

$$N_\mu^{(\alpha)}(dy) = \left(\sup_x \sup_{\mu^{-1} \leqslant \theta \leqslant \mu} \left| n^{(\alpha)}(x, \theta y) \right| \theta^d \right) dy$$

satisfies

$$\int |y|^\alpha \wedge 1 N_\mu^{(\alpha)}(dy) < \infty.$$

The methods of the proof of Theorem 1 are somewhat different from the methods of (Stroock, 1975) and (Komatsu, 1984). Namely, first, we consider the Cauchy problem for the operator L in Sobolev and Hölder spaces. Next, a priori estimates for the solution of the Cauchy problem together with the properties of the Green measures for the corresponding solution of the martingale problem allow us to derive L^p–estimates for the solutions of the martingale problem. Using these L^p–estimates, we obtain the statement of the theorem in a standard way.
The second group of assumptions deals with a degenerate operator L. In order to consider various partial cases, let us introduce some additional notation.
Let $(V, \mathcal{V})$ be a measurable space, $V = V_1 \cup V_2$, $V_1 \cap V_2 = \varnothing$, $V_1, V_2 \in \mathcal{V}, c : H \times V \to \mathbf{R}^d$ be a measurable function, $\Pi(t,x,dv)$ be a measurable function on H with values in $\mathcal{R}^+(V)$, $\varphi : \mathbf{R}^d \to \mathbf{R}^d$ be a twice differentiable function with bounded partial derivatives of the first and the second order, $\varphi(0) = 0$, $\varphi_{ix_j}(0) = \delta_{ij}$, where δ_{ij} is the Kronecker's symbol.
We shall assume that there exists a constant K such that for each $(t,x) \in H$

$$\int_{V_1} \left(|c(t,x,v)|^2 \wedge 1 + |\varphi(c(t,x,v))|^2 \right) \Pi(t,x,dv) +$$

$$+ \int_{V_2} |c(t,x,v)| \wedge 1 \Pi(t,x,dv) \leqslant K$$

and

$$\lim_{\delta \downarrow 0} \sup_{0 \leqslant t \leqslant 1} \int_{V_1 \cap \{|c(t,x,v)| \leqslant \delta\}} |c(t,x,v)|^2 \Pi(t,x,dv) = 0.$$

For $u \in C_0^\infty(\mathbf{R}^d)$, $(t,z) \in H$ let us introduce the operator

$$\hat{L}_{t,z} u(x) = \sum_{i,j=1}^d a_{ij}(t,z) u_{x_i x_j}(x) + \sum_{i=1}^d b_i(t,z) u_{x_i}(x) +$$

$$+ \int \left[u(x + c(t,z,v)) - u(x) - \sum_{i=1}^d u_{x_i}(x) \varphi_i(c(t,z,v)) \mathbf{1}_{V_1}(v) \right] \Pi(t,z,dv),$$

and for brevity denote

$$\hat{L}u = \hat{L}_t u(x) = \hat{L}_{t,x} u(x).$$

It is easy to see, that the operator $\hat{L}$ can be written as the (standard) operator L with b, π replaced by $\hat{b}, \hat{\pi}$, where

$$\hat{b} = b - \int \Big\{ \Big[\varphi(c(\cdot, v)) - c(\cdot, v) \mathbf{1}_{\{|c(\cdot,v)| \leqslant 1\}} \Big] \mathbf{1}_{V_1}(v) +$$

$$+ c(\cdot, v) \mathbf{1}_{\{|c(\cdot,v)| \leqslant 1\}} \mathbf{1}_{V_2}(v) \Big\} \Pi(\cdot, dv),$$

$$\hat{\pi}(\cdot, dy) = \Pi\left(\cdot, \{v \in V : c(\cdot, v) \in dy\}\right).$$

ASSUMPTION 5. *There exists a $(d \times m)$–matrix valued function $\sigma(t, x)$ on H such that $a = \sigma\sigma^*$ on H, and a constant K such that for each $t \in [0,1]$, $x, y \in \mathbf{R}^d$*

$$\|\sigma(t,x) - \sigma(t,y)\| + |b(t,x) - b(t,y)| + \int_{V_1} \Big[|c(t,x,v)|^2 \wedge 1 +$$

$$+ |\varphi(c(t,x,v))|^2 \Big] \, \big| \Pi(t,x,dv) - \Pi(t,y,dv) \big| +$$

$$+ \Big\{ \int_{V_1} \Big[|\varphi(c(t,x,v)) - \varphi(c(t,y,v))|^2 + |c(t,x,v) -$$

$$- c(t,y,v)|^2 \wedge 1 \Big] \Pi(t,x,dv) \Big\}^{1/2} + \int_{V_2} |c(t,x,v)| \wedge 1 \, \big| \Pi(t,x,dv) -$$

$$- \Pi(t,y,dv) \big| + \int_{V_2} \big| c(t,x,v) - c(t,y,v) \big| \wedge 1 \Pi(t,x,dv) \leqslant K|x-y|.$$

It is well–known that if a $(d \times d)$–matrix valued function a on H is twice differentiable in x and its second order derivatives are bounded, then $a = \sigma\sigma^*$, where σ is a $d \times d$–matrix valued function on H, satisfying Lipshitz condition in x uniformly with respect to $t \in [0,1]$.

Let $\varepsilon_0 \in (0,1]$, $(Z, \mathcal{Z})$ be a measurable space, α, β, γ be measurable functions on $H \times Z$ with values in $(0,2), (0,2), (0,1)$, respectively, and $\alpha \leqslant \beta \leqslant \alpha/2 + 1$ on $H \times Z$.

ASSUMPTION $A(\varepsilon_0, \alpha, \beta, \gamma)$. *We shall say that a measurable function $\mu(t,x,dy)$ on H with values in $\mathcal{R}^+(\mathbf{R}_0^d)$ satisfies the assumption $A(\varepsilon_0, \alpha, \beta, \gamma)$ if the following conditions are satisfied:*

a) *$\mu(t, x, \{|y| > \varepsilon_0\}) = 0$ for any $(t,x) \in H$,*

b) *μ has the following polar decomposition*

$$\mu(t,x,dy) = \int_Z R(t,x,z,dr) \mathbf{1}_{d\omega}(p(t,x,z)) S(t,x,dz),$$

where $y = r\omega$, $r = |y|$, $\omega = y/|y| \in S^{d-1}$, $(t,x) \in H$ and the functions R, S, p with values in $\mathcal{R}^+((0,\infty))$, $\mathcal{R}^+(Z)$, S^{d-1}, respectively, are measurable,

c) there exist constants $K_1 > 0$, $K_2 > 0$, a bounded measurable function $n : H \times Z \times (0,\infty) \to [0,\infty)$ and measurable function $\tilde{\pi}(t,x,dy)$ on H with values in $\mathcal{R}^+(\mathbf{R}_0^d)$ such that

$$R(t,x,z,[r,\varepsilon_0]) \leqslant K_1 \left[n(t,x,z,r)r^{-\beta(t,x,z)} + r^{-\gamma(t,x,z)}\right],$$

$$(\tilde{\pi} + \hat{\pi})(t,x,dy) \geqslant$$

$$\geqslant K_2 \int_Z n^2(t,x,z,r)\mathbf{1}_{d\omega}(p(t,x,z))r^{-1-\alpha(t,x,z)}dr\mathbf{1}_{(0,\varepsilon_0)}(r)S(t,x,dz),$$

$$\tilde{\pi}(t,x,R_0^d) + S(t,x,Z) + \int \Big[(1+\alpha(t,x,z)/2 - \beta(t,x,z))^{-1} +$$

$$+(1-\gamma(t,x,z))^{-1} + (2-\beta(t,x,z))^{-1}\Big] S(t,x,dz) \leqslant K_1.$$

ASSUMPTION 6. *There exist constants $\varepsilon_0 \in (0,1), K > 0$, a bounded measurable function $h : H \times (0,\infty) \to [0,\infty)$ and measurable function $\mu(t,x,dy)$ on H with values in $\mathcal{R}^+(\mathbf{R}_0^d)$ satisfying Assumption $A(\varepsilon_0,\alpha,\beta,\gamma)$ such that for each $(t,x) \in H$, $u \in C_0^\infty(\mathbf{R}^d)$*

$$\int\limits_{V_1 \cap \{|c(t,x,v)|<\varepsilon_0\}} \left|u(c(t,x,v)) - u(0) - \sum_{i=1}^d u_{x_i}(0)c_i(t,x,v)\right| |\Pi(t,x,dv) -$$

$$-\Pi(t,x+y,dv)| \leqslant K|y|\Bigg[\sup_{\mathbf{R}^d}(|\mathrm{grad}\, u| + |u|) + h(t,x,|y|)\sup_{\mathbf{R}^d}|u_{xx}| +$$

$$+ \int\limits_{|\xi|\leqslant\varepsilon_0} \left|u(\xi) - u(0) - \sum_{i=1}^d u_{x_i}(0)\xi_i\right| \mu(t,x,d\xi)\Bigg]$$

and $h(t,x,r) \to 0$ as $r \to 0$.

THEOREM 2. *Let Assumptions 5 and 6 (resp., 5, 6 and 4 with π replaced by $\hat{\pi}$) be satisfied. Then for each $(s,x) \in H$ there exists a unique solution $\mathbf{P}_{s,x}$ to the $(s,x,\hat{L})$-martingale problem, the function $(s,x) \to \mathbf{P}_{s,x}$ is measurable (resp., continuous) and $(X_t, (\mathcal{D}_{t+}), \mathbf{P}_{s,x})$ is strongly Markovian.*

The proof of this theorem is based on a certain extension of Oleinik's method, given in (Oleinik and Radkevič, 1973) (see also (Cancelier, 1986; Negoro and Tsuchiya, 1989)).
If the function $\Pi(t,x,dv)$ does not depend on x, then the trajectorial unicity holds and Theorem 2 can be proved using probabilistic methods (see (Jacod, 1979)).

We now give an example of conditions under which the Levy's measure of order $\beta \geqslant 1$ satisfies Assumptions 5 and 6.
Let $(Z, \mathcal{Z})$ be a measurable space, $Z = Z_1 \cup Z_2$, $Z_1 \cap Z_2 = \varnothing$, $Z_i \in \mathcal{Z}$, $i = 1, 2$, S be a finite non-negative measure on $(Z, \mathcal{Z})$,$\bar{\alpha} : Z_1 \to (0, 2)$, $\beta : Z \to [1, 2)$, $\bar{\gamma} : Z_2 \to (0, 1)$ be measurable functions such that $\bar{\alpha} \leqslant \beta < 1 + \bar{\alpha}/2$ on Z_1, $1 + \bar{\gamma} > \bar{\alpha}$ on Z_2 and

$$\int_Z (2 - \beta)^{-1} S(dz) + \int_{Z_1} (1 + \bar{\alpha}/2 - \beta)^{-1} S(dz) +$$

$$+ \int_{Z_2} (1 + \bar{\gamma} - \beta)^{-1} S(dz) < \infty.$$

Suppose that the function $\hat{\pi}$ has the following polar decomposition

$$\hat{\pi}(t, x, dy) = \int_Z \mathbf{1}_{d\omega}(p(t, x, z)) n(t, x, z, r) \frac{dr}{r^{1+\bar{\alpha}(z)}} S(dz),$$

where $y = r\omega$, $r = |y|$, $\omega = y/|y| \in S^{d-1}$, $p : H \times Z \to S^{d-1}$ and $n : H \times Z \times (0, \infty) \to (0, \infty)$ are bounded measurable functions, satisfying the following assumptions:

a) there exists a constant $N > 0$ such that $n(t, x, z, r) \geqslant N r^{\beta(z) - \bar{\alpha}(z)}$ on $H \times Z_1 \times (0, 1)$,

b) the function n is twice differentiable in x, the functions $n_{x_i} = 1, \dots, d$ are bounded on $H \times Z_2 \times (0, \infty)$ and there exists a constant N such that for any $t \in [0, 1]$, $x, x' \in \mathbf{R}^d$, $z \in Z_2$, $r, r' \in (0, \infty)$, $i = 1, \dots, d$

$$|n_{x_i}(t, x, z, r) - n_{x_i}(t, x, z, r')| \leqslant N |r - r'|^{\bar{\gamma}(z)},$$

$$|n_{x_i}(t, x, z, r) - n_{x_i}(t, x', z, r)| \leqslant N |x - x'|,$$

c) there exists a constant N such that for any $t \in [0, 1]$, $x, x' \in \mathbf{R}^d$, $z \in Z_1$, $r \in (0, \infty)$

$$|n(t, x, z, r) - n(t, x', z, r)| \leqslant N |x - x'|,$$

$$\int_Z |p(t, x, z) - p(t, x', z)|^2 S(dz) \leqslant N |x - x'|^2.$$

Using Oleinik's inequality (see (Oleinik and Radkevič, 1973, Lemma I.7.1)) it is easy to verify that if the Lévy's measure satisfies the conditions a)—c) then it satisfies the Assumptions 5 and 6.
Negoro and Tsuchiya in (Negoro and Tsuchiya, 1989) have considered the case: $Z_2 = \varnothing$, $Z_1 = S^{d-1}, p(t, x, z) = z$ or $Z_1 = \{1, \dots, m\}$ the conditions a)—c) are satisfied, the functions n, a, b are twice continuously differentiable in x and the function n is strictly positive.
The "mixed" case can be considered using Theorems 1 and 2 and the following statement.

THEOREM 3. *Suppose that for every $x \in \mathbf{R}^d$ one can find a neighbourhood $O = O(x)$ of the point x, an operator L^O, a measurable family of probability measures $(\mathbf{P}^O_{t,z})$ on $(D, \mathcal{D})$ such that $L^O_{t,z} = L_{t,z}$ for $(t,z) \in [0,1] \times O$ and $\mathbf{P}^O_{t,z}$ is a solution to the (t, z, L^O)-martingale problem for $(t,z) \in H$. Then the following statements hold:*

a) *if for each $(t,z) \in H$ and any neighbourhood $O(x)$, $x \in \mathbf{R}^d$, the solution to the (t, z, L^O)-martingale problem is unique, then for each $(t,z) \in H$ the solution to the (t, z, L)-martingale problem is unique,*

b) *if Assumption 4 is satisfied, then for each $(s,x) \in H$ the (s, x, L)-martingale problem has a solution $\mathbf{P}_{s,x}$ such that the function $(s,x) \to \mathbf{P}_{s,x}$ is measurable.*

A complete exposition with full proofs will appear elsewhere.

REFERENCES

Cancelier, C. (1986). Problèmes aux limites pseudo-différentiels donnant lieu au principe du maximum. *Commun. P.D.E.* **11(15)**, 1677–1726.

Jacod, J. (1979). Calcul stochastique et problèmes de martingales. *Lect. Notes in Math.* **714.**

Komatsu, T. (1984). Pseudo-differential operators and Markov processes. *J. Math. Soc. Japan* **36**, 387–418.

Negoro, A. and Tsuchiya, M. (1989). Stochastic processes and semigroups associated with degenerate Lévy generating operators. *Stochastics and Stochastic Rep.* **26**, 29–61.

Oleinik, O.A. and Radkevič, E.V. (1973). *Second order equations with non-negative characteristic form.* Amer. Math. Soc., Providence.

Stroock, D.W. (1975). Diffusion processes associated with Lévy generators. *Z. Wahrsch. verw. Geb.* **32**, 209–244.

Prob. Theory and Math. Stat., Vol. 2, pp. 176–180
B.Grigelionis *et al.* (Eds.)
1990 VSP/Mokslas

PREFERRED DIRECTIONS ON THE HYPERSPHERE

P. MILASEVIC

Institut de Mathématiques Appliquées SSP, BFSH–2, Université de Lausanne, CH–1015 Lausanne, Switzerland

ABSTRACT

Several problems involving directional data analysis are concerned with the identification of a "modal" or "preferred" direction for the distribution generating the data (see (Watson, 1983)). In this context, a widely used parametric model is the distribution on the unit hypersphere with density (with respect to the invariant measure) proportional to $\exp(k\ \theta' x)$ for some concentration coefficient $k > 0$ and some unit vector θ. This distribution was introduced in dimension three by P. Langevin in 1905 in a physical context and was for the first time investigated from a statistical point of view by K. J. Arnold in 1941. For further details see (Watson, 1983). The present note contains some remarks on the notion of preferred direction and on statistical estimation of that direction.

1. INTRODUCTION

Whatever the precise definition of a "preferred" direction may be, it must be θ in the Langevin model: the density is rotationally symmetric about θ, at which it attains its maximal value, and strictly decreases when one moves away from θ along great circles. As a consequence, a cap of any given size has maximal mass when and only when centered at θ. It would seem that the latter property better correspond to the concept of a "preferred" direction, which should somehow take into account the mass carried by the neighborhood of that direction and not only the fact of maximizing the density. The notion of preferred direction is the object of Section 3 below. If rotational symmetry is suspected it is naturally of interest to estimate its center, which provides a measure of location of the distribution, even if it is not a "preferred" direction. We shall present below some estimators of the center of symmetry and/or preferred direction and also some measures of concentration of the distribution about that direction. Although most of applications arise in dimension two or three we shall adopt, as in (Watson, 1983), a general viewpoint, which provides a better understanding of the various approaches.

Section 2 contains some remarks on more or less standard material and Section 3 is partly based on recent results due to Deborah Nolan and the author (Milasevic and Nolan, 1989).

2. THE MEAN AND THE MEDIAN

Consider the Euclidean unit (hyper)sphere $S = S^{p-1} = \{x \in \mathbf{R}^p;\ x'x = \|x\|^2 = 1\}$. Langevin's distribution is a member of the family of rotationally symmetric absolutely continuous distributions on S, namely absolutely continuous distributions which are invariant with respect to rotations with some given axis $\theta \in S$. Every member of this class has a density of the form $f(\theta' x)$ with respect to the invariant measure on S, for some f defined on $[-1,1]$ and some unit vector θ. Here θ' denotes the transpose of θ and $\theta' x$ is the standard euclidean inner product.

2.1. *The normalized mean*

Consider a distribution P on S with density of the form $f(\theta' x)$ and assume that its mean value $\mathbf{E}(P) = \int_S x P(dx)$ is not zero. It then follows that $\mathbf{E}(P) = \lambda\theta$ for some $\lambda \in (-1,1)$ which we may assume without loss of generality to be positive. Given independent random elements $X_1, \ldots, X_n$ distributed according to P, a natural estimator of θ is then given by the normalized empirical mean

$$T_{1,n} = \|\mathbf{E}(P_n)\|^{-1}\mathbf{E}(P_n),$$

where P_n is the empirical measure $\frac{1}{n}\sum \delta_{X_i}$. It turns out that, under very mild regularity assumptions on f, $T_{1,n}$ is the maximum likelihood estimator of θ if and only if P is a Langevin distribution (see (Watson, 1983)). In that sense Langevin's distribution is analogous to the normal distribution on the line.

2.2. *The spatial median*

The normalized mean is based on $\mathbf{E}(P)$, which is the unique minimizing vector of $\int_S \|x - \mu\|^2 P(dx)$. One would intuitively think that an estimator based on the minimizing vector of $\phi(\mu) = \int_S \|x - \mu\| P(dx)$ would be better protected against outliers, since less weight is given to "distant" points. This indeed turns out to be true in some situations (see (Ducharme and Milasevic, 1987)) but we will not deal further with that issue here.

Now, given a probability measure P on S, there is exactly one vector minimizing ϕ provided P is not concentrated on two antipodal points (see (Milasevic and Ducharme, 1987)). This vector $\mu(P)$ is called the spatial median of P.

For P as in 2.1, we have $\mu(P) = \lambda'\theta$ for some $\lambda' \in (-1,1), \lambda' \neq 0$, which we may again assume to be positive. This suggests to estimate θ by the normalized empirical spatial median

$$T_{2,n} = \|\mu(P_n)\|^{-1}\mu(P_n).$$

Note that because of the uniqueness property of the spatial median and the absolute continuity of P, $\mu(P_n)$ is almost surely unique for $n \geqslant 3$.

2.3. *Asymptotic behaviour*

Both estimators above converge almost surely to θ. Interestingly enough, it turns out that they have a limiting distribution of the same form. More precisely, we have, as $n \to \infty$,

$$\sqrt{n}(T_{j,n} - \theta) \to \mathcal{N}(0, \sigma_j^2(I_p - \theta'\theta))$$

for $j = 1, 2$, where I_p denotes the $p \times p$ identity matrix and σ_1, σ_2 are expressions depending on P which can be explicitly given and consistently estimated from the data (see (Watson, 1983) for the mean and (Ducharme and Milasevic, 1987) for the median).

Since $(I_p - \theta'\theta)$ is the projection on the orthogonal complement of θ, we see that for both estimators, the component orthogonal to θ determines the $\sqrt{n}$ rate of convergence and the limiting distribution. The component parallel to θ has a n rate of convergence and a χ^2 limiting distribution.

2.4. *Measures of concentration*

Both approaches provide us with a measure of concentration of P about θ, respectively given by $\|\mathbf{E}(P)\|$ and $\|\mu(P)\|$, which can be estimated respectively by $\|\mathbf{E}(P_n)\|$ and $\|\mu(P_n)\|$. Asymptotic normality of these estimates readily follows from the result stated in 2.3.

2.5. *Remark.* There are situations where rotationally symmetric distributions with zero expectation are of interest (see (Watson, 1983)). An example of such a distribution is the Scheidegger—Watson distribution, whose density is proportional to $\exp[k(\theta'x)^2]$ for some real k and some $\theta \in S$. Here we are interested in the axis supporting θ and the normalized mean and median cannot be used. An approach for estimating this "preferred" axis, based on the inertia moment of the sample, is proposed and investigated in (Watson, 1983).

3. PREFERRED DIRECTIONS AND MINIMAL CAPS

Given a unit vector $u \in S$ and $t \in [-1, 1]$ we shall denote by $C(u,t)$ the spherical cap on S with center u and half-angle $\arccos(t)$, i.e., $C(u,t) = \{x \in S;\ u'x \geqslant t\}$. Consider a continuous density on S which has exactly one local maximum θ. It then follows that it strictly decreases when one moves away from θ along great circles. If furthermore the density is rotationally symmetric (necessarily about θ), the center of symmetry can unequivocally be interpreted as the preferred direction. Moreover, for any fixed "size" t, the cap $C(\theta, t)$ is the (only) one containing maximal mass, and, at first sight, it would seem that this property could be reasonably used as a definition for a preferred direction. Consider however a continuous rotationally symmetric density with unique local maximum θ and support $C(\theta, t_0)$ for some $t_0 > -1$. Although θ is intuitively the preferred direction of such a distribution the preceding property clearly does not hold and thus appears to be too stringent.

3.1. *α-preferred directions*

We shall adopt a different approach in which it is not the size of the cap which is held fixed but the mass put by the distribution on that cap. More precisely, given $\alpha \in (0, 1)$, we shall consider the smallest cap on S with mass at least α. Such a

cap is called a *minimal α-cap* of the distribution and we shall call its center an *α-preferred direction* of the distribution.
Generally, a distribution P has more than one α-preferred direction. There is a unique such direction u_0 if and only if P satisfies the following property:

$$\text{For every } u \neq u_0 \text{ and every } t \text{ such that } P(C(u,t)) \geqslant \alpha,$$
$$\text{there exists } t' > t \text{ with } P(C(u_0,t')) \geqslant \alpha.$$

Even in case of nonuniqueness, the class of α-preferred directions can provide some information on the distribution. For example, for the density proportional to $\exp[k(\theta' x)^2]$ - the Scheidegger—Watson distribution, there are two α-preferred directions if $k > 0$, yielding the axis supporting θ, while if $k < 0$, the set of α-preferred directions is the $(p-2)$-dimensional unit sphere orthogonal to θ, the "girdle" along which the distribution stretches most of its mass. For the uniform distribution the class of α-preferred directions is the entire sphere S.

3.2. *A measure of concentration*

This approach also provides us with a natural measure of concentration of the distribution about an α-preferred direction. Indeed, if $C(u_0,t_0)$ is a minimal α-cap of P then t_0 is given by

$$t_0 = \sup\{t;\ \sup_{u \in S} P(C(u,t)) \geqslant \alpha\}.$$

Note that it is the $(1-\alpha)$-quantile of the variable $u_0'X$, where X has distribution P.
For example, if P is the Langevin distribution with density proportional to $\exp(k\theta' x)$, t_0 is not surprisingly a strictly increasing function of the concentration coefficient k appearing in the density. In the case $p = 3$ we have the relation $t_0 = k^{-1} \log[e^k - 2\alpha \sin h(k)]$.

3.3. *Preferred directions*

When α is close to one, the "separation power" of the caps is low and it is likely that the minimal α-cap be not unique, apart from very clear-cut cases like Langevin's distribution. On the other hand, one might expect from the "preferred" direction of a distribution to be the only α-preferred direction for all sufficiently small values of α. This suggests defining a unit vector θ as *the preferred direction* of the distribution P if θ is the unique α-preferred direction of P for each $\alpha \in (0,\alpha^*]$, $\alpha^* \in (0,1)$. Note that this definition implies uniqueness of the preferred direction when it exists.

3.4. *Estimation*

Consider independent random elements $X_1,\ldots,X_n$ from an absolutely continuous distribution P and denote by $C_n = C(u_n,t_n)$ an α-minimal cap of the empirical measure P_n, that is a cap with minimal opening which contains at least $n\alpha$ observations. An algorithm for the computation of such an empirical cap is given in (Milasevic and Nolan, 1989).

The coefficient t_n can be used as an estimator of the concentration coefficient t_0, and, if P has a unique α-preferred direction θ, it can be estimated by the empirical center u_n. Using the fact that the family of all spherical caps on S is a Vapnik—Červonenkis class of sets, it can be shown that t_n and u_n are strongly consistent estimators of the corresponding theoretical parameters (see (Milasevic and Nolan, 1989)).

Furthermore, under some mild regularity assumptions, the concentration t_n turns out to be asymptotically normally distributed (see (Milasevic and Nolan, 1989)). The limiting distribution of the empirical center has been investigated in (Milasevic and Nolan, 1989) under the additional assumption of rotational symmetry, which greatly simplifies the computations. Its behaviour is very different than the one encountered in 2.3. The component orthogonal to θ still determines the rate of convergence, but this rate is no longer $\sqrt{n}$ and (consequently) the limiting distribution is no longer normal. More precisely it is shown in (Milasevic and Nolan, 1989) that, under some regularity assumptions and rotational symmetry of P about its unique α-preferred direction θ, as $n \to \infty$,

$$n^{1/3}(I_p - \theta\theta')u_n \to x_{\max}$$

in distribution, where $x_{\max}$ denotes the almost surely unique minimizing vector of a certain gaussian random process with continuous sample paths. For details see (Milasevic and Nolan, 1989).

REFERENCES

Ducharme, G.R. and Milasevic, P. (1987). Spatial median and directional data. *Biometrika* **74**, 212—215.

Milasevic, P. and Ducharme, G.R. (1987). Uniqueness of the spatial median. *Ann. Statist.* **15**, 1332—1333.

Milasevic, P. and Nolan, D. (1989). Estimation on the sphere: a geometric approach. *Tech. Report.* No. 206, Dept. of Statistics, Univ. of California, Berkeley.

Watson, G.S. (1983). *Statistics on Spheres.* Wiley, New York.

Prob. Theory and Math. Stat., Vol. 2, pp. 181–192
B.Grigelionis *et al.* (Eds.)
1990 VSP/Mokslas

TWO–PARAMETER STRONG MARTINGALES: INEQUALITIES FOR QUADRATIC VARIATION AND SOME DECOMPOSITIONS

Yu.S.MISHURA and A.A.GUSHCHIN
Kiev University, 252601 Kiev, Ukraine, USSR
Steklov Mathematical Institute, Vavilova 42, 117966 Moscow GSP–1, USSR

ABSTRACT

The paper contains generalizations of Burkholder—Davis—Gundy inequalities for strong two parameter martingales with discrete and continuous parameter and decomposi-tions of strong martingales on a locally square–integrable part and a part of bounded variation.

Let $U = \mathbf{N}^2$ or $\mathbf{R}^2_+$, $t = (t_1, t_2) \in U$, $s \leqslant t$ if $s_1 \leqslant t_1$, $s_2 \leqslant t_2$; $s < t$ if $s_1 < t_1$, $s_2 < t_2$. Let $(\Omega, \mathcal{F}, \mathcal{F}_t, \mathbf{P}, t \in \mathbf{R}^2_+)$ be a stochastic basis,

$$\gamma^1_{t_1} = \bigvee_{t_2 \geqslant 0} \mathcal{F}_t, \quad \gamma^2_{t_2} = \bigvee_{t_1 \geqslant 0} \mathcal{F}_t, \quad \gamma_t = \gamma^1_{t_1} \vee \gamma^2_{t_2}.$$

Assume that $(\mathcal{F}_t, t \in \mathbf{R}^2_+)$ satisfies the conditions (F1)—(F4) of (Cairoli and Walsh, 1975).
Let

$$Q^{++}_t = \{s : s \geqslant t\}, \quad Q^{+-}_t = \{s : s_1 \geqslant t_1, s_2 < t_2\},$$

$$Q^{-+}_t = \{s : s_1 < t_1, s_2 \geqslant t_2\}, \quad Q^{--}_t = \{s : s < t\},$$

$(x_t, \mathcal{F}_t, t \in U)$ be an integrable $\mathcal{F}_t$–adapted stochastic process, $x_t = x_0$ on the axes. If $U = \mathbf{R}^2_+$ we suppose that trajectories of x_t are a.s. in the space $D = D(\mathbf{R}^2_+)$, i.e., for all $t, t_i \neq 0$, there exist four limits

$$x^{\pm\pm}_t = \lim_{s \to t \; s \in Q^{\pm\pm}_t} x_s, \quad x_t = x^{++}_t \quad (x \text{ is } D\text{–process}).$$

Let $x^*_t = \sup_{s \leqslant t} |x_s|$, $x^* = \sup_{s \in U} |x_s|$.

Definition 1 (Cairoli and Walsh, 1975). The process x is a *two–parameter martingale* if $\mathbf{E}(x_t / \mathcal{F}_s) = x_s, s \leqslant t$, and a *strong martingale* if $\mathbf{E}(x]s,t] / \gamma_s) = 0, s \leqslant \leqslant t$, $(x]s,t] = x_t - x_{t_1 s_2} - x_{s_1 t_2} + x_s)$.

Denote $\mathcal{M}_s^*$ the class of strong martingales with integrable supremum, $\mathcal{M}_s^p$, $p \geqslant 1$, the class of strong martingales in $\mathcal{L}_p$ and $\mathcal{B}$ the class of processes of bounded variation on each interval $[0,T], T \in \mathbf{R}_+^2$.

1. INEQUALITIES FOR QUADRATIC VARIATION

Let $U = \mathbf{N}^2$. We define the quadratic variation of x as the sum

$$[x]_t = \sum_{s<t} \Big(x]s, s+1]\Big)^2, \quad s+1 = (s_1+1,\ s_2+1),$$

$$[x] = \sum_{s \in \mathbf{N}^2} \Big(x]s, s+1]\Big)^2.$$

PROPOSITION 1. *Let $(x_t, \mathcal{F}_t, t \in \mathbf{N}^2)$ be a strong martingale.*

(i) *If $\sup\limits_{t \in \mathbf{N}^2} \mathbf{E}|x_t| < \infty$ then $[x]$ is finite* a.s.

(ii) *If $\Phi(x) : [0,+\infty) \to [0,+\infty)$ is a nondecreasing convex function, $\Phi(0) = 0$, $\Phi(2x) \leqslant C\Phi(x)$ for some $C > 0$ then there exist $C_1 > 0$ and $C_2 > 0$ such that*

$$C_1 \mathbf{E}\Phi\Big([x]_t^{1/2}\Big) \leqslant \mathbf{E}\Phi(x_t^*) \leqslant C_2 \mathbf{E}\Phi\Big([x]_t^{1/2}\Big), \quad t \in \mathbf{N}^2,$$

$$C_1 \mathbf{E}\Phi\Big([x]^{1/2}\Big) \leqslant \mathbf{E}\Phi(x^*) \leqslant C_2 \mathbf{E}\Phi\Big([x]^{1/2}\Big).$$

Proposition 1 for the case $\Phi(x) = x^p, p \geqslant 1$, was proved in (Mishura, 1982). For the general case the proof is essentially the same. Let $U = \mathbf{R}_+^2$, $T \in \mathbf{R}_+^2$ be fixed, $\{\lambda^T = \lambda^T_{n_1 n_2},\ (n_1, n_2) \in \mathbf{N}^2\}$ be an increasing sequence of grids on $[0,T]$ whose norms $|\lambda^T| = |\lambda^T_{n_1 n_2}|$ tend to zero, $\lambda^T = \lambda^{T_1}_{1,n_1} \times \lambda^{T_2}_{2,n_2}$, where

$$\lambda^{T_i}_{i,n_i} = \left\{0 = t^0_{i,n_i} < t^1_{i,n_i} < \cdots < t^{n_i}_{i,n_i} = T_i\right\}, \quad i = 1,2,$$

$$t_{jk} = \Big(t^j_{1,n_1}, \ t^k_{2,n_2}\Big), \quad t_{jk}(t) = t_{jk} \wedge t,$$

$$x_{jk} = x_{t_{jk}}, \quad x_{jk}(t) = x_{t_{jk} \wedge t},$$

$$\Delta^1_{jk} x = x_{j+1k} - x_{jk},$$

$$\square_{jk} x(t) = x]t_{jk} \wedge t, t_{j+1k+1} \wedge t],$$

$$\square_{jk} x = \square_{jk} x(T), \quad \square_t x = x_t^{++} - x_t^{+-} - x_t^{-+} + x_t^{--},$$

$$[x]_T^1 = \mathbf{P} - \lim_{|\lambda^T|\to 0} \sum_{j=0}^{n_1-1} \left(\Delta^1_{jn_2} x\right)^2,$$

$$[x]^t_{n_1 n_2} = \sum_{j,k} \left(\square_{jk} x(t)\right)^2.$$

PROPOSITION 2. *If $x \in \mathcal{M}_s^*$ then the sums*

$$\left\{ \left([x]^t_{n_1 n_2}\right)^{1/2}, \quad (n_1, n_2) \in \mathbf{N}^2 \right\}$$

are uniformly integrable.

Definition 2. The *quadratic variation* $[x]_t, t \leqslant T$, is the process

$$[x]_t = \mathbf{P} - \lim_{|\lambda^T|\to 0} \sum_{j,k} \left(\square_{jk} x(t)\right)^2,$$

if this limit exists and does not depend on $\lambda^T_{n_1 n_2}$.

Definition 3. A D-process is called $\square$*-continuous* if $\square_t x \equiv 0$.

THEOREM 1. *Let $x \in \mathcal{M}_s^*$. Then $[x]_t$ exists, is nondecreasing and has a D-modification. If x_t is continuous ($\square$-continuous, $|\square_t x| \leqslant C$) then $[x]_t$ has a continuous modification (a $\square$-continuous modification, a modification with $\square_t[x] \leqslant C^2$).*

Proof. Let $n_2 > l_2 \geqslant 1$, $\lambda^T = \lambda^T_{n_1 n_2}$, $\bar{\lambda}^T = \lambda^T_{n_1 l_2}$, $\varepsilon > 0$.
For all $C > 0$

$$\mathbf{P}\left\{ |[x]^T_{n_1 n_2} - [x]^T_{n_1 l_2}| > \varepsilon \right\} \leqslant \mathbf{P}\{x_T^* > C\} +$$

$$+\mathbf{P}\left\{ \Big| \sum_{j,k,k'} \square_{jk'} x \square_{jkk'} x^c \Big| > \frac{\varepsilon}{2} \right\},$$

where index k' corresponds to those points $t^{n_2}_{k',2} \in \lambda^{T_2}_{2,n_2}$ for which $t^{n_2}_{k,2} \leqslant t^{n_2}_{k',2} < t^{n_2}_{k+1,2}$,

$$\square_{jkk'} x_s = (x_{j+1k'} - x_{jk'} - x_{j+1k} + x_{jk})(s),$$

$$\square_{jkk'} x = \square_{jkk'} x_T, \ \alpha^c = (\alpha \wedge C) \vee (-C).$$

The process

$$\left\{ \zeta_s = \sum_{j,k,k'} \square_{jk'} x(s) \, \square_{jkk'} x_s^c, \ \mathcal{F}_s, s \in \lambda^T \right\}$$

is a strong martingale with discrete parameter. For $\Phi(x) = x$ from Proposition 1, (ii) we obtain

$$\mathbf{P}\left\{\zeta_T > \frac{\varepsilon}{2}\right\} \leqslant \frac{C_2}{\varepsilon}\mathbf{E}\Big(\sum_{j,k,k'} (\Box_{jk'}x)^2\ (\Box_{jkk'}x^c)^2\Big)^{1/2}. \qquad (1)$$

Let $\alpha > 0, \tau_r, r = \overline{1, R(\omega)}$ be such points that

$$|\Box x_{\tau_r}| > \frac{\alpha}{2}, \ \Box_{jkk'} =]t_{jk},\ t_{j+1\ k'}],$$

$$E_\lambda = \left\{\Box_{jk'} : \Box_{jkk'} \cap \left\{\bigcup_{r=\overline{1,R(\omega)}} \tau_r\right\} \neq \varnothing\right\},$$

$$\Omega_{\lambda T} = \left\{\omega : \Big(\exists \Box_{jk'} \notin E_\lambda : |\Box_{jkk'}x| > \alpha\Big)\right\}.$$

For a D-process x

$$\mathbf{P}(\Omega_\lambda) \to 0, \quad |\lambda^T| \to 0. \qquad (2)$$

Then using Proposition 1, (ii) we obtain

$$\lim_{|\bar\lambda^T|\to 0}\ \sup_{n_2 > l_2} \mathbf{E}\Big(\sum_{j,k,k'} (\Box_{jk'}x)^2(\Box_{jkk'}x^c)^2\Big)^{1/2} \leqslant$$

$$\leqslant \alpha C_1 \mathbf{E}x_T^* + C \lim_{|\bar\lambda^T|\to 0}\ \sup_{n_2 > l_2} \mathbf{E}\Big(\sum_{j,k,k'} (\Box_{jk'}x)^2 I\left\{\Box_{jk'} \in E_\lambda\right\}\Big)^{1/2} +$$

$$+ C \lim_{|\bar\lambda^T|\to 0}\ \sup_{n_2 > l_2} \mathbf{E}\left[\Big(\sum_{j,k,k'} (\Box_{jk'}x)^2\Big)^{1/2} I_{\Omega_\lambda}\right]. \qquad (3)$$

It follows from Proposition 2 and (2) that the last term is zero. The second term is also zero, because

$$\Big(\sum_{j,k,k'} (\Box_{jk'}x)^2 I\left\{\Box_{jk'} \in E_\lambda\right\}\Big)^{1/2} \to 0$$

as $|\bar\lambda^T| \to 0$ for an arbitrary D-process x and they are uniformly integrable (Proposition 2).
The first term is small when α is small. Choosing $C > 0$ sufficiently large we deduce from (1), (2) and (3) that

$$\lim_{|\bar\lambda^T|\to 0}\ \sup_{n_2 > l_2} \mathbf{P}\left\{\left|[x]^T_{n_1 n_2} - [x]^T_{n_1 l_2}\right| > \varepsilon\right\} = 0.$$

The difference $\left|[x]^T_{n_1 n_2} - [x]^T_{l_1 n_2}\right|$ may be considered similarly. Changing T by an arbitrary $t \leqslant T$ we obtain that the sums $[x]^t_{n_1 n_2}$ are $\mathbf{P}$-fundamental, therefore the limit

$$\mathbf{P}-\lim_{|\lambda^T|\to 0}[x]^t_{n_1 n_2}=[x]_t$$

exists.
The properties of trajectories of $[x]_t$ are established by similar estimates.

THEOREM 2. *Let $x \in \mathcal{M}^*_s$. Then for all $p \geqslant 1$ there exist $C_p > 0$ and $D_p > 0$ such that*

$$C_p\mathbf{E}[x]_t^{p/2} \leqslant \mathbf{E}(x_t^*)^p \leqslant D_p\mathbf{E}[x]_t^{p/2}. \qquad (4)$$

Proof. Using Proposition 1, (ii) with $\Phi(x) = x^p, p \geqslant 1$, and Theorem 1 we obtain

$$\mathbf{E}[x]_t^{p/2} \leqslant \varliminf_{|\lambda^T|\to 0} \mathbf{E}\left([x]^t_{n_1 n_2}\right)^{p/2} \leqslant \widetilde{C}_p\mathbf{E}(x_t^*)^p.$$

Thus the left hand side of (4) is true. The right–hand side will be proved only for $p = 1$ and $p = 2$. The estimates for other p are similar but more complicated.

(a) $p = 1$. Since x is a D–process it has a separable modification. Let $Z \subset [0, T]$ be the separante of x, and $\lambda^T_{n_1 n_2}$ be such sequence of grids that

$$\bigcup_{n_1,n_2} \lambda^T_{n_1 n_2} = Z.$$

Then using Proposition 1 (ii) we obtain

$$\mathbf{E}x_T^* \leqslant \lim_{|\lambda^T|\to 0} \mathbf{E}\sup_{z\in\lambda^T}|x_z| \leqslant C \lim_{|\lambda^T|\to 0}\mathbf{E}\Big([x]^T_{n_1 n_2}\Big)^{1/2}.$$

The sequence $\Big([x]^T_{n_1 n_2}\Big)^{1/2}$ is uniformly integrable and

$$[x]^T_{n_1 n_2} \xrightarrow{P} [x]_T$$

as $|\lambda^T| \to 0$, so the right–hand side of (4) is true for $p = 1$.

(b) $p = 2$. Let $A > 0$. Consider the stopping point

$$\nu = \nu_\lambda^A = \inf\left\{ l : \sum_{j=0}^{n_1-1}(\Delta^1_{jl}x)^2 > A\right\} \bigwedge n_2.$$

Since

$$\sum_{j=0}^{n_1-1}(\Delta^1_{jl_2}x)^2 = \sum_{j=0}^{n_1-1}\sum_{k=0}^{l_2-1}(\Box_{jk}x)^2+$$

$$+2\sum_{j=0}^{n_1-1}\sum_{k=0}^{l_2-1}\Box_{jk}x\Delta^1_{jk}x,$$

we have

$$\max_{1\leqslant l_2\leqslant\nu}\Big(\sum_{j=0}^{n_1-1}(\Delta^1_{jl_2}x)^2\Big)\leqslant\Big(\sum_{j,k}(\Box_{jk}x)^2\Big)+$$

$$+2\max_{1\leqslant l_2\leqslant n_2}\Big|\sum_{j=0}^{n_1-1}\sum_{k=0}^{\nu\wedge l_2-1}\Box_{jk}x\Delta^1_{jk}x\Big|, \tag{5}$$

where

$$\Big\{\sum_{j=0}^{n_1-1}\sum_{k=0}^{\nu\wedge l_2-1}\Box_{jk}x\Delta^1_{jk}x,\ 1\leqslant l_2\leqslant n_2\Big\}$$

is a one-parameter martingale with respect to $\{\gamma^2_{l_2}\wedge\nu, 1\leqslant l_2\leqslant n_2\}$. Using the Burkholder inequality for one-parameter martingales and the usual Cauchy—Schwartz inequality we obtain

$$\mathbf{E}\max_{1\leqslant l_2\leqslant n_2}\Big|\sum_{j=0}^{n_1-1}\sum_{k=0}^{\nu\wedge l_2-1}\Box_{jk}x\Delta^1_{jk}x\Big|\leqslant$$

$$\leqslant C\mathbf{E}\Big(\sum_{j,k}(\Box_{jk}x)^2\max_{1\leqslant l_2\leqslant\nu}\Big[\Big(\sum_{j=0}^{n_1-1}(\Delta^1_{jk}x)^2\Big)\bigwedge A\Big]\Big)^{1/2}, \tag{6}$$

and from (5) and (6) we deduce

$$\mathbf{E}x^\lambda\leqslant\mathbf{E}\Big(\Big(\sum_{j,k}(\Box_{jk}x)^2\Big)\bigwedge A\Big)+$$

$$+C\mathbf{E}\Big([x]_Tx^\lambda\Big)^{1/2}+C\mathbf{E}\Big(\Big|\sum_{j,k}(\Box_{jk}x)^2-[x]_T\Big|x^\lambda\Big)^{1/2},$$

where

$$x^\lambda=\Big(\max_{1\leqslant l_2\leqslant n_2}\sum_{j=0}^{n_1-1}(\Delta^1_{jl_2}x)^2\Big)\bigwedge A.$$

By Proposition 2 and Theorem 1

$$\mathbf{E}\Big(\Big|\sum_{j,k}(\Box_{jk}x)^2-[x]_T\Big|x^\lambda\Big)^{1/2}\to 0,\quad |\lambda^T|\to 0$$

and

$$\mathbf{E}\Big(\Big(\sum_{j,k}(\Box_{jk}x)^2\Big)\wedge A\Big)\to\mathbf{E}\Big([x]_T\wedge A\Big).$$

Therefore

$$\overline{\lim_{|\lambda^T|\to 0}} \mathbf{E}x^\lambda \leqslant \mathbf{E}\Big([x]_T \wedge A\Big) +$$

$$+C\left(\mathbf{E}[x]_T \overline{\lim_{|\lambda^T|\to 0}} \mathbf{E}x^\lambda\right)^{1/2} \leqslant C\mathbf{E}[x]_T. \tag{7}$$

From (7) we have

$$\overline{\lim_{|\lambda^T|\to 0}} \mathbf{E}\left[\left(\sum_{j=0}^{n_1-1}(\Delta^1_{jn_2}x)^2\right) \wedge A\right] \leqslant C\mathbf{E}[x]_T.$$

By properties of one-parameter martingales

$$\lim_{|\lambda^T|\to 0} \mathbf{E}\left[\left(\sum_{j=0}^{n_1-1}(\Delta^1_{jn_2}x)^2\right) \wedge A\right] = \mathbf{E}\Big([x]^1_T \wedge A\Big),$$

and

$$\mathbf{E}[x]^1_T \leqslant \lim_{A\to\infty} \mathbf{E}\Big([x]^1_T \wedge A\Big) \leqslant C\mathbf{E}[x]_T. \tag{8}$$

By the Burkholder inequality for one-parameter martingales with continuous parameter there exists $C > 0$ such that

$$\mathbf{E}x^2_T \leqslant C\mathbf{E}[x]^1_T, \tag{9}$$

and by the Doob maximal inequality for two-parameter martingale (Cairoli and Walsh, 1975)

$$\mathbf{E}(x^*_T)^2 \leqslant 16\mathbf{E}(x_T)^2. \tag{10}$$

From (8)—(10)

$$\mathbf{E}(x^*_T)^2 \leqslant C\mathbf{E}[x]_T,$$

where $C > 0$ is a fixed constant which does not depend on t and x_t.

2. DECOMPOSITIONS OF STRONG MARTINGALES

In this section we shall use the notions of a weakly predictable two-parameter process and a weakly predictable $\widehat{\Pi}x$ and a dual weakly predictable $x^{\hat{\pi}}$ projection of x. These notions were introduced and discussed in (Gushchin, 1982). In particular, $\widehat{\Pi}x$ exists if x is non-negative and bounded and satisfies the equality

$$\mathbf{E}\int x_t dA_t = \mathbf{E}\int \widehat{\Pi}x_t dA_t \tag{11}$$

for an arbitrary increasing integrable weakly predictable process $\{A_t,\ t \in \mathbf{R}^2_+\}$. Let $x = x^+ - x^-$. Define $\widehat{\Pi}x$ as $\widehat{\Pi}x^+ - \widehat{\Pi}x^-$. Then (11) is true at least for those A which satisfy

$$\mathbf{E}\int \widehat{\Pi}|x_t|dA_t < \infty.$$

THEOREM 3. *Let $x \in \mathcal{M}_s^*$. Then $\widehat{\Pi}(\Box x) = 0$, i.e.,*

$$\mathbf{P}\left\{\omega : \left\{\exists t \in \mathbf{R}_+^2 : \left(\widehat{\Pi}(\Box x)_t \neq 0\right)\right\}\right\} = 0.$$

Proof. We give only the main ideas of the proof.

(a) The theorem is true if $\mathbf{E}\int \Box x_t dA_t = 0$ for an arbitrary integrable increasing weakly predictable A_t such that $\mathbf{E}\int \widehat{\Pi}|\Box x_t|dA_t < \infty$.

(b) The theorem is true if $\mathbf{E}\int \Box x_t dA_t = 0$ for an arbitrary integrable increasing weakly predictable process A_t such that $\mathbf{E}x_T^* A_T < \infty$.

(c) Let $T \in \mathbf{R}_+^2$ be fixed, $\lambda_{n_1 n_2}^T$ be a sequence of partitions of $[0,T]$, A be as in (b), $\widehat{\mathcal{F}}_{jk} = \mathcal{F}_{jk+1} \vee \mathcal{F}_{j+1k}$, and

$$B_T^\lambda = \sum_{j,k} \mathbf{E}\left(\Box_{jk} A/\widehat{\mathcal{F}}_{jk}\right).$$

Lemma 14 in (Gushchin, 1982) states that B_T^λ weakly converges to A_T as $|\lambda^T| \to 0$. It is easy to prove that

$$\lim_{c\to\infty} \lim_{|\lambda^T|\to 0} \mathbf{E}x_T(B_T^\lambda \wedge c) = \mathbf{E}x_T A_T = \mathbf{E}\int_{[0,T]} x_s dA_s. \tag{12}$$

On the other hand,

$$\mathbf{E}x_T(B_T^\lambda \wedge c) = \mathbf{E}S_1^\lambda - \mathbf{E}S_2^{c,\lambda} + \mathbf{E}S_3^{c,\lambda}, \tag{13}$$

where

$$\mathbf{E}S_1^\lambda = \mathbf{E}\sum_{j,k}(x_{jk+1} + x_{j+1k} - x_{jk})\Box_{jk}A \to$$

$$\to \mathbf{E}\int_{[0,T]} (x_s^{+-} + x_s^{-+} - \dot{x}_s^{--})dA_s, \quad |\lambda^T| \to 0, \tag{14}$$

$$\mathbf{E}S_2^{c,\lambda} = \mathbf{E}\sum_{j,k}\left(x_{jk+1} + x_{j+1k} - x_{jk}\right)\Box_{jk}B^\lambda I\left\{B_T^\lambda > C\right\},$$

$$\mathbf{E}S_3^{c,\lambda} = \mathbf{E}\sum_{j,k}\left(x_{jk+1} + x_{j+1k} - x_{jk}\right)\Box_{jk}\left(B^\lambda \wedge C\right)I\left\{B_T^\lambda > C\right\},$$

and by simple but tedious calculations we can obtain that

$$\lim_{c\to\infty} \lim_{|\lambda^T|\to 0} \mathbf{E}S_i^{c,\lambda} = 0, \quad i = 2,3. \tag{15}$$

Equalities (12)—(15) imply that

$$\mathbf{E}\int_{[0,T]} \Box M_t dA_t = 0$$

for all A as in (b) and the theorem is proved.
Introduce now the following functional of integral type

$$((x,z)\circ y)_t = \mathbf{P} - \lim_{|\lambda^T|\to 0} \sum_{j,k} x_{j+1k}(t) z_{jk+1}(t) \Box_{jk} y(t),$$

if this limit exists and does not depend on $\lambda^T = \lambda^T_{n_1 n_2}$.

Definition 4. A stochastic process $\{x_t,\ t\in \mathbf{R}^2_+\}$ is called a $\tilde{D}$–process if its trajectories a.s. satisfy the condition: for all $t \in \mathbf{R}^2_+$ there exist limits x_t^{++}, $\lim_{s\to t} x_s$, $\lim_{s\to t} x_s$ with $s_1 < t_1$, $s_2 > t_2$.

PROPOSITION 3. *Let $\{x_t, z_t, \mathcal{F}_t\}$ be a bounded $\tilde{D}$–processes.*

(i) *If $\{y_t, \mathcal{F}_t, t\in \mathbf{R}^2_+\}$ belongs to $\mathcal{M}^*_s$ then the integral $\Big((x,z)\circ y\Big)_t$ exists, is of class $\mathcal{M}^1_s$ and*

$$\Big[\Big((x,z)\circ y\Big)\Big]_t = \Big(\Big((x)^2,(z)^2\Big)\circ y\Big)_t.$$

(ii) *If $y_t \in \mathcal{B}$ then $((x,z)\circ y)_t$ exists and belongs to class $\mathcal{B}$.*

Definition 5. A weakly predictable random set $D = D(\omega)$ is called a stopping set if $[0,t]\subset D(\omega)$ for $t\in D(\omega)$.

Definition 6. A two–parameter process x is called a locally–square integrable strong martingale ($x \in l\mathcal{M}^2_s$) if there exists a sequence (x^n, D^n) where $x^n \in \mathcal{M}^2_s$, D^n is a stopping set, $D^n \subset D^{n+1}$, $\bigcup_n D^n = \mathbf{R}^2_+$ and $(x-x^n)_t I_{D^n}(t) = 0$. We say that (x^n, D^n) is a localizating sequence for x. We shall use also the notions of optional Ox and i–optional $O^i x$ projections of x (Bakry, 1981).

THEOREM 4. *Let $x \in \mathcal{M}^*_s$ and assume has bounded jumps: $|\Box_t x| \leqslant C$. Then $x \in l\mathcal{M}^2_s$.*

Proof. By Theorem 1, $[x]_t$ exists and has bounded jumps: $\Box_t[x] \leqslant C^2$. Therefore there exists a sequence E_n of stopping sets such that $E_n \subset E_{n+1}$, $\bigcup_n E_n = \mathbf{R}^2_+ \times \Omega$,

$$\mathbf{E}\int I_{E_n}(t) d[x]_t < \infty$$

and a dual weakly predictable projection $\langle x\rangle_t = [x]^{\star}_t$. If $\tilde{D}$–processes (y_t, z_t) are bounded and non–negative and their O^i–projections are also $\tilde{D}$–processes, then

$$\mathbf{E}\Big((O^2 y, O^1 z)\circ [x]\Big)_t = \mathbf{E}\Big((O^2 y, O^1 z)\circ \langle x\rangle\Big)_t =$$

$$= \mathbf{E}\Big((z, y) \circ \langle x\rangle\Big)_t, \quad t \in \mathbf{R}^2_+. \tag{16}$$

Let $\langle x\rangle_\infty < \infty$ (otherwise consider $\langle x\rangle_{t_m}$, $t_m \to (\infty, \infty)$),

$$\tau_1^n = \inf\{t_1 : \langle x\rangle_{t_1\infty} > n\},$$

$$\tau_2^n = \inf\{t_2 : \langle x\rangle_{\infty t_2} > n\}, \quad \xi_t^{i,n} = I_{[0,\tau_i^n]}(t).$$

There exists $\widehat{\Pi}\xi_t^{i,n}$ and by Theorem 5 (Gushchin, 1982), $\widehat{\Pi}\xi^{i,n} = O\xi^{i,n} = O^j\xi^{i,n}$. By Theorems 4 (Bakry, 1981) and 4.2 (Millet, and Sucheston, 1981) $O^j\xi^{i,n}$ is a $\tilde{D}$-process, $i = 1, 2$, $j = 3 - i$.
It is easy to see that the processes

$$\eta^{i,n} = I_{\left\{\widehat{\Pi}\xi^{i,n} > \frac{1}{2}\right\}} + 2\widehat{\Pi}\xi^{i,n} I_{\left\{\widehat{\Pi}\xi^{i,n} \leqslant \frac{1}{2}\right\}}$$

are also $\tilde{D}$-processes, $i = 1, 2$. So there exists $x^n = \Big((\eta^{1,n},\ \eta^{2,n}) \circ x\Big) \in \mathcal{M}_s^1$ and from (16)

$$\mathbf{E}[x^n]_t = \mathbf{E}\Big((\eta^{1,n},\ \eta^{2,n}) \circ [x]\Big)_t \leqslant$$

$$\leqslant 8\mathbf{E}\Big((\widehat{\Pi}\xi^{1,n}, \widehat{\Pi}\xi^{2,n}) \circ [x]\Big)_t =$$

$$= 8\mathbf{E}\Big((\xi^{1,n},\ \xi^{2,n}) \circ [x]\Big)_t \leqslant 16n + 8C^2.$$

From (4) $x^n \in \mathcal{M}_s^2$.
We have

$$\alpha_t^n = ((\eta^{1,n}, \eta^{2,n}) \circ [x - x^n])_t;$$

$$\alpha_t^n = (((\eta^{1,n})^2, (\eta^{2,n})^2) \circ [x])_t - 2(((\eta^{1,n})^2, (\eta^{2,n})^2) \circ [x, x^n])_t +$$
$$+ (((\eta^{1,n})^2, (\eta^{2,n})^2) \circ [x^n])_t = ((\nu^{1,n}, \nu^{2,n}) \circ [x])_t,$$

where

$$\nu^{i,n} = 2\widehat{\Pi}\xi^{i,n}\Big(2\widehat{\Pi}\xi^{i,n} - 1\Big)^2 I_{\{\widehat{\Pi}\xi^{i,n} \leqslant \frac{1}{2}\}}.$$

Let

$$r_t^n = \Big((\eta^{1,n}, \eta^{2,n}) \circ (x - x^n)\Big)_t; \ r^n \in \mathcal{M}_s^2,$$

$$D^n = \left\{t : \ \widehat{\Pi}\xi_s^{i,n} > \frac{1}{2}, \ \forall s \leqslant t, \ i = 1, 2\right\}.$$

Then D^n is a stopping set, and

$$\mathbf{E}\Big(\int I_{D^n}(t)dr_t^n\Big)^2 = \mathbf{E}\int I_{D^n}(t)d[r^n]_t = 0,$$

since $[r^n]I_{D^n} = \alpha^n I_{D^n} = 0$. Consequently $\int I_{D^n}(t)dr_t^n = 0$ and $r_t^n I_{D^n}(t) = 0$, i.e.,

$$\left((\eta^{1,n}, \eta^{2,n}) \circ (x - x^n)\right)_t I_{D^n}(t) = 0.$$

But

$$\left((\eta^{1,n}, \eta^{2,n}) \circ (x - x^n)\right)_t I_{D^n}(t) = (x - x^n)_t \, I_{D^n}(t),$$

and

$$(x - x^n)_t \, I_{D^n}(t) = 0.$$

Note that $D^n = D_1^n \cap D_2^n$, where

$$D_i^n = \left\{t : \widehat{\Pi}\xi_s^{i,n} > \frac{1}{2}, \ \forall s \leqslant t\right\}.$$

Then

$$\mathbf{P}\{t \notin D_1^n\} = \mathbf{P}\left\{\inf_{s \leqslant t} \widehat{\Pi}\xi_s^{i,n} \leqslant \frac{1}{2}\right\} = \mathbf{P}\left\{\sup_{s\leqslant t} \widehat{\Pi} I_{]\tau_1^n,\infty[}(s) > \frac{1}{2}\right\} =$$

$$= \mathbf{P}\left\{\sup_{s\leqslant t} O^2 I_{]\tau_1^n,\infty[}(s) > \frac{1}{2}\right\} \leqslant 4\mathbf{E} \sup_{s_2 \leqslant t_2} \left[\sup_{s_1 \leqslant t_1} O^2 I_{]\tau_1^n,\infty[}(s)\right]^2.$$

The process

$$\zeta_{s_2} = \sup_{s_1 \leqslant t_1} O^2 I_{]\tau_1^n,\infty[}(s)$$

is a submartingale. By Doob's inequality

$$4\mathbf{E} \sup_{s_2 \leqslant t_2} \zeta_{s_2}^2 \leqslant 16\mathbf{E}\left[\sup_{s_1 \leqslant t_1} \mathbf{E}(I_{]\tau_1^n,\infty[}(s_1)/\gamma_{t_2}^2)\right]^2 \leqslant$$

$$\leqslant 16\mathbf{E} \sup_{s_1 \leqslant t_1} I_{]\tau_1^n,\infty[}(s_1) = 16\mathbf{P}\{\tau_1^n < t_1\} \to 0$$

as $n \to \infty$, and $\mathbf{P}\{t \in D_1^n\} \to 1$ as $n \to \infty$. Similarly $\mathbf{P}\{t \in D_2^n\} \to 1$ as $n \to \infty$. Thus $\mathbf{P}\{t \in D^n\} \to 1$ as $n \to \infty$ and $\bigcup_n D^n = \mathbf{R}_+^2 \times \Omega$. Stopping sets D^n and $x^n \in \mathcal{M}_s^2$ localize x and $x \in l\mathcal{M}_s^2$.

THEOREM 5. *Let $x \in \mathcal{M}_s^*$ and*

$$\mathbf{E}\sum_{s\leqslant t} |\Box_s x| I_{\{|\Box_s x| > a\}} < \infty \text{ for some } a > 0$$

and all $t \in \mathbf{R}_+^2$.

Then $x = x^1 + x^2$, where

$$x^1 \in l\mathcal{M}_s^2 \cap \mathcal{M}_s^*, \quad x^2 \in \mathcal{B} \cap \mathcal{M}_s^*.$$

Proof. Consider the process

$$A_t = \sum_{s \leqslant t} \Box_s x I_{\{|\Box_s x| > a\}} \in \mathcal{B}.$$

Since $\mathbf{E}A_t < \infty$ there exists a dual weakly predictable projection $A^{\hat{*}}$ and $x^2 = A - A^{\hat{*}} \in \mathcal{B} \cap \mathcal{M}_s^*$. Thus $x^1 = x - x^2 \in \mathcal{M}_s^*$. The left-hand side of the equality $x - A = x^1 - A^{\hat{*}}$ has bounded jumps: $|\Box_t(x - A)| \leqslant a$. From Theorem 3 $\widehat{\Pi}(\Box x^1) = 0$, therefore

$$a \geqslant |\widehat{\Pi}(\Box(x^1 - A^{\hat{*}}))_t| = |\widehat{\Pi}(\Box A^{\hat{*}})_t| = |\Box_t A^{\hat{*}}|.$$

Hence $|\Box_t x^1| \leqslant 2a$ and by Theorem 4 $x^1 \in l\mathcal{M}_s^2$. The decomposition $x = x^1 + x^2$ completes the proof.

REFERENCES

Bakry, D. (1981). Limites "quadrantales" des martingales. *Lect. Notes in Math.* **863**, 40–49.

Cairoli, R. and Walsh, J.B. (1975). Stochastic integrals in the plane. *Acta Mathem.* **134**, 111–183.

Gushchin, A.A. (1982). On the general theory of random fieldes on the plane. *Russian Math. Surveys* **37**, 55–80.

Millet, A. and Sucheston, L. (1981). On regularity of multiparameter amarts and martingales. *Z. Wahr. verw. Geb.* **56**, 21–45.

Mishura, Yu.S. (1982). On properties of the quadratic variation of two-parameter strong martingales. *Theory Probab. and Math. Statist.* **25**, 99–106.

Prob. Theory and Math. Stat., Vol. 2, pp. 193–206
B.Grigelionis *et al.* (Eds.)
1990 VSP/Mokslas

EXACT RESOLUTION THRESHOLDS FOR CLOSE FREQUENCIES

G.M. MOLCHAN
Institute of Physics of the Earth USSR Academy of Sciences,
123810 Moscow, USSR

ABSTRACT

The location problem of peaks in the time series spectral density is considered. Two frequencies ω_1 and ω_2 are called to be close with order $\gamma > 0$, if $\omega_i = \omega + (-1)^i h N^{-\gamma}$, $i = 1, 2$, where N is sample size. The maximum resolution capability $\sup\{\gamma(\mu), \mu \in M\}$ is determined for separate frequencies resolution methods μ, and these methods are considered.

INTRODUCTION

The mathematical problem of time series spectral analysis consists of building a better estimate in some sense of the power spectral density. However, in the presence of limited observations, as occurs in real situations, the investigator has to focus on the estimation of certain features of the spectrum, for example, on its main peaks. In this situation those methods are preferable which can accurately locate the peak frequencies in the spectrum, even if other spectrum details are determined worse.

First attempts to create such methods were made in application to geophysics in the 1970's. This resulted in a number of practical methods, namely "high–resolution capability" methods: the Burg Maximum Entropy Method, the Pisarenko Harmonic Decomposition method and his Non–Linear Spectral Estimations, etc (Kay and Marple, 1981). The problem emerges of how to estimate the resolution capability of certain methods and the spectral analysis on the whole. The engineering literature displays mainly simulation results for the comparison of different methods. Traditionally the hidden periodicity model

$$x_n = \sum_{1 \leqslant k \leqslant p} \alpha_k \exp(i\omega_k n) + \sigma w_n, \quad n = 0, 1, \ldots, N-1 \tag{1}$$

is taken as an object for comparison. Here, $\{w_n\}$ is a standard complex Gaussian white noise, $\theta = \{\sigma, \alpha_k, \omega_k\}$ are unknown signal parameters and p is the known order of the model.

The notions "short time series" and "close frequencies" are interrelated. A pair of frequencies ω_1, ω_2 are considered below as close with the proximity order $\gamma > 0$, if

$$\omega_i = \omega + (-1)^i h N^{-\gamma}, \quad i = 1, 2, \tag{2}$$

where (ω, h) are new frequency parameters. If $\gamma > 1$, a series is traditionally considered as "short" for frequencies resolution (2).
Let us say that a method μ resolves frequencies (2) if μ allows to construct such consistent estimations $\hat{\omega}_i$ that $\hat{\omega}_1 - \hat{\omega}_2 = o_p(|\omega_1 - \omega_2|)$, i.e.,

$$\mathbf{P}_\theta \left\{ |\hat{\omega} - \omega|_* > uN^{-\gamma}, \ |\hat{h} - h| > u \right\} \to 0, \quad N \to \infty. \tag{3}$$

Here $|\cdot|_*$ is the distance on the circle S^1.
The resolution is uniform with respect to $\theta \in F$ if (3) is uniform with respect to $\theta \in F$.
The upper bound $\gamma_\theta^*(\mu) = \sup \gamma$ of those orders γ for which the resolution is possible characterizes the resolution of the method at a point θ. Put

$$\gamma_G^*(\mu) = \inf\{\gamma_\theta^*(\mu), \ \theta \in G\}.$$

The value

$$\gamma_G^{**} = \sup\{\gamma_G^*(\mu), \ \mu \in M\}$$

determines the maximum resolution capability of the spectral analysis in the problem (1, 2) for a class of methods M.
Our goal is to determine γ^{**} and $\gamma^*(\mu)$ for separate methods. We show that $\gamma^{**} = 5/4$ for the generic parameter θ and $\gamma^*(\mu) = 1/6$ for the Harmonic Decomposition (HD) method. It seems that the bound of frequency resolution $\gamma^* = 1/6$ is typical for all correlation methods of spectral analysis which use a finite number of correlations.
The HD–method is interesting also from an other viewpoint. Here an unusual effect similar to phase transition is observed: the limit distribution for the estimate of the parameter h is normal for $\gamma < \gamma_0$ and is X_1^2 for $\gamma \in (\gamma_0, \gamma^*)$. At the critical point $\gamma_0 = 1/8$ these two distributions mix. Under certain conditions this situation is typical for statistical problems with converging parameters.

1. ABSOLUTE BOUNDARY OF THE FREQUENCY RESOLUTION γ^{**}

Consider the problem (1, 2) with two harmonics ($p = 2$). V. Pisarenko in (Pisarenko, 1976) and V.Gertzik (1983, unpublished) analyzed this model for Rex_n and continuous time n. They studied properties of the estimates of θ, obtained by the Least Square method (LS). This leads up to non–trivial estimates of γ^{**}:

$$7/6 \leqslant \gamma^{**} \leqslant 5/4, \text{ if } \omega \neq 0, \ \text{Im } \alpha_1 \bar{\alpha}_2 \neq 0 \text{ (Gertzik)}, \tag{4}$$

$$\gamma^{**} \leqslant 7/6, \text{ if } \omega \neq 0, \ \text{Im } \alpha_1 \bar{\alpha}_2 = 0 \text{ (Pisarenko)}. \tag{5}$$

More precisely, the authors calculated the upper thresholds of those γ for which the inverse Fischer matrix of the problem (1, 2) does not converge to zero. These

thresholds are given above as upper estimates of γ^{**}. The lower estimate of γ^{**} in (4) corresponds to the accurate upper bound of the values γ for which in problem (1, 2) the condition of uniform local asymptotic normality is satisfied (Ibragimov and Khasminskii, 1981). Under these conditions Gertzik applied a slightly modified Ibragimov—Khasminskii theory to the LS–estimates of θ, found their optimal properties and implicitly proved property (3). Similar analysis leads to the following improvement of (5): $9/8 \leqslant \gamma^{**} \leqslant 7/6$, if $\mathrm{Im}\ \alpha_1\bar{\alpha}_2 = 0$, since $9/8$ is the upper bound of the uniform LAN–condition in the case $\mathrm{Im}\ \alpha_1\bar{\alpha}_2 = 0$.
Below, by direct methods we show that γ^{**} coincides with the upper estimates in (4, 5).
Though the case $\mathrm{Im}\ \alpha_1\bar{\alpha}_2 = 0$ is degenerate, it is of particular interest for the analysis of resolution capability of correlation spectral methods. The point is that the majority of these methods asymptotically lose initial phases of harmonics, $\arg \alpha_i$, and, consequently, they should be compared on the model with $\mathrm{Im}\ \alpha_1\bar{\alpha}_2 = 0$.
V.Pisarenko and V.Gertzik analyzed the LS–estimates for a priori bounded parameters $\{\sigma, \alpha_1, \alpha_2, \omega, h\}$. This condition is not natural for the parameter $2h = |\omega_1 - \omega_2|N^\gamma$, which is related to the microscale of inter–frequency distances. Thus, the condition $h < h_0$ requires larger a priori information regarding the value $|\omega_1 - \omega_2|$. Below we shall analyze the least square method under the conditions $h < N^\alpha$, $0 < \alpha < \gamma$; corresponding LS–estimates will be called the LS(α)–estimates.
Let $\Theta = \Theta' \times \Theta''$ be the parametric space, where

$$\Theta' = \{\omega, h : \omega \in S^1, h > 0\},$$

$$\Theta'' = \{\sigma, \alpha_1, \alpha_2 : \sigma > 0,\ \alpha_i \neq 0,\ \alpha_1 + \alpha_2 \neq 0,\ \alpha_i \in \mathbf{C}^1\}.$$

In $\mathbf{C}^N$ we define a scalar product by $(x, y) = \sum x_i\bar{y}_i/N$, and an operator A by $(Ax)_i = \sum a_{ij}x_j/N$.

THEOREM 1. 1) *In the problem* (1, 2) *the* LS(α)*–method has uniform frequency resolution on any compact set* $F = F' \times F''$, $F^i \in \Theta^i$, *if*

(a) $F'' \subset \Theta'' \cap \{\mathrm{Im}\ \alpha_1\bar{\alpha}_2 \neq 0\} = \Theta''_1$, $1 < \gamma < 5/4$, $0 < \alpha < (\gamma - 1)/3$,

(b) $F'' \subset \Theta'' \cap \{\mathrm{Im}\ \alpha_1\bar{\alpha}_2 = 0\} = \Theta''_2$, $1 < \gamma < 7/6$, $0 < \alpha < (\gamma - 1)/4$.

2) *Let* $\hat{h}$ *be a consistent estimate of the parameter* h *in the problem* (1, 2) *with* $\gamma > 1$ *for any* θ *from an open set* $G \subset \Theta' \times \Theta''_i$. *Then* $\gamma < 5/4$, *when* $\mathrm{Im}\ \alpha_1\bar{\alpha}_2 \neq 0$ *and* $\gamma < 7/6$, *when* $\mathrm{Im}\ \alpha_1\bar{\alpha}_2 = 0$.

Sketch of the proof. 1). Equations for the LS–estimates of θ are linear in amplitudes α_i. If we exclude them, we get the variation principle for the LS(α)–estimates of frequencies:

$$\{\hat{\omega}_1, \hat{\omega}_2\} = \arg \max_{\omega_1, \omega_2} \sum (x, e^{(i)})\left[(e^{(i)}, e^{(j)})\right]^{-1}(e^{(j)}, x),$$

where $x = (x_0, \dots, x_{N-1})$, $e^{(j)} = \{\exp(i\omega_j n),\ n = 0, \dots, N-1\}$, $j = 1, 2$.
Let $P_{\theta'}$ be the projection operator in $\mathbf{C}^N$ to the plane $L_{\theta'} = \{ue^{(1)} + ve^{(2)}\}$.

Then

$$\{\hat{\omega},\hat{h}\} = \arg\{\max \|P_{\theta'}x\|,\ \omega\in S^1,\ 0<h<N^2\},$$

where $\theta' = (\omega, h)$ are equivalent parameters for ω_1, ω_2 (see (2)).
Let $P_0 = P_{\theta_0}$, where θ_0 is the true parameter. It is easy to see that (3) follows from the condition: for any $u > 0$

$$\mathbf{P}_{\theta_0}\left\{\min_{\Theta_u(\alpha)}\left(\|P_0x\|^2 - \|P_\theta x\|^2\right)\leqslant 0\right\}\to 0,\ N\to\infty, \tag{6}$$

where $\Theta_u(\alpha) = \{(\omega,h)\colon |\omega-\omega_0|_* \geqslant uN^{-\gamma},\ |h-h_0|>u,\ -\pi<\omega\leqslant\pi,\ h<N^\alpha\}$.
The frequency resolution is uniform in $\theta_0\in F$, if the convergence (6) is uniform in $\theta_0\in F$.
Consider the random function

$$\|P_0x\|^2 - \|P_{\theta'}x\|^2 = \|s_\theta^\perp\|^2 + 2\sigma\mathrm{Re}(s_\theta^\perp, w) + \sigma^2\Big(w,(P_0-P_\theta)w\Big) =$$

$$= \|s_\theta^\perp\|^2\Big(1+2\sigma\mathrm{Re}\eta_N(\theta)/(\|s_\theta^\perp\|\sqrt{N}) + \sigma^2\xi_N(\theta)/(\|s_\theta^\perp\|\sqrt{N})\Big)^2, \tag{7}$$

where $w = (w_0,\ldots,w_{N-1})$, $s = x-\sigma w$ is the useful signal, $s_{\theta'}^\perp = s - P_{\theta'}s$ is the component of s orthogonal to the plane L_θ. Our goal is to show that $\|s_{\theta'}^\perp\|^2 > c(N,u) > 0$ on $\Theta_u(\alpha)$, and the multiplier in parenthesis in (7) is $1+o_p(1)$ as $N\to\infty$. From this (6) follows.

LEMMA 1. *Let $\Delta = hN^{-\gamma} < \pi/2$, then for any $\beta>1$*

$$\|P_{\theta'} - P_{\tilde{\theta}'}\| \leqslant cN^\beta\Big(|\omega-\tilde{\omega}|_* + |\Delta-\tilde{\Delta}|\Big).$$

The proof follows from elementary estimates of $\|\partial/\partial\omega P_{\theta'}\|$ and $\|\partial/\partial\Delta P_\theta\|$; $P_\theta = [p_{n,m}^{(\theta)}]$, $n,m = 0,\ldots,N_1$, where

$$p_{n,m}(\omega,h) = 2\exp((i\omega(n-m))[\cos(t_n-t_m)\delta -$$

$$-\phi_N(\delta)\cos(t_n-t_m)\delta](1-\phi_N^2(\delta))^{-1} \tag{8}$$

and

$$t_n = (n-(N-1)/2)/N,\ \delta = N^{1-\gamma}h,$$

$$\phi_N(x) = \sin x/(N\sin x/N).$$

LEMMA 2. *Let $\theta_0', \theta'\in(-\pi,\pi]\times(0,\pi N^{\gamma/2})$. Then, $\forall\varepsilon>0$*

$$\sup_{\theta'}|\xi_N(\theta')| = \sup_{\theta'}|(w,(P_0-P_{\theta'})w)|N = O_p(N^\varepsilon) \tag{9}$$

uniformly in θ_0 and N.

Proof. Since $P_0 - P_{\theta'}$ is self–adjoint and w is normal we have

$$\xi_N(\theta') = \sum_{1\leqslant i\leqslant 4} \lambda_i|\varepsilon_i|^2,$$

where $|\lambda_i| \leqslant 1$, and ε_i are standard Gaussian independent variables. Hence

$$\mathbf{E}|\xi_N(\theta')|^m \leqslant \|P_0 - P_{\theta'}\|^m a_m \leqslant a_m,$$

and

$$\mathbf{E}|\Delta\xi_N(\theta')|^m \leqslant a_m\|\Delta P_{\theta'}\|^3,\ m > 3.$$

Using Lemma 1 we have

$$\mathbf{E}|\Delta\xi_N(\theta')|^m \leqslant c a_m N^{3\beta}|\Delta\theta'|^3.$$

Then, according to (Ibragimov and Khasminskii, 1981), we get

$$\mathbf{E}\sup_{\theta}|\xi_N(\theta')| \leqslant c_m N^{3\beta/m}.$$

Due to the choice of m, we have $3\beta/m < \varepsilon$. The Chebyshev inequality completes the proof of the lemma.

LEMMA 3. *Let $0 < \alpha < \gamma$. Then,* $\forall(\varepsilon, \varepsilon', u) > 0$

$$\sup\Big(|\eta_N(\theta')|: \theta' \in \Theta_u(\alpha)\Big) = O_p\Big(N^{\varepsilon'}\|s\|^{\varepsilon}/\inf\left(\|s_{\theta'}^{\perp}\|^{\varepsilon}: \theta' \in \Theta_u(\alpha)\right)\Big)$$

uniformly in θ_0' and N.

Proof. To estimate the supremum of the Gaussian function

$$\eta_N(\theta') = \sqrt{N}(s_{\theta'}^{\perp}, w)/\|s_{\theta'}^{\perp}\|,\ \mathbf{E}\eta_N(\theta') = 0,\ \mathbf{E}(\eta_N)^2 = 1,$$

it is sufficient to estimate the supremum of $\psi^2 = \mathbf{E}|\Delta\eta_N(\theta')|^2$, $\theta' = (\omega, hN^{-\gamma})$ for $|\Delta\theta'| < \delta$. Let $n(\theta) = \|s_{\theta'}^{\perp}\|$, $\mu \in (0,1)$ and

$$n_{u,\alpha} = \inf\left\{\|s_{\theta'}^{\perp}\|: \theta' \in \Theta_u(\alpha)\right\}.$$

We have

$$\psi = \|\Delta(s_{\theta}^{\perp}/n(\theta))\| \leqslant 2\|\Delta s_{\theta}^{\perp}\|/n(\theta) = 2\|(\Delta P)s\|/n(\theta) \leqslant 2\|s\|\|\Delta P\|^{\mu}/n_{u,\alpha}.$$

Taking into account Lemma 1, we get

$$\psi \leqslant c\|s\|(N^{\beta}\delta)^{\mu}/n_{u,\alpha}.$$

On the other hand, $\psi < 2$. Connecting estimates with powers ε and $1-\varepsilon$ we have

$$\psi \leqslant A_N\delta^{\varepsilon_1},\ \varepsilon_1 = \varepsilon\mu,$$

where

$$A_N = c\|s\|^\varepsilon N^{\varepsilon_2}/n^\varepsilon_{u,\alpha}, \ \varepsilon_2 = \beta\mu\varepsilon.$$

The Fernique result (Fernique, 1975, Theorem 4.1.1) leads to the final estimate

$$\sup\left\{|\eta_N(\theta')|: \theta' \in \Theta_u(\alpha)\right\} = O_p(A_N L^{\varepsilon_1}\varepsilon_1^{-1/2}),$$

which is uniform in N and θ'_0. Here L is the linear size of the parametrical set $\{\omega, \Delta = hN^{-\gamma}: |\omega| \leqslant \pi, \ \Delta < \pi/2\}$, i.e., $L = \pi$.

LEMMA 4. *Let $1 < \gamma < 1 + 1/k$ and on a compact set $F \in \Theta$*

$$\inf\left\{\|s^\perp_{\theta'}\|^2, \theta' \in \Theta_u(\alpha)\right\} \geqslant c_F N^{(1-\gamma)k}, \ c_F > 0, u > 0, \theta_0 \in F. \tag{10}$$

Then the LS(α)*–method resolves frequencies uniformly in $\theta_0 \in F$.*

Proof. Substituting estimates (9, 10) into (7) we get

$$\|P_0 x\|^2 - \|P_{\theta'} x\|^2 = \|s^\perp_\theta\|^2\left(1 + 2\sigma O_p(\|s\|^\varepsilon V_N^{-(1+\varepsilon)/2})\right) + \sigma^2 O_p(V_N^{-1}),$$

where

$$V_N = N^{1-\varepsilon}\inf\left\{\|s^\perp_{\theta'}\|^2, \ \theta' \in \Theta_u(\alpha)\right\}.$$

It follows from the a priori estimate (10), that

$$V_N \geqslant c_F N^{(k+1-\gamma k)-\varepsilon} = c_F N^{\varepsilon/2}$$

if $\varepsilon = (k + 1 - \gamma k)/2$.
On F the values $\sigma, \alpha_1, \alpha_2$ and, hence, $\|s\|$ are bounded. Thus,

$$\|P_0 x\|^2 - \|P_{\theta'} x\|^2 \geqslant c_F N^{(1-\gamma)}(1 + o_p(1)).$$

So (6) is true.

LEMMA 5. *Let $1 < \gamma < 5/4$ and $F_1 = \left\{\theta: \max(|\alpha_i|, |\alpha_1 + \alpha_2|^{-1}, h_0, h_0^{-1}) \leqslant f\right\}$. Then the estimate* (10) *is true if*

$$k = 4, \ F = F_1 \cap \{|\mathrm{Im}\alpha_1\bar{\alpha}_2| \geqslant \varepsilon\}, \ \alpha < (\gamma - 1)/3,$$

and if

$$k = 6, \ F = F_1 \cap \{\mathrm{Im}\alpha_1\bar{\alpha}_2 = 0, |\alpha_1\alpha_2| \geqslant \varepsilon\}, \ \alpha < (\gamma - 1)/4.$$

Proof. The norm $\|s^\perp_\theta\|$ will not change if we assume $\omega_0 = 0$, and

$$s = \alpha_1\exp(-i\delta_0 t) + \alpha_2\exp(i\delta_0 t), \ \delta_0 = N^{1-\gamma}h_0,$$

where t runs through the lattice $\{t_n\}$ (see (8)).
Let $0 < h < N^\alpha$, $0 \leqslant \alpha < \gamma - 1$, then $\delta = N^{1-\gamma}h = o(1)$. This allows to use the expansions of $\|s_0\|^2$ and $P_{\theta'}$ by small parameters δ and δ_0 to analyze

$$\|s_{\theta'}^{\perp}\|^2 = \|s_0\|^2 - \|P_{(0,h)}s_0\|^2.$$

We have

$$\|s_0\|^2 = |\alpha_1 + \alpha_2|^2 + 2\mathrm{Re}(\alpha_1\bar{\alpha}_2)\Big(-\frac{\delta_0^3}{3!} + \frac{\delta_0^4}{5!} - \frac{\delta_0^6}{7!} + rN^{(1-\gamma)6}\Big), \ |z| < c_f,$$

and the matrix elements of $P_{\theta'}$ can be written as

$$p_{n,m}(0,h) = \sum_{0\leqslant k\leqslant 3} u_k(t_n,t_m)(-\delta^2)^k/(2k)! + r_1\delta^8 + N^{-2}r_2,$$

where $|r_i(n,m,\delta)| < c_f$; $u_k(x,y)$ are polynomials of the form:

$$u_0 = 1 + 12xy,$$
$$u_1 = 2S[4x^3y + x^2 - 3/5xy - 1/12],$$
$$u_2 = S[24/5x^5y + 8x^3y^3 + 2x^4 + 6x^2y^2 - 24/5x^3y - 2x^2 + 66/125xy + 1/15]$$

and S is the symmetrization symbol: $Sf(x,y) = [f(x,y) + f(y,x)]/2$.
From the above expansions we consequently find estimates of the form $\|s_{\theta'}^{\perp}\| \geqslant c_f N^{-\beta}$, where

$$\beta = 0, \text{ if } |\omega N| \geqslant 2\pi,\ hN^{-\gamma} < \pi/2,$$
$$\beta = \gamma - 1, \text{ if } 2N^{-\mu} \leqslant |\omega N| \leqslant 2\pi,\ h < N^{\alpha},\ \mu > 0,$$
$$\beta = 4(\gamma - 1 - \alpha), \text{ if } N^{1-\gamma+\alpha} < |\omega N| = o(N^{-1}),\ h = 0(N^{\alpha}).$$

At last, for $h < N^{\alpha}$, $|\omega N| < N^{-\alpha_1}$, $c(\gamma) \leqslant \alpha_1 \leqslant \gamma - \alpha - 1$, $\max\{|\alpha_i|, h_0\} < f$ we have

$$\|s_{\theta'}^{\perp}\|^2 = \frac{1}{6!}|L_2 - L_0\delta^2|^2 + \\ + \begin{cases} o(N^{4(1-\gamma)}), & c(\gamma) = 2(\gamma-1)/3, \\ \varphi(\omega,\delta,\delta_0)/8! + o(N^{6(1-\gamma)}), & c(\gamma) = 3(\gamma-1)/4, \end{cases} \tag{11}$$

where

$$\varphi = \frac{2}{5}|L_3 - L_1\delta^2|^2 - \frac{2}{3}\mathrm{Re}\Big[(3L_4 - 7L_2\delta^2 + 4L_0\delta^4)(\bar{L}_2 - L_0\delta^2)\Big]$$

and

$$L_k = \alpha_1(\omega N - \delta_0)^k + \alpha_2(\omega N + \delta_0)^k.$$

The second asymptotics in (11) is necessary to estimate $\|s_{\theta'}^{\perp}\|^2$ in the case $\mathrm{Im}\alpha_1\bar{\alpha}_2 = 0$. In this case (11) takes the form:

$$15\cdot 8!\|s_{\theta'}^{\perp}\|^2 = 15\cdot 41|M_2|^2 + 6|M_3|^2 + |15M_2 - M_4|^2 + 40|M_2|^2\delta^2 + o(N^{6(1-\gamma)}),$$

where $M_k = L_k - L_{k-2}\delta^2$.
This leads to the estimate

$$\|s_{\theta'}^{\perp}\|^2 \geqslant c(|M_2|^2 + |M_3|^2) + o(N^{6(1-\gamma)}).$$

This estimate and the first one in (11) are easily analyzed and they lead (10).
2). Let $\mathrm{Im}\alpha_1\bar{\alpha}_2 \neq 0$, $\theta = (\sigma, A_+, A_-, \omega, h)$, $A_\pm = \alpha_2 \cdot \pm \alpha_1$ be the new parametrization of signal (1). Lét $\hat{h}_N$ be a sequence of estimates of h, consistent for certain compact vicinity $\theta_0 \subset \Theta \cap \{\mathrm{Im}\alpha_1\bar{\alpha}_2 \neq 0\}$, where $h < H$. The new sequence $\hat{h}_N^* = \hat{h}_N[|\hat{h}_N| < H] + H[|\hat{h}_N| > H]$ preserves the consistency property and is uniformly bounded.
The smoothness of the likelihood function $p_N(x|\theta)$ in θ implies the smoothness of $m_N(\theta) = \mathbf{E}_\theta \hat{h}_N^*$. Applying the Rao—Cramer inequality to $\hat{h}^*$ we have

$$\mathbf{E}_0|\hat{h}^* - h|^2 \geqslant |m(\theta) - h|^2 + \nabla^* m(\theta) I_N^{-1}(\theta) \nabla \bar{m}(\theta), \tag{12}$$

where

$$\nabla = \{\nabla_i\} = (\partial/\partial A_+,\ \partial/\partial A_-,\ \partial/\partial \bar{A}_+,\ \partial/\partial \bar{A}_-,\ \partial/\partial\omega,\ \partial/\partial h)'$$

and $I_N^{-1}(\theta)$ is the Fischer matrix for parameters (A_+, A_-, ω, h)

$$I_N(\theta) = \sigma^{-2} N\Big[(\nabla_i s, \nabla_j s) + (\nabla_i \bar{s}, \nabla_j \bar{s})\Big],\ i, j = 1, \ldots, 6.$$

The inequality (12) leads to a contradiction if $\gamma \geqslant 5/4$ (see below).
First we shall study asymptotic behaviour of $I_N^{-1}(\theta)$. The idea of asymptotic analysis of the inverse Gramm's matrices which are close to degenerate, goes back to V.Gertzik.
Again we can receive signal s with the accuracy of the multiplier $Z_N, |Z_N| = 1$.

$$s = (A_+ \cos\delta t + iA_- \sin\delta t)\exp(i\omega N t),\ \delta = hN^{1-\gamma},$$

where t runs through the lattice t_n on $[-1/2, 1/2]$. Let us expand $\cos\delta t$ and $\sin\delta t$ up to the order $O(\delta^3)$.
Then

$$\nabla s = D_1\{AD_2 e + O(\delta^3)\},$$

where

$$e = \big(e_0(t), e_1(t), e_2(t)\big),\ e_p(t) = (it)^p \exp(i\omega N t),$$

$$D_1 = \mathrm{diag}(1,1,1,1,N^\gamma h^{-1}, h^{-1}),\ D_2 = \mathrm{diag}(1, \delta, \delta^2),$$

$$A = \begin{pmatrix} 1 & 0 & 0 & 0 & 0 & 0 \\ 0 & 1 & 0 & 0 & A_+ & A_- \\ -1/2 & 0 & 0 & 0 & A_- & A_+ \end{pmatrix}'.$$

Similarly,

$$\nabla \bar{s} = D_1\{BD_2\bar{e} + O(\delta^3)\},$$

where

$$B = \begin{pmatrix} 0 & 0 & 1 & 0 & 0 & 0 \\ 0 & 0 & 0 & 1 & \bar{A}_+ & \bar{A}_- \\ 0 & 0 & -1/2 & 0 & \bar{A}_- & \bar{A}_+ \end{pmatrix}' .$$

Thus

$$\sigma^2 I_N(\theta) = N D_1 [C D \Gamma_N D C^* + R_N] D_1,$$

where

$$C = (A|B),\ D = \mathrm{diag}(D_2, D_2),\ \Gamma_N = \mathrm{diag}(G_N, G_N)$$

are 6×6 block–matrices and G_N is Gramm's matrix

$$G_N = \Big[(e_i, e_j)\Big]_{i,j=0,1,2} = \begin{pmatrix} 1 & 0 & -1/12 \\ 0 & 1/12 & 0 \\ -1/12 & 0 & 1/80 \end{pmatrix} +$$

$$+ O(N^{-1}) = \Gamma + O(N^{-1}).$$

The matrix $R_N = \delta^3(CDE + E^*DC^*) + O(\delta^6)$, where E has uniformly bounded elements. Matrices D and Γ_N are evidently invertible. Let us show that C is also invertible.

The cyclic transposition of lines (2, 3) and columns (2, 3, 5, 4) transforms C to a triangle–block form:

$$C_1 = \begin{pmatrix} I & 0 & \frac{1}{2}I \\ 0 & I & 0 \\ 0 & \alpha & \beta \end{pmatrix} = (2,3)C(2,3,5,4),$$

where

$$I = \begin{pmatrix} 1 & 0 \\ 0 & 1 \end{pmatrix}, \quad \alpha = \begin{pmatrix} A_+ & \bar{A}_+ \\ A_- & \bar{A}_- \end{pmatrix}, \quad \beta = \begin{pmatrix} A_- & \bar{A}_- \\ A_+ & \bar{A}_+ \end{pmatrix}.$$

Hence,

$$C^{-1} = (2,3,5,4) \begin{pmatrix} I & \beta^{-1}\alpha/2 & -\beta^{-1}/2 \\ 0 & I & 0 \\ 0 & -\beta^{-1}\alpha & \beta^{-1} \end{pmatrix} (2,3).$$

Let $\varphi = \sigma N^{-1/2} D^{-1} C^{-1} D_1^{-1}$, then

$$\varphi I_N(\theta) \varphi^* = \Gamma_N + (E C^{*-1} D^{-1} + D^{-1} C^{-1} E)\delta^3 + O(\delta^2) = \Gamma + O(N^{-1}) + O(\delta).$$

The residual terms are uniform in parameters, if $|\mathrm{Im}\alpha_1 \bar{\alpha}_2| > \varepsilon$ and

$$\max\{|\alpha_i|, h_0, h_0^{-1}\} < f.$$

Thus $\varphi^{*-1} I_N^{-1} \varphi^{-1} = \Gamma^{-1}(I + o(1)) \succ \Gamma_0$ as $N \to \infty$, where Γ_0 is the matrix which has all elements equal to zero, except $g_{33} = g_{66} = 1$.
Let us return to the proof of the second part of Theorem 1. We have

$$I_N^{-1}(\theta) \succ \varphi^* \Gamma_0 \varphi = \sigma^2 h^{-4} D_1^{-1} \Big[(3,2) U (3,2)\Big] D_1^{-1} N^{4\gamma - 5}, \tag{13}$$

where

$$U = \Big[0, \beta^{-1}\alpha, -\beta^{-1}\Big]^* \Big[0, \beta^{-1}\alpha, -\beta^{-1}\Big].$$

Let $\gamma \geqslant 5/4$. Then, using (13), we can extend the Rao—Cramér inequality

$$\mathbf{E}_\theta |\hat{h}^* - h|^2 \geqslant |m(\theta) - h|^2 + c^2 \partial^* m(\theta) U \partial \bar{m}(\theta), \tag{14}$$

where

$$c^2 = \sigma^2 h^{-4} N^{4\gamma-5} > c_0, \text{ if } h^{-1} > \varepsilon,\ \sigma > \varepsilon$$

and the operator

$$\partial = (3,2) D_1^{-1} \nabla = \Big(\partial/\partial A_+,\ \partial/\partial \bar{A}_+,\ \partial/\partial A_-,\ \partial/\partial \bar{A}_-,\ hN^{-\gamma}\partial/\partial\omega,\ h\partial/\partial h\Big)'.$$

We can easily check that the matrix U has the eigen–vector f with the eigen–value $\lambda \geqslant 1$, so that $U \succcurlyeq f \cdot f^*$, where

$$f = \Big(0, 0, -(\bar{A}_+ u + \bar{A}_- v), -(A_+ u + A_- v), u, v\Big)$$

and $\varphi = (u, v)$ is the solution of the equation

$$(I + \alpha\alpha^*)\varphi = \lambda \beta\beta^* \varphi$$

with maximal λ.
If $\mathrm{Im}\alpha_1\bar{\alpha}_2 \neq 0$ and $\mathrm{Re}\alpha_1\bar{\alpha}_2 \neq 0$, then $A_+u + A_-v$ and v are not equal to zero. The eigenvalue λ can have multiplicity 2 only on the set $(A_+, \bar{A}_+, A_-, \bar{A}_-)$ of the codimensionality 1. So, outside it, φ is a smooth function of $A_\pm, \bar{A}_\pm$.
Let us extend the inequality (14):

$$\mathbf{E}_\theta |\hat{h}^* - h|^2 \geqslant |m(\theta) - h|^2 + c_0 |Dm(\theta)|^2, \tag{15}$$

where

$$D = f^* \partial = -2\mathrm{Re}(A_+ u + A_- v)\partial/\partial A_- + uN^{-\gamma} h\partial/\partial\omega + vh\partial/\partial h.$$

Let G be an open subset of Θ where $\hat{h}_N^*$ are consistent, $\mathrm{Re}A_+\bar{A}_- \neq 0, \mathrm{Im}A_+\bar{A}_- \neq 0$ and λ have multiplicity 1. Then the unique characteristic θ_t of the operator D goes through $\theta \in G$.
Let θ_t, $0 \leqslant t < \tau$, belong to G. On the characteristic,

$$Dm(\theta_t) = \frac{d}{dt} m(\theta_t).$$

Since $\hat{h}^*$ is consistent and bounded, we have $\mathbf{E}_\theta|\hat{h}^* - h|^2 \to 0, N \to \infty$.
Due to (15), $dm(\theta_t)/dt \to 0$, $m(\theta_t) \to h(t)$. Due to the uniform boundedness of $dm(\theta_t)/dt$ in N and t, we have

$$\lim \int_0^\tau Dm(\theta_t)dt = \int_0^\tau \lim Dm(\theta_t)dt = 0, \quad N \to \infty.$$

On the other hand, $\lim[m(\theta_\tau) - m(\theta_0)] = h(\tau) - h(0)$.
But $\dot{h} = vh$ and $v \not\equiv 0$, thus, with an appropriate τ the last limit is not equal to zero. Hence we arrived at a contradiction having assumed $\gamma \geqslant 5/4$.
The case $\mathrm{Im}\alpha_1\bar{\alpha}_2 = 0$. There are only two differences from the preceding analysis. The signal is represented in the form:

$$s = (\tilde{A}_+ \cos \delta t + i\tilde{A}_- \sin \delta t)\exp(iN\omega t + i\varphi),$$

where $\tilde{A}_+ = |\alpha_1 + \alpha_2|$ and $\varphi = \arg A_+$. To analyze the asymptotics of the Fischer matrix we expand the trigonometric functions up to the order $O(\delta^4)$, i.e., s is expanded into basic functions $e_p = (it)^p \exp(i\omega Nt + i\varphi)$, $0 \leqslant p \leqslant 3$.

2. HARMONIC DECOMPOSITION METHOD

Any correlation function $B(k)$, $|k| \leqslant r$, has such extension to all $k \in \mathbf{Z}^1$, which coincides with the correlation function of a series (1) with uncorrelated amplitudes α_i, $\mathbf{E}\alpha_i = 0$, $1 \leqslant i \leqslant p$ (Grenander and Szego, 1958). An extension with $p \leqslant r$ is unique. It is constructed in the following way: σ^2 is the minimal eigen-value of the Toeplitz matrix $B_r = \{B(k-l),\ k,l = 0,\dots,r\}$; $p = \mathrm{rank}\{B_r - \sigma^2 I\}$; $\omega_i =$ $= \arg z_i$, where z_i are the zeroes of the polynomial $P_f(z) = f_0 z^p + \dots + f_p$ and $f = (f_0,\dots,f_p)'$ is the eigen-vector of B_p with the eigen-value σ^2. The powers $\mathbf{E}|\alpha_i|^2$ evidently depend linearly on $\{B(k), |k| \leqslant p\}$.
The described extension of the empiric correlations

$$\hat{B}(k) = N^{-1} \sum_{0 \leqslant n \leqslant N-k} x_{n+k}\bar{x}_n,\ \hat{B}(-k) = \overline{\hat{B}(k)},\ k = 0,\dots,p,$$

is called the Signal harmonic decomposition (Pisarenko, 1976). The HD-method selects several peaks in the spectrum, its remaining details are levelled in a noise component. Thus the model of "hidden periodicity" is most appropriate for the demonstration of this method's frequency resolution capability.
If in the described algorithm we substitute the matrix B_p by the empirical matrix with elements

$$\hat{B}_{ij} = N^{-1} \sum_{0 \leqslant n < N} x_{n+i}\bar{x}_{n+j},\ 0 \leqslant i,j \leqslant p,$$

then the vector f will define a plane $L{:}\, Xf^* = 0$ in $\mathbf{C}^{p+1}$ which minimizes the sum of the squared distances $X_n = (x_{n+p},\dots,x_n)$, $n = 0,\dots,N-1$ to L. Such modification of the HD-method is equivalent to the orthogonal regression method. The application of this method can be justified by two reasons: in the absence of

noise ($\sigma = 0$), model (1) satisfies the difference equation $X_n f^* = 0$ and the covariance matrix of noise in X_n is $\sigma^2 I$. The difference in methods manifests itself in the fact that in the first case (the HD–method) the zeroes of $P_f(t)$ lie on the unit circle, whereas in the second case it is true only asymptotically.
At first, let us consider a series of experiments corresponding to observations of the process (1) in time intervals of growing length $N \to \infty$. We assume that the parameters of the process do not unchange, i.e., $\sigma > 0$, $|\alpha_i| > 0$, $\omega_i \neq \omega_j$, $i = 1, \ldots, p$.

THEOREM 2. *As $N \to \infty$, the HD–estimates of $\{\sigma^2, \omega_1, \ldots, \omega_p\}$ are $\sqrt{N}$–consistent, asymptotically unbiased and asymptotically normal. That is, we have the following limits in probability*

$$\sqrt{N}(\hat{\omega}_k - \omega_k) \to |\sigma/\alpha_k|^2 \prod_{l \neq k} |e^{i\omega_k} - e^{i\omega_l}|^{-2} \eta_k,$$

$$\sqrt{N}(\hat{\sigma}^2 - \sigma^2) \to \sigma^2 \eta_0,$$

where $\eta = (\eta_0, \ldots, \eta_p)$ are Gaussian random variables with $\mathbf{E}\eta = 0$ and variances bounded from zero:

$$\sqrt{1/2} < \sigma(\eta_k) < 2^{2p-1}, \ k \geqslant 1; \ 1 < \sigma(\eta_0) < 2^p.$$

The vector η admits the following representation with $w(\omega)$ being the Wiener process

$$\sqrt{2\pi}\eta_k = -i \exp(-i\omega_k) \int_{-\pi}^{\pi} f(\omega) \bar{f}_k^{\perp}(\omega) dw(\omega), \ k \geqslant 1,$$

where

$$f(\omega) = \prod_{1 \leqslant n \leqslant p} (e^{i\omega} - e^{i\omega_n}) = f_k(\omega)(e^{i\omega} - e^{i\omega_k})$$

and $f_k^{\perp}(\omega_k)$ is the component of $f_k(\omega)$ orthogonal to $f(\omega)$ in $L^2(-\pi, \pi)$.

Comments. The accuracy of the frequency estimation by the HD–method has the usual decreasing order in observations volume, i.e., $N^{-1/2}$, and heavily depends on the geometry of all peak frequencies. The latter disadvantage can be compensated by the prior band filtration of the signal.
The optimal method in this situation (least–square method) is not sensitive to the frequency geometry. Its accuracy is in the linear (instead of quadratic) dependence upon the noise–to–signal ratio of the estimated frequency, but it has the non–standard decreasing order in N: $\sqrt{N}\sigma(\hat{\omega}_k) = c|\sigma/\alpha_k|N^{-1}$ (see (Ibragimov and Khasminskii, 1981)).

THEOREM 3. *In the problem (1, 2) with two harmonics the HD–method resolves frequencies iff $\gamma < 1/6$. The resolution is uniform in $\theta_0 \in F = \{\theta : 0 < c_1 < < (|\alpha_i|, \sigma, h) < c_2\}$.*

Under the conditions of frequency resolution there are the following limits in distribution:

$$N^{1/2-2\gamma}(\hat{\omega}-\omega) \to (2h)^{-2}B_{+}\sqrt{5/2}\,\xi,$$

$$N^{1/2-3\gamma}(\hat{h}-h) \to (2h)^{-2}B_{-}\sqrt{5/2}\,\xi,\ |\alpha_1| \neq |\alpha_2|,$$

where

$$B_{\pm} = 1/2\Big(|\sigma/\alpha_2|^2 \pm |\sigma/\alpha_1|\Big)^2.$$

When $|\alpha_1| = |\alpha_2|$ *the limit distribution of* $\hat{h}$ *undergoes a "phase transition"*

$$N^{1/2-2\gamma}(\hat{h}-h) \to (2h)^{-1}B_{+}\eta/3,\ 0<\gamma<1/8,$$

$$N^{1-6\gamma}(\hat{h}-h) \to (2h)^{-1}B_{+}\eta/3[\gamma=1/8] + (2h)^{-1}((2h)^{-2}B_{+}\sqrt{5/2}\,\xi)^2,\ \gamma \geqslant 1/8,$$

where ξ, η *are independent standard Gaussian random variables,* $[\gamma=\gamma_0]=1$ *for* $\gamma=\gamma_0$, *and* 0 *for* $\gamma \neq \gamma_0$.

The convergence is uniform in $\theta \in F \cap \{|\alpha_1-\alpha_2|>\varepsilon\}$.

Sketch of the proof. If $f=(f_0,f_1,f_2)$ is the eigen–vector of the correlation matrix B_p, $p=2$, with the minimal eigen–value, then

$$f_0z^2+f_1z+f_2 = f_0(z-e^{i\omega_1})(z-e^{i\omega_2}),$$

where

$$\omega_i = \omega + (-1)^i\rho, \quad \rho = hN^{-\gamma}.$$

Hence, we have the following parametrization of f:

$$f = \Big(\exp(i\omega), -2\cos\rho, \exp(-i\omega)\Big)$$

and we see that the estimation of f is connected with the estimation of (ω,h). In the HD–method the estimate of f minimizes the functional

$$Z = \langle \hat{B}_p f, f\rangle / \langle f,f\rangle - \hat{B}(0) =$$

$$= \mathrm{Re}\Big[\hat{B}(2)\exp(-2i\omega) - 4\cos\rho\hat{B}(1)\exp(-i\omega)\Big](2+\cos 2\rho)^{-1},$$

where $\langle\cdot\rangle$ is the quadratic form with the empirical matrix $\hat{B}_p$.
Thus, the HD–estimates of (ω,ρ) are

$$(\hat{\omega},\hat{\rho}) = \arg\inf Z(\omega,\rho),\ -\pi<\omega\leqslant\pi,\ 0<\rho<\pi/2.$$

Similar to (6), the frequency resolution for $\gamma<1/6$ follows from the condition:

$$\mathbf{P}_{\theta_0}\Big\{\min_{\Omega_u} Z(\omega,\rho) - Z(\omega_0,\rho_0) > c_F N^{-4\gamma}\Big\} \to 1,$$

where $\theta_0=(\omega_0,\rho_0)$, $\rho_0=h_0N^{-\gamma}$, are the true parameters,

$$\Omega_u = \Theta \cap \left\{ |\Delta\omega| > uN^{-\gamma},\ \Delta\rho > uN^{-\gamma} \right\}.$$

The estimate in (16) is obtained by direct calculations, using asymptotics $\hat{B}(k)$. Let the HD–method resolve frequencies. The HD–estimates satisfy equation $\nabla Z =$ $= 0$, since $\hat{\omega}, \hat{h}$ are consistent and (ω_0, h_0) is the inter point of Θ. In other words,

$$\operatorname{Im}\left[\hat{B}(2)e^{-2i\hat{\omega}} - 2\hat{B}(1)e^{-i\hat{\omega}}\cos\hat{\rho}\right] = 0, \tag{17}$$

$$\operatorname{Re}\left[\hat{B}(2)e^{-2i\hat{\omega}}\cos\hat{\rho} - \hat{B}(1)e^{-i\hat{\omega}}\cos 2\hat{\rho}\right] = 0 \tag{18}$$

or

$$\{(17),\ \hat{\rho} = 0\},\ \ \pi/2 < \hat{\omega} \leqslant \pi,\ \ 0 < \hat{\rho} < \pi/2.$$

The latter system contradicts the consistency of h. The system (17, 18) is easily analyzed due to the fact that $\hat{\omega} = \omega_0 + o_p(N^{-\gamma})$ and $\hat{\rho} = \rho_0 + o_p(N^{-\gamma})$. Thus we can pass to the equivalent linear system:

$$\begin{bmatrix} 1 + o_p(1) & 0 \\ \frac{\Delta\omega}{2\rho}(1 + O_p(\rho^2)) & 1 + O_p(\rho^2) \end{bmatrix} C_N \begin{bmatrix} \xi_N \\ \eta_N \end{bmatrix} = \begin{bmatrix} \rho^2\sqrt{N}\,\Delta\omega \\ \rho^2\sqrt{N}\,\Delta\rho \end{bmatrix}, \tag{19}$$

where

$$C_N = \frac{1}{4}\begin{pmatrix} B_+ & B_- \\ B_- & B_+ \end{pmatrix} + o_p(1),$$

$B_\pm$ are defined by (16), (ξ_N, η_N) is the centered asymptotically normal vector with the diagonal covariance matrix $\operatorname{diag}(5/8, 1/9)$, $\Delta\omega = \hat{\omega} - \omega$, $\Delta\rho = \hat{\rho} - \rho$.
If $\gamma > 1/6$ then $\rho^2\sqrt{N}\Delta\omega = \sqrt{N\rho^6}\cdot\Delta\omega/\rho = o(1)\cdot O_p(1)$. On the other hand, from the first equation (19) we have $\rho^2\sqrt{N}\Delta\omega \nrightarrow 0$. Hence $\gamma < 1/6$.
From (19) follow all the limit distributions of $\hat{\omega}, \hat{h}$.

REFERENCES

Fernique, X. (1975). Regularite des trajectoires des fonctions alcatoires gaussiennes. *Lect. Notes in Math.* **480.**

Grenander, V. and Szego, G. (1958). *Toeplitz Forms and their Application.* Univ. California Press, Berkeley.

Ibragimov, I.A. and Khasminskii, R.Z. (1981). *Statistical Estimation: Asymptotic Theory.* Springer, Berlin—Heidelberg etc.

Kay, S.M. and Marple (Jr.), S.L. (1981). Spectrum analysis–a modern perspective. *Proc. IEEE* **69**, 1380–1419.

Molchan, G. and Newman, W. (1988). On the resolution capability of the method of harmonic decomposition in spectral analysis. *Soviet Math. Dokl.* **298**, 1352–1355.

Pisarenko, V.F. (1976). Resolution capability of spectrum analysis. *Comput. Seismology* **9**, 185–196.

Prob. Theory and Math. Stat., Vol. 2, pp. 207–213
B.Grigelionis *et al.* (Eds.)
1990 VSP/Mokslas

ASYMPTOTIC PROPERTIES OF TWO STATISTICAL DECISION RULES BASED ON VALUES OF STUDENTIZED RANGE

A.V. NAGAEV
Institute of Mathematics Academy of Sciences,
ul. F.Khodzhaeva 29, 700143 Tashkent, USSR

ABSTRACT

The choice problem of one of two position–scale families is considered. The decision rule is Bayes likelihood ratio test corresponding to the measures, generated by standard distributions, over σ–algebra of position–scale invariant sets. Asymptotic properties of such a test for normal–uniform alternatives are established.

Let the density function $p_i(z)$ correspond to the distribution function $F_i(z)$, $i = 1, 2$. Denote by G_i the location–scale family generated by F_i, i.e.,

$$G_i = \Big\{F_i\Big((z-a)/\sigma\Big)\Big\}, \qquad -\infty < a < \infty, \qquad \sigma > 0.$$

Further let $x = (x_1, x_2, \ldots, x_n)$ be a random sample of size n from a distribution with the distribution function F. While testing the composite hypotheses $F \in G_1$ and $F \in G_2$ it is natural to use decision rules which are location and scale invariant. Consider two such procedures. Put for $i = 1, 2$

$$l_i(x) = \int_{-\infty}^{\infty}\int_{0}^{\infty} v^{n-2}\prod_{j=1}^{n} p_i(vx_j + u)dudv$$

and

$$L_i(x) = \max_{a,\sigma'} \sigma^{-n} p_i\Big((x-a)/\sigma\Big).$$

Introduce statistics

$$t(x) = l_1(x)/l_2(x), \quad T(x) = L_1(x)/L_2(x)$$

and the decision rules $d_t(x)$ and $d_T(x)$ based on these statistics such that for some $r > 0$

$$d_t(x) = \begin{cases} G_1, & \text{if} \quad t(x) > r \\ G_2, & \text{if} \quad t(x) < r; \end{cases}$$

$d_T(x)$ is determined in the same way.

For the sake of simplicity we suppose that

$$\mathbf{P}_i(T(x) = r) = \mathbf{P}_i(t(x) = r) = 0, \quad i = 1, 2.$$

Here P_i are measures defined on the σ–algebra $\mathfrak{F}$ of Borel sets which are generated by the distribution functions F_i. Let $\mathfrak{F}_0 \subset \mathfrak{F}$ be a σ–algebra which is invariant of location and scale transformations and P_i^0 be the restriction of P_i onto $\mathfrak{F}_0$.
It is known that (see, for example (Hajek and Sidak, 1967, II.2))

$$t = \frac{dP_i^0}{dP_2^0}. \tag{1}$$

Thus d_t is a Bayesian test corresponding to the case where $r = \pi_2 l_{21}/\pi_1 l_{12}$, (π_1, π_2) is the a priori distribution of G_1 and G_2, l_{ik}, $i, k = 1, 2$, $i \neq k$, are the losses caused by wrong decisions. Choosing properly the loss scale one can write down the Bayesian risk of d_t as

$$R_t = \mathbf{P}_1(d_t = G_2) + r\mathbf{P}_2(d_t = G_1). \tag{2}$$

The question arises how much will the risk grow if we replace d_t in (2) by more convenient d_T. Denote by R_T the risk corresponding to d_T. It is very difficult to compare R_t and R_T in the general case even as $n \to \infty$.
The example considered below shows that the study of the asymptotic behaviour of R_t/R_T is quite non–trivial even in the case when both families possess sufficient statistics.
Let $G_1 = G_N$ and $G_2 = G_U$ be the families of normal and uniform distributions, respectively. It is easily seen that in this case

$$t(x) = a_n\Big((x_{(n)} - x_{(1)})/s\Big)^{n-1} \tag{3}$$

and

$$T(x) = A_n\Big((x_{(n)} - x_{(1)})/s\Big)^{n}, \tag{4}$$

where

$$a_n = (2\pi e)^{n/2}, \quad A_n = n^{-(n-2)/2}\pi^{-(n-1)/2}\Gamma((n+1)/2),$$

$x_{(1)}, x_{(n)}$ are the extreme order statistics and s^2 is the sample variance. Thus both t and T are expressed in a similar way in terms of the studentized range $\Delta_n(x) = (x_{(n)} - x_{(1)})/s$ which is an invariantly sufficient statistic. Denote

$$\delta_n(x) = \Big(\Delta_n(x)\Big)^{-2} = s^2 \Big/ \Big(x_{(n)} - x_{(1)}\Big)^2. \tag{5}$$

In view of (2)—(5) we have

$$R_t = \mathbf{P}_N\Big(\delta_n(x) > z_n\Big) + r\mathbf{P}_U\Big(\delta_n(x) < z_n\Big) \tag{6}$$

and

$$R_T = \mathbf{P}_N\Big(\delta_n(x) > Z_n\Big) + r\mathbf{P}_U\Big(\delta_n(x) < Z_n\Big),$$

where

$$z_n = A_n^{2/(n-1)} r^{-2/(n-1)}, \quad Z_n = r^{-2/(n-1)}/(2\pi e) \tag{7}$$

and the subscripts N and U specify the case considered.
We have as $n \to \infty$

$$z_n = 1/(2\pi e) + \ln n/(\pi e n) + \big(\ln(\pi/e) - 2\ln r\big)/n + o(1/n) \tag{8}$$

and

$$Z_n = 1/(2\pi e) - 2\ln r/n + o(1/n). \tag{9}$$

Note that $\delta_n(x) \to 0$ or $\delta_n(x) \to 1/12$ in probability depending on $F \in G_N$ or $F \in G_U$. In view of (6)—(9) both risks represent the large deviations probabilities for the statistic $\delta_n(x)$.
Note that if $r = 1$, then

$$R_t = 1 - \sup_{A \in \mathfrak{F}_0} \big|\mathbf{P}_N(A) - \mathbf{P}_U(A)\big|. \tag{10}$$

It is easily seen that for any absolutely continuous distribution function $F(x)$

$$\mathbf{P}\Big(\delta_n(x) < z\Big) = n(n-1) \iint\limits_{-\infty<u<v<\infty} \Big(F(v) - F(u)\Big)^{(n-2)} \mathbf{P}\Big(\sum_{j=1}^{n-2} \xi_j^2 -$$

$$-\Big(\sum_{j=1}^{n-2} \xi_j + u + v\Big)^2 \Big/ n < nz(v-u)^2 - u^2 - v^2\Big) dF(u)dF(v), \tag{11}$$

where ξ_j are independent and

$$\mathbf{P}(\xi_j < z) = \mathbf{P}(x_1 < z | u < x_1 < v).$$

So the problem is reduced to the estimation of the probability of the event $\Big(\sum_{j=1}^{n-2} \xi_j, \sum_{j=1}^{n-2} \xi_j^2\Big) \in A$, where the distribution of summands (ξ_j, ξ_j^2) depends on the parameters u and v while the boundary of A depends also on z and n.
Before we formulate the large deviation theorem for $\delta_n(x)$ in the case $F \in G_N$, we introduce some notation. Put

$$\mu(t) = (1/4)t^{-2} \int_0^t y^2 e^{-y^2/2} dy \Big/ \int_0^t e^{-y^2/2} dy,$$

$$\lambda(t) = (1/4)t^{-1} e^{-t^2/2} \Big/ \int_0^t e^{-y^2/2} dy.$$

It is easy to understand that both these functions decrease in the interval $(0, \infty)$, so that $\mu(0) = 1/12$, $\lambda(0) = \infty$, $\mu(\infty) = 0$, $\lambda(\infty) = 0$. Further define

$$s(z) = 2^{-1}\lambda\Big(\mu^{-1}(z)\Big) \Big/ \Big(z + \lambda(\mu^{-1}(z))\Big), \quad v(z) = \mu^{-1}(z)\Big(1 + \lambda(\mu^{-1}(z))/z\Big)^{1/2},$$

$$\phi(z) = 2\int_0^{v(z)} e^{s(z)x^2}\,d\Phi(x)\exp\Big(-4zs(z)v^2(z)\Big),$$

$$d_1^2(z) = \int_0^{v(z)} x^4 e^{s(z)x^2}\,d\Phi(x)\Big/\int_0^{v(z)} e^{s(z)x^2}\,d\Phi(x) - 16z^2v^4(z).$$

THEOREM 1. *Under our assumptions as $n \to \infty$*

$$\mathbf{P}_N\Big(\delta_n(x) > z\Big) = A(z)\phi^n(z)n^{1/2}\Big(1 + o(1)\Big)$$

uniformly in $z \in [\varepsilon, 1/12 - \varepsilon]$, where

$$A(z) = 16z^2s^2(z)v^2(z)\Big(2\pi D(z)s^2(z)d_1^2(z)\Big(1 + 8z^2s(z)v(z)\Big)\Big)^{-1/2}$$

and

$$D(z) = 16z^2s(z)v^2(z)\Big(1 - 2s(z) + 8s^2(z)v^2(z)\Big/\Big(1 + 8zs(z)v^2(z)\Big)\Big)\Big(s(z) +$$

$$+ 8zs^2(z)v^2(z) + s(z)v^2(z)\Big(1 - 2s(z)\Big) + v(z)d_1^{-2}(z)\Big(1 + 4zs(z)v^2(z) - s(z)v^2(z)\Big)\Big).$$

Theorem 1 has the following

COROLLARY 1. *The probabilities of wrong acception of the uniformity hypothesis as $n \to \infty$ behave as*

$$\mathbf{P}_N\Big(d_t(x) = G_U\Big) = c_N r^\beta \rho_N^n n^{\gamma_N}\Big(1 + o(1)\Big),$$

and

$$\mathbf{P}_N\Big(d_T(x) = G_U\Big) = c'_N r^\beta \rho_N^n n^{1/2}\Big(1 + o(1)\Big),$$

where

$$\rho_N = \phi(z_0), \qquad \gamma_N = 1/2 - 4s(z_0)v^2(z_0)/(\pi e),$$
$$c_N = A(z_0)\exp\Big(-2s(z_0)v^2(z_0)\ln(\pi/e)/(\pi e)\Big),$$
$$c'_N = A(z_0), \qquad \beta = 8zs(z)v^2(z), \qquad z_0 = 1/(2\pi e).$$

If $F \in G_U$, the representation (11) becomes considerably simple. In this case

$$\mathbf{P}_U\Big(\delta_n(x) < z\Big) = \mathbf{P}\Big(\sum_{j=1}^{n-2} e_j^2 - \Big(\sum_{j=1}^{n-2} e_j\Big)^2 < nz - \frac{1}{2}\Big),$$

where independent variables e_j are distributed uniformly in $(-1/2,\ 1/2)$.

Let $\tau(z)$ be the solution of the equation

$$\int_0^{1/2} u^2 e^{\tau u^2} du \Big/ \int_0^{1/2} e^{\tau u^2} du = z.$$

It is easily seen that $\tau(z) < 0$ if $z < 1/12$.
Put

$$\psi(z) = 2e^{-z\tau(z)} \int_0^{1/2} e^{\tau(z)u^2} du,$$

$$d_2^2(z) = \int_0^{1/2} u^4 e^{\tau(z)u^2} du \Big/ \int_0^{1/2} e^{\tau(z)u^2} du - z^2,$$

$$B(z) = \Big(2\pi\Big(1 - 2z\tau(z)\Big)\Big)^{-1/2} e^{-\tau(z)(2z-1/2)} \psi^{-2}(z)\Big(d_2(z)\Big(-\tau(z)\Big)\Big)^{-1}.$$

THEOREM 2. *Under our assumptions as $n \to \infty$*

$$\mathbf{P}_U\Big(\delta_n(x) < z\Big) = B(z)\psi^n(z)n^{-1/2}\Big(1 + o(1)\Big)$$

uniformly in $z \in [\varepsilon, 1/12 - \varepsilon]$.

As the previous theorem this one has the obvious

COROLLARY 2. *Probabilities of wrong acception of the normality hypothesis corresponding to the decision rules d_t and d_T as $n \to \infty$, behave as*

$$\mathbf{P}_U\Big(d_t(x) = G_N\Big) = c_U r^{2\tau(z_0)} \rho_U^n n^{\gamma_U}\Big(1 + o(1)\Big)$$

and

$$\mathbf{P}_U\Big(d_T(x) = G_N\Big) = c'_U r^{2\tau(z_0)} \rho_U^n n^{-1/2}\Big(1 + o(1)\Big),$$

where

$$\rho_U = \psi(z_0), \qquad \gamma_U = -1/2 - \tau(z_0)/(\pi e),$$
$$c_U = B(z_0)\exp\big(-\tau(z_0)\ln(\pi/e)/(2\pi e)\big),$$
$$c'_U = B(z_0),$$

z_0 is the same as above.

From Corollaries 1 and 2 it follows that

$$R_t = c_N r^\beta \rho_N^n n^{\gamma_N}\Big(1 + o(1)\Big) + c_U r^{2\tau(z_0)} \rho_U^n n^{\gamma_U}\Big(1 + o(1)\Big)$$

and

$$R_T = c'_N r^\beta \rho_N^n n^{1/2}\Big(1 + o(1)\Big) + c'_U r^{2\tau(z_0)} \rho_U^n n^{-1/2}\Big(1 + o(1)\Big).$$

In particular, according to (10)

$$1 - \sup_{A\in\mathfrak{F}_0} |\mathbf{P}_U(A) - \mathbf{P}_N(A)| = c_N\rho_N^n n^{\gamma_N}\Big(1+o(1)\Big)+$$

$$+c_U\rho_U^n n^{\gamma_U}\Big(1+o(1)\Big).$$

The constants ρ_N, γ_N and ρ_U, γ_U, which play the crucial role in the behaviour of R_t and R_T are calculated by means of quite different formulas. The computer calculations show that

$$\max\left(\,|\,\rho_N-\rho_0\,|,\ |\,\rho_U-\rho_0\,|,\ |\,\gamma_N-\gamma_0\,|,\ |\,\gamma_U-\gamma_0\,|\,\right) < 10^{-8}$$

for

$$\rho_0 = 0.94127124, \quad \gamma_0 = 0.10471149.$$

Thus it seems to be likely that $\rho_N = \rho_U$, $\gamma_N = \gamma_U$. If we assume that it is really so, then

$$R_t = \Big(c_N r^\beta + c_U r^{2\tau(z_0)}\Big)\rho_N^n n^{\gamma_N}\Big(1+o(1)\Big)$$

and

$$R_t/R_T = n^{\gamma_N - 1/2}\Big((c_N/c'_N) + (c_U/c'_U)r^{2\tau(z_0)-1/2}\Big)\Big(1+o(1)\Big).$$

Thus the asymptotic properties of d_t and d_T are somewhat different.
Consider the problem of testing two simple alternatives representing families G_N and G_U. Naturally, in this situation the test d_t is in general worse than that based on the corresponding likelihood ratio.
Unfortuhately the size of this paper does not allow to present here the proofs of Theorems 1 and 2. We hope to do it elsewhere.
The following questions arise:

1. Which pair of alternatives representing G_N and G_U is the worst from the point of view of asymptotic distinguishability?

2. How much does the asymptotic risk of the corresponding likelihood ratio test differ from that of d_t?

It is easy to answer both these questions. Without loss of generality consider simple alternatives

$$F(x) = F_1(x) = \Phi(x/\sigma)$$

and

$$F(x) = F_2(x) = x + 1/2, \qquad x \in (-1/2, 1/2).$$

Let

$$d_\sigma(x) = \begin{cases} F_1, & \text{if } \|x\| > 1/2 \text{ or } \|x\| < 1/2 \text{ and } \sum_{j=1}^n \ln p(x_j) > \ln r, \\ F_2, & \text{otherwise,} \end{cases}$$

where $\|\,x\,\| = \max_{1\leqslant j\leqslant n} |\,x_j\,|$, $\quad p(x) = \sigma^{-1}\Phi'(x/\sigma)$.

Further let R_σ be the Bayesian risk corresponding to d_σ, i.e.,

$$R_\sigma = \mathbf{P}_1\Big(d_\sigma(x) = F_2\Big) + r\mathbf{P}_2\Big(d_\sigma(x) = F_1\Big).$$

Denote

$$c(\sigma) = \lim_{n\to\infty} \frac{-\ln R}{n}.$$

It is easy to show that $c(\sigma)$ has its maximum at $\sigma_0 = (2\pi e)^{-1/2}$. Moreover

$$R_\sigma = (2\pi n)^{-1/2} f^n(s_0) r^{1-s_0}\Big(s_0(1-s_0)d_0\Big)^{-1}\Big(1+o(1)\Big),$$

where

$$f(s) = s^{-1/2}(2\pi\sigma)^{(1-s)/2}\Big(2\Phi(s^{1/2}\sigma^{-1}2^{-1}) - 1\Big).$$

There s_0 is determined by the equation

$$f'(s_0) = 0,$$

d_0^2 is the value of the second logarithmic derivative of $f(s)$ at $s = s_0$. It is clear that $f(s_0) = \exp(-c(\sigma))$. Calculations on computer show that $\mid f(s_0) - \rho_0 \mid < 10^{-8}$. Thus it is very likely that Chernoff's indices of all three decision rules d_t, d_T and d_σ are equal.

In conclusion we make several remarks.

1. If indeed $\rho_N = \rho_U = f(s_0)$, then the question arises to what extent this property is universal.

2. It is unknown how to deal with asymptotics of R_t in more general situations.

3. Apparently the studentized range can be used for testing families G_1 and G_2 generated by symmetric distributions F_1 and F_2 such that F_1 has a compact support and the support of F_2 is $(-\infty, \infty)$.

4. Apparently using the Cramér's techniques one can obtain lower and upper bounds for the minimal sample size which guarantees a given risk (see (Nagaev and Jurovskaya, 1989)).

REFERENCES

Hajek, J. and Sidak, Z. (1967). *Theory of Rank Tests*. Academia, Prague.

Nagaev, A.V. and Jurovskaya, S.A. (1989). Remark on ε–distinquishability of alternatives. *Wiss. Z. TU Dresden*, **38**, 167–168.

Prob. Theory and Math. Stat., Vol. 2, pp. 214–226
B.Grigelionis *et al.* (Eds.)
1990 VSP/Mokslas

ON A NEW APPROACH TO THE STUDY OF THE DISTRIBUTION OF A NORM OF A RANDOM ELEMENT IN HILBERT SPACE

S.V. NAGAEV
Institute of Mathematics, 630090 Novosibirsk, USSR

ABSTRACT

The new formula for moments of the norm of Hilbert space–valued r.v. is deduced. By means of this formula the new bounds for moments of the sum of of independent Hilbert space–valued r.v. are obtained. Further the representation for Laplace transform of the norm is discussed. After all the bound for the error of the normal approximation is stated.

Let H be a separable real Hilbert space with a norm $|\cdot|$ and an inner product $\langle \cdot, \cdot \rangle$, S be a Borel σ–algebra in H, we shall denote r.v. taking values in H by X, Y, Z, ...
Let $\alpha = (\alpha_1, \alpha_2, \ldots)$ be a sequence of i.i.d. normal $\mathcal{N}(0,1)$ r.v. Put for any $x \in H$

$$\langle x, \alpha \rangle = \sum_{j=1}^{\infty} x_j \alpha_j,$$

where $x_j = \langle x, e_j \rangle$, e_j is an orthonormal basis in H. In other words,

$$\langle x, \alpha \rangle = \lim_{N \to \infty} \langle x, \widehat{\alpha}_N \rangle, \tag{1}$$

where $\widehat{\alpha}_N = (\alpha_1, \alpha_2, \ldots, \alpha_N)$ is a standard Gaussian r.v. in $\mathbf{R}^N$.
We say the form $\langle x, \alpha \rangle$ *is defined* for the pair (x, α), if the lim in (1) exists and is finite.
If x is fixed, then the bilinear form $\langle x, \alpha \rangle$ is defined for almost all α with respect to Kolmogorov's distribution generated by the sequence of r.v. $\alpha_1, \alpha_2, \ldots$.
Consider now

$$\langle X, \alpha \rangle = \lim_{N \to \infty} \langle X, \widehat{\alpha}_N \rangle,$$

where X does not depend on α.
Let A_x be a set of α such that the form $\langle x, \alpha \rangle$ is defined. As we noted already, for any x

$$\mathbf{P}(A_x) = 1. \tag{2}$$

Put $A = \bigcup_x \{x \times A_x\}$. In other words, A is a set of pairs (x, α) such that the form $\langle x, \alpha \rangle$ is defined. The set $A \subset S \times S'$, where S' is a σ-algebra generated by cylindrical sets in $\mathbf{R}^\infty$.
Therefore, the probability $Q(A) = \mathbf{P}(\langle X, \alpha \rangle \in A)$ is defined and

$$Q(A) = \int_H \mathbf{P}(A_x) F(dx),$$

where F is the distribution of r.v. X.
By (2)

$$Q(A) = 1$$

and therefore $\langle X, \alpha \rangle$ is a proper r.v. In the sequel, c and $c(t)$ will denote absolute and depending on t constants. We shall use also the notation

$$\mathbf{E}_X f(X, \alpha) = \mathbf{E}[f(X, \alpha) \mid \alpha],$$

$$\mathbf{E}_\alpha f(X, \alpha) = \mathbf{E}[f(X, \alpha) \mid X],$$

where f is a Borel function of (X, α).

THEOREM 1. *For any* $t > 0$

$$\mathbf{E} \mid X \mid^t = c(t) \mathbf{E}_\alpha (\mathbf{E}_X \mid \langle X, \alpha \rangle \mid^t), \tag{3}$$

where $c(t) = 1/\mathbf{E} \mid \alpha_1 \mid^t$.

Proof. For any fixed x the form $\langle x, \alpha \rangle$ is a Gaussian $\mathcal{N}(0, \mid X \mid)$ r.v. Consequently,

$$\mathbf{E} \mid \langle x, \alpha \rangle \mid^t = \mid x \mid^t \mathbf{E} \mid \alpha_1 \mid^t .$$

Hence,

$$\mathbf{E} \mid X \mid^t = \mathbf{E}_X \mathbf{E}_\alpha \mid \langle X, \alpha \rangle \mid^t / \mathbf{E} \mid \alpha_1 \mid^t .$$

Changing now the order of integration, we obtain the formula (3). Formula (3) gives a representation of the moment of a norm of r.v. taking values in H in terms of the mixture of moments of one-dimensional r.v.
We apply now the formula (3) to deduce a Rosenthal type inequality for the moment of a norm of the sum of H-valued i.i.d. r.v.
So let $X_1, X_2, \dots, X_n$ be H-valued independent r.v., $\mathbf{E}X_j = 0$, $j = 1, 2, \dots, n$, $S = \sum_{j=1}^n X_j$, $A_t = \sum_{j=1}^n \mathbf{E} \mid X_j \mid^t$. Let further B be a covariance operator of r.v. S, and Y be a Gaussian r.v. with a covariance operator B. In the sequel $\sigma_1^2 \geqslant \sigma_2^2 \geqslant \geqslant \dots \geqslant \sigma_n^2 \geqslant \dots$ denote eigen-values of B.

THEOREM 2. *For any* $t > 2$

$$\mathbf{E} \mid S \mid^t < c_1(t) A_t + c_2(t) \mathbf{E} \mid Y \mid^t / \mathbf{E} \mid \alpha_1 \mid^t, \tag{4}$$

where $c_1(t)$ and $c_2(t)$ are the same constants, as in one–dimensional Rosenthal type inequality.

Proof. Since the constants don't depend on distributions of r.v. $X_1, X_2, \ldots, X_n$, it is sufficient to prove (4) in the finite–dimensional case. In view of formula (3),

$$\mathbf{E} \mid S \mid^t = \mathbf{E}_\alpha \mathbf{E}_S \mid \langle S, \alpha \rangle \mid^t / \mathbf{E} \mid \alpha_1 \mid^t . \tag{5}$$

By one–dimensional Rosenthal inequality,

$$\mathbf{E}_S \mid \langle S, \alpha \rangle \mid^t \leqslant c_1(t) \sum_{j=1}^{n} \mathbf{E}_{X_j} \mid \langle X_j, \alpha \rangle \mid^t + c_2(t) \Big(\mathbf{E}_S (\langle S, \alpha \rangle)^2 \Big)^{t/2} . \tag{6}$$

In view of (3),

$$\mathbf{E}_\alpha \mathbf{E}_{X_j} \mid \langle X_j, \alpha \rangle \mid^t = \mathbf{E} \mid \alpha_1 \mid^t \mathbf{E} \mid X_j \mid^t . \tag{7}$$

Further $\mathbf{E}_S(\langle S, \alpha \rangle)^2 = (B\alpha, \alpha)$. It is easy to see that

$$\mathbf{E}_\alpha (B\alpha, \alpha)^{t/2} = \mathbf{E}_\alpha \left(\sum_{j=1}^{N} \sigma_j^2 \alpha_j^2 \right)^{t/2} ,$$

where N is a dimension, σ_j^2 are eigenvalues of the operator B. On the other hand, $\sum\limits_{j=1}^{N} \sigma_j^2 \alpha_j^2 = \mid Y^* \mid^2$, where $Y^* = DY$, D is an orthogonal transformation. Thus

$$\mathbf{E}_\alpha \Big(\mathbf{E}_S (\langle S, \alpha \rangle)^2 \Big)^{t/2} = \mathbf{E}_\alpha (B\alpha, \alpha)^{t/2} = \mathbf{E} \mid Y \mid^t . \tag{8}$$

The inequality (3) follows from (5)—(8).
Now we proceed to estimating of the difference

$$\nabla(q) = \mathbf{E} \mid S \mid^q - \mid Y \mid^q .$$

For this purpose we need some one–dimensional results.

THEOREM A. *If X_k are real r.v., then for any $0 < q \leqslant p \leqslant 3$*

$$\mid \nabla(q) \mid < c(p,q) A_p . \tag{9}$$

One may derive this bound from Theorem 18.1 in (Bhattacharya and Ranga Rao, 1976).

THEOREM B. *If X_k are real r.v., then for any $p > 3$, $0 < q \leqslant p$*

$$\mid \nabla(q) \mid < c(p,q) (A_p A_2^{(q-p)/2} + A_3 A_2^{(q-3)/2}) . \tag{10}$$

Proof. It follows from Theorem 2 in (Sakoyan, 1974) that for $p \geqslant 3$

$$| \mathbf{P}(S < A_2^{1/2}u) - \Phi(u) | < c(p)(L_3 + L_p)/(1+ | u |^p), \tag{11}$$

where

$$\Phi(u) = \frac{1}{\sqrt{2\pi}} \int\limits_{-\infty}^{u} e^{-v^2/2}\, dv, \quad L_p = A_p/A_2^{p/2}.$$

Fix any $x > 2b$ and put $\overline{X}_k = X_k \mathbf{1}_{\{|X_k|<x\}}$, $\overline{S} = \sum_{k=1}^{n} \overline{X}_k$, where $b^2 = A_2$.
Put

$$b_0^2 = \sum_{k=1}^{n} \{\mathbf{E}X_k^2;\ | X_k |< 2b\} - \left(\mathbf{E}(X_k;\ | X_k |< 2b)\right)^2.$$

Applying the bound (11), we have

$$| \mathbf{P}(\overline{S} < u) - \Phi((u - \overline{a})/\overline{b}) | < c(p)(\overline{A}_3 \overline{b}^{p-3} + \overline{A}_p)/(\overline{b}^p + | u - \overline{a} |^p). \tag{12}$$

Here $\overline{a} = \mathbf{E}\overline{S}$, $\overline{b}^2 = \mathbf{E}\overline{S}^2 - (\mathbf{E}\overline{S})^2$, $\overline{A}_p = \sum_{k=1}^{n} \mathbf{E} | \overline{X}_k - \mathbf{E}\overline{X}_k |^p$.
It is easy to see that

$$b^2 - b_0^2 = \sum_{k=1}^{n} \left(\mathbf{E}(X_k^2;\ | X_k |> 2b) + \left(\mathbf{E}(X_k;\ | X_k |> x)\right)^2\right) \leqslant$$

$$\leqslant 2\sum_{k=1}^{n} \mathbf{E}(X_k^2;\ | X_k |> 2b) \leqslant A_p/(2b)^{p-1}, \qquad p > 2.$$

If $b_0^2 \leqslant b^2/2$, then $A_p > b^p/4$. In according to Rosenthal inequality

$$\mathbf{E} | S |^q \leqslant (\mathbf{E} | S |^p)^{q/p} \leqslant c(p,q)(b^p + A_p)^{q/p} \leqslant c(p,q)(b^q + A_p^{q/p}).$$

Clearly,

$$A_p^{q/p} = A_p A_p^{(q-p)/p} < 4A_p b^{q-p}, \qquad b^q < 4A_3 b^{q-3}.$$

Thus, in the case $b_0^2 < b^2/2$

$$\mathbf{E} | S |^q \leqslant c(p,q)(A_3 b^{q-3} + A_p b^{q-p}).$$

Since

$$| \nabla(q) | \leqslant \mathbf{E} | S |^q + \mathbf{E} | Y |^q,$$

it leads us to (10). Now consider the case $b_0^2 \geqslant b^2/2$. Obviously,

$$\nabla(q) = q \int\limits_{-\infty}^{\infty} u^{q-1} \left(\Phi(u) - \mathbf{P}(S < u)\right) du. \tag{13}$$

Further,

$$| \mathbf{P}(S < u) - \Phi(u/b) | < | \mathbf{P}(\overline{S} < u) - \Phi((u - \overline{a})/\overline{b}) | +$$
$$+ | \Phi(u/\overline{b}) - \Phi((u - \overline{u})/\overline{b}) | + | \mathbf{P}(\overline{S} < u) - \mathbf{P}(S < u) | . \tag{14}$$

Putting in (12) $x = \max(2b, u)$, estimate

$$I_1 = \int_{-\infty}^{\infty} | u |^{q-1} \left| \mathbf{P}(\overline{S} < u) - \Phi((u - \overline{a})/\overline{b}) \right| du.$$

Clearly,

$$I_1 = \int_{|u| \leqslant 2b} + \int_{|u| > 2b} = I_{11} + I_{12}. \tag{15}$$

In view of (12),

$$I_{11} < c(p, q)(\overline{A}_3(2b)\overline{b}^{\,q-3} + \overline{A}_p(2b)\overline{b}^{\,p-q}).$$

It is easy to see that $\overline{A}_p \leqslant 2^p A_p$. Therefore

$$I_{11} < c(p, q)(A_3 b^{q-3} + A_p b^{q-p}). \tag{16}$$

Since

$$| \overline{a} | < \sum_{k=1}^{n} \mathbf{E}(| X_k |; | X_k | > 2b) < b/2, \tag{17}$$

then

$$I_{12} < c(p + 1) \int_{2b}^{\infty} u^{q-p-2}(\overline{A}_3 b^{p-2} + \overline{A}_{p+1}) \, du. \tag{18}$$

Further,

$$\int_{2b}^{\infty} u^{q-p-2} \, \overline{A}_{p+1} \, du = \sum_{k=1}^{n} \int_{2b}^{\infty} u^{q-p-2} \int_0^u v^{p+1} \, dF_k(v),$$

where $F_k(v) = \mathbf{P}(| X_k | < v)$.
Clearly,

$$\int_{2b}^{\infty} u^{q-p-2} \int_0^u v^{p+1} \, dF_k(v) = \int_0^{\infty} v^{p+1} \, dF_k(v) \int_{\max[2b, v]}^{\infty} u^{q-p-2} \, du =$$

$$= (1/(p + 1 - q)) \left((2b)^{q-p-1} \int_0^{2b} v^{p+1} \, dF_k(v) + \int_{2b}^{\infty} v^q \, dF_k(v) \right).$$

On the other hand,

$$b^{q-p-1} \int_0^{2b} v^{p+1} \, dF_k(v) < 2\mathbf{E} \, | X_k |^p \, b^{q-p},$$

$$\int_{2b}^{\infty} v^q \, dF_k(v) < (2b)^{q-p} \mathbf{E} \mid X_k \mid^p .$$

Therefore,

$$\int_{2b}^{\infty} u^{q-p-2} \, \overline{A}_{p+1} \, du < c(p,q) A_p b^{q-p}. \tag{19}$$

Further,

$$\int_{2b}^{\infty} u^{q-p-2} \, \overline{A}_3 \, du < 2A_3 (2b)^{q-p-1} / (p+1-q). \tag{20}$$

It follows from (16)—(20) that

$$I_{12} < c(p,q)(A_3 b^{q-3} + A_p b^{q-p}). \tag{21}$$

Combining (16) and (21), we get

$$I_1 \leqslant c(p,q)(A_3 b^{q-3} + A_p b^{q-p}). \tag{22}$$

Estimate now

$$I_2 = \int_{-\infty}^{\infty} \mid u \mid^{q-1} \left| \Phi((u - \overline{a})/\overline{b}) - \Phi(u/b) \right| dx.$$

Clearly,

$$\mid \Phi((u-\overline{a})/\overline{b}) - \Phi(u/\overline{b}) \mid < \mid \overline{a} \mid (2\pi \overline{b}^2)^{-1/2} (e^{-u^2/2\overline{b}^2} + e^{-(u-\overline{a})^2/2\overline{b}^2}). \tag{23}$$

Hence, in view of (17) and the bound $\overline{a} < A_3/4b^2$, we have

$$\int_{-\infty}^{\infty} \mid u \mid^{q-1} \mid \Phi((u - \overline{u})/\overline{b}) - \Phi(u/\overline{b}) \mid \, du < c(p,q) A_3 b^{q-3}.$$

On the other hand,

$$\int_{0}^{\infty} u^{q-1} (\Phi(u/\overline{b}) - \Phi(u/b)) \, du < c(q)(b^q - b_0^q) < c(q) A_3 b^{q-3}.$$

Consequently,

$$I_2 < c(p,q) A_3 b^{q-3}. \tag{24}$$

At last

$$\int_{0}^{\infty} \mid u \mid^{q-1} \left| \mathbf{P}(\overline{S} < u) - \mathbf{P}(S < u) \, du \right| \leqslant \int_{0}^{\infty} u^{q-1} \sum_{k=1}^{n} \mathbf{P}(\mid X_k \mid > \max[2b, u]) \, du =$$

$$= \int_0^{2b} + \int_{2b}^{\infty} < c(p,q)(A_3 b^{q-3} + A_p b^{q-p}). \tag{25}$$

Combining (13), (14), (22), (24) and (25), we get (10).

Remark 1. Our arguments in the proof of Theorem B are similar to those in (Petrov, 1965).

Remark 2. It follows from (12) and (14) that

$$| \mathbf{P}(S < u) - \Phi(u/A_2^{1/2}) | \leqslant c(p)(\overline{A}_3 \overline{A}_2{}^{(p-3)/2} + \overline{A}_p)/(\overline{A}_2^p + | u - \overline{a} |^p) +$$

$$+ | \Phi((u-\overline{a})/\overline{A}_2^{1/2}) - \Phi(u/A_2^{1/2}) | + \sum_{k=1}^{n} \mathbf{P}(| X_k | > \max(2b, u)).$$

In the case of identically distributed r.v. this bound sharpens that of (Osipov, 1967).

One may estimate the difference $\Phi((u-\overline{u})/A_2^{1/2}) - \Phi(u/A_2^{1/2})$, applying (23) and the inequalities

$$\Phi(u/\overline{b}) - \Phi(u/b) < u/(2\pi)^{1/2}(1/\overline{b} - 1/b)e^{-u^2/b^2}, \quad u > 0,$$

$$1/\overline{b} - 1/b = (b^2 - \overline{b}^2)/[b\overline{b}(b+\overline{b})] < (1/\overline{b}b^2) \sum_{j=1}^{n} \mathbf{E}(| X_j |^2; | X_j | > \max(2b, u)).$$

For any $s > 0$ define $N(s)$ as the least of integers N such that $N > 2s$. Let

$$f(B,s) = \begin{cases} \left(\prod_{j=1}^{2s} \sigma_j \right)^{-1} \ln(\sigma_{N-1}/\sigma_N), \ N = N(s), & \text{if } s > 0, \ N(s) = 2s+1, \\ \left(\sigma_N^{2s-N} \prod_{j=1}^{N} \sigma_j \right)^{-1}, \ N = N(s), & \text{if } s > 0, \ N(s) \neq 2s+1, \\ 1, & \text{if } s = 0, \end{cases}$$

$$g(B,s) = \begin{cases} f(B,s), & \text{if } s \geqslant 0, \\ \mathbf{E} | Y |^{-s}, & \text{if } s \leqslant 0. \end{cases}$$

THEOREM 3. *For any* $0 < q \leqslant p < \infty$

$$\left| \mathbf{E} | S |^q - \mathbf{E} | Y |^q \right| <$$

$$< c(p,q) \begin{cases} A_p f(B,(p-q)/2), & \text{if } p \leqslant 3, \\ A_p f(B,(p-q)/2) + A_3 g(B,(q-3)/2), & \text{if } p > 3. \end{cases} \tag{26}$$

Proof. Define

$$A_t(\alpha) = \sum_{j=1}^{n} \mathbf{E} | \langle X_j, \alpha \rangle |^t .$$

According to Theorem B and the formula (3) for any $0 < q < p,\ p > 3$,

$$| \mathbf{E} \mid S \mid^q - \mathbf{E} \mid Y \mid^q | < c(p,q)(\mathbf{E}_\alpha A_3(\alpha)\langle B\alpha, \alpha\rangle^{(q-3)/2} +$$

$$+ \mathbf{E}_\alpha A_p(\alpha)\langle B\alpha, \alpha\rangle^{(p-q)/2}). \tag{27}$$

Define $\xi = \sum\limits_{j=1}^{\infty} \sigma_j^2 \alpha_j^2$.

Lemma 1. *For any* $1 \leqslant k < \infty$

$$\mathbf{P}(\xi < r) < (er^{k/2}) / \prod_{j=1}^{k} \sigma_j. \tag{28}$$

Proof. It is known that

$$g(t) = \mathbf{E}\exp\{-t\xi\} = \prod_{j=1}^{\infty}(1 + 2\sigma_j^2 t)^{-1/2}. \tag{29}$$

Apply the unequality

$$\mathbf{P}(\xi < r) < e\, g(1/r) \tag{30}$$

to estimate $\mathbf{P}(\xi < r)$. It is easily seen that for any $1 \leqslant k \leqslant N$

$$g(1/r) < r^{k/2} / \prod_{j=1}^{k} \sigma_j. \tag{31}$$

The bounds (30) and (31) yield the assertion of the lemma. Note that

$$g(1/r) > r^{k/2} \Big/ \left(3^N \prod_{j=1}^{k} \sigma_j\right),$$

provided $\sigma_{k+1}^2 \leqslant r \leqslant \sigma_k^2$.

Lemma 2. *For any* $s > 0$

$$\mathbf{E}\xi^{-s} < c(s) f(B, s). \tag{32}$$

Proof. Clearly,

$$\mathbf{E}\xi^{-s} = (s+1)\int_0^\infty u^{-s-1}\mathbf{P}(\xi < u)\, du = (s+1)\left(\int_0^{\sigma_N^2} + \sum_{k=1}^{N-1} \int_{\sigma_{k+1}^2}^{\sigma_k^2} + \int_{\sigma_1^2}^{\infty}\right). \tag{33}$$

By Lemma 1,

$$\int_{\sigma_{k+1}^2}^{\sigma_k^2} < 2e\Big/\sigma_{k+1}^{2s-k}\prod_{j=1}^{k}\sigma_j(2s-k)$$

provided $k < 2s$. If $k = 2s$, then

$$\int_{\sigma_{k+1}^2}^{\sigma_k^2} < e\Big(\prod_{j=1}^{k}\sigma_j\Big)^{-1}\ln\left(\sigma_k/\sigma_{k+1}\right).$$

Further,

$$\int_{0}^{\sigma_N^2} < 2e/(2s-N)\sigma_N^{2s-N}\prod_{j=1}^{N}\sigma_j, \quad \int_{\sigma_1^2}^{\infty} < 1/(s+1)\sigma_1^{2s}.$$

Combining (33) and the above bounds, we obtain the assertion of Lemma 2.
Define $\eta^2 = \sum_{i=1}^{N(s)}\alpha_i^2$.

LEMMA 3. *For any* $t > 0,\ s > 0$

$$\mathbf{E}(\eta^t/\xi^s) < c(t,s)\,f(B,s). \tag{34}$$

Proof. First of all,

$$\eta^t < c(s,t)\sum_{j=1}^{N(s)} |\,\alpha_j\,|^t. \tag{35}$$

It is easily seen that

$$\mathbf{E}(|\,\alpha_j\,|^t/\xi^s) = c(t)\mathbf{E}\Big(\sum_{i\neq j}\alpha_i^2\sigma_i^2 + \xi_j^2\Big)^{-s}, \tag{36}$$

where ξ_j does not depend on $\alpha_i,\ i\neq j$, and has the density

$$|\,x\,|^t\,e^{-x^2/2\sigma_j^2}\big/\sigma_j\sqrt{2\pi}\,c(t), \quad c(t) = \mathbf{E}\,|\,\alpha_1\,|^t.$$

Defining

$$\mu = \sum_{i\neq j}\alpha_i^2\sigma_i^2 + \xi_j^2,$$

we have

$$\mathbf{E}\,e^{-\mu u} = \sigma_j^t/(1+2\sigma_j^2u)^{-(t+1)/2}\prod_{i\neq j}(1+2\sigma_i^2u)^{1/2}.$$

Hence,

$$\mathbf{E}\,e^{-\mu u} \leqslant \prod_{i=1}^{\infty}(1+2\sigma_i^2u)^{-1/2}.$$

We see that $\mathbf{E}\, e^{-\mu u}$ has the same bound as $\mathbf{E}\, e^{-\xi u}$. It means that the bound (32) is valid for $\mathbf{E}\, \mu^{-s}$. It remains to apply (35) and (36).
Continue to prove Theorem 3.
Put

$$\alpha_1^m = \{\alpha_1, \alpha_2, \ldots, \alpha_m, 0, 0, \ldots\}, \quad \alpha_2^m = \alpha - \alpha_1^m.$$

Clearly,

$$A_p(\alpha) \leqslant 2^{p-1}\big(A_p(\alpha_1^m) + A_p(\alpha_2^m)\big), \quad \langle B\alpha, \alpha\rangle \geqslant \langle B\alpha_1^m, \alpha_1^m\rangle.$$

Therefore,

$$2^{1-p}\mathbf{E}_\alpha A_p(\alpha)\langle B\alpha, \alpha\rangle^{(q-p)/2} \leqslant \mathbf{E}_\alpha A_p(\alpha_1^m)\langle B\alpha_1, \alpha_1\rangle^{(q-p)/2} +$$

$$+\mathbf{E}_\alpha A_p(\alpha_2^m)\, \mathbf{E}\langle B\alpha_1^m, \alpha_1^m\rangle^{(q-p)/2}. \tag{37}$$

It is easily seen, using the formula (3), that

$$\mathbf{E}_\alpha A_p(\alpha_1^m)\langle B\alpha_1^m, \alpha_1^m\rangle^{(q-p)/2} =$$

$$= \sum_{j=1}^{n} \mathbf{E}_{X_j} \mathbf{E}_\alpha \mid \langle X_j, \alpha_1^m\rangle \mid^p \mid \langle B\alpha_1^m, \alpha_1^m\rangle \mid^{(q-p)/2}.$$

Letting $m = N(s)$ and applying Lemma 3, we have

$$\mathbf{E}_\alpha \mid \langle X_j, \alpha_1^m\rangle \mid^p \langle B\alpha_1^m, \alpha_1^m\rangle^{(q-p)/2} \leqslant$$

$$\leqslant \mid X_j \mid^p \mathbf{E}_\alpha \left(\sum_{j=1}^{m} \alpha_j^2 \right)^{p/2} \langle B\alpha_1^m, \alpha_1^m\rangle^{(q-p)/2} \leqslant c(p,q) \mid X_j \mid^p f(B, (p-q)/2).$$

Thus,

$$\mathbf{E}_\alpha A_p(\alpha_1^m)\langle B\alpha_1^m, \alpha_1^m\rangle^{(q-p)/2} < c(p,q)\, A_p\, f(B, (p-q)/2), \quad m = N(s). \tag{38}$$

According to formula (3)

$$\mathbf{E}_\alpha \mathbf{E}_{X_j} \mid \langle X_j, \alpha_2^m \mid^p = \mathbf{E}_{X_j} \left(\sum_{i=m+1}^{\infty} \langle e_i, X_j\rangle^2 \right)^{p/2} \leqslant \mathbf{E} \mid X_j \mid^p. \tag{39}$$

By Lemma 2, for $m = N(s)$

$$\mathbf{E}_\alpha \langle B\alpha_1^m, \alpha_1^m\rangle^{(q-p)/2} \leqslant c(p,q)\, f(B, (p-q)/2). \tag{40}$$

It follows from (37)—(40) that

$$\mathbf{E}_\alpha A_p(\alpha)\langle B\alpha, \alpha\rangle^{(q-p)/2} \leqslant c(p,q)\, A_p\, f(B, (p-q)/2). \tag{41}$$

Proceed now to estimating $\mathbf{E}_\alpha A_3(\alpha)\langle B\alpha, \alpha\rangle^{(q-3)/2}$. If $q \leqslant 3$, then by (41)

$$\mathbf{E}_\alpha A_3(\alpha)\langle B\alpha, \alpha\rangle^{(q-3)/2} \leqslant c(q)\, A_3\, f(B, (3-q)/2). \tag{42}$$

The case $q > 3$ is more complicated.

LEMMA 4. *Let ξ_j, $j = 1,2,\dots,m$, be real Gaussian r.v., $\mathbf{E}\xi_j = 0$, $j = 1,2,\dots,m$. Then for any integer k*

$$\mathbf{E}\prod_{j=1}^{m}\xi_j^k = \begin{cases} \sum^1 a_1 \prod_{i,k} r_{ik}^{\mu_{ik}}, & \text{if } km \text{ is even,} \\ 0, & \text{if } km \text{ is odd,} \end{cases}$$

where $r_{ik} = \mathbf{E}\xi_i\xi_k$, $\sum^1$ is the summation over all matrix $\{\mu_{ik}\}$ such that μ_{ik} are integers, $\sum_{l=1}^{m}\mu_{jl} + \sum_{l=1}^{m}\mu_{lj} - \mu_{jj} = 2$, $1 < j \leqslant m$, coefficients a_1 depend on $\{\mu_{ik}\}$.

Proof. Let

$$\xi = \sum_{j=1}^{m}\lambda_j\xi_j, \quad \varphi(s) = \mathbf{E}\, e^{s\xi}.$$

It is not difficult to see that $\mathbf{E}\prod_{j=1}^{m}\xi_j^k$ is equal to the coefficient at $\prod_{j=1}^{m}\lambda_j^k$ in the polynomial $\mathbf{E}\xi^{km}$. On the other hand,

$$\mathbf{E}\xi^l = \frac{d^l}{ds^l}\varphi(s)\mid_{s=0} = (-1)^l H_l(0)(\mathbf{E}\xi^2)^{l/2}, \tag{43}$$

where H_l is the Hermite polynomial of the degree l.
Clearly,

$$\mathbf{E}\xi^2 = \sum_{i,k}\lambda_i\lambda_k r_{ik}. \tag{44}$$

It remains to insert (44) into (43), put $l = km$ and calculate the coefficient at $\prod_{j=1}^{m}\lambda_j^k$. Note that $\mathbf{E}\,\xi^l \equiv 0$, provided l is odd.
For any integer $k > t > 0$

$$\begin{aligned} \mathbf{E}_\alpha\mathbf{E}_{X_j} \mid \langle X_j,\alpha\rangle \mid^3 \langle B\alpha,\alpha\rangle^t &= \mathbf{E}_{X_j}\mathbf{E}_\alpha \mid \langle X_j,\alpha\rangle \mid^3 \langle B\alpha,\alpha\rangle^t \leqslant \\ &\leqslant \left(\mathbf{E}_{X_j}(\mathbf{E}_\alpha \mid \langle X_j,\alpha\rangle \mid^{3k/(k-t)})^{(k-t)/k}\right)\left(\mathbf{E}_\alpha\langle B\alpha,\alpha\rangle^k\right)^{t/k}. \end{aligned} \tag{45}$$

Clearly,

$$\mathbf{E}_\alpha \mid \langle X_j,\alpha\rangle \mid^{3k/(k-t)} = \mid X_j \mid^{3k/(k-t)} \mathbf{E} \mid \alpha_1 \mid^{3k/(k-t)}.$$

Hence,

$$\begin{aligned} &\mathbf{E}_{X_j}(\mathbf{E}_\alpha \mid \langle X_j,\alpha\rangle \mid^{3k/(k-t)})^{(k-t)/k} \leqslant \\ &\leqslant \mathbf{E} \mid X_j \mid^3 (\mathbf{E} \mid \alpha_1 \mid^{3k/(k-t)})^{(k-t)/k}. \end{aligned} \tag{46}$$

Next

$$\mathbf{E}_\alpha\langle B\alpha,\alpha\rangle^k = \mathbf{E}_\alpha(\mathbf{E}_Y\langle Y,\alpha\rangle^2)^k = \mathbf{E}_{Y_1,Y_2,\dots,Y_k}\mathbf{E}_\alpha\prod_{j=1}^{k}\langle Y_j,\alpha\rangle^2, \tag{47}$$

where Y_j are independent copies of Y. Applying Lemma 4, we obtain the bound

$$\mathbf{E}_\alpha \prod_{j=1}^{k} \langle Y_j, \alpha\rangle^2 < c(k)(\mathbf{E} \mid Y \mid^2)^k.$$

From (45)—(47) we get, letting $k = [t] + 1,\ t = (q-3)/2$,

$$\mathbf{E}_\alpha \mathbf{E}_{X_j} \mid \langle X_j, \alpha\rangle \mid^3 \langle B\alpha, \alpha\rangle^{(q-3)/2} < c(q)\,\mathbf{E} \mid X_j \mid^3 (\mathbf{E} \mid Y \mid^2)^{(q-3)/2}, \tag{48}$$

provided $q > 3$.
Combining (27), (41), (42) and (48), we come to the assertion of Theorem 3 in the case $p > 3$. In the case $p \leqslant 3$ we are able to use Theorem A and the bound (41).

Remark 3. In the case of i.d. r.v. X_k the bounds for $\left|\mathbf{E} \mid S \mid^q - \mathbf{E} \mid Y \mid^q \right|$ were obtained recently in (Zalessky et al., 1988a). It is rather complicated to compare in detail our results with those of (Zalessky et al., 1988a). However, it is clear that our bounds can not be derived from results obtained there even in the case of i.d. r.v. X_k.
Now deduce, in addition to (3), another formula concerning the distribution of r.v. $\mid X \mid$.
First for any complex z

$$e^{z|X|^2} = \mathbf{E}_\alpha \exp\left\{\sqrt{2z}\langle X, \alpha\rangle\right\}.$$

Hence,

$$\mathbf{E}\, e^{z|X|^2} = \mathbf{E}_X \mathbf{E}_\alpha \exp\left\{\sqrt{2z}\langle X, \alpha\rangle\right\}.$$

If the change of the order of integration is admissible, then

$$\mathbf{E}\, e^{z|X|^2} = \mathbf{E}_\alpha \mathbf{E}_X \exp\left\{\sqrt{2z}\langle X, \alpha\rangle\right\}. \tag{49}$$

Let X is Gaussian, $\mathbf{E} X = 0$ and b_j^2 be eigenvalues of the covariance operator of X. Then

$$\mathbf{E}_X \exp\left\{\sqrt{2z}\langle X, \alpha\rangle\right\} = \exp\left\{z \sum_{j=1}^{\infty} b_j^2\, \alpha_j^2\right\},$$

$$\mathbf{E}_\alpha \exp\left\{z \sum_{j=1}^{\infty} b_j^2\, \alpha_j^2\right\} = \prod_{j=1}^{\infty} (1 - 2\, z\, b_j^2)^{-1/2},$$

if $\operatorname{Re} z < 1/(2b^2),\ b^2 = \max b_j^2$.
Thus the formula (49) holds for Gaussian r.v. provided $\operatorname{Re} z < 1/(2b^2)$.
Clearly,

$$\mathbf{E}_X \mathbf{E}_\alpha \mid \exp\left\{\sqrt{2z}\langle X, \alpha\rangle\right\} \mid \leqslant \sum_{k=0}^{\infty} \mathbf{E}_X \mathbf{E}_\alpha \mid \langle X, \alpha\rangle \mid^k (2 \mid z \mid)^{k/2}/k!\,.$$

It means, by the formula (3), that the representation (49) is valid for any z, provided r.v. X is bounded.

In the general case the formula (49) holds in some generalized sense. Putting in (49) $z = it$, $X = S$, we have

$$\mathbf{E}\, e^{it|S|^2} = \mathbf{E}_\alpha \prod_{j=1}^{n} \mathbf{E}_{X_j} \exp\left\{\sqrt{2it}\langle X_j, \alpha\rangle\right\}.$$

Using this representation of a characteristic function, it is possible to estimate the error of the normal approximation for the distribution of $| S - a |$.

THEOREM 4. *If X_j are i.d., then*

$$\left|\mathbf{P}(| S - a |< r) - \mathbf{P}(| Y - a |< r)\right| < c A_3 \Big((\mathbf{E} \mid S \mid^2)^{3/2} + \mid a \mid^3 \Big) \Big/ \prod_{j=1}^{6} \sigma_j.$$

Remark 4. This result was obtained in the slightly different form in (Zalessky et al., 1988b, 1989), using the different approach.

Remark 5. The formula similar to (49) for $\operatorname{Im} z = 0,\ z < 0$ is used in (Csörgő, 1975) in more special terms.

REFERENCES

Bhattacharya, R.N. and Ranga Rao, R. (1976). *Normal approximation and asymptotic expansions.* Wiley, New York.

Csörgő, S. (1975). Asymptotic expansion for the Laplace transform of the von Mises w^2-test. *Theory Probab. Appl.* **20**, 158–162.

Osipov, L.V. (1967). Asymptotic expansions in the central limit theorem. *Leningrad Univ. Vestnik* **19**, 45–62.

Petrov, V.V. (1965). One estimate of the deviation of the sum of independent random variables from the normal law. *Soviet Math. Dokl.* **160**, 1013–1015.

Sakoyan, S.K. (1974). *Some Estimates of Distributions of Sums of Independent Random Variables, Random Processes and Statistical Inference.* Fan, Tashkent, pp. 136–151.

Zalessky, B.A., Sazonov, V.V. and Ulyanov, V.V. (1988a). Normal approximation in Hilbert space. I, II. *Theory Probab. Appl.* **33**, No 2 225–245; No 3 508–521.

Zalessky, B.A., Sazonov, V.V. and Ulyanov, V.V. (1988b). A precise estimate of accuracy of normal approximation in Hilbert space. *Theory Probab. Appl.* **33**, 753–754.

Zalessky, B.A., Sazonov, V.V. and Ulyanov, V.V. (1989). A precise estimate of accuracy of normal approximation in Hilbert space. *Preprint* N 6, Steklov Math. Inst.

Prob. Theory and Math. Stat., Vol. 2, pp. 227–233
B.Grigelionis *et al.* (Eds.)
1990 VSP/Mokslas

RECORDS FOR NONIDENTICALLY DISTRIBUTED RANDOM VARIABLES

V.B. NEVZOROV
Leningrad University, Bibliotechnaya square 2, 198904 Leningrad, USSR

ABSTRACT

Let $X_1, X_2, \ldots$ be a sequence of independent random variables (r. v.) with distribution functions (d. f.) $F_1, F_2, \ldots, L(n)$ be the record times, i.e., $L(1) = 1$ and $L(n+1) = \min\{j : X_j > X_{L(n)}\}$ $(n = 1, 2, \ldots)$, and $X(n) = X_{L(n)}$ $(n = 1, 2, \ldots)$ be the corresponding record values. Beginning from the early fifties (Chandler, 1952) there are now more than 210 papers on records (see, for example, the reviews (Galambos, 1978, Chapter 6) and (Nevzorov, 1987), but only a few of them consider record times and record values in nonstationary situations (see (Nevzorov, 1987a, §9)). In (Nevzorov, 1985, 1986, 1987a) the so-called F^α-scheme was studied. The idea of this scheme was taken from (Yang, 1975). A generalization of the F^α-scheme including the case of dependent r.v. $X_1, X_2, \ldots$ was suggested in (Ballerini and Resnick, 1987).

1. THE F^α-SCHEME

We consider the F^α-scheme, that is a sequence of independent r.v. $X_1, X_2, \ldots$ with d.f. $F_k(x) = (F(x))^{\alpha_k}$ $(k = 1, 2, \ldots)$, where $F(x)$ is a continuous d.f. and $\alpha_1; \alpha_2, \ldots$ are positive constants (it's natural to assume that $\alpha_1 = 1$). The F^α-scheme is a generalization of the scheme of sequences of independent identically distributed random variables.

Define

$$S(n) = \alpha_1 + \cdots + \alpha_n, \quad A(n) = \sum_{1 \leqslant k \leqslant n} \alpha_k / (\alpha_1 + \cdots + \alpha_k)$$

and

$$B(n) = \sum_{1 \leqslant k \leqslant n} \alpha_k^2 / (\alpha_1 + \cdots + \alpha_k)^2 \ (n = 1, 2, \ldots).$$

Let $S(n) \to \infty$ $(n \to \infty)$. Then $P(L(n) < \infty) = 1$, $P(X(n) < \infty) = 1$ $(n = 1, 2, \ldots)$ and $A(n) \to \infty$ $(n \to \infty)$.

The following lemma which represents a generalization of Renyi's lemma for i.i.d. sequences is the key result for the F^α-scheme.

LEMMA 1(Nevzorov, 1985). *For the F^α-scheme define the following sequence of r.v. $\xi_1, \xi_2, \ldots : \xi_k = 1$, if X_k is a record, and $\xi_k = 0$, otherwise.*

The r.v. $\xi_1, \xi_2, \dots$ are independent and

$$\mathbf{P}(\xi_k = 1) = 1 - \mathbf{P}(\xi_k = 0) = \alpha_k/(\alpha_1 + \dots + \alpha_k) \qquad (k = 1, 2, \dots).$$

Remark 1. Let $N(n)$ be the number of records among the r.v. $X_1, X_2, \dots, X_n$. Then

$$N(n) = \xi_1 + \dots + \xi_n$$

and

$$\mathbf{P}(L(n) > m) = \mathbf{P}(N(m) < n) = \mathbf{P}(\xi_1 + \dots + \xi_m < n).$$

The latter equality makes it possible to use the well-known limit theorems for sums of independent r.v. for investigation of the record times $L(n)$.

2. RECORDS AND MARTINGALES

In (Nevzorov, 1987a, 1987b, 1988) the martingale approach was offered for studying record times for i.i.d. sequences. It's known that a generalization of record times —the so-called k-th record times $L^{(k)}(n)$ are defined as $L^{(k)}(1) = k$ and $L^{(k)}(n+1) = = \min\{j > L^{(k)}(n) : X_j > X_{j-k,j-1}\}$ $(n = 1, 2, \dots)$, where $X_{1,n} \leqslant \dots \leqslant X_{n,n} (n = 1, 2, \dots)$ are the corresponding order statistics. The martingale method has allowed to obtain some interesting results for r.v. $L^{(k)}(n)$. Theorem 1 and Theorem 2 are formulated for i.i.d. random variables $X_1, X_2, \dots$ with a continuous d.f.

THEOREM 1. *For any $k = 1, 2, \dots$ and $\beta < k$ the sequence*

$$W_n = \frac{(k-\beta)^n \Gamma(L^{(k)}(n)+1)}{k^n \Gamma(L^{(k)}(n) - \beta + 1)} \quad (n = 1, 2, \dots),$$

where $\Gamma(s) = \int_0^\infty x^{s-1} e^{-x} dx$, is a martingale.

COROLLARY 1. *If $\alpha > 0$ then*

$$\mathbf{E}\Big(L^{(k)}(n)\Big)^{k-\alpha} = \frac{\Gamma(k)}{\Gamma(\alpha)}\left(\left(\frac{k}{\alpha}\right)^n + \frac{(\alpha-k)(\alpha-k+1)}{2\alpha}\left(\frac{k}{\alpha+1}\right)^n\right) + $$
$$+ O\left(\left(\frac{k}{\alpha+2}\right)^n\right) \quad (n \to \infty).$$

THEOREM 2. *For any $k = 1, 2, \dots$ and $\gamma > 1$ the sequence*

$$U_n = \gamma^n B\Big(L^{(k)}(n) + 1,\ k(\gamma - 1)\Big) \quad (n = 1, 2, \dots),$$

where $B(a,b)=\int_0^1 x^{a-1}(1-x)^{b-1}dx$, is a martingale.

COROLLARY 2. *For any $k=1,2,\dots$ the following equality for the generating functions $P_n^{(k)}(s)$ of r.v. $L^{(k)}(n)$ holds:*

$$P_n^{(k)}(s)=E\ s^{L^{(k)}(n)}=H(-\ln(1-s)),$$

where

$$H(v)=\frac{k^n}{\Gamma(n)}\int_0^v x^{n-1}e^{-kx}(1-e^{-(v-x)})^{k-1}dx.$$

This result for $k=1$ was obtained in (Renyi, 1962):

$$P_n^{(1)}(s)=\frac{1}{\Gamma(n)}\int_0^{-\ln(1-s)} x^{n-1}e^{-x}dx.$$

The martingale method can be also used for the F^α–scheme.

THEOREM 3. *For the F^α–scheme the sequences*

$$R_n=A(L(n))-n \quad (n=1,2,\dots)$$

and

$$T_n=(A(L(n))-n)^2-A(L(n))+B(L(n)) \quad (n=1,2,\dots)$$

are martingales.

THEOREM 4. *The sequence*

$$V_n=(1+c)^n\Big/\prod_{1\leqslant k\leqslant L(n)}\Big(1+c\alpha_k\big/(\alpha_1+\cdots+\alpha_k)\Big) \quad (n=1,2,\dots)$$

is a martingale for any $c>0$.

Below we give some useful corollaries of Theorems 3 and 4.

COROLLARY 3. $\mathbf{E}\big(A(L(n))\big)=n \quad (n=1,2,\dots)$.

COROLLARY 4. *$D(A(L(n))) = n - \mathbf{E}(B(L(n)))$ $(n = 1, 2, \ldots)$ and $D(A(L(n))) = n - B + o(1)$ $(n \to \infty)$ if $B = B(\infty) < \infty$.*

COROLLARY 5. *For any $\gamma > 0$ and $\delta = (e/\alpha_1)^{\gamma-1}$ the inequalities*

$$\mathbf{E}\big(S(L(n))\big)^{1-\gamma} \leqslant \delta\gamma^{-n} \quad (n = 1, 2, \ldots)$$

hold.

These corollaries may be interesting because of the following inequalities.

THEOREM 5. *If $\sum_{k=1}^{\infty} \alpha_k^2/(\alpha_1 + \cdots + \alpha_k)^2 < \infty$, then*

$$\sup_x \left|\mathbf{P}\big(A(L(n)) - n < xn^{1/2}\big) - \Phi(x)\right| \leqslant C/n^{1/2}$$

and

$$\sup_x \left|\mathbf{P}\big(\ln\ S(L(n)) - n < xn^{1/2}\big) - \Phi(x)\right| \leqslant C/n^{1/2},$$

where

$$\Phi(x) = (2\pi)^{-1/2} \int_{-\infty}^{x} \exp(-t^2/2)dt$$

and C is a constant.

The main idea of the proofs of Theorems 1—4 consists in substituting Markov times into certain martingales.
We know, for example, that

$$V_1(n) = \sum_{i=1}^{n} \xi_i - \sum_{i=1}^{n} \mathbf{E}\xi_i =$$

$$= N(n) - \mathbf{E}N(n) = N(n) - \sum_{k=1}^{n} \alpha_k/(\alpha_1 + \cdots + \alpha_k) \quad (n = 1, 2, \ldots)$$

is a martingale as a sequence of centered sums of independent random variables. It isn't difficult to see that $L(n)$ $(n = 1, 2, \ldots)$ is a sequence of Markov times. After verifying some conditions we can obtain that

$$V_1\big(L(n)\big) = N\big(L(n)\big) - \sum_{k=1}^{L(n)} \alpha_k/(\alpha_1 + \cdots + \alpha_k) = n - A\big(L(n)\big) \quad (n = 1, 2, \ldots)$$

is a martingale. We used here the obvious equality $N(L(n)) = n$. So

$$R_n = -V_1(L(n)) \ (n = 1, 2, \ldots)$$

is also a martingale.
To prove the second part of Theorem 3 we consider the martingale

$$V_2(n) = \left(N(n) - \mathbf{E}N(n)\right)^2 - D\left(N(n)\right) \quad (n = 1, 2, \ldots)$$

and the Markov times $L(n)$ $(n = 1, 2, \ldots)$.
Theorems 1, 2 and 4 can be proved by substituting the Markov times $L^{(k)}(n)$ or $L(n)$ into the martingale

$$V_3(n) = \gamma^{N(n)}/\mathbf{E}\gamma^{N(n)} \quad (n = 1, 2, \ldots)$$

with some $\gamma > 1$.

3. RECORDS AND CHARACTERIZATIONS

Let $M_n = \max\{X_1, \ldots, X_n\}$ and $\xi_1, \xi_2, \ldots$ be the random variables from lemma 1. It follows from (Ballerini and Resnick, 1987) that the vectors $M = (M_1, \ldots, M_n)$ and $\xi = (\xi_1, \ldots, \xi_n)$ $(n = 1, 2, \ldots)$ are independent for the F^α-scheme. The following characterization theorem is connected with this result.

THEOREM 6. *Let the d.f. $F_1, F_2, \ldots, F_n$ of independent r.v. $X_1, X_2, \ldots, X_n$ be continuous and $0 < F_i(a) < F_i(b) < 1$ $(1 \leqslant i \leqslant n)$ for some a and $b(-\infty < a < < b < \infty)$. If M and ξ are independent then there exist such positive constants α_i $(2 \leqslant i \leqslant n)$ that $F_i(x) = (F_1(x))^{\alpha_i}$ $(i = 2, 3, \ldots, n)$ for $-\infty < x < \infty$ and the r.v. $\xi_1, \xi_2, \ldots, \xi_n$ are independent.*

It's interesting to compare Theorem 6 with the following result, which also characterizes the F^α-scheme.

THEOREM 7 (Nevzorov, 1986). *Let $X_1, X_2, \ldots, X_n$ be independent r.v. having d.f. $F_1, F_2, \ldots, F_n$ and densities $f_1, f_2, \ldots, f_n$, such that $\prod_{i=1}^{n-1} f_i(x) \neq 0$ for all x from some interval $(\alpha, \beta)(-\infty \leqslant \alpha < \beta \leqslant \infty)$ and let the vector $\xi = (\xi_1, \ldots, \xi_{n-1})$ and the r.v. ξ_n be independent for any density function f_n. There exist such positive constants α_i $(2 \leqslant i \leqslant n-1)$ that $F_i(x) = (F_1(x))^{\alpha_i}$ $(2 \leqslant i \leqslant n-1)$ for $-\infty < x < \infty$ and the components $\xi_1, \xi_2, \ldots, \xi_{n-1}$ of ξ are independent.*

4. LIMIT THEOREMS FOR RECORD VALUES IN THE F^α–SCHEME

Above we have given some asymptotic results for the record times (Theorem 5) in the F^α-scheme. Now we are going to obtain limit theorems for record values in this scheme. We consider first the F^α-scheme with the d.f. $F(x) = \max\{0, 1 - e^{-x}\}$. In this case the following theorem is valid.

THEOREM 8. *Let $p_n = \alpha_n/(\alpha_1 + \cdots + \alpha_n) \to p \quad (n \to \infty)$, where $0 \leqslant p < 1$. Then*

$$\mathbf{P}\Big(X(n) - an < bxn^{1/2}\Big) \to \Phi(x) \quad (n \to \infty),$$

where $a = b = 1$, if $p = 0$, and $a = (-\ln(1-p))/p$, $b = a(1-p)^{1/2}$, if $0 < p < 1$ and $\sum_{n=1}^{\infty} |p_n - p| < \infty$.

Theorem 8, the Smirnov transformation and standard methods used in the i.i.d. case help to find limit distributions for suitable normalized r.v. $X(n)$ in the F^{α}–scheme with any continuous d.f.F. It can be proved that the set of possible asymptotic distributions for the F^{α}–scheme under the conditions on $\alpha_1, \alpha_2, \ldots$ given in Theorem 8 coincides with the set of asymptotic distributions in the stationary scheme with $\alpha_1 = \alpha_2 = \cdots = 1$. This set includes the following three types of limit distribution functions (Resnick, 1973):

$$H_i(x) = \Phi(g_i(x)) \quad (i = 1, 2, 3),$$

where

$$g_1(x) = x \quad (-\infty < x < \infty);$$

$$g_2(x) = \alpha \ln x, \quad \text{if} \quad x > 0, \quad \text{and} \quad g_2(x) = -\infty, \quad \text{if} \quad x < 0 \quad (\alpha > 0);$$

$$g_3(x) = -\alpha \ln(-x), \quad \text{if} \quad x < 0, \quad \text{and } g_3(x) = \infty, \quad \text{if} \quad x > 0 \quad (\alpha > 0).$$

Remark 2. The limit laws for record values $X(n)$ in the F^{α}–scheme, but only with random norming, were considered in (Ballerini and Resnick, 1987).

Remark 3. In (Nevzorov, 1987a) it was proposed to extend the F^{α}–scheme and to consider sequences of d.f. $F_1, F_2, \ldots$, which can be represented as

$$F_k(x) = \int_0^{\infty} (F(x))^{\alpha} H_k(d\alpha) \quad (-\infty < x < \infty).$$

Here $F(x)$ is any continuous d.f. and $H_k(x) \quad (k = 1, 2, \ldots)$ are d.f.'s or even functions of bounded variation such that $H_k(\infty) - H_k(0) = 1$. This approach makes it possible to find the asymptotic distributions of $X(n)$ in more general than the F^{α}–scheme nonstationary situations.

REFERENCES

Ballerini, R. and Resnick, S.I. (1987). Embedding sequences of successive maxima in extremal processes, with applications. *J. Appl. Probab.* **24**, 827–837.

Chandler, K.N. (1952). The distribution and frequency of record values. *J.Royal Statist. Soc.* B **14**, 220–228.

Galambos, J. (1978). *The asymptotic theory of extreme order statistics.* Wiley, New York.

Nevzorov, V.B. (1985). On record times and inter-record times for sequences of nonidentically distributed random variables. *Notes Sci. Seminars Leningrad Branch of Math. Inst.* **142**, 109–118.

Nevzorov, V.B. (1986). Two characterizations using records. *Lect. Notes Math.* **1233**, 79–85.

Nevzorov, V.B. (1987a). Records. *Theory Probab. Appl.* **32**, 219–251.

Nevzorov, V.B. (1987b). Moments of some random variables connected with records. *Vestnik Leningrad State University* **8**, 33–37.

Nevzorov, V.B. and Stepanov, A.V. (1988). Records: martingale approach to finding of moments. In: *Rings and modules. Limit theorems of probability theory,* **2**, pp. 171–181. Leningrad University, Leningrad.

Renyi, A. (1976). On the extreme elements of observations. In: *Selected Papers of Alfred Renyi,* **3**, pp. 50–65. Akademiai Kiado, Budapest.

Resnick, S.I. (1973). Limit laws for record values. *Stochastic Processes and Appl.* **1**, 67–82.

Yang, M.C.K. (1975). On the distribution of the inter-record times in an increasing population. *J. Appl. Probab.* **12**, 148–154.

Prob. Theory and Math. Stat., Vol. 2, pp. 234–243
B.Grigelionis *et al.* (Eds.)
1990 VSP/Mokslas

LOCAL CHERNOFF EFFICIENCY OF LINEAR RANK TESTS

Ya.Yu. NIKITIN
Faculty of Mathematics and Mechanics, Leningrad University,
198904 Leningrad, USSR

ABSTRACT

Large deviation asymptotics of one–sample rank statistics under the alternative is found under some regularity conditions imposed on scores and on the distribution of observations. This enables to evaluate local Chernoff and Hodges—Lehmann indices of these statistics. It turns out that they coincide up to a constant factor with corresponding Pitman and Bahadur efficacies.

1. INTRODUCTION

The notion of asymptotic relative efficiency (ARE) is frequently used for the asymptotic comparison of statistical tests. The most known and studied is the Pitman ARE (see, e.g., (Kendall and Stuart, 1973; Pratt and Gibbons, 1981)). Two other types of efficiency are the Bahadur ARE (Bahadur, 1967; 1971) and the Hodges—Lehmann ARE (Hodges and Lehmann, 1956; Raghavachari, 1982). They are also important and have some indisputable merits. An objection for their use is that they are in some sense "unbalanced", because in their computation one of the two probabilities of errors is kept fixed and the other goes to zero.
The Chernoff efficiency introduced in (Chernoff, 1952) doesn't possess this drawback as both probabilities of the first–kind and of the second–kind error tend to zero. The theory of Chernoff ARE was developed by Kallenberg (Kallenberg, 1982) who computed it for certain parametric tests. Meanwhile the Chernoff efficiency of nonparametric tests is not studied well enough. The reason for that is probably connected with the difficulty of asymptotic evaluation of large deviation probabilities of test statistics under the alternative when their "distribution–free" property fails. For instance the attempt in (Hwang, 1978) to find the Chernoff ARE of the Wilcoxon two–sample test was not successful.
In this paper we find the local Chernoff efficiency of linear rank tests in one– and two–sample problem under some regularity conditions on the score function and the structure of the alternative distribution. To obtain the large deviation probability under the alternative we use the Chernoff—Savage representation of linear rank statistics and the Sanov principle (Sanov, 1957; Groeneboom et al., 1979). Then the problem of the minimization of the Kullback—Leibler information

on a special set of distributions is solved by means of variational calculus and nonlinear analysis. This method was for the first time applied to nonparametric statistics in (Mogulski, 1977), then it was developed by the author in a series of papers (Nikitin, 1979, 1980, 1982 etc.). Some results on the Chernoff efficiency for simplest statistics for testing symmetry were obtained by another method in (Nikitin, 1989).

Our results as a by-product imply also expressions for the local Hodges—Lehmann efficiency of linear rank statistics (partly announced in (Nikitin, 1987c)). Moreover we find that the local Chernoff ARE coincides in the class of linear rank statistics with the local Bahadur, intermediate and Pitman ARE known before (possibly under some other regularity conditions). Hence the local ordering of linear rank tests by their ARE under general conditions doesn't depend on the type of ARE. It is emphasized that for other test statistics (e.g., Kolmogorov—Smirnov and Cramér—von Mises statistics) the situation is completely different (Nikitin, 1987a; 1987b).

We shall give only brief proofs of results mentioned above. Complete proofs will appear elsewhere.

2. TESTING OF SYMMETRY

Let $X_1, X_2, \ldots, X_n$ be i.i.d. observations with a common distribution function (d.f.) $F(x)$. We are testing the hypothesis of symmetry

$$H_0 : 1 - F(x) - F(-x) = 0 \quad \text{for all} \quad x \in \mathbf{R}^1$$

against the parametric alternative

$$A_0 : F(x) = G(x;\theta) \quad \text{for some} \quad \theta > 0,$$

where $\{G(x;\theta)\}$ is a parametric family of d.f.'s satisfying some regularity conditions stated below. Specifically we require that the d.f. $G(x;\theta)$ is symmetric only for $\theta = 0$.

Let R_i, $i = 1, \ldots, n$, be the rank of the observation $|X_i|$ among $|X_1|, \ldots, |X_n|$. We consider the linear rank statistic

$$S_n = \frac{1}{n}\sum_{j=1}^{n} J\left(\frac{R_j}{n+1}\right) s(X_j)$$

where J is some non-constant smooth function called the score function, and $s(x) = 1$ if $x > 0$, $s(x) = 0$ otherwise. As usually we suppose that J is non-negative and that $J(0) = 0$. One requires also that the score function J is two times continuously differentiable on $[0, 1]$. It will be convenient to suppose that J is normalized in such a way that

$$\frac{1}{4}\int_0^1 J^2(u)\,du = 1.$$

We denote for brevity $J_0 = \frac{1}{2} \int\limits_0^1 J(u)\,du$.

Concerning the family $\{G(x;\theta)\}$ we assume that the d.f.'s $G(x;\theta)$ are defined for $x \in R^1$ and $\theta \in [0, \theta_0]$, $\theta_0 > 0$, and that $G(x;\theta) \neq G(x;\theta')$ for $\theta \neq \theta'$ and all x. It is also supposed that the d.f.'s $G(x;\theta)$ are absolutely continuous (in x) for all θ, the corresponding densities $g(x;\theta)$ being strictly positive. Denote by $G^{-1}(x;0)$ the inverse d.f. of $G(x;0)$ and introduce the auxiliary d.f. $r(x;\theta)$ defined on $[0,1]$ by

$$r(x;\theta) = G\Big(G^{-1}(x;0);\theta\Big). \tag{1}$$

We shall assume that for all x and θ the function $r(x;\theta)$ possesses mixed derivatives up to the second order in x and up to the first order in θ which are jointly continuous in x and θ. The derivatives in x will be denoted r', r'', the derivative in θ and mixed derivatives will be denoted r_θ, r'_θ and r''_θ.

We introduce one more quantity

$$\sigma = -\int\limits_{1/2}^{1} J'(2t-1)\Big(r_\theta(t;0) + r_\theta(1-t;0)\Big)\,dt$$

and require that $\sigma > 0$. This condition ensures the consistency of the test based on S_n for small θ.

It can be deduced from (Ho, 1973) and (Groeneboom, 1980) that under some regularity conditions (less restrictive than above) the following large deviation result holds under H_0 : as $a \downarrow 0$

$$\lim_{n\to\infty} n^{-1} \ln \mathbf{P}(S_n - J_0 \geqslant a) = -\frac{a^2}{2}(1 + o(1)). \tag{2}$$

As a corollary we obtain for the local Bahadur slope $c_S(\theta)$ of the sequence $\{S_n\}$ that as $\theta \to 0$

$$C_S(\theta) \sim \sigma^2\theta^2. \tag{3}$$

It is well-known that the Pitman efficacy of the sequence $\{S_n\}$ is also equal to σ^2 (Pratt and Gibbons, 1981; Hettmansperger, 1984).

To obtain a large deviation result under the alternative we shall use the well-known representation

$$S_n = \int\limits_0^\infty J\Big(\frac{n}{n+1}\Big(F_n(x) - F_n(-x)\Big)\Big)\,dF_n(x),$$

where F_n is the empirical d.f. of the sample $X_1, \ldots, X_n$. In order to apply large deviation results for empirical measures we introduce for any absolutely continuous d.f. f on $[0,1]$ the entropy functional

$$I(f;r) = \int\limits_0^1 f' \ln \frac{f'}{r'}\,dt,$$

and for any set Ω of such d.f.'s put

$$I(\Omega;r)=\inf\left\{I(f;r):\ f\in\Omega\right\}.$$

We shall consider for any $a>0$ the following set of d.f.'s connected with the statistic S_n:

$$\Omega_a^S=\left\{f:\int_{1/2}^{1}J\left(f(x)-f(1-x)\right)f'dx-J_0\leqslant a\right\}.$$

THEOREM 1. *Under the regularity conditions imposed above on the score function J and the parametric family $\{G(x;\theta)\}$, for sufficiently small a and θ, such that $\sigma\theta>a$, the function $a\to I(\Omega_a^S;r)$ is continuous. Moreover, if $a\to 0$ and $\theta\to 0$ so that $\sigma\theta>a$, then*

$$I(\Omega_a^S;r)=\frac{1}{2}(\sigma\theta-a)^2+o(\theta^2+a^2).$$

Applying the Smirnov transformation and Corollary 3.3 from (Groeneboom et al., 1979) we obtain the following result directly connected with the large deviation problem.

THEOREM 2. *Suppose that the alternative A_0 and all the regularity conditions stated above are fulfilled. Then for any sequence $\{\gamma_n\}$, $\gamma_n\to 0$, and sufficiently small a and θ, such that $\sigma\theta>a$, we have*

$$\lim_{n\to\infty}n^{-1}\ln P(S_n-J_0\leqslant a+\gamma_n)=-I(\Omega_a^S;r).$$

The proof of Theorem 1 is rather long and of technical nature. It can be divided into several steps.
Step 1 consists in proving that we commit a negligible error minimizing the entropy I on the subset of Ω_a^S containing only d.f.'s with the densities uniformly separated from zero.
This fact enables us at Step 2 to use the Lagrange principle (see (Alexeev et al., 1979, §3.2)) in calculating $I(\Omega_a^S)$. Putting $h=f-r$, we obtain the following Euler—Lagrange equation with a Lagrange multiplier $\mu\geqslant 0$:

$$h''-\frac{r''}{r'}h'+\mu J'\big(v(1-x)\big)(h'+r')\big(h'(1-x)+r'(1-x)\big)=0,\quad 0\leqslant x\leqslant 1/2,$$
$$h''-\frac{r''}{r'}h'+\mu J'(v)(h'+r')(h'(1-x)+r'(1-x))=0,\quad 1/2\leqslant x\leqslant 1, \tag{4}$$

where we denote $v(x)=h(x)-h(1-x)+r(x)-r(1-x)$.
Equations (4) should be considered together with the boundary conditions

$$h(0)=h(1)=0 \tag{5}$$

and the norming condition

$$\int_{1/2}^{1} J(v)(h' + r')\,dx = a + J_0. \tag{6}$$

It is easy to show that the problem (4)—(6) for $a = \theta = 0$ admits only trivial solution $h \equiv 0$, $\mu = 0$. So one has to construct a "perturbation" of this solution for non-null a and θ.
Step 3 consists in considering (4)—(5) as an implicit operator

$$A(h; \theta, \mu) = 0 \tag{7}$$

acting from $C^2[0,1] \times \mathbf{R}^2$ to $C[0,1]$. Clearly $A(0;0,0) = 0$. We can use the implicit operator theorem (Alexeev et al., 1979, §2.3.4) since the conditions of this theorem are fulfilled due to the regularity conditions imposed above. It follows that there exists a smooth solution $h(t;\mu,\theta)$ such that as $\theta \to 0$ and $\mu \to 0$

$$h(t; \mu, \theta) = h_{10}(t)\,\mu + o\left(\sqrt{\mu^2 + \theta^2}\right), \tag{8}$$

where

$$h_{10}(t) = \begin{cases} \frac{1}{2}\int\limits_0^t J(1-2s)ds, & 0 \leqslant t \leqslant 1/2, \\ \frac{1}{2}\int\limits_t^1 J(2s-1)ds, & 1/2 \leqslant t \leqslant 1. \end{cases}$$

At step 4 we substitute (8) in the equation (6) and use again the implicit function theorem. It follows that for sufficiently small θ and a there exists a smooth solution $\mu(\theta, a)$ such that as $\theta \to 0$ and $a \to 0$ (but $\sigma\theta > a$ as $\mu > 0$)

$$\mu(\theta, a) = \sigma\theta - a + o\left(\sqrt{\theta^2 + a^2}\right). \tag{9}$$

Returning to the extremal d.f. f and using (9) we obtain

$$f(t) = r(t) + h_{10}(t)(\sigma\theta - a) + o\left(\sqrt{\theta^2 + a^2}\right). \tag{10}$$

It remains in step 5 to put (10) into the entropy functional. After some calculations this gives the conclusion of Theorem 1.
Theorems 1 and 2 are valid of course for $a = 0$. In this case we obtain from them the local Hodges—Lehmann index $d_S(\theta)$ of the sequence of statistics $\{S_n\}$. As in (3) we obtain

$$d_S(\theta) \sim \sigma^2\theta^2, \quad \theta \to 0.$$

As to the local Chernoff index it is sufficient to use Proposition 2 of (Kallenberg, 1982) to link Theorems 1 and 2 with (2). Choosing $a = \sigma\theta(1 + o(1))/2$ we obtain for the Chernoff index $\rho_S(\theta)$ of the sequence $\{S_n\}$ as $\theta \to 0$

$$\rho_S(\theta) \sim \sigma^2\theta^2/4.$$

Hence the Hodges—Lehmann and Chernoff efficacies locally coincide with the Bahadur and Pitman efficacies (up to a constant factor).
As an example take $J(u) = \sqrt{12}\,u$ that corresponds to the Wilcoxon signed rank test and suppose that $G(x;\theta) = G(x-\theta)$ where G is a regular d.f. with density g. Obviously

$$\sigma^2 = 12\Big(\int_{-\infty}^{\infty} g^2(x)\,dx\Big)^2$$

in accordance with the classic result ascending to (Pitman, 1949).

3. TWO–SAMPLE CASE

Let $X_1, \dots, X_m$ and $Y_1, \dots, Y_n$ be two independent samples of i.i.d. observations with continuous d.f.'s F_1 and F_2, correspondingly. It is required to test the homogeneity hypothesis $H_1 : F_1 \equiv F_2$ of coincidence of these d.f.'s. We suppose that under the alternative A_1 the first sample has the d.f. $G(x;0)$, whereas the second sample has the d.f. $G(x;\theta)$ with some $\theta > 0$. The parametric family $\{G(x;\theta)\}$ satisfies exactly the same conditions as in Section 2. The d.f. $r(x;\theta)$ is defined as in (1).
Let R_i, $i = 1, \dots, m$, be the rank of the observation X_i in the combined sample and $N = m + n$. We shall consider the class of statistics defined by

$$T_N = N^{-1}\sum_{i=1}^{m} J\Big(\frac{R_i}{N+1}\Big), \tag{11}$$

where J is again a non–constant smooth score function on $[0,1]$ satisfying the conditions of section 2. We may assume that

$$\int_0^1 J(u)\,du = 0, \qquad \int_0^1 J^2(u)\,du = 1.$$

Suppose that the sizes of the samples m and n tend to infinity in such a way that

$$\lim_{m,n\to\infty} \frac{m}{m+n} = \rho_1, \qquad \lim_{m,n\to\infty} \frac{n}{m+n} = \rho_2,$$

where ρ_1 and ρ_2 are some positive constants.
Let

$$\tau = \rho_1\,\rho_2 \int_0^1 J'(t)\; r_\theta(t;\,0)\,dt$$

and suppose that $\tau > 0$.
It is known (Chernoff and Savage, 1958) that the Pitman efficiency of the sequence of statistics $\{J_N\}$ is equal to $(\rho_1\rho_2)^{-1}\tau^2$. Large deviations under H_1 of this sequence were studied in (Woodworth, 1970; Kremer, 1979; Ledwina, 1987). It was shown that under minimal regularity conditions imposed on J as $a \to 0$

$$\lim_{N\to\infty} N^{-1}\ln \mathbf{P}(J_N \geqslant a) = -\frac{a^2}{2\rho_1\rho_2}\Big(1 + o(1)\Big). \tag{12}$$

Kremer used (12) to obtain the local Bahadur exact slope $c_T(\theta)$ of the sequence (11). Under special regularity conditions he proved that as $\theta \to 0$

$$c_T(\theta) \sim (\rho_1 \rho_2)^{-1} \tau^2 \theta^2. \tag{13}$$

So the local Bahadur ARE coincides for a large class of linear rank statistics with the Pitman ARE. The same is true also for the intermediate efficiency (see (Kallenberg, 1983)).
To obtain the large deviation asymptotics of the sequence of statistics (11) under the alternative we introduce for any absolutely continuous d.f.'s f and g on $[0,1]$ the entropy functional

$$I_\rho(f,g;r) = \rho_1 \int_0^1 f' \ln f' dt + \rho_2 \int_0^1 g' \ln \frac{g'}{r'} dt.$$

We put for any set Ω of pairs of d.f.'s on $[0,1]$

$$I_\rho(\Omega; r) = \inf \left\{ I_\rho(f,g;r) : (f,g) \in \Omega \right\}.$$

The following set Ω_a^T is closely connected with our problem:

$$\Omega_a^T = \left\{ (f,g) : \rho_1 \int_0^1 J(\rho_1 f + \rho_2 g) f' dt \leqslant a \right\}.$$

THEOREM 3. *Under the conditions on the score function J and the d.f. (1) stated above, for sufficiently small a and θ, such that $\tau\theta > a$, the function $a \to I_\rho(\Omega_a^T; r)$ is continuous. Moreover, if $\theta \to 0$ and $a \to 0$ so that $\tau\theta > a$, then*

$$I_\rho(\Omega_a^T; r) = (2\rho_1\rho_2)^{-1} (\tau\,\theta - a)^2 + o(\theta^2 + a^2). \tag{14}$$

The next theorem may be deduced from Theorem 3 using Corollary 3.3 from (Groeneboom et al., 1979) as in Section 2.

THEOREM 4. *Suppose the alternative A_1 is true and the regularity conditions stated above hold. Then any real sequence $\gamma_N \to 0$ and sufficiently small a and θ, such that $\tau\theta > a$, we have*

$$\lim_{N \to \infty} N^{-1} \ln P(T_N \leqslant a + \gamma_N) = -I_\rho(\Omega_a^T; r). \tag{15}$$

The proof of Theorem 3 is similar to the proof of Theorem 1. After the Smirnov transformation of the initial samples we use the Chernoff—Savage representation of T_N in the form

$$T_N = \frac{m}{N} \int_0^1 J\left(\frac{N H_N(t) + 1}{N+1} \right) dF_m(t),$$

where $H_N(t) = (m/(m+n))F_m(t) + (n/(m+n))G_n(t)$, F_m and G_n are the empirical d.f.'s of the transformed samples.
Using the Lagrange principle to minimize the entropy functional on the set Ω_a^T, we obtain the following system of nonlinear equations for the extremal functions $x(t) = f(t) - t$ and $y(t) = g(t) - r(t)$ and the Lagrange multiplier $\mu \geqslant 0$

$$\rho_1 x'' + \mu\,(x'+1)\,(y'+r')\,J'\,(z+h) = 0,$$

$$\rho_2 y'' r' - \rho_2 y' r'' - \mu\,(x'+1)\,(y'+r')\,J'\,(z+h)\,r' = 0, \tag{16}$$

where for brevity we put $z = \rho_1 x + \rho_2 y$, $h = \rho_1 t + \rho_2 r$.
The system (16) should be considered together with the boundary conditions

$$x(0) = x(1) = 0, \qquad y(0) = y(1) = 0$$

and the norming condition

$$\rho_1 \int_0^1 J\,(z+h)\,(x'+1)\,dt = a.$$

The regularity conditions enable us to use twice the implicit operator Theorem and to obtain after some calculations as $\theta \to 0$, $a \to 0$:

$$f(t) = t + x_{10}(t)\,(\tau\theta - a) + o\left(\sqrt{\theta^2 + a^2}\right), \tag{17}$$

$$g(t) = r(t) + y_{10}(t)\,(\tau\theta - a) + o\left(\sqrt{\theta^2 + a^2}\right),$$

$$\mu = \tau\theta - a + o\left(\sqrt{\theta^2 + a^2}\right),$$

where

$$x_{10}(t) = \rho_1^{-1} \int_0^1 \Big(\min(t,s) - ts\Big)\,J'(s)\,ds,$$

$$y_{10}(t) = -\rho_2^{-1} \int_0^1 \Big(\min(t,s) - ts\Big)\,J'(s)\,ds.$$

Substituting (17) into the entropy functional, we obtain the conclusion of Theorem 3.
The Hodges—Lehmann and Chernoff local indices $d_T(\theta)$ and $\rho_T(\theta)$ of $\{T_N\}$ can be obtained combining Theorems 3 and 4 as in Section 2. We have putting $a = 0$ in (14) and (15)

$$d_T(\theta) \sim (\rho_1\rho_2)^{-1}\tau^2\theta^2, \quad \theta \to 0.$$

If we choose $a = \tau\theta\,(1 + o(1))/2$, then from Proposition 2 of (Kallenberg, 1982) we obtain

$$\rho_T(\theta) \sim (8\rho_1\rho_2)^{-1}\tau^2\theta^2, \quad \theta \to 0.$$

4. CONCLUDING REMARKS

Let us discuss briefly the regularity conditions introduced above. The smoothness condition imposed on J was used in a number of papers on limiting behaviour of the linear rank statistics, especially under an alternative. On the contrary, the conditions on the regularity of the "comparison d.f." (1) are quite restrictive. Still they are often satisfied. For instance the logistic location family has a very simple comparison d.f.

$$r(x;\theta) = x\Big(x + (1-x)e^{-\theta}\Big)^{-1}$$

which satisfies all our conditions. The same is true also for the Cauchy family and for many other well–known families of distributions.

REFERENCES

Alexeev, V.M., Tikhomirov, V.M. and Fomin, S.V. (1979). *Optimal Control.* Nauka, Moscow.

Bahadur, R.R. (1967). Rates of convergence of estimates and test statistics. *Ann. Math. Stat.* **38**, 303–324.

Bahadur, R.R. (1971). *Some Limit Theorems in Statistics.* SIAM, Philadelphia.

Chernoff, H. (1952). A measure of asymptotic efficiency for tests of a hypothesis based on sums of observations. *Ann. Math. Stat.* **23**, 493–507.

Chernoff, H. and Savage, I.R. (1958). Asymptotic normality and efficiency of certain nonparametric tests. *Ann. Math. Stat.* **29**, 972–994.

Groeneboom, P., Oosterhoff, J. and Ruymgaart, F.H. (1979). Large deviation theorems for empirical probability measures. *Ann. Probab.* **7**, 553–586.

Groeneboom, P. (1980). Large deviations and asymptotic efficiencies. *Math. Centrum Tracts* **118**, Amsterdam.

Hettmansperger, T. (1984). *Statistical Inference Based on Ranks.* Wiley, New York.

Ho, Nguen van (1973). Asymptotic efficiency in the Bahadur sense for the signed rank tests. In: *Proc. Prague Symp. Asymp. Stat.* **2**, pp. 127–156.

Hodges, J. and Lehmann, E. (1956). On some nonparametric competitors of the t–test. *Ann. Math. Stat.* **27**, 324–335.

Hwang, Tea–Yuan (1978). The Chernoff efficiency of the Wilcoxon rank test. *Commun. Statist.* **A7**, 543–555.

Kallenberg, W.C.M. (1982). Chernoff efficiency and deficiency. *Ann. Statist.* **10**, 583–594.

Kallenberg, W.C.M. (1983). Intermediate efficiency, theory and examples. *Ann. Statist.* **11**, 170–182.

Kendall, M. and Stuart, A. (1973). *The Advanced Theory of Statistics.* **2** (3–rd ed.), Hafner, New York.

Kremer, E. (1979). Approximate and local Bahadur efficiency of linear rank tests in the two–sample problem. *Ann. Statist.* **7**, 1246–1255.

Ledwina, T. (1987). An expansion of the index of linear rank statistics. *Statist. Probab. Letters* **5**, 247–248.

Mogulski, A. (1977). Notes on large deviations of ω^2–statistic. *Theory Probab. Appl.* **22**, 170–175.

Nikitin, Ya.Yu. (1979). Large deviations and asymptotic efficiency of integral type statistics. I. *Notes Sci. Seminars Leningrad Branch of Math. Inst.* **85**, 175–187.

Nikitin, Ya.Yu. (1980). Large deviations and asymptotic efficiency of integral type statistics. II. *Notes Sci. Seminars Leningrad Branch of Math. Inst.* **97**, 151–175.

Nikitin, Ya.Yu. (1982). Asymptotic efficiency of integral symmetry tests. *Notes Sci. Seminars Leningrad Branch of Math. Inst.* **119**, 181–194.

Nikitin, Ya.Yu. (1987a). On asymptotic Hodges—Lehmann efficiency of nonparametric goodness-of-fit and homogeneity tests. *Theory Probab. Appl.* **32**, 82–91.

Nikitin, Ya.Yu. (1987b). Hodges—Lehmann efficiency of nonparametric tests. In: *Proc. IV Vilnius Confer. on Probab. Theory and Math. Stat.* VNU Sci. Press, **2**, pp. 391–408.

Nikitin, Ya.Yu. (1987c). On Hodges—Lehmann indices of nonparametric tests. In: *Proc. First World Congress of the Bernoulli Soc.* VNU Sci. Press, **2**, pp. 275–278.

Nikitin, Ya.Yu. (1989). Chernoff efficiency of the sign and Wilcoxon tests for testing symmetry. *Notes Sci. Seminars Leningrad Branch of Math. Inst.* **177**, 101–107.

Pitman, E. (1949). *Lecture Notes on Nonparametric Statistical Inference.* Columbia University.

Pratt, J. and Gibbons, J.D. (1981). *Concepts of Nonparametric Theory.* Springer, Berlin—Heidelberg etc.

Sanov, I.N. (1957). On the probability of large deviations of random variables. *Sov. Math. Sb.* **42(84)**, 11–44.

Woodworth, G. (1970). Large deviations and Bahadur efficiency of linear rank statistics. *Ann. Math. Stat.* **41**, 251–283.

Prob. Theory and Math. Stat., Vol. 2, pp. 244–256
B.Grigelionis *et al.* (Eds.)
1990 VSP/Mokslas

EXTENDED VECTOR STOCHASTIC INTEGRAL IN SOBOLEV SPACES OF WIENER FUNCTIONALS

N.V. NORIN

Moscow Institute of Radio Engineering, Electronics and Automation, Chair of Higher Mathematics, Vernadskogo 78, 117454 Moscow, USSR

ABSTRACT

An abstract analog of Ito's integral with a variable upper limit — the extended vector stochastic integral (vector ESI) — is defined and studied in this paper. A linear stochastic integral equation with vector ESI is defined and theorems of existence and uniqueness of the solution are proved. Also the commutation relations formulae are established.

1. INTRODUCTION

The theory of (non–vector) ESI was developed in (Hitsuda, 1972; Daletsky and Paramonova, 1973; Skorokhod, 1975, etc). Recently in (Gaveau and Trauber, 1982) the important connection between the ESI and the Malliavin calculus (Malliavin, 1978) was discovered. The ESI is a useful tool in the investigating the of smoothness of distribution of the Wiener functionals (cf. (Nualart and Zakai, 1986)). Many authors investigated the ESI as a self–contained mathematical object. It is interesting because of the interpretation of the ESI as the logarithmic derivative of a measure along a vector field (see (Norin, 1986; 1987a; Daletskii and Maryanin, 1985), where measures may be non–Gaussian). The vector ESI is the logarithmic derivative of a Gaussian measure along the operator field (for a definition see (Norin, 1987a)) of some special form. Since the ESI is a differential operator of the first order, it is natural to investigate this operator in the Sobolev spaces of Wiener functionals (Sugita, 1985; Watanabe, 1984). In this framework we shall study the vector ESI which is p–summable for some $1 < p < \infty$.
The results of this paper were briefly announced in (Norin, 1989).

2. SOBOLEV SPACES OF WIENER FUNCTIONALS

Let X be a locally convex space, X' its adjoint space, $\mathcal{B}(X)$ be the σ–field of Borel subsets of X and γ be a mean zero Gaussian Radon measure on X. Recall that a measure $\mu : \mathcal{B}(X) \to \mathbf{R}$ is called differentiable (Averbuch et al., 1971) along a vector $h \in X$ if there exists the limit

$$d_h\mu(B) = \lim_{t\to 0} t^{-1}\Big(\mu(B+th) - \mu(B)\Big)$$

for any $B \in \mathcal{B}(X)$. The set of such vectors h is a linear subspace of X.

PROPOSITION 1. *The subspace H of differentiability of a Gaussian measure γ is isometric to the closure of X' in $L_2(X, d\gamma)$. The space H is separable and has the Hilbert structure with the inner product*

$$\langle h, g\rangle = \int \rho_h \rho_g d\gamma,$$

where ρ_h is the Radon—Nikodym derivative of $d\gamma$ relative to γ.

Remark 1. The function ρ_h is called the logarithmic derivative of the measure γ along the vector h. If $h \in X'$, then $\rho_h(x) = -\langle h, x\rangle$, where $\langle\cdot,\cdot\rangle$ is the duality form between X' and X. For $h \in H$ we shall write also $\rho_h(x) = -\langle h, x\rangle$.
Let $\bar{H}^m$ be the m–fold noncompleted Hilbert tensor product of H, $(X')^m$ the subspace of H, E a separable Hilbert space and $V_m \in E\bar{\otimes}(X')^m$ ($\bar{\otimes}$ denotes the noncompleted Hilbert tensor product). The measure γ is infinitely differentiable along H and we can correctly define the E–valued derivative $d_{V_m}\gamma$ of γ along V_m (if $V_m = e \otimes h_1 \otimes \ldots \otimes h_m$, then $d_{V_m}\gamma = e d_{h_1} d_{h_2} \ldots d_{h_m}\gamma$, and for $V_m \in E\bar{\otimes}(X')^m$ we can define $d_{V_m}\gamma$ by linearity). Of course, the measure $d_{V_m}\gamma$ is absolutely continuous relative to γ and the Radon—Nikodym derivative will be denoted by $J_{E,m}(V_m)$.

Remark 2. The function $J_{E,m}(V_m, x)$ for $V_m \in E\bar{\otimes}(X')^m$ is an E–valued continuous Hermite polynomial. We shall consider the following representation

$$P_E = \left\{\sum_{m=0}^{N} J_{E,m}(V_m),\ V_m \in E\bar{\otimes}(X')^m\right\}$$

of the set P_E of all E–valued continuous cylindrical polynomials on X.
In this terms we can formulate an analog of the Ito—Wick decomposition. Let $1 < p < \infty$, $L_{p,E} = L_p(X, E, d\gamma)$ with the norm $\|\cdot\|_{p,E}$, the norm in the complete Hilbert tensor product $E \otimes H^m$ be $|\cdot|_{m,E}$, S^m be the symmetrization operator in H^m, $H^{m_s} = S^m H^m$.

THEOREM 1. (Ito—Wick). *Let $f \in L_{2,E}$. Then there exists a sequence $\{V_m\}$, $m = 0, 1, \ldots$, $V_m \in E \otimes H^m$ such that*

$$\sum_{m=0}^{\infty} m!|S^m V_m|^2_{m,E} < \infty \text{ and } f = \sum_{m=0}^{\infty} J_{E,m}(V_m)$$

(the last equality is in $L_{2,E}$). This representation is unique if $V_m \in E \otimes H^{m_s}$.

Definition 2. Let $\{\varphi(m)\}$, $m = 0, 1, \ldots$, be a sequence of real numbers. The linear operator

$$A_\varphi : P_E \to P_E : f = \sum_{m=0}^{N} J_{E,m}(V_m) \to \sum_{m=0}^{N} \varphi(m) J_{E,m}(V_m)$$

is called a multiplication operator.
Let $p > 1$, then

$$C_{p,k} = \exp(kt_0)\{1 + (\exp(kt_0) - 1)/(\exp t_0 - 1)\},$$

where $t_0 = 2^{-1}\ln(p-1)$ if $p > 2$; $t_0 = 0$ if $p = 2$ (then $C_{2,k} = 1 + k$) and $t_0 = -2^{-1}\ln(p-1)$ if $1 < p < 2$.

THEOREM 2. *If there exists a real function $h(x)$ analytic in some neighbourhood U of zero and $\alpha > 0$ is such that $\varphi(m) = h(m^{-\alpha})$, $m = m_0, m_0 + 1, \ldots$, then the multiplication operator A_φ is a continuous operator in $L_{p,E}$ and*

$$\|A_\varphi\|_{p\to p} \leqslant \left\{ C_{p,k} h(k^{-\alpha}) + (\exp(kt_0) - 1)/(\exp(t_0) - 1) \max_{0 \leqslant m < k} |\varphi(m)| \right\} \beta_p, \quad (1)$$

where k is any positive integer greater then m_0 such that $k^{-\alpha} \in U$, and β_p is the constant from the Khinchine inequality (see (Vachanya et al., 1985, p. 250)).

This theorem is due to Meyer and Shigekawa (see (Watanabe, 1984)), but the inequality (1) is new. The proof of (1) can be obtained by appropriate calculations at all steps of the proof of this theorem in (Watanabe, 1984).
Let

$$f = \sum_{m=0}^{N} J_{E,m}(V_{m_s}) \in P_E,$$

where

$$V_{m_s} \in E\bar{\otimes}(X')^{m_s}.$$

Definition 3. The stochastic derivative of a functional f is (Skorokhod, 1975)

$$Df = \sum_{m=1}^{N} m\ J_{E\otimes H,\ m-1}(V_{m_s}).$$

The stochastic derivative of order $n > 1$ is defined by induction.
Let $C : P_E \to P_E$ be the multiplication operator with $\varphi(m) = \sqrt{m}$.

THEOREM 3. *Let $f \in P_E$. There exist two positive constant C_p and C'_p non depending on E such that*

$$C'_p \|Cf\|_{p,E} \leqslant \|Df\|_{p,E\otimes H} \leqslant C_p \|Cf\|_{p,E}.$$

Let $s \in \mathbf{R}$, $(I + C)^s$ be the multiplication operator with $\varphi(m) = (1 + \sqrt{m})^s$.

Definition 4. The Sobolev space $D_{p,s}(E)$ of Wiener functionals is the completion of P_E by the norm

$$\|f\|_{p,s,E} = \|(I+C)^s f\|_{p,E}.$$

(The further properties of $D_{p,s}(E)$ see in (Sugita, 1985; Watanabe, 1984).) Let $D_{p,+\infty}(E) = \bigcap_{s\in\mathbf{R}} D_{p,s}(E)$ be the projective limit.

THEOREM 4. *The stochastic derivative D is a linear continuous operator from $D_{p,r}(E)$ to $D_{p,\,r-1}(E\otimes H)$ and the constant $d_p>0$ in the estimate*

$$\|Df\|_{p,\,r-1,\,E\otimes H} \leqslant d_p\|f\|_{p,r,E}$$

may be chosen independently of E and r.

Proof. Let S be the multiplication operator with

$$\varphi(m) = (1+\sqrt{m-1})/(1+\sqrt{m}).$$

By Theorem 2 the operator S is continuous in $L_{p,E}$ with

$$h(x) = (x+\sqrt{1-x^2})/(1+x),\ \alpha = 1/2.$$

We can compute an estimate for the operator norm of S^{r-1} with help of (1):

$$\|S^{r-1}\|_{p\to p} \leqslant$$

$$\leqslant \left\{C_{p,k}h^{r-1}(k^{-1/2}) + (\exp(kt_0)-1)/(\exp t_0 - 1)\max_{0\leqslant m<k}|\varphi(m)|^{r-1}\right\}\beta_p \leqslant$$

$$\leqslant \left\{C_{p,k} + (\exp(kt_0)-1)/(\exp t_0 - 1)\right\}\beta_p = a_p$$

(t_0 depends only on p and k depends on S). Note that the norm $\|C(I+ +C)^{-1}\|_{p\to p} = \delta_p$ depends on p only and by Theorem 3 we have

$$\|Df\|_{p,r-1,E\otimes H} = \|(I+C)^{r-1}Df\|_{p,E\otimes H} = \|D(I+C)^{r-1}S^{r-1}f\|_{p,E\otimes H} \leqslant$$

$$\leqslant C_p\|C(I+C)^{r-1}S^{r-1}f\|_{p,E} = C_p\|C(I+C)^{-1}(I+C)^r S^{r-1}f\|_{p,E} \leqslant$$

$$\leqslant C_p\delta_p\|S^{r-1}(I+C)^r f\|_{p,E} \leqslant C_p\delta_p a_p\|(I+C)^r f\|_{p,E} = d_p\|f\|_{p,r,E}$$

with $d_p = C_p\cdot\delta_p\cdot a_p$.

LEMMA 2. *Let A be a multiplication operator from $D_{p,s}(E)$ to $D_{q,r}(E)$ with $\varphi(m)>0$ for all $m=0,1,\ldots$ Then* $\mathrm{Ker}(A)=\{0\}$.

Proof. Let $p^{-1}+(p')^{-1}=1, q^{-1}+(q')^{-1}=1$. The operator

$$A^*\colon D_{q',-r}(E)\to D_{p',-s}(E)$$

is a multiplication operator with the same $\varphi(m)$. Moreover, A^* maps P_E on P_E. But P_E is tight in any Sobolev space, hence $\mathrm{Ker}(A) = \perp \mathrm{Im}(A^*) = \{0\}$.

3. THE EXTENDED VECTOR STOCHASTIC INTEGRAL

Let $Q: E \to E \otimes H$ be a linear continuous operator with the operator norm $q > 0$ and let G be a separable Hilbert space. Let $e \otimes g \in E \otimes G$ and define $Q(e \otimes g) = (Qe) \otimes g \in E \otimes H \otimes G$. By linearity we can define Q: $E \otimes G \to E \otimes H \otimes G$ with the same operator norm q. Indeed, if

$$b = \sum b_{ij} e_i \otimes g_j \in E \otimes G \text{with} |b|^2_{E \otimes G} = \sum b^2_{ij}$$

then

$$|Qb|^2_{E\otimes H \otimes G} = \sum_j \left| Q\Big(\sum_i b_{ij} e_i\Big)\right|^2_{E \otimes H} \leqslant q^2 \sum b^2_{ij} = q^2 |b|^2_{E \otimes G}$$

and we see that $\|Q\| \leqslant q$. Now let $\varepsilon > 0$ and $e \in E$ be such that $|Qe|_{E\otimes H} > (q-\varepsilon)|e|_E$. Then

$$|Q(e \otimes g)|_{E \otimes H \otimes G} = |g|_G |Qe|_{E \otimes H} \geqslant (q - \varepsilon)|g|_G |e|_E$$

and $\|Q\| \geqslant q - \varepsilon$. Thus $\|Q\| = q$.
We can also correctly define $Q^n: \; E \otimes G \to E \otimes H^n \otimes G$ with the operator norm q_n independent of G.

Definition 5. The linear continuous operator $Q: \; E \to E \otimes H$ is called quasinilpotent if

$$\lim_{n \to \infty} q_n^{1/n} = 0. \tag{2}$$

Remark 3. The operator Q maps one space into another space and therefore this definition is different from the generally accepted definition of a quasinilpotent operator. But we can construct the new Hilbert space F_- and extend Q to F_- so that this Q will be a usual quasinilpotent operator.
Let $F = \oplus_{m=0}^{\infty} E \otimes H^m$ be the "vector Fock space"

$$\left(b = \sum V_m \in F \Leftrightarrow \|b\|^2_F = \sum m! |V_m|^2_{E \otimes H^m} < \infty\right)$$

and let F_- be the completion of F by the norm

$$\|b\|^2_- = \sum_{m=0}^{\infty} m! \Big(\sum_{i=0}^{m}{}' A^i_m\Big)^{-1} |V_m|^2_{E \otimes H^m}, \; A^i_m = m!/(m-i)!.$$

LEMMA 3. *There exist the natural extension $\hat{Q}$ of $Q: \; E \to E \otimes H$ which maps F_- into F_- with the operator norm $\|\hat{Q}\| = q$. The operator Q is quasinilpotent in the sense of the Definition 5 iff $\hat{Q}$ is a usual quasinilpotent operator.*

Proof. The operator $\hat{Q}$ on $E \otimes H^m$ is defined as at the beginning of this section $(G = H^m)$ and on F_- by linearity. This operator is continuous since

$$\|\hat{Q}^n b\|_-^2 = \sum_{m=0}^{\infty}(m+n)!\Big(\sum_{i=0}^{m+n} A_{m+n}^i\Big)^{-1} |Q^n V_m|^2_{E\otimes H^{m+n}} \leqslant$$

$$\leqslant q_n^2 \sum_{m=0}^{\infty}(m+n)!\Big(\sum_{i=0}^{m+n} A_{m+n}^i\Big)^{-1} |V_m|^2_{E\otimes H^m} \leqslant q_n^2\|b\|_-^2. \qquad (3)$$

It follows from (3) and the equality $\|\hat{Q}V_0\|_- = |QV_0|_{E\otimes H}$, that $\|\hat{Q}\| = q$. The last assertion of the lemma is true because

$$\|\hat{Q}^n V_0\|_-^2 = n!\Big(\sum_{i=0}^{n} A_n^i\Big)^{-1} |Q^n V_0|^2_{E\otimes H^n} \geqslant e^{-1}|Q^n V_0|^2_{E\otimes H^n}$$

and $q_n e^{-1/2} \leqslant \|\hat{Q}^n\| \leqslant q_n$.

LEMMA 4. *There exist the natural extension of $Q : E \to E\otimes H$ to $D_{p,r}(E\otimes G)$. This extension is also denoted by Q and is a linear continuous operator from $D_{p,r}(E\otimes G)$ to $D_{p,r}(E\otimes H\otimes G)$ with the operator norm q.*

THEOREM 5. *The bilinear form $(Qf, Dg)_{2,E\otimes H\otimes G}$ is continuous on the Cartesian product $D_{p,r}(E\otimes G)\times D_{q,1-r}(E\otimes G)$, where $p > 1, p^{-1}+q^{-1} = 1, r \in \mathbf{R}$.*

Proof. By Theorem 4 and Lemma 4

$$|(Qf,\ Dg)_{2,E\otimes H\otimes G}| = \left|\int \langle\{(I+C)^r\ Qf\},\ \{(I+C)^{-r}Dg\}\rangle_{E\otimes H\otimes G} d\gamma\right| \leqslant$$

$$\leqslant \|Qf\|_{p,r,E\otimes H\otimes G}\|Dg\|_{q,-r,E\otimes H\otimes G} \leqslant q d_p \|f\|_{p,r,E\otimes G}\|g\|_{q,1-r,E\otimes G}.$$

Definition 6. The linear continuous operator $I_Q : D_{p,r}(E\otimes G) \to D_{p,r-1}(E\otimes G)$ such that

$$(I_Q f,\ g)_{2,E\otimes G} = (Qf,\ Dg)_{2,E\otimes H\otimes G} \qquad (4)$$

$\big(f \in D_{p,r}(E\otimes G),\ g \in D_{q,1-r}(E\otimes G)\big)$ and its value $I_Q f$ will be called the extended vector stochastic integral (vector ESI).
Clearly this definition is correct and $\|I_Q\| \leqslant q d_p$.

Example 1. Let $X = C[0,1]$ be the space of all continuous functions $x : [0,1] \to \to \mathbf{R}$, which vanish at zero. Let γ be the Wiener measure on X. Then the subspace $H \subset X$ of differentiability of the measure γ consists of all absolutely continuous functions $h \in X$ with $h' \in L_2([0,1])$ and the scalar product in H is

$$\langle h_1,\ h_2\rangle = \int_0^1 h_1'(\tau)h_2'(\tau)d\tau.$$

Hence the operator

$$U: Uh(t) = \int_0^t h'(\tau)d\tau$$

is an isometry from $L_2([0,1])$ to H. Let $U_2 = U \otimes U$ be the tensor square of U. Then $U_2: L_2([0,1]^2) \to H^2$ is an isometry and

$$H^2 = \left\{ g(\cdot,\cdot) \middle| \exists \tilde{g} \in L_2([0,1]^2),\ g(t,s) = \int_0^t \int_0^s \tilde{g}(u,v)dvdu \right\}.$$

Let $Q: H \to H^2$ be any linear continuous operator and

$$\tilde{Q} = U_2^{-1}QU : 0L_2([0,1]) \to L_2([0,1]^2).$$

In the special case when $Qh'(\tau, s) = \chi_{[0,\tau)}(s)h'(s)$ ($h \in H$, hence $h' \in L_2([0,1])$) and $f'(x,s)$ is a nonanticipating summable in square functional ($f = Uf'$), the equality

$$[I_Q f(x)](t) = \int_0^t \int_0^1 \chi_{[0,\tau)}(s) f'(x,s)dx(s)d\tau =$$

$$= \int_0^t \int_0^\tau f'(x,s)dx(s)d\tau = U\Big(\int_0 f'(x,s)dx(s)\Big)(t)$$

is true.
This example is a clue to the connection between the vector ESI and the Ito's integral with a variable upper limit.
We also have the following decomposition of the vector ESI (it is an analogy of the well-known decomposition of the usual ESI).

LEMMA 5. *Let $f = \sum J_{E\otimes G,m}(V_m) \in D_{2,r}(EG)$. Then*

$$I_Q f = \sum_{m=0}^{\infty} J_{E\otimes G,\ m+1}(QV_m) \tag{5}$$

(this equality is in $D_{2,r-1}(E \otimes G)$).

LEMMA 6. *Let $f \in D_{p,r}(E \otimes G)$, $g \in D_{q,n-r}(E \otimes G)$. Then*

$$(I_Q^n f, g)_{2,E\otimes G} = (Q^n f,\ D^n g)_{2,E\otimes H^n \otimes G}. \tag{6}$$

LEMMA 7. *Let $f \in D_{p,r}(E \otimes G)$: Then*

$$[D, I_Q](f) \stackrel{d}{=} D(I_Q f) - I_Q(Df) = Qf \tag{7}$$

(this equality is in $D_{p,r-2}(E \otimes G \otimes H)$).

Proof. Let

$$f = \sum_{m=0}^{N} J_{E\otimes G,m}(V_{m_s}) \in P_{E\otimes G}.$$

Noting that V_{m_s} is a $E \otimes\ G$–valued m–linear symmetric form on H, we have

$$Df(\cdot) = \sum_{m=1}^{N}\sum_{k=1}^{m} J_{E\otimes G\otimes H,m-1}(V_{m_s}(x_1,\ldots,x_{k-1},\cdot,x_{k+1},\ldots,x_m))$$

and

$$I_Q Df(\cdot) = \sum_{m=1}^{N}\sum_{k=1}^{m} J_{E\otimes G\otimes H,m}(QV_{m_s}(x,x_1,\ldots,x_{k-1},\cdot,x_{k+1},\ldots,x_m)).$$

On the other hand we also see that

$$I_Q f = \sum_{m=0}^{N}(m+1)\Big[J_{E\otimes G,m+1}(QV_{m_s}(x,x_1,\ldots,x_m))+$$

$$+\sum_{k=1}^{m} J_{E\otimes G,m+1}(QV_{m_s}(x_k,x_1,\ldots,x_{k-1},x,x_{k+1},\ldots,x_m))\Big]$$

and

$$DI_Q f(\cdot) = \sum_{m=0}^{N}\Big[J_{E\otimes G\otimes H,m}(QV_{m_s}(\cdot,x_1,\ldots,x_m))+$$

$$+\sum_{k=1}^{m} J_{E\otimes G\otimes H,m}(QV_{m_s}(x_k,x_1,\ldots,x_{k-1},\cdot,x_{k+1},\ldots,x_m))\Big].$$

Finally,

$$DI_Q f - I_Q(Df) = \sum_{m=0}^{N} J_{E\otimes G\otimes H,m}(QV_{m_s}(\cdot,x_1,\ldots,x_m)) = Qf$$

(in the last equality we identify $E\otimes G\otimes H$ and $E\otimes H\otimes G$). From the tightness of $P_{E\otimes G}$ in $D_{p,r}$ $(E\otimes G)$ we obtain the assertion of the lemma.

Definition 7. Let $f \in D_{p,r}(G)$, $Q:\ H \to H^2$. The operator

$$\mathcal{L}_Q:\ f \to I_Q(Df) :: D_{p,r}(G) \to D_{p,r-2}(G\otimes H)$$

will be called the vector Ornstein—Uhlenbeck operator.

LEMMA 8. *Let $f \in D_{p,r}(G)$, $Q:\ H \to H^2$. Then*

$$[D,\mathcal{L}_Q] = Q(Df) \tag{8}$$

(the equality is in $D_{p,r-3}(G \otimes H^2)$).

Proof. By Lemma 7

$$[D, \mathcal{L}_Q](f) = D(I_Q(Df)) - I_Q(DDf) = [D, I_Q](Df) = Q(Df).$$

THEOREM 6. (cf. (Norin, 1987b)). *Let $f \in D_{p,s}(H \otimes G)$, $Q: H \to H^2$. Then*

$$[\mathcal{L}_Q, I_Q](f) = I_Q(Qf) \tag{9}$$

(the equality is in $D_{p,r-3}(G \otimes H^2)$).

Proof. By Lemma 7

$$[Z_Q, I_Q](f) = I_Q(DI_Qf) - I_Q(I_Q(Df)) = I_Q([D, I_Q](f)) = I_Q(Qf).$$

Recall that the Ornstein—Uhlenbeck semigroup T_t is defined on P_E as a multiplication operator with $\varphi(m) = \exp(-mt)$. It is interesting that we can not extend T_t on to $L_{p,E}$ by Theorem 2. This example shows that Theorem 2 gives us only a sufficient condition. Of course, T_t has the strong contraction property known as hyper–contractivity. Since we couldn't find Nelson's theorem for the case $L_{p,E}$ with a Hilbert space E, we have proved this theorem in the following form which is sufficient for us.

THEOREM 7. (cf. (Simon, 1974, Theorem I. 17′)). *Let $1 < p < q < \infty$. Then there exists $t_0 = t_0(p, q) > 0$ such that for $t \geqslant t_0$ the operator T_t is a contraction from $L_{p,E}$ to $L_{q,E}$. Furthermore,*

$$\lim_{q \downarrow p} t_0(p, q) = 0.$$

The proof of this theorem is essentially the same as in (Simon, 1974).

LEMMA 9. *Let $1 < p \leqslant q < \infty$, $f \in D_{p,r}(E)$. Take $t > 0$ such that $T_t : L_{p,E} \to$ $\to L_{q,E}$ is a contraction. Then*

$$T_t I_Q f = \exp(-t) I_Q(T_t f) \tag{10}$$

(the equality is in $D_{q,r-1}(E)$).

Proof. Let $f = \sum J_{E,m}(V_m) \in P_E$. Then

$$T_t I_Q f = \exp(-t) \sum \exp(-mt) J_{E,m+1}(QV_m).$$

On the other hand,

$$I_Q(T_t f) = \sum \exp(-mt) J_{E,m+1}(QV_m) = \exp(t) T_t I_Q f.$$

LEMMA 10. *Let $1 < p \leqslant q < \infty$. Take $t > 0$ such that $T_t: L_{p,E} \to L_{q,E}$ is a contraction. Then $T_t: D_{p,r}(E) \to D_{q,r}(E)$ has a trivial kernel.*

Proof follows easily from Lemma 2.

Definition 8. The following equation

$$f = \lambda I_Q f + g \tag{11}$$

will be called the abstract stochastic integral equation.

Example 1 (continued). Let $\tilde{Q}h'(\tau, s) = K(\tau, s)h'(s)$, where $\|K(\tau, \cdot)\|_\infty \leqslant R(\tau)$, $R \in L_2([0,1])$. Then we can write (11) in the form (cf. (Shevlyakov, 1979))

$$f(x,t) = \lambda \int_0^t \int_0^1 K(\tau, s) f'(x, s) dx(s) d\tau + g(x,t).$$

Define now the space $F_{p,+\infty}(E) \subseteq D_{p,+\infty}(E)$, which is the class of uniqueness and existence of the solution of the abstract stochastic integral equation (11).
Let A be the set of all quickly–decreasing (for all $C > 0$ the series $\sum \alpha_n C^n$ absolutely converges) sequences $\alpha = \{\alpha_n, n = 0, 1, \ldots\}$ with $\alpha_n > 0$,

$$P_{p,r,\alpha}(g) = \sum \alpha_n \|g\|_{p,n+s,E} \quad (g \in D_{p,+\infty}(E)) \tag{12}$$

and

$$F_{p,+\infty}(E) = \{g \in D_{p,+\infty}(E) |\ \forall r \in \mathbf{R},\ \forall \alpha \in A,\ P_{p,r,\alpha}(g) < \infty\}.$$

THEOREM 8. i) *The space $F_{p,+\infty}(E)$ is a locally convex space with a noncountable base of topology.*

ii) *The inclusion $F_{p,+\infty}(E) \subseteq D_{p,+\infty}(E)$ is continuous and tight.*

iii) *For $p \geqslant 2$ the spaces $F_{p,+\infty}(E)$ and $D_{p,+\infty}(E)$ do not coincide as sets.*

Proof. i) Follows from the definitions.

ii) The norms $P_{p,r,\alpha}$ have the following properties:

1) if $\alpha \leqslant \beta$ ($\alpha \leqslant \beta$, if $\alpha_n \leqslant \beta_n$ for all $n = 0, 1, \ldots$), then $P_{p,r,\alpha}(g) \leqslant P_{p,r,\beta}(g)$;

2) if $1 < p < q$, then $P_{p,r,\alpha}(g) \leqslant P_{q,r,\alpha}(g)$;

3) if $s < t$, then $P_{p,s,\alpha}(g) \leqslant P_{p,t,\alpha}(g)$.

Since the sets of norms in both spaces are ordered, it is sufficient to obtain the inequality

$$\|g\|_{p,r,E} \leqslant \sum \alpha_n \|g\|_{p,s+n,E} = P_{p,s,\alpha}(g),$$

where $\alpha = \{1, \alpha_1, \ldots\}$. Now let us prove that $P_E \subset F_{p,+\infty}(E)$ (since P_E is tight in $D_{p,+\infty}(E)$, then ii) will be proved). Let $f = \sum J_{E,m}(V_m) \in P_E$. Then there exists $C > 0$ such that

$$\|f\|_{p,s+n,E} = \left\| \sum (1 + \sqrt{m})^{s+n} J_{E,m}(V_m) \right\|_{p,E} \leqslant$$

$$\leqslant \sum (1+\sqrt{m})^{s+n} \|J_{E,m}(V_m)\|_{p,E} \leqslant C^{s+n}$$

and

$$P_{p,s,\alpha}(f) = \sum \alpha_n \|f\|_{p,s+n,E} \leqslant \sum \alpha_n C^{s+n} < \infty.$$

iii) We shall construct functional $\varphi \in [D_{p,+\infty}(E)] \backslash [F_{p,+\infty}(E)]$ (for $p \geqslant 2$). Remark that the sequence $\beta_m = \exp[-m^2/2], m = 0, 1, \dots$ has the following two properties:

1) for all $k \in N$ and $t > 0$, the series $\sum m^k \exp(2mt)\beta_m$ converges;

2) there is no such constant $C > 0$ that $F(k) = \sum m^k \beta_m \leqslant C^k$ for all $k \in N$. We shall prove the property 2). Let us estimate $F(k)$:

$$F(k) \geqslant \int\limits_0^{+\infty} x^k \exp(-x^2/2) dx - k^{k/2} \exp(-k/2) \geqslant$$

$$\geqslant k^{k/2} \exp(-k/2) \left\{ (1+1/k)^{k/2} (\pi/e)^{1/2} \exp[\Theta/(6(k+1))] - 1 \right\},$$

where $0 < \Theta < 1$. But

$$(1+1/k)^{k/2} (\pi/e)^{1/2} \exp[\Theta/(6(k+1))] \to \sqrt{\pi} \text{ as } k \to \infty.$$

Hence for sufficiently large k ($k > k_0$) we have

$$F(k) \geqslant (ke^{-1})^{k/2} [(1-\varepsilon)\sqrt{\pi} - 1],$$

where $0 < \varepsilon < 1/5$. If 2) is true, then $(kC^{-2}e^{-1})^{k/2}[(1-\varepsilon)\sqrt{\pi} - 1] \leqslant 1$ for $k > k_0$ and we have a contradiction.
Now let $E = \mathbf{R}$, $h_m \in H$, $m = 1, 2, \dots$, with $|h_m| = [\exp(m/4)(m!)^{1/(2m)}]^{-1}$. Let

$$\varphi_t = \sum \exp(mt) J_m(V_m)$$

with $V_m = h_m \otimes \cdots \otimes h_m$. It is easy to check that $\varphi_t \in D_{2,s}$ for all $s \in \mathbf{R}$. Let $\varphi = T_t \varphi_t$ (φ doesn't depend on t). For $p > 2$ take $t_0 > 0$ such that $T_{t_0}\colon L_2 \to L_p$ is a contraction. Then

$$\|\varphi\|_{p,s} = \|T_{t_0} \varphi_{t_0}\|_{p,s} \leqslant \|\varphi_{t_0}\|_{2,s} < \infty,$$

hence $\varphi \in D_{p,+\infty}$. Let $\alpha_0 = 1$, $\alpha_k = (2e/k)^{k/4}$, $k = 1, 2, \dots$. Then $\alpha = \{\alpha_k\} \in$ $\in A$ but

$$P_{p,0,\alpha}(\varphi) \geqslant \|\varphi\|_{2,0} + \sum (2e/k)^{k/4} \|\varphi\|_{2,k} \geqslant \sum_{k=k_0+1}^{\infty} \left[(1-\varepsilon)\sqrt{\pi} - 1 \right] = \infty$$

and $\varphi \notin F_{p,+\infty}(\mathbf{R})$.

THEOREM 9. *The abstract stochastic integral equation* (11) *has no more than one solution in* $D_{p,+\infty}(E)$.

Proof. It is enough to show that the equation

$$f = \lambda I_Q f \tag{13}$$

has only a trivial solution. If $p \geqslant 2$, then $D_{p,+\infty}(E) \subseteq D_{2,+\infty}(E)$. Moreover, the value $I_Q f$ doesn't depend on p (if it exists). The end of the proof is similar to a well-known reasoning (see (Skorokhod, 1975)).
Let $1 < p < 2$. Take $t > 0$ such that $T_t: L_{p,E} \to L_{2,E}$ is a contraction. Then T_t is a contraction from $D_{p,s}(E)$ to $D_{2,s}(E)$. By Lemma 9, applying T_t to both sides of (13), we have

$$T_t f = \lambda T_t I_Q(f) = \exp(-t)\lambda I_Q(T_t f).$$

Since $T_t f \in D_{2,s}(E)$, it follows from the first part of the proof that $T_t f = 0$. By Lemma 10, $f = 0$ and the result follows.

Remark 4. In this theorem $Q: E \to E \otimes H$ may be arbitrary (not necessarily quasinilpotent).

THEOREM 10. *Let Q be quasinilpotent, $g \in F_{p,+\infty}(E)$. Then the abstract stochastic integral equation* (11) *has a solution $f \in F_{p,+\infty}(E)$. This solution continuously depends on g.*

Proof. It is enough to show that the series $f = \sum \lambda^n I_Q^n g$ converges in $F_{p,+\infty}(E)$. Let $p^{-1} + q^{-1} = 1$, $\psi \in D_{q,-s}(E)$. Since, by Lemma 6 and Theorem 4,

$$\left|(f,\psi)_{2,E}\right| \leqslant \sum |\lambda|^n \left|(Q^n g, D^n\psi)_{2,E\otimes H^n}\right| \leqslant$$

$$\leqslant \sum |\lambda|^n q_n \|g\|_{p,n+s,E} \|D^n\psi\|_{q,-n-s,E\otimes H^n} \leqslant$$

$$\leqslant \sum |\lambda|^n q_n \|g\|_{p,n+s,E} (d_q)^n \|\psi\|_{q,-s,E} < \infty$$

(the sequence $\beta_n = |\lambda|^n q_n (d_q)^n$ is quickly decreasing and $g \in F_{p,+\infty}(E)$). Let $\alpha = \{\alpha_k\} \in A$. Then

$$P_{p,s,\alpha}(f) \leqslant \sum \gamma_m \|g\|_{p,s+m,E} < \infty,$$

where $\gamma_m = \sum_{i=0}^{m} \alpha_{m-i}\beta_i > 0$ is a quickly decreasing sequence.
We also proved that $P_{p,s,\alpha}(f) < P_{p,s,\gamma}(g)$. Hence the theorem is proved.

REFERENCES

Averbuch, V.I., Smolyanov, O.G. and Fomin, S.V. (1974). Generalized functions and differential equations in linear spaces. I. *Trans. Moscow Math. Soc.* (1971). **24**, Amer. Math. Soc. Providence.

Daletskii, Yu.L. and Maryanin, B.D. (1985). Smooth measures on infinitedimensional manifolds. *Soviet Math. Dokl.* **285**, 863–866.

Daletskii, Yu.L. and Paramonova, S.N. (1973). Stochastic integrals with respect to a normally distributed additive set function. *Soviet Math. Dokl.* **14**, 96–100.

Gaveau, B. and Trauber, P. (1982). L'integrale stochastique comme operateur de divergence dans l'espace fonctionnel. *J. Funct. Anal.* **46**, 230–238.

Hitsuda, M. (1972). Formula for Brownian partial derivatives. In: *Proc. Second Japan—USSR Symp. on Probab. Theory,* **2**, pp. 111–114.

Malliavin, P. (1978). Stochastic calculus of variations and hypoelliptic operators. *Proc. Intern. Symp. on SDE's. Kyoto,* 1976, pp. 195–263. Kinokuniya—Wiley, Tokyo.

Norin, N.V. (1986). An extended stochastic integral for non–Gaussian measures in a locally convex space. *Russian Math. Surveys* **41**, 229–230.

Norin, N.V. (1987a). Stochastic integrals and differentiable measures. *Theory Probab. Appl.* **32**, 107–116.

Norin, N.V. (1987b). Two commutation's theorems with extended stochastic integral for non–Gaussian measures. In: *All–Union School on ergodic theory of Markov processes.* Kyzyl, p. 43.

Norin, N.V. (1989). Extended vector stochastic integral in Sobolev spaces of Wiener functionals. In: *Abstr. Fifth Int. Vilnius Conf. on Probab. Theory and Math. Statist.,* **4**, pp. 109–110.

Nualart, D and Zakai, M. (1986). Generalized stochastic integrals and the Malliavin calculus. *Probab. Theor. Rel. Fields* **73**, 255–280.

Shevlyakov, A.Yu. (1979). On a class of stochastic integral equations. *Theory of Random Processes* **7**, 118–127.

Simon, B. (1974). *The $P(\varphi)_2$ Euclidean (Quantum) Field Theory.* Prinston Univ. Press, Prinston.

Skorokhod, A.V. (1975). On a generalization of a stochastic integral. *Theory Probab. Appl.* **20**, 219–233.

Sugita, H. (1985). Sobolev spaces of Wiener functionals and Malliavin calculus. *J. Math. Kyoto Univ.* **25**, 31–48.

Vachanya, N.N., Tarieladze, B.I. and Chobanyan, S.A. (1985). *Probability Distributions on Banach Spaces.* Nauka, Moscow.

Watanabe, S. (1984). *Lectures on SDE's and Malliavin Calculus.* Springer, Berlin—Heidelberg etc.

Prob. Theory and Math. Stat., Vol. 2, pp. 257–274
B.Grigelionis *et al.* (Eds.)
1990 VSP/Mokslas

WEIGHTED MULTIVARIATE EMPIRICAL PROCESSES AS RANDOM MEASURES

R. NORVAIŠA
Department of Mathematics, University of Vilnius,
Naugarduko 24, 232006 Vilnius, Lithuania

ABSTRACT

Strong laws of large numbers and laws of the iterated logarithm for multivariate empirical processes with heavy weights are proved. Old and some new results are derived from ones known for Banach space valued random variables. The method is based on the representation of weighted point mass *via* signed measure.

1. INTRODUCTION

This note is devoted to some limit theorems for supremum of the suitably weighted multivariate empirical processes based on the uniform sample from the unit cube. Our approach is based on the viewpoint to the weighted empirical process as a sum of appropriately restricted σ–finite measures on a class of sets instead of a sum of point measures indexed by functions (weights). This allows us to use operator methodology developed in the context of the probability theory on Banach spaces. In order to be more explicit let us specify our setup and introduce some notation. Throughout $(\mathfrak{X}, \mathcal{A}, \mu)$ will denote the measurable space considered as a sample space and $(\mathfrak{X}^{\mathbf{N}}, \mathcal{A}^{\mathbf{N}}, \mu^{\mathbf{N}})$ will be its infinite product. We define the canonical probability space $(\Omega, \mathcal{F}, \mathbf{P})$ to be the product of this infinite product with a rich enough probability space such as $([0,1], \mathcal{B}, \mathbf{m})$. If $X_i : \Omega \to \mathfrak{X}$, $i \in \mathbf{N}$, are the natural projections of Ω onto the i–th copy of $\mathfrak{X}$, then they are i.i.d. with law μ. The n–th empirical measure and centered empirical measure are defined by

$$\mu_n := n^{-1} \sum_{i=1}^{n} \delta_{x_i}, \quad \nu_n := \mu_n - \mu, \ n \geqslant 1,$$

where δ_x denotes the point mass at $x \in \mathfrak{X}$.
Let $\mathcal{C} \subset \mathcal{A}$ be a class of sets. We will consider the weighted empirical process indexed by $\mathcal{C}$. Given a non–negative function $w \in \mathbf{D}((0,1])$, i.e., $w(t+) = w(t)$, for $t \in (0,1)$, which will be termed by the weight function (w.f.), define this process by

$$\nu_n^w(C) := w\Big(\mu(C)\Big)\nu_n(C) = w\Big(\mu(C)\Big)n^{-1}\left[\sum_{i=1}^{n} \delta_{x_i}(C) - \mu(C)\right], \quad C \in \mathcal{C}. \quad (1.1)$$

When $\mathcal{C}$ is

$$D^d := \Big\{[0,y]:\ y \in [0,1]^d\Big\}, \quad d \in \mathbb{N}, \tag{1.2}$$

and μ is Lebesgue measure on $[0,1]^d$, then ν_n^w becomes the weighted multivariate empirical process $w(|t|)[F_n(t) - |t|]$, where $|t| = t_1 \cdot \ldots \cdot t_d$. Let us consider for the moment the most simple case, i.e., $\mathcal{C} = D^1$ and μ is Lebesgue measure $\mathbf{m}$ or, in other words, uniform distribution. Then it is easy to see that the weighted point mass can be represented in the following manner

$$w(y)\delta_x\Big([0,y]\Big) = w(x)\delta_x\Big([0,y]\Big) - \mathbf{m}^w_{|[x,1]}\Big([0,y]\Big), \quad 0 \leqslant x \leqslant y \leqslant 1, \tag{1.3}$$

where $\mathbf{m}^w$ is the Lebesgue—Stieltjes measure, i.e., $\mathbf{m}^w([x,y]) = w(x) - w(y)$. This means that the weighted (univariate) empirical process ν_n^w can be considered as a random measure on $\mathcal{B}$. This fact among others was used in (Andersen et al., 1988) to prove Marcinkiewicz—Zygmund laws of large numbers in this setting (thanks to E. Giné for communicating this reference to us).

We show in Section 3 that measure representation extending (1.3) remains possible in the multivariate case. The technique used here suggest us to consider weighted empirical processes indexed by sets with Vapnik—Červonenkis property (defined below) incorporating appropriate measure representation condition. More precisely, given a class $\mathcal{C} \subset \mathcal{A}$ and a w.f. $w \in \mathsf{D}((0,1))$, define for $x \in \mathfrak{X}$ the set

$$G_x := \cap\{C \in \mathcal{C}:\ x \in C\}$$

and the envelope

$$F_w(x) := w\Big(\mu(G_x)\Big) = \sup\Big\{w(\mu(C))\delta_x(C):\ C \in \mathcal{C}\Big\}.$$

We may assume G_x is measurable, if not, then replace it throughout by a measurable envelope $G_x^* \supset G_x$ with $\mu(G_x^*) = \mu^*(G_x)$. Let $\mathcal{M}(\mathfrak{X}, \mathcal{A})$ be the Banach space of signed measures of finite total variation norm $\|\cdot\|_{\mathrm{var}}$. As has been mentioned earlier it may be quite reasonable to assume that for μ–a.a. $x \in \mathfrak{X}$

$$w \circ \mu\delta_x \in \mathcal{M}(\mathfrak{X}, \mathcal{A}) \quad \text{and} \quad \|w \circ \mu\delta_x\|_{\mathrm{var}} \leqslant CF_w(x), \tag{RM}$$

for some constant $C = C(\mathcal{C})$.

Letting $\mathcal{C}$ to be D^1 and μ to be the uniform distribution on $[0,1]$ we get that $G_x = [0,1]$ and (RM) is satisfied invoking the (1.3) and the obvious inequality

$$\|w \circ \mu\delta_x\|_{\mathrm{var}} \leqslant 2w(x), \quad \forall x \in (0,1), \tag{1.4}$$

if w is non–increasing on (0, 1).

Now let us describe some functional spaces generated by the weighted empirical process in consideration. The advantage is the possibility to use results from the theory of probabilities on Banach spaces. Following (Dudley, 1978) and (Gaenssler, 1983) we construct the space $\mathsf{D}(\mathcal{C},\ \mu)$ which constitutes a measurable structure. Recall that $(\mathsf{B},\ \mathcal{B}, \|\cdot\|)$ is a linear measurable space, where B denotes a real vector space, $\mathcal{B}$ a σ–algebra of subsets of B and $\|\cdot\|$ a semi–norm on B if (i) addition, scalar multiplication and $\|\cdot\|$ are $\mathcal{B}$ measurable operations on $\mathbf{B}$, (ii) there exists a set $\mathcal{F}$ of the $\mathcal{B}$ measurable linear functionals on $\mathbf{B}$ such that

$$\|x\| = \sup_{f\in\mathcal{F}} \left|f(x)\right|, \quad x \in \mathbf{B}.$$

Define the pseudo–metric e_μ on $\mathcal{C}$ by

$$e_\mu(C_1, C_2) := \mu(C_1 \Delta C_2), \quad C_1, C_2 \in \mathcal{C},$$

$C_1\Delta C_2$ denoting the symmetric difference between C_1 and C_2. For given $\mathcal{C} \subset \mathcal{A}$ let $\mathbf{C}_b(\mathcal{C},\ \mu)$ be the space of all bounded real functions on $\mathcal{C}$ continuous for e_μ. Let $\mathbf{D}_0(\mathcal{C}, \mu)$ be the linear space of all functions $\phi_1 + \phi_2$, where $\phi_1 \in \mathbf{C}_b(\mathcal{C}, \mu)$ and $\phi_2 = \sum_{i=1}^{n} f_i\delta_{x_i}$, $x_i \in \mathfrak{X}$, $f_i \in \mathbf{D}((0,1))$. Finally, let $\mathbf{D} = \mathbf{D}(\mathcal{C}, \mu)$ be the closure of $\mathbf{D}_0(\mathcal{C}, \mu)$ in $\ell^\infty(\mathcal{C})$ for the sup norm $\|\cdot\|_{\mathcal{C}}$. Note that we use the functions f_i in our construction instead of the constants as in previous works. The proof of the following statement is easy and we omit it.

PROPOSITION 1.1. *Let the class $\mathcal{C} \subset \mathcal{A}$ be such that*

$$\mathcal{B}_b(\mathbf{D},\ \mu) \subset \mathcal{K}(\mathbf{D}), \tag{1.5}$$

with $\mathcal{B}_b(\mathbf{D}, \mu)$ being the σ–algebra generated by the open e_μ–balls in $\mathbf{D}$, and $\mathcal{K}(\mathbf{D})$ being Kolmogorov's σ–algebra, i.e., $\mathcal{K}(\mathbf{D}) = \sigma(\{\pi_c : C \in \mathcal{C}\})$, with $\pi_c : \mathbf{D} \to \mathbf{R}$ being defined by $\pi_c(\phi) = \phi(C)$, $C \in \mathcal{C}$. Then the triple $(\mathbf{D}, \mathcal{K}(\mathbf{B}), \|\cdot\|_{\mathcal{C}})$ is the linear measurable space.

Condition (1.5) on $\mathcal{C}$ will be kept throughout the article. This is the measurability assumption, which turns out to be satisfied in important cases of interest. The proof of the following statement is the same as the proof of Lemma 20 in (Gaenssler, 1983) and we also omit it.

PROPOSITION 1.2. *Let the class $\mathcal{C} \subset \mathcal{A}$ has a countable subclass $\mathcal{C}_0$ such that for any $C \in \mathcal{C}$ there is the sequence $\{C_n, n \geqslant 1\} \subset \mathcal{C}_0$ with*

i) $\delta_x(C_n) \to \delta_x(C)$, as $n \to \infty$, *for μ–a.a. $x \in \mathfrak{X}$,*

ii) $C_n \supset C$, $\forall\, n \geqslant 1$,

then (1.5) *holds.*

Remark. The assumption ii) is needed because the functions $f_i, i = 1, \ldots, n$, in the definition $\mathbf{D}_0(\mathcal{C}, \mu)$ belong to $\mathbf{D}((0,1))$ only and this one can be dropped in the case $f_i \in \mathbf{C}((0,1))$, $i = 1, \ldots, n$. Moreover, under the assumption ii) the reverse inclusion in (1.5) holds also.

The control of the tail of the distribution will be produced using an operator methodology. Define the inclusion map $\mathcal{J} : \mathcal{M}(\mathfrak{X}, \mathcal{A}) \to \ell^\infty(\mathcal{C})$ by

$$\mathcal{J}(\phi) = \Big(\phi(C)\Big)_{C\in\mathcal{C}}, \quad \phi \in \mathcal{M}(\mathfrak{X}, \mathcal{A}). \tag{1.6}$$

If the class $\mathcal{C}$ is not too large then this operator allows to control the average given by $\| \sum r_i \phi_i \|_{\mathcal{C}}$ where (r_i) is the Rademacher sequence and $(\phi_i) \subset \ell^\infty(\mathcal{C})$. Here we consider the case when $\mathcal{C}$ constitutes a Vapnik—Červonenkis or VČ class, i.e., the class with the property

$$\sup\Big\{\#\{F \cap C : C \in \mathcal{C}\} : \#F = n, F \subset \mathfrak{X}\Big\} < 2^n$$

for some $n \geqslant 1$. Note that the class D^d is the most simple VČ class.

THEOREM 1.3 (Pisier, 1984). *For the class $\mathcal{C} \subset \mathcal{A}$ the inclusion map $\mathcal{J}$ is of type 2, i.e., there exists a constant $C = C(\mathcal{C})$ such that the inequality*

$$\mathbf{E}\Big\|\sum_{i=1}^{n} r_i \mathcal{J}(\phi_i)\Big\|_{\mathcal{C}} \leqslant C\Big(\sum_{i=1}^{n} \|\phi_i\|_{\mathrm{var}}^2\Big)^{1/2} \tag{1.7}$$

holds for any $\phi_1, \ldots, \phi_n \subset \mathcal{M}(\mathfrak{X}, \mathcal{A})$ and all $n \geqslant 1$ iff $\mathcal{C}$ is VČ class.

Comment. The type 2 inequality for the operator $\mathcal{J}$ when $\mathcal{C}$ is D^d was proved in (Zaman, 1988) also. This is all what is needed for the multivariate empirical process considered here. More general fact comparing with the Theorem 1.3 is proved in (Kolchinskii, 1988).
We do not use here the type 2 property explicitly. We need a weaker property expressed *via* the notion of type (p, q) which means that the ℓ_2^n norm of $(\|\phi_i\|_{\mathrm{var}})_{i=1}^n$ in the RHS of (1.7) is replaced by the $\ell_{p,q}^n$ semi-norm $|\cdot|_{p,q}$. Recall that

$$\Big|\{a_i, i \geqslant 1\}\Big|_{p,q} = \begin{cases} \Big(\sum_i [i^{1/p} a_i^*]^{q i-1}\Big)^{1/q}, & \text{if } q < \infty, \\ \sup_i i^{1/p} a_i^*, & \text{if } q = \infty, \end{cases}$$

where $\{a_i^*, i \geqslant 1\}$ is the non-increasing rearrangement of $\{a_i, i \geqslant 1\}$. Note that $\ell_{p,p}^n = \ell_p^n$. We also use a slightly more general property than type (p, ∞). To formulate this one we need the following quantity

$$\{Z\}_{R,\infty} = \sup_{\lambda > 0} R(\lambda)\mathbf{P}\Big(\{Z > \lambda\}\Big), \tag{1.8}$$

for some function $R : \mathbf{R}^+ \to \mathbf{R}^+$. If $R(\lambda) = \lambda^p, p > 0$, then (1.8) is equivalent to the semi-norm of the Marcinkiewicz space $L_{p,\infty}$.

LEMMA 1.4. *Let the linear operator $T : (\mathbf{B}_1, \|\cdot\|_1) \to (\mathbf{B}_2, \|\cdot\|_2)$ be of type 2 and let R be a regularly varying function with exponent $p \in (1, 2)$, i.e., $R(\lambda) = \lambda^p L(\lambda)$, for some slowly varying function L. Then there exist a constant $C = C(T, R)$ such that*

$$\Big\{\Big\|\sum_{i=1}^{n} TY_i\Big\|_2\Big\}_{R,\infty} \leqslant C \sum_{i=1}^{n} \Big\{\Big\|Y_i\Big\|_1\Big\}_{R,\infty}$$

for all zero mean $\mathbf{B}_1$*–r.v.'s* $Y_1, \ldots, Y_n$ *and any* $n \geqslant 1$.

Proof. The standard argument infer that we may assume r.v.'s $Y_1, \ldots, Y_n$ to be symmetric. Moreover, type 2 property of the operator T and Fubini theorem yields that

$$\mathbf{E}\left\|\sum_{i=1}^{n} TY_i\right\|_2^2 \leqslant C(T)\sum_{i=1}^{n}\mathbf{E}\|Y_i\|^2$$

for some constant $C(T)$. Thus we get an inequality

$$\mathbf{P}\left(\left\{\left\|\sum_{i=1}^{n} TY_i\right\|_2 > \lambda\right\}\right) \leqslant \sum_{i=1}^{n}\mathbf{P}\left(\left\{\|Y_i\|_1 > \lambda\right\}\right)+$$

$$+C(T)\lambda^{-2}\sum_{i=1}^{n}\int_0^{\lambda}\mathbf{P}\left(\left\{\|Y_i\|_1 > t\right\}\right)dt^2$$

for all $\lambda > 0$. Hence the standard calculation of an integral containing slowly varying function (see e.g., Theorem 2.1 in (Seneta, 1976)) gives

$$\int_0^{\lambda}\mathbf{P}\left(\left\{\|Y\| > t\right\}\right)dt^2 \leqslant 2\left\{\|Y\|\right\}_{R,\infty}\int_0^{\lambda}\left(t^{1-p}/L(t)\right)dt =$$

$$= 2(1+o(1))\left\{\|Y\|\right\}_{R,\infty}\lambda^{2-p}/(2-p)L(\lambda), \text{ as } \lambda \to \infty. \tag{1.9}$$

Combining above estimates yields

$$R(\lambda)\mathbf{P}\left(\left\{\left\|\sum_{i=1}^{n} TY_i\right\|_2 > \lambda\right\}\right) \leqslant C\sum_{i=1}^{n}\left\{\|Y_i\|_1\right\}_{R,\infty},$$

for some constant C and all λ, thereby proving the proposition.
Now we are ready to state and prove limit results for weighted empirical processes which will be done in the following section.

2. LIMIT THEOREMS

This section is devoted to the limit theorems for the supremum of the weighted multivariate uniform empirical processes. Non–standardness of this results constitutes conditions expressed *via* variation of w.f. . For an explanation of the assumptions used here assume for the moment the class $\mathcal{C}$ to be $\mathcal{D}^1$.
Let us consider the function $w \in \mathbf{D}((0,1))$ with the finite variation on any interval $[x,1]$, i.e., var $(w|_{[x,1]}) < \infty, \forall\, x \in (0,1)$. Taking into account (1.3) we see that the necessary inequality in the assumption (RM) is satisfied if for some constant c

$$\operatorname{var}\left(w|_{[x,1]}\right) \leqslant cw(x),\ \forall\, x \in (0,1).$$

As noted in the introduction this condition is trivially satisfied when w is non-increasing on (0, 1). Optimal conditions for (RM) in the multivariate case are more complex (see Corollary 3.2). But the usual monotonicity assumption is still sufficient in this case also (see the remark at the end of this note also).
The current remark concerns the characterization of the weight "obeying" the Zygmund—Marcinkiewicz type law of large numbers (Corollary 2.3 below). The similar extension of the class of weights can be done in the CLT and LIL for the weighted empirical process also. But straightforward application of this approach (see e.g., Theorem 1 in (Giné and Zinn, 1986)) gives square integrability condition of the w.f. only.
In other strong limit theorems stated below, traditional power weighting (see e.g., Chapter 4 in (Einmahl, 1987)) is replaced by more general one including slowly varying functions. The similar refinements related to some weak limit theorems is done in (Csörgő and Horváth, 1988).
Now we propose the aforementioned Marcinkiewicz—Zygmund type laws of large numbers. This result complements and extends to the multivariate case those one being stated in the Theorem 7.3 in (Andersen et al., 1988). According to our scheme we start with the general result in the operator setting. For given linear measurable space $(\mathbf{B}, \|\cdot\|)$ we denote the space of classes equivalencies of symmetric (*sic!*) $\mathbf{B}$–r.v.'s by $L_0(\mathbf{B})$. We recall the definition of the Lorentz—Marcinkiewicz spaces

$$L_{p,q}(\mathbf{B}) = \Big\{Y \in L_0(\mathbf{B}) : \int_0^\infty \Big[\mathbf{P}(\{\|Y\| > \lambda\})\Big]^{q/p} d\lambda^q < \infty, \text{ if } q < \infty,$$

and

$$\sup_{\lambda>0} \lambda^p \mathbf{P}\Big(\{\|Y\| > \lambda\}\Big) < \infty, \text{ if } q = \infty\Big\};$$

$$L^0_{p,\infty}(\mathbf{B}) = \Big\{Y \in L_0(\mathbf{B}) : \lim_{\lambda\to\infty} \lambda^p \mathbf{P}\Big(\{\|Y\| > \lambda\}\Big) = 0\Big\}.$$

Moreover, for given i.i.d. $\mathbf{B}$–r.v.'s $Y, Y_1, \ldots, Y_n$ we denote

$$S_{n,p}(Y) = n^{-1/p}(Y_1 + \ldots + Y_n).$$

THEOREM 2.1 (Norvaiša, 1984). 1. *Assume that a linear operator T maps a normed space $\mathbf{B}_1$ into a linear measurable space $\mathbf{B}_2$. For given $p \in [1,2)$ and $q \in [p, \infty]$ the following statements are equivalent:*

i) *T is of type (p, q);*

ii) *the sequence*

$$\Big\{S_{n,p}(TY) : n \geqslant 1\Big\} \tag{2.1}$$

converges (to zero) in $L_{p,q}(\mathbf{B}_2)$ if

$$Y \in \begin{cases} L_{p,q}(\mathbf{B}_1), & \text{for } q < \infty, \\ L^0_{p,\infty}(\mathbf{B}_1), & \text{for } q = \infty; \end{cases}$$

or (in the case $q = \infty$) is bounded in $L_{p,\infty}(\mathbf{B}_2)$ if $Y \in L_{p,\infty}(\mathbf{B}_1)$.
2. *Let $\mathbf{B}$ be the linear measurable space, then for given $p \in [1,2)$ and $q \in [p,\infty]$ the following statements holds:*

i) *the sequence (2.1) converges (to zero) a.s. iff converges (to zero) in $L_p(\mathbf{B})$;*

ii) *the sequence (2.1) converges (to zero) in probability iff converges (to zero) in $L_{p,\infty}(\mathbf{B})$;*

iii) *the sequence (2.1) is stochastically bounded iff is bounded in $L_{p,\infty}(\mathbf{B})$;*

iv) *the "only if" part holds true in the statement 1 ii) (with T being identity map).*

Remark. The Theorem 2.1 is not stated in the present form in (Norvaiša, 1984), but it is actually proved. The difference is an assumption used early that $\mathbf{B}, \mathbf{B}_1$ and $\mathbf{B}_2$ must be separable Banach spaces. But this was used for measurability only and the Theorem 2.1 remains true in the setting presented here.
Now we apply this theorem to the case $\mathbf{B}_1 = \mathcal{M}(\mathfrak{X}, \mathcal{A})$, $\mathbf{B} = \mathbf{B}_2 = \mathbf{D}(\mathcal{C}, \mu)$ and T let be an inclusion map $\mathcal{J}$ given by (1.6). Moreover, for given w.f. w we define the function

$$Y(C,x) := w(\mu(C))\big[\delta_x(C) - \mu(C)\big], \quad x \in \mathfrak{X}, C \in \mathcal{C}, \tag{2.2}$$

and put

$$Y_i(C) := Y(C, X_i), \quad i = 1, \dots, n.$$

We assume throughout and, following (Alexander, 1987b), say that $\mathcal{C}$ is half–bounded, i.e., $\mu(C) \leqslant 1/2$ for all $C \in \mathcal{C}$. Then due to the definition of the envelope F_w we get

$$\mu\Big(\big\{x : F_w(x) > 2\lambda\big\}\Big) \leqslant P\Big(\big\{\|Y(\cdot, X)\|_{\mathcal{C}} > \lambda\big\}\Big) \leqslant \mu\Big(\big\{x : F_w(x) > \lambda\big\}\Big), \tag{2.3}$$

for all sufficiently large $\lambda > 0$.

Corollary 2.2. *Let $\mathcal{C} \subset \mathcal{A}$ be a half bounded VČ class and the assumption (RM) holds. If $p \in [1,2)$ and $q \in [p,\infty]$ then the convergence*

$$n^{1-1/p}\big\|\nu_n^w\big\|_{\mathcal{C}} \to 0, \quad as \quad n \to \infty,$$

takes place in the form:

i) *a.s. iff $F_w \in L_p(\mathfrak{X}, \mu)$;*

ii) *in $L_{p,q}(\Omega, \sum, \mathbf{P})$ iff*

$$F_w \in \begin{cases} L_{p,q}(\mathfrak{X}, \mu), & if \quad q < \infty, \\ L^0_{p,\infty}(\mathfrak{X}, \mu), & if \quad q = \infty; \end{cases}$$

iii) *in probability iff $F_w \in L^0_{p,\infty}(\mathfrak{X}, \mu)$.*
Moreover, the sequence

$$\Big\{n^{1-1/p}\big\|\nu_n^w\big\|_{\mathcal{C}} : \quad n \geqslant 1\Big\}$$

is stochastically bounded (and/orin$L_{p,\infty}(\Omega, \sum, \mathbf{P})$) iff

$$F_w \in L_{p,\infty}(\mathfrak{X}, \mu).$$

Proof. As usual, we can randomize and consider the sequence

$$\left\{ \left(w(\mu(C)) \sum_{i=1}^{n} r_i \delta_{x_i}(C) \right)_{C \in \mathcal{C}} : n \geqslant 1 \right\}.$$

Taking into account (2.3) we see that the statement follows immediately from Theorem 2.1.
Let us reformulate the corollary for the case when $\mathcal{C} = \mathcal{D}^1, d \in \mathbb{N}$, and μ is Lebesgue measure $\mathbf{m}$ on $\mathfrak{X} = [0,1]^d$. In this case sufficient conditions given in Corollary 3.2 below implies the assumption (RM) to hold. Moreover, conditions on the distribution of the envelope function we are able to reduce to the conditions on w.f. Induction argument shows that

$$\mathbf{m}^{(d)}(t) := \mu\Big(\{x \in \mathfrak{X} : \mu(G_x) \leqslant t\}\Big) = \mathbf{m}\Big(\{x \in [0,1]^d : |x| \leqslant t\}\Big) =$$

$$= t \sum_{k=0}^{d-1} \ln^k(1/t)/k! \sim t\ln^{d-1}(1/t)/(d-1)! \text{ as } t \to 0. \tag{2.4}$$

Without notation abuse we denote by $\mathbf{m}^{(d)}$ the measure on $([0,1/2], \mathcal{B})$ with the distribution function $\mathbf{m}^{(d)}$. Note that $\mathbf{m}^{(1)}$ is the Lebesgue measure. This establishes the next corollary.

COROLLARY 2.3. *Let $w \in W^d; d \in \mathbb{N}$ (see (3.2)) and (3.5) holds. If $p \in [1,2)$ and $q \in [p, \infty]$ then the convergence*

$$n^{1-(1/p)} \sup\left\{ w(|t|) \Big| F_n(t) - |t| \Big| : t \in [0,1]^d, |t| \leqslant 1/2 \right\} \to 0 \quad \text{as} \quad n \to \infty,$$

takes place in the form:
i) a.s. *iff $w \in L_p([0,1/2], \mathbf{m}^{(d)})$;*
ii) *in $L_{p,q}(\Omega, \sum, \mathbf{P})$ iff*

$$w \in \begin{cases} L_{p,q}([0,1/2], \mathbf{m}^{(d)}), & \text{if } q < \infty, \\ L^0_{p,\infty}([0,1/2], \mathbf{m}^{(d)}), & \text{if } q = \infty; \end{cases}$$

iii) *in probability iff $w \in L^0_{p,\infty}([0,1/2], \mathbf{m}^{(d)})$.*
Moreover, the sequence

$$\left\{ n^{1-(1/p)} \sup\left\{ w(|t|) \Big| F_n(t) - |t| \Big| : t \in [0,1]^d, |t| \leqslant 1/2 \right\} : n \geqslant 1 \right\}$$

is stochastically bounded (and/or in $L_{p,\infty}(\Omega, \sum, \mathbf{P})$) *iff* $w \in L_{p,\infty}([0,1/2], \mathbf{m}^{(d)})$.

Remarks. 1. If the w.f. w is non–increasing then (3.5) holds and (e.g.,) the condition $w \in L^0_{p,\infty}([0,1/2], \mathbf{m}^{(d)})$ can be rewritten by

$$w^p(t) t \ln^{d-1}(1/t) \to 0, \quad \text{as} \quad t \to 0,$$

and the second one $w \in L_p([0,1/2], \mathbf{m}^{(d)})$ takes form

$$\int_0^{1/2} w^p(t) d\Big[t \ln^{d-1}(1/t)\Big] < \infty.$$

2. If $p = 1$ and w is non–increasing, then the statement i) follows from Theorem 2 in (Mason, 1982). The statements i), iii) and the stochastic boundedness are proved in (Andersen et al., 1988) when $d = 1$.
The second limit theorem for the weighted empirical processes concerns the law of large numbers for random variables with heavy tails (or heavy weights). The general form of this result can be stated by

THEOREM 2.4 (Mikosch and Norvaiša, 1986). *Let* $(\mathbf{B}, \|\cdot\|)$ *be a measurable linear space,* $b_n \to \infty$ *and* $Y, Y_1, \dots, Y_n$ *be i.i.d.* $\mathbf{B}$*–r.v.'s such that*

$$\sup\Big\{\mathbf{E}\|Y\|^2 \mathbf{1}\Big(\{\|Y\| \leqslant \lambda\}\Big) \Big/ \lambda^2 \mathbf{P}\Big(\{\|Y\| > \lambda\}\Big) : \lambda \geqslant 1\Big\} < \infty. \tag{2.5}$$

Then

$$\text{a.s.} - \overline{\lim_{n\to\infty}}\, b_n^{-1} \Big\| \sum_{i=1}^n Y_i \Big\| = 0 \quad or \quad = +\infty,$$

according as

$$\sum_{n=1}^{\infty} \mathbf{P}\Big(\{\|Y\| > b_n\}\Big) < +\infty \quad or \quad = +\infty.$$

Moreover, for the first case we require in addition

$$\textit{in probability} - \overline{\lim_{n\to\infty}}\, b_n^{-1} \Big\| \sum_{i=1}^n Y_i \Big\| = 0. \tag{2.6}$$

Remark. This theorem is proved in the referred paper under assumptions that $\mathbf{B}$ is a separable Banach space and the sequence $\{b_n : n \geqslant 1\}$ is increasing. But this is superfluous as noted in (Norvaiša, 1988).
We show that condition (2.6) is satisfied for the weighted empirical processes making use of the Lemma 1.4.

COROLLARY 2.5. *Let* $\mathcal{C} \subset \mathcal{A}$ *be a half-bounded* VČ *class and the assumption* (RM) *holds. Suppose that the tail function*

$$R(\lambda) = \mu\Big(\{x \in \mathfrak{X} : F_w(x) > \lambda\}\Big)$$

is regularly varying with exponent $p \in (1,2)$. Then

$$\text{a.s.} - \overline{\lim_{n\to\infty}} \frac{n}{b_n} \left\| \nu_n^w \right\|_{\mathcal{C}} = 0 \text{ or } = +\infty,$$

according as

$$\sum_{n=1}^{\infty} R(b_n) < +\infty \quad \text{or} \quad = +\infty.$$

Moreover, for the first case we require in addition that the sequence $\{nR(b_n) : n \geqslant 1\}$ is non-increasing.

Proof. Using the setup given by (2.2) and (2.3) we must verify (2.5) and (2.6). Due to the regular variation of the tail function R the calculation similar to (1.9) implies (2.5). Now using Lemma 1.4 we get

$$\mathbf{P}\left(\left\{\left\|\sum_{i=1}^{n} Y_i\right\|_{\mathcal{C}} > \varepsilon b_n\right\}\right) \leqslant R(\varepsilon b_n)\left\{\left\|\sum_{i=1}^{n} Y_i\right\|_{\mathcal{C}}\right\}_{1/R,\infty} \leqslant$$

$$\leqslant c\varepsilon^p nR(b_n)n\left\{\|Y\|_{\text{var}}\right\}_{1/R,\infty} \leqslant$$

$$\leqslant c\varepsilon^p nR(b_n)\left\{F_w\right\}_{1/R,\infty} = o(1) \quad \text{as} \quad n \to \infty.$$

This completes the proof.
Let us reformulate the corollary for the case when $\mathcal{C} = \mathcal{D}^1, d \in \mathbb{N}$, and μ is Lebesgue measure $\mathbf{m}$ on $\mathfrak{X} = [0,1]^d$. For the regular variation of the tail function R we must take the w.f. to be of the special form. For regularly varying with exponent $1/p$ function b, i.e., $b(\lambda) = \lambda^{1/p}L(\lambda)$, we put

$$w_p(t) = b(1/t) = t^{-1/p}L(1/t), \quad \text{for} \quad t \in (0, 1/2]. \tag{2.7}$$

W.l.o.g. we will assume (see property 1.5.4° in (Seneta, 1976)) that $b(\lambda)$ is increasing as $\lambda \to \infty$. Then according to (2.4) and (2.7) we get

$$R(\lambda) \sim b^{\#}(\lambda) \ln^{d-1}\left[1/b^{\#}(\lambda)\right], \quad \text{as} \quad \lambda \to \infty,$$

where the function $b^{\#}$ varies regularly with exponent $-p$ and such that $b^{\#}(b(\lambda)) \sim \lambda^{-1}$, as $\lambda \to \infty$. The existence of the function $b^{\#}$ can be proved as the property 1.5.5° in (Seneta, 1976). The next corollary summarizes this.

COROLLARY 2.6. *Let $p \in (1,2), d \in \mathbb{N}$ and $\{a_n : n \geqslant 1\}$ be a sequence. Then*

$$\text{a.s.} - \overline{\lim_{n\to\infty}}\, nb^{-1}(a_n^{-1}) \sup\Big\{w_p(|t|)\Big|F_n(t) - |t|\Big| :$$

$$t \in [0,1]^d,\ |t| \leqslant 1/2\Big\} = 0 \quad \text{or} \quad = +\infty,$$

according as

$$\sum_{n=1}^{\infty} a_n \log^{d-1}\left(1/a_n\right) < +\infty \quad or \quad = +\infty.$$

Moreover, for the first case we require in addition that the sequence $\{na_n : n \geqslant 1\}$ *is non–increasing.*

Remark. The statement of this corollary has been proved in (Einmahl, 1987) for $b(\lambda) = \lambda^p$ and by (Norvaiša, 1988) for $d = 1$. Under the additional assumptions Einmahl in (Einmahl, 1989) extends this corollary to other w.f. when $d = 1$.
The last limit theorem for the weighted empirical processes in this note concerns the law of the iterated logarithm for the random variables with heavy tails (or heavy weights). The general form of this result can be stated by

THEOREM 2.7 (Norvaiša, 1989). *Let* $(\mathbf{B}, \|\cdot\|)$ *be a measurable linear space and* $\{\|\cdot\|_n, n \geqslant 1\}$ *be a non–decreasing sequence of semi–norms not exceeding the initial one. Assume* $Y, Y_1, \ldots, Y_n$ *to be i.i.d.* $\mathbf{B}$*–r.v.'s such that for some non–decreasing sequence* $\{b_n, n \geqslant 1\}$

$$\|Y\|_n \leqslant b_n, \quad \mathbf{E}\|Y\|_n^2 \leqslant C_Y b_n^2 (\log_2 n)/n, \quad \forall\, n \geqslant 1, \tag{2.8}$$

for some constant C_Y *and*

$$in\ probability\!-\lim_{n\to\infty} \left\|\sum_{i=1}^{n} Y_i\right\|_n (b_n \log_2 n)^{-1} = 0. \tag{2.9}$$

Then

$$\text{a.s.} - \varlimsup_{n\to\infty} \left\|\sum_{i=1}^{n} Y_i\right\|_n (b_n \log_2 n)^{-1} \leqslant 2C, \tag{2.10}$$

where the constant C *is defined by* $C[\log(1 + C/2C_Y) - 1] > 2$.

Remark. The proof of this statement follows the lines of the proof of Theorem 2 in (Norvaiša, 1989). The difference is the use here of the sequence of semi–norms instead of the procedure of truncation in the referred paper. As the result we get better bound in (2.10) but still not attaining one known for the empirical processes.
Continuing this translation process we restate Theorem 2.7 for the weighted empirical process using (2.2) and the w.f. w_p defined by (2.7). Moreover, we choose the norming sequence $\{b_n, n \geqslant 1\}$ using the same regularly varying with exponent $1/p$ function b as for the w.f. w_p. So, we put

$$b_n = b(n/\log_2 n) \sim (n/\log_2 n)^{1/p} L(n) \quad \text{as} \quad n \to \infty.$$

COROLLARY 2.8. *Let* $\mathcal{C} \subset \mathcal{A}$ *be a half–bounded VČ class and the assumption* (RM) *holds. Suppose that the tail function*

$$R(\lambda) = \mu\Big(\{x \in \mathfrak{X} : F_w(x) > b(\lambda)\}\Big) \sim \lambda^{-1} \text{ as } \lambda \to \infty, \ . \tag{2.11}$$

for some $p \in (1.2)$. *Then*

$$\text{a.s.} - \overline{\lim_{n\to\infty}} \sup\left\{\frac{n\nu_n(C)w_p(\mu(C))}{b_n \log_2 n} : C \in \mathcal{C},\right.$$

$$\left.\mu(C) \geqslant \frac{\log_2 n}{n}\right\} b_n^{-1} \log_2^{-1} n \leqslant 2C \tag{2.12}$$

with the same constant C *as in* (2.10) *and* C_Y *given by*

$$C_Y = 2\left(1 + \frac{2}{2-p}\left\{F_{w_p}\right\}_{1/b\#,\infty}\right). \tag{2.13}$$

Remark. This statement complements Theorem 3.1 in (Alexander, 1987) where the w.f. $w(t) \sim 1/\sqrt{t}$ as $t \to 0$, is considered. It must be noted that the bound given in the referred paper is exact.

Proof. We precede the proof by denoting

$$\mathcal{C}_n = \{C \in \mathcal{C} : \mu(C) \geqslant \log_2 n/n\},$$

$$a_n = b_n \log_2 n, \quad F = F_{w_p}.$$

We use the sequence of semi–norms $\{\|\cdot\|_n,\ n \geqslant 1\}$ given by

$$\|\phi\|_n = \|\phi\|_{\mathcal{C}_n}, \quad \phi \in \ell^\infty(\mathcal{C}).$$

According to (2.2)

$$\frac{1}{a_n}\|n\nu_n w_p \circ \mu\|_{\mathcal{C}_n} = \frac{1}{a_n}\left\|\sum_{i=1}^n Y_i\right\|_n.$$

Hence it is sufficient to verify the conditions (2.8) and (2.9). Due to (2.2) we have for all $x \in \mathfrak{X}$

$$\left\|Y(\cdot,x)\right\|_n \leqslant b_n \vee \sup\left\{\mu(C)w(\mu(C)) : C \in \mathcal{C}_n, x \notin C\right\} = b_n,$$

at least for sufficiently large $n \geqslant 1$. This implies an estimate

$$\mathbf{E}\|Y\|_n^2 \leqslant b_n^2 \mathbf{P}\Big(\{\|Y\| > b_n\}\Big) + \mathbf{E}\|Y\|^2 \mathbf{1}\Big(\{\|Y\| \leqslant b_n\}\Big). \tag{2.14}$$

Using the theorem on the uniform convergence of the slowly varying functions (see Theorem 1.1 in (Seneta, 1976)) and (2.11), we get

$$\mathbf{P}\Big(\{\|Y\| > b(\lambda)\}\Big) \sim \mu\Big(\{x \in \mathfrak{X} : F(x) > b(\lambda)\}\Big) \sim \lambda^{-1} \text{ as } \lambda \to \infty. \tag{2.15}$$

Hence standard calculation of the integral containing the slowly varying function (see e.g., Theorem 2.1 in (Seneta, 1976)) gives

$$\mathbf{E}\|Y\|^2 \mathbf{1}\Big(\{\|Y\| \leqslant b_n\}\Big) = 2\int_0^{b_n}\int_0^{\lambda} t\,dt\,d\mathbf{P}\Big(\{\|Y\| \leqslant \lambda\}\Big) \leqslant$$

$$\leqslant 2\int_0^{b_n}\lambda\mathbf{P}\Big(\{\|Y\|>\lambda\}\Big)d\lambda \leqslant 2\{\|Y\|\}_{1/b\#,\infty}\int_0^{b_n}\lambda b(\lambda)d\lambda \sim$$

$$\sim\frac{2}{2-p}\{\|Y\|\}_{1/b\#,\infty}b_n^2\log_2 n/n \text{ as } n\to\infty.$$

Combining this with (2.14) and (2.15) we obtain (2.8), with C_Y as in (2.13) at least for sufficiently large n. Now it remains to establish (2.9) only. Standard randomization argument implies

$$a_n\mathbf{P}\left(\left\{\Big\|\sum_{i=1}^n Y_i\Big\|_n > a_n\right\}\right) \leqslant 2\varepsilon^{-1}\mathbf{E}\Big\|\sum_{i=1}^n r_i w\circ\mu\delta_{X_i}\Big\|_n. \tag{2.16}$$

For the remainder estimation it is necessary to split the RHS of (2.16) into the two terms by $\mathbf{1}(\{F(X_i)>b_n\})$ and $\mathbf{1}(\{F(X_i)\leqslant b_n\}), i=1,\dots,n$. Let Z be non-negative r.v. and $q\in(1,p)$, then for some constant c

$$\mathbf{E}Z\leqslant(\mathbf{E}Z^q)^{1/q}\leqslant\Big(1+c\{Z\}_{1/b\#,\infty}\Big)^{1/q}.$$

Using this we obtain for the first term

$$b_n^{-1}\mathbf{E}\Big\|\sum_{i=1}^n r_i w\circ\mu\delta_{X_i}\mathbf{1}\Big(\Big\{F(X_i)>b_n\Big\}\Big)\Big\|_n \leqslant$$

$$\leqslant\left(1+c\left\{\Big\|\sum_{i=1}^n r_i b_n^{-1}w\circ\mu\delta_{X_i}\mathbf{1}\Big(\{F(X_i)>b_n\}\Big)\Big\|\right\}_{1/b\#,\infty}\right)^{1/q}\leqslant$$

$$\leqslant\left(1+cn\sup_{x>0}\mathbf{P}\Big(\{\mathbf{1}(\{F(X)>b_n\})>x\}\Big)\right)^{1/q}\leqslant$$

$$\leqslant\Big(1+cn\mu\Big(\{x:F(x)>b_n\}\Big)\Big)^{1/q}.$$

Therefore

$$a_n^{-1}\mathbf{E}\Big\|\sum_{i=1}^n r_i w\circ\mu\delta_{X_i}\mathbf{1}\Big(\{F(X_i)>b_n\}\Big)\Big\|_n\leqslant$$

$$\leqslant\Big((\log_2 n)^{-q}+cn(\log_2 n)^{-q}\mu(\{x:F(x)>b_n\})\Big)^{1/q}=o(1),\text{ as } n\to\infty,$$

because $q>1$. Once again the classical calculation of the integral with the slowly varying function (see e.g., Theorem 2.1 in (Seneta, 1976)) will provide the needed bound of the second term

$$a_n^{-1}\mathbf{E}\Big\|\sum_{i=1}^n r_i w\circ\mu\delta_{X_i}\mathbf{1}\Big(\{F(X_i)\leqslant b_n\}\Big)\Big\|_n\leqslant$$

$$\leqslant c\Big[na_n^{-2}\mathbf{E}\|w\circ\mu\delta_X\|_{\mathrm{var}}^2\mathbf{1}(\{F(X)\leqslant b_n\})\Big]^{1/2}\leqslant$$

$$\leqslant c\Big[na_n^{-2}\int_{\mathfrak{X}} F^2(x)\mathbf{1}\Big(\{x : F(x) \leqslant b_n\}\Big)\mu(dx)\Big]^{1/2} \leqslant$$

$$\leqslant c\Big[n(\log_2 n)^{-2}\mu(\{x : F(x) > b_n\})\Big]^{1/2} = o(1), \text{ as } n \to \infty.$$

This completes the proof of the corollary.
When $\mathcal{C} = \mathcal{D}^1$ and μ is uniform on $[0,1]$ (the one dimensional empirical process), we have

$$\mu\Big(\{x \in \mathfrak{X} : F_{w_p}(x) > b(\lambda)\}\Big) \sim \mu\Big(\{x \in \mathfrak{X} : \mu(G_x) \leqslant \lambda^{-1}\}\Big) = \lambda^{-1} \text{ as } \lambda \to \infty,$$

so (2.11) is valid. Hence the following statement holds.

COROLLARY 2.9. *Let $p \in (1,2)$, then*

$$\text{a.s.} - \overline{\lim_{n\to\infty}} \sup\Big\{w_p(t)|F_n(t) - t| : t \in \Big[\frac{1}{n}\log_2 n, \frac{1}{2}\Big]\Big\}\Big/ b_n \log_2 n \leqslant 2C,$$

with the same constant C as in (2.10) and C_Y given by $2(1 + 2/(2-p))$.

3. APPENDIX

It is the purpose of this section to extend representation (1.3) to the case when $\mathcal{C} = \mathcal{D}^d, d \in \mathbb{N}$ and μ is Lebesgue measure on $\mathfrak{X} = [0,1]^d$. Let us outline the case $d = 2$ before giving the full story.
For given two vectors x, y of $[0,1]^2$ with $x \leqslant y$, we can write

$$w(\mathbf{m}([0,y]))\delta_x([0,y]) \equiv [w(y_1 \cdot y_2) - w(x_1 \cdot y_2) - w(y_1 \cdot x_2) + w(x_1 \cdot x_2)] -$$

$$-[w(x_1 \cdot x_2) - w(x_1 \cdot y_2)] - [w(x_1 \cdot x_2) - w(y_1 \cdot x_2)] + w(x_1 \cdot x_2) =$$
$$= Q^2([x,y]) - Q^1(\{x_1\} \times [x_1, y_2]) - Q^1([x_1, y_2] \times \{x\}) + Q^0(\{x\}),$$

where Q^l are the set functions on the indicated face of the quadrate $[x,y]$. For the case when $l = 0$, $Q^0(\{x\})$ is the point mass at x, i.e., $w(x_1 \cdot x_2)\delta_x$. Under the variation boundedness assumption on the w.f. w, the set function Q^l will be the appropriately dimensioned Lebesgue—Stieltjes measure generated by w. We formulate this condition more precisely in the general case (see (3.2) below).
We now introduce the necessary notation to permit us to give above representation for an arbitrary positive integer d. For given $1 \leqslant l \leqslant k \leqslant d$, let $I(k,>)(I(k,<))$ be any k–tuple of distinct decreasing (respectively, increasing) positive integers from $\Delta = \{1,\ldots,d\}$.
Let I be any k–tuple from Δ, with $k = \#I$. Put $I' = \Delta/I$. It is convenient to introduce the projection $\mathrm{pr}_{I'}$ acting on the cube $[x,y]$ to the $(d-k)$–dimensional face. For given two vectors x, y of $[0,1]^d$, i.e., $x = (x_1,\ldots,x_d)$ and $y = (y_1,\ldots,y_d)$, with $x \leqslant y$ we define

$$\mathrm{pr}_{I'}([x,y]) = \mathop{\times}_{j=1}^{d} [x_j, I(y_j)],$$

where

$$I(y_j) = \begin{cases} x_j, & \text{if } j \in I, \\ y_j, & \text{if } j \notin I. \end{cases}$$

Now on any such $(d-k)$–dimension face $\mathrm{pr}_{I'}([x,y])$ we define (cf. (Shilov and Gurevich, 1967, p. 68)) the σ–additive set function $Q^{d-k} = Q^{d-k,w}$ by

$$Q^{d-k}\Big(\mathrm{pr}_{I'}([x,y])\Big) = \sum_{l=0}^{d-k}(-1)^{l+k} \sum_{I(k,<)\subset I'} w\Big(\prod_{j\in I(l)} x_j \prod_{j\notin I(l)} y_j\Big), \tag{3.1}$$

where $I(l) = I \cup I(l,<)$. This set function becomes appropriately dimensioned Lebesgue—Stieltjes measure generated by the function w if its variation is bounded, i.e., for the face $B = \mathrm{pr}_{I'}([x,y])$ and any its partition π given by

$$B_p = \prod_{j\in I}\Big\{x_j\Big\} \times \prod_{j=1}^{d-k}\Big[z_j^{p-1}, z_j^p\Big], \quad p = 0,\ldots,m,$$

the following quantity is bounded

$$V_B\Big(Q^{d-k}\Big) = \sup_{\pi}\sum_{p=0}^{m}\Big|Q^{d-k}(B_p)\Big| =$$

$$= \sup_{\pi}\sum_{p=0}^{m}\Big|\sum_{l=0}^{d-k}(-1)^{l+k} \sum_{I(l,<)\subseteq I'} w\Big(\prod_{j\in I} x_j \prod_{j\in I(l,<)} z_j^{p-1} \prod_{j\notin I(l)} z_j^p\Big)\Big| < \infty.$$

Recall that in our applications x is fixed and y varies in $[x,1]$. For given function w and any k–tuple I with $k = \#I$, define the function $V^{I,w}$ by

$$V^{I,w}(x) = V_{\mathrm{pr}_{I'}([x,1])}(Q^{d-k,w}), \quad x \in (0,1]^d.$$

Now we are able to define the initial class of the w.f. w which is suggested by the method used in this note. For given positive integer $d \geqslant 1$, put

$$W^d = \Big\{w : [0,1] \to \mathbf{R}^+ \Big| w(0) = 0,\ w \in \mathbf{D}((0,1))$$

$$\text{and } V^{I,w}(x) < \infty,\ \forall\, x \in (0,1]^d,\ \forall\, I \subset \Delta\Big\}. \tag{3.2}$$

An assumption $w \in \mathbf{D}((0,1))$, i.e., $w(t+) = w(t)$, was used by us to prepare the measurability structure (see definition of the space $\mathbf{D}(\mathcal{C},\mu)$). Here, i.e., in the context of construction of measures, this assumption guaranties (see (Shilov and Gurevich, 1967, p.74)) uniqueness of the set function Q generated by w in the sense that equality

$$\int f dQ = 0,\ \forall f \in \mathbf{C}([0,1]^d)$$

implies $Q \equiv 0$.

THEOREM 3.1. *Assume that $w \in W^d, d \in \mathbb{N}$ and $x, y \in [0,1]^d$, with $x \leqslant y$. Then*

$$w\Big(\mathbf{m}([0,y])\Big)\delta_x([0,y]) =$$

$$= \sum_{k=0}^{d}(-1)^k \sum_{I(k,>)\subset\Delta} Q^{d-k}\Big(\mathrm{pr}_{I(k,>)'}([x,y])\Big), \tag{3.3}$$

where the measure Q^{d-k} is defined by (3.1).

Proof. Fix $l = 0, 1, \ldots, d$. Let N_l be the number of

$$w\Big(\prod_{j\in I(l)} x_j \cdot \prod_{j\notin I(l)} y_j\Big)$$

in the RHS of (3.3). Note that $N_0 = 1$, i.e., there is exactly one quantity $w(|y|)$. Examination of the RHS of (3.3) gives

$$N_l = \sum_{m=0}^{l}\Big(-1\Big)^m \sum_{I(m,>)\subset\Delta} \quad \sum_{I(l-m,<)\subset\Delta/I(m,>)} 1 = \sum_{m=0}^{l}\Big(-1\Big)^m A_m,$$

where

$$A_0 = \sum_{I(l,<)\subset\Delta} 1, \quad A_l = \sum_{I(l,>)\subset\Delta} 1$$

and

$$A_m = \sum_{I(m,>)\subset\Delta} \quad \sum_{I(l-m,<)\subset\Delta/I(m,>)} 1 =$$

$$= \sum_{I(m,>)\subset\Delta/\{1\}} \quad \sum_{I(l-m,<)\subset\Delta/I(m,>),i_{m+1}\in\{1,\ldots,i_m-1\}} 1+$$

$$+ \sum_{I(m,>)\subset\Delta} \quad \sum_{I(l-m,<)\subset\Delta/I(m,>),i_{m+1}>i_m} 1 = B_m + C_m,$$

for $m = 1, \ldots, l-1$. We show that $B_m = C_{m+1}$ for $m = 1, \ldots, l-2$. Indeed

$$B_m = \sum_{I(m,>)\subset\Delta/\{1\}} \quad \sum_{i_{m+1}\in\{1,\ldots,i_m-1\}} \quad \sum_{I(l-m-1,<)\subset\Delta/I(m+1,>),i_{m+2}>i_{m+1}} 1 =$$

$$= \sum_{I(m+1,>)\subset\Delta} \quad \sum_{I(l-m-1,<)\subset\Delta/I(m+1,>),i_{m+2}>i_{m+1}} 1 = C_{m+1}.$$

Hence

$$N_l = \sum_{m=0}^{l}(-1)^m(B_m + C_m) = A_0 - C_1 \mp B_{l-1} \pm A_l, \tag{3.4}$$

where the sign $-$ or $+$ depends on an integer l is even or odd. Note that

$$C_1 = \sum_{i_1 \in \Delta} \sum_{I(l-1,<) \subset \Delta/\{i_1\}} 1 = \sum_{I(l,<) \subset \Delta} 1 = A_0$$

and

$$B_{l-1} = \sum_{I(l-1,>) \subset \Delta/\{1\}} \sum_{i_l \in \{1,\ldots,i_{l-1}-1\}} 1 = \sum_{I(l,>) \subset \Delta} 1 = A_l.$$

This and (3.4) gives $N_l = 0$, for $l = 1, \ldots, d$. This establishes Theorem 3.1.

COROLLARY 3.2. *Let $w \in W^d$ and there is a constant c such that*

$$\max_{I \subset \Delta} V^{I,w}(x) \leqslant cw(|x|), \ \forall\, x \in (0,1]^d, \tag{3.5}$$

then an assumption (RM) *is satisfied by the sample* $(\mathfrak{X}, \mathcal{A}, \mu) = ([0,1], \mathcal{B}, \mathbf{m})$, *the subclass* $\mathcal{C} = \mathcal{D}^d$ *and the w.f.* w.

Remarks. 1. It is easy to see that (3.5) holds if the w.f. w is non–increasing.

2. In (Alexander, 1987b) there is no conditions on the w.f. w. But assumptions are expresses *via* the envelope F_w. If we had lefted conditions expressed *via* the variation norm then we should have got analoguous results. But an estimate in the assumption (RM) or (3.5) excludes this.

REFERENCES

Alexander, K.S. (1987a). Rates of grows and sample moduli for weighted empirical processes indexed by sets. *Probab. Th. Rel. Fields* **75**, 379–423.

Alexander, K.S. (1987b). The law of the iterated logarithm for empirical processes on Vapnik–Červonenkis classes. *Preprint.*

Andersen, N.T., Giné, E. and Zinn, J. (1988). Central limit theorem for empirical processes under local conditions: the case of Radon infinitely divisible limits without Gaussian component. *Trans. Amer. Math. Soc.* **308**, 603–636.

Csörgő, M. and Horváth, L. (1988). On the distributions of L_p norms of weighted uniform empirical and quantile processes. *Ann. Probab.* **16**, 142–161.

Dudley, R.M. (1978). Central limit theorems for empirical measures. *Ann. Probab.* **6**, 899–929.

Einmahl, J.H.J. (1987). *Multivariate Empirical Processes.* CWI Tract, **32.**

Einmahl, J.H.J. (1989). The a.s. behaviour of the weighted empirical process and the LIL for the weighted tail empirical process. *Univ. Limburg, Report* **25.**

Gaenssler, P. (1983). *Empirical Processes.* IMS Lecture Notes, **3.**

Giné, E. and Zinn, J. (1986). A remark on the central limit theorem for random measures and processes. In: *Proc. Vilnius Conf. Probab. Theory and Math. Statist,* **1**, pp. 483–487.

Kolchinskii, V.I. (1988). A type p operators and metric entropy. *Probab. Theory and Math. Stat.* **38**, 69–76.

Mason, D.M. (1982). Some characterizations of almost sure bounds for weighted multidimensional empirical distributions and a Glivenko–Cantelli theorem for sample quantiles. *Z. Wahr. verw. Geb.* **59**, 505–513.

Mikosch, T. and Norvaiša, R. (1986). On almost sure behaviour of sums of independent Banach space valued random variables. *Math. Nachr.* **125**, 217–231.

Norvaiša, R. (1984). The law of large numbers for identically distributed random variables with values in Banach spaces. *Liet. Matem. Rink.* **24**, 133–150.

Norvaiša, R. (1988). Remark on some stability result for empirical processes. *Probab. Theory Appl.* (to appear).

Norvaiša, R. (1989). The law of the iterated logarithm for random variables with heavy tails and empirical processes. *Liet. Matem. Rink.* (to appear).

Pisier, G. (1984). Remarques sur les classes de Vapnik–Červonenkis. *Ann. Inst. H. Poincaré* **20**, 287–298.

Shilov, G.V. and Gurevich, B.L. (1967). *Integral, Measure and Derivative. General Theory.* Nauka,, Moscow.

Seneta, E. (1976). Regularly varying functions. *Lect. Notes in Math.* **508.**

Zaman, A. (1988). A type 2 inequality for functions of bounded variation. *J. Mult. Anal.* **24**, 53–55.

Prob. Theory and Math. Stat., Vol. 2, pp. 275–277
B.Grigelionis *et al.* (Eds.)
1990 VSP/Mokslas

REMARKS ON THE RANDOM MULTIPLICATIVE ERGODIC THEOREM

V.I. OSELEDEC
III Frunzenskaya str., 1—83, 119270 Moscow, USSR

ABSTRACT

Our aim here is to present a version of the multiplicative ergodic theorem (Oseledec, 1968) adapted for applications to the study of deterministic and stochastic dynamical systems in a homogeneous random media.
For example we take the kinematic dynamo problem (see (Zeldovich, 1984; Molchanov et al., 1984, 1985)). According to Ya.B. Zeldovich and V.I. Arnold (Zeldovich, 1984, p. 42–45), this is a stability problem for the Lyapunov exponents in the limit of small magnetic diffusivity. In (Molchanov et al., 1984, 1985) on a stochastic model of the kinematic dynamo the Lyapunov exponents are studied, but their existence in the general case is not proved. This will follow from a version of the multiplicative ergodic theorem presented here.
All measure spaces in this paper are assumed to be Lebesgue spaces.

Let $\{p_t, t \in \mathbf{T}\}$ be a measurable measure preserving flow on a probability measure space $\mathbf{M}$ with a measure ϱ. Here $\mathbf{T}$ denotes either the integers $\mathbf{Z}$ or the reals $\mathbf{R}$. The multiplicative ergodic theorem gives information about the dynamical structure of a cocycle A, or a linear skew product $\{\bar{p}_t, t \in \mathbf{T}\}$, over the base space $\mathbf{M}$. The flow $\bar{p}_t$ is defined by

$$\bar{p}_t(m, e) = (p_t m, A(t, m)e),$$

$m \in \mathbf{M}$, $e \in \mathbf{R}^d$, where A is a measurable separable cocycle with range $\mathrm{GL}(d, \mathbf{R})$, that is

$$A(t + s, m) = A(s, p_t m) A(t, m).$$

Assume that

$$\int_M \sup\{\ln^+ \|A(t, m)\|,\ |t| \leqslant 1\} d\varrho(m) < \infty.$$

The multiplicative ergodic theorem asserts that there is a measurable splitting of the vector bundle $\mathbf{M} \times \mathbf{R}^d$ into $\bar{p}_t$–invariant measurable subbundles and the trajectories $\bar{p}_t(m, e)$ in any of these subbundles have strong Lyapunov exponents, i.e. almost everywhere

$$\lambda(m, e) = \lim_{t \to \infty} t^{-1} \ln \|A(t, m)e\|.$$

For a random measure preserving flow $q_t(w)$ on a probability space $\mathbf{X}$ with a measure μ over a measure preserving flow r_t on a probability space Ω with a measure ν we have that q_t is a measurable cocycle over r_t, that is

$$q_{t+s}(\omega) = q_s(r_t\omega)q_t(\omega).$$

The random linear skew product flow $\bar{q}_t(\omega)$ over a random flow q_t on $\mathbf{X}$ is defined by

$$\bar{q}_t(\omega)(x, e) = (q_t(w)x, A(t, \omega, x)e),$$

$x \in \mathbf{X}$, $e \in \mathbf{R}^d$, where A is a measurable separable cocycle with range $\mathrm{GL}(d, \mathbf{R})$, that is

$$A(t+s, \omega, x) = A(s, r_t\omega, q_t(\omega)x)A(t, \omega, x).$$

The random multiplicative ergodic theorem asserts in particular that the trajectories $\bar{q}_t(\omega)(x, e)$ have strong Lyapunov exponents (almost everywhere), if

$$\sup\{\ln^+ \|A(t, \omega, x)e)\|, \ |t| \leqslant 1\}$$

is integrable.

Now put $\mathbf{M} = \Omega \times \mathbf{X}$, $\varrho = \nu \times \mu$. A flow p_t on $\mathbf{M}$ is defined by the formula

$$p_t m = p_t(\omega, x) = (r_t\omega, q_t(\omega)x).$$

Then for the random linear skew product flow $\bar{q}_t$ over a random flow q_t on $\mathbf{X}$ we have

$$\bar{p}_t(m, e) = (p_t m, A(t, m)e) = (r_t\omega, \bar{q}_t(\omega)(x, e)).$$

The multiplicative ergodic theorem for the flow $\bar{p}_t$ on $\mathbf{M} \times \mathbf{R}^d$ allows us to conclude that the random multiplicative ergodic theorem is valid for the random linear skew product flow $\bar{q}_t(\omega)$ on $\mathbf{X} \times \mathbf{R}^d$.

This idea of applying a "nonrandom" ergodic theorem to a product space is a recipe which has furnished numerous random ergodic theorems.

The following ergodic construction illustrates that the condition $\mu(\mathbf{X}) = 1$ is in fact not necessary when some other conditions are fulfilled, so we assume now that μ is a σ–finite measure on $\mathbf{X}$.

Let G be a locally compact separable group. Recall that a dynamical system with time group G and invariant measure μ is a measurable family $\{S_g, \ g \in G\}$ of μ–preserving transformations of $\mathbf{X}$

$$S_{g_1 g_2} = S_{g_1} S_{g_2}, \quad S_e = id.$$

We also assume that there is a dynamical system $\{N_g, g \in G\}$ on Ω with invariant measure ν such that the following four conditions hold

(i) $p_t(N_g \times S_g) = (N_g \times S_g)p_t$,

(ii) $A(t, (N_g \times S_g)m) = A(t, m)$,

(iii) $\mathbf{X} = \{S_g x,\ g \in G\}$ for each $x \in \mathbf{X}$,

(iv) $\int\limits_{\Omega} \sup\{\ln^+ \|A(t,\omega,x)\|,\ |t| \leqslant 1\} d\nu(\omega) < \infty$.

Then for the cocycle A we have a random multiplicative ergodic theorem. We note that all the above conditions are satisfied in the stochastic model of the kinematic dynamo mentioned in the beginning of this paper. In the simplest case when the magnetic diffusivity is 0 we have that $G = \mathbf{R}^d = \mathbf{X}$, $S_g x = x - g$ and Ω is a space of functions $\omega = \omega(x)$, $x \in \mathbf{X}$ with G-action $N_g: \omega(x) \longmapsto \omega(S_g x)$ and $A(t,\omega,x)$ is the Jacoby cocycle $\partial_x q_t(\omega)$ for the cocycle $q_t(\omega)$. If the flow r_t is trivial that is $r_t = id$ for each $t \in \mathbf{T}$, then $q_{t+s}(\omega) = q_s(\omega) q_t(\omega)$ which means that the random cocycle $q_t(\omega)$ is in this case a random flow on $\mathbf{R}^d$. For example, let $q_t: x \longmapsto x_t$, where $dx_t = v(\omega, x_t)dt$, $x_0 = x$ and where $v(\omega,x)$ is a random homogeneous vector field on $\mathbf{R}^d$. In order to take into account a finite magnetic diffusivity it is sufficient to introduce the Ito equation

$$dx_t = v(\omega, x_t)dt + \varepsilon dW_t, \quad x_0 = x,$$

where W_t is the Wiener process on $\mathbf{R}^d$.

The proof of our main statements is based on a simple ergodic construction. Namely, let $\mathbf{M}_0$ be the space of ergodic components of the dynamical system $\{N_g \times S_g,\ g \in G\}$ on $\mathbf{M} = \Omega \times \mathbf{X}$ with invariant measure $\varrho = \nu \times \mu$. Conditions (iii), (i) imply that the measure ϱ induces on $\mathbf{M}$ a probability measure ϱ_0 invariant with respect to the flow p_t on $\mathbf{M}_0$ defined by

$$p_t m_0(m) = m_0(p_t m),$$

where $m_0(m)$ is the ergodic component with $m \in m_0$. Condition (i) implies that this definition is correct.

Now the cocycle A can be defined on $\mathbf{M}_0$ by the formula

$$A(t, m_0(m)) = A(t,m).$$

It follows from condition (ii), that this definition is also correct. Now condition (iv) implies that the random multiplicative ergodic theorem is valid in our case. It is clear that condition (iii) can be weakened but for this we must modify condition (iv). It is easy to see that the existence of an invariant probability measure ϱ_0 is a necessary condition and then we must assume, that $\sup\{\ln^+ \|A(t,m_0)\|, |t| \leqslant 1\}$ is integrable.

REFERENCES

Molchanov, S.A., Ruzmaikin, A.A. and Sokolov, D.D. (1984). A dynamo theorem. *Geophys. Astrophys. Fluid Dynamics* **30**, 242–259.

Molchanov, S.A., Ruzmaikin, A.A. and Sokolov, D.D. (1985). Kinematic dynamo in random flow. *Uspekhi Phys. Nauk* **145**, 594–628.

Oseledec, V.I. (1968). A multiplicative ergodic theorem. Lyapunov characteristic exponents for dynamical systems. *Proc. Moscow Math. Soc.* **19**, 119–210.

Zeldovich, Y.B. (1984). *Selected papers. Chemical physics and hydrodynamics.* Nauka, Moscow.

Prob. Theory and Math. Stat., Vol. 2, pp. 278–287
B.Grigelionis *et al.* (Eds.)
1990 VSP/Mokslas

PROBABILITY DISTRIBUTIONS AND BOREL MEASURES UNIQUELY DETERMINED BY THEIR RESTRICTIONS TO A HALF–SPACE

I.V. OSTROVSKII and A.M. ULANOVSKII
Low Temperature Physics and Engineering Institute,
Ukr. SSR Acad. Sci., 47 Lenin ave., 310164 Kharkov, USSR

ABSTRACT

The paper is a survey of results on classes of Borel measures in the space $\mathbf{R}^m$ that are uniquely determined by restrictions to a half–space. In particular, classes of convolutions and mixtures of probability distributions are considered.

This paper is devoted to the branch of research which studies the probability distributions and (complex–valued) Borel measures uniquely determined by their restrictions to sufficiently massive sets. That branch was initiated by V.M. Zolotarev in the middle of the 70–ies and a plenty of papers are devoted to it. We shall not try to give the complete history of the results under consideration.
The aim of this paper is to give the results which are, to the best of our knowledge, the strongest ones.
Further, we use the following abbreviations: f.m. — the finite (generally speaking, complex–valued) Borel measure; p.d. — the probability distribution; i.d.p.d. — the infinitely divisible probability distribution. The standard Gaussian distribution in $\mathbf{R}^m$ will be denoted by Φ_m.

1. CLASSES OF F.M. WHICH ARE UNIQUELY DETERMINED BY THEIR RESTRICTIONS TO A HALF–SPACE

In the 50ths A.N. Kolmogorov proposed the following question. Let μ be a i.d.p.d. on the line $\mathbf{R}^1$ and the restrictions of μ and Φ_1 to some half–line $(-\infty, a)$ coincide: $\mu|_{(-\infty,a)} = \Phi_1|_{(-\infty,a)}$. Is it true that $\mu \equiv \Phi_1$? The affirmative answer was obtained in (Rossberg et al., 1985). So the Gaussian distribution Φ_1 is uniquely determined in the set of all i.d.p.d. by its restriction to a half–line. In 1977 I.A. Ibragimov proved that a wide class of i.d.p.d. possesses the same property. Let us consider a measure μ which satisfies the following conditions:

(A) μ is i.d.p.d.;

(B) $\inf \operatorname{supp} \mu = -\infty$;

(C) $\lim_{r\to\infty} r^{-1}\ln\Big(1/\mu((-\infty,-r))\Big) = \infty.$

Then the following implication holds

$$\big(\nu \text{ is i.d.p.d.}, \ \exists\, a : \nu|_{(-\infty,a)} = \mu|_{(-\infty,a)}\big) \Rightarrow (\nu = \mu).$$

Definition. Let @ be a class of f.m. in $\mathbf{R}^m$. We say that the class @ possesses the property of the unique determination by restrictions to a half-space (and write @ $\in \mathrm{UDRS}_m$) if

$$(\nu,\mu \in @, \exists a : \nu|_{\mathbf{R}^m_a} = \mu|_{\mathbf{R}^m_a}) \Rightarrow (\nu = \mu),$$

where $\mathbf{R}^m_a = \{(x_1,\ldots,x_m) \in \mathbf{R}^m : x_1 < -a\}$ is the half-space.

It's obvious that the Ibragimov's theorem mentioned above is equivalent to the statement that $@_\infty \in \mathrm{UDRS}_1$, where $@_\infty$ is the class of measures μ in $\mathbf{R}^1$ satisfying the conditions (A), (B), (C). The class $@_\infty$ appeared to be the first example of a non-trivial class of measures possessing the property UDRS_1. As a result the interest to find and to study such classes naturally arises. It was found later that the condition (A) on the Ibragimov's theorem could be weakened in two directions: a) by passing to the f.m. whose Fourier transforms have "few zeros" in some sense (see (Ostrovskii, 1979; Blank, 1982)); b) by replacing the condition of infinitely divisibility by the condition of representability in the form of the n-th power convolution, $n \geqslant 2$, or in the form of the mixture of such convolutions (see (Titov, 1981; Blank, 1984; Ostrovskii, 1985; Ulanovskii, 1988a; Kruglov and Titov, 1986; Kruglov and Ulanovskii, 1987)). The first direction was initiated in (Ostrovskii, 1979), the second one in (Titov, 1981).

a) Let $\mathfrak{L}$ be the class of f.m. in $\mathbf{R}^1$ satisfying the condition (B) and the following condition

(D) the Fourier transform $\widehat{\mu}(t)$ of the measure μ has the non-vanishing analytic extension into the half-plane $\{\mathrm{Im} t > 0\}$ which is continuous on $\{\mathrm{Im} t \geqslant 0\}$.

It's known that the conditions (A) and (C) imply the condition (D). Hence, $@_\infty \subset \mathfrak{L}$ and it is clear that $@_\infty \neq \mathfrak{L}$. We extend the class $\mathfrak{L}$ by introducing the class $\mathfrak{L}'$ of finite linear combinations of the measures from $\mathfrak{L}$.

THEOREM 1 (Ostrovskii, 1979). $\mathfrak{L}' \in \mathrm{UDRS}_1$.

Some variant of the Theorem 1 for f.m.'s whose Fourier transforms have non-vanishing analytic extensions to the strip $\{0 < \mathrm{Im}\ t < H\}$ and satisfy some additional assumptions was obtained in (Blank, 1982).

b) Denote by $@(n,m)$, $n \geqslant 2$, $m \geqslant 1$, the class of f.m. μ in $\mathbf{R}^m$ which satisfy the following conditions:

(A_n) the measure μ is representable by the n-th power convolution, $\mu = \nu^{n*}$, where ν is some f.m.

(B_m)supp $\mu \cap \mathbf{R}^m_r \neq \varnothing$, $\forall r$;

(C_m) $\lim_{r\to\infty} r^{-1}\ln\big(1/|\nu|((x_1,\ldots,x_m) : x_1 < 0; x_1^2 + \ldots + x_m^2 > r^2)\big) = \infty.$

Let $@_+(n,m)$ be a subclass of $@(n,m)$ such that the measure ν in (A_n) is p.d. It's easy to see that $@_+(n,m)$ consists of p.d.'s μ satisfying the conditions (B_m), (A_n) with ν being a p.d. and

$$\lim_{r\to\infty} r^{-1}\ln\big(1/\mu((x_1,\ldots,x_m) : x_1 < 0; x_1^2 + \ldots + x_m^2 > r^2)\big) = \infty.$$

Hence, $@_\infty \subset @(n,1)$, $n = 2,3,\dots$. Moreover, it's easy to see that

$$@_\infty = \bigcap_{n=2}^{\infty} @_+(n,1).$$

THEOREM 2 (Ulanovskii, 1988a). *$@(n,m) \in \mathrm{UDRS}_m$ for all $n \geqslant 3$, $m \geqslant 1$.*

The assumption $n \geqslant 3$ in this theorem is essential since $@_+(2,m) \notin \mathrm{UDRS}_m$. Let $@^0(2,m), m \geqslant 1$, be a class of measures $\mu \in @(2,m)$ satisfying the additional assumption:

$$(\mathrm{E}_m) \quad \lim_{r\to\infty} (r \ \ln \ r)^{-1} \ln \left(1/|\mu(\mathbf{R}_r^m)|\right) = \infty.$$

THEOREM 3 (Ulanovskii, 1988a). *$@^0(2,m) \in \mathrm{UDRS}_m$ for all $m \geqslant 1$.*

It's interesting to note that the proofs of Theorems 2 and 3 even for the classes $@_+(n,m)$ and $@_+^0(2,m)$ instead of $@(n,m)$, $@^0(2,m)$, require the respective one–dimensional theorems (Ostrovskii, 1985) for f.m. but not for p.d.
The Theorems 2 and 3 are exact in the following sense. The classes $@(n,m)$, $@^0(2,m)$ can't be extended by weakening one of the conditions (B_m), (C_m) or (E_m). The condition (B_m) can't be weakened due to a trivial reason: any two different p.d. is ν_1 and ν_2 concentrated on $[0,\infty)$ satisfy $\nu_1^{n*}|_{(-\infty,0)} = \nu_2^{n*}|_{(-\infty,0)}$.

Remark 1. The paper (Ostrovskii, 1983) contains a description of a wide class of p.d. μ possessing the following property. There exist infinitely many different i.d.p.d. coinciding with μ outside some finite ball. For instance, all the spherically symmetric stable i.d.p.d.'s with the characteristic exponent $0 < \alpha < 2$ possess this property. This gives the negative answer to I.A. Ibragimov's question (Ibragimov, 1977): whether the stable i.d.p.d. in $\mathbf{R}^1$ can be uniquely determined by their restrictions to a half–line.

Remark 2. There exist f.m. ν_1, ν_2 in $\mathbf{R}^1$ such that

$$\nu_1^{(1+2k)*}|_{(-\infty,0)} = \nu_2^{(1+2k)*}|_{(-\infty,0)}, \quad k = 1,2,\dots; \quad \nu_1|_{(-\infty,0)} \neq 0,$$

but $\nu_1 \neq \nu_2$. For any natural number N there exist i.d.p.d. ν_1, ν_2 in $\mathbf{R}^1$ satisfying (B) and such that

$$\nu_1^{k*}|_{(-\infty,0)} = \nu_2^{k*}|_{(-\infty,0)}, \quad k = 1,\dots,N,$$

but $\nu_1 \neq \nu_2$. Moreover, the following theorem is true.

THEOREM 4. *Let ν_1, ν_2 be f.m. in $\mathbf{R}^m$, $m \geqslant 1$, and*

$$\nu_1|_{\mathbf{R}_0^m} \neq 0; \quad \nu_1^{(1+n_k)*}|_{\mathbf{R}_0^m} = \nu_2^{(1+n_k)*}|_{\mathbf{R}_0^m}, \quad k = 0,1,\dots, \quad n_0 = 0,$$

where $1 \leqslant n_1 < n_2 < \dots$ is the infinite sequence of numbers whose greatest common divisor equals to 1. Then $\nu_1 = \nu_2$.

Theorems 2 and 3 are exact in the sense that the measures from the classes $@(n,m)$, $@^0(2,m)$ (and their subclasses $@_+(n,m), @_+^0(2,m)$) for $m \geqslant 2$ are not determined by their restrictions to the sets which are "more narrow" than the half-space $\mathbf{R}_a^m$ sets. Namely, the following theorem holds.

THEOREM 5 (Ulanovskii, 1988a). *Let $M \subset \mathbf{R}^m$ be a convex set which does not contain any subspace $\mathbf{R}_a^m$. For any $n \geqslant 2$ there exist μ, $\nu \in @_+(n,m)$ ($\mu, \nu \in$ $\in @_+^0(2,m)$ for $n = 2$) such that $\nu|_M = \mu|_M$, $\quad \nu \neq \mu$.*

A.I.Il'inskii (1985) has proved that if $K \subset \mathbf{R}^m$ is a convex cone with nonempty interior and ν is an i.d.p.d. in $\mathbf{R}^m$, then the equality $\nu|_K = \Phi_m|_K$ implies $\nu = \Phi_m$. He also (communication) showed that the condition, ν is an i.d.p.d., can't be substituted by the condition, $\exists\ n \geqslant 2$, $\nu = \nu_1^{n*}$, where ν_1 is p.d. It's unknown whether it's possible to replace the Gaussian distribution Φ_m in the result of (Il'inskii, 1985) by an arbitrary i.d.p.d. μ satisfying (B_m) and (C_m).
Note that the proofs of Theorems 2 and 3 are based on one of the deepest results at the theory of functions, namely, the second main theorem of the Nevanlinna value distribution theory. The factorization theory in Hardy's classes also was substantially used. This establishes the close connection between the branch considered here and the complex analysis. The following applications of Theorems 2 and 3 also indicate this connection.

(i) *The generalization of the Titchmarsh convolution theorem.* The well-known Titchmarsh theorem states that if f.m.'s ν_1, ν_2 satisfy the condition

$$\inf \operatorname{supp} \nu_j > -\infty, \ j = 1, 2,$$

then

$$\inf \operatorname{supp}(\nu_1 * \nu_2) = \inf \operatorname{supp} \nu_1 + \inf \operatorname{supp} \nu_2. \tag{1}$$

The equality (1) is not true for all f.m. ν_1 and ν_2 because there exist f.m.'s ν_1, ν_2 such that

$$\inf \operatorname{supp} \nu_1 = -\infty, \quad \inf \operatorname{supp}(\nu_1 * \nu_2) > -\infty.$$

The question arises whether the condition $\inf \operatorname{supp} \nu_j > -\infty$ can be replace by the condition of sufficiently rapid decreasing of $|\nu_j|((-\infty, -r))$ as $t \to +\infty$? The affirmative answer is given by the following theorem which is the consequence of Theorem 3.

THEOREM 6 (Ostrovskii, 1985). *The equality* (1) *is true for all f.m.'s ν_1, ν_2 satisfying the condition* (E_1).

The example in which the Fourier transforms are

$$\widehat{\nu}_j(t) = \exp\left\{(-1)^j e^{-it}\right\}, \quad j = 1, 2,$$

shows that the condition (E_1) in Theorem 6 can't be weakened. Recently A.A. Borichev (report at the Conference on Differencial Equations and Complex Analysis, Chernogolovka, 1989) had extended Theorem 6 to the case of non-finite measures in $\mathbf{R}^1$. Note that such theorems have applications in the theory of invariant subspaces (see, for instance (Domar, 1983)).

(ii)*Laurent series determined by their principal parts.* Let $f_1(Z), f_2(Z)$ be analytic functions in the punctured unit disk $\{|Z| < 1, Z \neq 0\}$ and let f_1 have an essential singularity in $Z = 0$. Then Theorem 2 implies

THEOREM 7. *Let the function $f_1^n - f_2^n$, $n \geqslant 3$, has a pole or a removable singularity at the origin. Then $f_1^n \equiv f_2^n$.*

This theorem was strengthened by Zaidenberg (Zaidenberg, 1988) who investigated the following more general question. Let $\eta(u, v)$ be a polynomial. What can we say about the function $\eta(f, g)$ if it has a pole or a removable singularity at the origin?

In (Kruglov and Ulanovskii, 1987; Kruglov and Titov, 1986) the following question was studied. Is the property UDRS possessed by the classes of mixtures of p.d. of the form $\sum_{k=1}^{\infty} a_k \mu^{k*}$ where μ is p.d. or of the form $\int_{[0,\infty)} \mu^{u*}\chi(du)$, where μ is an i.d.p.d. and μ^{u*} is the i.d.p.d. with Fourier transform $\widehat{\mu}^u$?

Denote by $@(\{a_k\}_1^{\infty})$ the class of p.d. representable in the form

$$\sum_{k=1}^{\infty} a_k \mu^{k*}, a_k \geqslant 0, \ k = 1, 2, \ldots,$$

where μ is a p.d. satisfying (B), (C). Denote by $@(\chi)$ the class of p.d. representable in the form $\int_{[0,\infty)} \mu^{u*}\chi(du)$ where μ is an i.d.p.d. satisfying (B), (C) and χ is a p.d. concentrating on $[0, \infty)$.

THEOREM 8 (Kruglov and Ulanovskii, 1987).

(i) *Let $\{a_k\}$ be the sequence of non-negative numbers such that*

$$\sum_{k=3}^{\infty} a_k > 0; \quad \lim_{k\to\infty} k^{-1} \ln(1/a_k) = \infty.$$

Then $@(\{a_k\}_1^{\infty}) \in \mathrm{UDRS}_1$.

(ii) *Let χ be a p.d. such that*

$$\chi((-\infty, 0]) = 0; \quad \lim_{r\to\infty} r^{-1} \ln\Big(1/\chi((r, \infty))\Big) = \infty.$$

Then $@(\chi) \in \mathrm{UDRS}_1$.

The papers (Kruglov and Titov, 1986; Kruglov and Ulanovskii, 1987) deal with the problem of the description of a p.d. ν which does not admit a representation of the form

$$\nu = \sum_{k=0}^{\infty} a_k \mu^{k*}, \ a_k \geqslant 0,$$

on any half-line $(-\infty, a)$, where μ is p.d.

THEOREM 9 (Kruglov and Ulanovskii, 1987). *Let*

$$\left(\sum_{k=1}^{\infty} a_k \mu^{k*}\right)\Big|_{(-\infty,a)} = \Phi_1\Big|_{(-\infty,a)},$$

where μ *is a p.d.,* $a \in \mathbf{R}^1$, $a_k \geqslant 0$, $\sum_{k=2}^{\infty} a_k > 0$, $\sum_{k=1}^{\infty} a_k = 1$. *Then there exists a number* N *such that* $a_N = 1$, $a_k = 0$ *for* $k \neq N$ *and* $\mu^{N*} = \Phi_1$.

2. MEASURES WHOSE CONVOLUTIONS COINCIDE WITH THE GAUSSIAN DISTRIBUTION ON A HALF-LINE

In 1984 V.M. Kruglov posed the following question. Let μ be a p.d. in $\mathbf{R}^1$ and let $\mu^{(a)} = \mu^{(a)}(E) = \mu(a, E)$, $a > 0$. Is it true that

$$\left(\mu^{(a_1)} * \ldots * \mu^{(a_n)}|_{(-\infty,0)} = \Phi_1|_{(-\infty,0)}\right) \Rightarrow \left(\mu^{(a_1)} * \ldots * \mu^{(a_n)} = \Phi_1\right)?$$

For $a_1 = \ldots = a_n$ the implication was proved in (Titov, 1981). The affirmative answer is contained in the following theorem.

THEOREM 10 (Ulanovskii, 1988c). *Let a f.m.* μ *satisfies the condition* (C_1). *Then*

$$\left(\mu^{(a_1)} * \ldots * \mu^{(a_n)}|_{(-\infty,0)} = \Phi_1|_{(-\infty,0)}\right) \Rightarrow \left(\mu^{(a_1)} * \ldots * \mu^{(a_n)} = \Phi_1 + c\delta\right),$$

where c *is a constant with* $|c| = 1$ *or* $c = 0$ *and* δ *is a p.d.,* supp $\delta = \{0\}$. *If* μ *is a p.d. then* $c = 0$.

The theorem ceases to be true if the convolution $\mu^{(a_1)} * \ldots * \mu^{(a_n)}$ is replaced by $\mu_1 * \ldots * \mu_n$, where μ_j are different f.m.

Example 1. Let χ be a f.m., supp $\chi \subset [0, \infty)$ and let μ be a f.m. with the Fourier transform

$$\widehat{\mu}(t) = \exp\{-t^2/2\} - \widehat{\chi}^n(t), \quad n \geqslant 2.$$

It's clear that $\mu = \mu_1 * \ldots * \mu_n$, where $\mu_j = \Phi_1^{(\sqrt{n})} - \exp\{2\pi i j/n\}\chi$ and $\mu|_{(-\infty,0)} = \Phi_1|_{(-\infty,0)}$, but $\mu \neq \Phi_1$. If $n = 2$ then the measure χ can be chosen in such a way that μ_1 and μ_2 are p.d.'s.

In connection with this let us formulate the fact which can be considered as a version of the Cramer's theorem on a half-line.

THEOREM 11 (Ulanovskii, 1988c). *Let* $\mu_1, \ldots, \mu_n$ *be p.d.'s,* $n \geqslant 2$, *and let* $a_1, \ldots, a_n$ *be positive numbers not equal to* 1. *Then*

$$\left(\mu_1 * \ldots * \mu_n|_{(-\infty,0)} = \mu_1^{(a_1)} * \ldots * \mu_n^{(a_n)}|_{(-\infty,0)} = \Phi_1|_{(-\infty,0)}\right) \Rightarrow \left(\mu_1 * \ldots * \mu_n = \Phi_1\right).$$

The condition $\mu^{(a_1)} * \ldots * \mu^{(a_n)}|_{(-\infty,0)} = \Phi_1|_{(-\infty,0)}$ in Theorem 10 can be replaced by $\mu_1 * \ldots * \mu_n|_{(-\infty,0)} = \Phi_1|_{(-\infty,0)}$ when the measures $\mu_1, \ldots, \mu_n$ are "mutually close".

THEOREM 12. *Let $\mu_1, \ldots, \mu_n$ be f.m.'s, $n \geqslant 2$, satisfying the condition* (C_1). *Then the conditions*

$$\mu_1\big|_{(-\infty,0)} = \ldots = \mu_n\big|_{(-\infty,0)}; \quad \mu_1 * \ldots * \mu_n\big|_{(-\infty,0)} = \Phi_1\big|_{(-\infty,0)} \tag{2}$$

imply that there exists a f.m. χ, $\operatorname{supp}\chi \subset [0,\infty)$, *such that* $\mu_j = \Phi_1^{(\sqrt{n})} - \varepsilon_j\chi$, *where* $\{\varepsilon_j\}_1^n$ *are all the n-th roots of* 1.

Thus the structure of all the collections of f.m. $\mu_1, \ldots, \mu_n$ satisfying (2) is the same as in Example 1.

COROLLARY 1 (Ronkin and Ulanovskii, 1986).

(i) *Let $\mu_1, \ldots, \mu_n$ be f.m.'s satisfying the conditions of Theorem 12. Then* $\mu_1 + \ldots + \mu_n = n\,\Phi_1^{(\sqrt{n})}$.

(ii) *Let $n \geqslant 3$, $\mu_1, \ldots, \mu_n$ be p.d.'s satisfying* (2). *Then* $\mu_1 = \ldots = \mu_n = \Phi_1^{(\sqrt{n})}$.

Example 1 shows that the condition $n \geqslant 3$ in the part (ii) of this corollary can't be omitted.

In Theorem 12 and Corollary 1 the "mutually closeness" of the measures is understood as their coincidence on a half-line $(-\infty, 0)$. In this case it is natural to call the measure $\mu = \mu_1 * \ldots * \mu_n$ the n-th power "half-line convolution". In Corollary 1 the restriction of n-th power "half-line convolution" to a half-line $(-\infty, 0)$ determines the sum $\mu_1 + \ldots + \mu_n$, while the measures μ_j are determined only under additional assumptions. The similar situation takes place in more general cases.

THEOREM 13 (Ronkin and Ulanovskii, 1987a). *Let μ_k and ν_k be p.d.'s, $k = 1, \ldots, n$, $n \geqslant 2$, and let a p.d. μ_1 satisfies the conditions* (B) *and* (E_1). *Then the following implication holds*

$$\left.\begin{array}{r} \mu_1|_{(-\infty,0)} = \ldots = \mu_n|_{(-\infty,0)}; \\ \nu_1|_{(-\infty,0)} = \ldots = \nu_n|_{(-\infty,0)}; \\ \mu_1 * \ldots * \mu_n|_{(-\infty,0)} = \nu_1 * \ldots * \nu_n|_{(-\infty,0)} \end{array}\right\} \Rightarrow \mu_1 + \ldots + \mu_n = \nu_1 + \ldots + \nu_n.$$

If we assume additionally that one of the following conditions holds:

(i) $n \geqslant 3$, $\mu_1 = \mu_2 = \mu_3$;

(ii) $\mu_1 = \nu_1$,

then the measures μ_j can be renumerated in such a way that $\mu_j = \nu_j$, $j = 1, \ldots, n$. Note that the proofs of Theorems 10, 11 are based on the deep theorems of the function theory devoting to the theory of division of generalized quasipolynomials (Gordon and Levin, 1971), i.e., the functions admitting the representation

$$\sum_{k=1}^{N} f_k(t)e^{\lambda_k t},$$

where f_k are analytic functions in an angle which have the "slow growth". The character of these theorems is following: if the quotient of two generalized quasi-polynomials is analytic in a sufficiently large angle then it is a generalized quasi-polynomial. Moreover, the proof of Theorem 13 uses theorems of the Beurling's type on the connection between increasing and decreasing of an analytic function (Beurling, 1949).
Note that in some of the theorems from Sections 1 and 2 the condition of coincidence of the measures to a half-line can be replaced by a more weak condition of the asymptotic "proximity" when $x \to -\infty$ (Rossberg, 1985; Kruglov and Ulanovskii, 1987; Riedel, 1975; Blank, 1987; Ronkin and Ulanovskii, 1987b).

3. MEASURES COINCIDING ON THE SETS WHICH DIFFER FROM A HALF-LINE

The question arises as to whether the theorems from Sections 1, 2 may be strengthened by replacing the condition of coincidence of the measures to a half-line (or a half-space) with weaker condition of their coincidence on some "sufficiently large" set. In this direction we have the following result.

THEOREM 14 (Skaskiv, 1986). *Let $\mathfrak{L}(\mathbb{Z}^-)$ be the class of f.m.'s which are concentrated on the set $\mathbb{Z}^-$ of nonpositive integers and satisfy the conditions* (B) *and* (D). *Let*

$$M = \mathbb{Z}^- \setminus \bigcup_{k=1}^{\infty} \{-q_k\}$$

where q_k are positive integers such that

$$\exists\, \varepsilon > 0 : q_k^{-1} \ln\ k(\ln\ln\ k)^{2+\varepsilon} \to 0, \quad k \to \infty. \tag{3}$$

Then the following implication holds

$$(\mu, \nu \in \mathfrak{L}(\mathbb{Z}^-), \quad \mu|_M = \nu|_M) \Rightarrow (\mu = \nu).$$

This result is connected with the Nevanlinna value distribution theory for meromorphic functions in the following way. The classical Picard theorem ("small") is equivalent to the impossibility of the equality $g_1 + g_2 = 1$, where g_1 and g_2 are non-constant and non-vanishing entire functions. Theorem 14 is equivalent to the impossibility of the equality

$$g_1 + g_2 = \sum_{k=0}^{\infty} a_k t^{q_k},$$

where g_1 and g_2 non-constant and non-vanishing entire functions, the numbers q_k satisfy (3).
In (Ulanovskii, 1988b) the same question was studied for the measures from the classes $@_+(n,1)$. It was shown that any p.d. $\mu \in @_+(n,1)$, $n \geqslant 3$, is uniquely determined by its restriction to a sufficiently wide system (depending on μ) of intervals $M \subset (-\infty, 0)$. Particularly, the following theorem is true.

THEOREM 15 (Ulanovskii, 1988b). *Let $n \geqslant 3$, μ be a p.d.,*

$$M = \bigcup_{k=1}^{\infty} (-x_{2k+1}, -x_{2k}),$$

where $0 < x_1 < x_2 < \ldots$ be the sequence satisfying the condition $x_{2k+1}/x_{2k} \to \to \infty, k \to \infty$. Then

$$(\mu^{n*}|_M = \Phi_1|_M) \Rightarrow (\mu = \Phi_1^{(\sqrt{n})}).$$

This theorem was proved with the help of studying the stability of the property UDRS for the classes $@_+(n,1)$ in the following sense. If a p.d. ν satisfies (B), (C) and the difference $\mu^{n*} - \nu^{n*}$, $n \geqslant 3$, is "small" on a half-line $(-\infty, 0)$ "to within ε" in some metrics then the difference $\mu - \nu$ is "small" on a line $(-\infty, \infty)$ "to within $\delta(\varepsilon)$" in the same metrics. Moreover the estimate depending on ε and p.d. V was given for $\delta(\varepsilon)$.

THEOREM 16 (Ulanovskii, 1988b). *Let μ be a p.d. and*

$$\left|(\mu^{n*} - \Phi_1)((-\infty, x))\right| \leqslant \varepsilon, \quad x \in (-\infty, 0)$$

for some $n \geqslant 3$. Then

$$\left|(\mu - \Phi_1^{(\sqrt{n})})((-\infty, x))\right| \leqslant C \frac{\left[\ln(1 + \ln 1/\varepsilon)\right]^{3/4}}{\left[\ln 1/\varepsilon\right]^{1/4}}, \quad x \in (-\infty, \infty),$$

where $C > 0$ depends only on n.

REFERENCES

Beurling, A. (1949). Some theorems on boundedness of analytic functions. *Duke Math. J.* **16**, 355–359.

Blank, N.M. (1982). On the uniqueness conditions for functions of bounded variation and for distribution functions with given values on a half-line. *Theory Probab. Appl.* **27**, 784–787.

Blank, N.M. (1984). On distributions the convolutions of which coincide on a half-line. In: *Theory of Funct., Funct. Analysis and Appl.* **41**, Kharkov, pp. 17–25.

Blank, N.M. (1987). On unique determination of the infinitely divisible distribution functions by their asymptotic behavior for $x \to -\infty$. In: *Stability Problems of Stoch. Models,* Kuibyshev, pp. 19–24.

Domar, Y. (1983). Extension of the Titchmarsh convolution theorem with applications in the theory of invariant subspaces. *Proc. London Math. Soc.* **46**, 288–300.

Gordon, A.Ya. and Levin, B.Ya. (1971). On division of quasipolynomials. *Funct. Anal. Appl.* **5**, 22–29.

Ibragimov, I.A. (1977). On determining of infinitely divisible distribution by its values on a half-line. *Theory Probab. Appl.* **22**, 393–399.

Il'inskii, A.I. (1985). Normality of multidimensional infinitely divisible distribution which coincides with a normal distribution on cone. *Lect. Notes in Math.* **1155**, 47–59.

Kruglov, V.M. and Titov, A.N. (1986). Mixtures of probability distributions. *Lect. Notes in Math.* **1233**, 41–56.

Kruglov, V.M. and Ulanovskii, A.M. (1987). Mixtures of probability distributions uniquely determined by their behaviour on a half-line. *Theory Probab. Appl.* **32**, 670–678.

Ostrovskii, I.V. (1979). On a certain class of functions of the bounded variation on a line determined by its values on a half-line. In: *Notes Sci. Seminars Leningrad Branch of Math. Inst.* **92**, pp. 220–229.

Ostrovskii, I.V. (1983). Measures with infinitely many infinitely divisible extensions. *Sankhya,* **45**, 257–270.

Ostrovskii, I.V. (1985). Generalization of the Titchmarsh convolution theorem and the complex-valued measures uniquely determined by their restriction to a half-line. *Lect. Notes in Math.* **1155**, 256–283.

Riedel, M.N. (1975). On the one-sided tails of infinitely divisible distributions. *Math. Nachr.* **70**, 155–163.

Ronkin, A.L. and Ulanovskii, A.M. (1986). Distributions the convolutions of which coincide on a half-line with a normal distribution. In: *Dokl. Acad. Nauk Ukr. SSR. Ser. A* **4**, 106–110.

Ronkin, A.L. and Ulanovskii, A.M. (1987a). On the determination of the sum of mutually closed distributions by their convolutions. *Theory Probab. Appl.* **32**, 800–804.

Ronkin, A.L. and Ulanovskii, A.M. (1987b). On the determination of the distribution functions by values of their convolution for $x \to -\infty$. In: *Stability Problems of Stoch. Models,* Kuibyshev, pp. 78–86.

Rossberg, H.-J., Jesiak, B. and Siegel, G. (1985). *Analytic Methods of Probability Theory.* Akademie-Verlag, Berlin.

Skaskiv, O.B. (1986). Generalization of the Picard's small theorem. In: *Theory of Funct., Funct. Anal. and Appl.* **46**, Kharkov, pp. 90–100.

Titov, A.N. (1981). On the determination of the convolution of equal distribution functions by their values on a half-line. *Theory Probab. Appl.* **26**, 610–611.

Ulanovskii, A.M. (1988a). On unique determination of the convolution of measures in $\mathbf{R}^m$, $m \geqslant 2$, by their restrictions to the sets. In: *Theory of Funct., Funct. Anal. and Appl.* **49**, Kharkov, pp. 86–90.

Ulanovskii, A.M. (1988b). On the determination of the n-th power convolution of the distribution function by its values on the system of intervals. *Dokl. Akad. Nauk Ukr. SSR, Ser. A* **3**, 32–35.

Ulanovskii, A.M. (1988c). On functions of bounded variation that are determined by restrictions to a semiaxis. *Math. USSR Sb.* **60**, 427–436.

Zaidenberg, M.G. (1988). Transcendental solutions of the diophant equations. *Funct. Anal. Appl.* **22**, 78–81.

Prob. Theory and Math. Stat., Vol. 2, pp. 288-297
B.Grigelionis *et al.* (Eds.)
1990 VSP/Mokslas

DISCRETE TIME QUANTUM STOCHASTIC FLOWS, MARKOV CHAINS AND CHAOS EXPANSIONS

K.R. PARTHASARATHY
Indian Statistical Institute, Delhi Centre 7, S.J.S.
Sansanwal Marg, New Delhi-110016

ABSTRACT

Within the framework of a tensor product of a $\star$ unital algebra and copies of a finite dimensional matrix algebra a completely elementary account of discrete time quantum stochastic flows and their Isobe—Sato type chaos expansions is given. Well-known classical models like the hypergeometric urn scheme, Ehrnfest's dog-flea model and Polya urn scheme are realised in terms of independent spin half observables.

1. INTRODUCTION

The idea of a continuous time quantum diffusion was introduced through an example in (Hudson and Parthasarathy, 1983, 1984) using the language of quantum stochastic differential equations in Fock space. In a series of papers (Evans, 1989; Evans and Hudson, 1987; Hudson, 1987; Meyer, 1989). M.P. Evans and R.L. Hudson outlined a general approach to the construction of a class of quantum diffusion processes which are more appropriately called quantum stochastic flows. Their method involves the notion of structure maps of a (phase space) $\star$ unital algebra $\mathcal{B}$ and a quantum stochastic differential equation involving them. Recently, V.P. Belavkin pointed out to me that the structure maps of an Evans—Hudson flow are better described through a natural $\star$ homomorphism from $\mathcal{B}$ into its tensor product with a certain matrix algebra equipped with a special involution. In an attempt to understand the Belavkin homomorphism it seems appropriate to examine Evans—Hudson flows in discrete time and then make a passage to the continuous time limit. However, such a programme is far from complete even though a beginning was made in (Lindsay and Parthasarathy, 1988).
In a short note (Meyer, 1989) P.A. Meyer indicated how certain finite Markov chains in continuous time can be expressed as Evans—Hudson flows and their Isobe—Sato type chaos expansions can be arrived at by following a suggestion of Ph. Biane. The scope of Meyer's construction was enlarged in (Mohari and Sinha, 1989; Parthasarathy and Sinha, 1989) by constructing structure maps of an Evans—Hudson flow through certain group actions. It is curious that in spite of several interesting remarks made by P.A. Meyer concerning observables in a baby Fock space (Meyer, 1986; 1987; 1988a; 1988c; 1989a) much attention has not

been paid to the construction of Evans—Hudson flows in discrete time and the realization of classical Markov chains in terms of quantum mechanical observables. Within the framework of a tensor product of a $\star$ unital algebra $\mathcal{B}$ and a finite number of copies of a matrix algebra $\mathcal{M}_d$ we present here a completely elementary account of discrete time Evans—Hudson flows and develop their Isobe—Sato type chaos expansions (Isobe and Sato, 1984). Such a flow is determined by a $\star$ homomorphism from $\mathcal{B}$ into $\mathcal{B} \otimes \mathcal{M}_d$. We believe that in the passage to the continuous time limit this homomorphism 'becomes' the Belavkin homomorphism. As a spin off from our elementary approach a large class of classical Markov chains is realized in terms of quantum mechanical observables. Such a realization leads to an interesting connection between concrete classical models like the hypergeometric urn scheme, Ehrnfest's dog–flea model and Polya's urn scheme on the one hand and discrete time martingales obeying the Lévy condition on the other.

2. DISCRETE TIME FLOWS

Let $\mathcal{B}$ be a $\star$ unital algebra and let $\mathcal{M}_d$ be the algebra of all $(d \times d)$–matrices with complex entries, considered as operators in the Hilbert space $\mathbf{C}^d$. The standard basis in $\mathbf{C}^d$ is expressed as $e_0, e_1, \ldots, e_{d-1}$ where e_j is the column vector with 1 in the j–th position and 0 elsewhere for each $j = 0, 1, 2, \ldots, d-1$. Fix a positive integer N and denote

$$\widetilde{\mathcal{B}} = \mathcal{B} \otimes \mathcal{M}_d(1) \otimes \mathcal{M}_d(2) \otimes \ldots \otimes \mathcal{M}_d(N) \tag{2.1}$$

where $\mathcal{M}_d(n) = \mathcal{M}_d$ is the n–th copy of $\mathcal{M}_d$ and $\otimes$ denotes the algebraic tensor product. In any algebra with identity we denote by 1 its identity element and $\lambda 1 = \lambda$ for any scalar λ. Let $\mathcal{B}_{n]} \subset \widetilde{\mathcal{B}}$ denote the subalgebra of all elements of the form $X \otimes 1$ where $X \in \mathcal{B} \otimes \mathcal{M}_d(1) \otimes \ldots \otimes \mathcal{M}_d(n)$ and $1 \in \mathcal{M}_d(n+1) \otimes \ldots \otimes \mathcal{M}_d(N)$. Then $\{\mathcal{B}_{n]},\ 0 \leqslant n \leqslant N\}$ is an increasing sequence of subalgebras with $\mathcal{B}_{N]} = \widetilde{\mathcal{B}}$. It is called the ***standard filtration*** in $\widetilde{\mathcal{B}}$.
The unique linear map $\mathcal{E}_{n]} \colon \widetilde{\mathcal{B}} \to \mathcal{B}_{n]}$ satisfying

$$\mathbf{E}_{n]} X \otimes M_1 \otimes \ldots \otimes M_N = X \otimes M_1 \otimes \ldots \otimes M_n \otimes \prod_{i=n+1}^{N} \langle e_0, M_i e_0 \rangle$$

for all $X \in \mathcal{B}$, $M_j \in \mathcal{M}_d$, $j = 1, 2, \ldots, N$ is called the ***vacuum conditional expectation*** from $\widetilde{\mathcal{B}}$ onto $\mathcal{B}_{n]}$. $\mathbf{E}_{n]}$ is a positive, identity preserving and linear map satisfying

$$\mathbf{E}_{m]}\, \mathbf{E}_{n]} = \mathbf{E}_{n]}\, \mathbf{E}_{m]} = \mathbf{E}_{m]} \quad \text{for } m \leqslant n.$$

A sequence $\{X_n,\ 0 \leqslant n \leqslant N\}$ in $\widetilde{\mathcal{B}}$ is called ***adapted*** if $X_n \in \mathcal{B}_{n]}$ for each n. It is called a ***martingale*** if $\mathbf{E}_{m]}\, X_n = X_m$ for $m \leqslant n$. For any $X \in \widetilde{\mathcal{B}}$ the sequence $\{\mathbf{E}_{n]} X,\ 0 \leqslant n \leqslant N\}$ is a martingale. Let $M_i^j \in \mathcal{M}_d$ be the matrix with 1 as the ij–th entry and 0 elsewhere. Define

$$M_i^j(n) = 1 \otimes M_i^j \otimes 1, \quad n = 1, 2, \ldots, N,$$

where the first 1 is in $\mathcal{B} \otimes \mathcal{M}_d(1) \otimes \ldots \otimes \mathcal{M}_d(n-1)$ and the last 1 is in $\mathcal{M}_d(n+1) \otimes \ldots \otimes \mathcal{M}_d(N)$. Put

$$B^i_j(0) = 0, \; B^i_j(n) = \sum_{r=1}^{n} M^i_j(r), \; n = 1, 2, \ldots, N. \tag{2.2}$$

Each sequence $B^i_j = \{B^i_j(n),\ 0 \leqslant n \leqslant N\}$ is a martingale for $i + j > 0$, $0 \leqslant i, j \leqslant d-1$. Furthermore

$$\mathbf{E}_{0]}\, B^i_j(n) = 0,$$

$$\mathbf{E}_{m]}\, \{B^i_j(n) - B^i_j(m)\}^* \, \{B^k_l(n) - B^k_l(m)\} = \begin{cases} n - m, & \text{if } j = l, \; i = k = 0, \\ 0, & \text{otherwise} \end{cases}$$

for all $i + j > 0$, $0 \leqslant i, j \leqslant d-1$. The set $\{B^i_j,\ i + j > 0,\ 0 \leqslant i, j \leqslant d-1\}$ is called the family of *basic martingales*.

Definition 2.1. A family $\{j_n,\ 0 \leqslant n \leqslant N\}$ of maps from $\mathcal{B}$ into $\widetilde{\mathcal{B}}$ is called a *quantum stochastic flow* or, simply, a *flow* if the following conditions are fulfilled:

(i) $j_0(X) = X \otimes 1$ for all $X \in \mathcal{B}$, where $1 \in \mathcal{M}_d(1) \otimes \ldots \otimes \mathcal{M}_d(N)$;

(ii) j_n is an identity preserving $\star$ homomorphism from $\mathcal{B}$ into $\widetilde{\mathcal{B}}$ such that $j_n(X) \in \mathcal{B}_{n]}$ for every $X \in \mathcal{B}$, $n = 0, 1, 2, \ldots, N$;

(iii) There exist maps $\theta^i_j : \mathcal{B} \to \mathcal{B}$ such that

$$j_n(X) = \sum_{0 \leqslant i,k \leqslant d-1} j_{n-1}\left(\theta^i_k(X)\right) M^k_i(n) \tag{2.3}$$

for all $X \in \mathcal{B}$.

$\{\theta^i_j,\ 0 \leqslant i, j \leqslant d-1\}$ is called the set of *structure maps* of the flow $\{j_n\}$.

PROPOSITION 2.2. *A set $\{\theta^i_k,\ 0 \leqslant i, k \leqslant d-1\}$ of maps from $\mathcal{B}$ into itself is a family of structure maps of a flow $\{j_n\}$ in $\widetilde{\mathcal{B}}$ if and only if the map*

$$X \to \left((\theta^i_j(X))\right)_{0 \leqslant i,j \leqslant d-1}$$

is an identity preserving $\star$ homomorphism from $\mathcal{B}$ into $\mathcal{B} \otimes \mathcal{M}_d$.

Proof. Necessity is immediate from the identity

$$j_1(X) = \sum_{i,k} \theta^i_k(X) M^k_i(1) = \left((\theta^i_k(X))\right) \in \mathcal{B} \otimes \mathcal{M}_d(1).$$

Sufficiency follows by induction on n and the relation

$$M^i_j M^k_l = \delta^i_l M^k_j, \; 0 \leqslant i, j, k, l \leqslant d-1.$$

COROLLARY 1. *The structure maps $\{\theta^i_k\}$ determine the associated flow $\{j_n\}$ uniquely. Furthermore, $\{j_n\}$ satisfies the Markov property*

$$\mathbf{E}_{m]} j_n(X) = j_{n-m}\left(T^m(X)\right), \; X \in \mathcal{B}, \; m \leqslant n, \tag{2.4}$$

where $T = \theta_0^0$.

Proof. The first part is immediate from definitions. When $m = 1$, (2.4) follows from (2.3) and the property $\langle e_0, M_i^k e_0\rangle = 0$ whenever $i + k > 0$.
Now induction completes the proof.

COROLLARY 2. *Let $\{\theta_k^i\}$ be structure maps in $\mathcal{B}$ determining a flow $\{j_n\}$ in $\widetilde{\mathcal{B}}$. Define*

$$T = \theta_0^0,\ \mu_k^i = \theta_k^i \quad \text{if } i \neq k,$$
$$\mu_i^i = \theta_i^i - T \quad \text{if } i \geqslant 1.$$

Then

$$j_n(X) = T^n(X) + \sum_{m=1}^{n} I_m^{(n)}(X) \tag{2.5}$$

where

$$I_m^{(n)}(X) = \sum_{\substack{1\leqslant t_1<t_2<\ldots<t_m\leqslant n\\ i_j+k_j>0,\ j=1,2,\ldots,m}} T^{t_1-1}\mu_{k_1}^{i_1}T^{t_2-t_1-1}\mu_{k_2}^{i_2}\ldots T^{t_m-t_{m-1}-1}\mu_{k_m}^{i_m}\times$$

$$\times T^{n-t_m}(X)M_{i_1}^{k_1}(t_1)\ldots M_{i_m}^{k_m}(t_m) \tag{2.6}$$

with the convention that $Z\otimes 1$ is denoted Z.

Proof. This is by straightforward induction starting from (2.3).

Remark. The expansion of the flow j_n given by (2.5) and (2.6) is the discrete time version of chaos expansion or Isobe—Sato formula where $T^n(X) = \mathbf{E}_{0]}j_n(X)$ and $I_m^{(n)}(X)$ is m–th order multiple stochastic sum (instead of integral) with respect to the d^2-1 basic martingales $\{B_j^i,\ i+j>0\}$ over the interval $[0,n]$. The elements $I_m^{(n)}(X)$ obey the orthogonality relations:

(i) $\mathbf{E}_{0]}I_m^{(n)}(X) = 0$,

(ii) $\mathbf{E}_{0]}I_m^{(n)}(X)^* I_{m'}^{(n)}(Y) = 0, \quad \text{if } m \neq m'$,

(iii) $\mathbf{E}_{0]}I_m^{(n)}(X)^* I_m^{(n)}(Y) =$

$$= \sum_{\substack{1\leqslant t_1<\ldots<t_m\leqslant n\\ i_j>0,\ j=1,2,\ldots,m}} \{T^{t_1-1}\mu_0^{i_1}T^{t_2-t_1-1}\mu_0^{i_2}\ldots T^{t_m-t_{m-1}-1}\mu_0^{i_m}T^{n-t_m}(X)\}^*\times$$

$$\times\{T^{t_1-1}\mu_0^{i_1}T^{t_2-t_1-1}\ldots\mu_0^{i_m}T^{n-t_m}(Y)\}$$

for all $X, Y \in \mathcal{B}$.

PROPOSITION 2.3. *Suppose $\mathcal{B}$ is commutative and $\{j_n\}$ is a flow in $\widetilde{\mathcal{B}}$ determined by the structure maps $\{\theta_k^i\}$. Then for any $X, Y \in \mathcal{B}$, $0 \leqslant m, n \leqslant N$*

$$[j_m(X),\ j_n(Y)] = 0.$$

Proof. By (2.3) induction gives

$$j_n(Y) = \sum_{\substack{0\leqslant i_1,\ldots,i_{n-m}\\ k_1,\ldots,k_{n-m}\leqslant d-1}} j_m(\theta^{i_1}_{k_1}\ldots\theta^{i_{n-m}}_{k_{n-m}}(Y))M^{k_1}_{i_1}(m+1)\ldots M^{k_{n-m}}_{i_{n-m}}(n) \tag{2.7}$$

for $n > m$. Since j_m is a homomorphism for the commutative algebra $\mathcal{B}$ we have $[j_m(X), j_m(Z)] = 0$ for all $X, Z \in \mathcal{B}$. Since $j_m(X) \in \mathcal{B}_{m]}$ the required result is immediate from (2.7).

3. CHAOS EXPANSIONS OF DISCRETE TIME MARKOV CHAINS

Let S be a measurable space, $\phi_j: S \to S$, $0 \leqslant j \leqslant d-1$ be measurable maps and let $p_j: S \to [0,1]$, $0 \leqslant j \leqslant d-1$ be measurable functions satisfying the identity $\sum_{j=0}^{d-1} p_j(x) = 1$ for all $x \in S$. Suppose $\mathcal{B} = \mathcal{B}(S)$ is the $\star$ unital and commutative algebra of all bounded complex valued measurable functions on S. Define the map $T: \mathcal{B} \to \mathcal{B}$ by

$$Tf = \sum_{j=0}^{d-1} p_j f \circ \phi_j \tag{3.1}$$

where $\circ$ denotes composition. Then T is the transition operator of a Markov chain with state space S for which the state changes in one step from x to one of the states $\phi_0(x), \phi_1(x), \ldots, \phi_{d-1}(x)$ with respective probabilities $p_0(x), p_1(x), \ldots, p_{d-1}(x)$. However, it is possible that $\phi_i(x) = \phi_j(x)$ for some $i \neq j$. If S is a finite set of cardinality k it follows from the arguments in (Lindsay and Parthasarathy, 1988) that every Markov transition operator T on $\mathcal{B}$ is of the form (3.1) with $d \leqslant k^2 - k + 1$. In many practically interesting models d is small. (See Example 1—3 at the end.)

Consider a unitary $(d \times d)$–matrix valued function U on S for which $U = ((u_{ij}))$, $0 \leqslant i,j \leqslant d-1$, $u_{0j} = \sqrt{p_j}$ for all j. For example one may choose the orthogonal matrix

$$U = \begin{pmatrix} p_0^{1/2} & \vdots & p_1^{1/2} & \cdots & p_{d-1}^{1/2} \\ \cdots & \cdots & \cdots & \cdots & \cdots \\ -p_1^{1/2} & \vdots & & & \\ \vdots & \vdots & & I-Q & \\ -p_{d-1}^{1/2} & \vdots & & & \end{pmatrix},$$

where $Q = ((q_{ij}))$, $q_{ij} = (p_ip_j)^{1/2}\left(1 + p_0^{1/2}\right)^{-1}$, $i,j \geqslant 1$. Define the identity preserving $\star$ homomorphism

$$f \to ((\theta^i_j(f))) = U \begin{pmatrix} f\circ\phi_0 & & & 0 \\ & f\circ\phi_1 & & \\ & & \ddots & \\ 0 & & & f\circ\phi_{d-1} \end{pmatrix} U^\star$$

from $\mathcal{B}$ into $\mathcal{B} \otimes \mathcal{M}$ so that

$$\theta^i_j(f) = \sum_{r=0}^{d-1} u_{ir}\bar{u}_{jr} f \circ \phi_r, \quad \theta^0_0(f) = Tf$$

where T is given by (3.1). By Proposition 2.2 and 2.3 there exists a unique flow $\{j_n,\ 0 \leqslant n \leqslant N\}$ in $\widetilde{\mathcal{B}}$ with structure maps $\{\theta^i_j\}$ for which

$$[j_m(f),\ j_n(g)] = 0,$$
$$\mathbf{E}_{n-1]} j_n(f) = j_{n-1}(Tf)$$

for all $f, g \in \mathcal{B}(S)$. Thus for any $x \in S$, in the (quantum theoretical) state $\delta_x \otimes \otimes e_0 \otimes e_0 \otimes \ldots \otimes e_0$ of the $\star$ algebra $\widetilde{\mathcal{B}}$ the sequence of observables $j_0(f) = f, j_1(f), \ldots, j_N(f)$ has the same distribution as the sequence of random variables $f(\xi_0(x,w)),\ f(\xi_1(x,w)), \ldots, f(\xi_N(x,w))$, where $\{\xi_n(x,w)\}$ is a discrete time classical Markov chain with state space S, $\xi_0(x,w) = x$ and transition operator T, f being any element of $\mathcal{B}$. In other words $\{j_n\}$ is a classical Markovian stochastic flow with transition operator T expressed in terms of the $d^2 - 1$ basic martingales $\{B^i_k,\ i + k > 0,\ 0 \leqslant i, k \leqslant d-1\}$. An application of Corollary 2 to Proposition 2.2 enables us to expand this flow as a polynomial in the variables $M^i_k(n),\ 0 \leqslant i, k \leqslant d-1,\ 0 \leqslant n \leqslant N$.

We shall now examine the case $d = 2$ in some detail. To this end we write $a_n = M^1_0(n),\quad a'_n = M^0_1(n)$ and observe that $M^1_1(n) = a'_n a_n,\ a_n a'_n + a'_n a_n = 1$, $[a_n, a_m] = [a_n, a'_m] = 0$ whenever $m \neq n$. We express the transition operator T of (3.1) as

$$Tf = p\ f \circ \phi + q\ f \circ \psi \tag{3.2}$$

where $\phi_0 = \phi,\ \phi_1 = \psi,\ p_0 = p_1,\ p_1 = q$ so that $p + q = 1$. Assume that $(pq)^{-1}$ is bounded on S. Consider the orthogonal matrix valued function

$$U = \begin{pmatrix} \sqrt{p} & \sqrt{q} \\ -\sqrt{q} & \sqrt{p} \end{pmatrix}$$

and the identity preserving $\star$ homomorphism from $\mathcal{B}$ into $\mathcal{B} \otimes \mathcal{M}_2$ given by

$$((\theta^i_j(f))) = U \begin{pmatrix} f \circ \phi & 0 \\ 0 & f \circ \psi \end{pmatrix} U^\star =$$
$$= \begin{pmatrix} Tf & Lf \\ Lf & Tf + ((p-q)/\sqrt{pq})Lf \end{pmatrix}, \quad 0 \leqslant i, j \leqslant 1, \tag{3.3}$$

where

$$Lf = \sqrt{pq}(f \circ \psi - f \circ \phi). \tag{3.4}$$

The flow $\{j_n\}$ induced by the structure maps $\theta^i_j,\ 0 \leqslant i, j \leqslant 1$, satisfies

$$j_n(f) = j_{n-1}(Tf) + j_{n-1}(Lf)(a_n + a'_n) + j_{n-1}\left(\frac{p-q}{\sqrt{pq}} Lf\right) a'_n a_n \tag{3.5}$$

for all $f \in \mathcal{B}$, $1 \leqslant n \leqslant N$.
Since j_n is a homomorphism and by assumption $(pq)^{-1/2}(p-q) \in \mathcal{B}$ it follows that (3.5) can be expressed as

$$j_n(f) = j_{n-1}(Tf) + j_{n-1}(Lf)D_n, \quad n \geqslant 1, \tag{3.6}$$

where

$$D_n = a_n + a_n' + j_{n-1}((pq)^{-1/2}(p-q))a_n' a_n \tag{3.7}$$

for each $1 \leqslant n \leqslant N$. Define $D_0 = 0$ and observe that $D_n = D_n^{\star}$.

PROPOSITION 3.1. *Let $\{j_n\}, \{D_n\}$ be defined by (3.6), (3.7). Then $\{j_r(f), D_s, 0 \leqslant r \leqslant N,\ 0 \leqslant s \leqslant N,\ f \in \mathcal{B}\}$ is a commutative family in $\widetilde{\mathcal{B}}$.*

Proof. By Proposition 2.3, $\{j_r(f),\ 0 \leqslant r \leqslant N,\ f \in \mathcal{B}\}$ is a commutative family. Since $j_r(f) \in \mathcal{B}_{r]}$ it follows from (3.7) that $[j_r(f),\ D_s] = 0$ if $r < s$. This together with (3.6) implies $[j_s(f),\ D_s] = 0$. Thus

$$[j_r(f),\ D_s] = 0 \text{ whenever } r \leqslant s. \tag{3.8}$$

Now suppose that $\{j_r(f), D_s,\ 0 \leqslant r \leqslant n,\ 0 \leqslant s \leqslant n,\ f \in \mathcal{B}\}$ is a commutative family. Then by (3.7)

$$[D_r, D_{n+1}] = \begin{cases} [D_r, j_n(pq)^{-1/2}(p-q)]a_{n+1}' a_{n+1}, \\ 0, \quad \text{if } r \leqslant n. \end{cases}$$

Now (3.8) and induction complete the proof.

PROPOSITION 3.2. *Let $\{j_n\}$, $\{D_n\}$ be as in (3.6), (3.7). Then*

$$\mathbf{E}_{n-1]}D_n = 0, \quad \mathbf{E}_{n-1]}D_n^2 = 1 \quad \text{for} \quad n \geqslant 1.$$

Proof. This is immediate from the identity $\langle e_0,\ M_j^i e_0\rangle = 0$ for $0 \leqslant i,j \leqslant 1$, $i + j > 0$ and (3.7).

COROLLARY. *Let $M_n = D_0 + D_1 + \ldots + D_n$, $n = 0,1,2,\ldots,N$. Then $\{M_n\}$ is a commutative self-adjoint martingale in $\widetilde{\mathcal{B}}$ satisfying the Lévy property:*

$$\mathbf{E}_{n-1]}(M_n - M_{n-1})^2 = 1,\ n \geqslant 1.$$

Furthermore, the Markovian stochastic flow $\{j_n\}$ admits the chaos expansion

$$j_n(f) = T^n f + \sum_{m=1}^{n} \sum_{1 \leqslant t_1 < t_2 < \ldots < t_m \leqslant n} T^{t_1 - 1} L T^{t_2 - t_1 - 1} L \ldots T^{t_m - t_{m-1} - 1} \times$$
$$\times L T^{n - t_m}(f) D_{t_1} D_{t_2} \ldots D_{t_m} \tag{3.9}$$

for all $f \in \mathcal{B}$, $n = 0,1,2,\ldots,N$.

Proof. The first part is immediate from Proposition 3.1, 3.2. The second part is immediate from (3.6) and induction.

Remark. Note that the eigen-values of the matrix

$$a_1 + a_1' + (pq)^{-1/2}(p-q)a_1'a_1 = \begin{pmatrix} 0 & 1 \\ 1 & (pq)^{-1/2}(p-q) \end{pmatrix}$$

are $(p/q)^{1/2}$ and $-(p/q)^{-1/2}$. Equations (3.6) and (3.7) yield the following classical description of the stochastic flow: let $\xi_0(x,\cdot) \equiv x$. Suppose random variables $\xi_1(x,\cdot), \ldots, \xi_{n-1}(x,\cdot)$ with values in S have been defined. Given these, choose a binomial random variable $D_n(x,\cdot)$ assuming the values $(p/q)^{1/2}(\xi_{n-1})$ and $-(p/q)^{-1/2}(\xi_{n-1})$ with respective probabilities $q(\xi_{n-1})$, $p(\xi_{n-1})$. Define

$$\xi_n(x,\cdot) = \begin{cases} \phi(\xi_{n-1}(x,\cdot), & \text{if } D_n = -(p/q)^{-1/2}(\xi_{n-1}), \\ \psi(\xi_{n-1}(x,\cdot), & \text{if } D_n = (p/q)^{1/2}(\xi_{n-1}). \end{cases}$$

Then $\{\xi_n(x,\cdot)\}$ is a Markov chain for each x with transition operator T. If $M_n = D_0 + D_1 + \cdots + D_n$, where $D_0 = 0$ then $\{M_n(x,\cdot)\}$ is a martingale for each x satisfying the Lévy conditions

$$\mathbf{E}M_n(x,\cdot) = 0,\ \mathbf{E}\{(M_n - M_{n-1})^2 | \xi_0, \xi_1, \ldots, \xi_{n-1}\} = 1.$$

For any $f \in \mathcal{B}$, $n \geqslant 1$, the following chaos expansion holds:

$$f(\xi_n(x,\cdot)) = (Tf)(\xi_{n-1}(x,\cdot)) +$$
$$+ (Lf)(\xi_{n-1}(x,\cdot))(M_n(x,\cdot) - M_{n-1}(x,\cdot)) = (T^n f)(x) +$$
$$+ \sum_{m=1}^{n} \sum_{1 \leqslant t_1 < t_2 < \ldots < t_m \leqslant n} (T^{t_1-1} L T^{t_2-t_1-1} L \ldots T^{t_m - t_{m-1} - 1} \times$$
$$\times L T^{n-t_m} f)(x) \prod_{j=1}^{m} (M_{t_j} - M_{t_j - 1})(x,\cdot).$$

It may be of interest to construct continuous time analogues of our method in terms of ordinary as well as quantum stochastic differential equations and examine the role of martingales satisfying the Lévy conditions.

Example (hypergeometric model). Consider an urn with a white balls and b black balls. Draw a ball at random successively without replacement. The state at any time is denoted (x,y) where x is the number of white balls and y is the number of black balls. Then

$$S = \{(x,y)\colon\ 0 \leqslant x \leqslant a;\ 0 \leqslant y \leqslant b\}.$$

Define maps ϕ, ψ on S by

$$\phi(x,y) = \begin{cases} (x-1,y), & \text{if } x > 0, \\ (0,y-1), & \text{if } x = 0, y > 0, \\ (0,0), & \text{if } x = y = 0, \end{cases}$$

$$\psi(x,y) = \begin{cases} (x, y-1), & \text{if } y > 0, \\ (x-1, 0), & \text{if } y = 0, \\ (0,0), & \text{if } x = y = 0. \end{cases}$$

Define

$$p(x,y) = \begin{cases} x/(x+y), & \text{if } x > 0, \\ 1/2, & \text{if } x = 0. \end{cases}$$

Then

$$(Tf)(x,y) = \begin{cases} \frac{x}{x+y} f(x-1,y) + \frac{y}{x+y} f(x,y-1), & \text{if } x > 0, y > 0, \\ f(x-1,0), & \text{if } x > 0, y = 0, \\ f(0,y-1), & \text{if } x = 0, y > 0, \\ f(0,0), & \text{if } x = y = 0. \end{cases}$$

Further $(pq)^{-1}$ is a bounded function on S.

Example 2 (Ehrnfest's dog–flea model). There are two urns with a and b balls so that $a + b = c$. One of the c balls is chosen at random and shifted from its urn to the other. The state of the system is the number of balls in the first urn. Then

$$S = \{0, 1, 2, \ldots, c\}.$$

Define the maps ϕ, ψ on S by

$$\phi(x) = \begin{cases} x - 1, & \text{if } x > 0, \\ 1, & \text{if } x = 0, \end{cases}$$

$$\psi(x) = \begin{cases} x + 1, & \text{if } x < c, \\ c - 1, & \text{if } x = c. \end{cases}$$

Define

$$p(x) = \begin{cases} c^{-1}x, & \text{if } 0 < x < c, \\ 1/2, & \text{if } x = 0 \text{ or } c. \end{cases}$$

Then $(pq)^{-1}$ is a bounded function on S and

$$(Tf)(x) = \begin{cases} c^{-1}xf(x-1) + (1 - c^{-1}x)f(x+1), & \text{if } 0 < x < c, \\ f(1), & \text{if } x = 0, \\ f(c-1), & \text{if } x = c. \end{cases}$$

Example 3 (Polya's urn scheme). Consider an urn with a white balls and b black balls where $a > 0$, $b > 0$. Draw a ball at random, replace it and add c balls of its colour. The state of the system can be described by (x, y) where x and y are the number of white and black balls, respectively.
Thus $S = \mathbb{N} \times \mathbb{N}$ where $\mathbb{N} = \{1, 2, \ldots\}$. Define

$$\phi(x,y) = (x + c, y), \quad \psi(x,y) = (x, y + c),$$

$$p(x,y) = x(x+y)^{-1}, \quad q(x,y) = y(x+y)^{-1}.$$

Then

$$(Tf)(x,y) = (x+y)^{-1}\{xf(x+c,y) + yf(x,y+c)\}.$$

In this case $(pq)^{-1}$ is not bounded. It is possible to modify the urn scheme in many ways to obtain flows for which $(pq)^{-1}$ is a bounded function.

Acknowledgement
I thank V.P. Belavkin and the participants of the Moscow Steklov Institute seminar on quantum probability for several stimulating conversations during June 89 which led me to write this paper.

REFERENCES

Evans, M.P. (1989). Existence of quantum diffusion. *Probab. Theory. Rel. Fields* **81**, 473—483.

Evans, M.P. and Hudson, R.L. (1987). Multidimensional diffusions. *Lect. Notes in Math.* **1303**, 69—88. Springer, Berlin.

Hudson, R.L. and Parthasarathy, K.R. (1983). Quantum diffusions. *Lect. Notes Control and Inform. Sci.* **49**, 111—121. Springer, Berlin.

Hudson, R.L. and Parthasarathy, K.R. (1984). Constraction of quantum diffusions. *Lect. Notes in Math.* **1055**, 173—198. Springer, Berlin.

Hudson, R.L. (1987). Quantum diffusions and cohomology of algebras. In: *Proc. First World Congress of Bernoulli Soc.*, **I**, pp. 479—485. VNU Scienc. Press, Utrecht.

Isobe, and Sato, K. (1984). Wiener—Hermite expansion of a process generated by an Ito stochastic differential equation. *J. Appl. Probab.* **20**, 754—765.

Lindsay, J.M. and Parthasarathy, K.R. (1988). The passage from random walk to diffusion in quantum probability. II. *Sankhya, Ser. A* **50**, 151—170.

Meyer, P.A. (1986). Elements de probabilites quantiques. *Lect. Notes Math.* **1204**, Springer, Berlin.

Meyer, P.A. (1987). Elements de probabilites quantiques. *Lect. Notes Math.* **1247**, Springer, Berlin.

Meyer, P.A. (1988a). Elements de probabilites quantiques. *Lect. Notes Math.* **1321**, Springer, Berlin.

Meyer, P.A. (1988b). *Diffusions Quantique, D'apres Evans et Hudson.* Strasbourg *(Preprint).*

Meyer, P.A. (1988c). *Un Cas de Representations Chaotique Discrete.* Strasbourg *(Preprint).*

Meyer, P.A. (1989a). *Fock spaces in classical and non-commutative probability..* Strasbourg *(Notes).*

Meyer, P.A. (1989b). *Chaines de Markov Finies et Representations Chaotique.* Strasbourg *(Preprint).*

Mohari, A. and Sinha, K.B. (1989). Quantum stochastic flows with infinite degrees of freedom and countable state Markov chains. *Indian Statist. Inst.* Delhi Centre *(Preprint).*

Parthasarathy, K.R. and Sinha, K.B. (1989). Markov chains as Evans—Hudson diffusions in fock space. *Indian Statist. Inst.* Delhi Centre *(Preprint).*

Prob. Theory and Math. Stat., Vol. 2, pp. 298–308
B.Grigelionis *et al.* (Eds.)
1990 VSP/Mokslas

THE ZERO–LEVEL SET OF THE BROWNIAN BRIDGE AS A MODEL OF DISORDERED PHYSICAL SYSTEMS

D. PETRITIS*
Institut de Recherche mathématique, Université de Rennes I, 35042 Rennes, France

ABSTRACT

A model of disordered systems is discussed, whose partition function is given by the random Hausdorff measure of the zero–level set of the Brownian bridge. The existence of phase transition at a critical temperature is established. Above the critical temperature the partition function does not vanish the state is unique and the free energy has the self–averaging property. Below the critical temperature the partition function is zero, and there are at least two states at low temperatures.

1. INTRODUCTION

Random walks are thoroughly studied in many areas of physics and applied mathematics. A challenging problem that remains open is the construction of the measure for the self–avoiding walk in low dimension (i.e., d =2 or 3). A promising method towards that aim is the study of weakly self–avoiding random walks by suppressing walks with many self–intersections (Varadhan, 1969; Westwater, 1980; 1981; 1982). Counting the self–intersections of a walk led to many new concepts; the most useful one seems to be that of local time. Intuitively, if X_s, $s \in [0, \infty[$ is a random process with values in **R**, its local time on $x \in \mathbf{R}$ can be defined *formally* as

$$L_t(x) = \int_0^t \delta(X_s - x) ds.$$

For Brownian motion, this concept can be given a rigorous meaning (Levy, 1948) and it turned out that it is intimately connected to the Hausdorff measure of the x–level sets of the Wiener process, i.e., of the random sets of instants where the Brownian motion attains the value x (Karatzas and Shreve, 1987). The study of the weakly self–avoiding walks and their intersections was the starting point of this work.

Another interesting problem – apparently completely disconnected from random walks – is the statistical mechanics of disordered systems, like random field Ising model, spin glasses etc. These physical systems are very difficult to study from

* Work supported by the Swiss National Science Foundation.

a rigorous point of view and only few partial results are known (Fröhlich and Zegarliński, 1988; Bricmont and Kupiainen, 1988; Ruelle, 1987). Since a general mathematical setting for these systems is still lacking, it seems interesting to study some simplified models, sharing the main features of the "real" systems, in order to grasp the mathematical difficulties. The purpose of this paper is to present such a model that arose in the study of the astonishingly different problem of the level sets of the Brownian bridge.

The paper is organized as follows. In Section 2, the definition of the zero-level set of the Brownian bridge and a random recursive construction of this set are given. In Section 3 the random Hausdorff measure of the zero-level set is shown to be equal to the partition function of a disordered system of the Derrida's GREM-type. Standard martingale arguments are used to prove the vanishing of the partition function at low temperature.

The main results are postponed in Section 4. Namely, it is shown in that Section that at high temperature there is a unique probability measure and at low temperature there are many states. Moreover, the existence of a specific free energy is proved at every temperature with a self-averaging property at high temperature and lack of self-averaging at low temperature. Section 5 is devoted to the study of the critical point. A characterization of the probability measure is given in terms of the *exact* Hausdorff measure of the zero-level set of the Brownian bridge. Some open problems are discussed in Section 6. Most of the proofs are omitted since they are published in (Petritis, 1989).

In (Koukiou, 1988), rigorous probabilistic methods are used to study the phase transition of a REM-type model; for GREM-type models, rigorous results, reminiscent of the ones presented here, are obtained for the first time for a model of random multiplications on the c-adic tree in (Collet and Koukiou, 1989). Some of their powerful methods are used here for the study of the high temperature region, different methods are used for the low temperature region.

Although the notation used is standard and defined before it is used, it is useful to remember that, in general, small Greek letters are used to denote measures; especially λ is always the Lebesque measure and π and ρ are always probability measures. Script capital Latin letters are used for σ-fields; exceptionally, the symbol $\mathcal{H}$ is used to denote the Hausdorff outer measure and not a σ-field.

2. THE ZERO-LEVEL SET OF THE BROWNIAN BRIDGE

2.1 *Definition of the zero-level set*

A one-dimensional Brownian bridge over [0, 1] is a centered, Gaussian random process $B_t(\omega)$, $t \in [0,1]$, over a probability space $(\Omega, \mathcal{F}, \rho)$ and taking values in $\mathbf{R}$, with covariance function $\mathbf{E}(B_s B_t) = s(1-t)$ for $0 \leqslant s \leqslant t \leqslant 1$. Obviously, $B_0 = B_1 = 0$. We are interested in the level sets of B_t, i.e., the random sets of constant B. More specifically we have the following

Definition 2.1. The random set

$$z_\omega = \{t \in [0,1] \ : \ B_t(\omega) = 0\}$$

is called the zero-level set of the Brownian bridge.

This set has a Cantor–like structure as stated in the following.

LEMMA 2.2. *The zero–level set of the Brownian bridge almost surely*

i) *has Lebesgue measure zero,*

ii) *is closed,*

iii) *has accumulation points at* $t = 0$ *and* $t = 1$,

iv) *has no isolated points in*]0,1[.

The previous lemma establishes that z_ω is not a trivial set. Although it has zero Lebesgue measure, it is closed and dense in itself. To get some insight into the structure of this set it is natural to look at its Hausdorff dimension.
Given some non–negative number s and some subset $A \subset R$, define

$$\mathcal{H}^s_\delta(A) = \inf \sum_{i=1}^{\infty} |U_i|^s$$

where the infimum is over all δ-covers of A. Now, define an exterior measure by taking

$$\mathcal{H}^s(A) = \lim_{\delta \downarrow 0} \mathcal{H}^s_\delta(A).$$

Notice that $\mathcal{H}^s_\delta(A)$ does not decrease when δ decreases, therefore the above limit always exists (may be infinite). $\mathcal{H}^s(A)$ is the Hausdorff measure of order s as originally defined by Hausdorff. In the special case of the metric space $R = \mathbf{R}$ with distance $d(x,y) = |x - y|$ the cover can be formed by intervals and the outer Hausdorff measure can be written as

$$\mathcal{H}^s(A) = \lim_{\delta \downarrow 0} \left(\inf \sum_{i=1}^{\infty} \lambda(U_i)^s \right),$$

where $\lambda(U_i)$ is the Lebesque measure of the interval U_i.

Definition 2.3. The Hausdorff dimension, $d_H(A)$, of the set A is given by

$$d_H(A) \equiv \inf\{s \ : \ \mathcal{H}^s(A) = 0\} = \sup\{s \ : \ \mathcal{H}^s(A) = \infty\}.$$

In general, it is very difficult to compute the Hausdorff measure of a set just by applying the general Definition 2.3 since it is not a constructive one. In the following, a recursive random construction of the set z_ω, similar to the one introduced in (Graf et al., 1988), is given that allows the computation of its Hausdorff measure by an appropriate limiting procedure.

2.2. THE RANDOM RECURSIVE CONSTRUCTION

Denote by $M = \{1,2\}$ and by $M^* = \bigcup_{m=0}^{\infty} M^m$. Every element $s \in M^*$ is in an one–to–one correspondence with a branch of the rooted dyadic tree in an obvious way. Therefore, branches of the dyadic tree are also denoted by s. Denote by $|s|$ the integer–valued length of the branch s and by $s|_n$, $n \leqslant |s|$, the restriction of s to the n first ancestors. Given two branches $s,t \in M^*$, we denote by $s \vee t$ the concatenation of t to s with length $|s \vee t| = |s| + |t|$. Notice that $s \vee t \neq t \vee s$ in general. By abuse of notation, write $s \vee 1$ (resp., $s \vee 2$) for the concatenation of branch (1) (resp., (2)) to s. Finally, use the symbol () to denote the empty branch of length zero.

Now, given the stochastic process of the Brownian bridge B_t over $(\Omega, \mathcal{F}, \rho)$ we construct a family of random, finite, closed intervals $\{K_s(\omega)\}$, $s \in M^*$, having the infinite, rooted, dyadic tree as index set. Denote by $a_s(\omega)$ and $b_s(\omega)$ the random edges of these intervals, i.e., $K_s(\omega) = [a_s(\omega), b_s(\omega)]$ and by $m_s(\omega) = (a_s(\omega)+b_s(\omega))/2$ their middle points. The recursive construction goes as follows:

Start from $K_{()}(\omega) = [0,1] = K$.

Define

$a_{(1)} = a_{()} = 0,$

$b_{(1)} = \inf\{t \geqslant m_{()} \ : \ B_t(\omega) = 0\},$

$a_{(2)} = \sup\{t \leqslant m_{()} \ : \ B_t(\omega) = 0\},$

$b_{(2)} = b_{()} = 1,$

i.e., $K_{(1)} = [a_{(1)}, b_{(1)}]$ and $K_{(2)} = [a_{(2)}, b_{(2)}]$. The set z_ω being closed by Lemma 2.2, B_t is by construction a Brownian bridge both over $K_{(1)}$ and $K_{(2)}$. Thus,

Repeat ad infinitum: Rescale and define general

$a_{s\vee 1} = a_s,$

$b_{s\vee 1} = \inf\{t \geqslant m_s \ : \ B_t(\omega) = 0\},$

$a_{s\vee 2} = \sup\{t \leqslant m_s \ : \ B_t(\omega) = 0\},$

$b_{s\vee 2} = b_s.$

Obviously, we have for the zero–level set of the Brownian bridge

$$z_\omega = \bigcap_{m=1}^{\infty} \bigcup_{s\in M^m} K_s(\omega).$$

The family, $\{D_s\}$, of deterministic intervals indexed by the dyadic tree are also used in the sequel. For every $s \in M^k$ there is exactly one interval D_s, of length 2^{-k}, and belonging to the family $\{D_s\}$ that contains K_s.

Now it is possible to express the Hausdorff measure of z_ω in a form convenient for computations by virtue of the following.

LEMMA 2.4. *Let*

$$Z_{\omega,k}(\beta) = \sum_{s\in M^k} (\lambda(K_s(\omega)))^\beta$$

with λ denoting the Lebesque measure. Then, the limit $\lim_{k\uparrow\infty} Z_{\omega,k}(\beta) = Z_\omega(\beta)$ *always exists in* $\mathbf{R} \cup \{+\infty\}$ *and is comparable to the Hausdorff measure of order* β *of the zero-level set* z_ω, *i.e.,* $\mathcal{H}^\beta(z_\omega) \leqslant Z_\omega(\beta) \leqslant 6\mathcal{H}^\beta(z_\omega)$.

Proof. If the limit exists, it is comparable to $\mathcal{H}^\beta(z_\omega)$ by the theorem on comparable net measures (see (Rogers, 1970; Chapter 2)). The problem that remains is the existence of the limit that is postponed to the Theorem 3.2 of the next section.

Remark. Notice that the estimate $Z_{\omega,k}(\beta)$ of the Hausdorff measure can be written as the partition function of a disordered system in the form

$$Z_{\omega,k}(\beta) = \sum_{s\in M^k} \exp \beta \log \lambda(K_s).$$

The Hausdorff measure of order β of z_ω appears as the thermodynamic limit of $Z_{\omega,k}(\beta)$.

3. THERMODYNAMICS

It is pointed out in the last section that the recursive construction of z_ω allows to express the Hausdorff measure of this set as a statistical sum (partition function) of a disordered system. This analogy is not a new one. It has been already used for the Hausdorff measure of deterministic fractals (Badii and Politi, 1985).
The question that arises here is whether the disordered statistical systems that naturally appears in this random construction has any physical relevance. In fact, at the level k there are 2^k terms appearing in the statistical sum and each such term is non-negative, hence it can be written as $\exp(-\beta\epsilon_s)$ where $\epsilon_s = -\log \lambda(K_s)$ is a random energy. Moreover, for $s, s' \in M^k$ the corresponding energies are correlated and $\mathbf{E}[\epsilon_s\epsilon_{s'}]-\mathbf{E}[\epsilon_s]\mathbf{E}[\epsilon_{s'}] = c[k-d(s,s')]$ where d is the ultrametric distance defined for every $s, s' \in M^k$ as

$$d(s,s') = k - \max\{i \ : \ s\mid_i = s'\mid_i\}.$$

Thus, the model studied here is reminiscent of the Generalized Random Energy Model (GREM) introduced in (Derrida, 1985) and studied in (Ruelle, 1987). There is nevertheless an important difference: contrary to Derrida's approach of the GREM, here the hamiltonian is bounded from below, guaranteeing thus the stability of the model. However, in view of the special construction used, the model must be viewed merely as a model providing some understanding of the real case.

LEMMA 3.1. *Let* $\phi(\beta) = \mathbf{E}\big[(\lambda(K_{s\vee 1})^\beta + \lambda(K_{s\vee 2})^\beta)/\lambda(K_s)^\beta\big]$. *Then,*

i) ϕ *is independent of* $s \in M^*$,

ii) ϕ *is strictly decreasing in* β,

iii) $\phi(1/2) = 1$.

It is possible now to state the main result about the thermodynamic limit of the partition function.

THEOREM 3.2. *The sequence of random variables $Z_k(\beta)$ converges almost surely to a random variable $Z(\beta)$ such that*

i) for $\beta > \beta_c = 1/2$, $Z(\beta) = 0$ almost surely,

ii) for $\beta < \beta_c = 1/2$, $Z(\beta) = \infty$ almost surely and the limit $\lim_{k\uparrow\infty} Z_k(\beta)/\phi(\beta)^k$ is a.s. strictly positive.

Proof. i) Consider the σ-algebra $\mathcal{F}_k$ generated by the random variables

$$X_{s,i} = \lambda(K_{s|_{i+1}})/\lambda(K_{s|_i})$$

for $s \in M^k$ and $i = 0, \ldots, k-1$. The sequence $\{\mathcal{F}_k,\ k \in \mathbb{N}\}$ is a filtration. Moreover, due to the rescaling of the Brownian bridge, the variables $X_{s,i}$ are i.i.d. Now, write

$$Z_k(\beta) = \sum_{s\in M^k} \lambda(K_s)^\beta = \sum_{s\in M^k} \prod_{i=0}^{k-1} X_{s,i}^\beta$$

and take the conditional expectations

$$\mathbf{E}(Z_k(\beta)|\mathcal{F}_{k-1}) = \sum_{s\in M^k} \mathbf{E}\left(\frac{\lambda(K_s)^\beta}{\lambda(K_{s|_{|s|-1}})^\beta}\right) \prod_{i=0}^{k-2} X_{s,i}^\beta = \phi(\beta) Z_{k-1}(\beta).$$

Defining the sequence of random variables $W_k(\beta) = Z_k(\beta)/\phi(\beta)^k$, one sees immediately that $\{W_k, \mathcal{F}_k\}_{k\in\mathbb{N}}$ is a positive martingale, hence it converges to a random variable $W(\beta)$ with $0 \leqslant W(\beta) < \infty$. The statement follows trivially because $\phi(\beta) < 1$ for $\beta > 1/2$.

ii) Following (Falconer, 1986), denote by $M^k(\beta, \delta)$ the random set

$$M^k(\beta,\delta) = \{s \in M^k \ :\ \lambda(K_s)^\beta \geqslant \delta^k\}$$

for $0 \leqslant \delta \leqslant 1$. It is easy to establish the inequalities

$$\delta^k \mathbf{E}(\mathrm{card} M^k(\beta,\delta)) \leqslant \phi(\beta)^k \leqslant k \sup_{0\leqslant\delta\leqslant 1} \delta^{k-1}\mathbf{E}(\mathrm{card} M^k(\beta,\delta)).$$

Therefore, if $\beta < \beta_c$, $\phi(\beta) > 1$ and it is possible to find some integer k and some $\delta \in [0,1]$ such that $\delta^k\mathbf{E}(\mathrm{card}M^k(\beta,\delta)) > 1$. Then using that $z_\omega \neq \emptyset$, it follows that $\rho(\{Z_\infty(\beta) = \infty\}) = 1$ for $\beta < \beta_c$,. Now, by the standard theorem on branching process, stating that either $W = 0$ almost surely or $\{W > 0\} = \{Z = \infty\}$ almost surely (Asmussen and Hering, 1983), we establish that $W > 0$ a.s.

Using the identification of $Z(\beta)$ with $\mathcal{H}^\beta(z_\omega)$ as explained in the previous section one has the

COROLLARY 3.3. *The Hausdorff dimension of the zero-level set of the Brownian bridge is 1/2 a.s.*

The previous theorem, showing that the partition function vanishes at low temperature, is the first step towards proving the existence of a phase transition.

Actually, if there is a phase transition the partition function vanishes, by Lee and Yang theorem. However, vanishing of the partition function is not a sufficient condition since models can be imagined where the partition function vanishes but which do not present any phase transition; it must be proved that there are in fact more than one states at low temperature. The existence of many states at low temperature in the thermodynamic limit is shown in the next section. Here, the definition of the state, adapted from (Collet and Koukiou, 1989), is given.

Definition 3.4. The "finite-volume" state is a random probability measure π on the measurable space $([0,1],\ \mathcal{B}([0,1]))$ defined for every $A \in \mathcal{B}([0,1])$ by

$$\pi_{\beta,N}(A) = Z_N(\beta)^{-1} \sum_{s \in M^N} \lambda(K_s(\omega) \cap A)^{\beta}.$$

The problem as usual is the thermodynamic limit ($N \to \infty$). As far as N is finite $\pi_{\beta,N}$ is well defined (unique). In the limit $N \to \infty$ the probability measure π_β remains well defined if $Z(\beta) > 0$ (i.e., at high temperature) but it may fail to exist if $Z(\beta) = 0$.
Instead of exploring the whole Borel σ-field $\mathcal{B}([0,1])$ it is enough to study $\pi_{\beta,N}(D_s)$ for every $s \in M^k$ with $N \geqslant k$ and eventually take the limit $N \to \infty$. An important role in the intermediate technical steps is also played by the unnormalized random measure $\mu_{\beta,N}$ on $\mathcal{B}([0,1])$ defined for every $A \in \mathcal{B}([0,1])$ by

$$\mu_{\beta,N}(A) = \sum_{s \in M^N} \lambda(K_s(\omega) \cap A)^{\beta}.$$

4. STUDY OF THE FREE ENERGY AND THE STATES IN THE THERMODYNAMIC LIMIT

The finite-volume free energy is defined as usual to be the random variable

$$f_k(\beta) = \frac{1}{k} \log\ Z_k(\beta).$$

This section is devoted to the study of the limits $\lim_{k\uparrow\infty} f_k(\beta)$ and "$\lim_{k\uparrow\infty}$" $\pi_{\beta,k}$.
The main results are summarized in the following

PROPOSITION 4.1. (Strong self-averaging property). *For $\beta < \beta_c$, $f(\beta) = \lim_{k\uparrow\infty} f_k(\beta)$ is almost surely a constant equal to the annealed mean.*

PROPOSITION 4.2. (Unicity). *For $\beta < \beta_c$, there is an almost surely unique random probability measure π_β on $\mathcal{B}([0,1])$, obtained as limit of $\pi_{\beta,k}$ for $k \uparrow \infty$.*

THEOREM 4.3. (Lack of self-averaging). *For β sufficiently larger than β_c,*

$$\limsup_{k\uparrow\infty} \frac{1}{k} \log \frac{Z_k(\beta)}{\phi(\beta)^k} < 0 \quad a.s.$$

THEOREM 4.4. (Non–Unicity). *Fix $A = [0, 1/2[\subseteq K$. For every $\varepsilon \in]0, 1/4[$ there is a $\beta_0 = \beta_0(\varepsilon) > \beta_c$ such that*

$$\left| \limsup_{k\uparrow\infty} \pi_{\beta,k}(A) - \liminf_{k\uparrow\infty} \pi_{\beta,k}(A) \right| \geqslant \varepsilon$$

for $\beta > \beta_0(\varepsilon)$.

In the sequel, only the proofs of Proposition 4.1 and Theorem 4.4 are given. The detailed proofs of the other assertions can be found in (Petritis, 1989).

Proof of Proposition 4.1. The annealed average is log $\phi(\beta)$ because $\mathbf{E}(Z_k(\beta)) = \mathbf{E}(Z_k(\beta)|\mathcal{F}_0) = \phi(\beta)^k$. Now, for $\beta < \beta_c$ the limit $\lim_{k\uparrow\infty} Z_k(\beta)/\phi(\beta)^k = W(\beta)$ is almost surely $0 < W(\beta) < \infty$ as shown in Theorem 3.2. Hence,

$$\lim_{k\uparrow\infty} \frac{1}{k} \frac{Z_k(\beta)}{\phi(\beta)^k} = 0 \text{ a.s.}$$

Before proving the main result, two particular subsets of M^* are introduced that can be easily visualized if one thinks of elements of M^* in terms of branches of a dyadic tree. For every $s \in M^*$, the set $\text{umb}(s) \equiv \{t \in M^* \ : \ t\ |_{|s|} = s\}$ is called the *umbra* of s and the set $\text{pen}(s) \equiv \{t \in M^* \ : \ \exists n$, with $n \leqslant |s|$ s.t. $\forall i \leqslant n, s\ |_i = t\ |_i\}$ is called the *penumbra* of s.

Proof of Theorem 4.4. Since the sequence of $\pi_{\beta,k}(A)$ is bounded, there is at least one converging subsequence. It will be shown that there are at least two subsequences converging to different limits. The proof uses essentially the upcrossing–downcrossing inequalities in the reverse way of the one used in the standard proof of the martingale convergence theorem.

Write

$$Z_N(\beta) = \sum_{t\in M^N} \lambda(K_t)^\beta =$$

$$= \sum_{t\in M^N \cap \text{umb}(A)} \lambda(K_t)^\beta + \sum_{t\in M^N,\ t\in \text{pen}(A)\setminus\text{umb}(A)} \lambda(K_t)^\beta + \sum_{t\in M^N \setminus \text{pen}(A)} \lambda(K_t)^\beta$$

where

$$\text{umb}(A) = \bigcup_{s:K_s\subset A} \text{umb}(s), \ \text{pen}(A) = \bigcup_{s:K_s\subset A} \text{pen}(s).$$

The particular choice of the set A is such that $\text{pen}(A) = \text{umb}(A)$ and the same holds for the set $A^c = K\setminus A$, hence the second term in the above sum vanishes and

$$Z_N(\beta) = \mu_{\beta,N}(A) + \mu_{\beta,N}(A^c).$$

Choose $a = 1/2 - \epsilon$. We show that $\pi_{\beta,k}(A) < a$ happens infinitely often. The difficulty here is that $\pi_{\beta,k-1}(A)$ and $\pi_{\beta,k}(A)$ are not independent random variables. Hence the standard Borel—Cantelli lemma is of no use. Consider therefore the sequence of events

$$A_k = \{\mu_{\beta,k}(A)(1-a) < \mu_{\beta,k}(A^c)a\}.$$

It is easy to see that $\{\pi_{\beta,k}(A) < a\} = A_k$ and moreover $A_k \in \mathcal{F}_k$. The sequence of random variables $Y_k = \sum_{n=1}^{k} \rho(A_n \mid \mathcal{F}_{n-1})$ is a submartingale. By the Doob's decomposition theorem, it can be written as $Y_k = M_k + I_k$ where M_k is a martingale and I_k is an increasing process. Now, Y_k fails to converge (in $\mathbf{R}$) on the set $\{I_\infty = = \infty\}$ and for sufficiently low temperature, this set occurs with at least positive probability. Then, by conditional Borel—Cantelli lemma, $\{Y_k, i.o.\} = \{I_\infty = \infty\}$ almost surely.

Finally, a result concerning the probability measure corresponding to the critical temperature is quoted here. In the Introduction, the notion of local time was given formally in terms of intuitive definitions. Here precise statements are made.

Definition 4.5. (Karatzas and Shreve, 1987). The occupation time for the Brownian bridge is the random measure $\Gamma_t(\cdot,\omega)$ on $\mathcal{B}(\mathbf{R})$ defined for every Borel set A by

$$\Gamma_t(A,\omega) = \int_0^t \mathbf{1}_A(B_s)ds = \lambda(\{s \in [0,t] \ : \ B_s \in A\})$$

for $0 \leqslant t \leqslant 1$.

Definition 4.6. If the random measure $\Gamma_t(\cdot,\omega)$ is absolutely continuous with respect to the Lebesgue measure λ, the local time L_t is defined as the Radon—Nikodym derivative $d\Gamma/d\lambda$.

THEOREM 4.7. *There is a positive constant* c *and a function* h *given by the formula* $h(y) \doteq (y\ \log|\log\ y|)^{1/2}$ *on* $\mathbf{R}^+$ *for which*

$$\mathcal{H}^{(h)}(z_\omega \cap [0,t]) = cL_t(0,\omega)$$

for $0 \leqslant t \leqslant 1$.

Some comments may be helpful to clarify the utility of this theorem. In the previous sections we used as *a priori* measure on the intervals $[a,b]$, $0 \leqslant a \leqslant b \leqslant 1$, the Lebesgue measure $\lambda([a,b]) = b - a$. If instead of λ we use some set function

$$m([a,b]) = (b-a)\ \log|\log(b-a)|$$

(m is not a measure) and construct the corresponding probability measure $\pi^{(m)}_{1/2,\omega}(\cdot)$ on $\mathcal{B}([0,1])$ by adapting the Definition 3.4, the previous theorem states that every interval $[a,b] \subset [0,1]$ (i.e., every configuration) has a probability

$$\pi^{(m)}_{1/2,\omega}([a,b]) = \frac{L_b(0,\omega) - L_a(0,\omega)}{L_1(0,\omega)}.$$

The interest of the previous formula stems from the fact that the local time can be computed by other methods (Rosen, 1985; Le Gall, 1986; Yor, 1986) thus it provides an ***explicit construction*** of the limit probability measure!

5. CONCLUSION

It is shown in this paper that the model of disordered system considered here presents a phase transition at a critical temperature. Above the critical temperature the partition function does not vanish, the state is unique and the free energy has the self-averaging property. Below the critical temperature it is established that the partition function is zero; this is the starting point for showing that there are many states in the thermodynamic limit and that the free energy has not any self-averaging property. The methods used here are probabilistic.

The results proved here are in agreement with the results in (Collet and Koukiou, 1989) for a closely related model of multiplicative chaos. However, in spite of the formal similarities between the two models there are also important mathematical differences and the method used here to prove the multiplicity of the states at low temperature is different from the methods used in (Collet and Koukiou, 1989).

It would be helpful to examine whether the methods used here can be applied to more realistic statistical models.

It is worth reminding some genuine problems that still remain open. It is proved that there are at least two states at low temperature. How many states exist actually? It is believed that there is an infinite number of them but this is not proved. More fundamentally, what can be interpreted as a physical phase is an extremal thermodynamic state. Is there a procedure for selecting extremal states in the limit? Some very preliminary results seem to indicate that for this particular model there are only finitely many extremal states. Other closely related models should admit – by the same arguments – infinitely many states. Work in this direction is in progress. In case it is firmly established that there are infinitely many extremal states, here or in similar models, there is still a remaining open problem, namely the Choquet theory of infinitely many extremal states.

The thermodynamic state that arose at the critical point is well characterized; the corresponding probability measure can be written in terms of local time. It should be interesting to obtain explicit expressions for the probability measures at other temperatures.

In summarizing, one can say that although some questions are rigorously answered in this work there are many important problems that still remain open.

Acknowledgments.

I should like to thank F. Koukiou for numerous useful discussions and for communicating to me the results in (Koukiou, 1988; Collet and Koukiou, 1989) prior to their publication.

REFERENCES

Asmussen, S. and Hering, H. (1983). *Branching Processes.* Birkhäuser, Basel.

Badii, R. and Politi, A. (1985). Statistical description of chaotic attractors. *J. Stat. Phys.* **40**, 725–750.

Bricmont, J. and Kupiainen, A.-J. (1988). Phase transition in the 3d random field Ising model. *Commun. Math. Phys.* **116**, 539–572.

Collet, P. and Koukiou, F. (1989). *Thermodynamics of the Multiplicative Chaos.* Preprint

Université de Lausanne.

Derrida, B. (1985). A generalization of the random energy model which includes correlations between energies. *J. de Physique-Lettres* **46**, 401–407.

Falconer, K.J. (1986). Random fractals. *Math. Proc. Camb. Phil. Soc.* **100**, 559–582.

Fröhlich, J. and Zegarliński, B. (1988). *Spin glasses and other lattice systems with long range interactions.* Preprint Eidgenössische Technische Hochschule Zürich.

Karatzas, I. and Shreve, S.E. (1987). *Brownian Motion and Stochastic Calculus.* Springer, Berlin–Heidelberg etc.

Koukiou, F. (1988). *A random covering interpretation of the Random Energy Model.* Preprint Université de Lausanne.

Koukiou, F., Pasche, J. and Petritis, D. (1988). The Hausdorff dimension of the two–dimensional Edwards' random walk. *J. Phys. A. Math. Gen.* (to appear).

Le Gall, J.-F. (1986). Sur les temps locaux d'intersection du mouvement Brownian et la méthode de renormalisation de Varadhan. *Lect. Notes in Math.* **1123**, 314–331.

Graf, S., Maudin, R.D. and Williams, S.C. (1988). The exact Hausdorff dimension in random recursive constructions. *Memoirs A.M.S.* **71**, No. 381.

Lévy, P. (1948). *Processus Stochastiques et Mouvement Brownien.* Gauthier–Villars, Paris.

Petritis, D. (1989). Thermodynamics for the zero–level set of the Brownian bridge. *Commun. Math. Phys.* (to appear).

Rogers, C.A. (1970). *Hausdorff Measures.* Cambridge University Press, Cambridge.

Rosen, J. (1985). A representation for the intersection local time of Brownian paths in space. *Ann. Probab.* **13**, 145–153.

Ruelle, D. (1987). A mathematical reformulation of Derrida's REM and GREM. *Commun. Math. Phys.* **108**, 225–239.

Varadhan, S.R.S. (1969). Appendix to the course by K.Symanzik. In: *Local Quantum Theory.* Academic Press, New York.

Westwater, J. (1980). On Edwards' model for long polymer chains. *Commun. Math. Phys.* **72**, 131–173.

Westwater, J. (1981). On Edwards' model for long polymer chains. *Commun. Math. Phys.* **79**, 53–73.

Westwater, J. (1982). On Edwards' model for long polymer chains. *Commun. Math. Phys.* **84**, 459–470.

Yor, M. (1986). Renormalisation et convergence en loi pour les temps locaux d'intersection du mouvement brownian dans $\mathbf{R}^3$ *Lect. Notes in Math.* **1123**, 350–365.

Prob. Theory and Math. Stat., Vol. 2, pp. 309–314
B.Grigelionis *et al.* (Eds.)
1990 VSP/Mokslas

SOME INEQUALITIES FOR MOMENTS OF SUMS OF INDEPENDENT RANDOM VARIABLES

V.V. PETROV
Faculty of Mathematics and Mechanics, Leningrad University,
198904 Leningrad, USSR

ABSTRACT

Some generalizations and strengthenings of well-known inequalities are proved for moments of sums of independent random variables. In particular, generalizations of Rosenthal's inequalities are found.

Let $X_1, X_2, \ldots, X_n$ be independent random variables, $S_n = \sum_{k=1}^{n} X_k$. In (Rosenthal, 1970; 1972) the following result is proved: if

$$\mathbf{E}|X_k|^p < \infty \qquad (k = 1, \ldots, n) \tag{1}$$

for some $p > 1$, then

$$\mathbf{E}|S_n|^p \leqslant 2^{p^2} \max\left\{M_{p,n}, \left(\sum_{k=1}^{n} \mathbf{E}|X_k|\right)^p\right\}, \tag{2}$$

where

$$M_{p,n} = \sum_{k=1}^{n} \mathbf{E}|X_k|^p. \tag{3}$$

If condition (1) is satisfied for some $p \geqslant 2$ and, moreover,

$$\mathbf{E}X_k = 0 \quad (k = 1, \ldots, n), \tag{4}$$

then

$$\mathbf{E}|S_n|^p \leqslant c(p) \max\left\{M_{p,n}, B_n^{p/2}\right\} \tag{5}$$

and

$$\mathbf{E}|S_n|^p \geqslant 2^{-p} \max\left\{M_{p,n}, B_n^{p/2}\right\}, \tag{6}$$

where

$$B_n = \sum_{k=1}^{n} \mathbf{E}X_k^2 \tag{7}$$

and $c(p)$ is a positive constant depending only on p and may be different in different formulas.
We shall prove some generalizations of inequalities (2) and (5).

1. Let G_0 be the set of non-negative even functions $g(x)$, non-decreasing on $(0, \infty)$ and such that $g(0) = 0$.

THEOREM 1. *Suppose that*

$$\mathbf{E}g(X_k) < \infty \qquad (k = 1, \dots, n) \tag{8}$$

for some $g \in G_0$ *and that* $0 < D_n < \infty$, *where*

$$D_n = \sum_{k=1}^{n} \mathbf{E}|X_k|. \tag{9}$$

Then

$$\mathbf{E}g(S_n) \leqslant \sum_{k=1}^{n} \mathbf{E}g(rX_k) + 2e^r \int_0^\infty \left(1 + \frac{x}{D_n}\right)^{-r} dg(x) \tag{10}$$

for every $r > 0$.

Taking $g(x) = |x|^p$, $p > 1$, we have

$$\mathbf{E}|S_n|^p \leqslant r^p \sum_{k=1}^{n} \mathbf{E}|X_k|^p + 2pe^r B(p,\ r - p) D_n^p \tag{11}$$

for every $r > p > 1$ where $B(x,\ y)$ is the beta-function. Therefore

$$\mathbf{E}|S_n|^p \leqslant c(p)\,(M_{p,n} + D_n^p)$$

if $p > 1$. This inequality differs from Rosenthal's inequality (2) only by constants.

THEOREM 2. *Suppose that conditions* (4) *and* (8) *are satisfied for some* $g \in G_0$. *If* $0 < B_n < \infty$, *where* B_n *is defined by* (7), *then*

$$\mathbf{E}g(S_n) \leqslant \sum_{k=1}^{n} \mathbf{E}g(rX_k) + 2e^r \int_0^\infty \left(1 + \frac{x^2}{rB_n}\right)^{-r} dg(x) \tag{12}$$

for every $r > 0$.

Putting here $g(x) = |x|^p$, $p \geqslant 2$, we have

$$\mathbf{E}|S_n|^p \leqslant r^p M_{p,n} + p\,e^r r^{p/2} B\left(\frac{p}{2},\ r - \frac{p}{2}\right) B_n^{p/2} \tag{13}$$

for every $p \geqslant 2$ and $r > p/2$. Therefore Rosenthal's inequality (5) is a consequence of (12).

If X is a random variable with the distribution function $F(x)$, then we put

$$\mathbf{E}^+ g(X) = \int_0^\infty g(x)\, dF(x).$$

THEOREM 3. *Suppose that*

$$\mathbf{E}^+ g(X_k) < \infty \qquad (k = 1, \ldots, n) \tag{14}$$

for some $g \in G_0$ *and that* $0 < D_n < \infty$ *where* D_n *is defined by* (9). *Then*

$$\mathbf{E}^+ g(S_n) \leqslant \sum_{k=1}^{n} \mathbf{E}^+ g\,(rX_k) + e^r \int_0^\infty \left(1 + \frac{x}{D_n}\right)^{-r} dg(x) \tag{15}$$

for every $r > 0$.

THEOREM 4. *Suppose that conditions* (4) *and* (14) *are satisfied for some* $g \in G_0$. *If* $0 < B_n < \infty$ *then*

$$\mathbf{E}^+ g(S_n) \leqslant \sum_{k=1}^{n} \mathbf{E}^+ g\,(rX_k) + e^r \int_0^\infty \left(1 + \frac{x^2}{rB_n}\right)^{-r} dg(x) \tag{16}$$

for every $r > 0$.

2. Proof of Theorems 1 and 3.

LEMMA 1. *Let* $y_1, \ldots, y_n$ *be positive numbers,* $y = \max\{y_1, \ldots, y_n\}$. *Then*

$$\mathbf{P}\,(S_n \geqslant x) \leqslant \sum_{k=1}^{n} \mathbf{P}\,(X_k \geqslant y_k) + \exp\left\{\frac{x}{y} - \frac{x}{y}\log\left(1 + \frac{x}{D_n}\right)\right\}, \tag{17}$$

$$\mathbf{P}(|S_n| \geqslant x) \leqslant \sum_{k=1}^{n} \mathbf{P}(|X_k| \geqslant y_k) + 2\exp\left\{\frac{x}{y} - \frac{x}{y}\log\left(1 + \frac{x}{D_n}\right)\right\} \tag{18}$$

for every $x > 0$.

This lemma is a consequence of more general inequalities stated in (Fuc and Nagaev, 1971).

LEMMA 2. *Let* X *be a random variable such that* $\mathbf{E}^+ g(X) < \infty$ *for some* $g \in G_0$. *Then*

$$\mathbf{E}^+ g(X) = \int_0^\infty \mathbf{P}\,(X \geqslant x)\, dg(x).$$

If $\mathbf{E}g(X) < \infty$ *for some* $g \in G_0$, *then*

$$\mathbf{E}g(X) = \int_0^\infty \mathbf{P}(|X| \geqslant x)\,dg(x).$$

To prove the first assertion note that if $\mathbf{E}^+g(X) < \infty$ then

$$\lim_{x \to +\infty} (1 - F(x))g(x) = 0,$$

where $F(x) = \mathbf{P}(X < x)$. This relation is a corollary of the inequalities

$$g(x)(1 - F(x)) \leqslant g(x) \int_x^\infty dF(y) \leqslant \int_x^\infty g(y)\,dF(y),$$

where $x > 0$. The right–hand side here tends to zero as $x \to +\infty$ since $\mathbf{E}^+g(X) < \infty$. Therefore,

$$\int_0^\infty \mathbf{P}(X \geqslant x)\,dg(x) = -\mathbf{P}(X \geqslant 0)\,g(0) + \int_0^\infty g(x)dF(x) = \mathbf{E}^+g(X)$$

since $g(0) = 0$. The second assertion of Lemma 2 can be proved in a similar way. To prove Theorem 3, we put in Lemma 1 $y_k = y = x/r$ $(k = 1, \ldots, n)$, where $x > 0$ and $r > 0$. Then we have

$$\mathbf{P}(S_n \geqslant x) \leqslant \sum_{k=1}^n \mathbf{P}\left(X_k \geqslant \frac{x}{r}\right) + \exp\left\{r - r\log\left(1 + \frac{x}{D_n}\right)\right\}$$

for every $x > 0$. Accordingly,

$$\int_0^\infty \mathbf{P}(S_n \geqslant x)\,dg(x) \leqslant T_1 + T_2,$$

where

$$T_1 = \sum_{k=1}^n \int_0^\infty \mathbf{P}\left(X_k \geqslant \frac{x}{r}\right) dg(x), \qquad T_2 = e^r \int_0^\infty \left(1 + \frac{x}{D_n}\right)^{-r} dg(x).$$

Taking into account Lemma 2, we conclude that $T_1 = \sum\limits_{k=1}^n \mathbf{E}^+g(rX_k)$. Theorem 3 is proved.

The proof of Theorem 1 is based on inequality (18) instead of (17) and the second part of Lemma 2.

If $g(x) = |x|^p$, $p > 1$, and $r > p$, then

$$T_2 = e^r D_n^p \int_0^\infty u^{p-1} (1+u)^{-r} \, du = e^r D_n^p B(p, \; r-p)$$

and inequality (11) follows.

3. The proof of Theorems 2 and 4 goes in a similar fashion. Instead of Lemma 1 it is possible to use consequences of more general probability inequalities which were proved in (Fuc and Nagaev, 1971) and (Nagaev and Pinelis, 1977) (see also (Petrov, 1987), Chapter 3, Lemma 13).

4. We now turn to another group of estimates of $\mathbf{E}|S_n|^p$, where $S_n = \sum_{k=1}^n X_k$ and $X_1, \ldots, X_n$ are independent random variables. If condition (4) holds and $p \geqslant 2$ then

$$\mathbf{E}|S_n|^p \leqslant c(p) \, n^{p/2-1} \sum_{k=1}^n \mathbf{E}|X_k|^p. \tag{19}$$

This inequality is well-known (see, for example, (Petrov, 1975b, p. 60), or (Petrov, 1987, pp. 89 and 281)). It is noted in (Petrov, 1987) that (19) is a simple consequence of the Marcinkiewicz—Zygmund inequality or the Rosenthal inequality (5).

By means of (5) and an inequality for moments of a random variable (Petrov, 1975a) it is possible to get some improvement of inequality (19). The following inequality was proved in (Petrov, 1975a) (see also (Petrov, 1987, p. 16)):

$$\beta_r^{\frac{1}{r}} \leqslant \gamma^{\frac{1}{r}-\frac{1}{s}} \, \beta_s^{\frac{1}{s}}, \tag{20}$$

where $0 < r < s$, $\beta_r = \mathbf{E}|X|^r$, X is a random variable, $\gamma = \mathbf{P}(X \neq 0)$. If $\gamma < 1$, i.e., $\mathbf{P}(X = 0) > 0$, then inequality (20) is stronger than Liapounoff's inequality $\beta_r^{1/r} \leqslant \beta_s^{1/s}$.

Consider a random variable Y with the distribution function $F(x) = \frac{1}{n} \sum_{k=1}^n F_k(x)$, where $F_k(x) = \mathbf{P}(X_k < x)$. Then $\mathbf{E}|Y|^r = \frac{1}{n} \sum_{k=1}^n \mathbf{E}|X_k|^r$ for any $r > 0$ and

$$\mathbf{P}(Y \neq 0) = \frac{1}{n} \sum_{k=1}^n \mathbf{P}(X_k \neq 0).$$

Applying inequality (20) to Y with $r = 2$ and $s = p \geqslant 2$, we have

$$B_n^{1/2} \leqslant \left(\sum_{k=1}^n \mathbf{P}(X_k \neq 0) \right)^{\frac{1}{2}-\frac{1}{p}} M_{p,n}^{1/p},$$

where B_n and $M_{p,n}$ are defined by (7) and (3), respectively. Using the Rosenthal inequality (5) we can state the following result.

THEOREM 5. *If conditions* (1) *and* (4) *hold for some* $p \geqslant 2$, *then*

$$\mathbf{E}|S_n|^p \leqslant c(p)\left[1+\left(\sum_{k=1}^{n}\mathbf{P}\,(X_k \neq 0)\right)^{p/2-1}\right] M_{p,n}, \tag{21}$$

where $M_{p,n}$ *is defined by* (3).

If the sum $\sum_{k=1}^{n}\mathbf{P}\,(X_k \neq 0)$ grows slower than n (for example, slower than cn^q with some $q<1$), than (21) gives a better estimate as compared to (19).
Applying inequality (20) to the same random variable Y with $r=1$ and $s=p>1$, we get the following result by applying the Rosenthal inequality (2):

THEOREM 6. *If condition* (1) *holds for some* $p>1$, *then*

$$\mathbf{E}|S_n|^p \leqslant c(p)\left[1+\left(\sum_{k=1}^{n}\mathbf{P}\,(X_k \neq 0)\right)^{p-1}\right] M_{p,n}.$$

Note that condition (4) is absent in Theorem 6.

REFERENCES

Fuc, D.H., Nagaev, S.V. (1971). Probability inequalities for sums of independent random variables. *Theory Probab. Appl.* **16**, No 4.

Gradstein, I.S., Ryzhik, I.M. (1963). *Tables of Integrals, Sums, Series and Products.* Fizmatgiz, Moscow.

Nagaev, S.V., Pinelis, I.F. (1977). Some inequalities for the distribution of sums of independent random variables. *Theory Probab. Appl.* **22**, No 2.

Petrov, V.V. (1975a). An inequality for moments of a random variable. *Theory Probab. Appl.* **20**, 391–392.

Petrov, V.V. (1975b). *Sums of Independent Random Variables.* Springer, Berlin–Heidelberg–New York.

Petrov, V.V. (1987). *Limit Theorems for Sums of Independent Random Variables.* Naùka, Moscow.

Rosenthal, H.P. (1970). On the subspaces of $L^p(p>2)$ spanned by sequences of independent random variables. *Israel J. Math.* **8**, 273–303.

Rosenthal, H.P. (1972). On the span in L^p of sequences of independent random variables. In: *Proc. 6th Berkeley Symp. on Math. Statist. and Probab.* Univ. Calif. Press, Berkeley and Los Angeles **2**, pp. 149–167.

Prob. Theory and Math. Stat., Vol. 2, pp. 315–324
B.Grigelionis *et al.* (Eds.)
1990 VSP/Mokslas

SOME MINIMAX ESTIMATION PROBLEMS

I. F. PINELIS
Novosibirsk Institute of Electrical Engineering,
Karl Marx avenue 20, 630092 Novosibirsk, USSR

ABSTRACT

Based upon the approach proposed earlier in connection with the abstract "max min= =min max" equality, we give duality theorems and equations for solutions of problems of the minimax linear estimation of a linear regression and of a value of a linear functional. This approach seems to be fruitful in various problems of the minimax estimation.

1. INTRODUCTION AND AN ABSTRACT MINIMAX THEOREM

Let us consider the minimax risk

$$\rho = \inf_{\delta \in \Delta} \sup_{P \in \mathbf{P}} R(P, \delta)$$

where $R(P, \delta) = \int L(P, \delta(\omega))P(d\omega)$, L is a loss function, Δ is a set of statistician's strategies and $\mathbf{P}$ is a family of probability measures. We shall also consider the dual risk

$$\rho^* = \sup_{P \in \mathbf{P}} \inf_{\delta \in \Delta} R(P, \delta)$$

and examine the situations when $\rho = \rho^*$ because in this case the search of the minimax decisions gets simplified and, often, ρ^* is much more easy to calculate than ρ (see, e.g., (Ibragimov and Hasminskii, 1984; 1987; Dodunekova and Hasminskii, 1987; Fedotov and Levchuk, 1986; Samarov, 1987)). When does the equality $\rho = \rho^*$ hold? The classical maximin–minimax theorems require the concavity of $R(P, \delta)$ w.r. to P and its convexity w.r. to δ. The last of these two requirements is indeed satisfied when the loss function L is convex w.r. to its second argument, but the concavity is usually violated.

To avoid this difficulty, the approach suggested in (Pinelis, 1986) (created originally for other purposes – see section 5), has proved to be convenient. To describe it, let us take the class $\mathbf{F}(X, Y)$ of all real functions $F(x, y)$, $x \in X$, $y \in Y$, which are convex w.r. to $y \in Y$ for each fixed $x \in X$; Y is any convex set, X is an arbitrary one. Define

$$\varphi_F(y) = \sup\{F(x, y) : x \in X\}$$

and consider the following problem: characterize the set $\Phi(Y)$ of all real convex functions $\varphi(y)$, $y \in Y$, such that for every (or almost every, in a certain sense) X

and $F \in \mathbf{F}(X,Y)$ the equality $\varphi = \varphi_F$ implies

$$\sup_{x \in X} \inf_{y \in Y} F(x,y) = \inf_{y \in Y} \sup_{x \in X} F(x,y).$$

We succeeded in finding a rather complete solution of this problem. For example, it turns out that almost every smooth convex function φ increasing at ∞ belongs to $\Phi(Y)$. The smoothness of φ may be checked either directly or by suitable means of the convex analysis.
As to the problems of the minimax estimation considered in this paper, the following simplest minimax theorem is already sufficient.

THEOREM 1.1. *Let F belong to $\mathbf{F}(X,Y)$, where X is a compact Hausdorff topological space, Y is a convex subset of some linear space E. Assume that $F(x,y)$ is upper-semicontinuous w.r. to x and there exists $y^* \in \arg\min\{\varphi_F(y) : y \in Y\}$. We define $X^* = \{x \in X : F(x,y^*) = \varphi_F(y^*)\}$. If $F(x_1,y) = F(x_2,y)$ for all $y \in Y$, $x_1 \in X^*$, $x_2 \in X^*$, then*

$$F(x,y^*) \leqslant F(x^*,y^*) \leqslant F(x^*,y), \quad x^* \in X^*, \ x \in X, \ y \in Y.$$

This theorem is an analogue of the result stated in (Pinelis, 1986). In this paper, some duality theorems for the minimax estimation of linear functionals and regression are given. For some proofs see the last section. Full proofs will be published elsewhere.

2. MINIMAX LINEAR ESTIMATION OF A VALUE OF A LINEAR FUNCTIONAL

Let $A \subseteq \mathbf{R}^d$ be any non–empty strictly convex compact set. We have to estimate the value $a'l$ on the basis of observations $(a+\xi)'m$, $m \in \mathbf{R}^d$, where $l \in \mathbf{R}^d$ is given, a is an unknown vector from A, ξ is a symmetrically distributed random vector in $\mathbf{R}^d$, $'$ denotes the transposition, the vectors from $\mathbf{R}^d$ are identified with the matrix–columns. The risk $R(P,\delta)$ here is equal to

$$R(a,m) = \mathbf{E}\lambda((a+\xi)'m - a'l)$$

where λ is a real, even, strictly convex differentiable function and $\mathbf{P}, \Delta$ are replaced by $A, \mathbf{R}^d$.

THEOREM 2.1. *If $A = -A$, then $\rho = \rho^*$.*

Analogous statements were obtained (Ibragimov and Hasminskii, 1984; 1987; Dodunekova and Hasminskii, 1987) for $\lambda(t) = t^2$ and Gaussian ξ; see also (Fedotov and Levchuk, 1986).
Some restrictions on A, viz., the finite dimensionality, the strictness of the convexity, the compactness, have been put only to evade cumbersome wordings. For example, similar results are true for A lying in an arbitrary linear space. Besides,

one may substitute the more general condition $R(a,m) = R(-a,m)$ for the symmetry of the distribution of ξ together with the evenness of λ. In particular, if $\lambda(t) = t^2$, then the symmetry of the distribution of ξ may be replaced by $\mathbf{E}\xi = 0$. Let us now examine some cases when $A \neq -A$, but, of course, assuming the other conditions given above. For this purpose, we fix some vector $e_0 \in \mathbf{R}^d$ with $|e_0| = 1$. Denote by H^0 and P^0 the orthogonal complement to e_0 and the orthoprojector from $\mathbf{R}^d$ to H^0 resp. Then for any $x \in \mathbf{R}^d$ we have $x = x_0 e_0 + x^0$ with $x_0 = e_0' x$, $x^0 := P^0 x$. Let us call our a priori given A a one-sided set if the implication

$$(a \in A,\ a_0 \leqslant 0) \Rightarrow -a \in A$$

holds. We call A a one-sided body of rotation if it is a one-sided set and, besides,

$$(\forall t \geqslant 0\ \exists r(t) \in \mathbf{R}) a_0 \geqslant 0 \Rightarrow [a \in A \Longleftrightarrow |a^0| \leqslant r(a_0)];$$

in this case, there exist r_A and $t_A \geqslant 0$ (the maximal radius and its abscissa) uniquely determined by the equalities

$$r_A = r(t_A) = \max\{r(t) : t \geqslant 0\},$$

as A is a strictly convex compact. It is convenient to exclude the trivial case $r_A = 0$ when card $A = 1$.
Let C be the covariance operator of ξ, i.e.,

$$l'Cm = \mathbf{E}\xi' l \xi' m,\ l \in \mathbf{R}^d,\ m \in \mathbf{R}^d.$$

Denote by T the restriction of the operator $T^0 = P^0 C P^0 + r_A^2 P^0$ to H^0. Then there exists T^{-1} because $T > 0$. (Here and in what follows, $T \geqslant 0 (> 0)$ means that T is (strictly) positive definite.) Thus the vector

$$h = -T^{-1} P^0 C l$$

is well-defined.

THEOREM 2.2. *Let A be a one-sided body of rotation and $\lambda(t) = t^2$. Then $\rho = \rho^*$ iff*

$$r_A t_A |h| \leqslant |e_0' C(l + h)|.$$

If $C = \sigma^2 I$, where I is the unity operator in $\mathbf{R}^d$, then the previous inequality takes the form

$$r_A t_A |l^0| \leqslant (r_A^2 + \sigma^2)|l_0|.$$

By a suitable linear transformation, one can obtain an analogue of Theorem 2.2 for sets A with the ellipsoidal sections in place of the circular ones.

3. MINIMAX LINEAR ESTIMATION OF A LINEAR REGRESSION

Suppose we have a vector of observations of the form

$$Y = Xa + \xi,$$

where $Y \in M_{n1}$, $X \in M_{nr}$, $a \in M_{r1}$, ξ is a random vector in M_{n1} with $\mathbf{E}\xi = 0$, $\operatorname{cov} \xi = \mathbf{E}\xi\xi' = V > 0$, M_{nk} is the set of all real $(n \times k)$–matrices (note that X and Y now are not the same as in Section 1).
Denote by $B_V Y$ the best linear unbiased estimator (BLUE) of a, i.e., $\mathbf{E}B_V Y = a$ and for any $B \in M_{rn}$ with $\mathbf{E}BY = a$ the inequality $\operatorname{cov} B_V Y \leqslant \operatorname{cov} BY$ holds; of course, the matrix inequality $S_1 \leqslant S_2$ with $S_i \in M_{rr}$ means $S_2 - S_1 \geqslant 0$ (i.e., $S_2 - S_1$ is positive definite, see Section 2). All the square matrices considered in the paper are supposed to be symmetric. To secure the existence of the BLUE, we demand that rank $X = r$ and $n \geqslant r$.
Assume that $\operatorname{cov} \xi = V$ is not known, but V belongs to a given class $\mathbf{V}$ of strictly positivé matrices. In such case one may apply the Rozanov's idea (Rozanov, 1969) of so–called pseudo–best estimator $B_W Y$ where an arbitrary $W \in \mathbf{V}$ substitutes for the unknown V. We come to the following definition of the risk:

$$R(V, W) = \mathbf{E}_V (B_W Y - a)' S (B_W Y - a)$$

where $S \in M_{rr}$, $S \geqslant 0$, $\mathbf{E}_V$ is the conditional expectation under the condition $\operatorname{cov} \xi = V$. Here $\mathbf{V}$ replaces both Δ and $\mathbf{P}$.

THEOREM 3.1. *If* $\mathbf{V}$ *is a convex set of matrices* $V \in M_{nn}$, $V > 0$ *with* $\mathbf{V}^* = \arg\max\{R(V, V) : V \in \mathbf{V}\} \neq \varnothing$, *then*

$$\rho = \rho^* = \max\{R(V, V) : V \in \mathbf{V}\}.$$

This theorem reduces the problem of finding minimax decisions to the problem of maximization of the function $\psi(V) = R(V, V), V \in \mathbf{V}$. It is easy to see that ψ is concave and its derivative $\psi'(V; \Delta)$ at a point $V > 0$ in a direction Δ is equal to tr $\Gamma\Delta$ where

$$\Gamma = \Gamma^V = V^{-1} X (X' V^{-1} X)^{-1} S (X' V^{-1} X)^{-1} X' V^{-1}$$

Thus, $\mathbf{V}^*$ coincides with the set of all solutions of the equation $\mathbf{T}(V) = 0$, where

$$\mathbf{T}(V) = \max\{\operatorname{tr}\Gamma\Delta : V + \Delta \in V\}.$$

Let us rewrite this equation in more explicit forms, which are more convenient for the following classes $\mathbf{V}$:

$$\mathbf{V}_{\text{contam}} = \{(1-\delta)V_0 + \delta U : \ U \geqslant 0, \operatorname{tr} U = 1\}, 0 < \delta < 1,$$

$$\mathbf{V}_p = \mathbf{V}_p(\varepsilon, V_0) = \{\|V - V_0\|_p \leqslant \varepsilon\},$$
$$\mathbf{V}_p^0 = \{V \in \mathbf{V}_p : \operatorname{tr} V = 1\}, \ \varepsilon > 0, \ 1 \leqslant p \leqslant \infty.$$

Here $V_0 > 0$, $\operatorname{tr} V_0 = 1$, $\|U\|_p = \left(\sum \lambda_i(U)^p\right)^{1/p}$ for $1 \leqslant p < \infty$, $\|U\|_\infty = \max|\lambda_i(U)|$, where $\lambda_i(U)$ are the eigen values of U, tr means trace. By Schur's theorem (see, e.g., (Marshall and Olkin, 1979)) one can obtain the inequality

$$\operatorname{tr} MN \leqslant \|M\|_p \|N\|_q,$$

where $M,\ N \in M_{nn}, q \in [1,\infty]$ is dual to p. When this inequality is actually an equality, we say that the pair (M,N) is a p–extremal one.
Let us give a simple criterion for this property: a pair (M,N) is p–extremal iff it satisfies, respectively, the equation

1) $M^{[p]}\|N\|_q^q = N^{[q]}\|M\|_p^p$, for $1 < p < \infty$,
2) $MN = \|N\|_\infty |M|$, for $p = 1$,
3) $NM = \|M\|_\infty |N|$, for $p = \infty$;

here we use the extension of the real functions $x^{[p]} = x^p \text{sign } x$ and $|x|$ to the functions of symmetric matrices (see, e.g., (Marshall and Olkin, 1979)).
We define the numbers $c(q) = c(q,\Gamma)$ by the relation

$$c(q) \in \arg\min\{\|\Gamma(c)\|_q : c \in R\},$$

where $\Gamma(c) = \Gamma - cI$ and Γ has been defined earlier. This definition is equivalent to the following conditions on $c = c(q)$:

1) $\mathrm{tr}\Gamma(c)^{[q-1]} = 0$, when $1 < q < \infty$,
2) $\forall\delta \in R\ \mathrm{tr}\Gamma(c-\delta)^{[0]}\delta^{[0]} \geqslant 0$, when $q = 1$,
3) $c = (\max\lambda_i(\Gamma) + \min\lambda_i(\Gamma))/2$, when $q = \infty$.

In this way, the value of $c(q)$ is uniquely determined when $1 < q \leqslant \infty$, but, in general, it is not so when $q = 1$. Rewriting the above equation $\mathbf{T}(V) = 0$ in the terms defined just now, one obtains

THEOREM 3.2. *The following statements are true.*
(i) $(\mathbf{V}_p^0)^* = \{V \in \mathbf{V}_p^0 : \|V - V_0\|_p = \varepsilon,\ (V - V_0, \Gamma_{c(q)})$ *is a p–extremal pair*$\} \neq \varnothing$ *if* $V_0 > \varepsilon I$.
(ii) $(\mathbf{V}_p)^* = \{V \in \mathbf{V}_p : \|V - V_0\|_p = \varepsilon,\ (V - V_0, \Gamma)$ *is a p–extremal pair*$\} \neq \varnothing$.
(iii) $(\mathbf{V}_{\mathrm{contam}})^* = \mathbf{V}_1(\delta, (1-\delta)V_0)^*$.

For the class $\mathbf{V}_2$ the equation characterizing the minimax decisions gets the simplest form:

$$(\mathbf{V}_2)^* = \{V \in \mathbf{V}_2 : V - V_0 = \varepsilon\Gamma/\|\Gamma\|_2\}.$$

Remark. Replacing the invariant matrix norm $\|U\|_p$ in the definition of the classes $\mathbf{V}_p, \mathbf{V}_p^0$ by $\left(\sum |U_{ij}|^p\right)^{1/p}$ for $1 \leqslant p < \infty$ and by $\max|U_{ij}|$ for $p = \infty$, it is not difficult to give another version of the last theorem (of course, U_{ij} are the elements of U), and, what is more, for this new norm one may obtain an analogue of the stated below Theorem 4.2′. If we consider only non–correlated observations, i.e., if V and V_0 are diagonal matrices then these two matrix norms coincide in the effect.

4. MINIMAX LINEAR ESTIMATION OF A LINEAR REGRESSION (AN ASYMPTOTIC VERSION)

Now we deal with an infinite number of observations and represent the regression equation from the previous section as

$$y_t = a'x_t + \xi_t, \quad t = 1, 2, \ldots,$$

where $x_t = (x_{t1}, \ldots, x_{tr})', (\xi_t)$ is a second-order stationary process having an unknown spectral density $f \in \mathbf{F} \subseteq \mathbf{G}_+ = \{f \in \mathbf{G} : f > 0\}$, $\mathbf{G}$ is the set of all even, continuous real functions on $[-\pi, \pi]$ equipped with the topology of uniform convergence (the evenness reflects that (ξ_t) is real), $\mathbf{F}$ is a given class.
To start with, let us define the normalized variant of the risk considered in Section 3, namely

$$R_n(f,g) = \mathbf{E}_f(a_n^g - a)' S_n (a_n^g - a) = \mathrm{tr} S \Sigma_n(f,g)$$

corresponding to the true spectral density f and to the pseudo-best estimator $a_n^g = B_{V(n,g)} Y^{(n)}$ constructed on the basis of the first n observations $Y^{(n)} = (y_1, \ldots, y_n)'$ by means of a "pseudo-best" spectral density $g \in \mathbf{G}_+$, i.e.,

$$V(n,g) = (V_{ts}(g))_{t,s=1}^n, \quad V_{ts}(g) = \int e^{i\lambda(k-l)} g(\lambda) d\lambda,$$

where, of course, i is the imaginary unit, $S_n = D_n S D_n, S \geqslant 0,\ S \in M_{rr}$, $D_n = \mathrm{diag}(d_{n1}, \ldots, d_{nr})$ is the diagonal matrix with $d_{nj} = (x_{1j}^2 + \ldots + x_{nj}^2)^{1/2}$ on its diagonal, and $\Sigma_n(f,g) = D_n(\mathrm{cov}\ a_n^g) D_n$; all the integrals in the paper are taken over $[-\pi, \pi]$ unless otherwise specified.
In (Rozanov and Kozlov, 1969; Ibragimov and Rozanov, 1970) the following result is obtained. Assume the Grenander conditions: $\max\{|x_{tj}| : t = 1, \ldots, n\} = o(d_{nj})$ and, for each integer s, there exists a matrix $C(s) = (C_{jk}(s)) \in M_{rr}$ such that $C(0) > 0$ and

$$\sum_{t=1}^n x_{(t+s)j} x_{tk} / d_{nj} d_{nk} \to C_{jk}(s), \quad j, k = 1, \ldots, r \quad \text{as} \quad n \to \infty.$$

Then one has for $f \in \mathbf{G}_+$, $g \in \mathbf{G}_+$ as $n \to \infty$

$$\Sigma_n(f,g) \to \Sigma(f,g) = J(g,g)^{-1} J(f,g) J(g,g)^{-1},$$

where

$$J(f,g) = 2\pi \int f g^{-2} d\mu, \quad \mu = \mathrm{Re}\ \eta (\geqslant 0),$$

and η is the spectral complex matrix measure for $C(s)$, i.e.,

$$C(s) = \int e^{is\lambda} \eta(d\lambda).$$

We can now formulate the asymptotic analogue of Theorem 3.1.

THEOREM 4.1. *If* $\mathbf{F}$ *is a convex subset of* $\mathbf{G}_+$ *with*

$$\mathbf{F}^* = \arg\max\{R(f,f) : f \in \mathbf{F}\} \neq \varnothing,$$

then

$$\rho = \rho^* = \max\{R(f,f) : f \in \mathbf{F}\}.$$

Here, similarly to Section 3, both Δ and $\mathbf{P}$ should be replaced by $\mathbf{F}$. Now we consider the following special classes $\mathbf{F}$ (cf. Section 3):

$$\mathbf{F}_{\text{contam}} = \{(1-\delta)f_0 + \delta g : g \geqslant 0,\ g \in \mathbf{G},\ \int g = 1\},$$

$$\mathbf{F}_p = \mathbf{F}_p(\varepsilon, f_0) := \{f \in \mathbf{G}_+ : \|f - f_0\|_p \leqslant \varepsilon\},$$

$$\mathbf{F}_p^0 = \{f \in \mathbf{F}_p : \int f = 1\},\ 0 < \delta < 1,\ \varepsilon > 0,\ 1 \leqslant p \leqslant \infty,$$

where $f_0 \in \mathbf{G}_+$, $\int f_0 = 1$, $\int f = \int f(\lambda)d\lambda$, $\|\cdot\|_p$ is the usual L^p–norm.
We say that $(f,g) \in \mathbf{G} \times \mathbf{G}$ is a p–extremal pair if $\int fg = \|f\|_p \|g\|_q$. It is easy to prove the following criterion. A pair $(f,g) \in \mathbf{G} \times \mathbf{G}$ is p–extremal iff it satisfies, respectively, the equation

1) $f^{[p]}\|g\|_q^q = g^{[q]}\|f\|_p^p$, for $1 < p < \infty$,
2) $fg = |f|\ \|g\|_\infty$, for $p = 1$,
3) $fg = \|f\|_\infty |g|$, for $p = \infty$.

In what follows, we suppose that there exists a continuous density

$$m(\lambda) = \mu(d\lambda)/d\lambda$$

of the spectral matrix measure μ defined earlier, and set $\gamma_c = \gamma - c$, $\gamma = b^2/f^2$, where $f \in \mathbf{G}_+$,

$$b(\lambda) = b^f(\lambda) = \left[\operatorname{tr} SJ(f,f)^{-1} m(\lambda) J(f,f)^{-1}\right]^{1/2}.$$

We define the numbers $c_q = c_q(\gamma)$ by the relation

$$c_q \in \arg\min\left\{\|\gamma_c\| : c \in R\right\},$$

which is equivalent to the following condition on $c = c_q$:

1) $\int \gamma_c^{[q-1]} = 0$, for $1 < q < \infty$,
2) $\forall \delta \in R$, $\int \gamma_{c-\delta}^{[0]} \delta^{[0]} \geqslant 0$, for $q = 1$,
3) $c_q = [s_+(\gamma) + s_-(\gamma)]/2$, for $q = \infty$,

where $s_\pm(\gamma) = \pm \max\{\pm\gamma(\lambda) : |\lambda| \leqslant \pi\}$. Note that here, in contrast to Section 3, the value of c_q is uniquely determined not only when $1 < q \leqslant \infty$, but also, when $q = 1$. This happens due to the continuity of γ. Moreover, $c_q > 0$ for every $q \in [1,\infty]$, since $\int m = C(0) > 0$ and, therefore, $b \not\equiv 0$, $\gamma \not\equiv 0$.
When considering $\mathbf{F}_p^0$, we shall assume, in addition, that $m(\lambda) > 0$ ($\forall \lambda \in [-\pi,\pi]$) (so $s_-(\gamma) > 0$) and write c instead of c_q for short.

THEOREM 4.2. *The following statements are true.*
(i) $(\mathbf{F}_p^0)^* = \{f \in \mathbf{F}_p^0 : (f - f_0, \gamma_c)$ *is a p–extremal pair and* $\|f - f_0\|_p = \varepsilon$ *if* $\gamma \neq \text{const}\} \neq \varnothing$.

(ii) $(\mathbf{F}_p)^* = \{f \in \mathbf{F}_p : (f - f_0, \gamma)$ *is a p–extremal pair and* $\|f - f_0\|_p = \varepsilon\} \neq \varnothing$.
(iii) $(\mathbf{F}_{\text{contam}})^* = (\mathbf{F}_1(\delta, (1-\delta)f_0))^*$.

Let us note that

$$\gamma_c(\lambda) = \alpha \Leftrightarrow f(\lambda) = b(\lambda)/(c+\alpha)^{1/2}.$$

Thus, one can rewrite the last theorem in terms of the following, more explicit, nonlinear functional equations.

THEOREM 4.2′. *Let* $\mathbf{F}$ *stands for any of the classes:* $\mathbf{F}_{\text{contam}}$, $\mathbf{F}_p$, $\mathbf{F}_p^0$. *Take* $f \in \mathbf{F}$. *Then* $f \in \mathbf{F}^*$ *iff, with* $s_\pm = s_\pm(\gamma)$:

$(\mathbf{F}_1)$ $f = f_0 \vee s_+^{-1/2} b$ and $\int f = 1 + \varepsilon$;
$(\mathbf{F}_1^0)$ $f = (f_0 \wedge s_-^{-1/2} b) \vee s_+^{-1/2} b$ and $(\gamma \neq \text{const} \Rightarrow \|f - f_0\|_1 = \varepsilon)$;
$(\mathbf{F}_\infty)$ $(f - f_0 - \varepsilon)\gamma = 0$;
$(\mathbf{F}_\infty^0)$ $f = ((f_0 + \varepsilon) \wedge c^{-1/2} b) \vee (f_0 - \varepsilon)$;
$(\mathbf{F}_p, 1 < p < \infty)$ $(f - f_0)\|\gamma\|_q^{q-1} = \varepsilon\gamma^{[q-1]}$;
$(\mathbf{F}_p^0, 1 < p < \infty)$ $(f - f_0)\|\gamma_c\|_q^{q-1} = \varepsilon\gamma_c^{[q-1]}$;
$(\mathbf{F}_{\text{contam}})$ $f = (1-\delta)f_0 \vee s_+^{-1/2} b$ and $\int f = 1$
where, in the parentheses, the classes are indicated to which the corresponding equations pertain,

$$(f \vee g)(\lambda) = \max[f(\lambda), g(\lambda)], \quad (f \wedge g)(\lambda) = \min[f(\lambda), g(\lambda)].$$

For the case of the contamination class $\mathbf{F}_{\text{contam}}$, the statements of Theorems 4.2, 4.2′ were obtained earlier in (Samarov, 1987) by means of a method adapted to this case, only for small δ, without indication how small δ should be.

5. A COMMENT ON HISTORY

Take the so–called function of deviations

$$\Lambda(x; \xi) = \inf_{t \geqslant 0} \log \mathbf{E} \exp t(\xi - x)$$

of a random variable (r.v.) ξ (Borovkov, 1967; Borovkov and Rogozin, 1965; Mogulskii, 1976). Its well–known probability sense is that under the Cramér condition

$$\Lambda(x; \xi) = \lim_{\varepsilon \downarrow 0} \lim_{n \to \infty} n^{-1} \log P\big(|n^{-1}\zeta_n - x| < \varepsilon\big),$$

where $\zeta_n = \xi_1 + \ldots + \xi_n$; $\xi, \xi_1, \xi_2, \ldots$ are i.i.d.r.v'.s.
Moreover, we have the obvious inequality

$$\mathbf{P}(\xi \geqslant x) \leqslant \exp \Lambda(x; \xi).$$

The approach declared in (Pinelis, 1986) (see the Introduction there), establishes the following

THEOREM 5.1 (Pinelis, 1984). *For any $x \in \mathbf{R}$*

$$\sup \Lambda(x; S_n) = \sup \inf \log \mathbf{E} \exp t(S_n - x) =$$

$$= \inf \sup \log \mathbf{E} \exp t(S_n - x) = \inf \left(- tx + \int_{-\infty}^{\infty} (e^{tu} - 1) dG(u) \right),$$

where $S_n = X_1 + \ldots + X_n$, sup is taken over all n and over all independent real r.v.*'s $X_1, \ldots, X_n$ satisfying the condition $\Sigma \mathbf{P}(X_j < u, X_j \neq 0) = G(u)$, $u \in \mathbf{R}$, and* inf *is taken over all $t \geqslant 0$.*

This result has been inspired by (Prokhorov, 1962). Basically we substituted $\Lambda(x; S_n)$ for the moments $\mathbf{E} S_n^{2m}$ (see also (Pinelis and Utev, 1984; 1989)).

6. SOME PROOFS

Proof of Theorem 1.1. We may assume E to be equipped with the strongest locally convex topology. Take $x^* \in X^*$. By Theorem 1.5 from (Levin, 1985, p. 96), $\partial\varphi(y^*) = \partial\varphi_{x^*}(y^*)$ where $\varphi_x(y) = F(x, y)$. Furthermore, $0 \in \partial\varphi(y^*)$ since y^* is a point where φ attains its minimum. This implies $0 \in \partial\varphi_{x^*}(y^*)$, i.e., y^* is a point where φ_{x^*} attains its minimum. It remains to take into account the construction of x^*.

Proof of Theorem 3.1. Fix an arbitrary $V^* \in \mathbf{V}^*$ and set

$$X_H = Z_H([-\delta, \delta]), \quad Y_H = Z_H(R)$$

where

$$Z_H(K) = \{V \in \mathbf{V} : V = V^* + tH, t \in K\}, \; K \subseteq R, \; H \in M_{nn}.$$

Let $\delta > 0$ be so small that X_H is compact. Set $F(W, V) = -R(V, W)$, $\varphi(V) = \sup\{F(W, V) : W \in X_H\}$. Using the identities

$$R(V, W) = \mathrm{tr} S^{1/2} B_W V B'_W S^{1/2}$$

and

$$(B_W - B_V)V(B_W - B_V)' = B_W V B'_W - B_V V B'_V, \; V > 0, \; W > 0,$$

one sees that (i) $R(V, W) \geqslant R(V, V)$, and (ii) if $R(V, W) = R(V, V)$, then $R(U, V) = R(U, W)$ for each $U > 0$. Now (i) implies $\varphi(V) = F(V, V)$, $V \in X_H$. Thus $\varphi(V^*) = \min\{\varphi(V) : V \in X_H\}$. The function φ is convex as the supremum of a family of linear functions $V \mapsto F(W, V)$, $W \in X_H$. Consequently,

$$\varphi(V^*) \leqslant \varphi(V) \quad \forall V \in Y_H.$$

Now, taking into account (ii) and Theorem 1.1, we obtain $F(V^*, V^*) \leqslant F(V^*, V)$ for every H and $V \in Y_H$, i.e., for all $V \in \mathbf{V}$. On the other hand, in view of (i), $F(W, V^*) \leqslant F(V^*, V^*)$, $W > 0$. This completes the proof.

Proof of Theorem 4.1 follows the same scheme and uses the identity

$$2\pi \int \Big[(f/g)J(g,g)^{-1} - J(f,f)^{-1}\Big](1/f)d\mu \times$$

$$\times \Big[(f/g)J(g,g)^{-1} - J(f,f)^{-1}\Big] = \Sigma(f,g) - \Sigma(f,f).$$

A similar, and even easier, proof does also for Theorem 2.1. The proofs of Theorems 2.2, 3.2, 4.2 are rather technical. The statements $(\mathbf{F}_p^0)^* \neq \emptyset$, $(\mathbf{F}_p)^* \neq \emptyset$ of Theorem 4.2 require especially cumbersome proofs.

REFERENCES

Borovkov, A.A. (1967). Boundary problems for random walks and large deviations in functional spaces. *Theory Probab. Appl.* **12**, 635–654.

Borovkov, A.A. and Rogozin, B.A. (1965). On the central limit theorem in the multidimensional case. *Theory Probab. Appl.* **10**, 61–69.

Dodunekova, R.D. and Hasminskii, R.Z. (1987). Estimators of the values of a linear functional based on indirect observations. *Problemy Pered. Inform.* **23**, 54–60.

Fedotov, A.M. and Levchuk, O.V. (1986). Optimal estimators of the linear functionals of the solutions of the 1st kind operator equations with random errors in the data. *Sov. Math. Dokl.* **287**, 63–66.

Ibragimov, I.A. and Hasminskii, R.Z. (1984). On the nonparametric estimation of a value of a linear functional in the Gaussian white noise. *Theory Probab. Appl.* **29**, 19–32.

Ibragimov, I.A. and Hasminskii, R.Z. (1987). On the estimation of linear functionals in a Gaussian noise. *Theory Probab. Appl.* **32**, 35–44.

Ibragimov, I.A. and Rozanov, Yu.A. (1970). *Gaussian Random Processes.* Nauka, Moscow.

Levin, V.L. (1985). *Convex Analysis in Spaces of Measurable Functions and Its Application in Mathematics and Economics.* Nauka, Moscow.

Marshall, A.W. and Olkin, I. (1979). *Inequalities: Theory of Majorization and Its Applications.* (Mathematics in Science and Engineering, 143). Academic Press, New York.

Mogulskii, A.A. (1976). Large deviations for trajectories of multidimensional random walks. *Theory Probab. Appl.* **21**, 309–323.

Pinelis, I.F. (1984). Some estimates for distributions of sums of independent random variables. *Theory Probab. Appl.* **29**, 188–189.

Pinelis, I.F. (1986). An approach to minimax theorems. In: *Abstr. First World Congr. Bernoulli Soc.*, Nauka, Moscow. **2**, pp. 902.

Pinelis, I.F. and Utev, S.A. (1984). Estimates of moments of sums of independent random variables. *Theory Probab. Appl.* **29**, 554–557.

Pinelis, I.F. and Utev, S.A. (1989). Exact exponential estimates for sums of independent random variables. *Theory Probab. Appl.* **4**, 384–390.

Prokhorov, Yu.V. (1962). Extremal Problems in Limit Theorems. In: *Proc. Sixth All-Union Meeting on Probab. Theory and Math. Statist.*, Vilnius, pp. 77–84.

Rozanov, Yu.A. (1969). On a new class of estimates. In: *Multivariate Analysis* **2**, Academic Press, New York, pp. 437–441.

Rozanov, Yu.A. and Kozlov, M.V. (1969). Asymptotically effective estimation of regression coefficients. *Sov. Math. Dokl.* **188**, 37–40.

Samarov, A.M. (1987). Robust spectral regression. *Ann. Statist.* **15**, 99–111.

Prob. Theory and Math. Stat., Vol. 2, pp. 325–329
B.Grigelionis *et al.* (Eds.)
1990 VSP/Mokslas

LARGE DEVIATIONS OF TRAJECTORIES IN THE CENTRAL LIMIT THEOREM FOR RANDOM PROCESSES: SOME NEW RESULTS

V.I. PITERBARG
Moscow State University, Department of Mathematics and Mechanics,
119899 Moscow, USSR

ABSTRACT

Some problems on large deviations and high excursions of trajectories of random processes closed to Gaussian ones are considered. The problems are connected with excursions of the random walks on large moving time period and with excursions of trajectories of a shot noise process. Solutions are given in classical terms of large deviations in the CLT.

The problem of large deviations in the central limit theorem in infinite dimensional spaces is one of most difficult problems in the modern probability theory. There are many more or less general results in this topic. My report contains some examples of evaluation of exact asymptotics and asymptotic decompositions of probabilities of large deviations in the uniform metric in the central limit theorem for some processes with simple specific structures.

The first group of the results is connected with random walks.

Let ξ_k, $k \geqslant 1$, be a sequence of i.i.d. random variables such that

(i) $\mathbf{E}\xi_1 = 0$, $\operatorname{Var}\xi_1 = 1$.

(ii) ξ_1 has a bounded and continuous density or ξ_1 has a lattice distribution.

(iii) (Cramér condition) $\mathbf{E}\exp(\alpha\xi_1) < \infty$ for all $|\alpha| \leqslant \alpha_0$, $\alpha_0 > 0$.

Consider the random process

$$\zeta_L(t) = \frac{1}{\sqrt{L}} \sum_{l=[tL]}^{[tL]+L-1} \xi_l, \quad 0 \leqslant t \leqslant T.$$

Note that $\zeta_L(\cdot) \xrightarrow{w} W(\cdot + 1) - W(\cdot)$ for bounded T, where $W(\cdot)$ is the Wiener process, and (Piterbarg, 1988)

$$\mathbf{P}\Big(\max_{0 \leqslant t \leqslant T} (W(t+1) - W(t)) > u\Big) =$$

$$= \frac{Tu}{\sqrt{2\pi}} \exp\Big(-\frac{u^2}{2}\Big)(1 + o(1)), \quad u \to \infty. \tag{1}$$

THEOREM 1. *If* $u \to \infty$, $L \to \infty$, $u = o(\sqrt{L})$, $Tu^2 \to \infty$, $T\exp(-au^2) \to 0$ *for some* $a < 1/2$ *(for example* T *is const), then*

$$\mathbf{P}\Big(\max_{0\leqslant t\leqslant T} \zeta_L(t) > u\Big) =$$

$$= \frac{Tu}{\sqrt{2\pi}} \exp\Big(- \frac{u^2}{2} + \frac{u^3}{\sqrt{L}}\lambda\Big(\frac{u}{\sqrt{L}}\Big)\Big)\Big(1 + o(1)\Big),$$

where $\lambda(\cdot)$ *is the Cramér series for* ξ_1.

The main idea of the proof of the theorem is the application of the "double-sum method" (Piterbarg, 1988). Let δ be some small positive number. Then

$$\Big(\frac{T}{\delta} + 1\Big)\mathbf{P}\Big(\max_{t\in\delta_0} \zeta_L(t) \geqslant u\Big) \geqslant \mathbf{P}\Big(\max_{[0,T]} \zeta_L(t) \geqslant u\Big) \geqslant$$

$$\geqslant \Big(\frac{T}{\delta} - 1\Big)\mathbf{P}\Big(\max_{t\in\delta_0} \zeta_L(t) \geqslant u\Big) - \sum_{k\neq l, 0\leqslant k,l\leqslant T/\delta+1} \mathbf{P}\Big(\max_{t\in\delta_k} \zeta_L(t) \geqslant u,\ \max_{t\in\delta_l} \zeta_L(t) \geqslant u\Big),$$

where $\delta_k = [k\delta, (k+1)\delta)$, $\quad k = 0, 1, \ldots$.
Next we shall try to choose the length δ of small intervals in such a way that single sums in the above equalities have the same order of vanishing and the double-sum has a greater one. The book contains the general description of this method for Gaussian processes and fields.
The proof of Theorem 1 is also based on next lemmas.

LEMMA 1. *Let* $S_k = \xi_1 + \ldots + \xi_k$, $k = 1, 2, \ldots$, $S_0 = 0$. *Then*

$$\mathbf{P}\Big(\max_{0\leqslant k\leqslant n} S_k > x\Big) \leqslant \exp\Big(- \alpha x + n \log \mathbf{E}e^{\alpha\xi_1}\Big)$$

for all $0 < \alpha < \alpha_0$.

LEMMA 2. *Let assumptions* (i)—(iii) *be fulfilled. Let* $L \to \infty$, $u \to \infty$, $\delta \to 0$, $\delta L \to \infty$, $u\sqrt{\delta} = A$, A *is some positive constant. Then*

$$\mathbf{P}\Big(\max_{t\in\delta_0} \zeta_L(t) \geqslant u\Big) = \frac{1}{\sqrt{2\pi}} h_A u^{-1} \exp\Big\{ - \frac{u^2}{2} + \frac{u^3}{\sqrt{L}}\lambda\Big(\frac{u}{\sqrt{L}}\Big)\Big\}\Big(1 + o(1)\Big),$$

$$h_A = A\exp\Big(- \frac{A^2}{2}\Big) \int_{-\infty}^{+\infty} e^{As}\mathbf{P}\Big(\max_{0\leqslant t\leqslant 1} \Big(W(t+1) - W(t)\Big) > s\Big)ds < \infty.$$

LEMMA 3. *Let* $\varepsilon_1, \ldots, \varepsilon_m$ *be independent random variables. Let the Laplace transforms*

$$f_i(t) = \log \mathbf{E}\exp(t\varepsilon_i)$$

be such that

$$f_i'(0) = 0, \quad f_i''(0) = \theta_i, \quad |f_i'''(0)| \leqslant M < \infty$$

for all $t \in [0, t_0]$, $i = 1, \ldots, m$, *where* t_0 *is some positive constant. Then*

$$\mathbf{E} \exp\Big(tm^{-1/2} \sum_{i=1}^{m} \varepsilon_i\Big) \leqslant \exp\Big(\frac{t^2}{2} \sum_{i=1}^{m} \theta_i + m^{-1/2} M |t|^3\Big)$$

for all $t \in [0, \sqrt{m} t_0]$.

Lemmas 1 and 3 are obvious, Lemma 2 is basic. Its proof follows from the next equalities

$$\mathbf{P}\Big(\max_{0 \leqslant t \leqslant \delta} \zeta_L(t) \geqslant u\Big) = \mathbf{P}\Big(\frac{1}{\sqrt{L}} \sum_{l=\delta L+1}^{L} \xi_l +$$

$$+ \max_{1 \leqslant k < \delta L} \frac{1}{\sqrt{L}} \sum_{l=k}^{k+\delta L-1} \xi_{l+L+\delta L+1} \geqslant u\Big) = \int \mathbf{P}\Big(\eta > u - v\Big) dF(v),$$

where

$$\eta = \frac{1}{\sqrt{L}} \sum_{\delta L+1}^{L} \xi_l$$

and F is the distribution function of

$$\max_{1 \leqslant k < \delta L} \frac{1}{\sqrt{L}} \sum_{l=k}^{k+\delta L-1} \xi_l.$$

The next lemma is a consequence of (1).

LEMMA 4.

$$\lim_{A \to \infty} A^{-2} h_A = 1.$$

The detailed proof of the theorem is too long for the volume of this publication. It will be published in the "Theory of probability and its applications".
Standard methods due to B.V. Gnedenko lead to the following result.

THEOREM 2. *If* $T \to \infty$, $L \to \infty$, $T/L \to \infty$, $L^{-1} \log T \to 0$, *then*

$$\mathbf{P}\Big(l_T\big(\max_{0 \leqslant t \leqslant T} \zeta_L(t) - l_T\big) < x\Big) =$$

$$= \exp\Big\{ - \exp\Big(- x + \frac{l_T^3}{\sqrt{L}} \lambda\Big(\frac{l_T}{\sqrt{L}}\Big)\Big)\Big\} + o(1),$$

where

$$l_T = \sqrt{2 \log\big(T/\sqrt{2\pi}\big)} + \frac{\log(2 \log T)}{2\sqrt{2 \log\big(T/\sqrt{2\pi}\big)}}. \tag{2}$$

It is necessary to note that at first the author proved Theorem 1 under more restrictive conditions (ξ_i's are sub–Gaussian). The proof of the present version was made jointly with V. Dovgalyuck.

The next results are connected with shot noise processes. Consider the Poisson point process $\pi_\lambda(\cdot)$ on $\mathbf{R}$ with intensity λ and a function f on $\mathbf{R}$ such that the stochastic integral

$$\xi_\lambda(t) = \int\limits_{\mathbf{R}} f(t-s)\pi_\lambda(ds)$$

exists for all t. We are interested in the probability

$$\mathbf{P}_{\lambda,f}(x) = \mathbf{P}\Big(l_T\Big(\lambda^{-1/2}(\xi_\lambda(t) - \lambda m) - l_T\Big) < x\Big)$$

as λ, $T \to \infty$. Here $m = \int f(t)dt$, i.e., λm is the expectation of $\xi_\lambda(t)$. The norming function l_T depends on the smooth properties of f.

Let us consider a measurable functional $\varphi_T(\cdot)$ on the space $\mathbf{R}^{[0,T]}$ such that the Lipschitz condition in the uniform metric is satisfied

$$|\varphi_T(x) - \varphi_T(y)| \leqslant \gamma_T \|x-y\|_C.$$

Introduce the sequences

$$\delta_n = \gamma_T \sqrt{\frac{n}{\lambda}} \sup_{0\leqslant t\leqslant 1} |\sqrt{n}(F_n(t)-t) - B_n(t)| \mathrm{Var} f,$$

$$\delta_n^1 = \gamma_T (S\lambda)^{-\varkappa} \|Z_n(t)\|_C + \gamma_T \frac{2m}{\sqrt{S}}|W|,$$

$$Z_n(t) = \int\limits_{t-\tau}^{t+\tau} f(t-s)W_n(ds), \quad B_n(t) = W_n(t) - tW_n(1),$$

where W is a standard Gaussian random variable, F_n and B_n are sequences from "Hungarian strong invariance principle" (see (Komlos et al., 1975)), that is, $\sqrt{n}(F_n(t)-t)$ are empirical processes for the uniform distribution on $[0,1]$ and B_n are the corresponding Brownian bridges on the same probability spaces, W_n are the corresponding Wiener processes, $\varkappa > 0$, $S = T + 2\tau$.

LEMMA 5. *Let the numbers $\lambda > 1$, $1/2 > \varkappa > 0$, $\delta > 0$, $\varepsilon > 0$ be such that for all integers $n \in [S\lambda - (S\lambda)^{1-\varkappa}, S\lambda + (S\lambda)^{1-\varkappa}]$ and $S > 1$ the inequality*

$$\mathbf{P}(\delta_n + \delta_n^1 > \delta) \leqslant \varepsilon$$

is fulfilled. Then

$$\mathbf{P}\Big(\varphi_T\Big(\lambda^{-1/2}\Big(\xi_\lambda(\cdot) - mn/S\Big)\Big) \leqslant x\Big) \leqslant \mathbf{P}\Big(\varphi_T\Big(Z_n(\cdot)\Big) \leqslant$$

$$\leqslant x+\delta\Big)+\varepsilon+\exp\Big(-(S\lambda)^{1-2\varkappa}\Big),$$

$$\mathbf{P}\Big(\varphi_T\Big(\lambda^{-1/2}\Big(\xi_\lambda(\cdot)-mn/S\Big)\Big)\leqslant x\Big)\geqslant\mathbf{P}\Big(\varphi_T\Big(Z_n(\cdot)\Big)\leqslant$$

$$\leqslant x-\delta\Big)-\varepsilon-2\exp\Big(-(S\lambda)^{1-2\varkappa}\Big).$$

The correctness and the proof of the lemma are based on the above mentioned Hungarian principle and on the famous fact that a fixed number of Poisson points in a fixed interval are uniform points.

This lemma implies a number of results on the probability $\mathbf{P}_{\lambda,f}(x)$. Consider the following conditions on f.

(A) f is absolutely continuous on $\mathbf{R}$;

(B) f is absolutely continuous on its support $[-\tau,\tau]$ and $|f(\tau)|+|f(-\tau)|=$ $=d>0$;

(C) f is differentiable on $\mathbf{R}$ and f' is absolutely continuous on $\mathbf{R}$.

THEOREM 3. *Let* supp$f\in[-\tau,\tau]$ *and* $\int f^2dt=1$.
If (A) *is fulfilled and*

$$l_T=\sqrt{2\log\left(\sqrt{\frac{C}{2\pi}}T\right)},\quad C=\int f'(t)^2dt, \tag{3}$$

then

$$\mathbf{P}_{\lambda,f}(x)\to\exp\{-e^{-x}\}$$

as $T\to\infty$, $\lambda/\log^3T\to\infty$.
If (B) *is fulfilled and* l_T *is defined by* (2), *then*

$$\mathbf{P}_{\lambda,f}(x+\log d)\to\exp\{-e^{-x}\}$$

as $T\to\infty$, $\lambda/\log^3T\to\infty$.
If (C) *fulfilled and* l_T *is defined by* (3) *then there exists* $\rho>0$ *such that*

$$\mathbf{P}_{\lambda,f}(x)=\exp\Big\{-e^{-x-\frac{x^2}{2l_T^2}}\Big\}+O(T^{-\rho}),$$

as $T\to\infty$, $\lambda/\log^5T\to\infty$.

REFERENCES

Komlos, J., Major, P. and Tusnady, G. (1975). An approximation of partial sums of independent random variables and sample distribution functions. I. *Z. Wahrsch. verw. Geb.* **32**, 111–131.

Piterbarg, V.I. (1988). *Asymptotic Methods in the Theory of Gaussian Processes and Fields.* Moscow Univ. Press, Moscow.

Prob. Theory and Math. Stat., Vol. 2, pp. 330–339
B.Grigelionis *et al.* (Eds.)
1990 VSP/Mokslas

HOMOGENIZATION AND VANISHING VISCOSITY

A.L. PYATNITSKY and S.M. KOZLOV
Lebedev Physical Institute, Leninsky prospect 53, 117333 Moscow, USSR
Moscow Civil Engineering Institute, Yaroslavskoe shosse 26,
127337 Moscow, USSR

ABSTRACT

We discuss the asymptotic behavior of the solution of the diffusion equation

$$\left(\mu\Delta + \frac{1}{\varepsilon}v\left(\frac{x}{\varepsilon}\right)\cdot\nabla\right)u^{\mu,\varepsilon} = 0, \quad x \in \mathbf{R}^n,$$

with two small parameters μ and ε. There $v(\cdot)$ is a periodic vector field.

INTRODUCTION

Our report is devoted to the diffusion ξ_t governed by the operator

$$A_\mu = \mu\Delta + v(x)\cdot\nabla, \quad x \in \mathbf{R}^n,$$

where Δ is the Laplacian, ∇ is the gradient and v is a periodic vector field. For fixed μ there is a well-known procedure of homogenization of such operator or equivalently, the method of proving the C.L.T. and calculating the covariance matrix $\sigma(\mu)$ of the ratio $(\xi_t - ct)/\sqrt{t}$ as $t \to \infty$. These methods do not take into account the smallness of μ and a possible influence of this smallness to the behaviour of the matrix $\sigma(\mu)$. The questions of interest are also the connections between μ and homogenization parameter ε, or between μ and the characteristic time T needed for the variable $(\xi_t - ct)/\sqrt{t}$ to become Gaussian.
Here we discuss two possible formulations of the corresponding problem. In the first one we are interested in asymptotics of the covariance matrix $\sigma(\mu)$ when $\mu \to 0$. This is the case when the homogenization parameter $\varepsilon \ll \mu$. In the second formulation we assume that $\mu = \varepsilon^\alpha$, $\alpha > 0$, and consider the Dirichlet problem in a bounded domain. In this report we formulate and discuss the results in these directions which were obtained recently with the help of the probabilistic method which was worked out in our previous paper (Kozlov and Pyatnitsky, 1990). Section 1 also contains a brief review of known results on effective diffusion. As in (Kozlov and Pyatnitsky, 1990), we use some symmetry conditions on the vector field v. It's necessary to remark that exact algebraic symmetry is not

principal here, but it gives a sufficient condition which allows us to control the behavior of the solution of the Dirichlet problem.
Section 2 is devoted to the contact problems. It includes a complete proof of the result on such a problem stated in (Kozlov and Pyatnitsky, 1990) and contains a new visual interpretation of diffusion in such media for several possible cases of contact. As in (Kozlov and Pyatnitsky, 1990), our approach is based on the method derived in (Wentzell and Freidlin, 1979) for analysis of small random perturbations of dynamical system. It allows us to reduce the trajectory of diffusion to a Markov chain with a countable number of states.

1. DIFFUSION IN CUBICAL SYMMETRIC MEDIA

Suppose that vector field $v(x)$ is periodic with respect to $\mathbf{Z}^n, v(x) \in C^2(\mathbf{R}^n)$. Let us denote by ξ_t the following diffusion in $\mathbf{R}^n$:

$$d\xi_t = v(\xi_t)\,dt + \sqrt{2\mu}\,dw_t,$$

where w_t is the standard Wiener process in $\mathbf{R}^n$, and μ is the diffusion coefficient. Under these assumptions, we have weak convergence on the space of trajectories:

$$\varepsilon(\xi_{t\varepsilon^{-2}} - ct/\varepsilon^2) \to \sqrt{2\sigma(\mu)}\,w_t$$

when $\varepsilon \to 0$, for some nondegenerate matrix $\sigma(\mu)$, which is called the effective diffusion, and for some constant vector $c \in \mathbf{R}^n$ (cf. (Kozlov and Pyatnitsky, 1990)). Here we discuss the situation when $c = 0$ due to some restrictions on $v(x)$. For example, it is always so in the potential case $v(x) = -\nabla U(x)$, where $U(x)$ is a $\mathbf{Z}^n$-periodic function, otherwise we introduce symmetry conditions guaranteeing that $c = 0$. Let us discuss first the behaviour of $\sigma(\mu)$ when $\mu \to 0$. Here we have the following result (cf. (Kozlov, 1989)).

THEOREM 1. *Suppose that U has only one local minimum, $\min U = 0$ and $U_z = \inf_{x(\cdot)} \sup_t U(x(t))$, where inf is taken over all paths which connect the origin with $z \in \mathbf{Z}^n \setminus \{0\}$, and $U_z = U_0$ does not depend on z. Then $\lim_{\mu \to 0} \mu \ln \sigma(\mu) = -U_0$.*

This theorem was improved in (Kozlov, 1989), where it was shown in an other way that for any Morse periodic function $U(x)$ with U_z independent of z, the exact asymptotic holds:

$$\sigma(\mu) = c\exp(-U_0/\mu)(1 + O(\mu)),$$

where the constant c may be calculated (Kozlov, 1989) (it depends on global topological properties of U and local characteristics of U in the minimum point and the saddle points where inf for U_z is attained). We stress that here no other assumptions on critical points of U are made.
In spite of similarity of the results of (Kozlov and Pyatnitsky, 1990; Kozlov, 1989), there is a deep difference in the behavior of trajectories in the special case of Theorem 1 and in general. In this direction we have the following hypothesis: If the local minimum point of U is unique then the convergence

$$\lim_{t\to\infty} \mathbf{E}\left(\frac{\xi_t^2}{t} - \sigma(\mu)\right) = 0$$

holds uniformly in μ.
If the potential has more than one local minima then usually there is no uniformity in μ.
The reason for this are the jumps of the particle between unessential 'pockets' (local minima). Jumps between them occur in exponential time scale, but they not take part in the formation of the effective diffusion.
Analytically, this phenomena leads to different structures of spectrum. In the first case the distance between the first and the second eigenvalues of the diffusion operator $A_\mu = \mu\Delta + v(x) \cdot \nabla$ (Δ is the Laplacian) is bounded from below by a constant which does not depend on μ. In the second case the difference is exponentially small (cf. (Wentzell and Freidlin, 1979)). The probabilistic method presented here has some important advantages. To demonstrate them, let us present a generalization of Theorem 1 to the non-potential case.

THEOREM 2. *Suppose that v is periodic and cubically symmetric, and has asymptotically stable attractive points at the integer points of $\mathbf{R}^n$. Suppose that there exist domain such that*
(i) *∂G is piecewise smooth,*
(ii) *G is attracted to the origin by $v(x)$,*
(iii) $\bigcup_{k\in Z^n} (\overline{G} + k) = \mathbf{R}^n$.
Then

$$\lim_{\mu\to 0} \mu \ln \sigma(\mu) = -U_0,$$

where

$$U_0 = \inf_{y\in\partial G} \inf_{t} \inf_{\{x(\cdot),x(0)=0,x(t)=y\}} \int_0^t |\dot{x}(s) - v(x(s))|^2 \, ds.$$

The proof of this theorem is quite similar to the proofs in (Kozlov and Pyatnitsky, 1990).
Starting from here, we discuss the stationary diffusion problem:

$$\left(\mu\varepsilon\Delta + v\left(\frac{x}{\varepsilon}\right) \cdot \nabla\right) u_\varepsilon = 0, \quad u_\varepsilon\Big|_{\partial\Omega} = \varphi, \quad \mu = \varepsilon^\alpha, \tag{1}$$

where Ω is a bounded domain and $\varphi(x)$ is a given boundary density, $\varphi \in C^1(\partial\Omega)$. The vector field $v(x)$ is Z^n-periodic but not necessarily potential. We suppose that v has a stable point at the origin and is cubically symmetric. Moreover, we assume the following nonexplicit condition:
(TM) If $K_R = \{x \in \mathbf{R}^n : |x| < R, \, |x - z| > r \text{ for all } z \in Z^n\backslash\{0\}\}$ then for arbitrary $x \in I^n$ the inequality holds

$$\mathbf{P}\{|\xi_\tau(x)| < R\} > \gamma > 0,$$

where τ is the exit time from K_R, $\xi_0(x) = x$, $I^n = (-1/2, 1/2)^n$. We have (cf. (Kozlov and Pyatnitsky, 1990)).

THEOREM 3. *The solution of* (1) *converges uniformly in* Ω *as* $\varepsilon \to 0$ *to the harmonic function* u_0 *with the same boundary density* $\varphi(x)$.

Condition (TM) was discussed in (Kozlov and Pyatnitsky, 1990), where a number of vector fields for which (TM) is true was given also. Here we present one counter example. What can happen if (TM) is not true? Suppose that in the cube I^3 there are three cylinders C_1, C_2, C_3 whose axes are parallel to the respective coordinate ones, and which do not intersect pairwise. Let's define the potential $U(x) = = \text{const}$, $x \in I^3 \backslash (C_1 \cup C_2 \cup C_3)$. Inside of C_i, $U(x)$ depends only on the respective radial variable and has only one minimum on the axe of the cylinder. If we consider the problem (1) with the vector field $v(x) = -\nabla U(x)$ then we'll discover the following behaviour of the solution. Inside of the cylinders C_i, u_ε will be approximately equal to a function which is linear along the axis of C_i and which takes two given values of $\varphi(x)$ at the boundary. Away from these cylinders, u_ε tends to some function $u_0(x/\varepsilon)$. The function u_0 can be obtained as an appropriate linear combination of the periodic harmonic functions $\psi_i(x)$ in $I^3 \backslash (C_1 \cup C_2 \cup C_3)$ with $\psi_i \Big|_{\partial C_j} = \delta_{ij}$, $i, j = 1, 2, 3$.

2. CONTACT PROBLEMS

Suppose that there are two cubically symmetric potentials $U_1, U_2 \in C^2(\mathbf{R}^n)$. We assume the following conditions on U_i, $i = 1, 2$:

A. Points of $\mathbf{Z}^n$ are the minimum points of U_i;

B. The potential U_i attracts $I^n = (-1/2, 1/2)^n$ to the origin;

C. U_i attains its minimum in ∂I^n at an inner point of an $(n-1)$-dimensional face of the cube I^n;

D. $\min\limits_{\partial I^n} U_1 < \min\limits_{\partial I^n} U_2$.

From A—D it follows that in I^n there is only one critical point of U_i, which is the minimum. All other critical points of the potentials are on ∂I^n.

If there are two media which are described by U_1 and U_2 respectively then the question of the properties of the composed media (i.e. the media which consists of the two parts) depends crucially on the character of the contact layer. It is convenient to assume that the first media occupies the half space $y_1 < -1/2 + \delta$ and the second the half space $y_1 > 1/2 - \delta$ for some δ, $0 < \delta < 1/2$.

Therefore we put:

$$U = \begin{cases} U_1, & y_1 < -1/2 + \delta, \\ U_2, & y_1 > 1/2 - \delta. \end{cases}$$

Moreover, the potential U assumed to be is $\mathbf{Z}^n$-periodic and $(n-1)$-dimensional cubically symmetric in $y_2, \ldots, y_n$.

We are interested in the problem:

$$\left(\mu\varepsilon\Delta+v\left(\frac{x}{\varepsilon}\right)\cdot\nabla\right)u_\varepsilon=0,\quad \mu=\varepsilon^\alpha,$$
$$u_\varepsilon\Big|_{\partial\Omega}=\varphi(x). \tag{2}$$

This is the so-called problem of contact of two periodical media.
Consider first the case of the following behaviour of the contact layer:

(i) The open cube I^n is attracted by the origin which is the minimum point of $U(x)$.

(ii) The point of the minimum of U in ∂I^n has the first coordinate equal to $-1/2$.

THEOREM 4. *The solution of* (2) *tends to the solution of the following problem:*

$$u_0=\begin{cases}u_0^+, & x_1\geqslant 0,\\ u_0^-, & x_1\leqslant 0,\end{cases}$$

$$\Delta_0^+=0,\quad x\in\Omega\cap\{x_1>0\},$$
$$\Delta_0^-=0,\quad x\in\Omega\cap\{x_1<0\},$$

$$u_0\Big|_{\partial\Omega}=\varphi(x),\qquad u_0^+=\Big|_{x_1=0}=u_0^-\Big|_{x_1=0},\qquad \frac{\partial}{\partial x_1}u_0^-\Big|_{x_1=0}=0.$$

Remark. Theorem 4 says that in order to find the limit of u_ε it is necessary:

1) to determine harmonic function in $\Omega\cap\{x_1<0\}$ with the Dirichlet boundary condition on $\partial\Omega\cap\{x_1<0\}$ and with the homogeneous von Neuman boundary condition on the plane $\{x_1=0\}$;

2) to determine harmonic function in $\Omega\cap\{x_1>0\}$ with the Dirichlet boundary condition consisting of the original condition $\varphi(x)$ on $\partial\Omega\cap\{x_1>0\}$ and the boundary values at $\partial\Omega\cap\{x_1=0\}$ of the harmonic function determined in 1).

Proof. After the transformation $y=x/\varepsilon$, from (2) we obtain

$$A_\mu u_\varepsilon=(\mu\Delta+v(y)\cdot\nabla)u_\varepsilon(\varepsilon y)=0,$$
$$u_\varepsilon\Big|_{\partial(\varepsilon^{-1}\Omega)}=\varphi(\varepsilon y). \tag{3}$$

Let's denote by $\xi_t^\mu(y)$ the diffusion governed by the operator A_μ. For $\rho>0$ according to (Kozlov and Pyatnitsky, 1990, Lemma 2) we have

$$|u_\varepsilon(\varepsilon y)-u_\varepsilon(\varepsilon z)|\leqslant ce^{-\kappa/\mu},\ \kappa>0, \tag{4}$$

for $y\in Q_\rho^z\{y\in\mathbf{R}^n:|y-z|<\rho\}$, $z\in\mathbf{Z}^n$.
Let $B^z=\{y\in\mathbf{R}^n:(y-z)\in 3I^n\}$, and let τ be the exit time for ξ_t^μ, $\xi_0^\mu=z$, from $\left(B^z\setminus\bigcup\limits_{k\neq z}Q_{\rho_1}^k\right)$.

PROPOSITION 1.
a) *For* $k : |k - z| = 1, z_1 \neq 0,$

$$\left|\frac{1}{2n} - \mathbf{P}\{\xi^\mu_\tau \in \partial Q^k_{\rho_1}\}\right| \leqslant ce^{-\beta/\mu}, \quad \beta > 0;$$

b)

$$\mathbf{P}\left\{\xi^\mu_\tau \in \partial B^z \cup \bigcup_{|k-z|>1} \partial Q^k_{\rho_1}\right\} \leqslant ce^{-\beta/\mu}.$$

Proof. Suppose that first coordinate $|z_1| > 1$. Then the symmetry leads us to the implication b) $\Rightarrow$ a). To prove b), let us introduce the following sequence of markovian moments:

$$\sigma_1 = \inf\left\{t : \xi^\mu_t \notin \bigcup_k Q^k_{\rho_0}\right\},$$

$$\tau_1 = \inf\left\{t > \sigma_1 : \xi^\mu_t \notin B^z \backslash \bigcup_k Q^k_{\rho_1}\right\},$$

$$\sigma_i = \inf\left\{t > \tau_{i-1} : \xi^\mu_t \notin \bigcup_k Q^k_{\rho_0}\right\},$$

$$\tau_i = \inf\left\{t > \sigma_i : \xi^\mu_t \notin B^z \backslash \bigcup_k Q^k_{\rho_1}\right\},$$

$$\dots\dots\dots\dots\dots\dots\dots\dots\dots\dots\dots$$

$$0 < \rho_1 < \rho_0.$$

According to (Wentzell and Freidlin, 1979),

$$\exp\left(-\frac{U^i_0 + \gamma}{\mu}\right) \leqslant \mathbf{P}\left\{\xi^\mu_{\tau_1} \in \bigcup_{|k-z|=1} \partial Q^k_{\rho_1}\right\} \leqslant \exp\left(-\frac{U^i_0 - \gamma}{\mu}\right) \tag{5}$$

for all sufficiently small μ, where

$$U^i_0 = \min_{\partial I^n} U_i - U_i(0),$$

and similarly,

$$\exp\left(-\frac{U^i_1 + \gamma}{\mu}\right) \leqslant \mathbf{P}\left\{\xi^\mu_{\tau_1} \in \partial B^z \cup \bigcup_{|k-z|>1} \partial Q^k_{\rho_1}\right\} \leqslant \exp\left(-\frac{U^i_1 + \gamma}{\mu}\right), \tag{6}$$

where $U^i_1 > U^i_0$ and γ can be chosen sufficiently small if we take suitable ρ_0 and ρ_1. It's clear that with probability 1 the Markov moment τ is equal to τ_k for some random k. The event $\left\{\xi^\mu_\tau \in \partial B^z \cup \bigcup_{|k-z|>1} \partial Q^k_{\rho_1}\right\}$ can be written as:

$$\left\{\xi^\mu_{\tau_1} \in \partial B^z \cup \bigcup_{|k-z|>1} \partial Q^k_{\rho_1}\right\} \cup \bigcup_{m=1}^{\infty} \left(\bigcap_{d=1}^{m} \{\xi^\mu_{\tau_d} \in \partial Q^k_{\rho_1}\} \cap \right.$$
$$\left. \cap \left\{ \xi^\mu_{\tau_{m+1}} \in \partial B^z \cup \bigcup_{|k-z|>1} \partial Q^k_{\rho_1} \right\}\right).$$

From the estimates (5), (6) and strong Markov property of ξ^μ_t we obtain

$$\mathbf{P}\left\{\xi^\mu_\tau \in \partial B^z \cup \bigcup_{|k-z|>1} \partial Q^k_{\rho_1}\right\} \leqslant \exp\left(-\frac{U^i_1-\gamma}{\mu}\right)\left(1+\right.$$
$$\left. + \sum_{m=1}^{\infty}\left(1-\exp\left(-\frac{U^i_0+\gamma}{\mu}\right)\right)^m\right) = \exp\left(-\frac{U^i_1-U^i_0-2\gamma}{\mu}\right).$$

Now it is enough to choose $\beta = \min_{i=1,2}(U^i_1 - U^i_0 - 2\gamma)$. If $z_1 = \pm 1$, the symmetry condition breaks down and a) does not follow from b). To prove a), take the cube B_δ in B^z with the same center z and the side length $(1+2\delta)$. In this cube symmetry condition holds and the exit probabilities for its faces are equal. Moreover, with probability exponentially close to 1 the exit occurs at an inner point of $\partial B_\delta \cap \{I^n + k\}$, $|k-z| = 1$. From condition (i) we conclude now that the transitions probabilities in $Q^k_{\rho_1}$, $|k-z|=1$ can differ from each other only by exponentially small quantities.
Then we see as earlier that a) follows from b). Proposition 1 is proved.
Using the probabilistic representation together with estimate (4) and Proposition 1, now we have

$$u_\varepsilon(\varepsilon z) = \frac{1}{2n} \sum_{|k-z|=1} u_\varepsilon(\varepsilon k) + O(e^{-\beta/\mu}). \tag{7}$$

Let's consider a point $z \in \mathbf{Z}^n$ with $z_1 = 0$. For such points we have

PROPOSITION 2. *For the event $C_\mu = \{\xi^\mu_\tau \in \partial Q^j_{\rho_1}\}$, $j_1 = -1$, $j_s = z_s$, $s > 1$, the estimate holds:*

$$1 - \mathbf{P}\{C^\mu\} \leqslant c e^{-\beta/\mu}, \quad \beta > 0.$$

Proof. Let B^z be the cube with the center at z and $\{\sigma_k, \tau_k\}_{k=1}^\infty$ be the same Markov moments as above. As in Proposition 1 but with ii) taken in to account, we get

$$\exp\left(-\frac{\overline{U}_0+\gamma}{\mu}\right) \leqslant \mathbf{P}\{\xi^\mu_{\tau_1} \in \partial Q^j_{\rho_1}\} \leqslant \exp\left(-\frac{\overline{U}_0-\gamma}{\mu}\right),$$

where $U_0 = \min_{\partial I^n} U - U(0)$. From ii) we know that $\min_{\partial I^n} U$ is attained on the set $\{y \in \partial I^n, y_1 = -1/2\}$. From hypothesis i), ii) and (Wentzell and Freidlin, 1979, Lemma 2.1) it follows that for some $\overline{U}_1 > \overline{U}_0$, for the probability of exit in any other direction we have

$$\exp\left(-\frac{\overline{U}_1+\gamma}{\mu}\right) \leqslant \mathbf{P}\left\{\xi^{\mu}_{\tau_1} \in \partial B^z \cup \bigcup_{k\neq z,j} \partial Q^k_{\rho_1}\right\} \leqslant \exp\left(-\frac{\overline{U}_1-\gamma}{\mu}\right).$$

To prove the estimate of Proposition 2 we'll represent the event

$$\overline{C}^{\mu} = \left\{\xi^{\mu}_{\tau} \in \partial B^z \cup \bigcup_{k\neq j} \partial Q^k_{\rho_1}\right\}$$

opposite to C^{μ} in the form:

$$\overline{C}^{\mu} = \left\{\xi^{\mu}_{\tau_1} \in \partial B^z \cup \bigcup_{k\neq z,j} \partial Q^k_{\rho_1}\right\} \cup$$

$$\cup \bigcup_{m=1}^{\infty}\left(\bigcap_{d=1}^{m}\{\xi^{\mu}_{\tau_d} \in \partial Q^k_{\rho_1}\} \cap \left\{\xi^{\mu}_{\tau_{m+1}} \in \partial B^z \cup \bigcup_{k\neq z,j} \partial Q^k_{\rho_1}\right\}\right).$$

Taking into account the strong Markov property we find that

$$\mathbf{P}(\overline{C}^{\mu}) \leqslant \exp\left(-\frac{\overline{U}_1-\gamma}{\mu}\right)\left(1+\sum_{m=1}^{\infty}\left(1-\exp\left(-\frac{\overline{U}_0+\gamma}{\mu}\right)\right)^m\right) \leqslant$$

$$\leqslant c\exp\left(-\frac{\overline{U}_1-\overline{U}_0-2\gamma}{\mu}\right).$$

Proposition 2 is proved.

COROLLARY. *For any $z : z_1 = 0$ the solution u_ε of the problem* (3) *satisfies the relation:*

$$u_\varepsilon(\varepsilon z) = u_\varepsilon(\varepsilon j) + O(e^{-\beta/\mu}), \quad j = (-1, z_2, \dots, z_n). \tag{8}$$

The question of boundary values of the solution u_ε can be treated in the same way as in the proof of the Theorem 3 (cf. (Kozlov and Pyatnitsky, 1990, Propositions 7, 8). The cited result states that

$$u_\varepsilon(\varepsilon z) = \varphi(x(\varepsilon z)) + o(1), \quad \rho(\varepsilon z, \partial\Omega) < 2\sqrt{n}\varepsilon, \tag{9}$$

where $x(\varepsilon z)$ is the point on the boundary which is the nearest to εz.
Equations (7)—(9) are the standard difference scheme for the following problem:

$$\begin{gathered}\Delta u_0^- = 0, \quad x \in \Omega \cap \{x_1 < 0\},\\ \left.\frac{\partial u_0}{\partial x_1}\right|_{x_1=0} = 0, \quad u_0^-\Big|_{\partial\Omega\cap\{x_1<0\}} = \varphi(x).\end{gathered} \tag{10}$$

Let us find the limit of u_ε in the domain $\Omega \cap \{x_1 > 0\}$. Equations (7), (9) derived above hold also true in this part of Ω. So we must only obtain the boundary condition on the contact plane $\{x_1 = 0\}$. From the convergence of u_ε to the solution of (10) which was proved above, we know that

$$\lim_{\varepsilon\to 0}|u_\varepsilon(\varepsilon z) - u_0^-(\varepsilon z)| = 0,\ z_1 = 0.$$

Now from Proposition 1 for all $z \in \{z \in \mathbb{Z}^n : z_1 = 1\}$ we have

$$u_\varepsilon(\varepsilon z) = \frac{1}{2n}\left(\sum_{k\neq z,\,|k-z|=1} u_\varepsilon(\varepsilon k) + u_0^-(0, \varepsilon z_2, \ldots, \varepsilon z_n)\right) + o(1). \tag{11}$$

This equation is equivalent to the Dirichlet boundary condition $u_\varepsilon\Big|_{x_1=0} = u_0^-\Big|_{x_1=0}$. The equations (7), (9), (11) are constitute a difference scheme for the Dirichlet problem:

$$\Delta u_0^+ = 0, \quad x \in \Omega \cap \{x_1 > 0\},$$

$$u_0^+\Big|_{\partial\Omega\cap\{x_1>0\}} = \varphi(x), \quad u_0^+\Big|_{x_1=0} = u_0^-\Big|_{x_1=0}.$$

So by a rather standard reference to the convergence theorem for difference scheme (cf. (Samarskii, 1983)) we obtain the result requested. Theorem 4 is proved.

The problem under consideration can be treated as a diffusion of charged particles which are attracted or repelled at the layer of the contact, according to the sign of their charge. Until now, we have discussed the case when the potential in the contact layer is some sort of average between the potentials of the contacting media.

Now we consider the opposite situation when the properties of the contact boundary differ essentially from the properties of the contacting medias. This situation is interpreted as repelling or trapping of the particle by the contact layer depending in its charge. These cases are treated mathematically with the same methods as above. We'll restrict ourselves with the formulation of the properties of the contact potential which lead to the picture of diffusion described above.

THEOREM 5. *Suppose that* A—D *and* i) *are satisfied. Assume, moreover,*
ii′) *points of the minimum of U in ∂I^n have the absolute value of the first coordinate less than 1/2.*
Then u_ε tends to the solution of the following problem:

$$u_0(x) = \begin{cases} u_0^+(x), & x_1 \geqslant 0, \\ u_0^-(x), & x_1 \leqslant 0, \end{cases}$$

$$\Delta u_0^+ = 0, \quad x \in \Omega \cap \{x_1 > 0\},$$
$$\Delta u_0^- = 0, \quad x \in \Omega \cap \{x_1 < 0\},$$

$$u_0\Big|_{\partial\Omega} = \varphi(x), \quad u_0\Big|_{x_1=0} = u_0^0(x),$$

$$\sum_{i=2}^n \frac{\partial^2}{\partial x_i^2} u_0^0(x) = 0, \quad u_0^0\Big|_{\partial\Omega\cap\{x_1=0\}} = \varphi\Big|_{x_1=0}.$$

THEOREM 6. *Suppose that condition* A—D *are satisfied. If the minimum of the potential* U *in the set* $\{y \in \partial I^n : y_1 = \pm 1/2\}$ *is less than the minimum of* U *in* $\{y_1 = 0\} \cup \{y \in \partial I^n : |y_1| < 1/2\}$, *then* u_ε *tends to the solution of the following problem:*

$$u_0(x) = \begin{cases} u_0^+(x), & x_1 > 0, \\ u_0^-(x), & x_1 < 0, \end{cases}$$

$$\Delta u_0^+ = 0, \quad x \in \Omega \cap \{x_1 > 0\},$$

$$\Delta u_0^- = 0, \quad x \in \Omega \cap \{x_1 < 0\},$$

$$u_0\Big|_{\partial\Omega} = \varphi(x), \quad \frac{\partial}{\partial x_1} u_0^\pm\Big|_{x_1=0} = 0.$$

So, in the case of Theorem 5, it is necessary first to find the solution of the $(n-1)$-dimensional Dirichlet problem on $\Omega \cap \{x_1 = 0\}$, and then using the found value of u_0 on $\{x_1 = 0\}$, to solve the equations in $\Omega \cap \{x_1 \gtrless 0\}$. In the situation described in Theorem 6, there are two independent limiting equations in $\Omega \cap \{x_1 \gtrless 0\}$. Hence the limiting function $u_0(x)$ is in general discontinuous on $\{x_1 = 0\}$.

REFERENCES

Kozlov, S.M. and Pyatnitsky, A.L. (1990). Homogenization and vanishing viscosity. *Sov. Math. Sb.* (to appear).

Kozlov, S.M. (1989). Geometric aspects of homogenization. *Russian Math. Surveys* **44**, 79–120.

Wentzell, A.D. and Freidlin, M.I. (1979). *Fluctuations in Dynamic Systems with Small Random Perturbations.* Nauka, Moscow.

Samarskii, A.A. (1983). *The Theory of Difference Schemes.* Nauka, Moscow.

Prob. Theory and Math. Stat., Vol. 2, pp. 340–348
B.Grigelionis et al. (Eds.)
1990 VSP/Mokslas

SOME REMARKS ON ASYMPTOTIC EFFICIENCY IN BAHADUR SENSE OF MAXIMUM LIKELIHOOD ESTIMATOR

M. RADAVIČIUS
Institute of Mathematics and Cybernetics, Akademijos 4, 232600 Vilnius, Lithuania

ABSTRACT

An example of regular experiment (Ibragimov and Khas'minskii, 1981) for which maximum likelihood estimator is not efficient in the Bahadur sense is given and a general new type condition of its efficiency is formulated.

1. INTRODUCTION

Let $X_1, X_2, \ldots, X_n$ be i.i.d. random variables with common distribution density $f(\cdot;\theta) = d\mathbf{P}_0/d\mu$ in the measurable space $(\mathfrak{X}, \mathfrak{A}, \mu)$, where μ is some σ–finite measure. The unknown parameter θ belongs to the open, convex, and bounded set $\Theta \subset \mathbf{R}^d$. Let us suppose that the Fisher information matrix $I(\theta)$ exists and is positive defined.

Let us denote by $r(\theta)$ and $R(\theta)$ the lower bound and the upper bound of the (local) exponential rate of convergence of an estimator $\hat{\theta}$ respectively:

$$r(\theta) = \liminf_{\tau\to 0} \liminf_{n\to\infty} \Big(-\frac{2}{n\tau^2} \log \mathbf{P}_\theta \Big\{ |I^{1/2}(\theta)(\hat{\theta}-\theta)| \geqslant \tau \Big\} \Big),$$

$$R(\theta) = \limsup_{\tau\to 0} \limsup_{n\to\infty} \Big(-\frac{2}{n\tau^2} \log \mathbf{P}_\theta \Big\{ |I^{1/2}(\theta)(\hat{\theta}-\theta)| \geqslant \tau \Big\} \Big).$$

Bahadur suggested (Bahadur, 1960) to use the exponential rate as a criterion of asymptotic efficiency of estimators. This suggestion is based on so–called Bahadur inequality

$$R(\theta) \leqslant 1, \tag{1.1}$$

which is true for all consistent estimators under quite general condition (Bahadur, 1967; 1971). An estimator $\hat{\theta}$ is said to be efficient in the Bahadur sense if $r(\theta) = 1$ (c.f. (Bahadur, 1960; Fu, 1982)).

Bahadur in (Bahadur, 1967; 1971) also considered the efficiency of the maximum likelihood estimator (MLE) and proved it to be efficient under rather restrictive conditions. These results were further developed in (Fu 1973; 1975; 1984) and (Radavičius 1981; 1983)). The exponential rate for other estimators were investigated in the papers (Fu, 1975, Sievers, 1978, Rubin and Rukhin, 1983). Let us note the paper (Ibragimov and Khas'minskii, 1981) in which the MLE efficiency in the Bahadur sense have been shown without any restrictions on a moment generating function of derivative of the density logarithm. Nevertheless, a large gap remains between the conditions insuring the MLE efficiency and those which are sufficient for the exponential consistency of MLE (see (2.1) below). An attempt to fill this gap was made in (Bai and Fu, 1987), who proved the asymptotic efficiency of MLE for the location parameter of a Cauchy distribution.

Our treatment of this problem is based on the approach of (Ibragimov and Khas'minskii, 1981). It means that we suppose the standard conditions of the regular experiments to be fulfilled (see, Section 2). Despite the weakness of these conditions they insure the following MLE properties:

(a) The Bahadur inequality (1.1) holds for MLE without any additional restriction (Radavičius, 1983),

(b) MLE converges exponentially (see (2.1)),

(c) MLE is asymptotically normal and asymptotically efficient in the minimax sense (Ibragimov and Khas'minskii, 1981).

Are these conditions sufficient for the asymptotic efficiency of MLE in the Bahadur sense as well? In this paper we give a counter–example and propose a new type condition of the MLE efficiency. The results are formulated in the next section. The proofs are carried over to the Sections 3 and 4.

2. RESULTS

Everywhere below the following regularity conditions are supposed to be satisfied (see (Ibragimov and Khas'minskii, 1981, Chapter 3, §3)):

(R1) The function $g(x;\theta) = f^{1/2}(x;\theta)$ is continuous on Θ for almost all x with respect to the measure μ and continuously differentiable in $L_2(\mu)$, moreover,

$$0 < c \leqslant \inf_{|u|=1} \big(I(\theta)u, u\big) \leqslant \sup_{|u|=1} \big(I(\theta)u, u\big) \leqslant C < \infty,$$

where

$$I(\theta) = 4\int \frac{\partial}{\partial\theta} g(\theta) \Big[\frac{\partial}{\partial\theta} g(\theta)\Big]^T d\mu$$

is the Fisher information matrix.

(R2) For each $\delta > 0$ there exists $\varepsilon > 0$ such that

$$\int |g(\theta) - g(u)|^2 d\mu \geqslant \varepsilon$$

for all

$$u, \theta \in \Theta,\ |u - \Theta| \geqslant \delta.$$

(R3) For some $\nu > d$

$$\mathbf{E}_\theta \Big|\frac{\partial}{\partial\theta} \log f(X_1;\theta)\Big|^\nu \leqslant C < \infty.$$

Then it is known that MLE $\hat{\theta}$ is asymptotically normal, asymptotically efficient in the minimax sense (see (Ibragimov and Khas'minskii, 1981, Chapter 3, §3)) and converges exponentially, i.e., there exist positive constants δ_0, c_0, and C_0 such that inequality

$$\mathbf{P}_0\{|\hat{\theta}-\theta|>t\} \leqslant C_0 e^{-c_0 t^2} \tag{2.1}$$

holds for all $t>0$ and all $\theta \in \Theta$, $|\theta-\theta_0| \leqslant \delta_0$ (see (Ibragimov and Khas'minskii, 1981, Chapter 1, §§5, 8)).
The following example shows that conditions (R1)—(R3) are not sufficient for the MLE efficiency in the Bahadur sense.

PROPOSITION. *Let $\alpha>1$ and $\beta>1$ be some constants and $f(x;\theta)=f_0(x-\theta)$, where*

$$f_0(x)=\begin{cases} c(\alpha,\beta)(1-|x|)^{\alpha}\exp\{-|x|^{-\beta}\}, & \text{if } |x|<1,\\ 0, & \text{otherwise.}\end{cases}$$

Then conditions (R1)—(R3) *are fulfilled, but $R(\theta) \leqslant 0.25$.*

Thus, some additional condition for the MLE efficiency is needed. To formulate it we introduce some notations. Let

$$\eta_\theta(v,u)=f(X_1;\theta+I^{-1/2}(\theta)(v+u))/f(X_1;\theta+I^{-1/2}(\theta)v),$$

$$B_\varepsilon=\{u\in\Theta-\theta\colon |u|<\varepsilon\},$$

$$\lambda(v,u)=(u,v)/|u|^2,$$

and let $\chi(A)$ be a indicator function of an event A.

THEOREM. *Let us assume that there exist some positive constants ε, C, and $\gamma<1$, and a positive monotone function $\varrho(s)$, $s>0$, tending to zero as $s\to 0$, such that inequality*

$$\mathbf{E}_\theta(\eta_\theta(v,u)^{\lambda(v,u)}\chi(\log\eta_\theta(v,u)>\varrho(|v|)) \leqslant C|v|^2\varrho(|v|) \tag{2.2}$$

holds for all $v\in B_\varepsilon$ and all $u\in H_v$, where H_v is given by

$$H_v=\{u\in\Theta-\theta-v\colon (v,u)\geqslant\gamma|v||u|,\ |v|\varrho(|v|)\leqslant|u|\leqslant\varepsilon|v|\}.$$

Then MLE $\hat{\theta}$ is efficient in the Bahadur sense.

When $d=1$ the inequality (2.2) is easy to prove for any logconcave density function. Indeed, we have

$$\eta_\theta(v,u)^{v/u}f(X_1;\theta)\leqslant f(X_1;\theta+v/\sqrt{I(\theta)}),$$

provided $vu>0$. Moreover, any density function satisfying

$$\sup_{u\in B_\varepsilon}\mathbf{E}_\theta\exp\left\{\varkappa\left|\frac{\partial}{\partial\theta}\log f(X_1;\theta+u)\right|\right\}\leqslant C$$

with some positive ε, $\varkappa$, and C satisfies (2.2) too. The following corollary contains a more interesting result.

COROLLARY. *If $f(\cdot;u)/f(\cdot;v) \leqslant C$ μ–almost everywhere for some positive ε and all $u, v \in \theta + B_\varepsilon$, then* MLE *is efficient in the Bahadur sense.*

A Cauchy distribution with unknown location parameter obviously satisfies the corollary condition.

3. PROOF OF THE PROPOSITION

The proof is rather technical. Therefore only a sketch of the proof will be given. Suppose $I = I(\theta)$ and

$$\mathcal{E} = \{ \sup_{|v|<\tau} Z_n(v) \leqslant \sup_{\tau\leqslant|u|<2\tau} Z_n(u)\},$$

$$\mathcal{E}_1 = \{ \sup_{|v|<2\tau} (Z_n(v)/\zeta_j(v)) < \exp\{\nu\tau^2 n\} \quad \text{for all } j = 1,\ldots,n\}.$$

Here

$$Z_n(v) = \prod_{j=1}^n \zeta_j(v), \ \zeta_j(v) = f_0(X_j - v)/f_0(X_j), \ j = 1,\ldots,n.$$

From Lemma 3.3.2 and Theorem 1.A.19 in (Ibragimov and Khas'minskii, 1981) we conclude that

$$\mathbf{E}_0 \sup_{|v|<2} Z_n^{1/\nu}(\tau v) \leqslant C\tau\sqrt{n}$$

holds for some constant $C > 0$. Consequently,

$$\mathbf{P}_0(\mathcal{E}_1^c) \leqslant Cn^{3/2}\tau\epsilon^{-\tau^2 n}. \tag{3.1}$$

Given $0 < \tau_0 < \tau < \tau_1$ and $h > 0$ we denote $t_k = \tau_0 + kh$, $k = 0,1,\ldots,m$, $m = [(\tau - \tau_0)/h] + 1$ and $\Delta_k = [t_{k-1}, t_k)$, $k = 1,\ldots,m$, $\Delta = [-\tau_1, \tau_1]$. Let us introduce the following events

$$A = \bigcap_{k=1}^m \Big(\Big\{X_{2k-1} \in \Delta_k\Big\} \bigcap \Big\{-X_{2k} \in \Delta_k\Big\}\Big), \quad B = \bigcap_{k=2m+1}^n \Big\{X_k \notin \Delta\Big\}.$$

Let $h = (2\nu n\tau^2)^{-1/\beta}$ and $\tau_0 = o(\tau)$, $h = o(\tau_0)$ when $n \to \infty$ and $\tau \to 0$. Then for sufficiently large n and small τ the event $\{X_1 \in \Delta_k\}$ implies $\zeta_1(v) \leqslant \exp\{-\nu\tau^2 n\}$ for all $v \in \Delta_k$. Hence

$$\mathcal{E} \bigcap \mathcal{E}_1 \bigcap A = \mathcal{E}_0 \bigcap \mathcal{E}_1 \bigcap A, \tag{3.2}$$

where $\mathcal{E}_0$ is given by

$$\mathcal{E}_0 = \{ \sup_{|v|<\tau_0} Z_n(v) \leqslant \sup_{1\leqslant|u|<2} Z_n(\tau u)\}.$$

It follows from (3.1) and (3.2) that

$$\mathbf{P}_0\{\,|\,\hat\theta\,|\geqslant \tau\} \geqslant \frac{n!}{(n-2m)!}\mathbf{P}_0\Big(\mathcal{E}_0\bigcap A\bigcap B\Big) - Cn^{3/2}\tau e^{-\tau^2 n}. \tag{3.3}$$

The likelihood ratio $Z_n(v)$ may be represented as the product $Z_n^{(1)}(v)Z_n^{(2)}(v)$ where

$$Z_n^{(1)}(v) = \prod_{j=1}^{2m}\zeta_j(v),\ Z_n^{(2)}(v) = \prod_{j=2m+1}^{n}\zeta_j(v).$$

Choosing $v_0 = \tau + \tau_0 + 2h$, we have

$$A\bigcap\mathcal{E}_0 \supset \Big\{\sup_{|v_1|<\tau_0} Z_n^{(1)}(v_1)\sup_{|v_2|<\tau_0} Z_n^{(2)}(v_2) \leqslant Z_n(v_0)\Big\}\bigcap A \supset$$

$$\supset\Big\{\sup_{|v_2|<\tau_0} Z_n^{(2)}(v_2) \leqslant Z_n^{(2)}(v_0)\Big\}\bigcap A, \tag{3.4}$$

because $\sup\limits_{|v_1|<\tau_0} Z_n^{(1)}(v_1) < Z_n^{(1)}(v_0)$.

Further, let us suppose the event B takes place. Then $Z_n^{(2)}(u)$ is strictly log–concave on Δ. Therefore,

$$\Big\{\sup_{|v|<\tau_0} Z_n^{(2)}(v) \leqslant Z_n^{(2)}(v_0)\Big\}\bigcap B = \{Z_n^{(2)}(\tau_0) \leqslant Z_n^{(2)}(v_0)\}\bigcap B. \tag{3.5}$$

From (3.4) and (3.5) we obtain that

$$\mathbf{P}_0(\mathcal{E}_0\bigcap A\bigcap B) \geqslant \prod_{k=1}^{m} p_k^2\mathbf{P}_0\Big(\{Z_n^{(2)}(\tau_0) \leqslant Z_n^{(2)}(v_0)\}\bigcap B\Big), \tag{3.6}$$

where p_k denotes $\mathbf{P}_0\{X_1 \in \Delta_k\}$, $k = 1,\ldots,m$. Similarly as in the proof of Lemma 1, one can easy verify that

$$\frac{1}{n-2m}\log\mathbf{P}_0\Big(\{Z_n^{(2)}(\tau_0) \leqslant Z_n^{(2)}(v_0)\}\bigcap B\Big) = -I\tau^2/8 + o(\tau^2), \tag{3.7}$$

provided $\tau_0 = o(\tau)$, $\tau_1 = (3|\log\tau|)^{-1/\beta}$, and $v_0 = \tau + o(\tau)$ as $\tau \to 0$ (see also (Bahadur, 1971)). It remains to check that $m = o(n)$ and

$$S = \frac{1}{n}\log\left(\frac{n!}{(n-2m)!}\prod_{k=1}^{m} p_k^2\right) = o(\tau^2),\ n\to\infty,\ \tau\to 0.$$

Indeed, insertion of these relations together with (3.6) and (3.7) into (3.3) gives the desired result. Direct computations show that

$$\frac{m}{n} \leqslant \frac{\tau}{nh} = O\big(\tau^{(\beta+2)/\beta}n^{(1-\beta)/\beta}\big) = o(1)$$

and

$$| S | \leqslant C\Big(\frac{\tau}{nh}(1+ | \log nh |) + \frac{\tau_0^{1-\beta}}{nh} + \frac{1}{n^2}\Big) = o(\tau^2),$$

provided $h = \left(2\nu n\tau^2\right)^{-1/\beta} = o(\tau_0)$ and $\tau_0 = o(\tau)$ as $n \to \infty$ and $\tau \to 0$. The proof is completed.

4. PROOF OF THE THEOREM

We start with auxiliary lemmas. Define the rescaled likelihood process

$$Z_n(u) = \prod_{j=1}^{n} \zeta_j(u), \;\; \zeta_j(u) = f(X_j; \theta + I^{-1/2}(\theta)u)/f(X_j;\theta), \;\; j = 1, \ldots, n.$$

LEMMA 1. *Let $v \in B_\varepsilon$, $u \in H_v$ and $a \in \mathbf{R}^1$. Then for all positive $t \leqslant \lambda(v,u)$ and all integer $n > 0$*

$$\frac{1}{n} \log \mathbf{P}_\theta \Big\{ Z_n(u+v) \geqslant Z_n(v) e^{a|v|^2 n} \Big\} \leqslant$$

$$\leqslant -\big((u,v) + a|v|^2\big)t + |u|^2 \frac{t^2}{2} + (1+t^2)|v|^2 \varrho_1(|v|),$$

where $\varrho_1(s)$ is some positive and monotone function, tending to zero as $s \to 0$.

Proof. Let $\varrho_0(s)$ be some function, tending to zero as $s \to 0$ and satisfying $\varrho_0(s) \geqslant \varrho(s)$. Let us denote

$$A_v = \big\{ \,|\log \zeta_1(v)| \leqslant \varrho_0(|v|)\big\}.$$

Similarly as in the proof of local asymptotic normality (see (Ibragimov and Khas'-minskii, 1981, Theorem 2.1.1)) we obtain, that if $v \to 0$, $u = O(v)$, then

$$\mathbf{P}_\theta(A_v) = 1 + o(|v|^2), \tag{4.1}$$

$$\mathbf{E}_\theta \log \zeta_1(v) \chi(A_v) = -\frac{1}{2} \,|\, v\,|^2 + o(|v|^2), \tag{4.2}$$

$$\mathbf{E}_\theta \big(\log \eta_\theta(v,u)\big)^2 \chi\big(A_v \bigcap A_{v+u}\big) = |\, u\,|^2 + o(|v|^2), \tag{4.3}$$

provided ϱ_0 tends to zero sufficiently slowly. Because of (4.1) and condition (2.2) for all $v \in B_\varepsilon$ and $u \in H_v$ and all positive $t \leqslant \lambda(v,u)$ the inequality

$$\mathbf{E}_\theta |\log \eta_\theta(v,u)|^3 \eta_\theta^t(v,u) \chi\big(A_v \bigcap A_{v+u}\big) \leqslant C \varrho_0(|v|)|v|^2 \tag{4.4}$$

holds. Applying (2.2) and (4.1) once more, we get, as $v \to 0$,

$$\frac{1}{n} \log \mathbf{P}_\theta \Big\{ Z_n(v+u) \geqslant Z_n(v) e^{a|v|^2 n} \Big\} \leqslant$$

$$\leqslant \log\Big(\mathbf{E}_\theta \eta_\theta^t(v,u)\chi(A_v \bigcap A_{v+u})\Big) - a|v|^2 t + o(|v|^2). \tag{4.5}$$

To complete the prove we expand the exponent in (4.5) up to the third term and insert in it relations (4.1)—(4.4).

LEMMA 2. *Let $\delta(s) = \max(2\varrho(s),\ \varrho_1^{1/2}(s))$ and let $a(s)$ be some positive and monotone function, $a(s) \to 0$ and $a(s)/\delta(2s) \to \infty$ as $s \to 0$. If $D_w \subset \Theta - \theta$ is any class of sets with $w \in D_w$ and diameter $h = o(\delta(|w|)|w|)$, $w \to 0$, then for $v = (1 - \delta(|w|)w$ there exist constants $c_1 > 0$ and $C_1 > 0$, depending only on d and ν, such that*

$$\mathbf{P}_\theta\Big\{\sup_{y\in D_w} Z_n(y) > \max\Big(1, Z_n(v)e^{a(|v|)|v|^2 n}\Big)\Big\} \leqslant$$

$$\leqslant C_1 \exp\Big\{-c_1 n|v|^2 \frac{a(|v|)}{\delta(2|v|)}\Big\}.$$

Proof. If $|w|$ is sufficiently small $u = y - v \in H_v$ for all $y \in D_w$. Applying Lemma 1 with $t = \delta^{-1}(2|v|)$ yields for small $|v|$

$$\mathbf{P}_\theta\Big\{Z_n(y) > Z_n(v)e^{a(|v|)|v|^2 n}\Big\} \leqslant \exp\Big\{-\frac{a(|v|)}{2\delta(2|v|)}|v|^2 n\Big\}. \tag{4.6}$$

The rest of the proof is similar to that of Theorem 1.5.1 in (Ibragimov and Khas'-minskii, 1981), except that instead of inequality (1.5.2) used in the Theorem 1.5.1 the stronger inequality (4.6) must be used. We omit the details.

Proof of the Theorem. Let us take C_2 so large to satisfy

$$\mathbf{P}_\theta\{|I^{1/2}(\theta)(\hat\theta - \theta)| > C_2\tau\} \leqslant C_0 \exp\{-\tau^2 n\}. \tag{4.7}$$

It is possible in view of (2.1). Let $h = o(\tau\delta(\tau))$ as $\tau \to 0$, and let Δ_{kj}, $k = 1, \dots, m$, $j = 1, \dots, m$, be a partition of

$$D = B_{C_2\tau}\backslash B_\tau = \{v \in \Theta - \theta;\ \tau \leqslant |v| < C_2\tau\}$$

with the maximal diameter less then h and such that $m = \big[(C_2 - 1)\tau/h\big] + 1$, $m_k \leqslant C(k + \tau/h)^{d-1}$, and

$$\Delta_k = \{v \in \Theta - \theta\colon \tau_{k-1} \leqslant |v| < \tau_k\} = \bigcup_{j=1}^{m_k} \Delta_{kj},$$

where $\tau_k = \tau + kh$, $k = 1, \dots, m$. Let δ and a be the same as in Lemma 2. Given $w_{kj} \in \Delta_{kj}$ we take $v_{kj} = (1 - \delta(|w_{kj}|))w_{kj}$. Then

$$\mathbf{P}_\theta\Big\{\tau \leqslant |I^{1/2}(\theta)(\hat\theta - \theta)| < C_2\tau\Big\} \leqslant \mathbf{P}_\theta\Big\{\sup_{v\in D_0} Z_n(v) \leqslant \sup_{u\in D} Z_n(u)\Big\} \leqslant$$

$$\leqslant \sum_{k=1}^{m}\sum_{j=1}^{m_k}\Big(\mathbf{P}_\theta\Big\{\sup_{v\in D_{k-1}} Z_n(v)\leqslant Z_n(v_{kj})e^{a(\tau_k)\tau_k^2 n}\Big\}+$$

$$+\mathbf{P}\Big\{\sup_{u\in\Delta_{kj}} Z_n(u)>\max\Big(1, Z_n(v_{kj})e^{a(\tau_k)\tau_k^2 n}\Big)\Big\}\Big).$$

Here D_k denotes $D_0\bigcup(\bigcup_{l=1}^{k}\Delta_l)$, $k=1,\dots,m$, and $D_0=\{v\in\Theta-\theta\colon |v|<\tau\}$. According to Lemma 2 the last term for little τ does not exceed $C_1\exp\{-\tau_k^2\}$. Let us consider the first term. Let $\delta_1(s)$ be positive function tending to zero as $s\to 0$ slower then $a(s/2)$. Taking $u_{kj}=(1-\delta_1(\tau_k))v_{kj}$ and $t=(1-\delta_1(\tau_k))/\delta_1(\tau_k)$ we obtain by Lemma 1

$$\mathbf{P}_\theta\{Z_n(u_{kj})\leqslant Z_n(v_{kj})e^{a(\tau_k)\tau_k^2 n}\}\leqslant$$

$$\leqslant C\exp\Big\{-\frac{\tau_k^2 n}{2}\Big(1+O\Big(\frac{a(\tau_k)}{\delta_1(\tau_k)}+\frac{\varrho_1(\tau_k/2)}{\delta_1^2(\tau_k)}+\delta_1\Big(\frac{\tau_k}{2}\Big)\Big)\Big)\Big\}.$$

Thus,

$$\mathbf{P}_\theta\{\tau\leqslant |I^{1/2}(\theta)(\hat\theta-\theta)|<C_2\tau\}\leqslant C\Big(\frac{\tau}{h}\Big)^d\exp\Big\{-\frac{\tau^2 n}{2}\Big(1+o(1)\Big)\Big\}$$

and this inequality together with (4.7) completes the proof.

Proof of the Corollary. If ϱ_0 tends to zero sufficiently slowly we have by (4.1)

$$\mathbf{E}_\theta\big(\eta_\theta(v,u)\big)^{\lambda(v,u)}\chi\{\log\eta_\theta(v,u)>\varrho_0(|v|)\}\leqslant$$

$$\leqslant C^{\lambda(v,u)}\mathbf{P}_\theta\{\log\eta_\theta(v,u)>\varrho_0(|v|)\}\leqslant C^{\lambda(v,u)}|v|^2\varrho_2(|v|),$$

where $\varrho_2(s)\to 0$ as $s\to 0$. Now it is sufficient to assume

$$\varrho=\max\big(\varrho_0, 2/\log(1/\varrho_2)\big).$$

REFERENCES

Bahadur, R.R. (1960). On asymptotic efficiency of tests and estimates. *Sankhya* **22**, 229–252.

Bahadur, R.R. (1967). Rates of convergence of estimates and test statistics. *Ann. Math. Stat.* **38**, 303–324.

Bahadur, R.R. (1971). *Some Limit Theorems in Statistics.* SIAM, Philadelphia.

Bai, Z.D. and Fu, J.C. (1987). On the maximum likelihood estimator for the location parameter of a Cauchy distribution. *Canadian J.Statist.* **15**, 137–146.

Fu, J.C. (1973). On the theorem of Bahadur on the rate of convergence of point estimator. *Ann. Statist.* **1**, 745–749.

Fu, J.C. (1975). The rate of convergence of consistent points estimators. *Ann. Statist.* **3**, 234–242.

Fu, J.C. (1982). Large sample point estimation: A large deviation approach. *Ann. Statist.* **10**, 762–771.

Fu, J.C. (1984). A note on the interpretation of the Bahadur bound and the rate of convergence of the maximum likelihood estimator. *Statist. Probab. Lett.* **2**, 269–273.

Ibragimov, I.A. and Khas'minskii, R.Z. (1981). *Statistical Estimation: Asymptotic Theory.* Springer, Berlin–New York.

Radavičius, M. (1981). On probabilities of large and moderate deviations of maximum likelihood estimators. *Notes Sci. Seminars Leningrad Branch of Math. Inst.* **108**, 154–169.

Radavičius, M. (1983). On the probabilities of large deviations of maximum likelihood estimators. *Soviet Math. Dokl. Akad. Nauk SSSR* **27**, 268.

Rubin, H. and Rukhin, A.L. (1983). Convergence rates of large deviation probabilities for point estimators. *Statist. Probab. Lett.* **1**, 197–202.

Sievers, G. (1978). Estimators of location: A large deviations comparison. *Ann. Statist.* **6**, 610–618.

Prob. Theory and Math. Stat., Vol. 2, pp. 349–363
B.Grigelionis *et al.* (Eds.)
1990 VSP/Mokslas

ON DISTRIBUTION OF SUPREMUM TYPE FUNCTIONALS OF NONPARAMETRIC ESTIMATES FOR SPECTRAL DENSITY

R. RUDZKIS
Institute of Mathematics and Cybernetics,
Akademijos 4, 232600 Vilnius, Lithuania

ABSTRACT

Kernel type estimates of spectral density of a stationary sequence are considered. Accuracy of the estimates is characterized by a supremum type norm of the error. The asymptotics of the distribution of the norm is investigated when the number of observations tends to infinity.

1. INTRODUCTION

Let X_t be a zero mean stationary in widesense sequence with the spectral density f and let $\hat{f}$ be a nonparametric estimate based on the sample $X_1, \ldots, X_N$, $N \to \infty$. In this paper the asymptotic behaviour of

$$P(u) = \mathbf{P}\Big\{ \sup_{x \in K} q(x) \mid f(x) - \hat{f}(x) \mid < u \Big\}$$

is considered. Here q is some fixed positive function, $K = [x_1, x_2]$ is some fixed interval and $0 \leqslant x_1 < x_2 \leqslant \pi$. Define

$$\hat{f}(x) = \frac{1}{2\pi N} \int_{-\pi}^{\pi} W(x - \lambda) \Big| \sum_{t=1}^{N} X_t e^{it\lambda} \Big|^2 d\lambda$$

where W is some kernel function. In (Rudzkis, 1989a) the case $W(x) = \psi(x/h)/h$, where ψ is some finite function independent of N and $h = h(N) \to \infty$, is studied. In (Rudzkis, 1989a) the similar results for probability density estimates are presented too. The proofs of the results will appear in "Theory of Probability and its Applications".

In the present paper we deal with another important case

$$W(x) = \sum_{k=-n}^{n} g(k/n) e^{ikx},$$

where g is some fixed even function, $g(x) = 0$ for $|x| > 1$, and $n = n(N) \to \infty$. Denoting empirical covariance function by

$$\hat{R}(k) = \frac{1}{N} \sum_{t=1}^{N-|k|} X_t X_{t+|k|}$$

we have

$$\hat{f}(x) = \sum_{k=-n}^{n} g(k/n)\hat{R}(k)e^{ikx}. \tag{1}$$

The method of proofs is based on theorems obtained in (Rudzkis, 1989b) which deal with probabilities of large excursions of stochastic processes close to the Gaussian ones. We apply these theorems to the empirical process ξ generated by statistics $\hat{f}$. To define this process we denote

$$\bar{f}(x) = (\hat{f}(x) - \mathbf{E}\hat{f}(x))/\mathbf{D}^{1/2}\hat{f}(x)$$

and

$$x^* = \int_{x_1}^{x} \mathbf{D}^{1/2}\bar{f}'(\lambda)\, d\lambda.$$

Then we define

$$\xi(x^*) = \bar{f}(x), \quad x \in K. \tag{2}$$

For all $y \in [0, x_2^*]$ we have $\mathbf{E}\xi(y) = 0$ and $\mathbf{D}\xi(y) = \mathbf{D}\xi'(y) = 1$. Further,

$$P(u) = \mathbf{P}\{-v_1(y) < \xi(y) < v_2(y),\ 0 \leqslant y \leqslant x_2^*\}, \tag{3}$$

where the functions $v_j(y) = v_j(y, u, N)$ are given by

$$v_j(x^*) = \Big(u/q(x) + (-1)^j(f(x) - \mathbf{E}\hat{f}(x))\Big)/\mathbf{D}^{1/2}\hat{f}(x), \quad x \in K.$$

Note that $x_2^* \asymp n \to \infty$ as $N \to \infty$. Because of (3) the investigation of $P(u)$ is reduced to the analysis of the probability that the trajectory of the process ξ lies between the curvilinear bounds in the expanding interval. Let

$$G(u) = \prod_{j=1}^{2} \exp\left\{-\frac{1}{2\pi}\int_0^{x_2^*} dy \int_0^{\infty} \exp\left\{-\left(v_j^2(y) + (v_j'(y) + z)^2\right)/2\right\} z\, dz\right\}. \tag{4}$$

In (Rudzkis, 1989b) sufficient conditions for

$$\sup_{u \geqslant 0} |P(u) - G(u)| \to 0, \quad N \to \infty,$$

are presented. The proofs of the results given below include verification of these conditions and some reduction of (4).

2. MAIN RESULTS

Suppose that $\hat{f}$ is defined by (1) and the following conditions are satisfied:
1) there exists the square integrable derivative of the function g;
2) $N^c/n \to \infty$ for some constant $c < 1$;
3) $\sup\limits_{x\in K} f(x) > 0$;
4) there exist some constants C_1 and $\gamma \geqslant 1$ such that for any integer τ cumulants of the sequence X_t satisfy the inequality

$$\sum_{t_1,\dots,t_k=-\infty}^{\infty} \left|\mathrm{cum}\left\{X_{t_1},\dots,X_{t_k},\ X_\tau\right\}\right| \leqslant C_1^k (k!)^\gamma, \quad k = 2,3,\dots \,. \tag{5}$$

Denote $\| g \|_2^2 = \int\limits_{-1}^{1} g^2(x)\,dx$, $\sigma(x) = 2\pi \| g \|_2\, f(x)\sqrt{n/N}$,

$$H = \frac{n}{\| g \|_2} \left(\int\limits_{-1}^{1} g^2(x)x^2\,dx\right)^{1/2}.$$

THEOREM 1. *Let for some positive* α *and* C_2

$$\sum_{k=1}^{\infty} k^\alpha |R(k)| < C_2 \tag{6}$$

where $R(k) = \mathbf{E}X_0X_k$. *Then there exists an independent of* N *number* $\nu > 0$ *such that*

$$\sup_z \left|\mathbf{P}\left\{\sup_{x\in K} \frac{(\hat{f}(x) - \mathbf{E}\hat{f}(x))^2}{2\sigma^2(x)} \leqslant \log\frac{H(x_2 - x_1)}{\pi} + z\right\} - \exp\{-e^{-z}\}\right| = O(n^{-\nu}). \tag{7}$$

We say that a function $\varphi \in \mathrm{Lip}_\alpha$ if

$$\sup_{x\neq\lambda} \frac{|\varphi(x) - \varphi(\lambda)|}{|x - \lambda|^\alpha} < \infty.$$

THEOREM 2. *Let there exist the* r*–th derivative* $f^{(r)} \in \mathrm{Lip}_\alpha$ *and let* $q \in \mathrm{Lip}_{\alpha_0}$, *where* $0 < \alpha, \alpha_0 \leqslant 1$. *Let for some positive number* x_0

$$g(x) = 1/2\pi, \quad |x| \leqslant x_0.$$

If $\beta = r + \alpha > 1$ *and* $n^{-1} = O(N^{-1/(2\beta+1)})$ *then*

$$\sup_{u\geqslant 0}\left|P(u) - \prod_{j=1}^{2} \exp\left\{-\frac{H}{2\pi}\int\limits_K \exp\left\{-\mu_j^2(x,u)/2\right\}dx\right\}\right| \to 0, \tag{8}$$

where

$$\mu_j(x,u) = \Big(u/q(x) + (-1)^j b(x)\Big)/\sigma(x), \quad b(x) = f(x) - \mathbf{E}\hat{f}(x). \tag{9}$$

Denote $l(a) = 1 + \int\limits_0^{|a|} (\mid a \mid - y) \exp\{-y^2/2\}\, dy$.

THEOREM 3. *If the conditions of Theorem 2 are satisfied and $\beta \geqslant 2$ then there exist positive constants C and ν such that*

$$\sup_{u \geqslant 0} \left| P(u) - \prod_{j=1}^{2} \exp\left\{ -\frac{H}{2\pi} \int\limits_K \exp\left\{ -\mu_j^2(x,u)/2 \right\} l\Big(b'(x)/H\sigma(x)\Big)\, dx \right\} \right| = O(n^{-\nu}) \tag{10}$$

provided $n > C(N/\log N)^{1/(2\beta+1)}$.

Note that under conditions of Theorem 2 (see, for example, Theorems 1.2 and 3.1 in (Bentkus and Rudzkis, 1982)).

$$\mathbf{E}\Big(\hat{f}(x) - f(x)\Big)^2 = O(n^{-2\beta} + n/N). \tag{11}$$

The convergence rate in (11) is optimal when $n \asymp N^{1/(2\beta+1)}$. Further, (11) and Theorem 2 in (Rudzkis, 1978) yields

$$\mathbf{E} \sup_{x \in K} |\hat{f}(x) - f(x)| = O\Big(n^{-\beta} + \sqrt{n \log n / N}\Big).$$

Therefore when the deviation is considered in the uniform metric then it is natural to choose $n \asymp (N/\log N)^{1/(2\beta+1)}$.

3. LEMMAS

Below θ denotes a complex number with modulus less than one and C and ν denote positive values independent of N and x, moreover the same letter in different places designates not necessarily the same value. For some complex sequence $w(k)$ define

$$A(w) = \sum_{k=-\infty}^{\infty} w(k)\hat{R}(k), \quad \| w \|_\infty = \sup_k |w(k)|, \quad \| w \|_p = \Big(\sum_{k=-\infty}^{\infty} |w(k)|^p \Big)^{1/p}.$$

LEMMA 1. *Suppose (5) is true. Then for any sequences $w_1(\cdot), w_2(\cdot), \dots$, and $m = 2, 3, \dots$*

$$\Big| \mathrm{cum}\Big\{ A(w_1), \dots, A(w_m) \Big\} \Big| \leqslant$$

$$\leqslant C_1^{2m} \left((2m)!\right)^{\gamma} \Big(\frac{4}{N}\Big)^{m-1} \max_{j \in \{1,\dots,m\}} \frac{\| w_j \|_\infty}{\| w_j \|_1} \prod_{i=1}^{m} \| w_i \|_1 . \tag{12}$$

Lemma 2. *Let (5) and (6) be satisfied. Then for any sequences $w_1(\cdot)$ and $w_2(\cdot)$*

$$N\operatorname{cov}\left\{A(w_1),A(w_2)\right\} = 2\pi\int_{-\pi}^{\pi}(W_1(x)+W_1(-x))W_2(x)f^2(x)\,dx+$$

$$+\theta C \parallel w_1 \parallel_\infty \parallel w_2 \parallel_\infty \left(1+N^{-\alpha}\sum_{j=1}^{2}\parallel w_j \parallel_1 / \parallel w_j \parallel_\infty\right), \tag{13}$$

where $W_j(x)=\sum_k w_j(k)e^{ikx}$, $C=C(C_1,C_2,\gamma)$.

Suppose $\hat f$ is defined by (1) and conditions 1)—4) and (6) are fulfilled. Given $\varkappa>0$ let $\tilde K = K\cap(n^{-\varkappa},\pi-n^{-\varkappa})$.

Lemma 3. *For all $x\in\tilde K$ and for some positive number $\nu=\nu(\varkappa)$*

$$|\mathbf{D}\hat f(x)-\sigma^2(x)|<Cn^{1-\nu}/N \tag{14}$$

and

$$\frac{dx^*}{dx}=H(1+\theta C/n^\nu). \tag{15}$$

Now consider some properties of the process ξ defined in (2). Write $K^*=$ $=\{x^*:x\in\tilde K\}$.

Lemma 4. *The inequality*

$$\sup_{y\in K^*}\mathbf{D}\xi''(y)<C$$

is valid.

Lemma 5. *For any $m=2,3,\dots,$ $j=0,\dots,m$ and all $y_1,\dots,y_m\in K^*$*

$$\left|\operatorname{cum}\left\{\xi(y_1),\dots,\xi(y_j),\xi'(y_{j+1}),\dots,\xi'(y_m)\right\}\right|\leqslant C^m(m!)^{2\gamma}\left(\sqrt{n/N}\right)^{m-2}. \tag{16}$$

Lemma 6. *For some non-increasing function ρ independent of N with restrictions $\rho(x)<1$ for $x>0$ and such that*

$$\exists c>0;\ \lim_{x\to\infty}x^c\rho(x)=0$$

the estimate

$$\left|\operatorname{cov}\left\{\xi(y),\ \xi(z)\right\}\right|\leqslant\rho(\,|\,z-y\,|\,)+C/n^\nu,\quad y,z\in K^*,$$

is valid.

LEMMA 7. *Let $b(x)$ be defined by (9). If functions f and g satisfy conditions of Theorem 2 then for all $k = 0, 1, \ldots, r$*

$$\sup_x | b^{(k)}(x) | \leqslant \begin{cases} C(n^{k-\beta} + N^{k-\beta}), & \beta - k < 1, \\ C(n^{k-\beta} + \log N/N), & \beta - k = 1, \\ C(n^{k-\beta} + N^{-1}), & \beta - k > 1 \end{cases} \tag{17}$$

and

$$\sup_{x \neq \lambda} \left| b^{(r)}(x) - b^{(r)}(\lambda) \right| / | x - \lambda |^{\alpha} \leqslant C.$$

4. PROOFS

Define $\Pi = [-\pi, \pi]$, $a \wedge c = \min(a; c)$.

Proof of Lemma 1. Let $D = \{t_1, t_2, \ldots, t_{2m}\}$. For any set of integers $B = \{\tau_1, \ldots, \tau_k\}$ denote $S(B) = \text{cum}\,\{X_{\tau_1}, \ldots, X_{\tau_k}\}$. According to Leonov—Shiryaev formula (see (Leonov and Shiryaev, 1959)) we have

$$\text{cum}\Big\{A(w_1), \ldots, A(w_m)\Big\} = N^{-m} \sum_{D = D_1 + \cdots + D_q}^{*} \sum_{t_1, \ldots, t_{2m} = 1}^{N} w_1(t_1 - t_2) \times$$

$$\times w_2(t_3 - t_4) \ldots w_m(t_{2m-1} - t_{2m}) S(D_1) \ldots S(D_q) \tag{18}$$

where $\sum\limits_{D = D_1 + \cdots + D_q}^{*}$ means summation over all indecomposable partitions of set D into unordered subset consisting of two and more elements. An unordered collection of disjoint sets $D_1, \ldots, D_q$ is called a indecomposable partition of D, if

$$D = \bigcup_{k=1}^{q} D_k$$

and in the case $q > 1$ there exist numbers $k(2), k(3), \ldots, k(q)$ such that for any $j = 2, 3, \ldots, q$ one element from set $\{t_{2k(j)-1}; t_{2k(j)}\}$ belongs to

$$\bigcup_{i=1}^{j-1} D_i$$

and the an other one belongs to D_j. Let $D_1, \ldots, D_q$ be some fixed indecomposable partition. By virtue of (5) for any integers $\tau_2, \tau_3, \ldots, \tau_q$ the estimate

$$\sum_{\substack{t_1, \ldots, t_{2m} = 1, \ldots, N \\ t_{2k(j)-1} - t_{2k(j)} = \tau_j,\ j = 2, \ldots, q}} |S(D_1) \ldots S(D_q)| \leqslant N C_1^{2m} (r_1! \ldots r_q!)^{\gamma}$$

holds where $r_i = \text{card } D_i, \quad i = 1, \dots, q$. Therefore

$$\sum_{t_1,\dots,t_{2m}=1}^{N} \Big| w_1(t_1 - t_2) \dots w_m(t_{2m-1} - t_{2m}) S(D_1) \dots S(D_q) \Big| \leqslant$$

$$\leqslant N C_1^{2m} (r_1! \dots r_q!)^{\gamma} \sum_{\tau_2,\dots,\tau_q=1-N}^{N-1} |w_{k(2)}(\tau_2) \dots w_{k(q)}(\tau_q)| \times$$

$$\times \prod_{j \in \{1,\dots,m\} \setminus \{k(2),\dots,k(q)\}} \| w_j \|_{\infty} . \tag{19}$$

According to the proof of Lemma 2 in (Rudzkis, 1978)

$$\sum_{D = D_1 + \dots + D_q}^{*} (r_1! \dots r_q!)^{\gamma} \leqslant 4^{m-1} \big((2m)!\big)^{\gamma}. \tag{20}$$

From (18)—(20) we obtain (12).

Proof of Lemma 2. Denoting $R(k) = \mathbf{E} X_0 X_k$ we have

$$N \text{ cov} \Big\{ A(w_1),\ A(w_2) \Big\} = \sum_{\tau,t,u,v=1}^{N} w_1(\tau - t) w_2(u - v) \Big(R(\tau - u) R(t - v) +$$

$$+ R(\tau - v) R(t - u) + S(\tau, t, u, v) \Big) / N = Q_1 + Q_2 + Q_3. \tag{21}$$

By change of variables we obtain

$$Q_1 = \sum_{k,l,m=-\infty}^{\infty} w_1(k) w_2(l) R(m) R(k - l + m) \times$$

$$\times \left(1 + \theta \Big(\frac{|k| + |l| + |m|}{N} \wedge 1 \Big) \right), \quad \theta = \theta(k, l, m).$$

Since due to (6)

$$\sum_{j=-\infty}^{\infty} \Big(|j| \wedge N \Big) |R(j)| \leqslant C N^{1-\alpha}$$

it is easy to verify that

$$\Big| Q_1 - \sum_{k,l,m=-\infty}^{\infty} w_1(k) w_2(l) R(m) R(k - l + m) \Big| \leqslant$$

$$\leqslant C \Big(\| w_1 \|_1 \| w_2 \|_{\infty} + \| w_1 \|_{\infty} \| w_2 \|_1 \Big) / N^{\alpha} = \chi,$$

where $C = C(C_2)$. By virtue of Parseval identity

$$\left| Q_1 - 2\pi \int_{\Pi} W_1(x) W_2(x) f^2(x)\, dx \right| \leqslant \chi. \tag{22}$$

Similarly,

$$\left| Q_2 - 2\pi \int_{\Pi} W_1(-x) W_2(x) f^2(x)\, dx \right| \leqslant \chi. \tag{23}$$

Finally because of (5)

$$|Q_3| \leqslant 6^{\gamma} C_1^3 \parallel w_1 \parallel_{\infty} \parallel w_2 \parallel_{\infty} . \tag{24}$$

(21)—(24) imply (13).

Proof of Lemma 3. Fix $x \in K$. For $j = 0, 1, \ldots$ we have

$$\hat{f}^{(j)}(x) = A(w_j), \quad w_j(k) = (ik)^j g(k/n) e^{ikx}. \tag{25}$$

Applying Lemma 2 and using the periodicity of W and f we obtain

$$N\mathbf{D}\hat{f}^{(j)}(x) = 2\pi \int_{\Pi} \left(W^{(j)}(\lambda) + W^{(j)}(\lambda + 2x) \right) \times$$

$$\times W^{(j)}(\lambda) f^2(x + \lambda)\, d\lambda + \theta C n^{1+2j-\nu}, \quad j = 0, 1, \ldots . \tag{26}$$

Let

$$\psi(\lambda) = \int_{-1}^{1} g(t) e^{it\lambda}\, dt.$$

The function ψ and its derivatives are bounded and

$$\int_{-\infty}^{\infty} \left| \psi^{(j)}(\lambda) \right| \lambda^2\, d\lambda < \infty, \quad j = 0, 1, \ldots . \tag{27}$$

Further, it is easy to show that

$$W(\lambda) = \sum_{k=-\infty}^{\infty} \tilde{W}(\lambda + 2k\pi), \quad \tilde{W}(x) = n\psi(nx). \tag{28}$$

From (27) and (28) it follows

$$\int_{\Pi} \left| W^{(j)}(\lambda) - \tilde{W}^{(j)}(\lambda) \right|^2 d\lambda \leqslant C^{1+j} n^{2j-1}, \quad j = 0, 1, \ldots . \tag{29}$$

Because of (27)—(29) we get for all $x \in K$

$$\int\limits_{\Pi} \left| W^{(j)}(\lambda + 2x) W^{(l)}(\lambda) \right| d\lambda \leqslant C n^{j+l} / (x \wedge (\pi - x)), \quad j, l \in \{0, 1, \ldots\}. \qquad (30)$$

Since due to (6) $f^2(x+\lambda) = f^2(x) + \theta C \mid \lambda \mid^{\alpha}$, using (26)—(30) it is easy to obtain the following relations for $j = 0, 1, \ldots$

$$N \mathbf{D} \hat{f}^{(j)}(x) / n^{2j+1} = 2\pi f^2(x) \int\limits_{-\infty}^{\infty} |\psi^{(j)}(\lambda)|^2 \, d\lambda + \theta C n^{-\nu}, \quad x \in \tilde{K}. \qquad (31)$$

In view of Parseval identity and positivity of the function f in K the relation (31) implies (14) and the equality

$$N n^{-3} \mathbf{D} \hat{f}'(x) = (2\pi f(x))^2 \int\limits_{-1}^{1} t^2 g^2(t) dt + \theta C / n^{\nu}, \quad x \in \tilde{K}. \qquad (32)$$

Further,

$$x'^{*} = \frac{dx^{*}}{dx} = \frac{1}{\tilde{\sigma}(x)} \left(\mathbf{D} \hat{f}'(x) - \left(\tilde{\sigma}'(x) \right)^2 \right)^{1/2}, \quad \tilde{\sigma}(x) = \mathbf{D}^{1/2} \hat{f}(x). \qquad (33)$$

Since $\tilde{\sigma}'(x) = \operatorname{cov} \{\hat{f}(x), \ \hat{f}'(x)\} / \tilde{\sigma}(x)$ and $\int\limits_{-\infty}^{\infty} \psi(x) \psi'(x) dx = 0$, Lemma 2, (14), (25), (29) and (31) yield

$$\sup_{x \in \tilde{K}} \left| \tilde{\sigma}'(x) \sqrt{N / n^3} \right| \leqslant C / n^{\nu}. \qquad (34)$$

From (14) and (32)—(34) we obtain (15).

Proof of Lemma 4. By (26)—(29) for any $x \in K$

$$\mathbf{D} \hat{f}^{(j)}(x) \leqslant C n^{2j+1} / N, \quad j = 0, 1, 2. \qquad (35)$$

Further, differentiating the equality $\tilde{\sigma}(x) \xi(x^{*}) = \hat{f}(x) - \mathbf{E} \hat{f}(x)$ twice with respect to x and having in mind relations $\mathbf{E} \xi(y) \xi'(y) = \mathbf{E} \xi''(y) \xi'(y) = 0$ and $\mathbf{E} \xi(y) \xi''(y) = -1$ for $y \in K^{*}$, we have

$$\mathbf{D} \hat{f}''(x) = \left(\tilde{\sigma}''(x) \right)^2 - 2 \tilde{\sigma}''(x) \tilde{\sigma}(x) (x'^{*})^2 + \tilde{\sigma}^2(x) (x'^{*})^4 \mathbf{D} \xi''(x^{*}) +$$

$$+ \left(2 \tilde{\sigma}'(x) x'^{*} + \tilde{\sigma}(x) x''^{*} \right)^2 \geqslant \tilde{\sigma}^2(x) (x'^{*})^4 \left(\mathbf{D} \xi''(x^{*}) - 1 \right). \qquad (36)$$

The relations (14), (15) and (34)—(36) imply Lemma 4 and

$$\sup_{x \in \tilde{K}} \left| \frac{d^2 x^{*}}{dx^2} \right| \leqslant C n^2. \qquad (37)$$

This inequality will be useful later on.

Proof of Lemma 5. Let

$$B = \bigcup_{x \in K} \left\{ \hat{f}(x);\ \hat{f}'(x)/n \right\}.$$

By (25) and Lemma 1 for all $m = 2, 3, \ldots$ we have for any $Y_1, \ldots, Y_m \in B$

$$\left| \text{cum } \{Y_1, \ldots, Y_m\} \right| \leqslant C^m (m!)^{2\gamma} (n/N)^{m-1}. \tag{38}$$

Moreover,

$$\hat{f}'(x) - \mathbf{E}\hat{f}'(x) = \tilde{\sigma}'(x)\xi(x^*) + \tilde{\sigma}(x)x'^*\xi'(x^*). \tag{39}$$

Due to linear properties of cumulants from (14), (15), (34), (38) and (39) it follows (16).

Proof of Lemma 6. Applying Lemma 2 and representation (25) similarly as in the proof of (31) we obtain for $x, z \in \tilde{K}$

$$\text{cov } \left\{ \hat{f}(x),\ \hat{f}(z) \right\} \frac{N}{n} = 2\pi f^2(x) \int_{-\infty}^{\infty} \psi(\lambda)\psi(\lambda + n(x - z))d\lambda + \theta C/n^{\nu}. \tag{40}$$

Denote

$$r(\Delta) = \max_{\tau \geqslant \Delta} \left| \int_{-\infty}^{\infty} \psi(\lambda)\psi(\lambda + \tau)d\lambda \right| \Big/ \int_{-\infty}^{\infty} | \psi^2(\lambda) | \, d\lambda.$$

Then by (27)

$$r(\Delta) \leqslant C/\Delta^2. \tag{41}$$

(14), (15), (40) and (41) yield the estimate

$$\left| \mathbf{E}\xi(y)\xi(z) \right| \leqslant r(\, | \, y - z \, | \, /C) + \theta C/n^{\nu}, \quad y, z \in K^*. \tag{42}$$

Since $r(\Delta) < 1$ for $\Delta > 0$ the statement of Lemma 6 follows from (41) and (42).

Proof of Lemma 7. Let

$$\Phi_N(\lambda) = (2\pi N)^{-1} \sin^2(N\lambda/2)/\sin^2(\lambda/2).$$

Then

$$\mathbf{E}\hat{f}^{(k)}(x) = \int_{\Pi^2} \Phi_N(u) W(\lambda) f^{(k)}(x + u - \lambda) \, du \, d\lambda. \tag{43}$$

In view of (27) and (28)

$$\int_{\Pi} |W(\lambda)| d\lambda \leqslant C. \tag{44}$$

The estimate (17) follows from (43), (44) and the first part of the proof of Theorem 3.1 in (Bentkus and Rudzkis, 1982) with f replaced by $f^{(k)}$. Finally applying (43), (44) and equality $\int_{\Pi} \Phi_N(\lambda)d\lambda = 1$ we get

$$\left|b^{(r)}(x) - b^{(r)}(z)\right| \leqslant \left|f^{(r)}(x) - f^{(r)}(z)\right| + \sup_a \left|f^{(r)}(a+x) - f^{(r)}(a+z)\right| \times$$

$$\times \int_{\Pi^2} \Phi_N(u)W(\lambda)\,du\,d\lambda \leqslant C \mid x - z \mid^{\alpha}.$$

Proof of Theorem 1. Let

$$f^*(x) = \hat{f}(x) - \mathbf{E}\hat{f}(x),$$

$$V(x) = \mathbf{P}\left\{\sup_{x \in K} \frac{f^{*2}(x)}{2\sigma^2(x)} < \log \frac{H(x_2 - x_1)}{\pi} + z\right\} \tag{45}$$

and let $\tilde{V}(z)$ be defined similarly as $V(z)$ with K replaced by $\tilde{K}$. Because of (14) and (15)

$$\tilde{V}(z) = \mathbf{P}\left\{\sup_{y \in K^*} \xi^2(y)/2 < \log \frac{\mid K^* \mid}{\pi} + z + \theta C n^{-\nu}\left(1 + \mid z \mid\right)\right\} \tag{46}$$

where $\mid K^* \mid$ is length of the interval K^*. From Lemmas 4—6 it follows that the random process $\{\xi(y),\ y \in K^*\}$ for sufficiently large N satisfies the conditions of Theorem 2 in (Rudzkis, 1989b) with $\Lambda = C^{-1}\sqrt{N/n}$. Therefore

$$\sup_z \left|\mathbf{P}\left\{\sup_{y \in K^*} \xi^2(y)/2 < \log \frac{\mid K^* \mid}{\pi} + z\right\} - \exp\{-e^{-z}\}\right| = O(n^{-\nu}). \tag{47}$$

In view of (46) and (47) the relation

$$\sup_z \left|\tilde{V}(z) - \exp\{-e^{-z}\}\right| = O(n^{-\nu}) \tag{48}$$

holds. Let us estimate the difference $\tilde{V}(z) - V(z)$. By (48)

$$\tilde{V}(-\log n/3) = O(n^{-\nu}).$$

Since for sufficiently large N the inequality $3\log\left(H(x_2 - x_1)/\pi\right) > 2\log n$ holds we have

$$\sup_z \left|\tilde{V}(z) - V(z)\right| \leqslant \mathbf{P}\left\{\sup_{x \in K \setminus \tilde{K}} \mid f^*(x) \mid /\sigma(x) > \sqrt{2\log n/3}\right\} + O(n^{-\nu}). \tag{49}$$

Further denote

$$T_m = (K \setminus \tilde{K}) \cap \{2^{-m}j/n,\ \ j = 0, 1, \ldots\},\quad m = 0, 1, \ldots\,.$$

Because of the relation $\sigma^2(x) \asymp n/N$, for any $u > 0$ and some $C > 0$ we have

$$\mathbf{P}\Big\{ \sup_{x\in K\setminus\tilde K} \mid f^*(x) \mid /\sigma(x) > u\Big\} \leqslant \sum_{x\in T_0} \mathbf{P}\Big\{ \mid f^*(x) \mid > \frac{u}{C}\sqrt{\frac{n}{N}}\Big\} +$$

$$+ \sum_{m=1}^{\infty} \sum_{x\in T_{m-1}} \mathbf{P}\Big\{ \Big| f^*(x) - f^*(x + 2^{-m}/n)\Big| > \frac{u}{C}\sqrt{\frac{n}{2^m N}}\Big\}. \tag{50}$$

By virtue of (38) and Lemma 2.1 in (Bentkus and Rudzkis, 1980), for all $u > 0$ and $x, z \in K$ we have

$$\mathbf{P}\Big\{ \mid f^*(x) \mid > u\sqrt{n/N}\Big\} + \mathbf{P}\Big\{ \mid f^*(x) - f^*(z) \mid > u \mid x - z \mid n\sqrt{n/N}\Big\} \leqslant$$

$$\leqslant C \exp\Big\{ - C^{-1}\Big(u^2 \wedge (u^2 N/n)^{1/4\gamma}\Big)\Big\}. \tag{51}$$

Since card $T_m \leqslant 2^m n^{\varkappa}$ the inequalities (50) and (51) yield for any $a > 0$ the estimate

$$\mathbf{P}\Big\{ \sup_{x\in K\setminus\tilde K} \big(f^*(x)/\sigma(x)\big)^2 > a\Big\} \leqslant C n^{\varkappa} \exp\Big\{ - C^{-1}\big(a \wedge (aN/n)^{1/4\gamma}\big)\Big\}, \tag{52}$$

where C does not depend on $\varkappa$. Choosing $\varkappa > 0$ in the definition of $\tilde K$ sufficiently small from (49) and (52) we obtain

$$\sup_z \big|\tilde V(z) - V(z)\big| \leqslant C/n^{\nu}. \tag{53}$$

Now (7) is a consequence of (48) and (53).

Proof of Theorems 2 and 3. Take $n \geqslant CN^{1/(2\beta+1)}$ and suppose the conditions of Theorem 2 to be fulfilled. Let us consider the function

$$\tilde P(u) = \mathbf{P}\Big\{ \sup_{x\in\tilde K} q(x)\big|\hat f(x) - f(x)\big| < u\Big\}.$$

For applying Theorem 1 in (Rudzkis, 1989b) a smoothing of q is needed:

$$\tilde q(x) = \int_{\Pi} q\Big(x + \frac{\lambda}{n}\Big)\frac{\cos^2\lambda}{\pi}\,d\lambda.$$

Then for any $x \in K$

$$\tilde q(x)/q(x) = 1 + \theta C/n^{\alpha_0}, \quad \mid \tilde q^{(j)}(x) \mid \leqslant Cn^{j-\alpha_0}, \quad j = 1, 2. \tag{54}$$

Define the functions $v_j(y, u)$ by the relations

$$v_j(x^*, u) = \Big(u/\tilde q(x) + (-1)^j b(x)\Big)\Big/\sigma(x), \quad x \in K, \quad j = 1, 2,$$

and denote

$$F(u) = \mathbf{P}\Big\{ -v_1(y,u) < \xi(y) < v_2(y,u), \quad y \in K^* \Big\}.$$

Because of (54) we have

$$\tilde{P}(u) = F\big(u + \theta C n^{-\nu}(1+ \mid u \mid)\big). \tag{55}$$

Let

$$a_1 = 2\pi \parallel g \parallel_2 \min_{x \in K} q(x) f(x), \quad a_2 = 4\pi \parallel g \parallel_2 \max_{x \in K} q(x) f(x).$$

Let us consider the function $F(u)$ for

$$u \in B = \Big(a_1 \sqrt{n \log n / N}, \quad a_2 \sqrt{n \log n / N}\Big). \tag{56}$$

By Lemma 7, (15), (37) and (56) for $j = 1,2$ we have

$$\inf_{y \in K^*,\ u \in B} v_j(y,u) > \Big(\sqrt{\log n} + o(1)\Big) \Big/ C \tag{57}$$

and

$$\Big| v_j'(y,u) - v_j'(z,u) \Big| \leqslant C \mid y - z \mid^{\nu}, \quad v_j'(y,u) = \partial v_j(y,u) / \partial y.$$

Denote

$$\tilde{G}(u) = \prod_{j=1,2} \exp\Big\{ (2\pi)^{-1} \int_{K^*} dy \Big\{ \int_0^{\infty} \Big(v_j^2(y,u) + \big(v_j'(y,u) + z\big)^2 \Big) \Big/ 2 \Big\} z \, dz \Big\}.$$

Applying Theorem 1 of (Rudzkis, 1989b) to random process ξ and to functions v_1 and v_2 we obtain

$$\sup_{u \in B} \Big| F(u) - \tilde{G}(u) \Big| \to 0, \quad N \to \infty. \tag{58}$$

Let φ be a positive function and let ω be continuity modulus of φ'. Then for any interval J

$$\int_J \exp\Big\{ -\varphi^2(x)/2 \Big\} \mid \varphi'(x) \mid dx \leqslant$$

$$\leqslant 2 \exp\{-a^2/2\}/a + 3 \sqrt{\omega\Big(\frac{1}{a}\Big)} \int_J \exp\Big\{ -\varphi^2(x)/2 \Big\} dx,$$

where $a = \inf_{x \in J} \varphi(x)$. Therefore using (57) it is easy to get the estimate

$$\sup_{u \in B} \Big| \tilde{G}(u) - \tilde{G}_1(u) \Big| = o(1), \tag{59}$$

where

$$G_1(u) = \prod_{j=1}^{2} \exp\Big\{ -\frac{1}{2\pi} \int_{K^*} \exp\Big\{ -v_j^2(y,u)/2 \Big\} dy \Big\}.$$

Further denote

$$G(u)=\prod_{j=1}^{2}\exp\Big\{-\frac{H}{2\pi}\int_{K}\exp\Big\{-\mu_j^2(x,u)/2\Big\}dx\Big\},$$

where the functions μ_j are defined in (9). Because of (15), (17) and (54) we get

$$\sup_{u\in B}\Big|\tilde{G}_1(u)-G(u)\Big|=O(n^{-\nu}). \tag{60}$$

Note that

$$G\Big(a_1\sqrt{n\log n/N}\Big)=O(n^{-\nu}),\quad G\Big(a_2\sqrt{n\log n/N}\Big)=1+O(n^{-\nu}) \tag{61}$$

by virtue of (17). From (55) and (58)—(61) it follows that

$$\sup_{u\in B}\Big|\tilde{P}(u)-G(u)\Big|\to 0. \tag{62}$$

Finally choosing $\varkappa$ sufficiently small we have by means of (17) and (52) the relation

$$\sup_{u\in B}\Big|\tilde{P}(u)-P(u)\Big|=O(n^{-\nu}). \tag{63}$$

From (61)—(63) follows (8).

Let now $\beta\geqslant 2$ and $n>C_0(N/\log N)^{1/(2\beta+1)}$ where C_0 is a sufficiently large number independent of N. For sufficiently large N and $u\in B$ we have (57) and

$$\sup_{y\in K^{*},\ j=1,2}\left|\frac{\partial^2 v_j(y,u)}{\partial y^2}\right|\leqslant\inf_{y\in K^{*},\ j=1,2}v_j(y,u)/2,$$

$$\sup_{x\in\tilde{K},\ j=1,2}\Big|v_j'(x^*,u)-(-1)^j b'(x)/H\sigma(x)\Big|=O(n^{-\nu}) \tag{64}$$

because of (15), (17), (37) and (54). For $u\in B$ the functions $v_j(\cdot,u)$ and the process ξ satisfy conditions of Theorem 2 in (Rudzkis, 1989b), thus

$$\sup_{u\in B}\Big|F(u)-\tilde{G}(u)\Big|=O(n^{-\nu}). \tag{65}$$

Using (15), (17), (64), the equality

$$2l(z)=\int_{-\infty}^{\infty}\exp\Big\{-(z+\lambda)^2/2\Big\}\mid\lambda\mid d\lambda,$$

and formula (2.3) in (Rudzkis, 1985) we get

$$\sup_{u\in B}\Big|\tilde{G}(u)-G_1(u)\Big|=O(n^{-\nu}), \tag{66}$$

where

$$G_1(u) = \prod_{j=1}^{2} \exp\left\{ -\frac{H}{2\pi} \int_K \exp\left\{ -\mu_j^2(x,u)/2 \right\} l\Big(b'(x)/H\sigma(x)\Big)\,dx \right\}.$$

Since (61) remains valid when G is replaced by G_1, relations (55), (65) and (66) imply (10).

REFERENCES

Bentkus, R.J. and Rudzkis, R.A. (1982). On the distributions of some statistical estimates of spectral density. *Theory Probab. Appl.* **27**, 795-814.

Bentkus, R. and Rudzkis, R. (1980). On exponential estimates of the distribution of random variables. *Liet. Mat. Rink.* **20**, 30–37.

Leonov, V.P. and Shiryaev, A.N. (1959). On the calculus of semi-invariants. *Theory Probab. Appl.* **4**.

Rudzkis, R. (1978). Large deviations for estimates of spectrum of stationary time series. *Lith. Math. J.* **18**, 81–98.

Rudzkis, R. (1985). On probability of large excursion of a non-stationary Gaussian process. I. *Lith. Math. J.* **25**, 76–84.

Rudzkis, R. (1989a). On distributions of supremum type functionals of non-parametric estimates of probability density and spectral density. In: *Abstr. Fifth Intern. Vilnius Conf. Probab. Theory and Math. Statist.*, Inst. Math. Cybern., Vilnius **2**, 124–125.

Rudzkis, R. (1989b). Probabilities of large excursions of stochastic processes close to the Gaussian ones. *Liet. Mat. Rink.* **29**, 785–805.

Prob. Theory and Math. Stat., Vol. 2, pp. 364–371
B.Grigelionis *et al.* (Eds.)
1990 VSP/Mokslas

A CENTRAL LIMIT THEOREM FOR QUADRATIC FORMS IN INDEPENDENT RANDOM VARIABLES

Ju.M. RYZHOV
Kiev State University, 252017 Kiev, USSR

ABSTRACT

Using some ideas from the Feynman integral theory and a limit theorem for centered quadratic forms in Gaussian random variables, we prove the central limit theorem for the independent identically distributed random variables with finite moments of all orders.

In the present paper a method is proposed which permits to obtain the limit distribution of the centered quadratic forms in independent random variables having finite moments. The idea of this method is closely related to some result in (Daletsky and Fomin, 1983) referring to Feynman's integral theory.
First let us explain this idea. If a quasi-measure μ in some Hilbert space H corresponds to the random element ξ and A is a symmetric positive operator in H such that the quadratic form $(A\xi, \xi)$ is well-defined, then the characteristic function

$$\psi(\lambda) = \int_H \exp\left\{i\lambda(Az, z)\right\}\mu(dz) \tag{1}$$

defines the Feynman integral

$$\tilde{\int_H} \varphi(x)\nu_{-2i\lambda A}(dx), \tag{2}$$

where

$$\varphi(x) = \int_H \exp\left\{i(x, z)\right\}\mu(dz) \tag{3}$$

is the characteristic functional of ξ. But it is shown in (Daletsky and Fomin, 1983) that under some condition the integral (2) can be rewritten as an ordinary integral with respect to the Gaussian measure γ_A having the correlation operator A:

$$\tilde{\int_H} \varphi(x)\nu_{-2i\lambda A}(dx) = \int_H \varphi\left(\sqrt{|\lambda|}[1 - i\,\mathrm{sign}\lambda]z\right)\gamma_A(dz). \tag{4}$$

Comparing (1)—(4) we can expect that under some conditions the equation

$$\psi(\lambda) = \int_H \varphi\Big(\sqrt{|\lambda|}(1 - i\,\mathrm{sign}\lambda)z\Big)\gamma_A(dz) \tag{5}$$

holds. Hence, in order to investigate the limit distribution of quadratic forms, one can to find the conditions under which (4) is true and then it is sufficient to investigate the limit behaviour in the integral (5).

Let us realize this idea now.

Let H be a separable Hilbert space with a fixed orthonormal basis $\{e_k\}$ and let ξ be a generalized H-valued random element having the distribution determined by a quasi-measure μ with the characteristic functional

$$\varphi(z) = \int_H \exp\{i(z,x)\}\mu(dx) = \prod_{k=1}^{\infty} \chi((e_k, z)), \tag{6}$$

where $\chi(t)$, $t \in \mathbf{R}^1$, is the characteristic function of some distribution on $\mathbf{R}^1$. The equality (6) is understood in the product-measure sense, i.e., for P_n being the orthogonal projection on span$\{e_1, e_2, \dots, e_n\}$, we shall set

$$\varphi(P_n z) = \int_{P_n H} \exp\{i(z,x)\}\mu(dx) = \prod_{k=1}^{n} \chi((e_k, z))$$

(see (Skorokhod, 1975)). In this meaning we consider ξ as the sequence of the independent identically distributed random variables

$$\xi = \{\xi_1, \xi_2, \dots, \xi_n, \dots\},$$

where $\xi_k = (e_k, \xi)$, $k = 1, 2, \dots$, and ξ_k has the characteristic functional $\chi(t) = \mathbf{E}\exp\{it\xi_k\}$.

Suppose $\mathbf{E}\xi_k = 0$, $\mathbf{E}\xi_k^2 = 1$ and let A be a symmetric operator in H having the matrix representation $\{a_{kj} = (Ae_k, e_j)\}$.

Denote

$$S_n = (AP_n\xi,\, P_n\xi) - \mathbf{E}(AP_n\xi,\, P_n\xi) = \sum_{k,j=1}^{n} a_{kj}(\xi_k\xi_j - \delta_{kj}), \tag{7}$$

i.e., we have the centered quadratic forms for the elements $P_n\xi$.

PROPOSITION 1. *If the random variables ξ_k, $k = 1, 2, \dots$, have the finite moment $\mathbf{E}\xi_k^4 = m_4$ then the centered quadratic form*

$$S = \lim_{n\to\infty} S_n = \sum_{k,j=1}^{\infty} a_{kj}(\xi_k\xi_j - \delta_{kj})$$

is well-defined if and only if $\sum\limits_{k,j=1}^{\infty} a_{kj}^2 < \infty$, *i.e., A is a Hilbert—Schmidt operator.*

The proof follows easily from the relation

$$\mathbf{E}S_n^2 = (m_4 - 3)\sum_{k=1}^{n} a_{kk}^2 + 2\sum_{k,j=1}^{n} a_{kj}^2, \tag{8}$$

with

$$m_4 - 1 = \mathbf{E}\xi_k^4 - \left(\mathbf{E}\xi_k^2\right)^2 \geqslant 0,$$

$$\mathbf{E}S_n^2 \geqslant (m_4 - 1)\sum_{k=1}^{n} a_{kk}^2 + 2\sum_{k\neq j} a_{kj}^2,$$

$$\mathbf{E}S_n^2 \leqslant (2 + |m_4 - 3|)\sum_{k,j=1}^{n} a_{kj}^2. \tag{9}$$

If $m_4 = 1$ then $E\xi_k^4 = \left(E\xi_k^2\right)^2$ and $|\xi_k| = 1$ a.s., but this case is of no interest. For symmetric non-degenerate bounded operator A in H let us denote by A_+ and A_- its positive and negative parts, respectively, that is, the operators A_+ and A_- are positive (non-negative) symmetric and $A_+ - A_- = A$, $A_+A_- = A_-A_+ = 0$. Denote

$$\sqrt{-2i\lambda A} = \sqrt{|\lambda|}\Big[(1 - i\,\mathrm{sign}\lambda)A_+^{1/2} + (1 + i\,\mathrm{sign}\lambda)A_-^{1/2}\Big], \tag{10}$$

where $A_+^{1/2}$, $A_-^{1/2}$ are the non-negative square roots of the operators $A_\pm$. It is checked easily that

$$\sum_k \left(\sqrt{-2i\lambda A}\,z, z\right) = -2i\lambda\Big[(A_+z, z) - (A_-z, z)\Big] +$$

$$+4|\lambda|\left(A_+^{1/2}z, A_-^{1/2}z\right) = -2i\lambda(Az, z).$$

In view of the equality

$$\left(\sqrt{-2i\lambda A}\right)^2 = -2i\lambda A$$

we can write the formal relation

$$\left(\sqrt{-2i\lambda A}\,z,\ \sqrt{-2i\lambda A}\,z\right) = \left(\left(\sqrt{-2i\lambda A}\right)^2 z, z\right)$$

and in this meaning we shall consider $\sqrt{-2i\lambda A}$ as some symmetric map from H into its complex extension.

Now let A_n be a sequence of symmetric nuclear operators in H and let $g(z) = \exp\left\{-\frac{1}{2}(z, z)\right\}$. Then denoting $\dot{w}$ the Gaussian white noise (i.e., $(\dot{w}, e_k)$, $k = 1, 2, \ldots$, are i.i.d. (0, 1)–Gaussian) one can write

$$\mathbf{E}\exp\{i\lambda\big[(A_n\dot{w}, \dot{w}) - E(A_n\dot{w}, \dot{w})\big]\} =$$

$$= \exp\{-i\lambda \mathrm{tr} A_n\} \int_H \exp\{i\lambda(A_n z, z)\}\gamma_E(dz) =$$

$$= \exp\{-i\lambda \mathrm{tr} A_n\} \int_H g(\sqrt{-2i\lambda A_n} z)\gamma_E(dz).$$

It is the characteristic function of the centered quadratic form in independent (0, 1)–Gaussian random variables, and writing the main theorem of (Sevastyanov, 1961) in a suitable for us form we obtain the following

PROPOSITION 2. *In order that*

$$\lim_{n\to\infty} \exp\{-i\lambda \mathrm{tr} A_n\} \int_H g(\sqrt{-2i\lambda A} z)\gamma_E(dz) = \exp\Big\{-\frac{\sigma^2}{2}\lambda^2\Big\},$$

it is necessary and sufficient that the conditions

A. $\|A_n\| \to 0, \quad n \to \infty;$

B. $\|A_n\|_2^2 \to \sigma^2/2, \quad n \to \infty,$

are satisfied, where $\|\cdot\|$ *denotes the operator norm and* $\|\cdot\|_2$ *denotes the double-norm (Hilbert—Schmidt norm).*

As it follows from this proposition, the conditions A and B are necessary in the central limit theorem for the quadratic forms in independent not obligatory Gaussian random variables with finite moments. The task is to find the needed additional conditions. One of such condition we obtain from Proposition 1:

C. $\mathrm{tr}(A_n \otimes A_n) = \sum_k (A_n e_k, e_k)^2 \to \sigma_1^2, \quad n \to \infty,$

where $A \otimes A$ is defined by relation

$$(A \otimes A) = \sum_{kj} (Ae_k, e_j)^2 (x, e_k) e_j.$$

We recall that in this paper the considered Hilbert space H is equipped with some fixed basis $(e_k,\ k \geqslant 1)$.

Let us return now to the formula (4). Using the definition of $\sqrt{-2i\lambda A}$ in (10) let us consider the function

$$\tilde{\psi}(\lambda) = \int_H \varphi(\sqrt{-2i\lambda A}\, z)\gamma_E(dz) =$$

$$= \int_H \varphi\Big(\sqrt{|\lambda|}\Big[(1 - i\,\mathrm{sign}\lambda)A_+^{1/2} z + (1 + i\,\mathrm{sign}\lambda)A_-^{1/2} z\Big]\Big)\gamma_E(dz), \qquad (11)$$

where γ_E is canonical Gaussian quasi–measure corresponding to the characteristic functional $g(z) = \exp\big\{-\frac{1}{2}(z, z)\big\}$ and $\varphi(z)$ is the characteristic functional of some quasi–measure μ in H.

In order that $\tilde{\psi}(\lambda)$ in (11) will be well–defined it is necessary to extend the functional $\varphi(z)$ on the complex extension of the space H, i.e., the integral

$$\varphi(u+i\nu)=\int_H \exp\{(u+i\nu,x)\}\mu(dx) \tag{12}$$

has to be defined. In the case under consideration, (12) may be rewritten as

$$\int_H \exp\Big\{\sqrt{|\lambda|}\operatorname{sign}\lambda\Big((A_+^{1/2}-A_-^{1/2})z,x\Big)+i\sqrt{|\lambda|}\Big((A_+^{1/2}+A_-^{1/2})z,x\Big)\Big\}\mu(dx)$$

and it is bounded by

$$\int_H \exp\Big\{\sqrt{|\lambda|}\Big((A_+^{1/2}-A_-^{1/2})z,x\Big)\Big\}\mu(dx)\leqslant$$

$$\leqslant \int_H \exp\Big\{\|x\|\,\|z\|\sqrt{|\lambda|\|A\|}\Big\}\mu(dx).$$

If the condition A is assumed to be satisfied, the function (11) exists if

$$\int_{H\times H} \exp\{\varepsilon\|z\|\,\|x\|\}\mu(dx)\gamma_E(dx)<\infty \tag{13}$$

for sufficiently small $\varepsilon>0$. But

$$\int_H \mu(dx)\int_H \exp\Big\{i\Big(\sqrt{-2i\lambda A}\,z,x\Big)\Big\}\gamma_E(dx)=$$

$$=\int_H \exp\{i\lambda(Ax,x)\}\mu(dx)=\psi(\lambda)=\mathbf{E}\exp\left(i\lambda(A\xi,\xi)\right)$$

and by the Fubini theorem $\psi(\lambda)=\tilde{\psi}(\lambda)$, i.e.,

$$\psi(\lambda)=\int_H \varphi(\sqrt{-2i\lambda A}\,z)\gamma_E(dz). \tag{14}$$

PROPOSITION 3. *If a random element ξ in H corresponds to a quasi–measure μ having characteristic functional $\varphi(z)=\mathbf{E}\exp\{i(\xi,z)\}$ and*

$$\iint_{H\times H} \exp\{\varepsilon\|z\|\,\|x\|\}\mu(dx)\gamma_E(dz)<\infty$$

for sufficiently small $\varepsilon > 0$, where γ_E is canonical Gaussian quasi-measure, then

$$\int_H \varphi(\sqrt{-2i\lambda A}\, z)\gamma_E(dz) = \mathbf{E}\exp\{i\lambda(A\xi,\xi)\}. \tag{15}$$

Remark. Under the conditions of Proposition 3 the random variables $\xi_k = (\xi, e_k)$, $k \geqslant 1$, have the moments of arbitrary orders and the functions $\varphi(\lambda z)$ are analytical in complex variable λ.

Now let us consider the case

$$\varphi(z) = \prod_{k=1}^{\infty} \chi((z, e_k)),$$

and denote by $\varkappa_m$, $m = 1, 2, \ldots$, the semi-invariants of $\chi(t)$.

It is followed from (15) and symmetry of measure γ_E that the odd semi-invariants do not influence on the limit behaviour of the quadratic forms. Therefore we can put $\varkappa_{2m+1} = 0$ and then

$$\varphi(\sqrt{-2i\lambda A}\, z) = \exp\Big\{\sum_k \ln \chi((\sqrt{-2i\lambda A}\, z, e_k))\Big\} =$$

$$= \exp\Big\{\sum_k \sum_m \frac{\varkappa_{2m}(2i\lambda)^m}{(2m)!}\Big((A_+^{1/2} + i\,\mathrm{sign}\lambda\, A_-^{1/2})z, e_k\Big)^{2m}\Big\} =$$

$$= \exp\Big\{i\lambda \sum_k \Big((A_+^{1/2} + i\,\mathrm{sign}\lambda\, A_-^{1/2})z, e_k\Big)^2 - \frac{\varkappa_4\lambda^2}{2}\sum_k (Ae_k, e_k)^2 -$$

$$-\frac{\varkappa_4\lambda^2}{6}\sum_k \Big[\Big((A_+^{1/2} + i\,\mathrm{sign}\lambda\, A_-^{1/2})z, e_k\Big)^4 - 3(Ae_k, e_k)^2\Big] +$$

$$+\sum_{m=3}^{\infty}\sum_k \frac{\varkappa_{2m}(2i\lambda)^m}{(2m)!}\Big((A_+^{1/2} + i\,\mathrm{sign}\lambda\, A_-^{1/2})z, e_k\Big)^{2m}\Big\}.$$

Suppose for every λ

$$|\varphi(\sqrt{-2i\lambda A}\, z)| \leqslant \psi_\lambda(z)$$

and $\psi_\lambda(z)$ to be integrable with respect to γ_E. Then

$$\varphi(\sqrt{-2i\lambda A}\, z) = \exp\Big\{i\lambda \sum_k \Big((A_+^{1/2} + i\mathrm{sign}\lambda\, A^{1/2})z, e_k\Big)^2 -$$

$$-\frac{\varkappa_4\lambda^2}{2}\sum_k (Ae_k, e_k)^2\Big\} g_\lambda(z)$$

where, assuming $\|A\|_2^2 < \infty$,

$$|g_\lambda(z)| \leqslant \psi_\lambda(z)\exp\Big\{\frac{\varkappa_4\lambda^2}{2}\sum_k (Ae_k, e_k)^2\Big\} \leqslant \psi_\lambda(z)\exp\Big\{\frac{\varkappa_4\lambda^2}{2}\|A\|_2^2\Big\}.$$

Therefore $|g_\lambda(z)|$ is majorized by an integrable function. Hence, if $\mathbf{E}(\xi, e_k) = 0$, $\mathbf{E}(\xi, e_k)^2 = 1$ then

$$\exp\{-i\lambda \mathrm{tr} A_n\}\mathbf{E}\exp\{i\lambda(A_n\xi,\xi)\} \cong$$

$$\cong \exp\{-i\lambda\, \mathrm{tr} A_n\} \int_H \exp\Big\{i\lambda \sum_k \Big((A_{n+}^{1/2} + i\,\mathrm{sign}\lambda\, A_{n-}^{1/2})z, e_k\Big)^2 -$$

$$-\frac{\varkappa_4\lambda^2}{2}\sum_k (A_n e_k, e_k)^2\Big\}\gamma_E(dz) + \int_H |(g_\lambda(z)-1)|\gamma_E(dz) =$$

$$= \exp\Big\{-\frac{\varkappa_4\lambda^2}{2}\sum_k (A_n e_k, e_k)^2\Big\}\int_H \exp\Big\{i\lambda\big[(A_n z, z) - \mathrm{tr} A_n\big]\Big\}\gamma_E(dz) +$$

$$+\int_H |g_\lambda(z) - 1|\gamma_E(dz).$$

Under the conditions A—C it follows by the Proposition 2, that

$$\exp\Big\{-\frac{\varkappa_4\lambda^2}{2}\sum_k (A_n e_k, e_k)^2\Big\}\int_H \exp\Big\{i\lambda\big[(A_n z, z) - \mathrm{tr} A_n\big]\Big\}\gamma_E(dz) \to$$

$$\to \exp\Big\{-\frac{\lambda^2}{2}(\sigma^2 + \varkappa_4\sigma_1^2)\Big\},$$

where

$$\sigma^2 = 2\lim_{n\to\infty}\|A_n\|_2^2, \quad \sigma_1^2 = \lim_{n\to\infty}\mathrm{tr}(A\otimes A).$$

Next, let us consider $g_\lambda(\dot w)$, where $\dot w$ is the white noise (a random element with the distribution γ_E). We have the following inequalities

$$\mathbf{E}|\ln g_\lambda(\dot w)| \leqslant \sum_{m=3}^{\infty}\sum_k \frac{|\varkappa_{2m}|(2|\lambda|)^m}{(2m)!}\mathbf{E}\big((A_{n+}^{1/2} + i\,\mathrm{sign}\lambda\, A_{n-}^{1/2})\dot w, e_k\big)^{2m} +$$

$$+\frac{\lambda^2|\varkappa_4|}{6}\Big[\mathbf{E}\Big(\sum_k\Big\{\Big((A_{n+}^{1/2} + i\,\mathrm{sign}\lambda\, A_{n-}^{1/2})\dot w, e_k\Big)^4 - 3(A_n e_k, e_k)^2\Big\}\Big)^2\Big]^{1/2} \leqslant$$

$$\leqslant \sum_{m=3}^{\infty}\sum_k \frac{|\varkappa_{2m}|\,|\lambda|^m}{m!}\sum_k\Big((A_+ + A_-)e_k, e_k\Big)^m +$$

$$+\frac{\lambda^2|\varkappa_4|}{6}\Big[\sum_{k\neq j}\Big(72(A_n e_k, e_k)(A_n e_j, e_j)(A_n e_k, e_j)^2 + 24(A_n e_k, e_j)^4\Big)\Big]^{1/2}$$

and therefore

$$\mathbf{E}|\ln g_\lambda(\dot w)| \leqslant C\|A_n\|\Big\{\sum_{m=3}^{\infty}\frac{|\varkappa_{2m}|\,|\lambda|^m}{m!}\|A_n\|^{m-3} + \|A_n\|_2\Big\},$$

where $\|A_n\|\,\|A_n\|_2 \to 0$, because $\|A_n\| \to 0$, $\|A_n\|_2 \leqslant C < \infty$ by the conditions A, B. Consequently, if $\|A_n\| \to 0$ (the condition A) and for some sufficiently small $\delta > 0$ relation

$$\sum_{m=3}^{\infty} \frac{|\varkappa_{2m}|\delta^m}{m!} \leqslant c < \infty$$

holds, then $\ln g(\dot{w}) \to 0$ in probability and by the bounded convergence theorem we obtain

$$\int_H |g(z) - 1| \gamma_E(dz) \to 0.$$

Finally, we can formulate the main result.

THEOREM. *Let $\xi_1, \xi_2, \ldots, \xi_n, \ldots$ be a sequence of independent identically distributed random variables having the moments of all orders, $\mathbf{E}\xi_k = 0$, $\mathbf{E}\xi_k^2 = 1$ and $\varkappa_m$, $m = 3, 4, \ldots$, denotes the semi-invariants of ξ_k. Let the sequence of nuclear symmetric (infinite dimensional) matrices $A_n = \{a_{kj}^{(n)}\}$, $k, j = 1, 2, \ldots$, satisfies the following conditions:*

A. $\|A_n\| \to 0$ *($\|\cdot\|$ denotes the operator norm);*

B. $\|A_n\|_2^2 = \sum_{k,j} |a_{kj}^{(n)}|^2 \to \sigma^2/2$;

C. $\operatorname{tr}(A_n \otimes A_n) = \sum_k |a_{kk}^{(n)}|^2 \to \sigma_1^2$ *as* $n \to \infty$;

D. *for sufficiently small $\delta > 0$*

$$\sum_{m=3}^{\infty} \frac{|\varkappa_{2m}|\delta^m}{m!} \leqslant c < \infty;$$

E. $\mathbf{E}\exp\left\{\varepsilon\left[\sum_{k=1}^{n} w_k^2 \cdot \sum_{k=1}^{n} \xi_k^2\right]^{1/2}\right\} < \infty$, *where $w_1, w_2, \ldots$ is the sequence of independent (0, 1)–Gaussian random variables independing on the sequence $\{\xi_k\}$. Then*

$$\lim_{n\to\infty} \mathbf{E}\exp\left\{i\lambda \sum_{k,j} a_{kj}^{(n)}(\xi_k\xi_j - \delta_{kj})\right\} = e^{-\lambda^2(\sigma^2 + \varkappa_4\sigma_1^2)/2}$$

uniformly on λ in every bounded interval.

REFERENCES

Daletsky, Yu.L. and Fomin, S.V. (1983). *Measures and Differential Equations in Infinite Dimensional Spaces.* Nauka, Moscow.

Ramachandran, B. (1967). *Advanced Theory of Characteristic Functions.* Statist. Publ. Soc., Calcutta.

Sevastyanov, B.A. (1961). A class of limit distributions of quadratic forms in normal random variables. *Theory Probab. Appl.* **6**, 368–372.

Skorokhod, A.V. (1974). *Integration in Hilbert Spaces.* Springer, Berlin—Heidelberg etc.

Prob. Theory and Math. Stat., Vol. 2, pp. 372–382
B.Grigelionis *et al.* (Eds.)
1990 VSP/Mokslas

SUBORDINATION DEPENDING ON A PARAMETER

KEN–ITI SATO

Department of Mathematics, College of General Education,
Nagoya University, Nagoya, 464–01 Japan

ABSTRACT

A family of subordinators depending on a parameter θ is considered as a special case of exponential transformation of a Lévy process. By the use of these subordinators, Bochner's subordination of a semigroup of operators depending on a parameter θ is defined. Limit theorems for the subordinate semigroups as $\theta \to \infty$ are given. These are connected with Ichinose's results concerning convergence of relativistic model process to non–relativistic model process as the light velocity tends to infinity.

1. EXPONENTIAL TRANSFORMATION OF LÉVY PROCESSES

Let $\mathbf{R}^d$ be the n–dimensional Euclidean space, each element of which is a column vector. Let $\Omega = D([0,\infty),\mathbf{R}^d)$, the space of right–continuous functions $\omega(t)$ on $[0,\infty)$ into $\mathbf{R}^d$ with left limits, and let $X(t,\omega) = X_t(\omega) = \omega(t)$, $\mathcal{F}_t = \sigma(X_s : s \in \in [0,t])$, and $\mathcal{F} = \sigma(X_s : s \in [0,\infty))$. Suppose that, under a probability measure $\mathbf{P}$ on $\mathcal{F}$, $\{X_t : t \geqslant 0\}$ is a Lévy process with generator (A,ν,γ). This means that $\{X_t\}$ is a process with stationary independent increments with $X_0 = 0$ a.s. and the characteristic function of X_t is as follows:

$$\mathbf{E}e^{i\langle z, X(t)\rangle} =$$

$$= \exp\left[t\left\{-\frac{1}{2}\langle z, Az\rangle + \int \left(e^{i\langle z,x\rangle} - 1 - i\langle z,x\rangle \mathbf{1}_D(x)\right)\nu(dx) + i\langle\gamma, z\rangle\right\}\right] \quad (1.1)$$

for $z \in \mathbf{R}^d$, where A is a non–negative definite symmetric matrix, γ is in $\mathbf{R}^d$, $\mathbf{1}_D(x)$ is the indicator function of $D = \{x : |x| < 1\}$, and ν is a measure on $\mathbf{R}^d$ (called Lévy measure of the process) satisfying $\nu(\{0\}) = 0$ and $\int (1 \wedge |x|^2)\nu(dx) < \infty$.

LEMMA 1.1. *For $c \in \mathbf{R}^d$ the following three properties are equivalent:*

$$\mathbf{E}e^{\langle c, X(t)\rangle} < \infty \quad \text{for every} \quad t > 0. \quad (1.2)$$

$$\mathbf{E}e^{\langle c, X(t)\rangle} < \infty \quad \text{for some} \quad t > 0. \quad (1.3)$$

$$\int_{|x|\geqslant 1} e^{\langle c,x\rangle}\nu(dx) < \infty. \tag{1.4}$$

In fact, since $e^{\langle c,x\rangle}$ is (sub)multiplicative on $\mathbf{R}^d$, the equivalence is a consequence of a result of (Kruglov, 1970, 1972) and (Sato, 1973).
Now, for $p = 1, \dots, d$, let E_p be either a closed interval with an endpoint at 0 or a singleton $\{0\}$, and let $E = E_1 \times \dots \times E_d$. We fix this E and assume that (1.2) holds for every $c \in E$. Then we have the following lemma and theorem.

LEMMA 1.2. *If $w = (w_p)_{1\leqslant p\leqslant d} \in \mathbf{C}^d$ satisfies $(\mathrm{Re}\, w_p)_{1\leqslant p\leqslant d} \in E$, then one can define*

$$\Psi(w) = \frac{1}{2}\langle w, Aw\rangle + \int \left(e^{\langle w,x\rangle} - 1 - \langle w,x\rangle \mathbf{1}_D(x)\right)\nu(dx) + \langle \gamma, w\rangle \tag{1.5}$$

and prove that

$$\mathbf{E}e^{\langle w,X(t)\rangle} = e^{t\Psi(w)}. \tag{1.6}$$

Here we denote $\langle w, v\rangle = \sum_{p=1}^{d} w_p v_p$ for $w = (w_p)_{1\leqslant p\leqslant d}$, and $v = (v_p)_{1\leqslant p\leqslant d}$ in $\mathbf{C}^d$.

THEOREM 1.3. *For $c \in E$ and $B \in \mathcal{F}_t$ define*

$$\mathbf{P}_c(B) = e^{-t\Psi(c)}\mathbf{E}e^{\langle c,X(t)\rangle}\mathbf{1}_B. \tag{1.7}$$

(i) *If $B \in \mathcal{F}_{t_0}$, then the right-hand side of (1.7) coincides for all $t \geqslant t_0$.*

(ii) *The set function $\mathbf{P}_c$ defined on $\bigcup_t \mathcal{F}_t$ can be extended to a probability measure on $\mathcal{F}$. The extension is unique (and denoted by the same symbol $\mathbf{P}_c$).*

(iii) *Under $\mathbf{P}_c$, $\{X_t\}$ is a Lévy process with generator (A_c, ν_c, γ_c) defined by*

$$A_c = A, \tag{1.8}$$

$$\nu_c(dx) = e^{\langle c,x\rangle}\nu(dx), \tag{1.9}$$

$$\gamma_c = \gamma + Ac + \int_{|x|<1} x(e^{\langle c,x\rangle} - 1)\nu(dx). \tag{1.10}$$

(iv) *Denote the distribution of X_t under $\mathbf{P}$ by μ_t. Then, for any Borel set F in $\mathbf{R}^d$ and $z \in \mathbf{R}^d$,*

$$\mathbf{P}_c(X_t \in F) = e^{-t\Psi(c)}\int_F e^{\langle c,x\rangle}\mu_t(dx), \tag{1.11}$$

$$\mathbf{E}_c e^{i\langle z,X(t)\rangle} = e^{t(\Psi(iz+c)-\Psi(c))}, \tag{1.12}$$

where $\mathbf{E}_c$ *denotes expectation by* $\mathbf{P}_c$.

We call the transformation of $\mathbf{P}$ to $\mathbf{P}_c$ the exponential transformation with parameter c. When $\{X_t\}$ is the Brownian motion under $\mathbf{P}$, the transformation is a simple case of drift transformation. Also, the transformation of Ichinose—Tamura (Ichinose and Tamura, 1986) is reduced to the exponential transformation if the scalar potential is constant. The formula (1.11) shows that the distribution of X_t under $\mathbf{P}_c$ is the conjugate distribution of μ_t in the sense of (Khinchin, 1949). The rest of this section gives proof of Lemma 1.2 and Theorem 1.3.

Proof of Lemma 1.2. The property (1.4) holds for every $c \in E$ by Lemma 1.1. Hence the right-hand side of (1.5) is finite and $\Psi(w)$ can be defined by (1.5). Let $w = (w_p)_{1 \leqslant p \leqslant d}$. If $\mathrm{Re} w_p = 0$ for $p = 1, \ldots, d$, then the identity (1.6) is true, since it is no other than (1.1). Fix w_p for $p = 2, \ldots, d$ with $\mathrm{Re} w_p = 0$ and consider $f(w_1) = \mathbf{E}e^{\langle w, X(t)\rangle}$ and $g(w_1) = e^{t\Psi(w)}$ as w_1 runs in $\widetilde{E}_1 = \{w_1 \in \mathbf{C} : \mathrm{Re} w_1 \in E_1\}$. Then it is easy to see that $f(w_1)$ is continuous on $\widetilde{E}_1$. Moreover it is analytic in the interior of $\widetilde{E}_1$, since $\mathbf{E}e^{\langle w, X(t)\rangle}\mathbf{1}_{\{|x| \leqslant n\}}(X(t))$ is analytic there and converges pointwise to $f(w_1)$ as $n \to \infty$. Similarly $g(w_1)$ is continuous on $\widetilde{E}_1$ and analytic in the interior of $\widetilde{E}_1$. Since $f(w_1) - g(w_1) = 0$ on the boundary $\{\mathrm{Re} w_1 = 0\}$, $f(w_1) - g(w_1)$ has the analytic continuation over the boundary by the Schwarz principle of reflection. Hence $f(w_1) - g(w_1) = 0$ on $\widetilde{E}_1$ by the theorem of uniqueness. In the next step, fix $w_1 \in \widetilde{E}_1$ and w_p with $\mathrm{Re} w_p = 0$ for $p = 3, \ldots, d$. Consider $\mathbf{E}e^{\langle w, X(t)\rangle}$ and $e^{t\Psi(w)}$ as functions of w_2 in $\widetilde{E}_2 = \{w_2 \in \mathbf{C} : \mathrm{Re}\ w_2 \in E_2\}$. Then, by the same argument, we see that they are identical. Repeating this, we get $\mathbf{E}e^{\langle w, X(t)\rangle} = e^{t\Psi(w)}$ for all w satisfying $(\mathrm{Re} w_p)_{1 \leqslant p \leqslant d} \in E$.

Proof of Theorem 1.3. If $s < t$ and $B \in \mathcal{F}_s$, then

$$\mathbf{E}e^{\langle c, X(t)\rangle}\mathbf{1}_B = \mathbf{E}e^{\langle c, X(s)\rangle + \langle c, X(t) - X(s)\rangle}\mathbf{1}_B = $$

$$= e^{(t-s)\Psi(c)}\mathbf{E}e^{\langle c, X(s)\rangle}\mathbf{1}_B$$

by Lemma 1.2. This proves (i). The set function $\mathbf{P}_c$ on $\mathcal{F}_t$ is a probability measure and $\{X_s : s \in [0, t]\}$ defined on the probability space $(\Omega, \mathcal{F}_t, \mathbf{P}_c)$ is easily proved to have stationary independent increments. We have, for $s \in [0, t]$ and $z \in \mathbf{R}^d$,

$$\int_\Omega e^{i\langle z, X(s)\rangle}\mathbf{P}_c(d\omega) = e^{-t\Psi(c)}\mathbf{E}e^{\langle c, X(t)\rangle}e^{i\langle z, X(s)\rangle} =$$

$$= e^{-s\Psi(c)}\mathbf{E}e^{\langle c, X(s)\rangle}e^{i\langle z, X(s)\rangle} = e^{s(\Psi(c+iz) - \Psi(c))},$$

by Lemma 1.2. Define A_c, ν_c and γ_c by (1.8), (1.9), and (1.10). Then, using (1.4), we get

$$e^{s(\Psi(c+iz) - \Psi(c))} =$$

$$= \exp\left[s\left\{-\frac{1}{2}\langle z, A_c z\rangle + \int \left(e^{i\langle z, x\rangle} - 1 - i\langle z, x\rangle\mathbf{1}_D(x)\right)\nu_c(dx) + i\langle \gamma_c, z\rangle\right\}\right].$$

On the other hand, it is well-known that there exists a probability measure $\widetilde{\mathbf{P}}_c$ on $\mathcal{F}$ such that $\{X_t : t \in [0, \infty)\}$ is, under $\widetilde{\mathbf{P}}_c$, a Lévy process with generator

(A_c, ν_c, γ_c). It follows that $\mathbf{P}_c$ and $\widetilde{\mathbf{P}}_c$ coincide on $\mathcal{F}_t$. Hence $\widetilde{\mathbf{P}}_c$ is an extension of $\mathbf{P}_c$. The extension is unique since $\bigcup_t \mathcal{F}_t$ generates $\mathcal{F}$. Thus (ii) and (iii) are proved. The assertion (iv) is now obvious.

2. BOCHNER'S SUBORDINATION OF SEMIGROUPS OF OPERATORS

Let Ω, F, $\mathbf{P}, \{X_t\}$ and (A, ν, γ) be as in Section 1. Assume that $d = 1$ and that $\{X_t\}$ is a subordinator under $\mathbf{P}$. That is, $X_t(\omega)$ is increasing in t (meaning $X_t(\omega) \leqslant X_{t'}(\omega)$ for $t < t'$) almost surely. This is equivalent to assuming that

$$A = 0, \ \nu((-\infty, 0)) = 0, \quad \int_{(0,1)} x\nu(dx) \leqslant \gamma.$$

Let

$$\gamma_0 = \gamma - \int_{(0,1)} x\nu(dx).$$

Since $\mathbf{E}e^{-uX(t)} < \infty$ for $u \geqslant 0$, it follows from Lemma 1.2 that

$$\mathbf{E}e^{-uX(t)} = e^{t\Psi(-u)}, \tag{2.1}$$

$$\Psi(-u) = -\gamma_0 u + \int_{(0,\infty)} (e^{-us} - 1)\nu(ds) \tag{2.2}$$

for $u \geqslant 0$. Let $\mathbf{B}$ be a real or complex Banach space, and let $\{P_t : t \geqslant 0\}$ be a strongly continuous contraction semigroup of linear operators on $\mathbf{B}$ (contraction means $\|P_t\| \leqslant 1$). Let L be the infinitesimal generator of $\{P_t\}$ in the sense of Hille and Yosida. The domain of L is denoted by $\mathbf{D}(L)$. Bochner in (Bochner, 1949; 1960) introduced a transformation of $\{P_t\}$ to a new semigroup $\{Q_t\}$ and call it *subordination*. The semigroup $\{Q_t\}$ is said to be subordinate to $\{P_t\}$. The transformation is as follows.

THEOREM 2.1. *Define, for* $f \in \mathbf{B}$,

$$Q_t f = \int_{[0,\infty]} P_s f \, \mathbf{P}(X_t \in ds). \tag{2.3}$$

Then $\{Q_t : t \geqslant 0\}$ *is a strongly continuous contraction semigroup of linear operators on* $\mathbf{B}$. *Let* M *be its infinitesimal generator. Then* $\mathbf{D}(L)$ *is a core of* M *and*

$$Mf = \gamma_0 Lf + \int_{(0,\infty)} (P_s f - f)\nu(ds) \tag{2.4}$$

for $f \in \mathsf{D}(L)$.

We define $\int_{[0,n]} P_s f \, \mathbf{P}(X_t \in ds)$ in the Riemann sense in the strong topology, and its strong limit as $n \to \infty$ is defined to be the integral in the right-hand side of (2.3.). The integral in the right-hand side of (2.4) is defined similarly, which is possible by virtue of $\|P_s f - f\| = O(s)$ as $s \downarrow 0$, and $\int_{(0,\infty)} (1 \wedge s)\nu(ds) < \infty$.

Proof of Theorem 2.1. Phillips in (Phillips, 1952) proves that $\{Q_t\}$ is a strongly continuous contraction semigroup and that, if $f \in \mathsf{D}(L)$, then $f \in \mathsf{D}(M)$ and (2.4) holds. In order to show that $\mathsf{D}(L)$ is a core of M, it is enough to prove that

$$f \in \mathsf{D}(L) \quad \text{implies} \quad Q_t f \in \mathsf{D}(L) \quad \text{for} \quad t \geqslant 0, \tag{2.5}$$

and apply Lemma 2.2 of (Watanabe, 1968). Let $f \in \mathsf{D}(L)$ and $g = Q_t L f$. Since

$$P_\varepsilon Q_t f = \int_{[0,\infty)} P_{\varepsilon+s} f \, \mathbf{P}(X_t \in ds),$$

we see that

$$\|\varepsilon^{-1}(P_\varepsilon Q_t f - Q_t f) - g\| \leqslant \int_{[0,\infty)} \|\varepsilon^{-1}(P_{\varepsilon+s} f - P_s f) - P_s L f\| \mathbf{P}(X_t \in ds) \leqslant$$

$$\leqslant \|\varepsilon^{-1}(P_\varepsilon f - f) - Lf\| \to 0 \quad \text{as} \quad \varepsilon \downarrow 0.$$

Thus (2.5) follows.

By the formulas (2.2) and (2.4) the infinitesimal generator M of $\{Q_t\}$ is denoted by $\Psi(L)$. If $\{X_t\}$ is, under $\mathbf{P}$, a stable subordinator of index $\alpha \in (0,1)$ and $\Psi(-u) = -u^\alpha$, then $M = -(-L)^\alpha$, which is identical with the usual fractional power (see (Yosida, 1965)).

3. LIMIT THEOREMS FOR SUBORDINATE SEMIGROUPS DEPENDING ON A PARAMETER

We combine Bochner's subordination with exponential transformation. Namely we construct a family of subordinators from a subordinator by exponential transformation with parameter $-\theta$, and consider subordination by these subordinators depending on the parameter θ. Our aim is to get limit theorems when θ tends to infinity.

Let us keep the assumption on $\mathbf{P}$ and $\{X_t\}$ in Section 2. So $\{X_t\}$ satisfies (2.1) and (2.2). Noting that $\mathbf{E}e^{cX(t)} < \infty$ for $c = -\theta \leqslant 0$, make the exponential transformation of $\mathbf{P}$ with parameter $-\theta \leqslant 0$. We have

$$\mathbf{P}_{-\theta}(B) = e^{-t\Psi(-\theta)}\mathbf{E}e^{-\theta X(t)}1_B \quad \text{for} \quad B \in \mathcal{F}_t. \tag{3.1}$$

The process $\{X_t\}$ under $\mathbf{P}_{-\theta}$ is a subordinator satisfying

$$\mathbf{E}_{-\theta}e^{-uX(t)} = e^{t\Psi^{(\theta)}(-u)} \quad \text{for} \quad u \geqslant 0, \tag{3.2}$$

where

$$\Psi^{(\theta)}(-u) = \Psi(-u-\theta) - \Psi(-\theta) = -\gamma_0 u + \int\limits_{(0,\infty)} (e^{-us} - 1)e^{-\theta s}\nu(ds). \tag{3.3}$$

Given a strongly continuous contraction semigroup $\{P_t\}$ on B with infinitesimal generator L, let $\{Q_t^{(\theta)}\}$ be the semigroup obtained from $\{P_t\}$ through subordination by the subordinator $\{X_t\}$ under $\mathbf{P}_{-\theta}$. That is,

$$Q_t^{(\theta)} f = \int\limits_{[0,\infty)} P_s f \, \mathbf{P}_{-\theta}(X_t \in ds). \tag{3.4}$$

The infinitesimal generator of $\{Q_t^{(\theta)}\}$ is $\Psi^{(\theta)}(L)$ by Theorem 2.1. It follows from (2.4) and (3.3) that

$$\Psi^{(\theta)}(L)f = \gamma_0 Lf + \int\limits_{(0,\infty)} (P_s f - f)e^{-\theta s}\nu(ds) \tag{3.5}$$

for $f \in \mathsf{D}(L)$. We are interested in asymptotic behaviour of $\{Q_t^{(\theta)}\}$ as $\theta \to \infty$. Assume that

$$\gamma_0 = 0 \quad \text{and} \quad \int\limits_{(0,\varepsilon)} s\nu(ds) > 0 \quad \text{for every} \quad \varepsilon > 0. \tag{3.6}$$

Let $a(\theta) = \int\limits_{(0,\infty)} se^{-\theta s}\nu(ds)$. Then $0 < a(\theta) < \infty$ for $\theta > 0$ and $a(\theta) \to 0$ as $\theta \to \infty$.

THEOREM 3.1. *The following convergence of infinitesimal generators holds:*

$$\|a(\theta)^{-1}\Psi^{(\theta)}(L)f - Lf\| \to \infty \quad \text{as} \quad \theta \to \infty, \tag{3.7}$$

for $f \in \mathsf{D}(L)$. The corresponding convergence of semigroups is as follows:

$$\|Q^{(\theta)}_{t/a(\theta)} f - P_t f\| \to 0 \quad \text{as} \quad \theta \to \infty, \tag{3.8}$$

for $f \in \mathsf{B}$ and $t \geqslant 0$.

Proof. Let $f \in \mathsf{D}(L)$. Given $\varepsilon > 0$, choose $\delta > 0$ such that

$$\|s^{-1}(P_s f - f) - Lf\| < \varepsilon \quad \text{for} \quad s \leqslant \delta.$$

By (3.5) and $\gamma_0 = 0$,

$$a(\theta)^{-1}\Psi^{(\theta)}(L)f - Lf = a(\theta)^{-1}\int_0^\infty (s^{-1}(P_s f - f) - Lf)se^{-\theta s}\nu(ds).$$

Decompose the integral to the sum of two integrals on $(0,\varepsilon]$ and (ε,∞). Then

$$\|a(\theta)^{-1}\Psi^{(\theta)}(L)f - Lf\| \leqslant$$

$$\leqslant \sup_{s\leqslant\delta}\|s^{-1}(P_sf - f) - Lf\| + a(\theta)^{-1}\int_\delta^\infty se^{-\theta s}\nu(ds)(2\delta^{-1}\|f\| + \|Lf\|).$$

Use (3.6) and note that, for $\theta > 1$,

$$a(\theta)^{-1}\int_\delta^\infty se^{-\theta s}\nu(ds) \leqslant \int_\delta^\infty se^{-\theta s}\nu(ds)\Big/\int_0^{\delta/2} se^{-\theta s}\nu(ds) \leqslant$$

$$\leqslant e^{-\theta\delta}\int_\delta^\infty se^{-\theta(s-\delta)}\nu(ds)\Big/\left(e^{-\theta\delta/2}\int_0^{\delta/2} s\nu(ds)\right) \leqslant$$

$$\leqslant e^{-\theta\delta/2}\int_\delta^\infty se^{-(s-\delta)}\nu(ds)\Big/\int_0^\infty s\nu(ds),$$

which tends to 0 as $\theta \to \infty$. Hence,

$$\|a(\theta)^{-1}\Psi^{(\theta)}(L)f - Lf\| < 2\varepsilon$$

if θ is large enough. This proves (3.7). The operator $a(\theta)^{-1}\Psi^{(\theta)}(L)$ is the infinitesimal generator of the strongly continuous contraction semigroup $\{Q^{(\theta)}_{t/a(\theta)} : t \geqslant 0\}$. It is known that, in general, convergence of infinitesimal generator on a core of the limiting infinitesimal generator implies strong convergence of the generated semigroups (see (Ethier and Kurtz, 1986, p.28)). Hence (3.8) for $f \in \mathbf{B}$ follows from (3.7) for $f \in \mathbf{D}(L)$.

THEOREM 3.2. *Suppose that, for some $\beta \in (0,1)$, $a(\theta)^{-1} = o(\theta^\beta)$ as $\theta \to \infty$. Then we have the following convergence of infinitesimal generators:*

$$\|\Psi^{(\theta)}(a(\theta)^{-1}L)f - Lf\| \to 0 \quad \text{as} \quad \theta \to \infty, \tag{3.9}$$

for $f \in \mathbf{D}(L)$. The corresponding convergence of semigroups is that

$$\|R^{(\theta)}_t f - P_t f\| \to 0 \quad \text{as} \quad \theta \to \infty, \tag{3.10}$$

for $f \in \mathbf{B}$ and $t \geqslant 0$, where

$$R^{(\theta)}_t f = \int_{[0,\infty)} P_{s/a(\theta)}f\,\mathbf{P}_{-\theta}(X_t \in ds). \tag{3.11}$$

Proof. Since $a(\theta)^{-1}L$ is the infinitesimal generator of the semigroup $\{P_{t/a(\theta)} : t \geqslant 0\}$, the operator $\Psi^{(\theta)}(a(\theta)^{-1}L)$ is the infinitesimal generator of the semigroup $\{R_t^{(\theta)}\}$ defined by (3.11). Let $f \in \mathsf{D}(L)$. By (3.5) and $\gamma_0 = 0$,

$$\Psi^{(\theta)}(a(\theta)^{-1}L)f - Lf = \int_0^\infty (P_{s/a(\theta)}f - f)e^{-\theta s}\nu(ds) - Lf =$$

$$= \frac{1}{a(\theta)} \int_0^\infty \left(\frac{P_{s/a(\theta)}f - f}{s/a(\theta)} - Lf\right) se^{-\theta s}\nu(ds).$$

Now decompose the integral into two integrals on $(0, \theta^{-\beta}]$ and $(\theta^{-\beta}, \infty)$. Notice that

$$\left\|\frac{1}{a(\theta)} \int_0^{\theta^{-\beta}} \left(\frac{P_{s/a(\theta)}f - f}{s/a(\theta)} - Lf\right) se^{-\theta s}\nu(ds)\right\| \leqslant \sup_{t \leqslant \theta^{-\beta}/a(\theta)} \left\|\frac{P_t f - f}{t} - Lf\right\|,$$

which tends to 0 as $\theta \to \infty$ by the assumption $a(\theta)^{-1} = o(\theta^\beta)$. Since

$$\sup_t \left\|\frac{P_t f - f}{t} - Lf\right\| = c < \infty,$$

we have

$$\left\|\frac{1}{a(\theta)} \int_{\theta^{-\beta}}^\infty \left(\frac{P_{s/a(\theta)}f - f}{s/a(\theta)} - Lf\right) se^{-\theta s}\nu(ds)\right\| \leqslant$$

$$\leqslant \frac{c}{a(\theta)} \int_{\theta^{-\beta}}^\infty se^{-\theta s}\nu(ds) = \frac{c}{a(\theta)} e^{-\theta^{1-\beta}} \int_{\theta^{-\beta}}^\infty se^{-\theta(s-\theta^{-\beta})}\nu(ds) \leqslant$$

$$\leqslant \frac{c}{a(\theta)} e^{-\theta^{1-\beta}} \left(\int_0^1 s\nu(ds) + \int_1^\infty se^{-\theta(s-1)}\nu(ds)\right),$$

which also tends to 0 by virtue of $a(\theta)^{-1} = o(\theta^\beta)$. This proves (3.9). Strong convergence of the corresponding semigroups is expressed by (3.10).

Remarks 3.3. The identity (3.3) suggests that

$$\Psi^{(\theta)}(L) = \Psi(L - \theta I) - \Psi(-\theta)I, \tag{3.12}$$

where I is the identity operator. This is justified in general (without assuming (3.6)) as follows. The operator $L - \theta I$ is the infinitesimal generator of the strongly continuous contraction semigroup $\{e^{-\theta t}P_t : t \geqslant 0\}$. So $\Psi(L - \theta I)$ is an operator with a core $\mathsf{D}(L - \theta I) = \mathsf{D}(L)$ and

$$\Psi(L - \theta I)f = \gamma_0(L - \theta I)f + \int_{(0,\infty)} (e^{-\theta s}P_s f - f)\nu(ds)$$

for $f \in \mathsf{D}(L)$. On the other hand, $\Psi^{(\theta)}(L)$ is an operator with a core $\mathsf{D}(L)$ and $\Psi^{(\theta)}(L)f$ has the expression (3.5) for $f \in \mathsf{D}(L)$. Thus, using (3.3), we see that

$$\Psi^{(\theta)}(L)f = \Psi(L - \theta I)f - \Psi(-\theta)f$$

for $f \in \mathsf{D}(L)$. Since the both sides of (3.12) have $\mathsf{D}(L)$ as a core, (3.12) is true as an identity between two operators. Another identity essentially same as (3.12) is

$$Q_t^{(\theta)} f = e^{-t\Psi(-\theta)} \int_{[0,\infty)} e^{-\theta s} P_s f \ \mathbf{P}(X_t \in ds) \quad \text{for} \quad f \in \mathsf{B}, \tag{3.13}$$

which follows from (1.11) and (3.4).

Example 3.4. Let $\{X_t\}$ under $\mathbf{P}$ be a strictly stable subordinator of index α $(0 < \alpha < 1)$. We have

$$\Psi(-u) = -ku^\alpha = -\frac{k\alpha}{\Gamma(1-\alpha)} \int_0^\infty (e^{-us} - 1)s^{-1-\alpha} ds$$

with a constant $k > 0$. Assume that $k = 1$. The assumption (3.6) is satisfied and $a(\theta) = \alpha\theta^{\alpha-1}$. We have

$$\Psi^{(\theta)}(L) = \theta^\alpha I - (\theta I - L)^\alpha$$

by Remark 3.3. Since the condition in Theorem 3.2 is fulfilled, we obtain (3.7) and (3.9). But, letting $(\theta/\alpha)^{1/\alpha} = \theta'$, we have

$$a(\theta)^{-1}\Psi^{(\theta)}(L) = \alpha^{-1}\theta(I - (I - \theta^{-1}L)^\alpha) = \Psi^{(\theta')}(a(\theta')^{-1}L).$$

Hence, in this example, (3.7) and (3.9) are identical assertions. We note that the operator $\theta^{1/2}I - (\theta I - L)^{1/2}$ is already mentioned in (Bochner, 1949) and treated in (Ichinose and Tamura, 1986).

Example 3.5. Suppose that $\{X_t\}$ under $\mathbf{P}$ is the gamma process with parameter 1. This means that X_t has gamma distribution with parameters t and 1,

$$\mathbf{P}(X_t \in B) = \frac{1}{\Gamma(t)} \int_B s^{t-1} e^{-s} ds.$$

We have, for $u \geqslant 0$,

$$\mathbf{E}e^{-uX(t)} = (u+1)^{-1} = e^{t\Psi(-u)}, \quad \Psi(-u) = \int_0^\infty (e^{-us} - 1)e^{-s}s^{-1} ds.$$

This satisfies (3.6) and $a(\theta) = (\theta + 1)^{-1}$. Theorem 3.1 says that

$$a(\theta)^{-1}\Psi^{(\theta)}(L)f = (\theta + 1) \int_0^\infty (P_s f - f)e^{-(\theta+1)s}s^{-1} ds \to Lf$$

for $f \in \mathbf{D}(L)$ and that

$$Q^{(\theta)}_{t/a(\theta)} f = \frac{(\theta+1)^{(\theta+1)t}}{\Gamma((\theta+1)t)} \int_0^\infty P_s f e^{-(\theta+1)s} s^{(\theta+1)t-1} ds \to P_t f$$

for $f \in \mathbf{B}$. In this example (3.9) does not hold in general, because

$$\Psi^{(\theta)}\big(a(\theta)^{-1}L\big) f = \int_0^\infty (P_{(\theta+1)s} f - f) e^{-(\theta+1)s} s^{-1} ds =$$

$$= \int_0^\infty (P_s f - f) e^{-s} s^{-1} ds = \Psi(L) f$$

for $f \in \mathbf{D}(L)$.

Example 3.6. Consider the case where $\{X_t\}$ is, under $\mathbf{P}$, the Poisson process with parameter 1. Then $\gamma_0 = 0$ and ν is a probability measure concentrated at the point 1. Thus the assumption (3.6) is not satisfied. Neither (3.7) nor (3.9) holds in general, since

$$a(\theta)^{-1}\Psi^{(\theta)}(L) f = P_1 f - f$$

and

$$\Psi^{(\theta)}\big(a(\theta)^{-1}L\big) f = e^{-\theta}(P_{e^\theta} f - f) \to 0 \quad \text{as} \quad \theta \to \infty.$$

Let us specify $\mathbf{B}$ as the Banach space of real-valued continuous functions f on a space $\mathbf{E}$ vanishing at infinity with norm $\|f\| = \sup_x |f(x)|$. We suppose that $\mathbf{E}$ is a complete separable metric space which is locally compact. Assume that the strongly continuous contraction semigroup $\{P_t\}$ on $\mathbf{B}$ in the argument above is positive and has norm one. Then the semigroups $\{Q^{(\theta)}_{t/a(\theta)} : t \geqslant 0\}$ are also positive and with norm one. It is well-known that such semigroups give Markov process on $\mathbf{E}$ with sample functions being right-continuous and having left limits (see (Ethier and Kurtz, 1986, p. 169)). Let $\widetilde{\Omega} = D\big([0,\infty), \mathbf{E}\big)$ with the Skorokhod topology (see (Ethier and Kurtz, 1986, p. 118)). For each $x \in \mathbf{E}$ let $\widetilde{\mathbf{P}}_x$ and $\widetilde{\mathbf{P}}^{(\theta)}_x$ be the probability measures on $\widetilde{\Omega}$ induced by the Markov processes starting at x given by the semigroups $\{P_t\}$ and $\{Q^{(\theta)}_{t/a(\theta)}\}$, respectively. Then, for each x,

$$\widetilde{\mathbf{P}}^{(\theta)}_x \to \widetilde{\mathbf{P}}_x \quad \text{weakly as} \quad \theta \to \infty.$$

In fact, by a general theorem (Ethier and Kurtz, 1^86, Theorem 2.5 of Chapter 4, p. 167), the convergence of the probability measures on $\widetilde{\Omega}$ follows from the convergence (3.8) of the semigroups in Theorem 3.1.

Similarly, if $\widetilde{\mathbf{P}}'^{(\theta)}_x$ is the probability measure on $\widetilde{\Omega}$ corresponding to the Markov process starting at x defined by the semigroup $\{R^{(\theta)}_t\}$ in Theorem 3.2, then the convergence (3.10) of the semigroups to $\{P_t\}$ leads to the convergence

$$\widetilde{\mathbf{P}}'^{(\theta)}_x \to \widetilde{\mathbf{P}}_x \quad \text{weakly as} \quad \theta \to \infty.$$

A special case of the convergence of the probability measures on $\widetilde{\Omega}$ discussed here is Theorem 2.1 of (Ichinose, 1987).

Acknowledgment
The author expresses his hearty thanks to Akinobu Shimizu for calling his attention to the use of the book of Ethier and Kurtz for convergence of probability measures on the path space, and to Kikuji Matsumoto for his kind advice to the proof of Lemma 1.2.

REFERENCES

Bochner, S. (1949). Diffusion equation and stochastic processes. *Proc. Nat. Acad. Sci. USA* **35**, 368–370.

Bochner, S. (1960). *Harmonic Analysis and the Theory of Probability.* Univ. California Press, Berkeley.

Ethier, S.N. and Kurtz, T.G. (1986). *Markov Processes. Characterization and Convergence.* Wiley, New York.

Ichinose, T. (1987). The nonrelativistic limit problem for a relativistic spinless particle in an electromagnetic field. *J. Func. Anal.* **73**, 233–257.

Ichinose, T. and Tamura, H. (1986). Imaginary-time path integral for a relativistic spinless particle in an electromagnetic field. *Comm. Math. Phys.* **105**, 239–257.

Khinchin, A.I. (1949). *Mathematical Foundations of Statistical Mechanics.* Dover, New York.

Kruglov, V.M. (1970). A note on infinitely divisible distributions. *Theory Probab. Appl.* **15**, 319–324.

Kruglov, V.M. (1972). Integrals with respect to infinitely divisible distributions in a Hilbert space. *Math. Notes* **11**, 407–411.

Phillips, R.S. (1952). On the generation of semigroups of linear operators. *Pacific J. Math.* **2**, 343–369.

Sato, K. (1973). A note on infinitely divisible distributions and their Lévy measures. *Sci. Rep. Tokyo Kyoiku Daigaku, Sect. A.* **12**, 101–109.

Watanabe, S. (1968). A limit theorem of branching processes and continuous state branching processes. *J. Math. Kyoto Univ.* **8**, 141–167.

Yosida, K. (1965). *Functional Analysis.* Springer, Berlin—Heidelberg etc.

Prob. Theory and Math. Stat., Vol. 2, pp. 383–393
B.Grigelionis *et al.* (Eds.)
1990 VSP/Mokslas

GENERAL LEMMAS ON THE DISTRIBUTION DENSITY FOR LARGE DEVIATIONS OF A RANDOM VECTOR

L. SAULIS
Institute of Mathematics and Informatics, Lithuanian Academy of Sciences, Akademijos 4, 232600 Vilnius, Lithuania

ABSTRACT

Limit theorems dealing with the distribution densities of large deviations of random vectors (r.v.) were considered in (Chaganty and Sethuraman, 1985; Saulis, 1984; 1987a; 1987b; 1988). The aim of this paper is to obtain general lemmas about the distribution density of large deviations of an arbitrary r.v. X with cumulants of a regular growth. The cumulant method for investigating the large deviation probabilities was presented by V. Statulevičius in (Statulevičius, 1966) and developed in (Saulis, 1987a; 1987b; 1988; Saulis and Statulevičius, 1989).

1. NOTATIONS AND RESULTS

Consider a r.v. X in a k-dimensional Euclidean space $\mathbf{R}^k$ with mean $\mathbf{E}X = 0$ and unit covariance matrix (c.m.) E. Let $(\cdot,\cdot)$ be the inner product in $\mathbf{R}^k$, $|x| = |x_1| + \cdots + |x_k|$, $\|x\| = (x,x)^{1/2}$ be the Euclidean norm of $x \in \mathbf{R}^k$. Below θ (indexed or not) denotes a quantity with $|\theta| \leqslant 1$. Let $\varphi(x)$ be the density of the standard normal distribution

$$\varphi(x) = (2\pi)^{-k/2} \exp\left\{-\frac{1}{2}\|x\|^2\right\} \tag{1}$$

and $f_X(t) = \mathbf{E}\exp\{i(t,X)\}$ be the characteristic function of a r.v. X. Denote

$$\gamma_{j_1,\dots,j_r}(0) = \left.\frac{\partial^r \ln \mathbf{E}\exp\{(h,X)\}}{\partial h_{j_1}\dots\partial h_{j_r}}\right|_{h_1=0,\dots,\, h_r=0}, \tag{2}$$

where $j_1,\dots,j_r$ run over $1,2,\dots,k$. Then the l-th order cumulant $\Gamma_l\{(X,z)\}$ of the random variable (X,z) is defined by

$$\Gamma_l\{(X,z)\} = \left(\frac{\partial}{\partial t}, z\right)^l \ln \mathbf{E}\exp\{(X,t)\}\Big|_{t=0} = \sum_{j_1,\dots,j_l=1}^{k} \gamma_{j_1,\dots,j_l}(0) z_{j_1}\dots z_{j_l}. \tag{3}$$

The density of the distribution of a r.v. $X = (X_1, \dots, X_k)$ will be denoted by $p_X(x)$.
Suppose X is a r.v. with $\mathbf{E}X = 0$ and unit c.m. E. Assume that there exist $H > 0$, $\Delta > 0$ such that for any $z \in \mathbf{R}^k$ with $\|z\| = 1$

$$| \Gamma_l\{(X, z)\} | \leqslant H(l-2)!/\Delta^{l-2}, \; l = 3, 4, \dots, s+2, \tag{S*}$$

where s is even and satisfy the inequality $s \leqslant 2\Delta^2$.
Denote

$$\tilde{\varphi}_X(x) = \exp\Big\{ \sum_{l=2}^{s} \frac{1}{l!} \Gamma_l\{(X, z)\} \Big\}, \tag{4}$$

and let $h = h(x)$ be the solution of the equation

$$x = \tilde{\varphi}_X^{-1}(h) \operatorname{grad} \tilde{\varphi}_X(h), \tag{5}$$

$$c(k, H) = \max\{1, H\}(1 + kH), \tag{6}$$

$$T = \frac{\left(1 - 5\sqrt{2e}\, c(k, H)\|x\|/(4\Delta)\right)\sqrt{s}}{4\sqrt{2e}\, c(k, H)}. \tag{7}$$

LEMMA 1. *If the r.v. X with $\mathbf{E}X = 0$ and c.m. E satisfies condition (S*) and there exists its density $p_X(x)$, where $\sup_x p_X(x) < \infty$, then in the domain*

$$0 \leqslant \|x\| < 2\sqrt{s}/\left(5\sqrt{e}\, c(k, H)\right)$$

the relation

$$\frac{p_X(x)}{\varphi(x)} = \exp\{L(x)\}\Bigg(1 + \theta_1 f(k, H) \frac{\|x\| + 1}{(1 - 5\sqrt{2e}\, c(k, H)\|x\|/(4\Delta))\sqrt{s}} +$$

$$+\theta_2 \frac{2}{(2\pi)^{k/2}} \sum_{l=0}^{s} \frac{2^l}{l!} \int\limits_{|t| \geqslant T} \left| \left(\frac{\partial}{\partial t}, x\right)^l f_X(t) \right| dt \Bigg) \tag{8}$$

holds. Here

$$L(x) = \frac{1}{2}\|x\|^2 - (h(x), x) + \ln \tilde{\varphi}_X(h) = \sum_{\nu=3}^{\infty} Q_\nu(x), \tag{9}$$

where $Q_\nu(x)$ is a ν–th order multilinear form. The coefficients of the form Q_ν depend on the first r_ν cumulants of X, where $r_\nu = \min\{\nu, s\}$, $\nu \leqslant s - 3$. They coincide with the corresponding coefficients in the multidimensional analogue of the classical Cramér—Petrov series,

$$f(k, H) = \frac{3k^2 H}{4}\left(\frac{7}{4}\right)^{(k/2)-1} + \frac{32 \cdot 2^{5k/2}\Gamma\left((k+1)/2\right) c(k, H)}{\Gamma(k/2)} +$$

$$+\theta_1 2^{(k+5)/2} e\, c(k,H) T \exp\Big\{-\frac{1}{32k}T^2\Big\} + \theta_2 \frac{32\sqrt{s}\,T^k \exp\{-2\sqrt{2e}\, c(k,H)T/s^{1/4}\}}{\max\{1,H\} 2^{k/2}\Gamma(k/2+1)}. \tag{10}$$

Now suppose that the cumulant $\Gamma_l\{(X,z)\}$ *satisfies the condition: there exist* $H>0,\ \gamma \geqslant 0$ *and* $\Delta > 0$ *such that for any* $z \in \mathbf{R}^k$ *with* $\|z\|=1$,

$$|\Gamma_l\{(X,z)\}| \leqslant H(l!)^{1+\gamma}/\Delta^{l-2},\ l=3,4,\dots . \tag{S^*_γ}$$

Denote

$$\Delta(\gamma) = c(\gamma)\Delta^{1/(1+2\gamma)},\ c(\gamma) = (1/3\sqrt{2e})(\sqrt{2}/6)^{1/(1+2\gamma)},$$
$$\Delta^*(\gamma) = 1,9\,\Delta(\gamma)/\big(5\sqrt{e}\, c(k,H)\big). \tag{11}$$

LEMMA 2. *If a r.v.* X *with* $\mathbf{E}X=0$ *and c.m.* E *satisfies condition* (S^*_γ) *and there exists its density* $p_X(x)$ *where* $\sup_x p_X(x) < \infty$, *then in the domain* $0 \leqslant \|x\| < \Delta^*(\gamma)$ *relation* (8) *holds true, with* $\sqrt{s}$ *substituted by* $\Delta^*(\gamma)$ *and*

$$L(x) = \sum_{3\leqslant\nu<p} Q_\nu(x) + \theta(\|x\|/\Delta^*(\gamma))^3,\ p \leqslant \frac{1}{\gamma}+2. \tag{12}$$

2. PROOF OF LEMMA 1

According to (4)

$$\tilde\varphi_X(z) = 1 + \sum_{l=2}^{\infty}\frac{1}{l!}\ \tilde m_l\{(X,z)\}, \tag{13}$$

where $\tilde m_l\{(X,z)\} = \Big(\frac{\partial}{\partial t}, z\Big)^l \tilde\varphi_X(t)\big|_{t=0}$ and $\tilde m_l\{(X,z)\} = m_l\{(X,z)\} = \mathbf{E}(X,z)^l,\quad 1\leqslant l \leqslant s$. It is easy to understand that

$$\tilde m_l\{(X,z)\} = l!\sum_{j=1}^{[l/2]}\frac{1}{j!}\sum_{\substack{l_1+\dots+l_j=l\\ l_i=2,\dots,s}} \frac{\Gamma_{l_1}\{(X,z)\}\dots\Gamma_{l_j}\{(X,z)\}}{l_1!\dots l_j!} \tag{14}$$

and

$$\big|\tilde m_l\{(X,z)\}\big| \leqslant l!\|z\|^l \sum_{j=1}^{[l/2]}\frac{1}{j!}\Big(\sum_{i=2}^{s}\frac{1}{i(i-1)}\Big)^j \frac{1}{\Delta_H^{k-2j}}. \tag{15}$$

After rather simple calculations we obtain

$$\big|\tilde m_l\{(X,z)\}\big| \leqslant l!\|z\|^l/a^l(l), \tag{16}$$

where $a(l) = \min\{\sqrt{l/(2e)},\ \Delta_H/\sqrt{e}\}$.

Define the vector $h = h(x)$ as the solution of the equality

$$x = \tilde{m}(h) = \tilde{\varphi}_X^{-1}(h)\operatorname{grad}\tilde{\varphi}_X(h). \tag{17}$$

Let $\tilde{E}(h)\|e_{i,j}(h)\|_{i,j=\overline{1,k}}$ be the matrix with elements

$$e_{i,j}(h) = \partial^2 \ln\tilde{\varphi}_X(h)/(\partial h_i \partial h_j), \ i,j = \overline{1,k}.$$

Suppose, that $0 \leqslant \|h\| \leqslant \delta a, \ 0 < \delta < 1, \ a = \sqrt{s/(4e)}$. Taking into account that $s \leqslant 2\Delta^2$ we obtain $a \leqslant \Delta/\sqrt{2e}$ and $\|h\|/\Delta \leqslant \delta/\sqrt{2e}$. Then

$$\ln\tilde{\varphi}_X(h) = \frac{1}{2}\sum_{j_1,j_2=1}^{k} \gamma_{j_1,j_2}(o)h_{j_1}h_{j_2} + \theta H\|h\|^2 \sum_{l=3}^{s} \frac{1}{l(l-1)}(\|h\|/\Delta)^{l-2} =$$

$$= \frac{1}{2}(h,h)\big(1 + \theta(6H\delta/23)\big). \tag{18}$$

According to (17) we have

$$\tilde{m}_j(h) = \frac{\partial}{\partial h_j}\ln\tilde{\varphi}_X(h) = h_j + \theta H h_j \sum_{l=3}^{s}\frac{1}{(l-1)}\left(\frac{\|h\|}{\Delta}\right)^{l-2} =$$

$$= h_j\big(1 + \theta(9H\delta/23)\big). \tag{19}$$

Similarly

$$e_{i,j}(h) = \frac{\partial^2 \ln\tilde{\varphi}_X(h)}{\partial h_i \partial h_j} = \delta_{i,j} + \theta H \sum_{l=3}^{s}(\|h\|/\Delta)^{l-2} = \delta_{i,j} + \theta H\delta/23, \tag{20}$$

where

$$\delta_{i,j} = \begin{cases} 1, & i = j, \\ 0, & i \neq j. \end{cases}$$

The Jacobian of the equation

$$x = \tilde{m}(h) = \tilde{\varphi}_X^{-1}(h)\operatorname{grad}\tilde{\varphi}_X(h)$$

is $\det\tilde{E}(h)$, which is continuous in h and equals to 1 at $h = 0$, which is the inner point of the set

$$O = \{h: \tilde{\varphi}_X(h) < \infty\} = \left\{h: \|h\|/\Delta \leqslant \delta/\sqrt{2e}\right\}.$$

So $\det\tilde{E}(h) \neq 0$ in some neighbourhood of zero, and the equation $x = \tilde{m}(h)$ defines a one–to–one and continuous map, under which the inner part of the set O is mapped to some open set G containing point 0, which corresponds to the value $h = 0$. From (19) it follows that

$$x = h\big(1 + \theta(9H\delta/16)\big), \ \|h\| \leqslant \delta\Delta/\sqrt{2e}, \tag{21}$$

$$h = x\big(1 + \theta(9H\delta/37)\big), \ \|h\| \leqslant \delta\Delta_H/\sqrt{2e}. \tag{22}$$

Now introduce the following functions

$$g_h(y) = \sum_{l=0}^{s} \frac{1}{l!}(h,y)^l + \frac{(h,y)^2}{\|h\|^2} \sum_{l=s+1}^{\infty} \frac{1}{l!}\, \tilde{m}_l\{(X,h)\}, \tag{23}$$

$$F_h(y) = \tilde{\varphi}_X^{-1}(h) \int\limits_{u<y+\tilde{m}(h)} g_h(u)p_X(u)du =$$

$$= \tilde{\varphi}_X^{-1}(h) \int \dots \int\limits_{\mathcal{U}} g_h(u_1,\dots,u_k)p_X(u_1,\dots,u_k)du_1\dots du_k, \tag{24}$$

$$\mathcal{U} = \{u_1 < y_1 + \tilde{m}_1(h), \dots, u_k < y_k + \tilde{m}_k(h)\}.$$

The Fourier transform of $F_h(y)$ we denote by

$$f_h(t) = \tilde{\varphi}_X^{-1}(h)\exp\{-i(t,\tilde{m}(h))\} \int\limits_{\mathbf{R}^k} e^{i(t,u)} g_h(u)p_X(u)du. \tag{25}$$

Let

$$\tilde{f}_h(t) = \exp\{-i(t,\tilde{m}(h))\} \int\limits_{\mathbf{R}^k} g_z(u)p_X(u)du, \tag{26}$$

where $z = h + it$ and $g_z(u)$ is defined by (23). Using the fact that the c.m. is E we obtain

$$\tilde{f}_h(t) = \exp\{-i(t,\tilde{m}(h))\}\tilde{\varphi}_X^{-1}(h)\Big\{ \sum_{l=0}^{s} \frac{1}{l!} \int\limits_{\mathbf{R}^k} (z,u)^l p_X(u)du +$$

$$+ \sum_{l=s+1}^{\infty} \frac{1}{l!}\, \tilde{m}_l\{(X,z)\} \frac{1}{\|z\|^2} \int\limits_{\mathbf{R}^k} (z,u)^2 p_X(u)du \Big\} =$$

$$= \exp\{-i(t,\tilde{m}(h))\}\tilde{\varphi}_X(h+it)/\tilde{\varphi}_X(h). \tag{27}$$

From (24) it follows that

$$p_h(y) = \tilde{\varphi}_X^{-1}(h) g_h\big(y+\tilde{m}(h)\big) p_X\big(y+\tilde{m}(h)\big). \tag{28}$$

It is easy to obtain that $\int\limits_{\mathbf{R}^k} p_h(y)dy = 1$. Because $f_h(t)$ is the Fourier transform of.$F_h(y)$, we have

$$p_h(u) = \frac{1}{(2\pi)^k} \int\limits_{\mathbf{R}^k} \exp\big\{-i(t,u)\big\} f_h(t)dt.$$

From (28) since $h = h(x)$ is the solution of the equation $x = \tilde{m}(h)$, we obtain

$$p_X(x) = \frac{1}{(2\pi)^k}\tilde{\varphi}_X(h)g_h^{-1}(x)\int\limits_{\mathbf{R}^k} f_h(t)dt. \tag{29}$$

From here it follows that

$$p_X(x) = \frac{1}{(2\pi)^k}\tilde{\varphi}_X(h)g_h^{-1}(x)(I_1 + I_2), \tag{30}$$

where

$$I_1 = \int\limits_{|t|\leqslant\varepsilon} f_h(t)dt,\ I_2 = \int\limits_{|t|>\varepsilon} f_h(t)dt. \tag{31}$$

The quantity $\varepsilon > 0$ will be chosen later. To investigate the integral I_2 we will use the following statement.

STATEMENT 1. *If a r.v. X with $\mathbf{E}X = 0$ and c.m. E satisfies condition (S^*), the function $g_h(u)$ defined by (23), can be expressed as*

$$g_h(u) = \sum_{l=0}^{s}\frac{1}{l!}(h,u)^l\big(1+\theta(\delta/\sqrt{2})^{s-1}\big),\ 0<\delta<1, \tag{32}$$

where $\|h\| \leqslant \delta\Delta/\big(\sqrt{2e}\,c(k,H)\big)$ and

$$g_h(u) = e^{(h,u)}\big(1+\theta_1(\delta/\sqrt{2})^{s-1}\big)\big(1+\theta_2(\delta/4)\big) \tag{33}$$

for $u \in B = \{u\colon \|u\| < 4\Delta/(5\sqrt{2e}\,c(k,H))\}$, where $c(k,H) = \max\{1,H\}(1+$ $+kH)$.

A complete proof of this statement can be found in (Saulis, 1987a; 1987b; 1988). Taking into account (25) and employing expression (32) for the function $g_h(u)$ we obtain

$$|I_2| = \Bigg|\int\limits_{|t|>\varepsilon}\Big(e^{-i(t,\tilde{m}(h))}\tilde{\varphi}_X^{-1}(h)\int\limits_{\mathbf{R}^k} e^{i(t,u)}g_h(u)p_X(u)du\Big)dt\Bigg| \leqslant$$

$$\leqslant \tilde{\varphi}_X^{-1}(h)\int\limits_{|t|>\varepsilon}|f_h^*(t)|dt\Big(1+\theta(\delta/\sqrt{2})^{s-1}\Big), \tag{34}$$

where

$$f_h^*(t) = \sum_{l=0}^{s}\frac{1}{l!}\int\limits_{\mathbf{R}^k} e^{i(t,u)}(h,u)^l p_X(u)du. \tag{35}$$

Similarly

$$\int\limits_{\mathbf{R}^k} e^{i(t,u)}(x,u)^l p_X(u)du = \frac{1}{i^l}\Big(\frac{\partial}{\partial t},x\Big)^l f_X(t),$$

where $f_X(t) = \mathbf{E}\exp\{i(t, X)\}$ is the characteristic function of the X. Thus

$$|f_h^*(t)| \leqslant \sum_{l=0}^{s} \frac{2^l}{l!} \left| \left(\frac{\partial}{\partial t}, x\right)^l f_X(t) \right| dt. \tag{36}$$

Remark. If we suppose $s = \infty$, then

$$g_h(y) = \exp\{(h, y)\}, \ \tilde{\varphi}_X(h) = \int_{\mathbf{R}^k} e^{(h,y)} p_X(y) dy$$

and $f_h(t)$ is the characteristic function of the conjugate r.v. with the distribution density

$$e^{(h,u)} p_X(u) \Big/ \int_{\mathbf{R}^k} \exp\{(h, u)\} p_X(u) du. \tag{37}$$

Considered now the integral I_1. By (31) we have

$$I_1 = I_1^{(1)} + I_1^{(2)}, \tag{38}$$

where

$$I_1^{(1)} = \int_{|t| \leqslant \varepsilon} \tilde{f}_h(t) dt, \quad I_1^{(2)} = \int_{|t| \leqslant \varepsilon} (f_h(t) - \tilde{f}_h(t)) dt$$

and the function $\tilde{f}_h(t)$ is defined by (27). Using Statement 1, we find

$$g_h^{-1}(u) = \exp\{-(h, u)\}\left(1 + \theta\left(5\exp\left\{-\frac{s}{3}\right\}\right)\right) \tag{39}$$

if $\|u\| < 4\Delta/(5\sqrt{2e}\, c(k, H))$. Denote

$$\widetilde{L}(u) = (h, u) - \ln \tilde{\varphi}_X(h), \tag{40}$$

$$\widetilde{G}(t) = i(t, u) - \ln \tilde{\varphi}_X(h + it) + \ln \tilde{\varphi}_X(h). \tag{41}$$

By (30), (38) and (27) we have

$$\frac{\tilde{\varphi}_X(h)}{(2\pi)^k g_h(x)} \int_{|t| \leqslant \varepsilon} \exp\{-i(t, x)\}\left(\tilde{\varphi}_X(h + it)/\tilde{\varphi}_X(h)\right) dt =$$

$$= \frac{\exp\{-\widetilde{L}(x)\}}{(2\pi)^k} \int_{|t| \leqslant \varepsilon} \exp\{-\widetilde{G}(t)\} dt \left(1 + \theta\left(5\exp\left\{-\frac{s}{3}\right\}\right)\right), \tag{42}$$

if $\|x\| \leqslant 4\Delta/(5\sqrt{2e}\, c(k, H))$. Furthermore

$$\int\limits_{|t|\leqslant\varepsilon} \exp\{-\widetilde{G}(t)\}dt = \int\limits_{|t|\leqslant\varepsilon} \exp\Big\{-\frac{1}{2}\Big(t,\widetilde{E}(h)t\Big)\Big\}dt+$$

$$+\int\limits_{|t|\leqslant\varepsilon}\Big(\exp\{-\widetilde{G}(t)\} - \exp\Big\{-\frac{1}{2}\Big(t,\widetilde{E}(h)t\Big)\Big\}\Big)dt. \tag{43}$$

According to (42), we have

$$-\widetilde{G}(t) = \frac{1}{2}\Big(\frac{\partial}{\partial h}, it\Big)^2 \ln\tilde{\varphi}_X(h) + \frac{1}{3!}\Big(\frac{\partial}{\partial z}, t\Big)^3 \ln\tilde{\varphi}_X(z)\big|_{z=h+\theta it} =$$

$$= -\frac{1}{2}\Big(t,\widetilde{E}(h)t\Big) + \theta H|t|^3/(6\Delta(1-\|z\|/\Delta))^2, \tag{44}$$

if $\|z\| \leqslant \delta_2 a$, $\delta < \delta_2 < 1$, or if $\|t\| \leqslant (\delta_2-\delta)a$, because

$$\frac{\partial^3 \ln\tilde{\varphi}_X(h)}{\partial z_i \partial z_j \partial z_l} = \theta\frac{H}{\Delta}\sum_{l=3}^{s}(l-2)(\|z\|/\Delta)^{l-3} = \theta\frac{H}{\Delta}(1-\|z\|/\Delta)^{-2}.$$

By (20) we obtain

$$t'\widetilde{E}(h)t = \Big(t,\widetilde{E}(h)t\Big) = (t,t) + \theta\frac{H\|h\|/\Delta}{(1-\|h\|/\Delta)}|t|^2 \geqslant$$

$$\geqslant (t,t)(1 - Hk\Delta/(1-\|h\|/\Delta)) \geqslant \frac{9}{16}\|t\|^2, \tag{45}$$

if $\|h\| \leqslant \delta\Delta/(1+16kH/7)$. It is obvious, that $\Delta/(1+16kH/7) \geqslant \Delta/\big(\sqrt{2e}(1+kH)\big)$. Put

$$T = (\delta_2-\delta)a/(1+kH), \tag{46}$$

where $\delta_2 c_H = 1-(1-\delta c_H)/(2s^{1/4})$, $c_H = \max\{1,H\}$.
We shall keep this notation in what follows. With this δ_2 the inequality $\delta < \delta_2$ will be satisfied if $\delta c_H < 1$. Then $\delta_2 - \delta = (1-\delta c_H)(1-1/(2s^{1/4}))/c_H$.
Taking into account that $\|h\| \leqslant \delta\Delta/\big(\sqrt{2e}(1+kH)\big)$, and using (19), we obtain

$$\|x\| = \|h\|/\big(1+\theta(1/5)\big),\ \delta \leqslant 5\sqrt{2e}(1+kH)\|x\|/(4\Delta). \tag{47}$$

So $1-\delta c_H \geqslant 1-5\sqrt{2e}\,c(k,H)\|x\|/(4\Delta)$, where $c(k,H) = c_H(1+kH)$ and

$$T \geqslant \big(1-5\sqrt{2e}\,c(k,H)\|x\|/(4\Delta)\big)a/\big(2\sqrt{2e}\,c(k,H)\big). \tag{48}$$

Returning to (44) we have

$$\frac{H|t|^3}{6\Delta(1-\|z\|/\Delta)^2} \leqslant \frac{2ekH\|t\|^2}{(\sqrt{2e}-\delta_2)^2 6\Delta} \leqslant \frac{7|t|\|t\|^2}{T}. \tag{49}$$

Denote $A_p = \{t\colon \|t\| \leqslant p,\ t\in\mathbf{R}^k\}$ and $B_p = \{t\colon |t| \leqslant p,\ t\in\mathbf{R}^k\}$. It is easy to see that $B_p \subset A_p$. Relations (44), (49) and (45) imply

$$\Big|\exp\{-\widetilde{G}(t)\} - \exp\Big\{-\frac{1}{2}\Big(t, \widetilde{E}(h)t\Big)\Big\}\Big| \leqslant$$

$$\leqslant \exp\Big\{-\frac{1}{2}\Big(t, \widetilde{E}(h)t\Big)\Big\}\Big|\exp\{H|t|^3/(6\Delta(1-\|z\|/\Delta)^2)\} - 1\Big| \leqslant$$

$$\leqslant \exp\Big\{-\frac{1}{2}\Big(t, \widetilde{E}(h)t\Big)\Big\}\sum_{l=1}^{\infty}\frac{1}{l!}\Big(\frac{|t|}{T}\Big)^l\Big(\frac{7\|t\|^2}{32}\Big)^l \leqslant$$

$$\leqslant \frac{|t|}{T}\exp\Big\{-\frac{1}{2}\Big(t, \widetilde{E}(h)t\Big)\Big\}\Big(\exp\Big\{\frac{7}{32}\|t\|^2\Big\} - 1\Big) \leqslant$$

$$\leqslant \frac{2|t|}{T}\exp\Big\{-\frac{1}{16}\|t\|^2\Big\}, \tag{50}$$

if $|t| \leqslant T$. Now it follows that

$$\int\limits_{|t|\leqslant T}\Big|\exp\{-\widetilde{G}(t)\} - \exp\Big\{-\frac{1}{2}\Big(t, \widetilde{E}(h)t\Big)\Big\}\Big|dt \leqslant$$

$$\leqslant \frac{4\sqrt{k}2^k}{T}\int\limits_{\mathbf{R}^k}\|t\|\exp\Big\{-\frac{1}{4}\|t\|^2\Big\}dt = \frac{2^{2k+3}\pi^{k/2}\,\Gamma((k+1)/2)\sqrt{k}}{\Gamma(k/2)T}$$

and

$$\int\limits_{|t|\leqslant T}\exp\Big\{-\frac{1}{2}\Big(t, \widetilde{E}(h)t\Big)\Big\}dt =$$

$$= (2\pi)^{k/2}\big(\det \widetilde{E}^{-1}(h)\big)^{1/2} + \theta\int\limits_{|t|>T}\exp\Big\{-\frac{1}{2}\Big(t, \widetilde{E}(h)t\Big)\Big\}dt =$$

$$= (2\pi)^{k/2}\big(\det \widetilde{E}^{-1}(h)\big)^{1/2} + \theta\exp\Big\{-\frac{1}{32k}T^2\Big\}\int\limits_{\mathbf{R}^k}\exp\Big\{-\frac{1}{4}\|t\|^2\Big\}dt =$$

$$= (2\pi)^{k/2}\big(\det \widetilde{E}^{-1}(h)\big)^{1/2} + \theta 2^k\pi^{k/2}\exp\Big\{-\frac{1}{32k}T^2\Big\}. \tag{51}$$

Taking now into account relations (41), (43) and (51), we obtain

$$\frac{\tilde{\varphi}_X(h)}{(2\pi)^k g_h(x)}\int\limits_{|t|\leqslant T}e^{-i(t,x)}\tilde{\varphi}_X(h+it)/\tilde{\varphi}_X(h)dt =$$

$$= \exp\{-\widetilde{L}(x)\}\Big\{\frac{\big(\det \widetilde{E}^{-1}(h)\big)^{1/2}}{(2\pi)^{k/2}} + \frac{2^{2k+3}\Gamma((k+1)/2)\sqrt{k}}{\pi^{k/2}\Gamma(k/2)T} +$$

$$+\theta\frac{1}{\pi^{k/2}}\exp\Big\{-\frac{1}{32k}T^2\Big\}\Big\}\Big(1+\theta\Big(5\exp\Big\{-\frac{s}{3}\Big\}\Big)\Big) \tag{52}$$

where $\widetilde{L}(x)$ is defined by (40).

To estimate the integral $I_1^{(2)}$ in (38) we must to estimate $|f_h(t) \dot{-} \tilde{f}_h(t)|$, where $f_h(t)$ and $\tilde{f}_h(t)$ are defined by (25) and (27).

STATEMENT 2. *In the domain $|t| \leqslant T$, where*

$$T = \left(1 - \frac{5\sqrt{2e}\, c(k,H)\|x\|}{4\Delta}\right)\frac{\sqrt{s}}{4\sqrt{2e}\, c(k,H)},$$

the following estimate

$$\left|f_h(t) - \tilde{f}_h(t)\right| \leqslant \frac{32\sqrt{s}\exp\left\{-\frac{1}{2}\left(1 - 5\sqrt{2e}\, c(k,H)\|x\|/(4\Delta)\right)s^{1/4}\right\}}{\left(1 - 5\sqrt{2e}\, c(k,H)\|x\|/(4\Delta)\right)c_H} \tag{53}$$

is valid.

This statement is proved in (Saulis, 1987a, p. 542—545). Using Statement 2 and relations (30), (37)—(40) we obtain

$$\frac{\tilde{\varphi}_X(h)}{(2\pi)^k g_h(x)} \int\limits_{|t|\leqslant T} \left|f_h(t) - \tilde{f}_h(t)\right| dt \leqslant \exp\left\{-\widetilde{L}(x)\right\}\left(1 + \theta e^{-s/3}\right)\times$$

$$\times \frac{32\sqrt{s}\, T^k \exp\left\{-\frac{1}{2}\left(1 - 5\sqrt{2e}\, c(k,H)\|x\|/(4\Delta)\right)s^{1/4}\right\}}{2^k \pi^{k/2}\left(1 - 5\sqrt{2e}\, c(k,H)\|x\|/(4\Delta)\right)\Gamma(k/2+1)c_H}. \tag{54}$$

From (30), (31), (37), (42), (52) and (54) it follows that

$$\frac{p_X(x)}{\varphi(x)} = \exp\left\{L(x)\right\}\left\{\left(\det \widetilde{E}^{-1}(h)\right)^{1/2} + \right.$$

$$+\theta_1 \frac{8\cdot 2^{5k/2}\Gamma\left((k+1)/2\right)}{\Gamma(k/2)}\cdot\frac{1}{T} + \theta_2 2^{k/2}\exp\left\{-\frac{1}{32k}T^2\right\} +$$

$$+\theta_3 \frac{32\sqrt{s}\, T^k \exp\left\{-\frac{1}{2}\left(1 - 5\sqrt{2e}\, c(k,H)\|x\|/(4\Delta)\right)s^{1/4}\right\}}{\left(1 - 5\sqrt{2e}\, c(k,H)\|x\|/(4\Delta)\right)c_H 2^{k/2}\Gamma(k/2+1)} +$$

$$+\theta_4 \frac{2}{(2\pi)^{k/2}}\sum_{l=0}^{s}\frac{2^l}{l!}\int\limits_{|t|\geqslant T}\left|\left(\frac{\partial}{\partial t}, x\right)^l f_X(t)\right| dt\left(1 + \theta e^{-s/3}\right), \tag{55}$$

if $\|x\| \leqslant 4\Delta/\left(5\sqrt{2e}\, c(k,H)\right)$. Here

$$L(x) = \frac{1}{2}\|x\|^2 - (h,x) + \ln\tilde{\varphi}_X(h),\ h = h(x).$$

Now we are ready to consider $\det \widetilde{E}^{-1/2}(h)$.

STATEMENT 3. *If $k \geqslant 2$ and $\|h\| < \Delta/\left(1 + \sqrt{2e}\, kH\right)$, then*

$$\left(\det \widetilde{E}(h)\right)^{-1/2} = 1 + \frac{k}{2}\left(1 + \theta\varrho_1\right)^{(k/2)-1}\varrho_1, \tag{56}$$

where

$$\varrho_1 = (kH\|h\|/\Delta)/(1-(1+kH)\|h\|/\Delta) \leqslant (\sqrt{2e}-1)^{-1}.$$

Relation (56) is proved in (Saulis, 1987a, p. 547). In order to finish the proof of Lemma 1, it is enough to substitute $h(x)$ of the equation $x = \tilde{\varphi}_X^{-1}(h)\operatorname{grad}\tilde{\varphi}_X(h)$ into the expression for $L(x)$, i.e.,

$$L(x) = \|x\|^2/2 - (x, h(x)) + \ln\tilde{\varphi}_X(h(x)) = \sum_{\nu\geqslant 3} Q_\nu(x). \tag{57}$$

Here $Q_\nu(x)$ is a ν–th order multilinear form in x_l, $l=1,\dots,k$. The coefficients of the form Q_ν are functions of the first cumulants of X, where $r_\nu = \min\{\nu, s\}$. Besides, for $\nu \leqslant s-3$, the coefficients of the form Q_ν coincide with the corresponding coefficient in the multidimensional analogue of the Cramér–Petrov series.

Proof as Lemma 2. At first we shall prove that condition (S_γ^*) implies

$$|\Gamma_l\{(X,z)\}| \leqslant H(l-2)!/\Delta_s^{l-2},\ l=3,4,\dots,s+2,$$

where s is even and does not exceed $2\Delta_s^2$. For $s\geqslant 4$ we have

$$(l!)^{1+\gamma}/\Delta^{l-2} \leqslant (l-2)!/\Delta_s^{l-2},$$

where $\Delta_s = \Delta/(6(s+2)^\gamma)$. Hence for all even s, satisfying the inequalities

$$4 \leqslant s \leqslant \Delta^2/(18(s+2)^{2\gamma}), \tag{58}$$

condition (S_γ^*) implies (S^*). Take $s = 2\left[\frac{1}{2}(\Delta^2/18)^{1/(1+2\gamma)}\right]-2$, where $[m]$ is the integral part of m. Such s is even and satisfies inequalities (57). Immediately (57) imply

$$0,95(\sqrt{2}\Delta/6)^{1/(1+2\gamma)} < \sqrt{s} < (\sqrt{2}\Delta/6)^{1/(1+2\gamma)}$$

It is easy to see that $\Delta_s = \Delta/(6(s+2)^\gamma) \geqslant \sqrt{s/2}$. Now we obtain Lemma 2 by applying Lemma 1 in the domain $0\leqslant \|x\| < \Delta^*(\gamma)$, where $\Delta^*(\gamma)$ is defined by (11).

REFERENCES

Chaganty, N.R. and Sethuraman, J. (1985). Multi–dimensional large deviation. *Ann. Probab.* **13**, 97–114.

Saulis, L. (1984). On large deviations in Euclidean space $\mathbf{R}^k$. *Soviet Math. Dokl.* **29**, 443–446.

Saulis, L. (1987a). General large deviation lemmas for random vector with semiinvariants of regular growth. I. *Liet. Mat. Rink.* **27**, 535–549.

Saulis, L. (1987b). General large deviation lemmas for random vector with semiinvariants of regular growth. II. *Liet. Mat. Rink.* **27**, 747–758.

Saulis, L. (1988). General large deviation lemmas for random vector with semiinvariants of regular growth. III. *Liet. Mat. Rink.* **28**, 99–111.

Saulis, L. and Statulevičius, V. (1989). *Limit theorems for large deviations.* Mokslas, Vilnius.

Statulevičius, V. (1966). On large deviations. *Z. Wahrsch. verw. Geb.* **6**, 133–144.

Prob. Theory and Math. Stat., Vol. 2, pp. 394–410
B.Grigelionis *et al.* (Eds.)
1990 VSP/Mokslas

SPEED OF CONVERGENCE IN THE CENTRAL LIMIT THEOREM IN HILBERT SPACE UNDER WEAKENED MOMENT CONDITIONS

V.V. SAZONOV and V.V. ULYANOV
Steklov Mathematical Institute, 42 Vavilov str., 117966 Moscow, USSR
Faculty of Computing Mathematics and Cybernetics, Moscow State University, Lenin Hills, 119899 Moscow, USSR

ABSTRACT

A new estimate of the speed of convergence on balls in the CLT for i.i.d. Hilbert space valued random variables X_k, $k = 1, 2, \ldots$ is obtained. The estimate is of order $n^{(2-p)/2}$, $2 \leqslant p \leqslant 3$, if $\mathbf{E}|X_1|^p < \infty$ and depends only on the centre of the ball, $\mathbf{E}X_1$, $\mathbf{E}|X_1|^2$, $\mathbf{E}|X_1|^3$ and on the first six eigen-values of the covariance operator of X_1.

1. THE MAIN RESULT

Let X_i, $i = 1, 2, \ldots$, be independent identically distributed random variables with values in a separable Hilbert space H. Assume that $\mathbf{E} \mid X_1 \mid^2 < \infty$ and denote by V the covariance operator of $X_1 : (Vx, y) = \mathbf{E}(X_1 - \mathbf{E}X_1, x)(X_1 - \mathbf{E}X_1, y)$. Put

$$S_n = n^{-1/2}\sigma^{-1}\sum_{i=1}^{n}(X_i - \mathbf{E}X_i),$$

where $\sigma^2 = \mathbf{E} \mid X_1 - \mathbf{E}X_1 \mid^2$, and let Y be a $(0, \sigma^{-2}V)$ Gaussian H-valued random variable.

The central limit theorem implies that for any $a \in H$, $r \geqslant 0$

$$\Delta_n(a, r) = \left|\mathbf{P}(|S_n - a| < r) - \mathbf{P}(|Y - a| < r)\right| \to 0 \quad \text{as} \quad n \to \infty.$$

In this paper we will construct an estimate of $\Delta_n(a, r)$, which involves only $\beta_p = \mathbf{E}|X_1 - \mathbf{E}X_1|^p$, where p is a number such that $2 \leqslant p \leqslant 3$ (the moments of higher order may be infinite).

THEOREM. *For any p, $2 \leqslant p \leqslant 3$, there exists $c(p)$ such that for all $n \geqslant 1$*

$$\Delta_n(a, r) \leqslant c(p)\left(\prod_{i=1}^{6}\sigma_i\right)^{-p/3}\sigma^p\beta_p(1 + |a|^p)n^{-(p-2)/2}, \tag{1}$$

where $\sigma_1^2 \geqslant \sigma_2^2 \geqslant \dots$ are the eigenvalues of V.

Remark. In the case $p = 3$ the theorem was proved in (Sazonov et al., 1989). It was noted there that in this case our estimate is *precise*. If $p = 2$, then (1) is obviously true with $c(p) = 1$. Thus we may assume that $2 < p < 3$.
In what follows, we will assume that $\mathbf{E}X_1 = 0$, $\mathbf{E}|X_1|^2 = 1$. The general case is reduced to this one if we replace X_i by $\sigma^{-1}(X_i - \mathbf{E}X_i)$.
Below the following notation will be used: δ_x is the probability measure concentrated at x, i.e., $\delta_x(E) = 1$ if $x \in E$ and $\delta_x(E) = 0$ if $x \notin E$; E^c is the complement of a set E; for $a \in H$, $r \geqslant 0$ put $B_r(a) = \{x \in H : \ |x - a| < r\}$, $B_r = B_r(0)$; if Q is a measure, then Q^n is the n-fold convolution of Q with itself.

2. AUXILIARY MEASURES

Let P be the distribution of X_1. Fix n and put

$$P_1(E) = P(E \cap B_{n^{1/2}}) + P(B^c_{n^{1/2}})\delta_0(E),$$

$$Q(E) = P(E \cap B_L)/P(B_L),$$

assuming that $P(B_L) > 0$.

Denote by W the covariance operator of Q and let $\tau_1^2 \geqslant \tau_2^2 \geqslant \dots$ be its eigenvalues. It was shown in (Sazonov et al., 1989, Lemma 8) that

$$\tau_i^2 \leqslant p\sigma_i^2, \quad i = 1, 2, \dots, \tag{2}$$

where $p = 1/P(B_L)$, and that if for some positive integer l

$$\int_{B_L^c} |x|^2 P(dx) \leqslant \sigma_l^2/3, \tag{3}$$

then

$$\tau_i^2 \geqslant (2/3)\sigma_i^2, \quad i = 1, 2, \dots, l. \tag{4}$$

Define now

$$s_p = (3\sigma_4^{-2}\beta_p)^{1/(p-2)},$$

$$R_p = 3^{1/(p-2)}\Lambda^{2(3-p)/(p-2)}\sigma_6^{-2/(p-2)}\beta_p^{3/p}n^{(3-p)/p}, \tag{5}$$

where $\Lambda = [\prod_{i=1}^{6} \sigma_i]^{1/3}$. We will fix a p, $2 < p < 3$, and write for simplicity $s = s_p$, $R = R_p$, $\beta = \beta_p$.
Note that if

$$\Lambda^{-p}\beta(1 + |a|^p)n^{(2-p)/2} \leqslant 1, \tag{6}$$

then $s \leqslant R$ since

$$s^{p-2} = R^{p-2}\sigma_4^2\beta\Lambda^{2(p-3)}\sigma_6^2\beta^{3(2-p)/p}n^{(p-2)(p-3)/p} \leqslant$$

$$\leqslant R^{p-2}\Lambda^{2(p-3)}\beta^{2(3-p)/p}n^{(p-2)(p-3)/p} =$$
$$= R^{p-2}(\Lambda^{-p}\beta n^{(2-p)/2})^{2(3-p)/p} \leqslant R^{p-2}.$$

But $P(B_s^c) \leqslant 1/2$. Hence if (6) is satisfied, then

$$P_2 \geqslant P_2'/2, \tag{7}$$

where

$$P_2(E) = P(E \cap B_R)/P(B_R),$$
$$P_2'(E) = P(E \cap B_s)/P(B_s),$$

and if (6) is satisfied and $R \leqslant n^{1/2}$, then

$$P_1 \geqslant P_2/2. \tag{8}$$

Inequalities (7), (8) imply, respectively

$$P_2 = (P_2' + P_3')/2$$

and

$$P_1 = (P_2 + P_3)/2,$$

where P_3, P_3' are some probability measures.

In what follows, we will assume that s, R are defined by (5) and that condition (6) is satisfied (if (6) is violated, the theorem is obviously true). We will say that we have

$$\text{the case } A \text{ if } R \leqslant n^{1/2}$$

and

$$\text{the case } B \text{ if } R > n^{1/2}.$$

Note that if V_2' (resp., V_2) is the covariance operator corresponding to P_2' (resp., P_2) and $(\sigma_{21}')^2 \geqslant (\sigma_{22}')^2 \geqslant \ldots$ (resp., $\sigma_{21}^2 \geqslant \sigma_{22}^2 \geqslant \ldots$) are its eigenvalues, then

$$(\sigma_{2i}')^2 \leqslant 2\sigma_i^2,\ \ i = 1, 2, \ldots;\ \ (\sigma_{2i}')^2 \geqslant (2/3)\sigma_i^2,\ \ 1 \leqslant i \leqslant 4, \tag{9}$$

and

$$\sigma_{2i}^2 \leqslant 2\sigma_i^2,\ \ i = 1, 2, \ldots;\ \ \sigma_{2i}^2 \geqslant (2/3)\sigma_i^2,\ \ 1 \leqslant i \leqslant 6. \tag{10}$$

Indeed, as we observed above, $P(B_s^c) \leqslant 1/2$ and obviously (3) is true with $L = s$, $l = 4$. Hence we have (2) and (4) with $\tau_i = \sigma_{2i}'$ and (9) follows. Furthermore, (6) implies $s \leqslant R$ so that $P(B_R^c) \leqslant 1/2$; moreover (6) implies (3) with $L = R, l = 6$. Thus we have (2) and (4) with $\tau_i = \sigma_{2i}$ and (10) follows.

Let P_4 be the probability measure corresponding to $Z' = \xi(Z - \mathbf{E}Z) + Y' + \mathbf{E}Z$, where ξ is a bounded real random variable such that $\mathbf{E}\xi = 0$, $\mathbf{E}\xi^2 = 1/2$, $\mathbf{E}\xi^3 = 1$ (see (Sazonov, 1981, p. 84)), Z be distributed as P_2, Y' be a $(0, V_2/2)$ Gaussian random variable and ξ, Z, Y' be independent. Note that Z' has mean $\mathbf{E}Z$, covariance operator V_2 and for any $h_1, h_2, h_3 \in H$

$$\mathbf{E}(Z', h_1)(Z', h_2)(Z', h_3) = \mathbf{E}(Z, h_1)(Z, h_2)(Z, h_3)$$

(see (Sazonov, 1981, p. 85)), i.e., Z and Z' have the same moments of the first three orders. In the case A we put $P_5 = (P_4 + P_3)/2$. Denote by P_6 the distribution of $\xi(Z - \mathbf{E}Z) + \mathbf{E}Z$. Finally, define

$$P_1'(E) = P(E \cap B_R) + P(B_R^c)\delta_0(E)$$

and let $P_7 = p_1 P_2 + p_2 \delta_0$, where $p_1 = 2P(B_R) - 1$, $p_2 = 2P(B_R^c)$. Since by (6) $p_1 \geqslant 0$ and $p_1 + p_2 = 1$, P_7 is a probability measure. Note that $P_1' = (P_2 + P_7)/2$ and denote $P_8 = (P_4 + P_7)/2$.

In what follows, $X_1^{(k)}, X_2^{(k)}, \dots$ is a sequence of independent H-valued random variables with the same distribution P_k, $k = 1, \dots, 8$. The measures P_k and the random variables $X_i^{(k)}$ with $k = 3, 5$ are considered only in the case A. Denote $\bar{P}_k(\cdot) = P_k(n^{1/2}\cdot)$.

3. LEMMAS

LEMMA 1. *Let X_i, $i = 1, 2, \dots$, and S_n be the same as in the theorem and let $\chi_i(u)$ be the indicator function of the event $\{| X_i | < u\}$, $u \geqslant 0$. Denote $S_n(u) = n^{-1/2} \sum_{i=1}^{n} X_i \chi_i(u)$. Then for any real Borel measurable function f defined on H, any u_1, u_2, $0 \leqslant u_2 \leqslant u_1 \leqslant \infty$, and any H-valued random variable Z*

$$\Delta_1 = \Big| \mathbf{E}\big(f(S_n(u_1) + Z) - f(S_n(u_2) + Z\big) \Big| \leqslant$$

$$\leqslant n\mathbf{E} \Big| f(S_n(u_1) + Z) - f(S_n(u_2) + Z) \Big| (1 - \chi_1(u_2)),$$

provided the expectations exist.

Proof. We have

$$\Delta_1 = \Big| \mathbf{E}\Big(f(S_n(u_1) + Z) - f(S_n(u_2) + Z)\Big) \Big(\mathbf{1}_{\cap\{|X_i| < u_2\}} + \mathbf{1}_{\cup\{|X_i| \geqslant u_2\}}\Big) \Big| \leqslant$$

$$\leqslant \mathbf{E} \Big| f(S_n(u_1) + Z) - f(S_n(u_2) + Z) \Big| \sum_{i=1}^{n} (1 - \chi_i(u_2)) \leqslant$$

$$\leqslant n\mathbf{E} \Big| f(S_n(u_1) + Z) - f(S_n(u_2) + Z) \Big| (1 - \chi_1(u_2)).$$

LEMMA 2. *Let $q \geqslant 0$, $0 \leqslant n_k \leqslant n$, $k = 1, \dots, 6$. Then if $k = 1$*

$$\mathbf{E} \Big| n^{-1/2} \sum_{i=1}^{n_k} X_i^{(k)} \Big|^q \leqslant c(q). \tag{11}$$

In the case A (11) is true also if $k = 5$; furthermore, if $|n_2 - n_3 + n_4| \leqslant n^{3/(2p)}$, then

$$\mathbf{E} \Big| n^{-1/2} \sum_{k=2}^{4} \sum_{i=1}^{n_k} X_i^{(k)} \Big|^q \leqslant c(q) \tag{12}$$

and if $|n_6 - n_3| \leqslant n^{3/(2p)}$, *then*

$$\mathbf{E}\Big|n^{-1/2}\sum_{k=3,6}\sum_{i=1}^{n_k} X_i^{(k)}\Big|^q \leqslant c(q). \tag{13}$$

In the case B for any n', $0 \leqslant n' \leqslant n$,

$$\mathbf{E}\Big|n^{-1/2}\sum_{i=1}^{n'} X_i^{(k)}\Big|^{4p/(6-p)} \leqslant c(p), \quad k = 2,4,6,7,8. \tag{14}$$

Finally, if Z is a Gaussian H-valued random variable, then

$$\mathbf{E}|Z|^q \leqslant c(q)(\mathbf{E}\mid Z\mid^2)^{q/2}. \tag{15}$$

Proof. Relation (11) with $k = 1$ and in the case A with $k = 5$, as well as (15), are proved in (Sazonov et al., 1989, Lemma 2).
Furthermore, by (6) and the obvious inequality $\sigma_6^2 \leqslant \Lambda^{3-p}$, we have

$$|\mathbf{E}X_1^{(4)}| = |\mathbf{E}X_1^{(2)}| = P(B_R)^{-1}\Big|\int_{B_R^c} xP(dx)\Big| \leqslant 2R^{-1} \leqslant n^{-(3-p)/(2p)}.$$

Using this relation in the case A we obtain (12) and (13) in the same way as in (Sazonov et al., 1989) (19) and (20) were proved.

To prove (13) in the case B we use the inequality

$$\mathbf{E}\Big|n^{-1/2}\sum_{i=1}^{n'} X_i^{(k)}\Big|^q \leqslant c(q)n^{-q/2}\Big((n'|\mathbf{E}X_1^{(k)}|)^q +$$

$$+(\mathbf{E}|X_1^{(k)}|^2)^{q/2} + n'\mathbf{E}|X_1^{(k)}|^q\Big) \tag{16}$$

(see, e.g., (Pinelis, 1980)) with $q = 4p/(6-p)$. It is easy to see that $\mathbf{E}|X_1^{(2)}|^2 \leqslant 2$ and, since $R > n^{1/2}$,

$$|\mathbf{E}X_1^{(2)}| = P(B_R)^{-1}\Big|\int_{B_R^c} xP(dx)\Big| \leqslant 2\int_{B_R^c} \mid x\mid P(dx) \leqslant 2n^{-1/2}.$$

Furthermore,

$$\mathbf{E}|X_1^{(2)}|^{4p/(6-p)} = P(B_R)^{-1}\int_{B_R^c} \mid x\mid^{p+p(p-2)/(6-p)} P(dx) \leqslant$$

$$\leqslant 2R^{p(p-2)/(6-p)}\beta.$$

Using now (6) and observing that $\Lambda^3 \leqslant \sigma_6$, we have

$$n^{-2p/(6-p)} n' \mathbf{E}|X_1^2|^{4p/(6-p)} \leqslant 2n^{3(2-p)/(6-p)} R^{p(p-2)/(6-p)} \beta =$$

$$= 2 \cdot 3^{p/(6-p)} \left[\Lambda^{3-p} \sigma_6^{-1} \beta n^{-(p-2)/2}\right]^{2p/(6-p)} \leqslant$$

$$\leqslant 6 \left[\Lambda^{-p} \beta n^{-(p-2)/2}\right]^{2p/(6-p)} \leqslant 6. \tag{17}$$

This proves (14) if $k = 2$. Using (17), one can prove (14) with $k = 4, 6, 7, 8$ in the same way as in (Sazonov et al., 1989) (21) was proved for these k's.

LEMMA 3. *Let Z_j, $j = 1, 2, \ldots$, be independent H-valued random variables with the same distribution Q such that $Q = (Q_1 + Q_2)/2$, where Q_1, Q_2 are probability measures, $Q_1(B_L) = 1$ for some $L > 0$, and the covariance operator V' of Q_1 has trace $\mathrm{tr} V' \leqslant 2$. For $a > 0$ let Y be a $(0, aV)$ Gaussian H-valued random variable and let Z_0, Z_{0j}, $j = 1, 2, \ldots$, be independent H-valued random variables, which are independent of Z_j, $j = 1, 2, \ldots$, and Y. Finally, let l, m, n_0, n be positive integers such that $l \leqslant m$, $l + m \leqslant n$, $n_0 \leqslant n$. Define*

$$U_1 = n^{-1/2} \sum_{j=1}^{l+m} Z_j, \quad U_2 = n^{-1/2} \sum_{j=1}^{n_0} Z_{0j}.$$

Then for any $A > 0$, integer $r \geqslant 0$, any t if $l \leqslant L^2$, and $t : |\, t \,| \leqslant c(A) L^{-1} n \times \times (l \ln(L^{-2} l))^{-1/2}$ if $l > L^2$,

$$\left| \mathbf{E} \exp\{it|U_1 + U_2 + Z_0|^2\} (x, U_1 + U_2)^r \right| \leqslant$$

$$\leqslant c_1 K_1 \Big(\exp\{-c_2 l\} + c(A)(L^2/l)^A + h^{1/2}(c_3 t_1^2 lm/n^2,\ V') \Big),$$

where $c_j = c_j(r)$, $j = 1, 2, 3$,

$$K_1 = \mathbf{E}\Big|(x, U_1 + U_2)\Big|^r \quad \text{if} \quad l < r + 1,$$

$$K_1 = \sum_{j=1}^{r+1} \mathbf{E}\Big|(x, U_{1j} + U_{2j})\Big|^r \quad \text{if} \quad l \geqslant r + 1,$$

where $U_{1j} = n^{-1/2} \sum\limits_{k=k_j-1}^{k_j} Z_k$, $k_j = j[(l+m)/(r+1)]$, $j = 0, \ldots, r$, $k_{r+1} = l + m$ and U_{2j}, $j = 1, \ldots, r+1$, is a partition of the sum U_2 into $r + 1$ parts (some of them may be empty),

$$t_1 = \min\Big\{ \,|\, t \,|,\ L^{-1} n \big(m \ln(L^{-2} m)\big)^{-1/2} \Big\},$$

$$h(s, V') = \prod_{j=1}^{\infty} \big(1 + 2s(\sigma_j')^4\big)^{-1/2}$$

and $(\sigma_1')^2 \geqslant (\sigma_2')^2 \geqslant \dots$ are the eigenvalues of V'.
Moreover, for any t

$$\left|\mathbf{E}\exp\left\{it|Y+U_2+Z_0|^2\right\}(x,Y+U_2)^r\right| \leqslant c_4 K_2 h^{1/2}(c_5 t^2 a, V),$$

where $c_j = c_j(r),\quad j=4,5$ *and*

$$K_2 = \sum_{j=1}^{r+1} \mathbf{E}\left|(x, Y_j + U_{2j})\right|^r,$$

where $Y_j,\quad j=1,\dots,r+1$, *are* $(0,(a/(r+1))V)$ *Gaussian* H*–valued random variables independent of* Z_{0j} *and of each other,* U_{2j}, $j=1,\dots,r+1$, *is a partition of the sum* U_2 *into* $r+1$ *parts (some of them may be empty).*

The proof of Lemma 3 may be found in (Sazonov et al., 1989).

LEMMA 4. *For any* $A>0,\quad a>1/2$

$$2^{-n}\sum_{q:|2q-n-1|\geqslant n^a}\binom{n}{q} \leqslant c(A,a)n^{-A}.$$

Lemma 4 follows from exponential inequalities for the binomial distribution (see, e.g., (Petrov, 1987, p. 84)).

LEMMA 5. *For any* $p,\quad 2<p<3$,

$$\Delta_2 = \left|\mathbf{P}(|S_n' - a| < r) - \mathbf{P}(|S_n'' - a| < r)\right| \leqslant$$

$$\leqslant c(p)\Lambda^{-p}\beta(1+|a|^p)n^{-(p-2)/2},$$

where

$$S_n' = S_n(n^{1/2}),\quad S_n'' = n^{-1/2}\sum_{i=1}^n X_i^{(5)}\quad \text{in the case A},$$

$$S_n' = S_n(R),\quad S_n'' = n^{-1/2}\sum_{i=1}^n X_i^{(8)}\quad \text{in the case B}$$

(the definition of $S_n(u)$ *was given in Lemma 1).*

Proof. Note that the probability measures of S_n', S_n'' may be written as $[(\bar{P}_2+\bar{P}_3)/2]^n$, $[(\bar{P}_4+\bar{P}_3)/2]^n$ in the case A and as $[(\bar{P}_2+\bar{P}_7)/2]^n$, $[(\bar{P}_4+\bar{P}_7)/2]^n$ in the case B.

First let us consider the case B. We have

$$\Delta_2 = \left|\left([(\bar{P}_2+\bar{P}_7)/2]^n - [(\bar{P}_4+\bar{P}_7)/2]^n\right)(B_r(a))\right| \leqslant$$

$$\leqslant 2^{-n}(\Sigma_1+\Sigma_2)\binom{n}{q}\left|(\bar{P}_2^q-\bar{P}_4^q*\bar{P}_7^{n-q})(B_r(a))\right|=I_1+I_2, \tag{18}$$

where Σ_1 is the summation over all q such that

$$|q-(n+1)/2|<(1/2)n^{3/(2p)} \tag{19}$$

and Σ_2 is the summation over the remaining q, $0\leqslant q\leqslant n$.
By Lemma 4,

$$I_2\leqslant c(p)n^{-(p-2)/2}. \tag{20}$$

Furthermore, fix q and consider the corresponding summand in Σ_1. We have

$$\mu_q=((\bar{P}_2^q-\bar{P}_4^q)*\bar{P}_7^{n-q})(B_r(a))=$$

$$=\mathbf{P}\left\{\left|n^{-1/2}\Big(\sum_{i=1}^{q}X_i^{(2)}+\sum_{i=1}^{n-q}X_i^{(7)}\Big)-a\right|<r\right\}-$$

$$-\mathbf{P}\left\{\left|n^{-1/2}\Big(\sum_{i=1}^{q}X_i^{(4)}+\sum_{i=1}^{n-q}X_i^{(7)}\Big)-a\right|<r\right\},$$

where $X_i^{(7)}$, $i=1,2,\ldots$, are assumed to be independent of $X_i^{(2)}$, $i=1,2,\ldots$, and $X_i^{(4)}$, $i=1,2,\ldots$. The random variables $X_i^{(4)}$ can be written as

$$X_i^{(4)}=\xi_i\left(X_i^{(2)}-\mathbf{E}X_i^{(2)}\right)+Y_i+\mathbf{E}X_i^{(2)},\quad i=1,2,\ldots,$$

where ξ_i and Y_i are distributed as ξ and Y', respectively (see the definition of P_4). Hence

$$U=n^{-1/2}\Big(\sum_{i=1}^{q}X_i^{(4)}+\sum_{i=1}^{n-q}X_i^{(7)}\Big)=Y_1'+Z_1, \tag{21}$$

where Y_1', Z_1 are independent and Y_1' is a $(0,\ (q/(2n))V_2)$ Gaussian random variable. Since for any $b\in H$ the density function of $|Y_1'+b|^2$ is bounded by $c(q/n)^{-1}\sigma_{21}^{-1}\sigma_{22}^{-1}$ (see, e.g., (Sazonov, 1981), p.72), by (10), (19), the density function of $|U-a|^2$ is not greater than $c\sigma_1^{-1}\sigma_2^{-1}$. Applying the Esseen inequality (see, e.g., (Petrov, 1987, Theorem 2, §1, Chapter 5)) we have for any $T>0$

$$|\mu_q|\leqslant c\Big(\int_{|t|\leqslant T} g_n(t)\,|\,t\,|^{-1}\,dt+\sigma_1^{-1}\sigma_2^{-1}T^{-1}\Big), \tag{22}$$

where

$$g_n(t)=|f_{2n}(t)-f_{4n}(t)|,$$

$$f_{2n}(t)=\mathbf{E}\exp\left\{it\left|n^{-1/2}\Big(\sum_{i=1}^{q}X_i^{(2)}+\sum_{i=1}^{n-q}X_i^{(7)}\Big)-a\right|^2\right\},$$

$$f_{4n}(t) = \mathbf{E}\exp\left\{it\left|n^{-1/2}\left(\sum_{i=1}^{q} X_i^{(4)} + \sum_{i=1}^{n-q} X_i^{(7)}\right) - a\right|^2\right\}.$$

Take

$$T = c(3)(s^{-2}n)^{5/9}(\ln(s^{-2}n))^{-1/2},$$

where $c(3)$ is from Lemma 3 (here, as we agreed, $s = (3\sigma_4^{-2}\beta)^{1/(p-2)}$). Define also

$$T_1 = (\Lambda^{p-3}\sigma_6\beta^{-1}(1+\mid a\mid^p)^{-1}n^{(p-2)/2})^{1/2},$$

$$T_2 = c(3)(s^{-2}n(\ln(s^{-2}n))^{-1})^{1/2}.$$

First we estimate $J_1 = \int_{T_2\leqslant|t|\leqslant T} g_n(t)\mid t\mid^{-1} dt$. Obviously

$$J_1 \leqslant \int_{T_2\leqslant|t|\leqslant T} (\mid f_{2n}(t)\mid + \mid f_{4n}(t)\mid)\mid t\mid^{-1} dt = J_{11} + J_{12}. \tag{23}$$

From representation (21), Lemma 1 in (Sazonov, 1981, p. 82) and (10) it follows that

$$|f_{4n}(t)| \leqslant |\mathbf{E}\exp\{it|Y_1' + Z_1 - a|^2\}| \leqslant$$

$$\leqslant \prod_1^{\infty}(1 + ct^2\sigma_{2j}^4)^{-1/4} \leqslant \prod_{j=1}^{3}(1 + ct^2\sigma_j^4)^{-1/4}. \tag{24}$$

Now, observing that

$$\sigma_4^2 \geqslant (\sigma_4\sigma_5\sigma_6)^{2/3} \geqslant (\sigma_4\sigma_5\sigma_6)^{p/3},$$

$$(\sigma_1\sigma_2\sigma_3)^{p-2} \geqslant (\sigma_1\sigma_2\sigma_3)^{p/3},$$

by (6) we have

$$J_{12} \leqslant c(\sigma_1\sigma_2\sigma_3)^{-1}T_2^{-3/2} \leqslant c(\sigma_1\sigma_2\sigma_3)^{-1}sn^{-1/2} \leqslant$$

$$\leqslant c(\Lambda^{-p}\beta n^{(2-p)/2})^{1/(p-2)} \leqslant c\Lambda^{-p}\beta n^{(2-p)/2}. \tag{25}$$

To estimate J_{11} we apply Lemma 3 with $Q = P_1$, $Q_1 = P_2'$, $L = s$, $A = 3$, $r = 0$, $l \approx q(s^2n^{-1})^{1/9}$, $m = q - l$. Using (9) and observing that $s^2/n \leqslant 1$ we obtain for all t, $T_2 \leqslant\mid t\mid\leqslant T$,

$$|f_{2n}(t)| \leqslant c\left((s^2/n)^{8/3} + \prod_{j=1}^{3}\left(1 + cT_2^2(s^2/n)^{1/9}\sigma_j^4\right)^{-1/4}\right) \leqslant$$

$$\leqslant c\left(s^2/n + \left(1 + cT_2^{3/2}(s^2/n)^{1/12}\sigma_1\sigma_2\sigma_3\right)^{-1}\right). \tag{26}$$

Now in the same way as (25) we deduce

$$J_{11} \leqslant c\left(s^2/n + (1 + (T_2^{3/2}(s^2/n)^{1/12}\sigma_1\sigma_2\sigma_3)^{-1})\ln T\right) \leqslant$$

$$\leqslant csn^{-1/2}(\sigma_1\sigma_2\sigma_3)^{-1} \leqslant c\Lambda^{-p}\beta n^{(2-p)/2}. \tag{27}$$

Next we estimate $J_2 = \int\limits_{T_1\leqslant|t|\leqslant T_2} g_n(t) \mid t \mid^{-1} dt$ assuming that $T_1 \leqslant T_2$. We have $J_2 = J_{21} + J_{22}$, where (see (24))

$$J_{22} \leqslant \int\limits_{T_1\leqslant|t|\leqslant T_2} |f_{4n}(t)| \mid t \mid^{-1} dt \leqslant 2\int\limits_{T_1}^{\infty} t^{-1}\prod_{j=1}^{4}(1+ct^2\sigma_j^4)^{-1/4}dt \leqslant$$

$$\leqslant c\left(\prod_{j=1}^{4}\sigma_j\right)^{-1} T_1^{-2} \leqslant c\Lambda^{-p}\beta(1+\mid a\mid^p)n^{(2-p)/2}. \tag{28}$$

Moreover, by Lemma 3 with the same parameters as in (26) we except that now $l \approx q/2,\quad m = q-l$

$$J_{21} \leqslant \int\limits_{T_1\leqslant|t|\leqslant T_2} \mid f_{2n}(t) \mid\mid t \mid^{-1} dt \leqslant$$

$$\leqslant c\int\limits_{T_1}^{T_2} t^{-1}\left((s^2/n)^3 + \prod_{j=1}^{4}(1+ct^2\sigma_j^4)^{-1/4}\right)dt \leqslant$$

$$\leqslant c\left(s^2/n + \left(\prod_{j=1}^{4}\sigma_j\right)^{-1} T_1^{-2}\right) \leqslant c\Lambda^{-p}\beta(1+\mid a\mid^p)n^{(2-p)/2}. \tag{29}$$

Finally, let us estimate $J_3 = \int\limits_{|t|\leqslant T'} g_n(t) \mid t \mid^{-1} dt$, where $T' = \min(T_1, T_2)$. We have

$$J_3 \leqslant \int\limits_{|t|\leqslant T'} \mid t \mid^{-1} \left|\int \exp\{it|x-a|^2\}(\bar P_2^q - \bar P_4^q) * \bar P_7^{n-q}(dx)\right| dt \leqslant$$

$$\leqslant \sum_{j=0}^{q-1}\int\limits_{|t|\leqslant T'} \mid t \mid^{-1} \left|\iiint \exp\{it|x+y+z-a|^2\}\times\right.$$

$$\left.\times \bar P_2^j * \bar P_4^{q-j-1}(dx)(\bar P_2 - \bar P_4)(dy)\bar P_7^{n-q}(dz)\right| dt \leqslant$$

$$\leqslant \sum_{j=0}^{q-1}\int I_{3j}(z)\bar P_7^{n-q}(dz), \tag{30}$$

where

$$I_{3j}(z) = \int\limits_{|t|\leqslant T'} \mid t \mid^{-1} \left|\iint \exp\{it|x+y+z-a|^2\}\bar P_2^j * \bar P_4^{q-j-1}(dx)(\bar P_2 - \bar P_4)(dy)\right| dt.$$

If we prove that for all j, $0 \leqslant j \leqslant q-1$,

$$I_{3j}(z) \leqslant c(p)\Lambda^{-p}\beta(1+ \mid a \mid^p)n^{-p/2}(1+ \mid z \mid^{4p/(6-p)}), \tag{31}$$

then by (18), (20), (22), (23), (25), (27)–(31) and by (14) with $k=7$ the lemma will follow in the case B.

We will prove (31) when $j = q-1$. The same proof applies for all $j \geqslant (q-1)/2$. If $j < (q-1)/2$, $I_{3j}(z)$ is estimated exactly as I_{3j} in Lemma 7 in (Sazonov et al., 1989). Denote $b = a - z$, $f(\lambda) = \exp\{it|x + \lambda y - b|^2\}$ and using limited expansions represent $f(1)$ as

$$f(1) = \exp\{it(|x-b|^2 + 2(x-b,y))\}+$$

$$+it \mid y \mid^2 \int_0^1 \exp\{it(|x-b|^2 + 2(x-b,y) + \theta_1 \mid y \mid^2\} d\theta_1 = \cdot$$

$$= \exp\{it|x-b|^2\} + 2it(x-b,y)\exp\{it|x-b|^2\}+$$

$$+[2it(x-b,y)]^2 \int_0^1 \exp\left\{it(|x-b|^2 + 2\theta_2(x-b,y))\right\}(1-\theta_2)d\theta_2+$$

$$+it \mid y \mid^2 \int_0^1 \exp\left\{it(|x-b|^2 + 2(x-b,y) + \theta_1) \mid y \mid^2 \right\} d\theta_1.$$

Since P_2 and P_4 have the same first moments, we have for all $\theta_1, \theta_2 \in [0,1]$

$$I_{3,q-1}(z) \leqslant 4 \int_{|t| \leqslant T'} |t| \left| \iint (x-b,y)^2 \exp\left\{it\Big(|x-b|^2 + 2\theta_2(x-b,y)\Big)\right\}\times\right.$$

$$\left.\times \bar{P}_2^{q-1}(dx)(\bar{P}_2 - \bar{P}_4)(dy)\right| dt+$$

$$+\int_{|t| \leqslant T'} \left| \iint \mid y \mid^2 \exp\left\{it\Big(|x-b|^2 + 2(x-b,y) + \theta_1 \mid y \mid^2\Big)\right\}\times\right.$$

$$\left.\times \bar{P}_2^{q-1}(dx)(\bar{P}_2 - \bar{P}_4)(dy)\right| dt = 4I_{301} + M_1.$$

Put $q_0 = 1$, $q_1 = [(q-1)/3]$, $q_2 = 2q_1$, $q_3 = q-1$, $Z_i = n^{-1/2} \sum\limits_{j=q_{i-1}}^{q_i} X_j^{(2)}$ and let L_i be the distribution of Z_i, $i = 1,2,3$. Using this notation we may write

$$I_{301} \leqslant \int_{|t| \leqslant T'} |t| \left| \iint \Big((x_1+x_2+x_3,y)^2 - 2(b,y)(x_1+x_2+x_3,y)\Big)\times\right.$$

$$\times \exp\left\{it\Big(|x_1 + x_2 + x_3 - b|^2 + 2\theta_2(x_1 + x_2 + x_3 - b, y)\Big)\right\}\times$$

$$\times L_1(dx_1)L_2(dx_2)L_3(dx_3)(\bar{P}_2 - \bar{P}_4)(dy)\Big| dt +$$

$$+ \int\limits_{|t|\leqslant T'} |t| \left| \iint (b,y)^2 \exp\{it(|x-b|^2 + 2\theta_2(x-b,y))\}\times\right.$$

$$\left.\times \bar{P}_2^{q-1}(dx)(\bar{P}_2 - \bar{P}_4)(dy)\right| dt = I_{302} + M_2.$$

To estimate I_{302} it is enough to estimate integrals of the type

$$I(j,l,m) = \int\limits_{|t|\leqslant T'} |t| \left| \iint (x_1,y)^j (x_2,y)^l (b,y)^m \times\right.$$

$$\times \exp\left\{it\Big(|x_1 + x_2 + x_3 - b|^2 + 2\theta_2(x_1 + x_2 + x_3 - b, y)\Big)\right\}\times$$

$$\left.\times L_1(dx_1)L_2(dx_2)L_3(dx_3)(\bar{P}_2 - \bar{P}_4)(dy)\right| dt,$$

where j,l,m are non–negative integers, $j \leqslant 2$, $l \leqslant 2$, $m \leqslant 1$ and $j + l + m = 2$. Let $Z_i' = n^{-1/2} \sum\limits_{j=q_{i-1}}^{q_i} X_j^{(2)} \cdot \mathbf{1}_{\{|X_j^{(2)}|<n^{1/2}\}}$ and denote L_i' the distribution of Z_i', $i = 1,2$. Obviously to estimate $I(j,l,m)$ it is enough to estimate

$$I_1(j,l,m) = \int\limits_{|t|\leqslant T'} |t| \left| \iint (x_1,y)^j (x_2,y)^l (b,y)^m \times\right.$$

$$\times \exp\left\{it\Big(|x_1 + x_2 + x_3 - b|^2 + 2\theta_2(x_1 + x_2 + x_3 - b, y)\Big)\right\}\times$$

$$\left.\times L_3(dx_3)(L_1 - L_1')(dx_1)L_2(dx_2)(\bar{P}_2 - \bar{P}_4)(dy)\right| dt,$$

where $j \geqslant 1$ and

$$M_3(j,l,m) = \int\limits_{|t|\leqslant T'} |t| \left| \iint (x_1,y)^j (x_2,y)^l (b,y)^m \times\right.$$

$$\times \exp\left\{it\Big(|x_1 + x_2 + x_3 - b|^2 + 2\theta_2(x_1 + x_2 + x_3 - b, y)\Big)\right\}\times$$

$$\left.\times L_3(dx_3)L_1'(dx_1)L_2'(dx_2)(\bar{P}_2 - \bar{P}_4)(dy)\right| dt.$$

Note now that if $j = 1, 2$, we have, by (14) with $k = 2$,

$$\mathbf{E}|Z_1|^j \mathbf{1}_{\bigcup_{k=1}^{q} \{|X_k^{(2)}| \geqslant n^{1/2}\}} \leqslant \sum_{k=1}^{q_1} \mathbf{E}|Z_1|^j \mathbf{1}_{\{|X_k^{(2)}| \geqslant n^{1/2}\}} \leqslant$$

$$\leqslant 2q\mathbf{E}\Big(\Big|n^{-1/2}X_1^{(2)}\Big|^j + \Big|n^{-1/2}\sum_{k=2}^{q} X_k^{(2)}\Big|^j\Big)\mathbf{1}_{\{|X_1^{(2)}| \geqslant n^{1/2}\}} \leqslant$$

$$\leqslant c(p)n^{-3(p-2)/(6-p)}\mathbf{E}|X_1^{(2)}|^{4p/(6-p)} \leqslant$$

$$\leqslant c(p)\beta(Rn^{-1/2})^{p(p-2)/(6-p)}n^{(2-p)/2}. \tag{32}$$

Applying Lemma 3 in the same way as in proving (29) we obtain

$$\int_{|t| \leqslant T'} |t| \Big| \int \exp\Big\{it\Big(|x_1 + x_2 + x_3 - b|^2 + 2\theta_2(x_1 + x_2 + x_3 - b, y)\Big)\Big\} L_3(dx_3)\Big| dt \leqslant$$

$$\leqslant c \int_{|t| \leqslant T'} |t| \Big((s^2/n)^3 + \prod_{j=1}^{4}(1 + ct^2\sigma_j^4)^{-1/4}\Big) dt \leqslant \tag{33}$$

$$\leqslant c\Big(T_2^2(s^2/n)^3 + \Big(\prod_{j=1}^{4}\sigma_j\Big)^{-1}\ln(1 + T_1)\Big) \leqslant c(p)\Big(\prod_{j=1}^{4}\sigma_j\Big)^{-1} T_1^{6(p-2)/(6-p)}.$$

Furthermore,

$$\Big((Rn^{-1/2})^p T_1^6\Big)^{(p-2)/(6-p)} \leqslant c\Lambda^{3-p}\sigma_6^{-1}(1+ | a |)^{2-p}. \tag{34}$$

Combining (32)–(34) and using Lemma 1 we find

$$I_1(j, l, m) \leqslant c(p)\Lambda^{-p}\beta n^{-p/2}.$$

Thus

$$I_{3,q-1}(z) \leqslant c(p)\Lambda^{-p}\beta n^{-p/2} + M_1 + M_2 + \Sigma_3 M_3(j, l, m), \tag{35}$$

where M_i, $i = 1, 2, 3$, are defined above and Σ_3 is the summation over all non-negative integers j, l, m such that $j \leqslant 2$, $l \leqslant 2$, $m \leqslant 1$ and $j + l + m = 2$. We will estimate M_2. The summands M_1, M_3 are estimated similarly since for any $q \geqslant 0$

$$\mathbf{E}|Z_i'|^q \leqslant c(q)$$

(in the case B this follows from (16)).

Consider separately two cases according to the value of p.

I. $6(p-2)/(6-p) \geqslant 1$.

Taking a limited expansion of $f_1(\lambda) = \exp\{it(|x - b|^2 + 2\theta_2\lambda(x - b, y))\}$ we may write

$$f_1(1) = f_1(0) + f_1'(0) + \int_0^1 f_1''(\theta_3)(1 - \theta_3)d\theta_3,$$

where

$$f_1'(0) = f_1(0) \cdot 2it\theta_2(x-b,y),$$
$$f_1''(\theta_3) = f_1(\theta_3)(2it\theta_2)^2(x-b,y)^2.$$

Since $\bar{P}_2$ and $\bar{P}_4$ have the same moments of the first three orders we have

$$M_2 = \int\limits_{|t|\leqslant T'} |t| \left| \int (b,y)^2 \int \left(f_1(1) - f_1(0) - f_1'(0)\right) \bar{P}_2^{q-1}(dx)(\bar{P}_2 - \bar{P}_4)(dy) \right| dt.$$

The domain of integration in y we divide into two parts

$$\Omega_1 = \{y : |y| \leqslant 1/((1+|b|)(1+|t|))\} \text{ and } \Omega_2 = \Omega_1^c.$$

According to this division we have $M_2 = M_{21} + M_{22}$.

In the domain Ω_1 we replace $f_1(1) - f_1(0) - f_1'(0)$ by $\int\limits_0^1 f_1''(\theta_3)(1-\theta_3)d\theta_3$ and in Ω_2 $f_1(1)$, $f_1(0)$, $f_1'(0)$ are taken care of individually.

By Lemma 3, we have for all $\theta_3 \in (0,1)$

$$M_{21} = 4 \int\limits_{|t|\leqslant T'} |t|^3 \left| \int\limits_{\Omega_1} (b,y)^2 \int (x-b,y)^2 f_1(\theta_3) \bar{P}_2^{q-1}(dx)(\bar{P}_2 + \bar{P}_4)(dy) \right| dt \leqslant$$

$$\leqslant 4 \int\limits_{|t|\leqslant T'} |t|^3 \int\limits_{\Omega_1} (b,y)^2 (1+|b|^2) |y|^2 \times$$

$$\times \left((s^2/n)^3 + \left(\prod_{j=1}^4 \sigma_j \right)^{-1} t^{-2} \right) (\bar{P}_2 + \bar{P}_4)(dy)dt \leqslant$$

$$\leqslant 4\left(1+|b|^{4p/(6-p)}\right) \int\limits_{|t|\leqslant T'} |t|^{(5p-6)/(6-p)} \left((s^2/n)^3 + \left(\prod_{j=1}^4 \sigma_j \right)^{-1} t^{-2} \right) dt \times$$

$$\times \int |y|^{4p/(6-p)} (\bar{P}_2 + \bar{P}_4)(dy).$$

Note that $6(p-2)/(6-p) \leqslant 2$ if $p \leqslant 3$, and

$$T' \leqslant T_2 \leqslant c(s^2/n)^{-1/2}.$$

Furthermore, by (6) and the definition of s,

$$(s^2/n)^3 \leqslant T_2^{-4} s n^{-1/2} (3\sigma_4^{-2} \beta n^{(2-p)/2})^{1/(p-2)} \leqslant$$

$$\leqslant c(p) T_2^{-4} \Lambda^{-p} \beta (1+|a|^p) n^{(2-p)/2}. \tag{36}$$

Thus, by the definition of P_2, P_4, we have

$$M_{21} \leqslant c(p) n^{-p/2} \left(1+|z|^{4p/(6-p)}\right) \left(\Lambda^{-p} \beta \left(1+|a|^p\right) n^{(2-p)/2} + \right.$$

$$+\Big(\prod_{j=1}^{4}\sigma_j\Big)^{-1}\beta(Rn^{-1/2})^{p(p-2)/(6-p)}T_1^{6(p-2)/(6-p)}\Big(1+\mid a\mid^{4p/(6-p)}\Big)\Big). \tag{37}$$

Now by (34), (37), we get

$$M_{21}\leqslant c(p)\Lambda^{-p}\beta(1+\mid a\mid^{p})n^{-p/2}(1+\mid z\mid^{4p/(6-p)}). \tag{38}$$

Finally, by Lemma 3 and (34), (36),

$$M_{22}\leqslant\int\limits_{|t|\leqslant T'}\mid t\mid\int\limits_{\Omega_2}(b,y)^2\Big(\Big|\int(f_1(1)-f_1(0))\bar{P}_2^{q-1}(dx)\Big|+$$

$$+2\mid t\mid\Big|\int(x-b,y)f_1(0)\bar{P}_2^{q-1}(dx)\Big|\Big)(\bar{P}_2+\bar{P}_4)(dy)dt\leqslant$$

$$\leqslant c(p)(1+\mid b\mid^{4p/(6-p)})\int\limits_{|t|\leqslant T'}\mid t\mid^{(5p-6)/(6-p)}\Big((s^2/n)^3+\Big(\prod_{j=1}^{4}\sigma_j\Big)^{-1}t^{-2}\Big)dt\times$$

$$\times\int\mid y\mid^{4p/(6-p)}(\bar{P}_2+\bar{P}_4)(dy)\leqslant$$

$$\leqslant c(p)\Lambda^{-p}\beta(1+\mid a\mid^{p})n^{-p/2}(1+\mid z\mid^{4p/(6-p)}). \tag{39}$$

II. $6(p-2)/(6-p)<1$.
In this case we use the expansion

$$f_1(1)=f_1(0)+\int\limits_0^1 f_1'(\theta_4)d\theta_4,$$

where $f_1'(\theta_4)=f_1(\theta_4)2it\theta_2(x-b,y)$.
Since $\bar{P}_2$ and $\bar{P}_4$ have the same moments of the first two orders we may write

$$M_2=\int\limits_{|t|\leqslant T'}\mid t\mid\Big|\int(b,y)^2\int\Big(f_1(1)-f_1(0)\Big)\bar{P}_2^{q-1}(dx)(\bar{P}_2-\bar{P}_4)(dy)\Big|dt.$$

By the same arguments as in the case I we obtain

$$M_2\leqslant M_{21}+M_{22}\leqslant c(p)\Lambda^{-p}\beta(1+\mid a\mid^{p})n^{-p/2}(1+\mid z\mid^{4p/(6-p)}). \tag{40}$$

Together (35), (38)–(40) imply (31). This proves the Lemma in the case B. In the case A the proof is essentially the same as the proof of Lemma 7 in (Sazonov et al., 1989). The few necessary changes due to the parameter p were indicated above in the proof in the case B. In the case A we have inequalities (11)–(13) and the introduction of the measures L_i is not needed.

LEMMA 6. *Let Z be a $(0,W)$ Gaussian H-valued random variable. Assume that* tr$W\leqslant 2$ *and let $\tau_1^2\geqslant\tau_2^2\geqslant\ldots$ be the eigenvalues of W. Let G be a $(0,V)$*

Gaussian measure on H and denote $\bar{G}(\cdot) = G(n^{1/2}\cdot)$. Finally, define $F = \bar{P}_5 - \bar{G}$ in the case A *and $F = \bar{P}_8 - \bar{G}$ in the case* B. *Then for any $b \in H$, $r \geqslant 0$, if*

$$\rho = \int P(|Z + x - b| < r)F(dx),$$

then

$$|\rho| \leqslant c(p)\tau^{-p}\beta n^{-p/2}(1+ \mid b \mid^p),$$

where $\tau = \left(\prod_{i=1}^{6} \tau_i\right)^{1/3}$.

The proof of the lemma is essentially the same as the proof of Lemma 12 in (Sazonov et al., 1989). The main difference is that now as in the proof of Lemma 5 one has to use for $\int Q(x)F(dx)$, where $Q(x) = \sum_{j=0}^{2} Q_j(x) + \bar{Q}(x)$ and Q_j is a j–linear form, the following representation

$$\int Q(x)F(dx) = \int_{B^c_{\tau/(1+|a|)}} \bar{Q}(x)F(dx) +$$

$$+ \int_{B^c_{\tau/(1+|a|)}} Q(x)F(dx) - \sum_{j=0}^{2} \int_{B^c_{\tau/(1+|a|)}} Q_j(x)F(dx) + \int \left(\sum_{j=0}^{2} Q_j(x)\right)F(dx).$$

LEMMA 7. *In the notation of the theorem and Lemmas 5, 6*

$$|\mathbf{P}(|S''_n - a| < r) - \mathbf{P}(|Y - a| < r)| \leqslant c(p)\Lambda^{-p}\beta(1+ \mid a \mid^p)n^{(2-p)/2}.$$

The proof repeats the proof of Lemma 13 in (Sazonov et al., 1989)

4. PROOF OF THE THEOREM

Above it has been pointed out that we may assume condition (6) to be satisfied. Let $\Delta'(u, E)$ be equal to Δ_1 from Lemma 1 corresponding to the indicator function $f = \mathbf{1}_E$ of a Borel set E, $u_1 = \infty$, $u_2 = u$, $Z = 0$. By Lemma 1,

$$\Delta'(n^{1/2}, E) \leqslant nP(|X_1| \geqslant n^{1/2}) \text{ in the case A},$$

$$\Delta'(R, E) \leqslant nP(|X_1| \geqslant R) \leqslant nP(|X_1| \geqslant n^{1/2}) \text{ in the case B},$$

and obviously $nP(|X_1| \geqslant n^{1/2}) \leqslant \beta n^{(2-p)/2}$. Now the theorem follows from Lemmas 5 and 7.

REFERENCES

Petrov, V.V. (1987). *Limit Theorems for Sums of Independent Random Variables.* Nauka. Moscow.

Pinelis, I.F. (1980). Estimates of moments of infinite dimensional martingales. *Math. Notes* **28**, 953–958.

Sazonov, V.V. (1981). *Normal Approximation–Some Recent Advances.* Lect. Notes in Math., **879**, Springer, Berlin–Heidelberg etc.

Sazonov, V.V., Ulyanov, V.V. and Zalesskii, B.A. (1988). Normal approximation in Hilbert space. I; II. *Theory Probab. Appl.* **33**, 225–245; 508–521.

Sazonov, V.V., Ulyanov, V.V. and Zalesskii, B.A. (1989). Precise estimate of the rate of convergence in the central limit theorem in Hilbert space. *Math. USSR Sb.* **180**, 1587–1613.

Prob. Theory and Math. Stat., Vol. 2, pp. 411–423
B.Grigelionis et al. (Eds.)
1990 VSP/Mokslas

STATISTICAL FUNCTIONALS, STOPPING TIMES AND ASYMPTOTIC MINIMUM RISK PROPERTY

PRANAB KUMAR SEN

Departments of Biostatistics and Statistics, University of North Carolina, Chapel Hill, NC 27599–3260, USA

ABSTRACT

In a nonparametric estimation problem, for a statistical functional, the risk (defined suitably) depends on n, the sample size, as well as F, the underlying distribution function. Sequential estimators based on appropriate stopping times possess the (first order) asymptotic minimum risk property, although the picture is somewhat different for shrinkage versions of such sequential statistical functionals. In this context, the role of stopping times is explored in a systematic manner.

1. INTRODUCTION

Let $\{X_i; i \geqslant 1\}$ be a sequence of independent and identically distributed (i.i.d.) random vectors (r.v.) with a distribution function (d.f.) F, defined on E^r, for some $r \geqslant 1$. The functional form of F is not assumed to be given, and it is only assumed that $F \in \mathcal{F}$, where $\mathcal{F}$ is a suitable class of d.f.'s, defined on E^r. In this nonparametric setup, an estimable parameter θ is regarded as a (possibly vector valued) functional of the d.f. F, i.e., we set

$$\theta = (\theta_1, \ldots, \theta_p)' = \mathbf{T}(F) = \big(T_1(F), \ldots, T_p(F)\big)', \tag{1.1}$$

where $p \geqslant 1$, and $T_1(\cdot), \ldots, T_p(\cdot)$ are suitable functionals. Based on a sample $(X_1, \ldots, X_n)$ of size n, let

$$F_n(x) = n^{-1} \sum_{i=1}^{n} I(X_i \leqslant x), \;\; x \in E^r, \tag{1.2}$$

be the empirical (sample) d.f. Then, a natural nonparametric estimator of θ is

$$T_n = \mathbf{T}(F_n) = \big(T_{n1} = T_1(F_n), \ldots, T_{np} = T_p(F_n)\big)', \quad n \geqslant n_0. \tag{1.3}$$

In estimating θ by T_n, we conceive of a plausible loss function

$$L(T_n, \theta) = L\big(\|T_n - \theta\|\big), \tag{1.4}$$

where $L(t),\ t > 0$, is nonnegative and nondecreasing in t and $\|T_n - \theta\|$ stands for a suitable norm of $T_n - \theta$. For example, we may consider a quadratic error loss

$$L(a,b) = (a-b)'Q^{-1}(a-b) = \|a-b\|_Q^2,$$

where Q is a given positive definite (p.d.f) matrix. For $p = 1$, another possibility is to consider the absolute error loss $L(a,b) = |a-b|$, and this definition can also be extended for $p \geqslant 1$. Also, let $c > 0$ be the cost of sampling per unit observation, so that incorporating the cost of sampling along with the chosen loss function, we may formulate the following risk function for estimating θ by T_n :

$$\rho(n; c, F) = E_F\{L(T_n, \theta) + cn\},\ F \in \mathcal{F},\ c > 0,\ n \geqslant n_0. \tag{1.5}$$

Note that, in general, the risk function may depend on θ as well. But, by (1.1), $\theta = \mathbf{T}(F)$ is itself a functional of the d.f. F, and hence, the risk $\rho(n; c, F)$ is regarded as a function of (n, c) and a functional of the d.f. F. Invoking the consistency of T_n (requiring usually some mild regularity conditions on $T(\cdot)$ and F), we may argue that $\|T_n - \theta\|$ becomes stochastically smaller as n increases, so that incorporating appropriate integrability condition on $L(\|T_n - \theta\|)$, it may be quite reasonable to assume that

$$E_F L(T_n, \theta) = \nu_n(F) \text{ is monotone decreasing function in } n \geqslant n_0; \quad \forall\, F \in \mathcal{F}, \tag{1.6}$$

where $\nu_n(F)$ may depend in general on the unknown $F \in \mathcal{F}$. Thus, by (1.5) and (1.6), we have

$$\rho(n; c, F) = \nu_n(F) + cn,\ c > 0,\ n \geqslant n_0,\ F \in \mathcal{F}, \tag{1.7}$$

where $\nu_n(F)$ is decreasing in n, while cn is increasing in n. Therefore, there exists an n_c^0 (depending generally on c and F), such that

$$\rho(n_c^0; c, F) = \inf\{\rho(n; c, F) : n \geqslant n_0\}; \quad F \in \mathcal{F},\ c > 0. \tag{1.8}$$

If there are more than one such n_c^0, we choose the smallest one among them as the desired solution. In this formulation, n_c^0 stands for the optimal samples size (for a given F and c) and $\rho_c^0(F) = \rho(n_c^0; c, F)$ stands for the minimum risk attainable for the estimation of θ when one confines oneself to the class of nonparametric estimators

$$\mathbf{T} = \{T_n = \mathbf{T}(F_n); \quad n \geqslant n_0\}, \tag{1.9}$$

and adopt the risk function in (1.5).
As has been mentioned earlier, $\nu_n(F)$ in (1.6) generally depends on the unknown $F \in \mathcal{F}$. The current situation is more complex than a parametric model where $\mathcal{F}$ is of a given functional form (involving some unknown parameters appearing as algebraic constants), so that $\nu_n(F)$ is also of a known functional form (involving the same parameters (or a subset of them) as algebraic constants). In our nonparametric formulation, $\nu_n(F)$ is a functional of the unknown $F \in \mathcal{F}$, and it satisfies (1.6). It is therefore clear from the above discussion that n_c^0 generally depends on $F \in \mathcal{F}$ as well as on $c > 0$, so that for any chosen $n_c^0 = n^0$, say, T_{n^0}

may not have the minimum risk property, for all $F \in \mathcal{F}$. Thus, a fixed sample size estimator T_n may not be a minimum risk estimator (MRE) of θ, simultaneously for all $F \in \mathcal{F}$.

In the negation of the MRE property of fixed sample size estimators, it is of genuine interest to consider suitable sequential estimators $\{T_{N_c};\ c > 0\}$ based on appropriate stopping times $\{N_c;\ c > 0\}$, such that in some meaningful way (viz., asymptotically as $c \downarrow 0$), T_{N_c} has the desired MRE property. Or, in other words, we intend to formulate suitable stopping times $\{N_c;\ c > 0\}$, where for every $c > 0$, N_c is a positive integer valued r.v. such that

$$\lim_{c\downarrow 0} \{E_F[L(T_{N_c}, \theta) + cN_c]/\rho_c^0(F)\} = 1, \quad \forall\, F \in \mathcal{F}. \tag{1.10}$$

In the literature, (1.10) relates to the first order asymptotic minimum risk property, and any $\{T_{N_c}\}$ satisfying (1.10) is termed an asymptotically (first order) minimum risk estimator (AMRE) of θ.

Section 2 is devoted to the formulation of such AMRE of θ. In this context, asymptotic expansions for $\nu_n(F)$ and sequential estimation of the leading terms are of prime importance, and these will be discussed there. For $p \geqslant 3$, for any fixed n, T_n may not have the smallest risk. There exist suitable shrinkage (or Stein–rule) estimators which may dominate T_n (in an exact or asymptotic sense) with respect to the risk in (1.4). This Stein–phenomenon also holds in the sequential case (viz., Ghosh et al., 1987) dealing with a specific parametric model and (Sen, 1989), for a general nonparametric setup). Thus, based on the stopping time $\{N_c\}$ leading to the AMRE property of T_{N_c}, it may be possible to construct a suitable shrinkage version of T_{N_c} (say, $T_{N_c}^S$), such that the (asymptotic) risk of $T_{N_c}^S$ is smaller than (or equal to) that of T_{N_c}, uniformly in θ. This relative risk dominance picture is depicted in Section 3. In passing, it may be remarked that the usual shrinkage versions of the T_{N_c} may not belong to the class $\mathbf{T}$. As such, there remains some open questions regarding the (asymptotic) optimality of the stopping times in the context of such shrinkage estimators. The main objective of the current study is to focus on this issue.

2. ASYMPTOTICALLY OPTIMAL STOPPING TIMES FOR STATISTICAL FUNCTIONALS

Formulation of stopping times and sequential estimators of θ depends very much on the nature of $\nu_n(F)$ in (1.6). Let us assume that for some positive number q,

$$\nu_n(F) = n^{-q}\{\nu^{(1)}(F) + n^{-1}\nu^{(2)}(F) + \ldots\}, \tag{2.1}$$

where the $\nu^{(j)}(F)$ are suitable functionals of the d.f. F. Typically, if we consider the case of a quadratic error loss with a given (p.d.) Q, then

$$\nu_n(F) = \text{trace}\big(Q\ E_F[T_n - \theta][T_n - \theta]'\big),$$

so that invoking the usual asymptotic expansions for the second order moment of $n^{1/2}(T_n - \theta)$, we obtain that (2.1) holds with $q = 1$. Actually, for Hoeffding's

U–statistics (Hoeffding, 1948) and von Mises' functionals such an expansion follows readily from the so called Hoeffding–decomposition (into orthogonal components); for more general forms of such statistics, a Hoeffding–type decomposition has been neatly worked out in (van Zwet, 1984). For differentiable statistical functions, parallel results (up to the order n^{-2}) have been considered in (Sen, 1988). Note that by (1.7) and (2.1), we have

$$\rho(n;c,F) = cn + n^{-q}\nu^{(1)}(F) + n^{-q-1}\nu^{(2)}(F) + \dots . \tag{2.2}$$

Thus, if we define

$$n_c^* = \left[c^{-1}q\ \nu^{(1)}(F)\right]^{(1+q)^{-1}}, \quad c > 0, \tag{2.3}$$

then, it is easy to show that as $c \downarrow 0$,

$$\rho_c^0(F) = \left\{c^q\nu^{(1)}(F)\right\}^{(1+q)^{-1}}\left\{q^{(1+q)^{-1}} + q^{-q(1+q)^{-1}}\right\} + o\left(c^{q(1+q)^{-1}}\right); \tag{2.4}$$

$$\rho(n_c^*;c,F) - \rho_c^0(F) = o\left(c^{q(1+q)^{-1}}\right), \tag{2.5}$$

$$n_c^* - n_c^0 = o\left(c^{-(1+q)^{-1}}\right); \quad n_c^0 = \left(c^{-1}q\ \nu^{(1)}(F)\right)^{(1+q)^{-1}} + o\left(c^{-(1+q)^{-1}}\right). \tag{2.6}$$

Thus, for $c \downarrow 0$, n_c^* provides a first order approximation for n_c^0, and the asymptotic risk of $T_{n_c^*}$ satisfies the limit in (1.10). For this reason, in the sequel, we shall replace n_c^0 by n_c^* for our subsequent analysis. It is worth mentioning in this context that for this first order equivalence results, we may not even need an asymptotic expansion (as in (2.1)); it suffices to assume that there exist a positive q and a positive $\nu^{(1)}(F)$, such that as $n \to \infty$,

$$n^q\nu_n(F) \to \nu^{(1)}(F), \quad \forall\ F \in \mathcal{F}, \tag{2.7}$$

and this can be justified even under less stringent regularity conditions. Also, (2.3) clearly reveals the dependence of n_c^* (or n_c^0) on F through $\nu^{(1)}(F)$ (and the $\nu^{(j)}(F)$). As such, a sequential estimator of $\nu^{(1)}(F)$ can be incorporated in the formulation of an appropriate stopping rule.

In the simplest situation, we may assume that the functional form of $\nu^{(1)}(\cdot)$ is known, and define

$$V_n = \nu^{(1)}(F_n),\ n \geqslant n_0, \tag{2.8}$$

as the sample counterparts of $\nu^{(1)}(F)$; it is tacticaly assumed that

$$V_n \to \nu^{(1)}(F) \text{ almost surely (a.s) as } \quad n \to \infty. \tag{2.9}$$

Then (2.3) and (2.9) lead us consider a stopping time

$$N_c = \inf\left\{n \geqslant n_0 : n^{q+1} \geqslant (c^{-1}q)[V_n + n^{-h}]\right\}, \quad c > 0, \tag{2.10}$$

where $h > 0$ (specified), and n^{-h} protects against too early termination (if V_n is too small).

It may be remarked that

$$\left\{[F_n(x) - F(x)] : \ x \in E^r\right\},\ n \geqslant 1$$

(or $\{[F_n(x),\ x \in E^r],\ n \geqslant 1\}$) is a reverse martingale (process) and $\nu^{(1)}(\cdot)$ is nonnegative. So whenever $\nu^{(1)}(\cdot)$ is a convex functional, such that $\mathbf{E}\ \nu^{(1)}(F_n)$ exists for some $n = n_0 \geqslant 1$, then

$$\{V_n; n \geqslant n_0\} \text{ is a nonnegative reverse submartingale.} \tag{2.11}$$

Incorporating this reverse sub–martingale property along with appropriate moment convergence results for $\{V_n\}$, one may virtually repeat the steps in (Sen and Ghosh, 1981) and conclude that the AMRE property in (1.10) holds for $\{T_{N_c}\}$. Thus, in this sense, $\{N_c,\ c > 0\}$ is asymptotically (as $c \downarrow 0$) an optimal stopping time. In the negation of such a reverse sub–martingale property of the V_n, one may require more stringent moment convergence properties of the V_n for the derivation of the AMRE property of $\{T_{N_c}\}$. In dealing with rank based ($R-$) estimators of location, such an approach has been systematically explored in (Sen, 1980, 1981) (see also (Sen, 1984) for the multiparameter case), and similar results can be derived for other cases too.

In some situations (particularly in the multiparameter case), $\nu^{(1)}(F)$ may be of quite complicated form, so that the formulation of V_n may involve rather complex computational schemes. For example, suppose that $\mathbf{T}(\cdot)$ is differentiable at F, so that $\mathbf{T}(F_n)$ can be expanded around $\mathbf{T}(F)$. In this expansion, the first term is

$$\int T^{(1)}(F;x)dF_n(x) = n^{-1}\sum_{i=1}^{n} T^{(1)}(F;X_i),$$

where $T^{(1)}(F;x)$ is the so called influence function (at F). If we choose a quadratic error loss function (with a given p.d. matrix Q), then

$$\nu^{(1)}(F) = \text{trace}\big(Q\Sigma^*(F)\big),$$

where $\Sigma^*(F)$ is the dispersion matrix of $T^{(1)}(F;X_1)$. Usually, the functional form of $T^{(1)}(F;x)$ depends on the unknown F, so that for the estimation of $\Sigma^*(F)$ (or $\nu^{(1)}(F)$), one may require the estimation of $T^{(1)}(F;x)$ as well. Jackknifing methods can be used with advantage in such a case. From the base sample $(X_1,\ldots,X_n)$ of size n, we delete the i–th observation X_i and denote the corresponding empirical d.f. by $F_{n-1}^{(i)}$, so that $T_{n-1}^{(i)} = T(F_{n-1}^{(i)})$ is an estimator of θ, for $i = 1,\ldots,n$. Let then

$$T_{n,1} = n\,T_n - (n-1)T_{n-1}^{(i)}, \quad i = 1,\ldots,n; \tag{2.12}$$

$$T_n^* = \frac{1}{n}\sum_{i=1}^{n} T_{n,i}, \quad V_n^* = \frac{1}{n-1}\sum_{i=1}^{n}(T_{n,i} - T_n^*)(T_{n,i} - T_n^*)'. \tag{2.13}$$

The $T_{n,i}$ are so called pseudovariables, T_n^* is the classical jackknifed estimator of θ and V_n^* is the jackknifed estimator of $\Sigma^*(F)$. Under quite general regularity conditions (viz. (Sen, 1988)), we have

$$n\|T_n - T_n^*\| = O(1) \quad \text{a.s.,} \quad \text{as} \quad n \to \infty, \tag{2.14}$$

$$V_n^* \to \Sigma^*(F) \quad \text{a.s.,} \quad \text{as} \quad n \to \infty, \tag{2.15}$$

so that

$$\dot{V}_n = \text{trace}(Q\ V_n^*) \to \nu^{(1)}(F) = \text{trace}\big(Q\ \Sigma^*(F)\big) \quad \text{a.s.,} \quad \text{as} \quad n \to \infty. \tag{2.16}$$

Note that for every $n \geqslant n_0$, $T_{n,1}, \ldots, T_{n,n}$ are exchangeable r.v.'s, but their dependence structure changes with n. As such, in general, it may be difficult to obtain a reverse sub–martingale characterization for $\{V_n\}$. However, using the differentiability approach in (Sen, 1988), moment convergence properties of V_n can be studied under appropriate regularity conditions on $\mathbf{T}(\cdot)$ and F, and hence, the asymptotic optimality of the stopping time $\{N_c,\ c > 0\}$ can be established as in before.

We conclude this section with a remark that the asymptotic optimality of the stopping time $\{N_c, c \downarrow 0\}$ in (2.10) is relative to the AMRE property in (1.10). In this context it is tactily assumed that $\rho_c^0(F)$ is the minimum risk attainable for the class $\mathbf{T}$ of estimators in (1.9). For the multiparameter model (i.e., $p \geqslant 3$), it is possible to construct some shrinkage versions of T_n which have smaller (asymptotic) risk. This leads to the basic question: Is $N_c, c \downarrow 0$, an asymptotically optimal stopping time for shrunken estimators too? We study this problem in the next section.

3. STOPPING–TIMES AND SHRUNKEN STATISTICAL FUNCTIONALS

Shrunken versions of statistical functionals have been considered in (Sen, 1989) where the asymptotic dominance of sequential shrunken functionals over their classical counterparts have been discussed. To motivate our main results, we recapitulate briefly the basic results on shrunken functions.

Let us define T_n, V_n^* and Q as in earlier sections, and let

$$d_n \quad \text{be the smallest characteristic root of} \quad QV_n^*,\ n \geqslant n_0. \tag{3.1}$$

Side by side, let δ be the smallest characteristic root of $Q\Sigma^*(F)$, where $\Sigma^*(F)$ is the dispersion matrix of the asymptotic (multi–normal) distribution of $n^{1/2}(T_n - \theta)$. Then

$$d_n \to \delta \quad \text{a.s.,} \quad \text{as} \quad n \to \infty. \tag{3.2}$$

Next, note that Stein–rule (or shrinkage) estimators are basically testimators: An appropriate test statistic (say, L_n) is incorporated in the estimator itself, and this test statistic is primarily a measure of divergence of θ from the assumed pivot θ_0. Without any loss of generality, we may set $\theta_0 = 0$ (otherwise, subtract θ_0 from T_n and reduce the case to a null pivot). Then an appropriate L_n to test the adequacy of the pivot 0 is

$$L_n = nT_n' V_n^{*-1} T_n, \tag{3.3}$$

where under the null hypothesis ($\theta = 0$), L_n has asymptotically (central) chi–square distribution with p degrees of freedom (DF). Larger values of L_n depict the inadequacy of the assumed null pivot. We consider the following type of shrinkage functionals:

$$T_n^S = \{I - k\ d_n L_n^{-1} Q^{-1} V_n^{*-1}\} T_n, \tag{3.4}$$

where p is taken greater than 2, the shrinkage factor k is a positive number $\left(0 < k < 2(p-2)\right)$ and, in this context, the loss function $L(a,b) = (a-b)'Q(a-b)$ is quadratic with the p.d. matrix Q as its discriminant.

In the parametric (normal theory) model, T_n has a multinormal distribution, $(n-1)V_n^* \sim$ Wishart $(\Sigma, p, n-1)$, and T_n and V_n^* are independent. Thus, L_n has the Hotelling T^2–distribution (noncentral when $\theta \neq 0$). This enables one to compute the dispersion matrix of T_n^S and incorporate the same in the computation of the risk of T_n^S. But, in a nonparametric model, the exact distribution of L_n^{-1} (or the independence of T_n and V_n^*) may be extremely difficult to comprehend, and moment–convergence results for T_n^S may require quite stringent regularity conditions. This poses a serious problem in the computation of the (asymptotic) risk of T_n^S (with an unbounded loss function), and the problem becomes harder in the sequential case. In the normal theory model, the technique employed in (Ghosh et al., 1987) rests heavily on the stochastic independence of $\{T_n; n \geqslant 1\}$ and $\{V_n^*; n \geqslant 2\}$ and their joint sufficiency. Parallel results may not hold for the nonparametric case. Hence, the theory developed so far (Sen, 1989) relates to the asymptotic case only.

There are two easy ways eliminating the technical difficulty caused by the presence of L_n^{-1} in (3.4). First, note that

$$d_n Q^{-1} V_n^{*-1} = \{\text{largest root of } Q^{-1} V_n^{*-1}\}^{-1} Q^{-1} V_n^{*-1}, \tag{3.5}$$

so that the characteristic roots of $d_n Q^{-1} V_n^{*-1}$ are all less than or equal to 1 (these are nonnegative too). As such, on the set $\{L_n > k\}$, $\{I - k\, d_n L_n^{-1} V_n^{*-1}\}$ is positive semi–definite (actually p.d. with probability one) and has characteristic roots all nonnegative and bounded by 1. The picture can be quite different on the set $\{L_n \leqslant k\}$, particularly when L_n is close to zero. This prompts us to consider a "positive rule" version of T_n^S in (3.4). We denote by

$$U_n^* = \begin{cases} 0, & \text{if } L_n \leqslant k, \\ I - k\, d_n L_n^{-1} Q^{-1} V^{*-1}, & \text{if } L_n > k, \end{cases} \tag{3.6}$$

and consider the modified estimator

$$T_n^{S+} = U_n^* T_n. \tag{3.7}$$

Note that $T_n^{S+} = 0$ on the set $\{L_n \leqslant k\}$, while for $L_n > k$, $L_n^{-1} \leqslant k^{-1}$, so that moment–convergence results for T_n^{S+} may not require any stringent regularity conditions (pertaining otherwise to the integrability of L_n^{-1}). In fact, such a positive–rule version is known to possess better dominance properties than T_n^S itself (see (Sclove et al., 1972) for the normal theory model). The other possibility is to use the original estimator T_n^S in (3.4), but to adopt a modified definition of the asymptotic risk. Towards this, we may note that if the true parameter point θ is different from the pivot then under quite general regularity conditions, as $n \to \infty$

$$n^{-1} L_n \to \Delta = \theta' \left[\Sigma^*(F)\right]^{-1} \theta > 0, \quad \text{in probability}, \tag{3.8}$$

so that $L_n = O_p(n)$ when $\theta \neq 0$. In this context, by (3.4), we have

$$n \| T_n^S - T_n \|_Q^2 = k^2 n\, L_n^{-2} T_n' V_n^{*-1} Q^{-1} V_n^{*-1} T_n =$$

$$= (k^2 L_n^{-1})(n\ L_n^{-1})(T_n' V_n^{*-1} Q^{-1} V_n^{*-1} T_n) = O_p(n^{-1}), \quad \text{as} \quad n \to \infty. \tag{3.9}$$

Thus, $n^{1/2}(T_n^S - \theta)$ and $n^{1/2}(T_n - \theta)$ are stochastically equivalent and they share the common asymptotic properties. Or, in other words, there is no asymptotic improvement due to shrinkage if the actual θ (fixed) is different from the assumed pivot. The situation is different when θ is "close to" the assumed pivot. In that case, there is some improvement due to shrinkage, and this will be considered here. We take into account such a 'small neighborhood' of the pivot (in conjunction with $c \downarrow 0$) in our formulation of the "asymptotic risk" function. For simplicity, we confine ourselves to a quadratic error loss, so that as in Section 2, we have

$$n_c^0 = O(c^{-1/2}), \quad \text{as} \quad c \downarrow 0. \tag{3.10}$$

It will be seen later on that the class of stopping times $\{N_c; c > 0\}$ may also be so chosen that

$$N_c = O_p(c^{-1/2}), \quad \text{as} \quad c \downarrow 0. \tag{3.11}$$

As such, looking at (3.8) and (3.9) along with (3.10) and (3.11), we may conceive of a sequence $\{A(c); c > 0\}$ of nested neighborhoods

$$A(c) = \{\theta : \|\theta - 0\| \leqslant A\ c^{-1/4}\},\ c > 0, \tag{3.12}$$

where A, $0 < A < \infty$, is a fixed positive number (otherwise, arbitrary). Within the domain $A(c)$ and for N_c satisfying (3.11), $c^{1/4}[T_{N_c}^S - T_{N_c}]$ has a nondegenerate limit distribution (as $c \downarrow 0$), and hence, $T_{N_c}^S$ and T_{N_c} are not generally asymptotically equivalent, while outside $A(c)$, as $c \downarrow 0$, (3.9) holds (with $O_p(n^{-1})$ replaced by $o_p(1)$). As such, we may as well consider the asymptotic (as $c \downarrow 0$) distribution of $c^{-1/4}(T_{N_c} - \theta)$ and $c^{-1/4}(T_{N_c}^S - \theta)$, denote then G and $G^{(S)}$ respectively, and compute the asymptotic risk by reference to such asymptotic distributional risk (ADR) (see (Sen, 1987a, 1989), for example). The main advantage of using the ADR criterion is that we do not have to consider more stringent regularity conditions pertaining to the asymptotic value of the actual risk $\rho(n; c, F)$, and whenever such a limit exists, it would also agree with the ADR. For the positive rule estimator $T_{N_c}^{S+}$ the asymptotic risk and ADR would have a common value (under the usual regularity conditions). The important fact is that the results based on these ADR comparisons remain applicable to the ones based on the asymptotic risks, when additional regularity (i.e., integrability) conditions are met.

It follows along the lines of (Sen, 1989) that in the light of the ADR and adapted to the stopping time $\{N_c, c > 0\}$ in (2.10), $T_{N_c}^S$ dominates T_{N_c}. A very similar conclusion holds for $T_{N_c}^{S+}$(vs T_{N_c}). Thus, with respect to the asymptotically optimal stopping times pertaining to the classical case (treated in Section 2), further reduction in the asymptotic risk is possible due to shrinking the functionals towards a pivot, when the true parameter lies "close to" the pivot. This is in agreement with the general results on sequential shrinkage estimation for the normal theory case (Ghosh et al., 1987), where also the dominance is perceptible only in a "small" neighborhood of the pivot. However, our conclusions remain applicable to a much larger class of estimators and a wider class of nonparameter models.

With these results at hand, we now consider the final question regarding asymptotically optimal stopping times relating to shrinkage estimators. In this context, first, we provide an answer to the query posed at the end of Section 2. An affirmative answer to the asymptotic optimality of N_c, $c > 0$, in (2.10) for shrinkage estimators can be made when the true θ is away from the pivot. Or, in other words, if a pivot is so chosen that it is not likely to be "very close" to the actual parameter point θ, then shrinkage estimators fail to provide any perceptible improvement over their classical counterparts, and hence, their asymptotic risk equivalence entails the asymptotic optimality of the stopping time N_c, $c > 0$, in (2.10). However, we may have an altogether different answer when θ lies "close to" the assumed pivot. We explore this situation in the remaining of this section. Let us now confine ourselves to the case of $\theta \in A(c), c \downarrow 0$. Towards this, we consider a sequence $\{F_{(c)}, c > 0\}$ of d.f.'s such that

$$\mathbf{K}_c : \theta_c = \mathbf{T}(F_{(c)}) = c^{1/4}\lambda, \quad \lambda = (\lambda_1, \ldots, \lambda_p)' \text{ fixed}, \quad c > 0. \tag{3.13}$$

This is more effectively conceived by allowing $F_{(c)} \to F$ as $c \downarrow 0$, such that $\theta_0 = = T(F) = 0$ (the pivot), so that $\mathbf{K}_c$ corresponds to a sequence of local alternatives. Note that the dispersion matrix of the asymptotic distribution of $n^{1/2}(T_n - \theta)$ is not affected by $\{\mathbf{K}_c\}$ in (3.13), and hence, the ADR of T_{N_c} (when N_c is defined by (2.10)) does not depend on λ (or the pivot)). On the other hand, the distribution of $n^{1/2}(T_n^S - \theta)$ (even asymptotically) may depend on λ, and hence, the ADR of $T_{N_c}^S$ depends on λ as well.

We define by Γ the dispersion matrix of the asymptotic distribution of $n^{1/2}(T_n - \theta)$, and for error function $L(\cdot)$, we take a quadratic error loss (with a p.d. Q), so that as $c \downarrow 0$

$$\rho(n; c, F_{(c)}) = n^{-1} \text{ trace } (Q\Gamma) + cn, \tag{3.14}$$

whenever cn^2 is bounded from below by some positive ϵ. For every $n : n^2 c = O(1)$ (as $c \downarrow 0$), we denote by

$$\Delta_{nc} = (nc^{1/2})\lambda'\Gamma^{-1}\lambda \quad \text{and} \quad \Delta_{nc}^0 = (nc^{1/2})\lambda'\Gamma^{-1}Q^{-1}\Gamma^{-1}\lambda. \tag{3.15}$$

Then, following the lines of (Sen, 1987a,b), it can be shown that the ADR of T_n^S (whenever $n^2 c = O(1)$ and $n^2 \; c > 0$) is given by

$$cn + n^{-1}\Big[\text{tr}(Q\Gamma) - 2k(p-2)\delta \; \mathbf{E}\big(\chi_p^{-2}(\Delta_{nc})\big) + k^2\delta^2\big[\text{tr}(Q^{-1}\Gamma^{-1})\big]\mathbf{E}\big(\chi_{p+2}^{-4}(\Delta_{nc})\big) + $$
$$+ \Delta_{nc}^0\big[\text{tr}(Q\Gamma)\big]\mathbf{E}\big(\chi_{p+4}^{-4}(\Delta_{nc})\big)\Big], \tag{3.16}$$

where k is the shrinkage factor, $0 < k < 2(p-2)$, δ is defined after (3.1) and denoting by $H_q(x; \Delta)$ the non-central chi-square d.f. with q DF and noncentrality parameter Δ,

$$\mathbf{E}\big(\chi_q^{-2r}(\Delta)\big) = \int_0^\infty x^{-r} dH_q(x; \Delta), \quad r \geqslant 0, \quad \Delta \geqslant 0. \tag{3.17}$$

Since Δ_{nc} and Δ_{nc}^0 both depend on λ in (3.13), it is clear that (3.16) depends on $(c, n), Q$, Γ as well as λ. For any given c, n, Q and Γ, (3.16) attains a minimum at $\Delta_{nc} = \Delta_{nc}^0 = 0$ (i.e., $\lambda = 0$), and this minimum value is given by

$$cn + n^{-1}\Big\{\text{tr}(Q\Gamma) - k\delta\big[2 - p^{-1}(p-2)^{-1}k\delta \; \text{tr}(Q^{-1}\Gamma^{-1})\big]\Big\}, \tag{3.18}$$

where $\delta\ \mathrm{tr}(Q^{-1}\Gamma^{-1}) = \mathrm{tr}(Q^{-1}\Gamma^{-2})/\Lambda \leqslant p$, Λ is the largest characteristic root of $Q^{-1}\Gamma^{-1}$ and $0 < k < 2(p-2)$. Hence, (3.18) is less than $cn + n^{-1}\mathrm{tr}(Q\Gamma) = \rho(n; c, F)$, so that for $\lambda = 0$, the minimum value of (3.18) (over n) is given by

$$2\left\{\sqrt{c}\left[\mathrm{tr}(Q\Gamma) - k\delta\{2 - p^{-1}(p-2)^{-1}k\delta\mathrm{tr}(Q^{-1}\Gamma^{-1})\}\right]^{1/2}, \qquad (3.19)$$

which is smaller than $\rho_c^0(F) = 2\sqrt{c}\{\mathrm{tr}(Q\Gamma)\}^{1/2}$. Thus, if we consider a stopping time N_c^0, $c > 0$, defined by

$$N_c^0 = \inf\Big\{n > n_0 : n^2 > \frac{1}{c}\{\mathrm{tr}(QV_n^*) -$$

$$-kd_n[2 - p^{-1}(p-2)^{-1}kd_n\ \mathrm{tr}(Q^{-1}V_n^{*-1})]\}\Big\}, \qquad (3.20)$$

(for $c > 0$), then it follows by the same technique as in (Sen, 1989) that the asymptotic (as $c \downarrow 0$) risk of $T_{N_c^0}^S$, when $\theta = 0$, is given by (3.19), and moreover, defining N_c, $c > 0$, as in (2.10) (with $q = 1$),

$$\lim_{c\downarrow 0}(N_c^0/N_c) = 1 - \frac{k\delta}{\mathrm{tr}(Q\Gamma)}\left[2 - \frac{k\delta}{p(p-2)}\mathrm{tr}(Q^{-1}\Gamma^{-1})\right] < 1 \quad \text{a.s.} \qquad (3.21)$$

Thus, at the pivot ($\theta = 0$), the stopping time N_c^0 is stochastically smaller than N_c and at the same time leads to a smaller asymptotic risk for the shrunken estimator $T_{N_c^0}^S$ (compared to $T_{N_c}^S$) when $\lambda \neq 0$ (i.e., $\theta \neq 0$). To see this, we may note that under (3.13), the ADR of $T_{N_c^0}^S$ is given by

$$c^{1/2}\left\{\mathrm{tr}(Q\Gamma) - k\delta\left[2 - \frac{k\delta}{p(p-2)}\mathrm{tr}(Q^{-1}\Gamma^{-1})\right]\right\}^{1/2} +$$

$$+c^{1/2}\left\{\mathrm{tr}(Q\Gamma) - k\delta\left[2 - \frac{k\delta}{p(p-2)}\mathrm{tr}(Q^{-1}\Gamma^{-1})\right]\right\}^{-1/2}\Big\{\mathrm{tr}(Q\Gamma) -$$

$$-k\delta\left(2 - \frac{k\delta}{p(p-2)}\mathrm{tr}(Q^{-1}\Gamma^{-1})\right) + h(\delta, Q, \Gamma, \Delta_{n_c^0 c}, \Delta_{n_c^0 c}^0)\Big\}, \qquad (3.22)$$

where $n_c^0 = c^{-1/2}\left[\mathrm{tr}(Q\Gamma) - k\delta\{2 - p^{-1}(p-2)^{-1}k\delta\ \mathrm{tr}(Q^{-1}\Gamma^{-1})\}\right]$ and

$$h(\delta, Q, \Gamma, \Delta_{nc}, \Delta_{nc}^0) \quad \text{is given by} \quad (3.16) - (3.18)\ (\text{at } n); \qquad (3.23)$$

(3.23) is nonnegative and bounded from above by

$$k\delta\left[2 - \frac{k\delta}{p(p-2)}\mathrm{tr}(Q^{-1}\Gamma^{-1})\right],$$

where the upper bound is attained when $\|\lambda\| \to \infty$. At this stage we may note that the ADR of $T_{N_c}^S$ is given by

$$c^{1/2}\{\mathrm{tr}(Q\Gamma)\}^{1/2}\Big[1 + [\mathrm{tr}(Q\Gamma)]^{-1}\{\mathrm{tr}(Q\Gamma) - k\delta\left(2 - \frac{k\delta}{p(p-2)}\mathrm{tr}(Q^{-1}\Gamma^{-1})\right) +$$

$$+h(\delta, Q, \Gamma, \Delta_{n_c^* c}, \Delta^0_{n_c^{0*} c})\}\Big], \tag{3.24}$$

where $n_c^{0*} \sim c^{-1/2}\big(\mathrm{tr}(Q\Gamma)\big)^{1/2} > n_c^0$. By using (3.23) along with the upper bound for $h(\cdot)$, we obtain that (3.24) can not be larger than $2\sqrt{c}\{\mathrm{tr}(Q\Gamma)\}^{1/2} = \rho_c^0(F)$. Hence, it suffices to show that (3.22) can be larger than $\rho_c^0(F)$, for large values of $\|\lambda\|$. Towards this note that for $0 \leqslant d \leqslant b < a$,

$$(a-b)^{1/2} + (a-b)^{-1/2}(a-b+d) = (a-b)^{-1/2}(2a-2b+d) =$$

$$= (2a-b)(a-b)^{-1/2} - (b-d)(a-b)^{-1/2} =$$

$$= 2\sqrt{a}\left(1 - \frac{1}{2a}\right)\left(1 - \frac{b}{a}\right)^{-1/2} - (b-d)(a-b)^{-1/2}. \tag{3.25}$$

Since

$$(1 - b/2a)^2 = 1 - b/a + b^2/4a^2 > 1 - b/a, \ \forall\, b < a,$$

the right hand side of (3.25) can be made greater than $2\sqrt{a}$ for all d sufficiently close to b. Now as $\|\lambda\| \to \infty$,

$$h(\delta, Q, \Gamma, \Delta_{n_c^0 c}, \Delta^0_{n_c^0 c}) \to d\delta\{2 - p^{-1}(p-2)^{-1}k\delta\mathrm{tr}(Q^{-1}\Gamma^{-1})\},$$

by (3.22) and (3.25), we conclude that the ADR of $T^S_{N_c^0}$ can indeed exceed that of T_{N_c} (and hence, $T^S_{N_c}$) when θ is away from the pivot. This exhibits the lack of asymptotic optimality of $\{N_c^0,\ c > 0\}$, uniformly in $\lambda \in E^p$. A similar criticism applies to $\{N_c,\ c > 0\}$ in (2.10) when one intends to use shrinkage functional (as in a neighborhood of the pivot (i.e., for $\|\lambda\|$ small) (3.24) will be larger than (3.22)), although $\{N_c, c > 0\}$ has the nice property that the ADR of $T^S_{N_c}$ can't be larger than that of T_{N_c} (while that of $T^S_{N_c^0}$ can be so). This leads us to the question: Does there exist a stopping rule $\{N_c^*, c > 0\}$, such that

(a) N_c^* is stochastically smaller than N_c,

(b) the ADR of $T^S_{N_c^*}$ is not greater than the ADR of T_{N_c}, uniformly in λ?

We construct an adaptive stopping time to meet both these requirements. In this context, we may note that there are nice series expansions for $\mathbf{E}\big(\chi_q^{-2}(\Delta)\big)$ and $\mathbf{E}\big(\chi_{q+2}^{-4}(\Delta)\big)$, $q > 2$, $\Delta \geqslant 0$, in terms of Poisson weights, and hence, $h(d_n, Q, V_n^*, a_n, b_n)$ can be conveniently computed for suitable a_n, b_n, d_n, Q and V_n^*. Let us define then $N_c^0, c > 0$, as in (3.20). Also, for every $n \geqslant n_0$, let

$$a_n^* = \min\big\{\mathrm{tr}(QV_n^*) -$$

$$-kd_n\left[2 - \frac{kd_n}{p(p-2)}\mathrm{tr}(Q^{-1}V_n^{*-1})\right] + h(d_n, Q, V_n^*, a_{n1}, a_{n2}), \mathrm{tr}(QV_n^*)\big\}, \tag{3.26}$$

where

$$a_{n1} = \max\big\{0,\ n\, T_n' V_n^{*-1} T_n - p\big\},$$

$$a_{n2} = \max\big\{0,\ n\, T_n' V_n^{*-1} Q^{-1} V_n^{*-1} T_n - \mathrm{tr}(Q^{-1}V_n^{*-1})\big\}$$

and the other notations have already been introduced earlier. Let then

$$N_c^{(1)} = \left[c^{-1/2}(a^*_{N_c^0})^{1/2}\right] + 1, \quad c > 0, \tag{3.27}$$

and by iteration, we let

$$N_c^{(r)} = \left[c^{-1/2}(a^*_{N_c^{(r-1)}})\right] + 1, \; c > 0, \quad \forall\, r \geqslant 1. \tag{3.28}$$

Finally, we define

$$N_c^* = \lim_{r\to\infty} N_c^{(r)}. \tag{3.29}$$

It is clear from the definition in (3.26)—(3.29) that N_c^* is stochastically smaller than N_c. (Note that $a_n^* \leqslant \{\mathrm{tr}(QV_n^*)\}, \quad \forall\, n \geqslant n_0$, and hence $N_c^{(r)} \leqslant N_c \quad \forall\, r \geqslant 0$.) Further note that under $\{K_c\}$ in (3.13),

$$n\,T_n'V^{*-1}T_n - p = n\,c^{1/2}\lambda'V^{*-1}\lambda + 2\,nc^{1/4}\lambda'V_n^{*-1}(T_n - c^{-1/4}\lambda) +$$

$$+ n(T_n - c^{1/4}\lambda)'V_n^{*-1}(T_n - c^{1/4}\lambda) - p, \tag{3.30}$$

where for $n \leqslant n_c^*$, $nc^{1/2} < k < \infty$, the second term on the right hand side of (3.30) has asymptotically a normal distribution with zero mean and variance $4nc^{1/2}\lambda'\Gamma^{-1}\lambda$, while the third term has the central chi–squared d.f. with p DF. Thus, $n\,T_n'V_n^{*-1}T_n - p$ behaves like $\Delta_{nc} + Z_c + (\chi_p^2 - p)$, where $Z_c \sim \mathcal{N}(0, 4\Delta_{nc})$. A similar result holds for a_{n2}. Thus, a_{n1} and a_{n2} are not stochastically equivalent to Δ_{nc} and Δ_{nc}^0, respectively; they are subject to some random variation (due to Z_c and χ_p^2) of the same order. This creates additional trouble for finding out the ADR of $T_{N_c^*}^S$. Fortunately, this relates to the case, where

$$N_c^{(r)}/n_c^* \xrightarrow{P} (1-\beta) + U_r, \quad \text{as} \quad c \downarrow 0, \tag{3.31}$$

where $0 < \beta < 1$ and U_r has a nondegenerate distribution over $(0, \infty)$. This representation allows us to use the central limit theorem for random sample size (cf. (Billingsley, 1968, p.147)), and justifies the computation of the ADR of $T_{N_c^*}^S$ from its asymptotic distribution. It follows that the ADR of $T_{N_c^*}^S$ is less than (or equal to) that of T_{N_c}, for all λ. Thus, the adaptation of the stopping time N_c^*, $c > 0$, leads to a stochastically smaller stopping time for $T_{N_c^*}^S$ without violating the asymptotic risk dominance of $T_{N_c^*}^S$ over T_{N_c}. But, viewed from a practical point of view, when θ is not close to the pivot (even under (3.13)), $T_{N_c^*}^S$ and $T_{N_c}^S$ may be very close to each other (with respect to their ADR), and $\{N_c,\ c > 0\}$ signals a great simplification of the computation of ADR, without changing the stopping time. Hence, one may as well favor N_c, $c > 0$, as a working rule.

REFERENCES

Billingsley, P. (1968). *Convergence of Probability Measures*. Wiley, New York.

Ghosh, M., Nickerson, D.M. and Sen, P.K. (1987). Sequential shrinkage estimation. *Ann. Statist.* **15**, 817–829.

Hoeffding, W. (1948). A class of statistics with asymptotically normal distribution. *Ann. Math. Statist.* **19**, 293–325.

Sclove, S.L., Morris, C. and Radhakrishnan, R. (1972). Nonoptimality of preliminary test estimators for the mean of a multivariate normal distribution. *Ann. Math. Statist.* **43**, 1481–1490.

Sen, P.K. (1980). On nonparametric sequential point estimation of location based on general rank order statistics. *Sankhya, Ser.A* **42**, 201–219.

Sen, P.K. (1981). *Sequential Nonparametrics: Invariance Principles and Statistical Inference.* Wiley, New York.

Sen, P.K. (1984). On sequential nonparametric estimation of multivariate location. In: *Proc. Third Prague Conf. Asymp. Meth.* Academia, Prague, pp. 119–130.

Sen, P.K. (1987). Sequential Stein–rule maximum likelihood estimation: General asymptotics. In: *Statist. Dec.Th. and Rel. Top.IV,* **2**, pp. 195–208.

Sen, P.K. (1987). Sequential shrinkage U–statistics: General asymptotics. *Rev. Brasileira de Prob. Estatist.* **1**, 1–21.

Sen, P.K. (1988). Functional Jackknifing. Rationality and general asymptotics. *Ann. Statist.* **16**, 450–469.

Sen, P.K. (1989). Asymptotic theory of sequential shrunken estimation of statistical functionals. In: *Proc. Fourth Prague Con. Asymp. Meth.* Academia, Prague.

Sen, P.K. and Ghosh, M. (1981). Sequential point estimation of estimable parameters based on U–statistics. *Sankhya, Ser.A* **43**, 331–344.

van Zwet, W.R. (1984). A Berry—Esseen bound for symmetric statistics. *Z. Wahrsch. verw. Geb.* **66**, 425–440.

Prob. Theory and Math. Stat., Vol. 2, pp. 424–431
B.Grigelionis *et al.* (Eds.)
1990 VSP/Mokslas

THE CENTRAL LIMIT THEOREM FOR FINITELY DEPENDENT RANDOM VARIABLES

V.V. SHERGIN
Institute of Chem. Technol., F. Engelsa 7, 153460 Ivanovo, USSR

ABSTRACT

We consider a finite or countable set of the finitely dependent random variables. The purpose of the paper is to estimate the remainder term in the central limit theorem for such weakly dependent random variables.

1. INTRODUCTION

Let $\mathcal{J}$ be a nonempty finite or countable set. A nonempty family of random variables is said to be finitely dependent if for every nonempty finite subset A of $\mathcal{J}$ there exists another finite subset B, $A \subset B$, such that $(\xi_j,\ j \in A)$ is independent of $(\xi_j,\ j \in \mathcal{J}\backslash B)$ and

$$d = \sup_{A} \inf_{B} |B|/|A| < \infty,$$

where $|\cdot|$ denotes the cardinality of a set (see (Chen, 1978)). Let $B(A)$ be the minimum set (with respect to $|\cdot|$) for given A. The order of dependence D is defined to be the smallest integer not less than d. Then, in particular, $|B(A)| \leqslant D|A|$. Examples of finitely dependent families are

m–dependent random sequences ($\mathcal{J} = \mathbb{N}$, $D = 2m + 1$) and

m–dependent random fields ($\mathcal{J} = \mathbb{Z}^r$ for some $r \geqslant 2$, $D = (2m+1)^r$).

Conditions of convergence of distributions of sums of finitely dependent random variables to infinitely divisible distributions are established in (Chen, 1978). The purpose of the given paper is to estimate the remainder term in the central limit theorem for this class of weakly dependent random variables. In (Shergin, 1979; Sunklodas, 1986; Heinrich, 1985) and others, estimates of the rate of convergence in the central limit theorem for m–dependent random variables and fields were obtained, most of the estimates being optimal. But the methods used in the above mentioned papers cannot be applied in our paper because of some peculiarities of the class of finitely dependent random variables. For instance, let $B_0(A) \equiv A$, $B_1(A) \equiv B(A)$, $B_k(A) = B(B_{k-1}(A))$, $k = 1, 2, \ldots$. Then $|B_k(A)| \leqslant D^k|A|$ takes place for finitely dependent random variables, while $|B_k(A)| \leqslant (2km+1)|A|$ is related to m–dependent random fields. The outlined methods of transform of characteristic function of a sum of finitely dependent random variables make it possible to construct the estimations in the central limit theorem in the form of Lyapunov fractions and to eliminate some assumptions from (Sunklodas, 1986; Heinrich, 1985).

2. RESULTS

In this paper we assume, that the introduced family of finitely dependent random variables satisfies the following additional condition: for any finite sets $A_1, A_2 \subset \subset \mathcal{J}$,

$$B(A_1) \cap A_2 = \varnothing \text{ implies } B(A_2) \cap A_1 = \varnothing. \tag{1}$$

It is easy to see that (1) is true for m–dependent random processes and fields. We are going to discuss some possible forms of (1) in the forthcoming papers. In (Chen, 1978) some inequalities for moments of random variables were used instead of (1).

Suppose $\mathbf{E}\xi_j = 0$, $\mathbf{E}|\xi_j|^p < \infty$ for all j from $\mathcal{J}$ and some p, $2 < p < 3$. Let V be an arbitrary finite subset of $\mathcal{J}$,

$$\zeta_V = \sum_{j \in V} \xi_j, \qquad B_V^2 = \mathbf{E}\zeta_V^2 > 0,$$

$$\Delta_V(x) = |\mathbf{P}(\zeta_V < xB_V) - \Phi(x)|,$$

$$\Delta_V = \sup_x \Delta_V(x), \quad L_{p,V} = B_V^{-p} \sum_{j \in V} \mathbf{E}|\xi_j|^p,$$

where

$$\Phi(x) = (2\pi)^{-1/2} \int_{-\infty}^{x} e^{-t^2/2} dt.$$

THEOREM 1. *In these notations, the inequality*

$$\Delta_V \leqslant c_1 L_{p,V}$$

holds, where c_1 is a positive constant dependent only on D and p.

THEOREM 2. *If there exist positive constants M and n_0 such that for any $w \subset \mathcal{J}$, $n_0 < |w| < \infty$,*

$$\sum_{j \in w} \mathbf{E}\xi_j^2 \leqslant M\mathbf{E}\Big(\sum_{j \in w} \xi_j\Big)^2 \tag{2}$$

holds, then

$$\Delta_V(x) \leqslant c_2(1 + |x|)^{-p} L_{p,V},$$

where the constant $c_2 > 0$ depends on p, D, M and n_0.

Investigations of dependence of constants c_1, c_2 on D are of some interest. This is done in Theorems 3 and 4.

THEOREM 3. *Suppose that* $\mathbf{E}|\xi_j|^3 < \infty, \quad j \in \mathcal{J}$. *Then:*

1) $\Delta_V \leqslant a_1 D^{14/3} L_{3,V}$ *where* a_1 *is an absolute constant.*

2) *If there exist constants* $\alpha \geqslant D$, $\beta > 0$, *such that for any finite* $A \subset \mathcal{J}$ *and* $k = 1, 2, \ldots$

$$\left|B_k(A)\right| \leqslant \alpha k^{\beta} \left|A\right|$$

holds, then

$$\Delta_V \leqslant c_3 \alpha^{7/3} L_{3,V},$$

where $c_3 > 0$ *depends only on* β.

THEOREM 4. *Let* ξ_j, $j \in \mathbf{Z}^r$, *be an* m*-dependent random field. Then*

$$\Delta_V \leqslant c_4 (m+1)^{(p-1)r} L_{p,V},$$

where $c_4 > 0$ *depends only on* p *and* r.

For a sequence of m-dependent random variables Theorem 2 holds without (2).

THEOREM 5. *Let* $\xi_1, \ \xi_2, \ldots$ *be a sequence of* m*-dependent random variables,* $V = \{1, 2, \ldots, n\}$. *Then*

$$\Delta_V(x) \leqslant c_5 (1 + |x|)^{-p} (m+1)^{p-1} L_{p,V},$$

where $c_5 > 0$ *depends only on* p.

3. PRELIMINARIES TO THE PROOFS OF THEOREMS 1—5

Due to the standard method of reduction of arbitrary m to $m = 1$, Theorem 4 is a simple corollary of Theorem 1. Thus we allow $m = 1$ in the proof of Theorem 5. As for Theorems 1 and 2 we discuss them in detail. Let a, a_0, a_1, $\ldots$ be positive constants dependent only on D and p (also dependent on M and n_0 in Theorem 2). The results of (Shergin, 1988) make it possible to assume that $L_{p,V} \leqslant a_0$ holds and we can consider the truncated random variables

$$\bar{X}_j = \xi_j I(|\xi_j| \leqslant B_V), \quad X_j = \bar{X}_j - \mathbf{E}\bar{X}_j.$$

In this case $\beta_V^2 = \mathbf{E}\big(\sum_V X_j\big)^2 = B_V^2(1+\lambda)$, where $|\lambda| \leqslant a_1 L_{p,V}$. If $a_0 < 1/(2a_1)$, then $|\beta_V - 1| < 1/2$ and without loss of generality we may assume that $B_V = 1$ (see also (Bhattacharia and Ranga Rao, 1976; Petrov, 1972). Let

$$S_V = \sum_{j \in V} X_j, \quad \bar{\Delta}_V(x) = \left|\mathbf{P}(S_V < x) - \Phi(x)\right|,$$

$$Q_V = \sum_{j \in V} \mathbf{E}|X_j|^3, \quad f_V(t) = \mathbf{E}\exp(itS_V), \quad g(t) = \exp(-t^2/2).$$

Since $Q_V \leqslant a_1 L_{p,V}$, for the proof of Theorems 1 and 2 it is sufficient to check that $\bar{\Delta}_V(x) \leqslant a_2 Q_V$, and under condition (2), $\bar{\Delta}_V(x) \leqslant a_3(1+|x|)^{-3} Q_V$. By means of Lemmas 7 and 8 (Petrov, 1972, Chapter 5) we see that these inequalities follow from Lemma 1.

LEMMA 1. *There exists a constant $a > 0$ such that for all $t \in \mathbf{R}$, $1 \leqslant |t| \leqslant a Q_V^{-1}$,*

$$\left|f_V(t) - g(t)\right| \leqslant a_4 |t|^3 Q_V \exp\left(-t^2/4\right) \tag{3}$$

and, under condition (2),

$$\left|f_V^{(\mu)}(t) - g^{(\mu)}(t)\right| \leqslant a_5 |t|^3 Q_V \exp\left(-t^2/4\right), \quad \mu = 1,2,3. \tag{4}$$

When $|t| \leqslant 1$, the inequalities necessary for applying Lemmas 7 and 8 from (Petrov, 1972) are easy to prove using the results of (Shergin, 1988).

4. PROOF OF LEMMA 1

Let $[k,l]$ be a subset of $\mathbf{N}$, $n = |V|$. For convenience, let $V = [1,n]$. First let us estimate $f_V(t)$ when $t \in \mathbf{R}$ satisfies the conditions of Lemma 1 and the inequality

$$|t|^3 Q_V < a_1. \tag{5}$$

The constant a_1 will be specified later. For some random variable X, set

$$\bar{v}(X) = \exp(itX) - 1, \quad v(X) = \bar{v}(X) - \mathbf{E}\bar{v}(X).$$

Let m be the smallest integer such that $[m,n] \subset B(m)$ (discussing various sets $B(\cdot)$ we use the same notations for the element $m \in \mathcal{J}$ and for the set consisting of this one element),

$$E(k) = [k+1,n] \cap B(k), \quad H(k) = [k,n] \cap \left(B_2(k) \backslash B(k)\right),$$

$$Z_k = \sum_{E(k)} X_j, \quad W_k = \sum_{H(k)} X_j, \quad q_k = \mathbf{E}X_k^2 + 2\mathbf{E}X_k Z_k,$$

$$Y_k = \sum_{j=k}^{n} X_j, \quad Y_{2,k} = Y_{k+1} - Z_k - W_k,$$

$$f_k = \mathbf{E}\exp(itY_k), \quad \omega_k = v(X_k) v(Z_k), \tag{6}$$

$$R_k = \sum_{j \in B(k)} \mathbf{E}|X_j|^3, \quad R_{2,k} = \sum_{j \in B_2(k)} \mathbf{E}|X_j|^3,$$

$$\lambda_k = \mathbf{E}\left(\exp(itX_k)\right) + \mathbf{E}\omega_k. \tag{7}$$

Using the same formulas we define Y_m, f_m, R_m, $R_{2,m}$. Let $\lambda_0 = 1$, $\lambda_m = f_m$, $q_m = \mathbf{E}Y_m^2$,

$$\Lambda_l = \prod_{j=0}^{l} \lambda_j, \quad l = 0, \dots, m.$$

It is not difficult to prove that $f_k = f_{k+1}\lambda_k + M_k$, where

$$M_k = \mathbf{E}\big(\exp(itY_{2,k})V_k\big),$$

$$V_k = \bar{v}(W_k)\omega_k - \bar{v}(Z_k + W_k)\mathbf{E}\omega_k \tag{8}$$

and

$$f_V(t) \equiv f_1 = \Lambda_m + \sum_{j=1}^{m-1} \Lambda_{j-1} M_j. \tag{9}$$

Applying Taylor's formula in (7), together with moments inequalities and estimates of cardinality of sets $B(\cdot)$, one can define the constant a_1 in (5) such that the relations

$$t^2 \sum_{j\in G} |q_j| \leqslant 1 \quad \text{if } G \subset B_5(l), \quad \text{for an arbitrary} \quad l, \tag{10}$$

$|\lambda_k - 1| \leqslant 1/2$, and the equality

$$\lambda_k = 1 - \frac{t^2}{2} q_k + a_2\theta(t)t^3 R_k, \quad k = 1,\ldots,m, \tag{11}$$

hold (from here on, $\theta(t)$ denotes various complex functions such that $|\theta(t)| \leqslant 1$). From (11) and $q_1 + q_2 + \ldots + q_m = 1$ it follows that for $j = 1,\ldots,m-1$,

$$\Lambda_j = \exp\Big(-\frac{t^2}{2}\sum_{k=1}^{j} q_k + a_3\theta(t)t^3\sum_{k=1}^{j} R_k\Big), \tag{12}$$

$$\Lambda_m = g(t)\exp\big(a_4\theta(t)t^3 Q_V\big)$$

and, under condition (2),

$$|\Lambda_j^{(\mu)}| \leqslant a_5|t^3|\exp\Big(-\frac{t^2}{2}\sum_{k=1}^{j} q_k + a_6|t|^3\sum_{k=1}^{j} R_k\Big), \ \mu = 1,2,3.$$

Let us estimate $\Lambda_{j-1}M_j$, $j = 1,\ldots,m-1$, under condition (5). Set

$$P_0 = B_2(j) \cap [j,n], \quad C_0 = [j,n]\backslash P_0, \quad U_0 = V_j, \quad Y(C_0) = \sum_{C_0} X_l.$$

Then

$$M_j = \mathbf{E}\big(\exp(itY(C_0))U_0\big) \equiv \psi_0.$$

Let us define sets $C_0 \supset C_1 \supset \ldots, P_0 \subset P_1 \subset \ldots$ and random variables $U_1, U_2, \ldots$, having in mind that $\psi_l = \mathbf{E}\big(\exp(itY(C_l)U_l)\big)$ and ψ_{l+1} must be related. Let

$$p_0 = j, \quad p = p_1 \in C_0, \quad B_3(p) \cap P_0 = \varnothing,$$

$$E(p) = C_0 \cap \big(B(p)\backslash(p)\big), \quad H(p) = C_0 \cap \big(B_2(p)\backslash B(p)\big).$$

Let us define for the sets $E(p), H(p)$ variables Z_p, W_p, q_p, ω_p by analogy with (6). Let

$$Y_{p,0} = Y(C_0) - X_p, \quad Y_{p,1} = Y_{p,0} - Z_p, \quad Y_{p,2} = Y_{p,1} - W_p.$$

Then the random variables U_0 and $(X_p,\ Z_p,\ W_p)$ are independent and

$$\psi_0 = \lambda_p^* \mathbf{E}\big(\exp(itY_{p,0})U_0\big) + K_p,$$

where

$$\lambda_p^* = \mathbf{E}\big(\exp(itX_p)\big) + \mathbf{E}\omega_p,$$

$$K_p = \mathbf{E}\big(\exp(itY_{p,2})U_0 N_p\big), \quad N_p = \bar{v}(\dot{W}_p)\omega_p - \bar{v}(Z_p + W_p)\mathbf{E}\omega_p.$$

Note that $\mathbf{E}|N_p| \leqslant a|t|^3 R_{2,p}$. Hence we may define a_1 in (5) such that $\mathbf{E}|N_p| < 1/4$ and for λ_p^* the equation analogous to (11) holds. It is obvious, that only two possibilities exist:

$$|K_p| \leqslant 2|\psi_0|\mathbf{E}|N_p|, \tag{13}$$

$$|K_p| > 2|\psi_0|\mathbf{E}|N_p|. \tag{14}$$

If (13) holds, then $\psi_0 = \lambda_{p_1}\mathbf{E}\big(\exp(itY_{p,0})U_0\big)$, where $\lambda_{p_1} = \lambda_p^* + a_2\theta(t)t^3 R_{2,p}$. Let $C_1 = C_0 \backslash (p)$, $P_1 \equiv P_0$, $U_1 = U_0$. Then $\psi_0 = \psi_1 \lambda_{p_1}$. If (14) holds, then let

$$C_1 = C_0 \backslash B_2(p), \quad P_1 = P_0 \cup B_2(p),$$

$$U_1 = U_0 N_p / E|N_p|, \quad \lambda_{p_1} = \frac{1}{2}\theta(t).$$

Then necessarily $\psi_0 = \psi_1 \lambda_{p_1}$. The final result of these transformations is

$$\psi_0 = \psi_\nu \prod_{l=1}^{\nu} \lambda_{p_l},$$

where ν is defined from the condition $B_3(p) \cap P_\nu \neq \emptyset$ for all $p \in C_\nu$.
Let

$$G_1 = \big(p_l : \ (13) \text{ holds}\big), \quad G_2 = \big(p_l : l = 0 \text{ or } (14) \text{ holds}\big).$$

Then

$$P_\nu = \bigcup_{G_2} B_2(p)$$

and from condition (1) and definition of ν it is obvious that $H_\nu = P_\nu \cup C_\nu$ can be represented as

$$H_\nu = \bigcup_{G_2} A_p,$$

where $|A_p| \leqslant D^5$.
Interchanging of elements of $[j, n]$ leads to variables Z_p, q_p, defined here and previously in (6) being identical for all $p \in G_1$. Then

$$\frac{t^2}{2}\sum_{k=j}^{n} q_k \geqslant \frac{t^2}{2}\sum_{p \in G_1} q_p - \frac{1}{2}|G_2| \tag{15}$$

implies (10) and the equality $[j,n] = G_1 \cup H_\nu$. The random variables N_p, $p \in G_2$, are independent, hence $\mathbf{E}|U_\nu| \leqslant \mathbf{E}|U_0|$ and

$$|\psi_\nu| \leqslant \Big|\prod_{G_1} \lambda_p\Big| 2^{-|G_2|}\mathbf{E}|U_0|.$$

It is obvious that $\exp(x/2) < 2^x$ for $x \geqslant 1$ and therefore the inequality

$$|\Lambda_{j-1}M_j| \leqslant a_3|t|^3 R_j \exp\Big(-\frac{t^2}{2} + a_4|t|^3 Q_V\Big)$$

implies (12), (15). These inequalities give us the proof of the first part of Lemma 1.

Slightly more sophisticated calculations allow to estimate $\big(M_j\Lambda_{j-1}\big)^{(\mu)}$, $\mu = 1,2,3$, in the same way. Let

$$|t|^3 Q_V > a_1, \quad |t| Q_V < a_2. \tag{16}$$

In this case for any t it is enough to check that

$$\Big|f_V^{(\mu)}(t)\Big| \leqslant a_3 \exp(-t^2/4), \quad \mu = 0,\ldots,3. \tag{17}$$

LEMMA 2. *There exists a set* A_0 *such that* $\mathbf{E}\big(\sum_{A_0} X_j\big)^2 < 1/3$ *and if* $j \in V\setminus A_0$ *then* $\mathbf{E}X_k^2 \leqslant a_4 Q_V^2$ *holds for all* $k \in B_2(j)$.

The set A_0 is constructed in the following way: order the elements of V so that $\mathbf{E}X_1^2 \geqslant \mathbf{E}X_2^2 \geqslant \ldots$, and let

$$I_l = (p\colon B_2(p) \cap [1,l] \neq \varnothing), \quad \gamma_l = \mathbf{E}\Big(\sum_{I_l} X_j\Big)^2,$$

$$k = \min(l : \gamma_l < 1/3,\ \gamma_{l+1} \geqslant 1/3).$$

Then $A_0 = I_k$ (see also (Shergin, 1979, Lemma 9)). It is seen from Lemma 2 that by proper choosing of the constant in condition (16) formulae (10) and (11) are true if the indexes of variables in these formulae are included into the set $H = V\setminus A_0$. Giving reordered elements V so that H coincides with $[1,\nu]$, $\nu = |H|$, we have $q_1 + \ldots + q_\nu \geqslant 1 - 1/3 = 2/3$. Repeating the previous procedure and adding the condition $B_2(p_l) \cap A_0 = \varnothing$ for p_l, we come to (17), with $C_0 = V$, $U_0 = 1$, $P_0 = \varnothing$.

5. SOME WORDS ABOUT THEOREM 3

We can prove the first statement of this theorem if we calculate constants more precisely on the basis of Theorem 1. The main point is to investigate the constants in Lemma 2 and in the inequality (10). If $|B_r(A)| \leqslant \alpha r^\beta |A|$, $r = 1,2\ldots$, we have $|I_k| \leqslant \alpha 2^\beta k$ in Lemma 2, and $\mathbf{E}X_{k+1}^2 \leqslant C_1 \alpha^{8/3} Q_V^2$. Furthermore, in (10) $|G_2| \leqslant C_2\alpha$ holds and hence (10) is true if $|t| < C_3\big(\alpha^{7/3} Q_V\big)^{-1}$ (C_1, C_2, C_3 depend only on β). These inequalities are sufficient to end the proof of Theorem 3.

6. ON THEOREM 5

We must verify that inequality (4) in Lemma 1 is true for sequence of m–dependent random variables without condition (2). Let $m = 1,\ a,\ a_1, \dots$ be absolute positive constants. Let us return to the proof of Lemma 1, using the previous notations. Using the estimates

$$\delta_i = |EX_i(X_{i-1} + X_{i+1})| \leqslant a_1 Q_V^{2/3} \leqslant a_2,$$

we have

$$\Big|\sum_{j=1}^{i} q_j\Big| = \Big|1 - E\Big(\sum_{j=i+1}^{n} x_j\Big)^2\Big| \leqslant 1 + \Big(1 + 2\delta_i - \mathbf{E}\Big(\sum_{j=1}^{i} X_j\Big)^2\Big) \leqslant a_3.$$

Using this inequality instead of condition (2), we have (4) for $|t| \leqslant aQ_V^{-1/3}$. In the case $aQ_V^{-1/3} < |t| \leqslant a_1 Q_V^{-1}$ we use some ideas from (Shergin, 1979). We omit routine details for verification of (4).

REFERENCES

Bhattacharya, R.N. and Ranga Rao, R. (1976). *Normal Approximation and Asymptotic Expansions.* Wiley, New York.

Chen, L.H.Y. (1978). Two central limit problems for dependent random variables. *Z. Wahrsch. verw. Geb.* **43**, 223–243.

Heinrich, L. (1985). Non–uniform estimates, moderate and large deviations in the central limit theorem for m–dependent random variables. *Math. Nachr* **121**, 107–121.

Petrov, V.V. (1972). *Sums of Independent Random Variables.* Nauka, Moscow.

Shergin, V.V. (1979). On the rate of convergence in the central limit theorem for m–dependent random variables. *Theory Probab. Appl.* **24**, 781–794.

Shergin, V.V (1988). On the central limit theorem for finite–dependent random variables. In: *Rings and Modulus. Limit Theorems of Probability Theory,* Leningrad State Univ. Press, Leningrad. **2**, 204–210.

Sunklodas, J (1986). An estimate of the rate of convergence in the central limit theorem for weakly dependent random fields. *Liet. Matem. Rink.* **26**, 541–559.

Prob. Theory and Math. Stat., Vol. 2, pp. 432–436
B.Grigelionis *et al.* (Eds.)
1990 VSP/Mokslas

ON HARMONIC FUNCTIONS OF RANDOM WALKS ON GROUPS

M.G. SHUR
Moscow Institute for Electronic Engineering,
Bol.Vuzovskii proezd, 3/5, 109817 Moscow Z–28, USSR

ABSTRACT

A right spread–out random walk on a locally compact group is considered. According to the key result, each harmonic function of the random walk can be decomposed into a sum of two harmonic functions, the first (respectively, the second) of which is generated in a sense by absolutely continuous (singular) components of the random walk transition probabilities. Some topological properties of harmonic functions and compactness properties of families formed by harmonic functions are also investigated.

1. INTRODUCTION

In this paper we consider properties of harmonic functions corresponding to a spread–out random walk on a topological group $\mathcal{G}$. The set of all such functions contains two special families. Here their elements are called almost absolutely continuously generated and singularly generated harmonic functions respectively. Roughly speaking, the generation of functions of the first or the second family depends only on the absolutely continuous or singular components of convolution powers of the random walk law. Every finite harmonic function admits a unique decomposition into the sum of a function of the first kind and that of the second kind (see Theorem 1).

All functions of the second family are unbounded and in the case of Abelian group $\mathcal{G}$ they vanish almost everywhere with respect to the Haar measure (see Theorem 2 and Example 1).

The functions of the first family turn out to be lower semicontinuous on an "essential" part of the group (see Theorem 3 and Corollary). It is interesting to compare these assertions with the well–known property of continuity of bounded harmonic functions (Revuz, 1975).

We consider also properties of the compactness for families of harmonic functions (see Theorems 6 and 7). The corresponding statements resemble the known ones from the excessive function theory for Markov processes (Dellacherie and Meyer, 1987, Sections XII.92—XII.94) but are not immediate analogies of them. Some results concerning the compactness of families of harmonic functions for standard Markov processes are proposed in (Shur, 1970).

All proofs omitted here will be given in two papers by the author which have been accepted for publication in the journal "Theory of Probability and Its Applications".

2. TERMINOLOGY AND NOTATION

We suppose a Hausdorff locally compact separable group $\mathcal{G}$ with multiplication as the group operation to be fixed. Let e be the neutral element, ξ be the left Haar measure, $\mathcal{B}$ be the family of Borel sets in $\mathcal{G}$, μ be a fixed probability measure given on $\mathcal{B}$. On the group $\mathcal{G}$ we consider the right random walk with the law μ (Revuz, 1975), i.e., the homogeneous Markov chain with the transition probabilities $p(x, dy) = \mu(x^{-1}dy)$, $x \in \mathcal{G}$, on the measurable space $(E, \mathcal{B})$. The n-step transition probabilities are denoted by $p(n, x, dy)$ where $n \geqslant 1$, $x \in \mathcal{G}$. If P is the transition operator of the random walk, then for bounded from below functions f given on $\mathcal{G}$ we have

$$P^n f(x) = \int f(y)p(n, x, dy) = \int f(xy)\mu^n(dy), \tag{1}$$

where μ^n, $n \geqslant 1$, is the n-fold convolution of the measure μ.
By definition, our random walk on $\mathcal{G}$ is

(i) finitary if the measure μ has a compact support;

(ii) spread-out if for some n the convolution power μ^n is non-singular with respect to the Haar measure ξ:
If the measure μ^n has the Lebesgue decomposition $\mu^n = \alpha_n + \beta_n$, where α_n and β_n are the absolutely continuous and singular components of μ_n with respect to the Haar measure, then $\alpha_n(\mathcal{G})$ does not decrease when n increases. Moreover, $\beta_n(\mathcal{G})$ tends to zero at the geometrical rate as $n \to \infty$ if the random walk in question is spread-out (Revuz, 1975).
In what follows we shall consider only spread-out random walks without special reservations as a rule.
We regard a Borel function $f : \mathcal{G} \to [0, \infty]$ as belonging to the class $\mathcal{H}$ of harmonic functions for the given random walk, if $f < \infty$ on a everywhere dense set and $f \equiv Pf$. It is easy to establish the local summability of all harmonic functions relative to the measure ξ.

3. DECOMPOSITION OF HARMONIC FUNCTIONS

In order to state the first group of our results, it is necessary to construct transition operators Q_n^α and Q_n^β corresponding to the transition probabilities $p_n^\alpha(x, dy) = \alpha_n(x^{-1}dy)$ and $p_n^\beta(x, dy) = \beta_n(x^{-1}dy)$, $n \geqslant 1$, $x \in \mathcal{G}$ (cf. (1)). The following simple lemma is a key one.

LEMMA 1. *If $f \in \mathcal{H}$, then $Q_1^\alpha f \leqslant Q_2^\alpha f \leqslant \ldots$ on the set $D_f = \{x : f(x) < \infty\}$ and*

$$Q_1^\beta f \geqslant Q_2^\beta f \geqslant \ldots \tag{2}$$

everywhere on $\mathcal{G}$.

We prove the relations (2), for example. Considering the Lebesgue decompositions of measures μ^n, one can see that $Q^\beta_{n+1} f \leqslant Q^\beta_n Q^\beta_1 f$.
Hence

$$Q^\beta_{n+1} f(x) \leqslant \int \Big[\int f(xyz)\beta_1(dz) \Big] \beta_n(dy) \leqslant \int f(xy)\beta_n(dy) = Q^\beta_n f(x)$$

as the integral in the square brackets does not exceed the integral

$$\int f(xyz)\mu(dz) = Pf(xy) = f(xy) \quad \text{(see (1))}.$$

We emphasize that the set D_f (see Lemma 1) is polar (the last term is related to any set contained in the set of polars $\{x: g(x) < \infty\}$ for some harmonic function g). Besides Lemma 1 permits us to represent any harmonic function as the sum of the functions $Q^\alpha f = \lim_{n\to\infty} Q^\alpha_n f$ and $Q^\beta f = \lim_{n\to\infty} Q^\beta_n f$, the first of which being defined on the set D_f only.
We call a function $f \in \mathcal{H}$ almost continuously generated and write $f \in \mathcal{H}_a$, if $f = Q^\alpha f$ outside some polar set E_f. Usually this set can be chosen in different ways and, if it can be supposed to be empty, we write $f \in \mathcal{H}^*_a$ and call the function f continuously generated (in this case $f < \infty$). A harmonic function is singularly generated, if it is the least majorant of the function $Q^\beta f$ on the set D_f (this means that $g \geqslant f$ on D_f, whenever $g \in \mathcal{H}$ and $g \geqslant Q^\beta f$ on D_f). In this case we write $f \in \mathcal{H}_s$.
Let f be a harmonic function. If $\varphi = Q^\beta f$ on D_f and $\varphi = 0$ outside D_f, then the function $\psi = \lim_{n\to\infty} P^n\varphi$ is a singularly generated harmonic function. This fact leads to the following statement in which the function ψ can be chosen as f_2 (we remind that we consider spread–out random walks only).

THEOREM 1. *Let f be a harmonic function for the given random walk. Then $f = f_1 + f_2$, where $f_1 \in \mathcal{H}_a$, $f_2 \in \mathcal{H}_s$ and $Q^\alpha f_1 = f_1$ on D_f. This decomposition is unique up to the values of f_1 and f_2 outside D_f.*

It is clear that the class $\mathcal{H}_s$ consists only of the function $f \equiv 0$ whenever some measure μ^n is absolutely continuous relative to the Haar measure ξ. According to the following result this class is also rather poor in the case of random walks on Abelian groups.

THEOREM 2. *Let $\mathcal{G}$ be an Abelian group. A harmonic function f is singularly generated if and only if either of the two equivalent statements is valid:* (I) $f = Q^\beta f$ *almost everywhere (with respect to ξ)* or (II) $f = 0$ *almost everywhere.*

It would be very interesting to describe the class of non–Abelian groups for which this result is in force.

Example 1. We give an example $f \in \mathcal{H}_s$ of a random walk on the real line $\mathcal{G} = \mathbf{R}$ (with the addition as the group operation). Let the law μ of the random walk be equal to the sum $\mu_1 + \mu_2 + \mu_3$, where μ_1 is an absolutely continuous measure on [−1, 1] having the density equal to 1/4, and μ_2 and μ_3 are the point masses concentrated at the points −1 and 1 respectively. We suppose $\mu_2\{-1\}$ and $\mu_3\{1\}$ to be equal to 1/4. Then the function equal to $(2+\sqrt{3})^x$ for integers and to zero otherwise, is harmonic and by Theorem 2 singularly generated.

4. TOPOLOGICAL PROPERTIES OF HARMONIC FUNCTIONS

For the sake of simplicity of formulations, in this section we agree to assume that the exceptional set E_f (see the definition of the class $\mathcal{H}_a$) contains the set $\{x : f(x) = \infty\}$. The following Theorem 3 implies, in particular, the lower semi-continuity of the functions from $\mathcal{H}_a^*$.

THEOREM 3. *The restriction of any function $f \in \mathcal{H}_a$ to the subspace $\mathcal{G}\backslash E_f$ of the topological space $\mathcal{G}$ is lower semicontinuous on the above subspace. If $f \in \mathcal{H}_a$ and the set E_f is closed, then on $\mathcal{G}\backslash E_f$ the function f is equal to a function $F \leqslant f$ which is lower semicontinuous on $\mathcal{G}$ and belongs to the class $\mathcal{H}_a$.*

It is also possible to state conditions of continuity of harmonic functions. Here we give the following partial result.

THEOREM 4. *Let the random walk under consideration be finitary. A function $f \in \mathcal{H}_a^*$ is continuous if and only if* (i) *it is locally bounded and* (ii) *the integrals*

$$Q_n^\beta f(x) = \int f(xy)\beta_n(dy)$$

tend to zero uniformly on every compact set in $\mathcal{G}$ as $n \to \infty$.

Since $\beta_n(\mathcal{G}) \to 0$ at the geometrical rate, the condition (ii) for a finitary walk is valid automatically if the function f can increase only sufficiently slow.

5. COMPACTNESS OF FAMILIES OF HARMONIC FUNCTIONS

As it was pointed out in Section 2, every function which is harmonic for the considered random walk, belongs to the space L_{loc} of functions locally integrable with respect to the left Haar measure ξ. We shall say that a sequence of functions $\{f_n;\ n \geqslant 1\}$ tends to $f \in L_{\text{loc}}$ locally weakly, if

$$\lim_{n\to\infty} \int f_n w d\xi = \int f w d\xi$$

for every bounded Borel function w on $\mathcal{G}$ with compact support. If $f_n \in L_{\text{loc}}$ and

$$\lim_{n\to\infty} \int_K |f_n - f| d\xi = 0$$

for any compact K, where $f \in L_{\text{loc}}$, then we shall say that the sequence $\{f_n\}$ tends to f locally strongly. So we work with the two kinds of convergence in the space L_{loc} (later the locally uniform convergence will be considered too). A family of harmonic functions S is called compact in $\mathcal{H}$ (with respect to the chosen kind of convergence) if every sequence of functions of S contains a subsequence tending to a harmonic function.

THEOREM 5. *A family $S \subset \mathcal{H}$ is compact in $\mathcal{H}$ relative to the locally weak convergence if and only if the following condition is fulfilled:*

(A) *whatever compact $K \subset \mathcal{G}$ is, the family S is uniformly integrable with respect to the measure $\xi_K P$, where ξ_K is the restriction of the measure ξ to K and P is the transition operator of the random walk.*

It should be mentioned here that

$$\xi_K P(C) = \int_K p(y, C)\xi(dy), \quad C \in \mathcal{B}.$$

THEOREM 6. *Let either of the following conditions be realized:* (i) *the group $\mathcal{G}$ is Abelian;* (ii) *the law μ of the random walk is absolutely continuous with respect to the Haar measure. Then a family $S \subset \mathcal{H}$ is compact in $\mathcal{H}$ with respect to the locally strong convergence if and only if the condition* (A) *from Theorem* 5 *is satisfied.*

Passing to the case of locally uniform convergence we consider finitary random walks only (cf. Theorem 4).

THEOREM 7. *Let a given random walk be finitary and all functions of a family $S \subset \mathcal{H}_a^*$ be continuous. Then the family S is compact in $\mathcal{H}$ with respect to the locally uniform convergence if and only if conditions* (i) *and* (ii) *from Theorem* 4 *are satisfied uniformly with respect to all functions of S.*

REFERENCES

Dellacherie, C. and Meyer, P.-A. (1987). *Probabilités et potentiel. Chapitres XII—XVI.* Hermann, Paris.

Revuz, D. (1975). *Markov Chains.* North Holland, Amsterdam—New York.

Shur, M.G. (1970). A property of compactness of families of harmonic functions for a Markov process. *Math. Notes* 7, 109–115.

Prob. Theory and Math. Stat., Vol. 2, pp. 437–445
B.Grigelionis *et al.* (Eds.)
1990 VSP/Mokslas

EXPONENTS OF OPERATOR STABLE DISTRIBUTIONS IN BANACH SPACES

G. SIEGEL
Technische Universität, Dresden, DDR.

ABSTRACT

The existence of exponents of an operator stable full probability measure in a separable Banach space in proved. Some properties of the class of all such exponents are presented.

1. INTRODUCTION

An operator stable probability measure (p.m.) μ on a real separable Banach space $\mathcal{E}$ represents a limit distribution of operator normed and centered sums of i.i.d. random elements taking values in $\mathcal{E}$. The classical stable law on $\mathbf{R}^1$ and (more general) on $\mathcal{E}$ are special cases when the norming operators are multiplies of the identity operator. It is well-known that an operator stable law is always infinitely divisible (e.g., (Siegel, 1989, Theorem 3.1)). Thus, if $\widehat{\mu}$ is the characteristic functional of μ, then $\widehat{\mu}^t$ stands for the characteristic functional of an infinitely divisible p.m. μ^t for all $t > 0$. Sharpe (Sharpe, 1969) showed in the finite dimensional case dim $\mathcal{E} < \infty$ that for every full and operator stable p.m. there exists at least one linear operator A and a function $a : (0, \infty) \to \mathcal{E}$ such that for all $t > 0$

$$\mu^t = t^A \mu \ * \ \delta_{a(t)},$$

where

$$t^A = \exp(A \ \ln \ t) = \sum_0^\infty (k!)^{-1}(A \ \ln \ t)^k, \ t^A\mu = \mu(t^A)^{-1},$$

and $\delta_{a(t)}$ denotes the degenerate distribution concentrated at $a(t)$. The operator A is then called exponent of μ. The present paper proves the existence of exponents of μ in separable Banach spaces. In general, this exponent is not unique. Hence the class of all exponents for a fixed operator stable p.m. on $\mathcal{E}$ is also established. These results are direct extensions of those in (Holmes et al., 1982).

Many authors have studied properties and representations of exponents for a full and operator stable p.m. in a finite dimensional vector space, for instance (Holmes et al., 1982; Hudson and Mason, 1981; Jurek, 1980; Luczak, 1984; Sato, 1987). However, the general case is more complicated. So it was shown in (Siegel, 1990b) that each infinitely divisible distribution on a vector space with dim $\mathcal{E} = \infty$ can be represented as a limit law of certain operator normed sums of i.i.d. random elements. To derive more typical properties of operator stable laws it is therefore necessary to suppose additional restrictions on the norming operators. Krakowiak (see (Krakoviak, 1979)) was the first one to prove assertions of this type. He claimed that the semigroup of operators spanned by

$$\{A_k A_n^{-1} : k \geqslant n, \quad n = 1, 2, \ldots\}$$

is relatively compact in the norm topology. A new access to this subject of research discussed in (Siegel, 1989) leads to significantly weaker conditions, see also (Siegel, 1990a). An example is given in Theorem 1. In the present paper we use the same method. Thus it is possible to show the existence of exponents of an operator stable p.m. only under suitably chosen boundedness relations for norming operators.

2. PRELIMINARIES AND MAIN RESULTS

Let $\|\cdot\|$ be the norm and $\mathcal{E}'$ the dual space of $\mathcal{E}$. L denotes the set of linear and bounded operators and L_0 the subset of all invertible operators in L. In particular, $I \in L$ and $O \in L$ stand for the identity and the zero operator, respectively. The set of p.m. on $\mathcal{E}$ will be denoted by $\mathcal{P}$. Further, $\mathcal{P}_0$ stands for the subset of all full p.m. on $\mathcal{E}$. For $A \in L$, the image measure $\lambda = A\mu$ is defined by

$$\lambda(B) = (A\mu)(B) = \mu(A^{-1}(B)),$$

where B is a Borel set of $\mathcal{E}$. For $\mu_1, \mu_2 \in \mathcal{P}$ it is well-known that

$$A(\mu_1 \ * \ \mu_2) = A\mu_1 \ * \ A\mu_2 \quad \text{and} \quad A(\mu^n) = (A\mu)^n.$$

An important tool for treating operator stable laws is the strong operator topology; the sequence $T_n \in L$ strongly converges to $T \in L$ if $T_n x \to Tx$ for all $x \in \mathcal{E}$. Then we briefly write $T_n \xrightarrow{s} T$. Further we make use of the uniform topology on L and write $T_n \xrightarrow{u} T$ if $\|T_n - T\| \to 0$.
To characterize full and operator stable p.m. it is useful to introduce the following subset of L_0 :

$$F(t) = \{A \in L_0 : \mu = A\mu^t * \ \delta_x \quad \text{for some} \quad x \in \mathcal{E}\}, \quad t > 0.$$

Now we are able to state the following assertion.

THEOREM 1 (Siegel, 1989, Proposition 6.4). *Let $\mu \in \mathcal{P}_0$ and suppose that we have the weak convergence*

$$A_n \nu^n * \ \delta_{a_n} \xrightarrow{w} \mu, \quad n \to \infty, \tag{2.1}$$

for certain $A_n \in L_0$, $a_n \in \mathcal{E}$, and $\nu \in \mathcal{P}$. Then $F(t)$ is non-empty for every $t > 0$ provided that the operators A_n in (2.1) satisfy the following condition:
For every $m > 1$ there are numbers n_0, $K_0 \geqslant 1$ such that

$$\|A_k A_n^{-1}\| \leqslant K_0 \tag{2.2}$$

for all $n \geqslant n_0$ and all k with $n/m \leqslant k \leqslant nm$.

Let $\mu \in \mathcal{P}_0$ be a fixed operator stable p.m. Our main interest is concerned with the set $E(\mu)$ of exponents of μ. Basically we have:

THEOREM 2. *Let μ be a full and infinitely divisible p.m. Then $E(\mu)$ is nonempty if and only if (2.1) is satisfied for certain $\nu \in \mathcal{P}$, $a_n \in \mathcal{E}$, and some operators $A_n \in L_0$ having the property*

$$\sup_{k,n\geqslant 1,\ k\neq n} \left|kn^{-1} - 1\right|^{-1} \|A_k A_n^{-1} - I\| < \infty. \tag{2.3}$$

Obviously, (2.3) is sharper than (2.2). Moreover, $E(\mu) \neq \emptyset$ implies $F(t) \neq \emptyset$ for all $t > 0$ since $(1/t)^A = t^{-A} \in F(t)$. Note that t^A is always invertible for every $A \in L$. Indeed, from $1^A = I$ and from the continuity of t^A in the uniform topology we conclude that t^A is invertible in a neighborhood of $t = 1$ and, accordingly, for all $t > 0$. In finite dimensional spaces the converse assertion is also true, i.e., $F(t) \neq \emptyset$ implies $E(\mu) \neq \emptyset$. The question whether this remains true in the general case is unsolved so far. The next theorem gives an answer in an affirmative provided that the norming operators satisfy, in addition, a Lipschitz condition similar to the corresponding one in Theorem 1. Before formulating this result we introduce the notion of G–stability of μ for a certain group G of operators from L_0; μ is called G–stable if for every $A, B \in G$ there exist $C \in G$ and $c \in \mathcal{E}$ such that

$$A\mu * B\mu = C\mu * \delta_c.$$

THEOREM 3. *Let μ be a full and G–stable p.m. Then $E(\mu)$ is nonempty if and only if there are positive numbers t_n tending to one and operators $B_n \in F(t_n)$ such that*

$$\sup_n |t_n - 1|^{-1} \|B_n - I\| < \infty. \tag{2.4}$$

Note that $F(t) \neq \emptyset$, $t > 0$, implies the G–stability of μ with

$$G = \bigcup_{t>0} F(t). \tag{2.5}$$

Conversely, if μ is G–stable, then μ is clearly infinitely divisible; further, it is easy to verify that $F(r) \neq \emptyset$ at least for each rational number $r > 0$ (for more information see (Siegel, 1989)).
Now we deal with representations of $E(\mu)$. An important role for proving such statements plays the symmetry group $S(\mu)$ of μ, i.e., the set of all operators $A \in L_0$ with $\mu = A\mu * \delta_x$ for some $x \in \mathcal{E}$. In the special case $\dim \mathcal{E} < \infty, S(\mu)$ is compact if and only if μ is full (see (Billingsley, 1966)). Unfortunately this statement is not true in the general case, that is, $S(\mu)$ is not necessarily compact even if L is equipped with the strong operator topology. We introduce further the tangent space of certain group G of operators belonging to L_0. This definition depends on

the use of the special topology. $T_u(G)$ denotes the set of operators $A \in L$ which satisfy

$$d_n^{-1}(D_n - I) \xrightarrow{u} A$$

for some $D_n \in G$ and some real numbers $d_n \neq 0$ tending to zero. Similarly we define the subset $T_s(G)$ of operators A from L by

$$d_n^{-1}(D_n - I) \xrightarrow{s} A.$$

Note that $T_u \subseteq T_s$ and T_u is nonempty since $O \in T_s$.
If $\mathcal{E}$ is infinite dimensional and μ is G stable for certain group $G \subseteq L_0$, then in general it is not possible to show that G is strongly closed. However, if G_s denotes the set of all strong limit points of mutual sequences of G, then we may expect that $G_s \cap L_0 = G$. This is so, for instance, if $F(t) \neq \emptyset$ for all t and $G = \bigcup_{t>0} F(t)$, see Lemmas 1 and 2 below. More results of this type can be found in (Siegel, 1990a).

THEOREM 4. *Let μ be a full and G–stable p.m. and assume that H is a subgroup of G satisfying $H_s \cap L_0 = H$. Then the space $T_s(H)$ is linear and all strong limit points from sequences from H belong to H. Further, the function*

$$e(A) = \sum_{k=0}^{\infty} A^k/k!$$

maps $T_s(H)$ into H and $T(H) = T_s(H) = T_u(H)$. In particular,

$$T_s(G) = T_u(G) \quad \text{and} \quad T_s\big(S(\mu)\big) = T_u\big(S(\mu)\big).$$

In the following theorem we relate $E(\mu)$ with the tangent space of $S(\mu)$.

THEOREM 5. *Let μ be full and operator stable on $\mathcal{E}$. If $E(\mu)$ is nonempty, then*

$$E(\mu) = B + T\big(S(\mu)\big) \tag{2.6}$$

for any exponent B of μ.

COROLLARY 6. *Let us assume that μ is full and operator stable; further let $E(\mu)$ be nonempty. If $S(\mu)$ is discrete, then there is exactly one exponent $B \in E(\mu)$.*

Next we illustrate the assertions mentioned above by an example.

Example. Consider the Hilbert space $\mathcal{E} = l_2$ and let $e_1, e_2, \ldots$ be a basis in l_2. We introduce a random element X taking values in l_2 by

$$X = \sum_{k=1}^{\infty} X_k e_k$$

such that the real–valued random variables X_k are independent. We suppose that every X_k has a symmetric and stable distribution with characteristic exponent $\alpha_k, 0 < \alpha_k \leqslant 2$. We suppose further that the α_k are pairwise different. If μ is the distribution of X, then each operator $A \in S(\mu)$ admits the representation

$$A : x = \sum_{k=1}^{\infty} x_k e_k \to Ax = \sum_{k=1}^{\infty} \varepsilon_k x_k e_k,$$

where $\varepsilon_k \in \{-1, +1\}$. Since $S(\mu)$ is discrete, $T(S(\mu)) = \{O\}$, see Corollary 6.

i) Assume now that there is a number $\delta > 0$ such that $\alpha_k \geqslant \delta$ for every k. Then by Corollary 6 the exponent B of μ is unique and has the representation

$$B\, x = \sum_{k=1}^{\infty} \frac{x_k}{\alpha_k} e_k.$$

ii) If we assume that $\alpha_{n'} \to 0$ for some subsequence n', then there exists operators $A_t \in L$ such that $\mu^t = A_t \mu$ for $0 \leqslant t \leqslant 1$, and A_t is given by

$$A_t x = \sum_{k=1}^{\infty} t^{1/\alpha_k} x_k e_k.$$

Clearly, A_t is not invertible in L and strongly (but not uniformly) converges to I as $t \to 1$. It is interesting to note that $G = \{A_t, 0 \leqslant t \leqslant 1\}$ is a strongly continuous semigroup of operators. Finally we note that in this case $E(\mu) = \emptyset$ and $F(t) = \emptyset$ for all t.

3. PROOFS

We begin with an assertion which is concerned with the convergence of operator types. It is clear that the mapping

$$(A, \mu) \in (L, \mathcal{P}) \to A\mu \in \mathcal{P}$$

is continuous in the weak topology of $\mathcal{P}$ provided that L is connected with the strong topology (see (Linde and Siegel, 1989)). The converse assertion is deeper. We have:

LEMMA 1 (Linde and Siegel, 1989). *Let $\lambda_n = A_n \mu_n * \delta_{a_n}$ for some $A_n \in L$, $\mu_n \in \mathcal{P}$, $a_n \in \mathcal{E}$. Further, we assume that $\lambda_n \xrightarrow{w} \lambda$, $\mu_n \xrightarrow{w} \mu \in \mathcal{P}_0$. If $\sup_n \|A_n\| <$ $< \infty$, then the set $\{A_n,\ n \geqslant 1\}$ is relatively compact in the strong topology.*

Note that the additional condition $\sup_n \|A_n\| < \infty$ is needed only in the case $\dim \mathcal{E} = \infty$.

The following assertion is well–known (see, for instance (Linde, 1986)).

LEMMA 2. *Let $t_n \geqslant 0$ be some real numbers and suppose that $\mu \in \mathcal{P}$ is infinitely divisible and non-degenerate. Then the set $\{\mu^{t_n},\ n \geqslant 1\}$ is weakly relatively compact if and only if t_n is bounded. In particular, $\mu^{t_n} \xrightarrow{w} \mu_0$ for some $\mu_0 \in \mathcal{P}$ if and only if $t_n \to t_0$ for some $t_0 \geqslant 0$. In either case, $\mu_0 = \mu^{t_0}$.*

The proofs of the theorems are organized as follows:

Theorem 3 → Theorem 2 → Theorem 4 → Theorem 5.

Proof of Theorem 3.
We show this assertion step by step.
Step 1. Denote by S the set of all rational numbers $r > 1$, then there exists a function $C_x : S \to L_0$ such that

$$C_{x_1}C_{x_2} = C_{x_1x_2}, \quad x_1, x_2 \in S. \tag{3.1}$$

To show this, we first note that $B_n \in F(t_n)$ implies $A_n^{-1} \in F(1/t_n)$; further the operators $\widehat{B}_n = B_n^{-1}$ also satisfy (2.4) with $\widehat{t}_n = 1/t_n$ instead of t_n. Hence we may assume without loss of generality that $t_n > 1$. For every $x \in S$ we now introduce a sequence of integers $k_n = k_n(x)$ by writing

$$k_n \leqslant \frac{\ln x}{(t_n - 1)} < k_n + 1.$$

By (2.4) we get, for some $c > 0$,

$$\|B_n^{k_n}\| \leqslant (\|B_n - I\| + 1)^{k_n} \leqslant (c(t_n - 1) + 1)^{k_n} \leqslant$$
$$\leqslant \left(\frac{c\ \ln\ x}{k_n} + 1\right)^{k_n} \leqslant x^c, \tag{3.2}$$

i.e., $\{B_n^{k_n}\}$ is bounded for fixed x. On the other hand, it is well-known that $B_n \in F(t_n)$ implies $B_n^{k_n} \in F(t_n^{k_n})$ (see (Siegel, 1989, Proposition 3.2)). Using

$$t_n^{k_n} = ((t_n - 1) + 1)^{k_n} \to x \quad \text{as} \quad n \to \infty,$$

we see that the p.m.

$$\mu^{s_n} = B_n^{k_n}\mu * \delta_{a_n}, \quad \text{with} \quad s_n = t_n^{k_n} \quad \text{and some} \quad a_n \in \mathcal{E},$$

are weakly relatively compact, see Lemma 2. Thus we may apply Lemma 1 according to which $\{B_n^{k_n}\}$ has to be strongly relatively compact. By means of the diagonal selection method it is possible to find a subsequence $n' \to \infty$ such that for every $x \in S$

$$B_{n'}^{k_{n'}(x)} \xrightarrow{s} C_x, \text{ say.} \tag{3.3}$$

It is obvious that

$$k_n(x_1x_2) = k_n(x_1) + k_n(x_2) + l_n, \quad x_1, x_2 \in S, \tag{3.4}$$

where $l_n \in \{0,1\}$. From $B^{l_n} \xrightarrow{u} I$, (3.3) and (3.4) it follows that

$$B_n^{k_n(x_1x_2)} \xrightarrow{s} C_{x_1x_2} = C_{x_1}C_{x_2}.$$

Step 2. There exists an uniformly continuous function $C : (1,\infty) \to L$ having the property (3.1) for all $x > 1$.
By (3.1) and (3.2) we have

$$\|B_n^{k_n} - I\| \leqslant \|B_n - I\| \sum_{j=1}^{k_n-1} \|B_n\|^j \leqslant c(k_n - 1)(t_n - 1)x^c \leqslant c\, x^c \ln\, x.$$

The right-hand side is arbitrarily small for sufficiently small $(x-1) > 0$ and all n. Hence $\|C_x - I\|$ is also small on S. The closure of $\{C_x,\ x \in S\}$ in the uniform topology yields an uniformly continuous semigroup $\{C_x,\ x \geqslant 1\}$ satisfying (3.1).
Step 3. There exists a subgroup $\{C_x,\ x > 0\}$ of L_0 with $C_x \in F(x)$.
To see this we use $B_x \xrightarrow{u} I$. Hence $B_x^{-1} \in L_0$ exists in a neighborhood of $x = 1$ and, accordingly, for all $x > 1$. By

$$F(t_n^{k_n}) \ni B_n^{k_n} \xrightarrow{s} C_x$$

and by $t_n^{k_n} \to x$, $C_x \in F(x)$ follows for all $x \geqslant 1$, see Lemma 2. Next we put $C_{1/x} = C_x^{-1} \in F(1/x)$. Note that the last relation is well-known (see, e.g., (Siegel, 1989, Proposition 3.2)).
Step 4. There exists an operator $B \in L$ such that $x^B \in F(x)$, for each $x > 0$.
Putting $D_t = C_{e^t}$ for real t and using (3.1) for $x > 0$ we get

$$D_{t_1+t_2} = D_{t_1}D_{t_2}.$$

Clearly, D_t is also uniformly continuous. By a standard result (Hille and Phillips, 1948) it is well-known that there is $B \in L$ with $D_t = \exp\{tB\}$ and thus

$$F(x) \ni C_x = \exp(\ln xB) = x^B.$$

This implies $\mu^x = (1/x)^B\mu \,*\, \delta_{a(x)}$ for some $a(x) \in \mathcal{E}$ and for all $x > 0$, i.e., $-B \in E(\mu) \neq \varnothing$.

Proof of Theorem 2.
It suffices to show that the suppositions of Theorem 3 are fulfilled. Since (2.3) implies (2.2) we have $F(x) \neq \varnothing$, $x > 0$. For each rational number $r > 1$ we use sequences of natural numbers k_j, n_j tending to infinity such that $k_j/n_j \to r$. It is shown in (Siegel, 1989) that there is a subsequence j' such that

$$A_{k_{j'}}A_{n_{j'}}^{-1} \xrightarrow{s} B(r) \in F(r) \tag{3.5}$$

for each r. Next we select a fixed sequence of rational numbers $r_l \to 1$ and put $B_l = B(r_l)$. Then from (2.4) and (3.5) it follows that

$$\sup_l |r_l - 1|^{-1}\|B_l - I\| \leqslant \sup_{k,n\geqslant 1} |kn^{-1} - 1|\ \|A_lA_n^{-1} - I\|.$$

This completes the proof.

Proof of Theorem 4.

Step 1. Supposing $L \ni D_n \xrightarrow{s} D \in L$ *and* $k_n \to \infty$, *we get*

$$\left(\frac{1}{k_n} D_n + I\right)^{k_n} \xrightarrow{s} e(D).$$

This assumption can be easily verified by standard arguments using

$$\left(\frac{1}{k_n} D_n + I\right)^{k_n} \xrightarrow{u} e(D) \quad \text{and} \quad e(D_n) \xrightarrow{s} e(D).$$

Step 2. We have $e(A) : T_s(H) \to H$.

Let $A \in T_s(H)$, then there is a sequence $A_n \in H$ and real numbers $t_n > 0$ tending to zero such that

$$\frac{1}{t_n}(A_n - I) \xrightarrow{s} A.$$

Next we define integers k_n by $k_n \leqslant 1/t_n \leqslant k_n + 1$ and put $D_n = k_n(A_n - I)$. Thus $D_n \xrightarrow{s} A$ and, by Step 1,

$$H \ni A_n^{k_n} = \left(\frac{D_n}{k_n} + I\right)^{k_n} \xrightarrow{s} e(A) \in L_0.$$

Hence $e(A) \in H_s \cap L_0 = H$ which was to be proved.

Step 3. $T_s(H)$ *is linear and* $e(A) \in H$ *implies* $e(tA) \in H$ *for every real* t.

The proof of this assertion can be carried out in just the same way as that in the appendix of (Holmes et al., 1982) for finite dimensional spaces, where in addition $H_s \cap L_0 = H$ is needed.

Step 4. $T_u(H) = T_s(H)$.

It suffices to show that $T_s(H) \subseteq T_u(H)$. Let $A \in T_s(H)$, then by Steps 2 and 3, $e(A/n) \in H$ and thus

$$n\big(e(A/n) - I\big) \xrightarrow{u} A \in T_u(H).$$

Step 5. $T_s(H)$ *is strongly closed.*

Indeed, $T_s(H) \ni A_n \xrightarrow{s} A$ implies $H \ni e(A_n) \xrightarrow{s} e(A) \in H_s \cap L_0 = H$.

The proof is complete.

Proof of Theorem 5.

This proof is similar to the corresponding one in (Holmes et al., 1982) for the special case $\mathcal{E} = \mathbf{R}^l$. Hence only a sketch of the proof is given.

Consider the class $F(t)$ and introduce a homomorphism η by $\eta(A) = t$, if $A \in F(t)$. Further we put $g(A) = \ln\, \eta\big(e(A)\big)$, $A \in T(G)$. Our basic aim is to show that $g : T(G) \to \mathbf{R}$ is linear. To see this we have to show that $tg(A) = g(tA)$ first for each natural number $t = n \geqslant 2$, then for positive rational numbers and, since $T(G)$ is strongly closed by Theorem 4, also for all $t > 0$. Moreover, we have also to show that $g(A) + g(B) = g(A + B)$ first for all A, B in a certain neighborhood of $O \in T(G)$ and then for each $A, B \in T(G)$ (note that the function $e(tA)$ is analytic in t). The case $t < 0$ is trivial by definition of the tangent space. Finally we use the same arguments as in (Holmes et al., 1982) and observe that

$$T(S(\mu)) = \{A \in T(G) : g(A) = 0\} - \text{the kernel of} \quad g-$$

and

$$E(\mu) = \{A \in T(G) : g(A) = 1\}.$$

These both representations imply the required relation (2.6).
The proof is complete.

REFERENCES

Billingsley, P. (1966). Convergence in K-spaces. *Z. Wahrsch. verw. Geb.* **5**, 175–179.

Hille, E. and Phillips, R.S. (1948). *Functional Analysis and Semi-Groups.* Wiley, New York.

Holmes, J.P., Hudson, W.N. and Mason, J.D. (1982). Operator stable laws: Multiple exponents and elliptic symmetetry. *Ann. Probab.* **10**, 602–612.

Hudson, W.N. and Mason, J.D. (1981). Operator stable distribution on $\mathbf{R}^2$ with multiple exponents. *Ann. Probab.* **9**, 482–489.

Jurek, Z.J. (1980). Domains of normal attraction of operator stable measure on Euclidean spaces. *Bull. Acad. Polon. Sci., Ser. Math.* **28**, 397–409.

Krakowiak, W. (1979). Operator–stable probability measures on Banach spaces. *Collog. Math.* **41**, 313–326.

Linde, W. (1986). *Infinitely Divisible and Stable Measures in Banach Spaces.* Wiley, New York.

Linde, W. and Siegel, G. (1989). On the convergence of operator types for Radon probability measures in Banach spaces. In: *Probability in Banach spaces.* Birkhäuser, Basel—Stuttgart.

Luczak, A. (1984). Elliptic symmetry and characterization of operator stable and operator semistable measures. *Ann. Probab.* **12**, 1217–1223.

Sato, K.–I. (1987). Strictly operator–stable distributions. *J. Multivariate Analysis* **22**, 278–295.

Sharpe, M. (1969). Operator–stable probability distributions on vector groups. *Trans. Amer. Math. Soc.* **136**, 51–65.

Siegel, G. (1987). Operator stable distributions in stationary sequences defined on separable Banach spaces. In: *Limit Theorems in Probab. Theory and Rel. Fields.* Dresden, pp. 135–153.

Siegel, G. (1989). Operator–stable distributions in separable Banach spaces. *Theory Probab. Appl.* **34**, 552–560.

Siegel, G. (1990a). Properties of operator stable distributions in infinite dimensional spaces. *J. Theor. Probab* (to appear).

Siegel, G. (1990b). On the class of operator stable distributions in separable Banach spaces. *Probab. and Math. Statist.* (to appear).

Prob. Theory and Math. Stat., Vol. 2, pp. 446–450
B.Grigelionis *et al.* (Eds.)
1990 VSP/Mokslas

INFINITE DIMENSIONAL DISTRIBUTIONS AND SECOND QUANTIZATION IN STRING THEORIES

O.G. SMOLYANOV
Department of Mathematics and Mechanics, Moscow University, Lenin Hills, 117234 Moscow, USSR.

ABSTRACT

The method of quantization is proposed for the infinite dimensional Hamilton systems consisting of some number of equal interacting subsystems. For this purpose the space of infinite dimensional distributions and Feynman measures for some exponents are used.

For the classical Schrödinger quantization of (finite dimensional) Hamilton system having linear phase $E = Q \times P$ and a Hamilton function $H: E \to \mathbf{R}^1$ one must use pseudodifferential operators in the space $\mathcal{L}_2(Q)$ with the symbol (in the Weil or Jordan or some else sense) H as an Hamiltonian in Schrödinger equation. In the case when a phase space has infinite dimension there are two difficulties to overcome. The first one is the absence of the standard Lebesgue measure on the infinite dimensional space, i.e., σ–finite and locally finite Borel measure what is invariant under translations. The second one (linking with the first) is the fact that usual formulas for pseudodifferential operators has no sense without Lebesgue measure. To overcome the first difficulty we use space of measures or distributions (generalized measures) on Q instead of the space $\mathcal{L}_2(Q)$; to overcome the second difficulty we substitute so–called Feynman measures for some exponents.
We use an analogous methods in the case of so–called second quantization (of infinite dimensional Hamilton systems consisting of some number of equal interacting subsystems). The second quantization in the sense of this paper has two steps. We name the first one as a "nonlinear quantization"; as a result of this step we get an equation with "nonlinear pseudodifferential operators" what are Hamilton equations for some new Hamilton system (the latter has infinite dimension even in the case when the initial Hamilton system has only finite dimension).
The second step is the (usual) quantization of this Hamiltonian system. In the case when the subsystems of initial Hamilton system describe fields in space–time of dimension two (as in the theory of superstrings) one replace the term "second quantization" by the term "third quantization" (because one can regard field equations as a result of a (first) quantization of a finite dimensional Hamilton system). This remark explain the title of the paper. Below we regard not only common vector spaces over real or complex fields but so–called superspaces as well (exact definition will be formulated in the text).

1. SUPERANALYSIS

We shall generalize the method due to Vladimirov and Volovich, and use without explanations the terms from the paper (Vladimirov and Volovich, 1984). Throughout what follows $\Lambda = \Lambda_0 \oplus \Lambda_1$ is a commutative Banach superalgebra over the real field supposed $\{\lambda \in \Lambda\colon \lambda \cdot \lambda_1 = 0, \quad \forall\, \lambda_1 \in \Lambda_1\} = 0$, Λ^c is the complexification of Λ. Let else $T = T_0 \oplus T_1$, $S = S_0 \oplus S$, are topological supermodules over Λ.

A topological supermodule over Λ is defined to be a topological module $T_0 \oplus T_1$ over Λ under following conditions:

$$\forall i \in \{0,1\},\ a_i \in T_i,\ \lambda_i \in \Lambda_i \Rightarrow \lambda_i, \quad a_i \in T_j,$$

where $j = (i + j) \pmod 2$.

A (real) superspace over a pair of supermodules (T, S) is defined to be a topological vector space $T_0 \oplus S_i$, the definition of a complex superspace is analogous the role of Λ playing in its complexification. Given real superspaces G and E over pairs (T^0, S^0) and (T^1, S^1) of topological supermodules, a Λ-linear function $g\colon E \to G$ is defined to be the restriction of a certain Λ-linear map whose range is the module $T^0 \oplus S^0$ and whose image belongs to the module $T^1 \oplus S^1$.

We suppose that the vector space $\Lambda(E, G)$ of (all) Λ-linear maps from E to G is endowed by a suitable topology and a natural structure of a superspace. Therefore, one can define high derivatives by the induction. Proper method of differentiation of maps between topological vector spaces given a map $f\colon E \to G$ (where E and G are superspaces) is defined to be ($\tau-$) differentiable iff this map is ($\tau-$) differentiable as map between topological vector spaces E, G and its derivative $f'(x)$ belongs to $\lambda(E, G)$. Below we suppose, that all derivatives are so-called compact derivatives (Smolyanov, 1979; Daletsky and Fomin, 1983).

Example 1. Let $A = A_0 \oplus A_1$ be Z_2-graduated locally convex space, $T = \Lambda \otimes A_0$, $S = \Lambda \otimes A_1$; then

$$T_0 \oplus S_1 = (\Lambda_0 \otimes A_0) \oplus (\Lambda_1 \otimes A_1)$$

(here the symbol $\otimes$ means a tensor product endowed by a proper topology).

2. HAMILTON SYSTEMS IN SUPERSPACES AND IN COMMON SPACES

We shall say that a superspace is free if it is as in the previous example and we shall suppose that all superspaces are free.

A Hamilton system is a triplet (E, b, H), where E is a superspace (named the phase space of the system), $H\colon E \to \Lambda$ is a map, named the Hamilton function, b is a simplectic structure, i.e., a restriction to $E' \times E'$ of Λ-bilinear map $\bar{b}\colon E'_\Lambda \times E'_\Lambda \to \Lambda$ $(E_\Lambda = \Lambda \otimes A,\ E' = \lambda(E_\Lambda, \Lambda))$, $\bar{b}$ possessing the following properties:

(a) the restriction of $\bar{b}$ to $A'_0 \times A'_0$ is non-degenerate real antisymmetric bilinear form;

(b) the restriction of $\bar{b}$ to $A'_0 \times A'_1$ is a usual scalar product;

(c) the restrictions of $\bar{b}$ to $A'_0 \times A'_1$ and to $A'_1 \times A'_1$ are equal to zero. A corresponding Hamilton equation is the equation $f'(t) = I_b(H'(f(t)))$; here $f: R^1 \to E$ is a function, I_b is a linear map from E' to E, corresponding to the function b.

Example 2. Let H is a (common) Hilbert space over the complex field, H_r is a corresponding real Hilbert space (H_r is identic to H as an additive group), $b(x,z) = (ix, z)$ where $(\cdot,\cdot)$ is the scalar product, $x \longmapsto \bar{x}$ is the complex conjugation, B is a self adjoined operator in the H, commuting with complex conjugation, $H(x,z) = i(Bx, \bar{z})$. Then the usual Schrödinger equation is the Hamilton equation for the Hamilton system (H_r, b, H).

3. PSEUDODIFFERENTIAL OPERATORS

Throughout what follows Q, P are superspaces $Q = P$, $P = Q$, $E = Q \times P$, B is canonical simplectic structure on E, $\Phi_{E'}$ is "the Feynman superdistribution" on E', i.e., linear function whose domain is the space of Fourier transforms of supermeasures on E' and whose Fourier transform is the function $(q,p) \mapsto \exp i\langle p, q\rangle$ (a supermeasure is defined in (Smolyanov and Shavgulidze, 1989); the definition of Feynman supermeasures and Fourier transforms of supermeasures are given in the same paper). If τ belongs to $[0,1]$ and H is a complex or real function on E, then a pseudodifferential operator (ΨDO) in a space M of supermeasures on Q having the τ–symbol H is defined by the next formula (we use natural signs):

$$\hat{H}_\tau \nu(dq) = \int\int H(\tau q_1 + (1-\tau)q, p) e^{i\langle q_1, p\rangle} \nu(dq_1) \Phi_E(dq_1, dp)$$

and ΨDO in a space of Λ–functions on E (i.e., functions on E whose images belong to Λ) having the τ–symbol H is defined by the next formula:

$$(\hat{H}_\tau \varphi)(q) = \int\int H(\tau q_1 + (1-\tau)_{q,p}) e(q_1) e^{-i\langle q_1, p\rangle} \Phi_E(dq_1, dp).$$

4. QUANTIZATION

The (linear) τ–quantization ($\tau \in [0,1]$) of the Hamilton system (E, b, H) is the Hamilton system $(S_Q \times \Sigma_Q,\ b_Q, P_\tau)$, where Σ_Q is a (super) space of a Λ^c–functions on Q, S_Q is a (super) space of Λ–linear maps from Σ_Q to Λ^c (it is naturally to name the latter by the term the space of "generalized supermeasures" on Q): b_Q is a canonical simplectic structure on $S_Q \times \Sigma_Q$, $P_\tau(\nu, \varphi) = -i\langle \hat{H}_\tau \nu, \varphi\rangle$; here $\hat{H}_\tau$ is the ΨDO in M having τ–symbol H. "The Hamilton equation" of this Hamilton system is linear and in fact consists of two (linear) "Schrödinger equations"; the one involves functions (of a real variable) having their images in S_Q and another involves functions having their images in S_Q).

5. NONLINEAR QUANTIZATION

Let $n \in N$ and $\mathcal{E} = E \times \ldots \times E$ (n factors), $h: \mathcal{E} \to \Lambda$. It is naturally to name the object $(\mathcal{E}, b, H, h)$ by the term "a Hamilton system consisting of n identical Hamilton subsystems (these subsystems are described by a Hamilton function H) with an interaction" (describing by the function h). Nonlinear quantization of such system is defined to be the Hamilton system $(S_Q \times \Sigma_Q,\ b_Q, P_\tau^n)$, where P_τ^n is a Hamilton function of the latter,

$$P_\tau^n(\nu, \psi) = -i\langle \hat{H}_\tau \nu, \varphi\rangle - \frac{i}{n}\langle \hat{h}_\tau(\nu \otimes \ldots \otimes \nu,\ \varphi \otimes \ldots \otimes \varphi)\rangle.$$

"The Hamilton equation" of this system is the nonlinear system of Schrödinger equations:

$$i\dot{\nu}(\cdot, t) = \hat{H}_\tau \nu + \hat{h}_\tau\langle(\nu \otimes \ldots \otimes \nu),\ (\varphi \otimes \ldots \otimes \cdot)\rangle,$$
$$i\dot{\varphi}(\cdot, t) = \hat{H}_{1-\tau}\varphi + \langle(\nu \otimes \ldots \otimes \cdot), \hat{h}_{1-\tau}\ (\varphi \otimes \ldots \otimes \varphi)\rangle;$$

so using this quantization we go to a polynomial Hamiltonian having even degree. It is interesting to mention that in (Bogolyubov and Bogolyubov, 1984) it was stated that polynomial Hamiltonians of odd degree had no sense. But on the contrary we can construct polynomial Hamiltonian of odd degree using a (nonlinear) quantization of more sophisticated Hamilton systems with interactions (Smolyanov, 1988).

6. SECOND QUANTIZATION

Let ν be a smooth (Smolyanov,1979) measure on the space Q of the subsystem. Then (under some assumptions) for every map $f: S_Q \to S_Q$ there is a map $f_\nu: \mathcal{L}_2(\nu) \to \mathcal{L}_2(\nu)$ defined by the next formal equation: $f(\psi\sqrt{\nu}) = f_\nu(\psi)\sqrt{\nu}$. Then the nonlinear Schrödinger system transform into a (nonlinear) system that involves functions having images in $\mathcal{L}_2(\cdot)$. This system is a "Hamilton equation" of a Hamilton system with phase space $\mathcal{L}_2(\nu)$ and symplectic structure described in the example. So besides the method of nonlinear quantization described above we have another such method (the latter depends on the choice of measure ν).
One can define the second Schrödinger quantization of a Hamilton system consisting of interacting subsystems by one of four methods. In each of these methods the second quantization is a composition of two operations, the first one being a nonlinear quantization by one of described above two methods and the second one being a linear quantization by one of analogous two methods. It is necessary to emphasize that in the models connecting with string theories, it is impossible to use a Gaussian measure because this measure is not quasiinvariant under action of a group of smooth reparametrization of the string; an existence of such measure was proved by Shavgulidze. It is worth to note, that the quantization in the first sense (without fixed measure ν) uses distributions and functions that are solutions of Schrödinger equations (Hamilton equations) but have no sense of a common wave function; generally the second power of a distribution has no sense, and the second power of a function (on a space having the infinite dimension) is not a density of a measure (it is a consequence of the absence of Lebesgue measure). It is a

reason to use a Wigner measure in the phase space; this measure is an analogous of Wigner function on this space.

REFERENCES

Bogolyubov, N.N. and Bogolyubov, N.N. (1984). *Introduction to the Quantum Statistical Mechanics.* Nauka, Moscow.

Daletsky, Yu.L. and Fomin, S.V. (1983). *Measures and Differential Equations on Infinite Dimensional Spaces.* Nauka, Moscow.

Smolyanov, O.G. (1979). *Analysis in Topological Linear Spaces and Its Applications.* Moscow State Univ., Moscow.

Smolyanov, O.G. (1988). Nonlinear pseudodifferential operators in superspaces. In: *Abstr. Tartu Conf. Meth. of Algebra and Analysis,* pp.103–106.

Smolyanov, O.G. and Shavgulidze, E.T. (1989). Fourier transforms and pseudodifferential operators in superanalysis. *Soviet Math. Dokl.* **299**, 818–820.

Vladimirov, V.S. and Volovich, I.V. (1984). Superanalysis I. Differential calculus. *Theor. Math. Phys.* **59**, 3–27.

Prob. Theory and Math. Stat., Vol. 2, pp. 451–459
B.Grigelionis *et al.* (Eds.)
1990 VSP/Mokslas

DIFFERENTIAL CALCULUS ON THE CONFIGURATION SPACE: SURFACES, SURFACE MEASURES, GAUSS–OSTROGRADSKII FORMULA

N.V. SMORODINA
NPO Pigment, Nab. Mojki 58, 190000 Leningrad, USSR

ABSTRACT

The purpose of this paper is to investigate and to discuss some concepts concerning the differential calculus on the configuration space which was introduced by the author in (Smorodina, 1987; 1988) and, in a revised form, in the paper submitted to "Probability Theory and their Applications". The last paper includes also some details and proofs omitted in the present publication. Below we present some generalizations which may be useful to the problem of smoothness of distribution of stochastic functionals. More specificaly, we investigate here general Poisson measures and even "locally Poisson" ones instead of the Poisson measure with Lebesgue intensity discussed in the above mentioned papers.

CONFIGURATION SPACE

Let $(G, \mathcal{B})$ be a measurable space with measurable one–point subsets and $\mathcal{X}(G)$ be the configuration space on G. By configuration we mean as usual any locally finite subset of G. We allow multiple points in configurations. By the intersection of a configuration with a usual set we mean the collection of the corresponding points with full multiplicities. We can represent every configuration with the help of sequence which enumerate all points of it (the representative sequence). The points appear in the representative sequence with the corresponding multiplicities. The standard σ–algebra Σ in the space $\mathcal{X}(G)$ is defined as the minimal one for which the mappings of the form $X \mapsto |X \cap V|$ are measurable (see (Matthes et al., 1978)). Here V runs over all bounded sets in $\mathcal{B}$, $|X \cap V|$ is the cardinality of the finite configuration $X \cap V$.

LOCAL STRUCTURE OF THE CONFIGURATION SPACE

Let S be any complete separable metric space. We fix also an open set $I \subset \subset \mathbf{R}^N$ ($N \geqslant 1$) and consider it as the metric space with Euclidean metrics. We introduce further G as $S \times I$ with any metrics compatible with the structure of the Cartesian product and with a Borel σ–algebra $\mathcal{B}$. The presence of the factor S is

essential in some applications to stable measures, see (Smorodina, 1988; Smorodina and Lifshitz, 1989).
The next step is the definition of the structure of "flat Hilbertian manifold in the direction Γ" on the configuration space $\mathcal{X}(G)$. For this reason we consider the functional Hilbert space

$$H(X)=\left\{h:X\to\mathbf{R}^N\,\middle|\,\sum_{x\in X}|h(x)|^2<\infty\right\}.$$

The mapping $X\mapsto H(X)$ defines the measurable field $\mathfrak{h}$ of Hilbert spaces on $\mathcal{X}$ in the sense of (Dixmier, 1969). The space $H(X)$ can be considered as the tangential space along I in the point $X\in\mathcal{X}$. Indeed, for $h\in H(X)$ we may define (under obvious restrictions) the configuration $X+h\in\mathcal{X}$ as the collection of points $x+h(x)$ $(x\in X)$. We understand here the addition as the addition along I : if $x=(\theta,r)\in G$ $(\theta\in S,\ r\in I)$, then $x+a=(\theta,r+a)$.
Now we define the orthoprojector in $\mathfrak{h}$ corresponding to the Borel set $V\subset G$ as the collection $Q=Q(V)=\{Q(X):H(X)\to H(X)|X\in\mathcal{X}\}$ of the mappings of the form $h\mapsto \mathbf{1}_{X\cap V}h$. We call the orthoprojector $Q(V)$ finite–dimensional if V is bounded (note that in such case all spaces $QH(X)$ are finite–dimensional). An increasing sequence $Q_n=Q_n(V_n)$ of finite–dimensional projectors (and the sequence of corresponding indicators $\mathbf{1}_{V_n}$) is called approximative if all the sets V_n are open and $\bigcup_n V_n=G$.
We shall introduce also the smooth analogue of approximative sequence. To define it we choose firstly a suitable increasing sequence of bounded open sets

$$I_n\subset\mathbf{R}^N,\ \bigcup_n I_n=\mathbf{R}^N,$$

and construct the sequence of smooth functions $\varphi_n:\mathbf{R}^N\to[0,1]$ such that $\varphi_n\equiv 1$ on I_n and $\varphi_n\equiv 0$ on $\mathbf{R}^N\setminus I_{n+1}$ (we understand here the smoothness as the $\mathbf{C}^{(1)}$–one but this is not essential and can be varied). If we choose now any increasing sequence $\{B_n\}$ of bounded (open) sets converging to the S, we can form the sequence $g_n:G\to[0,1]$ of functions $g_n(x)=\mathbf{1}_{B_n}(\theta(x))\varphi_n(r(x))$.
Now we define a topology $\mathcal{O}$ in $\mathcal{X}$ which describes its properties "along Γ". We name a set $U\subset\mathcal{X}$ $\mathcal{O}$–open $(U\in\mathcal{O})$ if: for any configuration $X\in U$ and for any bounded open $V\subset G$ there exists $\varepsilon>0$ such that for any $h\in Q(V)H(X)$ with $\|h\|<\varepsilon$ the configuration $X+h$ belongs to U.
Finally we denote $\overline{U}$ (resp., ∂U) the closure (resp., the boundary) of an open set $U\in\mathcal{X}(G)$ in the space $\mathcal{X}(S\times\mathbf{R}^N)\supset\mathcal{X}(G)$. It is obvious that $\overline{U}\subset\mathcal{X}(S\times\bar{I})$, where $\bar{I}$ is the Euclidean closure of I. We shall say that U is an internal open set if $\overline{U}\subset\mathcal{X}(G)$.

DERIVATION ON THE CONFIGURATION SPACE

We give here two definitions of smoothness close to (Smorodina, 1987; 1988).

Definition 1. Let $U\in\mathcal{O}\cap\Sigma$, $V\subset G$ be a bounded open set, $Q=Q(V)$. We say that a measurable function $f:U\to\mathbf{R}$ is $\mathbf{C}^{(l)}_Q(U)$–smooth if for any $X\in U$ the

mapping $h \mapsto f(X+h)$ is l times continuously differentiable in some neighborhood of zero in $QH(X)$.
By definition, the derivatives are polylinear functionals $f_Q^{(j)}(X,h)$ on $QH(X)$. We omit the argument h if $h=0$.
The classes of $\mathbf{C}_Q^{(l)}(U)$–smooth vector fields can be defined almost analogously.

Definition 2. Let $U \in \mathcal{O} \cap \Sigma$ and $f : U \to \mathbf{R}$ be a measurable function. We call f $\mathbf{C}^{(l)}(U)$–smooth if $f \in \mathbf{C}_Q^{(l)}(U)$ for all finite–dimensional Q and there exists "global" measurable vector fields $f^{(j)} : U \to \mathcal{L}^j(\mathfrak{h})$ $(j = 1,\dots,l)$ such that for any Q and any $X \in U$ $f_Q^{(j)}(X)(Q\cdot,\dots,Q\cdot) = f^{(j)}(X)(Q\cdot,\dots,Q\cdot)$ $(j = 1,\dots,l)$.
Here $\mathcal{L}^j(\mathfrak{h})$ is the measurable field of "polylinear functionals" $X \mapsto \mathcal{L}^j(H(X))$ which is associated with the field $\mathfrak{h}$.

SURFACES IN THE CONFIGURATION SPACE

Definition 3. A set $\Gamma \subset \mathcal{X}$ will be called an elementary surface of codimension k (an elementary hypersurface if $k=1$) if there exist an open measurable set $U \supset \Gamma$ and a function $f = (f_1,\dots,f_k) \in \mathbf{C}^{(1)}(U,\mathbf{R}^k)$ so that $\Gamma = f^{-1}(0)$ and

$$\inf_{X\in U} \det(\langle Qn(X;f_i), \quad Qn(X;f_j)\rangle) > 0$$

for some finite–dimensional projector $Q = Q(V)$ (V is open). Further we name the set $\Gamma \subset \mathcal{X}$ will be called a surface of codimension k if there exists a countable collection $\{\Gamma_n\}$ of elementary surfaces of codimension k with the properties: 1) $\bigcup_n \Gamma_n = \Gamma$, and 2) every finite intersection of the sets Γ_n is again an elementary surface of the same codimension.

LEMMA 1. *Let $U \in \mathcal{O} \cap \Sigma$, $f \in \mathbf{C}^{(1)}(U)$, $t \in f(U)$. Then the set $f^{-1}(t) \cap \cap \{X | f'(X) \neq 0\}$ is a hypersurface.*

Proof. Let $\{g_n\}$ be some approximative sequence of functions. We define

$$D_n(X) = \sum_{x\in X} g_n(x)\left|f'(X)(x)\right|^2, \quad X \in U,$$

$$R_n = \left\{X \in U \middle|\ D_n(X) > 2^{-1}\left\|f'(X)\right\|^2 > 0\right\}.$$

The statement of the lemma is an immediate consequence of the representation

$$f^{-1}(t) \cap \{X|\ f'(X) \neq 0\} = \bigcup_{n=0}^{\infty} f^{-1}(t) \cap R_n.$$

For elementary hypersurface $\Gamma = f^{-1}(0)$ we define the normal to Γ as the vector field $X \mapsto n(X;f)$. It is obvious that this normal depends on the choice of f.

IMPORTANT REMARK

It is clear that Definitions 1—3 use neither the concrete form of the metrics in I, nor the concrete form of the norm in the spaces $H(X)$. As far as it concerns the spaces $H(X)$ we may simply suppose that there are some Hilbert spaces $H(X)$ with norms $|\cdot|_{H(X)}$ with the only property that the mapping $X \mapsto H(X)$ defines some measurable field of Hilbert spaces on $\mathcal{X}$. For example, we may take

$$|h|_{H(X)} = \left(\sum_{x \in X} \kappa(x)|h(x)|^2_{\mathbf{R}^N}\right)^{1/2},$$

where κ is any strictly positive measurable function on G.
Nevertheless we shall suppose below, unless stated otherwise, that the metrics in I is Euclidean and that the norm in $H(X)$ is the standard one ($\kappa \equiv 1$).

POISSON RANDOM MEASURE AND CORRESPONDING SURFACE MEASURES IN THE CONFIGURATION SPACE

Let us consider Poisson random measure ν on $(G, \mathcal{B})$ with intensity $\Pi = \pi \times \lambda$, where π is any measure on G which is finite on bounded sets and λ is the Lebesgue measure on I. As usual we can naturally identify the realizations of random measure ν with configurations in G. Thus we obtain a probability measure $\mathbf{P}$ on the space $(\mathcal{X}, \Sigma)$ or the distribution of the random measure ν. Without loss of generality we can suppose that ν is defined on the probability space $(\mathcal{X}, \Sigma, \mathbf{P})$ by the formula $\nu(V, X) = |X \cap V|$. The random variables $\nu(V)$ are Poissonian (with parameter $\Pi(V)$) and independent for disjoint V's.
Let us construct now surface measures on elementary surfaces in $\mathcal{X}$ corresponding to the measure $\mathbf{P}$. It will be clear from the construction that it can be generalized in a standard way to arbitrary surfaces. So let $\Gamma = f^{-1}(0)$ $(f \in \mathbf{C}^{(1)}(U, \mathbf{R}^k))$ be an elementary surface of codimension k, $Q = Q(V)$ be the corresponding finite–dimensional orthoprojector of Definition 3. We can suppose, without loss of generality, that the open set V is of the form $V = B \times \Delta$. Consider the mapping $v : \mathcal{X}(G) \to \mathcal{X}(G \setminus V)$ defined by the formula $v(X) = X \cap (G \setminus V)$. We have the representation $X = (X \cap V) \cup v(X)$ for every $X \in \mathcal{X}$.
Now we fix a configuration $X \in \Gamma$. Let $m = |X \cap V|$ be the corresponding cardinality (it is clear that $mN \geqslant k$) and $(\theta_1, r_1), \ldots, (\theta_m, r_m)$ be the corresponding representative sequence for $X \cap V$. Let us denote by $h_1, \ldots, h_m$ the representation of the restriction $h|_{X \cap V}$ for every $h \in QH(X)$. If we consider the N–dimensional vectors $h_1, \ldots, h_m$ as the coordinates in $(\mathbf{R}^N)^m$ we may identify this space and $QH(X)$. This identification and the implicit function theorem immediately give us that the level sets of the function $h \mapsto f(X + h)$ defined on $\{h \in QH(X) | X + h \in U,\ (X \cap V) + h \subset V\}$ are $\mathbf{C}^{(1)}$–manifolds of codimension k in $QH(X)$ and so there is the usual Lebesgue surface measure $\widetilde{\sigma}^m_{v(X), \theta_1, \ldots, \theta_m}$ on them. The mapping $h \mapsto (X \cap V) + h$ transforms this measure in a new measure $\sigma^m_{v(X), \theta_1, \ldots, \theta_m}$ on the space $\mathcal{X}(V)$.
Let us fix now $Z \in \mathcal{X}(G \setminus V)$ and define a measure σ_Z in the space $\mathcal{X}(V)$ by the

formula

$$\sigma_Z(dY) = e^{-\Pi(V)} \sum_{m \geqslant k/N} \frac{1}{m!} \mathbf{1}_{\{|Y|=m\}} \times$$

$$\times \int_{B^m} \pi(d\theta_1) \ldots \pi(d\theta_m) \mathbf{1}_U(Z \cup Y) \sigma^m_{Z,\theta_1,\ldots,\theta_m}(dY).$$

Now we can define the surface measure on Γ.

Definition 4. Let Γ be an elementary surface of codimension k. The surface measure S_Γ on Γ is defined by the formula

$$S_\Gamma(dX) = \mathbf{1}_\Gamma(X) \Big[\det(\langle n_i(X), n_j(X)\rangle)^k_{i,j=1} /$$

$$/ \det(\langle Qn_i(X), Qn_j(X)\rangle)^k_{i,j=1} \Big]^{1/2} \sigma_{v(X)}\big(d(X \cap V)\big) \mathbf{P}v^{-1}\big(dv(X)\big).$$

Here $\mathbf{P}v^{-1}$ is the image of the measure $\mathbf{P}$ under the mapping v, $n_i(X) = n(X; f_i)$. We omit standard finite–dimensional arguments which motivate the correctness of this definition.

Let U be an open measurable set, $f : U \in \mathbf{R}$ be a $\mathbf{C}^{(1)}(U)$–function such that f' doesn't vanish on U. According to Lemma 1 the level sets $f^{-1}(t) = \{X \in \in U | f(X) = t\}$ are hypersurfaces. Let us denote by S_t the surface measure on $f^{-1}(t)$.

THEOREM 1. *The measure* $\mathbf{P}$ *has a representation* $d\mathbf{P} = \|f'\|^{-1} dt\, dS_t$, *that is for any* $\mathbf{P}$*–integrable function* $\psi : U \to \mathbf{R}$,

1. *for almost all (with respect to the Lebesgue measure)* $t \in \mathbf{R}$ *the function* $X \mapsto \psi(X) \|f'(X)\|^{-1}$ *is* S_t*–integrable;*
2. *the function*

$$t \mapsto \int_{f^{-1}(t)} \psi(X) \|f'(X)\|^{-1} S_t(dX)$$

is integrable and

$$\int_{\mathbf{R}} dt \int_{f^{-1}(t)} \psi \|f'\|^{-1} dS_t = \int_U \psi \, d\mathbf{P}.$$

Proof. Let us fix an approximative sequence $Q_n = Q_n(V_n)$ of finite–dimensional orthoprojectors and define the functions $D_n : U \to \mathbf{R}$ as in the proof of Lem ma 1. We may take V_n in the form of the product $V_n = B_n \times \Delta_n$. Let us take also an auxiliary continuous increasing function $\rho : \mathbf{R} \to [0,1]$ such that $\rho(1/3) = 0$, $\rho(2/3) = 1$. Further on let us define $\widetilde{\lambda}_n(X) = \rho\Big(\|f'(X)\|^{-2} D_n(X)\Big)$ $(X \in U)$ and an identity partition $\{\lambda_n\}_{n=1}^\infty$ by the formulas $\lambda_1 = \widetilde{\lambda}_1$, $\lambda_2 = (1 - \widetilde{\lambda}_1)\widetilde{\lambda}_2$, $\lambda_3 = = (1-\widetilde{\lambda}_1)(1-\widetilde{\lambda}_2)\widetilde{\lambda}_3, \ldots$. Note that $f^{-1}(t) \cap \{\lambda_n \neq 0\}$ is an elementary hypersurface with Q_n the finite–dimensional projector of Definition 3. It is sufficient to prove our theorem for $\psi_n = \psi\lambda_n$ instead of ψ.

Let us consider the obvious identity

$$\int\limits_U \psi \, d\mathbf{P} = \sum_{m=1}^{\infty} \int\limits_{U_m} \psi \, d\mathbf{P},$$

where $U_m = \{X \in U | \, |X \cap V| = m\}$ (the term with $m = 0$ disappears because $\psi(X) = 0$ for $X \cap V = \varnothing$). We can now factorize the measure in every term separately:

$$\mathbf{P}(dX) = \mathbf{P}v^{-1}(dv(X)) \frac{(\Pi(V))^m}{m!} e^{-\Pi(V)} \frac{(d\theta_1)\dots(d\theta_m)dr_1\dots dr_m}{(\Pi(V))^m}.$$

The last factor is a measure in the space V^m of representative sequences. Thus

$$\int\limits_{U_m} \psi(X)\mathbf{P}(dX) = \frac{1}{m!} e^{-\Pi(V)} \int\limits_{\mathcal{X}(G\setminus V)} \mathbf{P}v^{-1}(dZ) \times$$

$$\times \int\limits_{B^m} \pi(d\theta_1)\dots\pi(d\theta_m) \int\limits_{\Delta^m} \mathbf{1}_{U_m}(X)\psi(X)dr_1\dots dr_m.$$

Here $X = Z \cup \{(\theta_1, r_1), \dots, (\theta_m, r_m)\}$, $Z = v(X)$. We can represent the internal (finite–dimensional!) integral with the help of the surface measure $\widetilde{\sigma}^m_{Z,\theta_1,\dots,\theta_m}$:

$$\int\limits_{\Delta^m} \mathbf{1}_{U_m}(X)\psi(X)dr_1\dots dr_m = \int\limits_{\mathbf{R}} dt \int \psi(X)\left|Qf'(X)\right|^{-1} \widetilde{\sigma}^m_{Z,\theta_1,\dots,\theta_m} dr_1 \dots dr_m$$

(the internal integral is taken over the level set $\{(r_1, \dots, r_m) \in \Delta^m | X \in U_m \cap \cap f^{-1}(t)\}$). By substituting the last integral into the previous formula we obtain easily

$$\int\limits_U \psi d\mathbf{P} = \int\limits_{\mathbf{R}} dt \int\limits_{\mathcal{X}(G\setminus V)} \mathbf{P}v^{-1}(dZ) \int\limits_{\mathcal{X}(V)} \mathbf{1}_{f^{-1}(t)}(Z \cup Y)\psi(Z \cup Y) \times$$

$$\times \left|Qf'(Z \cup Y)\right|^{-1} \sigma_Z(dY) = \int\limits_{\mathbf{R}} dt \int\limits_{f^{-1}(t)} \psi(X)\left|Qf'(X)\right|^{-1} \left|Qn(X; f)\right| S_t(dX) =$$

$$= \int\limits_{\mathbf{R}} dt \int\limits_{f^{-1}(t)} \psi(X)\left\|f'(X)\right\|^{-1} S_t(dX).$$

From this theorem we obtain immediately that the image of the measure $\mathbf{P}_V = = \mathbf{P}(\cdot \cap V)$ under f is absolutely continuous with respect to the Lebesgue measure and

$$\frac{d\mathbf{P}_V f^{-1}}{d\lambda}(t) = \int\limits_{f^{-1}(t)} \|f'\|^{-1} dS_t.$$

The last statement can be used for the proof of the absolute continuity of the distributions of many stochastic functionals (it is well-known that $(\mathcal{X}, \mathbf{P})$ is the canonical probability space for the "Poisson part" of infinitely divisible distributions). The only inconvenience of this theorem is the presence of the Lebesgue intensity. This can be achieved (see (Gikhman and Skorokhod, 1982)) by some suitable transformation. Nevertheless it is more convenient to generalize the notion of the surface measure. We give such a generalization for the case of "locally Poisson" measures (this class contains in particular Poisson measures with absolutely continuous intensities).

Now let $G = S \times I$, as above, but with any metrics in I generating the standard topology. Let us suppose that for every $X \in \mathcal{X}(G)$ the Hilbert space $H(X)$ is defined in such a way that the mapping $X \mapsto H(X)$ gives a measurable field of Hilbert spaces on $\mathcal{X}$. Let $\widetilde{\mathbf{P}}$ be any probability measure on $(\mathcal{X}, \Sigma)$ with the property: for some σ-finite measure π on S and for every finite-dimensional projector $Q = Q(V)$ there exists a measurable function $q_Q(Y, Z)$, $Y \in v(\mathcal{X})$, $Z \in \mathcal{X}(V)$, such that for $\widetilde{\mathbf{P}}v^{-1}$-almost every Y

$$\widetilde{\mathbf{P}}_Q^Y \ll \mathbf{P}_Q \quad \text{and} \quad \frac{d\widetilde{\mathbf{P}}_Q^Y}{d\mathbf{P}_Q}(Z) = q_Q(Y, Z).$$

Here $\widetilde{\mathbf{P}}_Q^Y$ is the conditional measure (for $\widetilde{\mathbf{P}}$) on $\mathcal{X}(V)$ corresponding to the condition $v(X) = Y$ and $\mathbf{P}_Q$ is the Poisson measure on $\mathcal{X}(V)$ with the intensity $\pi \times \lambda$.

We define now a surface measure $\widetilde{S}_\Gamma$ for any such measure $\widetilde{\mathbf{P}}$ and for any surface Γ. As it has been already mentioned it is sufficient to define $\widetilde{S}_\Gamma$ for elementary surfaces. Let firstly Γ be an elementary hypersurface, Q, f be the corresponding finite-dimensional projector and a function, $n(X) = f'(X)|f'(X)|_{H(X)}^{-1}$. For $Y \in v(\mathcal{X})$, let us denote by Γ^Y the section of Γ, i.e., $\Gamma^Y = \{Z \in \mathcal{X}(V) | Y \cup Z \in \Gamma\}$, and by S_{Γ^Y} the surface measure on Γ^Y corresponding to $\mathbf{P}_Q$. Further let us denote by $|\cdot|_{Q(X)}$ the norm in $QH(X)$ defined as

$$\Big(\sum_{x \in X \cap V} |h(x)|^2 \Big)^{1/2}.$$

Now, we can set

$$\widetilde{S}_\Gamma(dY, dZ) = \widetilde{\mathbf{P}}v^{-1}(dY) \; |Qn|_{Q(X)}^{-1} q_Q(Y, Z) S_{\Gamma^Y}(dZ).$$

For the case of codimension k, we can set analogously

$$\widetilde{S}_\Gamma(dY, dZ) = \widetilde{\mathbf{P}}v^{-1}(dY) \; (\det\langle n_i, n_j\rangle_{H(X)})^{1/2} \times$$

$$\times (\det\langle Qn_i, Qn_j\rangle_{Q(X)})^{-1/2} q_Q(Y, Z) S_{\Gamma^Y}(dZ).$$

It is not difficult to prove Theorem 1 for $\widetilde{\mathbf{P}}$ in the form $d\widetilde{\mathbf{P}} = \|f'\|^{-1} dt \; d\widetilde{S}_t$.

POISSONIAN SPACES

Now we return to the original assumptions about the metrics of I and the norm in $H(X)$ and to the case of Poisson measure $\mathbf{P}$ with Lebesgue intensity. For bounded G we have a natural topology in $\mathcal{X}(G)$ (all configurations are finite), but the situation is rather delicate in the unbounded case. We shall define below a convenient class of topologies (we call them Poissonian) on some subsets of full measure in $\mathcal{X}$.

Call two configurations $X, Y \in \mathcal{X}$ similar if there exists a bounded $V \subset G$ such that $v(X) = v(Y)$ (X equals Y outside V). A subset $\mathcal{X}_0 \subset \mathcal{X}$ will be called saturated if $\varnothing \in \mathcal{X}_0$ and $\mathcal{X}_0$ contains, with any configuration, also all similar ones. Let $\mathcal{X}_0$ be a saturated set and $V \subset G$ be bounded. Then we denote $v(\mathcal{X}_0) = \{X \in \mathcal{X}_0 | X \cap V = \varnothing\}$. The mapping $i_V : (Y, Z) \mapsto Y \cup Z$ is a bijection of $\mathcal{X}(V) \times v(\mathcal{X}_0)$ on $\mathcal{X}_0$. If τ is any topology on $\mathcal{X}_0$ then we denote $\tau(V)$ (resp., $v(\tau)$) the topology of subspace on $\mathcal{X}(V)$ (resp., on $v(\mathcal{X}_0)$).

Now we can formulate our definition:

Definition 5. Let $\mathcal{X}_0$ be any measurable saturated set. A topological space $(\mathcal{X}_0, \tau)$ will be said a Poissonian (relative to $\mathbf{P}$) if

1) $\mathcal{X}_0$ is the set of the full measure — $\mathbf{P}(\mathcal{X}_0) = 1$;

2) the measure $\mathbf{P}$ on $\mathcal{X}_0$ is tight;

3) the topology τ is finally compact (see (Engelking, 1977)) and measurable ($\tau \subset \Sigma$);

4) for any open bounded $V \subset G$ the topology $\tau(V)$ on $\mathcal{X}(V)$ is a natural one;

5) the topology τ is not stronger than $\mathcal{O}(\tau \subset \mathcal{O})$;

6) for any open bounded $V \subset G$ the mapping $i_V : \mathcal{X}(V) \times v(\mathcal{X}_0) \to \mathcal{X}_0$ is continuous (relative to $\tau(V) \times v(\tau)$ and τ).

GAUSS—OSTROGRADSKII FORMULA

Definition 6. Let $(\mathcal{X}_0, \tau)$ be Poissonian space and let U be an internal τ–open set. We say that U the set with a smooth boundary if for any $X \in \partial U$ there exists a τ–neighborhood U_X, a function $f \in \mathbf{C}^{(1)}(U_X)$ and a finite–dimensional projector Q such that

1) $U \cap U_X = \{Y \in U_X | f(Y) < 0\}$,

2) $\inf_{Y \in U_X} \|Qn(Y; f)\| > 0$.

As a simple consequence of finite compactness of τ we have that ∂U is a hypersurface. For any set with a smooth boundary the formulas $n(Y) = n(Y; f)$, $Y \in U_X$ define a field of external normal.

If U is a set with a smooth boundary and $K \in \mathbf{C}^{(1)}(\overline{U})$ is a vector field we may consider the identity

$$\int_U \operatorname{tr} K' \, d\mathbf{P} = \int_{\partial U} \langle K, n \rangle dS_{\partial U} \tag{GO}$$

as a version of the Gauss—Ostrogradskii formula. The next two theorems give some conditions for the validity of it.

THEOREM 2. *Let us suppose that* K *is finite–dimensional* ($Q_0K = K$ *for some finite–dimensional projector* Q_0), $\mathrm{tr}K' \in L_1(U, \mathbf{P})$, $\langle K, n\rangle \in L_1(\partial U,\ S_{\partial U})$. *Then* (GO) *is valid.*

THEOREM 3. *Let us suppose that* $K'(X) \in \mathcal{L}_{(1)}(H(X))$ *for any* $X \in U$ *and* $\int_U \|K'\|_1 d\mathbf{P} < \infty$. *If* $\|K\| \in \bigcup_{\delta>0} L_{1+\delta}(U, \mathbf{P})$ *and* $\|K\| \in L_1(\partial U, S_{\partial U})$ *then* (GO) *is valid.*

See also (Smorodina, 1989).

REFERENCES

Dixmier, J. (1969). *Les* C^**–Algèbres et leurs Représentations.* Gauthier–Villars, Paris.

Engelking, R. (1977). *General Topology.* PWN, Warszawa.

Gikhman, I.I. and Skorokhod, A.V. (1982). *Stochastic Differential Equations and their Applications.* Naukova Dumka, Kiev.

Matthes, K., Kerstan, J. and Mecke, J. (1978). *Infinitely Divisible Point Processes.* Wiley, New York.

Smorodina, N.V. (1987). Differential calculus on measurable spaces and conditions of a smoothness of densities of distributions of random variables. *Soviet. Math. Dokl.* **292**, 1053.

Smorodina, N.V. (1988). Differential calculus on configuration spaces and stable measures. *Theory Probab. Appl.* **33**.

Smorodina, N.V. (1989). The Gauss–Ostrogradskii formula for the configuration space. In: *Abstr. Fifth Intern. Vilnius Conf. on Probab. Theory and Math. Statist.* **4**, pp. 241–242.

Smorodina, N.V. and Lifshitz, M.A. (1989). On the distribution of the norm of a stable vector. *Theory Probab. Appl.* **34**.

Prob. Theory and Math. Stat., Vol. 2, pp. 460–465
B.Grigelionis *et al.* (Eds.)
1990 VSP/Mokslas

THE LOCAL THEOREMS FOR ADDITIVE ARITHMETIC FUNCTIONS

G. STEPANAUSKAS
Department of Mathematics, Vilnius University
Naugarduko 24, 232006 Vilnius, Lithuania

ABSTRACT

Two local limit theorems for integer–valued additive arithmetic functions are proved. The theorems are valid in some new regions of values of additive arithmetic functions. The classical analytic method (the theorem of residues) is used.

Let $f(n)$ be an additive integer–valued arithmetic function, a be an integer, p be a prime number, $N_x(a)$ be the number of the natural numbers $n \leqslant x$ for which $f(n) = a$.

In (Sathe, 1953; 1954; Selberg, 1954) the behaviour of $N_x(a)$ for the functions $\omega(n)$, $\Omega(n)$ (the number of the prime divisors of the number n and the number of the prime divisors of n counted with multiplicity respectively), when $1 \leqslant a \leqslant$ $\leqslant c_1 \log\log x$, was investigated. The positive constant c_1 depends on the function under consideration only. At present the bounds for a in the local laws for the above–mentioned functions are essentially extended (Balazard, 1987; Hildebrand and Tenenbaum, 1988). Generalizing the results of L.G. Sathe and A. Selberg for wider sets of functions many of authors (see (Kubilius, 1968; Skrabutėnas, 1978; Stepanauskas, 1982)) used probabilistic methods, therefore the bounds for a contracted to $\log\log x\big(1 + o(1)\big)$.

In our theorems a can accept the values from the wider intervals. We suppose that the function $f(n)$ satisfies the condition $f(p) = 1$ for all primes p with the exception of primes from a rare set. More precisely, we suppose that

$$\sum_{f(p)\neq 1} \frac{h(p)\log p}{p} \leqslant c_2, \quad \sum_{p^k, k\geqslant 2} \frac{h(p^k)\log p^k}{p^k} \leqslant c_3, \tag{1}$$

where the multiplicative function $h(n)$ will be defined later.

Further, let $\Gamma(r)$ be Euler's gamma–function and

$$G(r) = \frac{1}{\Gamma(r+1)} \prod_p \left(1 - \frac{1}{p}\right)^r \sum_{k=0}^{\infty} \frac{r^{f(p^k)}}{p^k}. \tag{2}$$

The addition in (1) and the multiplication in (2) are extended over all primes p. Finally, B denotes a quantity (not always the same) which is bounded by a constant.

Under various assumptions concerning the function $f(n)$ we obtain the following two theorems.

THEOREM 1. *Let $f(n)$ be the additive integer–valued arithmetic function and there exist positive constants c_4 and c_5, $c_4 < c_5$, $c_4 \leqslant 1 \leqslant c_5$, such that the conditions* (1) *are satisfied with*

$$h(p^k) = \max\left(c_4^{f(p^k)},\ c_5^{f(p^k)}\right).$$

Then for all $x \geqslant 3$ satisfying

$$c_4 \log\log x \leqslant a-1 \leqslant c_5 \log\log x$$

the estimate

$$N_x(a) = \frac{x(\log\log x)^{a-1}}{\Gamma(a)\log x}\left(G\Big(\frac{a-1}{\log\log x}\Big) + \frac{B}{\log\log x}\right)$$

holds. Here the constant bounding the value of B depends on c_2, c_3, c_4 and c_5 only.

THEOREM 2. *Let $f(n)$ be the non–negative additive integer–valued arithmetic function and there exists a constant $c_6 \geqslant 1$ such that the conditions* (1) *are satisfied with*

$$h(p^k) = c_6^{f(p^k)}.$$

Then for all $x \geqslant 16$ satisfying

$$\delta \log\log\log x \leqslant a-1 \leqslant c_6 \log\log x$$

with $\delta > 0$ the estimate

$$N_x(a) = \frac{x\big(\log\log x\big)^{a-1}}{\Gamma(a)\log x}\left(G\Big(\frac{a-1}{\log\log x}\Big) + \right.$$

$$\left. +B\left(\frac{a}{\big(\log\log x\big)^2} + \frac{(\log\log x)^{1-\delta/2}}{(\delta\log\log\log x)^{1/2}}\right)\right)$$

holds. Here the constant bounding the value of B depends on c_2, c_3 and c_6 only.

Proof of Theorem 2. The proofs of Theorems 1 and 2 are inspired by the research of M. Balazard (Balazard, 1987). Applying the theorem of residues we obtain

$$N_x(a) = \sum_{\substack{n\leqslant x\\ f(n)=a}} 1 = \frac{1}{2\pi i}\int_{\gamma_r} \frac{\sum\limits_{n\leqslant x} z^{f(n)}}{z^{a+1}}\,dz, \tag{3}$$

where γ_r is the circle with the radius r and the center at the origin. This equality shows that the consideration of the behaviour of $N_x(a)$ can be replaced by the consideration of the mean of multiplicative function $z^{f(n)}$.
The rest of the proof is based on the analytic method using the result of J. Kubilius which we formulate as the following lemma.

LEMMA (Kubilius, 1971). *Let $g(n)$ be the multiplicative complex-valued arithmetic function and there exists a number $\varkappa$, independing from p, such that $|\varkappa| \leqslant c_7$ and*

$$\sum_p |g(p) - \varkappa| \frac{\log p}{p} \leqslant c_8, \quad \sum_p |g(p)| \frac{\log p}{p} \leqslant c_9, \quad \sum_p \sum_{k=2}^{\infty} |g(p^k)| \frac{\log p^k}{p^k} \leqslant c_{10}.$$

Then for all $x \geqslant 20$ the estimate

$$\sum_{n \leqslant x} g(n) = \frac{x(\log x)^{\varkappa-1}}{\Gamma(\varkappa)} \prod_p \left(1 - \frac{1}{p}\right)^{\varkappa} \sum_{k=0}^{\infty} \frac{g(p^k)}{p^k} +$$

$$+BRx(\log x)^{|\varkappa|/2-1}$$

with

$$R = \begin{cases} (1-|\varkappa|)^{-1/2}, & \text{if } |\varkappa| < 1 - (\log\log x)^{-1}, \\ (\log\log x)^{1/2}, & \text{if } |\,|\varkappa| - 1\,| \leqslant (\log\log x)^{-1}, \\ (|\varkappa|-1)^{-1/2}(\log x)^{|\varkappa|/2-1/2}, & \text{if } |\varkappa| > 1 + (\log\log x)^{-1} \end{cases}$$

holds. The infinite product is absolutely convergent. The constant bounding B depends on c_7, c_8, c_9, c_{10} only.

A detailed proof of lemma can be found in (Kubilius, 1971).
Let us continue the proof of theorem 2. We put $r = (a-1)/\log\log x$. From the conditions (1) with $g(n) = z^{f(n)}$ and $\varkappa = z$ it follows that in our situation the lemma of J. Kubilius can be used. Now the equality (3) can be written in the form

$$N_x(a) = I + I_1, \tag{4}$$

where

$$I = \frac{1}{2\pi i} \int_{\gamma_r} \frac{x(\log x)^{z-1} F(z)}{z^{a+1}} dz$$

with

$$F(z) = \frac{1}{\Gamma(z)} \prod_p \left(1 - \frac{1}{p}\right)^z \sum_{k=0}^{\infty} \frac{z^{f(p^k)}}{p^k}$$

and

$$I_1 = \frac{1}{2\pi i} \int\limits_{\gamma_r} \frac{BR\,x(\log x)^{r/2-1}}{z^{a+1}} dz.$$

Firstly we shall calculate the integral I. Because $f(n)$ is non–negative, the function $G(z) = F(z)/z$ is analytic in the circle $|z| \leqslant r \leqslant c_6$. Hence

$$G(z) = G(r) + (z-r)G'(r) + (z-r)^2\Phi(z,r),$$

where

$$\Phi(z,r) = \int\limits_0^1 (1-u)G''\big(r + u(z-r)\big)du$$

is an analytic function in the above–mentioned circle too and, hence, bounded. Thus, we can write

$$I = \frac{1}{2\pi i}\frac{x}{\log x}G(r)\int\limits_{\gamma_r}\frac{(\log x)^z}{z^a}dz +$$

$$+\frac{1}{2\pi i}\frac{x}{\log x}G'(r)\int\limits_{\gamma_r}\frac{(\log x)^z}{z^a}(z-r)dz +$$

$$+\frac{1}{2\pi i}\frac{x}{\log x}\int\limits_{\gamma_r}\frac{(\log x)^z}{z^a}(z-r)^2\Phi(z,r)dz =$$

$$= I_0 + I_2 + I_3. \tag{5}$$

Applying the theorem of residues we obtain

$$I_0 = \frac{x}{\log x}G(r)\frac{(\log\log x)^{a-1}}{(a-1)!}$$

and

$$I_2 = \frac{x}{\log x}G'(r)\left(\frac{(\log\log x)^{a-2}}{(a-2)!} - r\frac{(\log\log x)^{a-1}}{(a-1)!}\right) = 0.$$

Further, because $e^{i\varphi} - 1 = B\varphi$ for $|\varphi| \leqslant \pi$,

$$I_3 = \frac{Bx}{\log x}\int\limits_{-\pi}^{\pi}\frac{(\log x)^{r\cos\varphi}}{r^a}r^3\varphi^2 d\varphi =$$

$$= \frac{Bx}{\log x}\frac{(\log x)^r}{r^{a-3}}\int\limits_0^{\pi} e^{r(\cos\varphi - 1)\log\log x}\varphi^2 d\varphi. \tag{6}$$

From the inequality $\varphi \leqslant \pi \sin \frac{\varphi}{2}$, which holds in the interval $0 \leqslant \varphi \leqslant \pi$, we can deduce that the last integral in (6) does not exceed

$$\int_0^{\pi} e^{-(2r/\pi^2)\varphi^2 \log\log x} \varphi^2 d\varphi =$$

$$= \frac{\pi^3}{(2r \log\log x)^{3/2}} \int_0^{\sqrt{2r\log\log x}} e^{-u^2} u^2 du = \frac{B}{(r \log\log x)^{3/2}}.$$

Taking into account the magnitude of r we can easy obtain

$$I_3 = \frac{Bx}{\log x} \frac{(\log\log x)^{a-3}}{(a-2)!}.$$

Further we shall estimate the integral I_1. At first let r belong to the interval $0 < r < 1 - (\log\log x)^{-1}$. If $a - 1 \geqslant \frac{1}{2} \log\log x$ then

$$I_1 = \frac{Bx}{\log x} \frac{(\log x)^{r/2}}{(1-r)^{1/2}} \frac{1}{r^a} =$$

$$= \frac{Bx}{\log x} \frac{(\log\log x)^a}{(a-1)!} \left(\frac{\log\log x}{e^{a-1}(a-1)(\log\log x - (a-1))} \right)^{1/2} =$$

$$= \frac{Bx}{\log x} \frac{(\log\log x)^a}{(a-1)!} \left(\frac{\log\log x}{e^{a-1}(a-1)} \right)^{1/2} =$$

$$= \frac{Bx}{\log x} \frac{(\log\log x)^a}{(a-1)!} \frac{1}{(\log x)^{1/4}}. \tag{7}$$

In the case when $\delta \log\log\log x \leqslant a - 1 < \frac{1}{2} \log\log x$ it can be easily seen that

$$\frac{\log\log x}{e^{a-1}(a-1)(\log\log x - (a-1))} = \frac{B}{e^{a-1}(a-1)} = \frac{B}{\delta (\log\log x)^{\delta} \log\log\log x}$$

and

$$I_1 = \frac{Bx}{\log x} \frac{(\log\log x)^{a-\delta/2}}{(a-1)!} \frac{1}{(\delta \log\log\log x)^{1/2}}.$$

If the inequality $|r-1| \leqslant (\log\log x)^{-1}$ holds, then

$$I_1 = \frac{Bx}{\log x} (\log x)^{r/2} (\log\log x)^{1/2} \frac{1}{r^a} =$$

$$= \frac{Bx}{\log x} \frac{(\log\log x)^a}{(a-1)!(\log x)^{1/2}}.$$

Finally, in the case $1 + (\log\log x)^{-1} < r \leqslant c_6$ we

$$I_1 = \frac{Bx}{\log x} \frac{(\log x)^{r-1/2}}{(r-1)^{1/2}} \frac{1}{r^a} =$$

$$= \frac{Bx}{\log x} \frac{(\log\log x)^a}{(a-1)!(\log x)^{1/2}}.$$

Substituting the estimates of the integrals I_0, I_2, I_3 into (5) and those of I, I_1 into (4) we complete the proof of Theorem 2.
The proof of Theorem 1 differs from the proof of Theorem 2 mainly in that, that the function $G(z)$ is analytic in the ring $c_4 \leqslant |z| \leqslant c_5$ only. Therefore in the case $r \geqslant c_4$ the estimate of the integral I_1 is simpler. Estimating this integral similarly as in (7) one can obtain

$$I_1 = \frac{Bx}{\log x} \frac{(\log\log x)^a}{(a-1)!} \frac{1}{(\log x)^{c_4/2}}.$$

The estimates of other integrals remain the same.
Finally let us note that the condition $f(p) = 1$ for "almost" all primes p is not essential. The results on mean values with good remainders for multiplicative functions belonging to wider classes would allow to prove analogous local laws for wider classes of additive functions. Probably the classes of additive functions from (Skrabutėnas, 1978; Stepanauskas, 1982) are worth attention.

REFERENCES

Balazard, M. (1987). Sur la repartition des valeurs de certaines fonctions arithmétiques additives. *Theése Doct.* Université De Limoges, Limoges.

Hildebrand, A. and Tenenbaum, G. (1988). On the number of prime factors of an integer *Duke Math. J.* **56**, 471—501

Kubilius, J (1968). On local theorems for additive number-theoretic functions. In *Abhandlungen aus Zahlentheorie und Analysis (zur Erinnerung an E Landau),* VEB Deutscher Verlag der Wissenschaften, Berlin, pp. 175—191

Kubilius, J. (1971). The method of Dirichlet generating series in the theory of distribution of additive arithmetic functions. I. *Liet. Matem. Rink.* **11**, 125—134

Sathe, L.G (1953,1954). On a problem of Hardy on the distribution of integers having a given number of prime factors. I, II, III, IV *J. Indian Math. Soc.* **17**, 63—82, 83-141, **18**, 27—42, 43—81

Selberg, A (1954). Note on a paper by L.G Sathe. *J. Indian Math. Soc.* **18**, 83—87

Skrabutėnas, R. (1978). Limit local distribution laws for one class of arithmetic functions. *Liet. Matem. Rink.* **18**, *N* 1, 187—202.

Stepanauskas, G (1982). Der lokale Satz grosser Abweichungen für additive Funktionen. *Liet. Matem. Rink.* **22**, *N* 3, 170—184.

Prob. Theory and Math. Stat., Vol. 2, pp. 466–473
B.Grigelionis *et al.* (Eds.)
1990 VSP/Mokslas

QUADRATIC FORMS WITH LONG–RANGE DEPENDENCE

N. TERRIN and M.S. TAQQU*
Carnegie Mellon University, Department of Statistics,
232 Baker Hall, Pittsburgh, PA 15213, USA
Boston University, Department of Mathematics,
111 Cummington Street, Boston, MA 02215, USA

ABSTRACT

This is a survey of recent results on the weak limit of normalized quadratic forms in random variables that exhibit long–range dependence. In order to define long–range dependence, it is best to focus first, on the asymptotic behavior of normalized sums of stationary random variables.

1. ASYMPTOTIC BEHAVIOR OF NORMALIZED SUMS

Consider a stationary Gaussian sequence $X_1, X_2, \ldots$ with zero mean and unit variance and suppose $G \in L^2(\mu)$, where μ is the standard normal measure. The process

$$Z_N(t) = \frac{1}{d_N} \sum_{i=1}^{[Nt]} G(X_i), \tag{1}$$

with $d_N = \sqrt{N}$, converges weakly to Brownian motion when the X_i are independent, and even when some level of dependence is present. (See (Breuer and Major, 1983) and (Giraitis and Surgailis, 1985), for example). However, the covariances $r(k) = \mathbf{E}X_1X_{k+1}$ may decay so slowly that $Z_N(t)$ does not converge to Brownian motion, and (depending on G) may not converge to a Gaussian process at all. For example, if $r(k)$ is regularly varying with exponent $\alpha - 1$, $0 < \alpha < 1$, i.e.,

$$r(k) \sim k^{\alpha-1} L(k),$$

L slowly varying at infinity, then

$$Z_N(t) = \frac{1}{N^{(\alpha+1)/2} L(N)} \sum_{i=1}^{[Nt]} X_i \tag{2}$$

converges weakly to $CB_H(t)$ (Taqqu, 1975). Here C is a positive constant and $\{B_H(t),\ t \geqslant 0\}$ is a fractional Brownian motion, the Gaussian process which has

* This research was supported at Boston University by the National Science Foundation grant DMS–88–05627 and by the AFOSR grant 89–0115.

stationary increments and self-similarity parameter $H = (\alpha + 1)/2$. Fractional Brownian motion has mean zero and covariance

$$\mathbf{E}B_H(t_1)B_H(t_2) = \frac{1}{2}\{|t_1|^{2H} + |t_2|^{2H} - |t_1 - t_2|^{2H}\}.$$

This form for the covariance follows from the stationarity of the increments and the self-similarity. (A process $Y(t)$ is *self-similar* with parameter H if any finite dimensional vector $(Y(at_1), \ldots, Y(at_k))$ has the same distribution as $(a^H Y(t_1), \ldots, a^H Y(t_k))$.

Since the process $B_H(t)$ is equal to Brownian motion if and only if $H = 1/2$, the process $Z_N(t)$ as defined in (2) does not satisfy a classical central limit theorem. In such a setting, we say that the sequence $\{X_i\}$ has ***long-range dependence***. Observe that the assumption on $r(k)$ implies $\sum_{k=-\infty}^{\infty} r(k) = \infty$, while central limit theorems typically require $\sum_{k=-\infty}^{\infty} r(k) < \infty$. Quoting (Giraitis and Surgailis, 1989a), "although there is no strict definition of a LR (long-range) process, usually the term characterizes (stationary) sequences of random variables, normalized sums of which converge in distribution, but either the limiting law or the normalization is different from "classical" ones. For example, in the case of finite variances, "classical" means Gaussian limit law and normalization $\sqrt{N}$."

The limit of the more general process defined in (1) is influenced by the amount of dependence in the sequence $G(X_i)$, and that, it turns out, is determined by the Hermite rank of G, defined below.

Recall that the Hermite polynomials $H_0, H_1, H_2, \ldots$ form an orthogonal basis in $L^2(\mu)$, where μ is standard normal. The Hermite rank of an L^2 function G is the degree of the first Hermite polynomial with nonzero coefficient in the expansion of G. A relation between the covariance structures of X_i and $G(X_i)$ that has important implications for LR processes was found in (Taqqu, 1975). He established that $\mathrm{Var}\left(\sum_{i=1}^{N} G(X_i)\right)$ is regularly varying with exponent $2H$, $1/2 < H < 1$, as $N \to \infty$, if $r(k)$ is regularly varying with exponent $(2H-2)/m$ as $k \to \infty$, where m is the Hermite rank of G. Var means variance. We illustrate this with an example. If $G(x) = H_2(x) = x^2 - 1$, then the Hermite rank m is 2. Hence, if $r(k) \sim k^{\alpha-1}$, $1/2 < \alpha < 1$, then $H = \alpha$ and $\mathrm{Var}\left(\sum_{i=1}^{N} G(X_i)\right) \sim CN^{2\alpha}$, where C is a constant. In this example the process

$$Z_N(t) = \frac{1}{N^\alpha}\sum_{i=1}^{[Nt]}(X_i^2 - 1) \tag{3}$$

has a non-Gaussian limit, called the Rosenblatt process (see (Rosenblatt, 1961) and (Taqqu, 1975)). This limit process is defined below. Because this limit is not Gaussian, we say that $\{G(X_i)\}$ satisfies a non-central limit theorem. Observe that for $G(X_i)$ to be sufficiently dependent for a non-CLT to hold, one needs an even greater degree of dependence in the original sequence. (We have been assuming X_i

Gaussian, although results exist for more general processes $\{X_i\}$. See (Surgailis, 1981; 1982; Giraitis and Surgailis, 1986 and Avram and Taqqu, 1987)).

2. ASYMPTOTIC BEHAVIOR OF NORMALIZED QUADRATIC FORMS

We have focused on normalized sums of stationary sequences. We shall now address the effect of long-range dependence on *quadratic forms* of stationary sequences. Suppose $X_1, X_2, \dots$ is Gaussian and stationary with mean zero and covariance function

$$r(k) = \mathbf{E}X_1X_{k+1} = 2\pi\sigma^2 \int_{-\pi}^{\pi} e^{ik\lambda}|\lambda|^{-\alpha}L_1(\lambda^{-1})d\lambda, \tag{4}$$

where L_1 is slowly varying at infinity and $\alpha < 1$. Note that $r(k)$ is regularly varying at infinity with exponent $\alpha - 1$, and hence it is either summable or unsummable, depending on the sign of α. Define constants a_k by

$$a_k = \int_{-\pi}^{\pi} e^{ik\lambda}|\lambda|^{-\beta}L_2(\lambda^{-1})d\lambda, \tag{5}$$

where L_2 is slowly varying at infinity and $\beta < 1$. Observe that the restriction $\alpha < 1$ and $\beta < 1$ is necessary for the $r(k)$'s and $a(k)$'s to be well-defined.
Consider now the normalized quadratic form

$$Y_N = \frac{1}{d_N}\left[\sum_{j=1}^{N}\sum_{k=1}^{N} a_{j-k}X_jX_k - \mathbf{E}\sum_{j=1}^{N}\sum_{k=1}^{N} a_{j-k}X_jX_k\right]. \tag{6}$$

It converges to a Gaussian random variable when

$$\alpha + \beta < 1/2 \text{ and } d_N = N^{1/2}$$

(Fox and Taqqu, 1987), and to a non-Gaussian random variable when

$$\alpha + \beta > 1/2 \text{ and } d_N = N^{\alpha+\beta}L_1(N)L_2(N) \tag{7}$$

(Terrin and Taqqu, 1989b). Thus, if the constants a_k are small enough to compensate for the dependence in the sequence, one obtains a central limit theorem. If they are too big, the quadratic form, it turns out, has an asymptotically non-Gaussian behavior.
Note that the slowly varying functions L_1 and L_2 are of importance because they provide a great flexibility in the choice of the $r(k)$'s and $a(k)$'s. They influence, however, only the normalization d_N in (7) and not the limit. In the sequel, we choose for simplicity $L_1 \equiv L_2 \equiv 1$.
We now want to characterize the non-Gaussian limit in (6). Set $\alpha + \beta > 1/2$ and suppose that the sums in (6) are taken from 1 to $[Nt]$, $0 \leqslant t \leqslant 1$. Then the process

$$Y_N(t) = \frac{1}{N^{\alpha+\beta}}\left[\sum_{j=1}^{[Nt]}\sum_{k=1}^{[Nt]} a_{j-k}X_jX_k - \mathbf{E}\sum_{j=1}^{[Nt]}\sum_{k=1}^{[Nt]} a_{j-k}X_jX_k\right]$$

converges weakly to a non–Gaussian limit $Y(t)$ representable as a Wiener—Ito integral.

More precisely, let G_N be the measure on $[-N\pi, N\pi]$ with density $|x|^{-\alpha}$, and let G_0 be the measure on $\mathbf{R}$ with density $|x|^{-\alpha}$. Given a measure G, let Z_G be the complex–valued Gaussian random measure satisfying

$$\mathbf{E}Z_G(A) = 0,\ \mathbf{E}|Z_G(A)|^2 = G(A),\ \overline{Z_G(A)} = Z_G(-A) \tag{8}$$

for any Borel set A of $\mathbf{R}$ with finite G measure. The quadratic form $Y_N(t)$ may be represented as the Wiener—Ito integral

$$Y_N(t) = 2\pi\sigma^2 \int''_{[-N\pi,N\pi]^2} K_{N,t}(x_1,x_2) Z_{G_N}(dx_1) Z_{G_N}(dx_2), \tag{9}$$

where

$$K_{N,t}(x_1,x_2) = \int_{-N\pi}^{N\pi} \frac{e^{it(x_1+u)}-1}{N(e^{i(x_1+u)/N}-1)} \frac{e^{it(x_2-u)}-1}{N(e^{i(x_2-u)/N}-1)} |u|^{-\beta} du.$$

The integral (9) is defined as the L^2 limit of sums of the form

$$\sum_{i=-M}^{M} \sum_{j=-M,\, j\neq\pm i}^{M} f(x_i,x_j) Z_{G_N}(\Delta_i) Z_{G_N}(\Delta_j),$$

where f is a step function approximating $K_{N,t}$, and $\Delta_i \times \Delta_j$ is a small rectangle on which f is constant. The symbol $\int''_{\mathbf{R}^2}$ signifies that integration on the diagonals $x_1 = \pm x_2$ is excluded. (For a more detailed description of Wiener—Ito integrals see (Major, 1981).)

One might guess from the integral representation (9) of $Y_N(t)$ that its limit in $D[0,1]$, as $N \to \infty$, is the non–Gaussian

$$Y(t) = 2\pi\sigma^2 \int''_{\mathbf{R}^2} K_t(x_1,x_2) Z_{G_0}(dx_1) Z_{G_0}(dx_2), \tag{10}$$

where

$$K_t(x_1,x_2) = \int_{-\infty}^{\infty} \frac{e^{it(x_1+u)}-1}{i(x_1+u)} \frac{e^{it(x_2-u)}-1}{i(x_2-u)} |u|^{-\beta} du.$$

Such a guess is often dangerous, but it is correct in this case because $\alpha + \beta >$ $> 1/2$. The convergence has been established in (Terrin and Taqqu, 1989b) using power counting arguments. (Power counting theorems are developed in (Terrin and Taqqu, 1989a).)

Consider the special case $\beta = 0$. Then $Y_N(t)$ reduces to $Z_N(t)$ of (3), which has as its limit

$$\int_{\mathbf{R}^2}'' \frac{e^{it(x_1+x_2)} - 1}{i(x_1 + x_2)} Z_{G_0}(dx_1) Z_{G_0}(dx_2), \tag{11}$$

called the Rosenblatt process. By comparing (10) to (11), one sees how the limit reflects the mixing effect of the constant in the quadratic form.

The parameter space corresponding to the central and non-central limit theorems for quadratic forms is best depicted in a figure. Recall that we must have $\alpha < 1$ and $\beta < 1$. In the region below the diagonal $\alpha + \beta = 1/2$ the limit is Gaussian. In the region above the diagonal, the limit is the non-Gaussian process represented in (10). The behavior of the limit near the boundary $\alpha + \beta = 1/2$ was explored in (Terrin and Taqqu, 1989d), where it was shown that when $\alpha > 0$ and $\beta > 0$,

$$(\alpha + \beta - 1/2)Y(1)$$

converges in distribution to a Gaussian random variable as $\alpha + \beta$ tends to $1/2$.

The preceding results have been expanded in two directions. First, convergence in both of the cases $\alpha + \beta < 1/2$ and $\alpha + \beta > 1/2$ has been obtained for linear processes (moving averages) that are not necessarily Gaussian. That is, the same results hold when X_i is defined as

$$X_i = \sum_{s=-\infty}^{\infty} b(i-s)\xi_s, \tag{12}$$

where

$$b(k) = \int_{-\pi}^{\pi} e^{ikx} |x|^{-\alpha/2} L_1^{1/2}(|x|^{-1}) dx$$

and $\ldots \xi_{-1}, \xi_0, \xi_1, \ldots$ is a sequence of i.i.d. random variables with $\mathbf{E}\xi_0 = 0$ and $\mathbf{E}\xi_0^2 = \sigma^2$. (We set again $L_1 \equiv 1$.) Convergence in the case $\alpha + \beta < 1/2$ is a consequence of results of (Giraitis and Surgailis, 1989b). Convergence in the case $\alpha + \beta > 1/2$ is established in (Terrin and Taqqu, 1989c).

The second extension concerns the asymptotic behavior of the more general form

$$\sum_{j=1}^{N} \sum_{k=1}^{N} a_{jk} G(X_j, X_k). \tag{13}$$

Some definitions are needed before further explanation can be given.

The *Wick power* $: Y_1 \ldots Y_k :$ of random variables $Y_1, \ldots, Y_k$ is defined recursively as follows: when $k = 0$, the Wick power equals 1, and when $k > 0$, $: Y_1 \ldots Y_k :$ is defined by

$$\mathbf{E} \; : Y_1 \ldots Y_k := 0,$$

$$\frac{\partial : Y_1 \ldots Y_k :}{\partial Y_i} =: Y_1 \ldots Y_{i-1} Y_{i+1} \ldots Y_k : \; .$$

The *multivariate Appell polynomials* $P_{n_1,\ldots,n_k}(Y_1,\ldots,Y_k)$, indexed by non–negative integers, are defined as

$$P_{n_1,\ldots,n_k}(Y_1,\ldots,Y_k) =: Y_1\ldots Y_1 Y_2\ldots Y_2\ldots Y_k\ldots Y_k :$$

where Y_i is repeated n_i times, $i = 1,\ldots,k$. The polynomials have mean zero and behave like powers in the sense that

$$\frac{\partial P_{n_1,\ldots,n_k}(Y_1,\ldots,Y_k)}{\partial Y_i} = n_i P_{n_1,\ldots,n_i-1,\ldots,n_k}(Y_1,\ldots,Y_k).$$

The polynomials depend on the joint distribution of $Y_1,\ldots,Y_k$ and in the event that $Y_1,\ldots,Y_k$ are Gaussian, the Appell polynomials are equivalent to the multivariate *Hermite polynomials* $H_{n_1,\ldots,n_k}(Y_1,\ldots,Y_k)$.
Since, in the Gaussian case, the asymptotic behavior of $Z_N(t)$ in (1) is determined by the asymptotic behavior of $\frac{1}{d_N}\sum_{i=1}^{N} H_m(Y_i)$, where m is the Hermite rank of G (Taqqu, 1975), one would expect knowledge concerning

$$\sum_{i=1}^{N}\sum_{j=1}^{N} a_{j-k} H_{m,n}(X_j,X_k) \tag{14}$$

to be useful in studying the limit behavior of (13), adequately normalized, when X_j is Gaussian. When X_j is not Gaussian, the Appell polynomials do not form an orthogonal basis. However, certain classes of analytic functions do have expansions in the Appell polynomials (see (Giraitis and Surgailis, 1986)), and hence the form

$$\sum_{j=1}^{N}\sum_{k=1}^{N} a_{j-k} P_{m,n}(X_j,X_k) \tag{15}$$

is still relevant to (13).
The asymptotic behavior of (15) has been studied in (Avram, 1988) and (Terrin and Taqqu, 1989c). Let X_i be moving averages as in (12). Avram showed that when X_i is Gaussian, the limit of (15), normalized by $N^{1/2}$, is also Gaussian if

$$\alpha \geqslant 0,\ 0 \leqslant \beta \leqslant 1/2,\ (m+n)\alpha + 2\beta < m+n-1, \tag{16}$$

and

$$\text{either}\quad m\alpha + 2\beta < m \quad\text{or}\quad n\alpha + 2\beta < n. \tag{17}$$

Observe that in Avram's case, expression (15) reduces to (14). Terrin and Taqqu in (Terrin and Taqqu, 1989c) do not suppose the X_i's necessarily Gaussian. They assume

$$(m+n)\alpha + 2\beta > m+n-1 \tag{18}$$

and

$$n\alpha > n-1 \quad\text{and}\quad m\alpha > m-1. \tag{19}$$

Setting

$$H = (m+n)(\alpha - 1)/2 + \beta + 1 \in (1/2, 1),$$

they show that, as $N \to \infty$,

$$\frac{1}{N^H} \sum_{j=1}^{N} \sum_{k=1}^{N} a_{j-k} P_{m,n}(X_j, X_k)$$

has the non-Gaussian random limit

$$(2\pi\sigma^2)^{(m+n)/2} \int''_{\mathbf{R}^{m+n}} K_0(x_1, \ldots, x_{m+n}) Z_{G_0}(dx_1) \ldots Z_{G_0}(dx_{m+n}), \tag{20}$$

where

$$K_0(x_1, \ldots, x_{m+n}) = \int_{-\infty}^{\infty} \Delta\left(\sum_{1}^{m} x_p + u\right) \Delta\left(\sum_{m+1}^{m+n} x_p - u\right) |u|^{-\beta} du,$$

$$\Delta(y) = \frac{e^{iy} - 1}{iy},$$

and G_0 is the measure on $\mathbf{R}^1$ with density $|x|^{-\alpha}$. Relation (20) represents a multiple Wiener—Ito integral of order $m + n$.

These results are best depicted in a figure (draw it in the case $m > n$). When (α, β) is in the region defined by (16) and (17), the limit is Gaussian (at least when X_j is Gaussian) and when (α, β) is in the region defined by (18) and (19), the limit is represented by (20). Research on the determination of the limits in the remaining regions is now in progress.

Related results can be found in (Giraitis, 1989). We point out that neither the central limit theorem of (Avram, 1988) nor the central limit theorem of (Giraitis and Surgailis, 1989b) require the covariance function to be regularly varying. Moreover, Giraitis and Surgailis (1989a) obtain a non-central limit theorem for $\sum_{i=1}^{N} P_m(X_i)$ assuming only the regular variation of $\mathrm{Var}\left(\sum_{i=1}^{N} X_i\right)$ and $\mathrm{Var}\left(\sum_{i=1}^{N} P_m(X_i)\right)$.

REFERENCES

Avram, F. (1988). On the central limit theorem for sums of Wick powers. *Preprint.*

Avram, F. and Taqqu, M.S. (1987). Noncentral limit theorems and Appell polynomials. *Ann. Probab.* **15**, 767–775.

Breuer, P. and Major, P. (1983). Central limit theorems for non-linear functionals of Gaussian fields. *J. Multivariate Analysis* **13**, 425–441.

Fox, R. and Taqqu, M.S. (1987). Central limit theorems for quadratic forms in random variables having long-range dependence. *Probab. Theory Rel. Fields* **74**, 213–240.

Major, P. (1981). *Multiple Wiener—Ito Integrals.* (Lect. Notes in Math. **849**). Springer, Berlin–Heidelberg etc.

Giraitis, L. (1989). Central limit theorem for polynomial forms. I. *Liet. Matem. Rink.* **29**, 266–289.

Giraitis, L. and Surgailis, D. (1985). CLT and other limit theorems for functionals of Gaussian processes. *Z. Wahr. verw. Geb.* **70**, 191–212.

Giraitis, L. and Surgailis, D. (1986). Multivariate Appell polynomials and the central limit theorem. In: *Dependence in Probability and Statistics.* Birkhäuser, Boston.

Giraitis, L. and Surgailis, D. (1989a). A limit theorem for polynomials of linear processes with long range dependence. *Liet. Matem. Rink.* **29**, 290–311.

Giraitis, L. and Surgailis, D. (1989b). CLT for quadratic forms in strongly dependent linear variables and application to asymptotical normality of Whittle's estimate. *Preprint.*

Rosenblatt, M. (1961). Independence and dependence. In: *Proc. Fourth Berkeley Sympos. Math. Statist. Probab.* Univ. Calif. Press, Berkeley, pp. 411–443.

Surgailis, D. (1981). Convergence of sums of nonlinear functions of moving averages to self-similar processes. *Soviet Math. Dokl.* **23**, 247–250.

Surgailis, D. (1982). Zones of attraction of self-similar multiple integrals. *Lith. Mat. J.* **22**, 327–340.

Terrin, N. and Taqqu, M.S. (1989a). Power counting theorem on $\mathbf{R}^n$. *Preprint.*

Terrin, N. and Taqqu, M.S. (1989b). A noncentral limit theorem for quadratic forms of Gaussian stationary sequences. *Preprint.*

Terrin, N. and Taqqu, M.S. (1989c). Convergence in distribution of sums of bivariate Appell polynomials with long-range dependence. *Preprint.*

Terrin, N. and Taqqu, M.S. (1989d). Convergence to a Gaussian limit as the normalization exponent tends to 1/2. *Preprint.*

Taqqu, M.S. (1975). Weak convergence to fractional Brownian motion and to the Rosenblatt process. *Z. Wahr. verw. Geb.* **31**, 287–302.

Prob. Theory and Math. Stat., Vol. 2, pp. 474–481
B.Grigelionis *et al.* (Eds.)
1990 VSP/Mokslas

BACKWARD LIMITS AND INHOMOGENEOUS REGENERATION

H. THORISSON
Department of Mathematics, Chalmers University of Technology,
41296 Göteborg, Sweden

ABSTRACT

Consider a time–inhomogeneous regenerative process starting from regeneration at time s. It is shown (under regularity conditions on the regeneration times) that, as the starting time s tends backward to $-\infty$, the process tends in total variation in any fixed time–interval $[t, \infty)$ to a two–sided limit process. Further, time–uniform rates of convergence are obtained, the analogue of the renewal measure is considered and a time–inhomogeneous key renewal theorem presented.

INTRODUCTION

"How are things now (and from now on) if they started long ago?"
The traditional probabilistic way to answer this loosely formulated question is to start a stochastic process at time 0, consider its distribution in a time interval $[t, \infty)$ and check whether it stabilizes as $t \to \infty$:

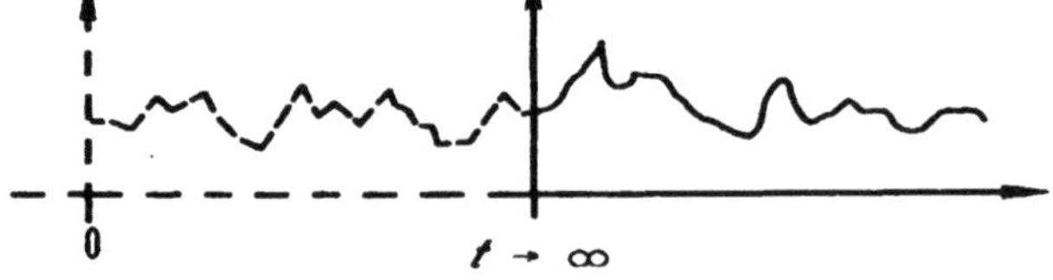

Fig. 1. *Forward limits*

For this to work the mechanism governing the development of the process (the transition probabilities if the process is Markovian) must be time–homogeneous, or asymptotically so in some sense.
Here a reverse approach is proposed: Start the process at an arbitrary time s and check what happens to its distribution in a *fixed* time interval $[t, \infty)$ as the *starting times* s goes *backward* to $-\infty$:

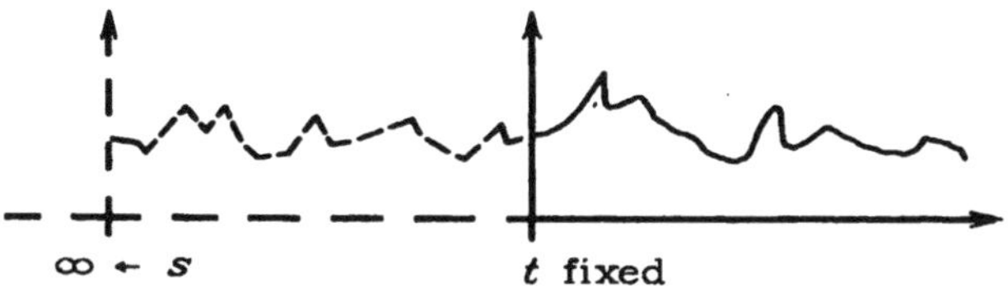

Fig. 2. *Backward limits*

As an answer to the above question this backward approach is even more natural than the forward one. Of course in the time–homogeneous case the two approaches are equivalent – the point is that unlike the forward one the backward approach also works for time–inhomogeneous processes and thus widely extends the class of processes admitting a limit law.

As far as I am aware, backward limits were first discussed in (Kolmogorov, 1936) dealing with Markov processes on a finite state space. Kolmogorov's results were elaborated in (Blackwell, 1945) but otherwise the idea seems to have passed more or less unnoticed. However, during the last decade Cohn has extensively studied Markov chains on finite and countable state space drawing attention to the tail σ–algebra; see (Cohn, 1974; 1982; 1989). See also (Loynes, 1962; Griffeath, 1978; Seneta, 1981; Sonin, 1987; 1988; Brandt et al., 1990).

In this paper we consider backward limits in the context of time–inhomogeneous regenerative processes; see (Thorisson, 1983; 1988).

An example of such a process is an inhomogeneous Markov chain with a recurrent state j; the regeneration times are the times of successive visits to j. See (Thorisson, 1986, Section 3).

Another example is the queue length process in a queueing system where the service time of a customer and the next interarrival time depend on the arrival instant; the regeneration times are the times of arrivals to an idle system. A relatively simple special case is the queue with nonstationary Poisson arrivals and i.i.d. service times. See (Thorisson, 1985; 1986, Section 2).

Observe in these examples that if the process regenerates at time x then the post–x process is independent of the pre–x process but has a distribution depending on the time of regeneration x, i.e., the regeneration is time–inhomogeneous.

The plan of the paper is as follows: We establish notation in Section 1, state the main backward limit theorem in Section 2 and prove it in Section 3. In Section 4 we give some results on time–uniform rates of convergence, in Section 5 consider the time–inhomogeneous analogue of the renewal measure and finally in Section 6 present a time–inhomogeneous key renewal theorem. The results of this paper rely heavily on the coupling results of (Thorisson, 1983).

1. TIME–INHOMOGENEOUS REGENERATIVE PROCESSES

Let $(Z_x)_{[s,\infty)}$ be a right–continuous stochastic process on the state space $(E, \mathcal{E})$

where E is Polish and $\mathcal{E}$ its Borel subsets. Let $S_0 < S_1 < \dots$ be an increasing sequence of random times in $[s,\infty)$ and $V_x = \inf\{S_n - x \geqslant 0\}$ the forward recurrence time at x. Call $(Z_x)_{[s,\infty)}$ a *time-inhomogeneous regenerative process* with *regeneration times* S_n if $\big((Z_{S_n+x}, V_{S_n+x})\big)_{[0,\infty)}$ is conditionally independent of $S_0, \dots, S_{n-1}$ given S_n and the conditional distribution does not depend on n. $(Z_x)_{[s,\infty)}$ is *time-homogeneous* if the conditional distribution does not depend on S_n. In that case $(S_n)_0^\infty$ is a renewal process on $[s,\infty)$ while in the general time-inhomogeneous case $(S_n)_0^\infty$ is a time-homogeneous Markov chain on $[s,\infty)$. For further details see (Thorisson, 1983, Section 1.3) and (Thorisson, 1988, Section 1); note that the process in (Thorisson, 1983) develops in the time-interval $[0,\infty)$ while our process develops in $[s,\infty)$ where $s \in (-\infty,\infty)$ is arbitrary.
For $-\infty < s \leqslant t < \infty$ let $\pi_{s,t}$ denote the distribution of $(Z_x)_{[t,\infty)}$ when $S_0 = s$; observe that $\pi_{s,t}$ is a probability measure on $(E^{[t,\infty)}, \mathcal{E}^{[t,\infty)})$, not on $(E^{[0,\infty)}, \mathcal{E}^{[0,\infty)})$. Let $\mathbf{P}_s$ be the distribution of $S_1 - S_0$ when $S_0 = s$; observe that $\mathbf{P}_s$ is a probability measure on $[0,\infty)$, not on $[s,\infty)$. In the time-homogeneous case $\mathbf{P}_s$ is independent of s, $\mathbf{P}_s = \mathbf{F}$ is the recurrence distribution of the renewal process $(S_n)_0^\infty$.

2. THE BACKWARD LIMIT THEOREM

THEOREM 1. *Suppose:*
the probability measures $\mathbf{P}_s$, $s \in (-\infty,\infty)$, have a common component that is absolutely continuous w.r.t. Lebesgue measure (1)
and

$$\int_0^\infty \sup_s \mathbf{P}_s[x,\infty)dx < \infty. \tag{2}$$

Then there exists a probability measure π on $(E^{(-\infty,\infty)}, \mathcal{E}^{(-\infty,\infty)})$ such that for all $t \in (-\infty,\infty)$

$$\pi_{s,t} \xrightarrow{\text{t.v.}} \pi_{-\infty,t} \quad as \quad s \downarrow -\infty \quad (\text{backward limit}) \tag{3}$$

where $\xrightarrow{\text{t.v.}}$ denotes convergence in total variation norm and

$$\pi_{-\infty,t}(A) = \pi(E^{(-\infty,t)} \times A), \quad A \in \mathcal{E}^{[t,\infty)}. \tag{4}$$

Remark 1. Conditions (1) and (2) are only simple sufficient conditions and far from being necessary. For time-homogeneous processes they reduce to the familiar condition that the recurrence distribution is non-singular and has finite first moment; in this case (3) is known (see (Thorisson, 1983, Corollary 1.1)) but is usually stated in the traditional *forward* form: The distribution of $(Z_{t+x})_{[0,\infty)}$ tends in total variation to the distribution of a stationary regenerative process $(Z_x^*)_{[0,\infty)}$ as $t \to \infty$. Clearly this cannot hold for time-inhomogeneous processes without the restrictive condition of asymptotic time-homogeneity; see (Thorisson, 1988, Section 5).

Remark 2. If we consider the regenerative process $\big((Z_x, V_x)\big)_{[s,\infty)}$ then π is the distribution of a process $\big((Z_x^*, V_x^*)\big)_{(-\infty,\infty)}$ and being the total variation limit of regenerative processes this process is easily seen to have the property that $(Z_x^*)_{[t,\infty)}$ is a regenerative process for each $t \in (-\infty, \infty)$ with the times of the successive visits of $(V_x^*)_{[t,\infty)}$ to 0 as regeneration times. In fact, $(Z_x^*)_{(-\infty,\infty)}$ itself is regenerative in the sense that for all $(V_x^*)_{(-\infty,\infty)}$ stopping times T^* satisfying $V_{T^*}^* = 0$ it holds that $\big((Z_{T^*+x}^*, V_{T^*+x}^*)\big)_{[0,\infty)}$ is conditionally independent of $(V_{T^*+x}^*)_{(-\infty,0]}$ given T^* and the conditional distribution of $\big((Z_{T^*+x}^*, V_{T^*+x}^*)\big)_{[0,\infty)}$ given $T^* = x$ depends only on x and not on the particular choice of T^*. (Actually, in order to prove this one must assume that the paths of Z are not only right–continuous but that they also have left–hand limits; see (Thorisson, 1988, Section 4).)

3. PROOF OF THE THEOREM

Let $(Z_x)_{[s,\infty)}$ and $(Z'_x)_{[s',\infty)}$ start from regeneration at s and s', respectively, let $s' \leqslant s$ and let V_x and V'_x be the forward recurrence times at x. Then $(Z_{s+x})_{[0,\infty)}$ and $(Z'_{s+x})_{[0,\infty)}$ are regenerative processes of the same type developing in the time–interval $[0, \infty)$, $(Z_{s+x})_{[0,\infty)}$ zero–delayed and $(Z'_{s+x})_{[0,\infty)}$ with delay V'_s, and Condition 1.1 in (Thorisson, 1983) is satisfied with pF_0 the common absolutely continuous component of the $\mathbf{P}'_s s$ and $\bar{\mathbf{F}}$ defined by $\bar{\mathbf{F}}[x, \infty) = \sup_s \mathbf{P}_s[x, \infty)$. Thus by (Thorisson, 1983, Lemma 2.6) V'_s is dominated in distribution by a random variables Y with distribution independent of s and s'. Now (Thorisson, 1983, Proposition 1.2 together with Theorem 1.2 and Lemma 1.1) yields the existence of a random variable $\bar{T}$, with distribution independent of s and s', such that for all $t \in [s, \infty)$

$$\|\pi_{s,t} - \pi_{s',t}\| \leqslant 2\, \mathbf{P}(\bar{T} > t - s) \tag{5}$$

where $\|\cdot\|$ denotes total variation norm. Taking supremum in s' satisfying $s' \leqslant s$ and letting $s \downarrow -\infty$ proves that $\pi_{s,t}$ is Cauchy convergent in total variation as $s \downarrow -\infty$. Since probability measures on a given measurable space form a complete metric space with respect to total variation the limit

$$\pi_{-\infty,t} = \lim_{s \downarrow -\infty} \pi_{s,t}$$

exists for each $t \in (-\infty, \infty)$ and is a probability measure on $(E^{[t,\infty)}, \mathcal{E}^{[t,\infty)})$. Define π on

$$\left\{E^{(-\infty,t)} \times A : t \in (-\infty, \infty), \quad A \in \mathcal{E}^{[t,\infty)}\right\}$$

through (4). Then, due to Kolmogorov's existence theorem and the fact that $(E, \mathcal{E})$ is Polish, π extends uniquely to a probability measure on $(E^{(-\infty,\infty)}, \mathcal{E}^{(-\infty,\infty)})$ and the proof is complete.

4. TIME–UNIFORM RATES OF CONVERGENCE

In (5) we can replace $\pi_{s',t}$ by the limit $\pi_{-\infty,t}$ and s by $t - x$ where $x \geqslant 0$ to obtain

$$\|\pi_{t-x,t} - \pi_{-\infty,t}\| \leqslant 2\, \mathbf{P}(\bar{T} > x).$$

Thus (3) with $s = t - x$ holds uniformly in t as $x \to \infty$. Further, by (Thorisson, 1983, Lemma 2.6) we have that the tail of the distribution of Y is of the same order as the tail of the finite measure $\bar{F}[x, \infty)dx$. Applying the moment results of (Thorisson, 1983, Proposition 1.2 and the discussion following Lemma 1.1) yields the following theorem.

THEOREM 2. *Under the conditions* (1) *and* (2) *the following results hold:* A) *time–uniform backward ergodicity*

$$\sup_t \|\pi_{t-x,t} - \pi_{-\infty,t}\| \to 0 \quad \text{as} \quad x \to \infty;$$

B) *time–uniform backward ergodicity of geometric order. If*

$$\exists\, \rho > 1 : \rho^x \sup_s \mathbf{P}_s[x, \infty) \to 0 \quad \text{as} \quad x \to \infty \tag{6}$$

then

$$\exists\, \rho > 1 : \rho^x \sup_t \|\pi_{t-x,t} - \pi_{-\infty,t}\| \to 0 \quad \text{as} \quad x \to \infty$$

C) *time–uniform backward ergodicity of order* φ. *If* $\varphi : [0, \infty) \to [0, \infty)$ *is increasing and satisfies*

$$x^{-1} \log \varphi(x) \quad \textit{decreases strictly to} \quad 0 \quad \text{as} \quad x \to \infty \tag{7}$$

and

$$\int_0^\infty \varphi(x) \sup_s \mathbf{P}_s[x, \infty)dx < \infty \tag{8}$$

then

$$\varphi(x) \sup_t \|\pi_{t-x,t} - \pi_{-\infty,t}\| \to 0 \quad \text{as} \quad x \to \infty.$$

5. THE TIME–INHOMOGENEOUS RENEWAL MEASURE

Consider the regeneration times S_n with $S_0 = s$. Let N be the point process associated with S_n, i.e.,

$$N(B) = \big|\{n \geqslant 0 : S_n \in B\}\big| = \big|\{x \in B : V_x = 0\}\big|, \quad B \in \mathcal{B}[s, \infty).$$

The intensity measure μ_s of N, defined by

$$\mu_s(B) = \mathbf{E}\big[N(B)\big], \quad B \in \mathcal{B}[s, \infty),$$

is the time–inhomogeneous analogue of the renewal measure. Let $(V_x^*)_{(-\infty,\infty)}$ be the limit process of $(V_x)_{[s,\infty)}$ and define N^* by

$$N^*(B) = \{x \in B : V_x^* = 0\}, \quad B \in \mathcal{B}(-\infty, \infty).$$

Then the intensity measure μ of N^*,

$$\mu(B) = \mathbf{E}\big[N^*(B)\big], \quad B \in \mathcal{B}(-\infty, \infty),$$

takes the place of Lebesque measure in traditional renewal theory. Now (Thorisson, 1983, Lemma 1.2 and Example 1.2) easily yields

$$\|\mu_{t-x} - \mu\|_{[t,t+a]} \leqslant 2\, C(a+1)\mathbf{P}(\bar{T} > x), \quad a \in [0, \infty),$$

where C is a finite constant independent of t, x and a, and $\|\cdot\|_B$ denotes total variation norm restricted to B. We thus obtain the following theorem.

THEOREM 3. *Suppose the conditions* (1) *and* (2) *hold and let* $a \in [0, \infty)$. *Then*

$$\mu_{t-x}\big([t, t+a]\cap\cdot\,\big) \xrightarrow{\text{t.v.}} \mu\big([t, t+a]\cap\cdot\,\big) \quad \textit{uniformly in} \quad t \quad \textit{as} \quad x \to \infty.$$

Further, if (7) *and* (8) *hold then*

$$\varphi(x) \sup_t \|\mu_{t-x} - \mu\|_{[t,t+a]} \to 0 \quad \textit{as} \quad x \to \infty.$$

From (Thorisson, 1983, Lemma 1.2) we also obtain

$$\|\mu_{t-x} - \mu\|_{[t,\infty)} \leqslant 2\, C \sum_{n=[x]}^{\infty} \mathbf{P}(\bar{T} > n)$$

which yields the following result.

THEOREM 4. *Suppose* (1) *holds and that*

$$\int_0^\infty x \sup_s \mathbf{P}_s[x, \infty)dx < \infty. \tag{9}$$

Then

$$\sup_s \|\mu_s - \mu\|_{[s,\infty)} < \infty \tag{10}$$

and

$$\mu_{t-x}\big([t, \infty)\cap\cdot\,\big) \xrightarrow{\text{t.v.}} \mu\big([t, \infty)\cap\cdot\,\big) \quad \textit{uniformly in} \quad t \quad \textit{as} \quad x \to \infty. \tag{11}$$

Further, if (6) *holds then*

$$\exists\, \rho > 1 : \rho^x \sup_t \|\mu_{t-x} - \mu\|_{[t,\infty)} \to 0 \quad \textit{as} \quad x \to \infty,$$

while if (7) *and* (8) *hold and* φ *has an increasing density* $\dot{\varphi}$ *then*

$$\dot{\varphi}(x) \sup_t \|\mu_{t-x} - \mu\|_{[t,\infty)} \to 0 \quad \textit{as} \quad x \to \infty.$$

Remark 3. More generally, consider time–inhomogeneous regenerative random measures ξ starting with regeneration at $S_0 = s$ and satisfying

$$\mathsf{E}[\xi[t,t+1)|S_1 > t] \leqslant C, \quad -\infty < s \leqslant t < \infty,$$

where C is a finite constant independent of s and t. Let ξ^* be the limit measure on $(-\infty,\infty)$. Then Theorems 3 and 4 hold with μ_s and μ replaced by the intensity measures $\mathbf{E}[\xi]$ and $\mathbf{E}[\xi^*]$, respectively, cf. (Thorisson, 1983, Section 1.6).

6. A TIME–INHOMOGENEOUS KEY RENEWAL THEOREM

Let $g : (-\infty,\infty) \to [0,\infty)$ be a measurable function satisfying

g is bounded; (12a)

$\int g \, d\mu < \infty$; (12b)

$g(s) \to 0$ as $s \downarrow -\infty$. (12c)

Further, let $f : (-\infty,\infty) \to (-\infty,\infty)$ be measurable and satisfy $|f| \leqslant g$. Then for any $t \geqslant s$

$$\Big| \int_{[s,\infty)} f \, d\mu_s - \int_{(-\infty,\infty)} f \, d\mu \Big| \leqslant \int_{(-\infty,s)} g \, d\mu + \int_{[s,t)} g \, d|\mu_s - \mu| + \int_{[t,\infty)} g \, d|\mu_s - \mu| \leqslant$$

$$\leqslant \int_{(-\infty,s)} g \, d\mu + \sup_{x \leqslant t} g(x) \int_{[s,t)} d|\mu_s - \mu| + \sup_{x} g(x) \int_{[t,\infty)} d|\mu_s - \mu| \leqslant$$

$$\leqslant \int_{(-\infty,s)} g \, d\mu + \sup_{x \leqslant t} g(x) \sup_{x} \|\mu_x - \mu\|_{[x,\infty)} + \sup_{x} g(x) \|\mu_s - \mu\|_{[t,\infty)}.$$

Assuming (1) and (9) to hold, letting $s \downarrow -\infty$ (applying (11), (12a) and (12b)) and then $t \downarrow -\infty$ (applying (10) and (12c)) yields the following uniform key renewal theorem.

THEOREM 5. *Suppose* (1) *and* (9) *hold. Then*

$$\sup_{|f| \leqslant g} \Big| \int f \, d\mu_s - \int f \, d\mu \Big| \to 0 \quad \text{as} \quad s \downarrow -\infty,$$

provided g satisfies (12a), (12b) *and* (12c).

As an immediate consequence we have the following

COROLLARY. *Suppose* (1) *and* (9) *hold and that g satisfies* (12a), (12b) *and* (12c). *Then*

$$\int g \, d\mu_s \to \int g \, d\mu \quad \text{as} \quad s \downarrow -\infty.$$

Remark 4. The above proof of Theorem 5 is similar to the first step in the proof of (Arjas et al., 1978, Theorem 1). In order to take the second step and replace the second order condition (9) by the first order condition (2) a closer look must be taken at the coupling epoch in (Thorisson, 1983, Theorem 1.2).

REFERENCES

Arjas, E., Nummelin, E. and Tweedie, R.L. (1978). Uniform limit theorems for non-singular renewal and Markov renewal processes. *J. Appl. Prob.* **15**, 112–125.

Blackwell, D. (1945). Finite non-homogeneous chains. *Ann. Math.* **46**, 594–599.

Brandt, A., Lisek, B. and Nerman, O. (1990). On stationary Markov chains and independent random variables. *Stoch. Proc. Appl.* (to appear).

Cohn, H. (1974). On the tail events of a Markov chain. *Z. Wahr. verw. Geb.* **29**, 65–72.

Cohn, H. (1982). On a class of non-homogeneous Markov chains. *Math. Proc. Camb. Phil. Soc.* **92**, 527–534.

Cohn, H. (1989). Products of stochastic matrices and applications. *Internat. J. Math. and Math. Sci.* **12**, 209–233.

Griffeath, D. (1978). Coupling methods for Markov processes. In: *Advances in Mathematics Supplementary Studies. Studies in Probability and Ergodic Theory* **2**, pp. 1–43.

Kolmogorov, A. (1936). Zur Theorie der Markoffschen Ketten. *Math. Ann.* **112**, 155–160.

Loynes, R.M. (1962). The stability of a queue with non-independent interarrival and service time. *Proc. Camb. Phil. Soc.* **58**, 494–520.

Sonin, I.M. (1987). A theorem on separation of jets and some properties of random sequences. *Stochastics* **21**, 231–250.

Sonin, I.M. (1988). The separation of jets and some asymptotic properties of random sequences. *Lect. Notes in Control. Inform.* **103**, 275–282.

Seneta, E. (1981). *Non-negative Matrices and Markov Chains.* Springer, Berlin—Heidelberg etc.

Thorisson, H. (1983). The coupling of regenerative processes. *Adv. Appl. Prob.* **15**, 531–561.

Thorisson, H. (1985). On regenerative and ergodic properties of the *k*-server queue with non-stationary Poisson arrivals. *J. Appl. Prob.* **22**, 893–902.

Thorisson, H. (1986). On non-time-homogeneity. In: *Semi-Markov Models: Theory and Applications.* Plenum Press, New York-London, pp. 351–368.

Thorisson, H. (1988). Backward limits. *Ann. Probab.* **16**, 914–924.

Prob. Theory and Math. Stat., Vol. 2, pp. 482–494
B.Grigelionis *et al.* (Eds.)
1990 VSP/Mokslas

ON THE NORMAL APPROXIMATION OF SUMS OF WEAKLY DEPENDENT HILBERT—VALUED RANDOM VARIABLES

A.N. TIKHOMIROV
Faculty of Physics and Mathematics, Syktyvkar State University,
Oktjabrskii pr. 55, 167001 Syktyvkar, USSR

ABSTRACT

Let $\{X_j\}_{j=1}^{\infty}$ be a stationary sequence of random variables (either m–dependent or φ–mixing) with values in a Hilbert space H. Denote: $Z_n = \sum_{j=1}^{n}(X_j - \mathbf{E}X_j)$, $\sigma_n^2 = \mathbf{E}\|Z_n\|^2$, $S_n = Z_n/\sigma_n$, $B_n = \operatorname{cov}(S_n)$. We investigate the rate of decreasing of $\Delta_n(a) = \sup_{r>0}|\mathbf{P}\{|S_n + a\| < r\} - \mathbf{P}\{\|Y_n + a\| < r\}|$, where Y_n is Gauss vector in H with $\mathbf{E}Y_n = 0$, $\operatorname{cov}(Y_n) = B_n$, with various conditions imposed on the moments of X_j. The resulting estimates are of the form $\Delta_n(a) \leqslant Cn^{-1/2}(1 + \|a\|)^3 \lg^p n$, $p > 0$.

1. INTRODUCTION. FORMULATIONS OF THE RESULTS

Let $X_1, X_2, \ldots$ be a stationary (in the strong sense) sequence of random variables with values in a Hilbert space H. The scalar product and the norm in H will be denoted by $\langle \cdot, \cdot \rangle$ and $\| \cdot \|$, respectively. We shall suppose that $\mathbf{E}X_j = 0$ and $\mathbf{E}\|X_j\|^2 < \infty$, so that the covariance operator for X_j and for sums of X_j's exist. We denote the random variable (r.v.) $n^{-1/2}(X_1 + \ldots + X_n)$ by Z_n, the covariance operator of Z_n by B_n, and the track of B_n by $\sigma_n^2 = \operatorname{Sp}B_n$.
Let Y_n be a Gauss H–valued r.v. with zero mean value and with the covariance operator $B_n/\operatorname{Sp}B_n$. In many papers had been considered the rate of the convergence to zero as $n \to \infty$ of the sequence

$$\Delta_n(a) = \sup_{r>0}\left|\mathbf{P}\{\|S_n + a\| < r\} - \mathbf{P}\{\|Y_n + a\| < r\}\right|,$$

where a is an arbitrary vector of H, and $S_n = Z_n/\sigma_n$. In the case of independent X_j's, the estimates , that are close to the best possible had been obtained (see, for example (Götze, 1979; Utev, 1989; Bradley, 1981; Zalesskii et al., 1988; Nagaev, 1985; Yurinskii, 1982)).
In the present paper we consider the problem of estimating the rate of the convergence of $\Delta_n(a)$ in the case of weakly dependent summands. We shall suppose from now on that one of the following variants of weak dependence holds:

a) the m–dependence condition: any two vectors $(X_{k-p}, X_{k-p+1}, \ldots, X_k)$ and $(X_l, X_{l+1}, \ldots, X_{l+q})$ with $k - l > m$ are independent;

b) the absolute regularity condition: as $n \to \infty$

$$\beta(n) = \mathbf{E} \sup |\mathbf{P}(A|\mathfrak{M}^0_{-\infty}) - \mathbf{P}(A)| \to 0$$

(here and below, $\mathfrak{M}^b_a$ denotes the σ–algebra, generated by X_j with $j \in [a, b]$);

c) uniform mixing condition: as $n \to \infty$

$$\varphi(n) = \sup |\mathbf{P}(AB) - \mathbf{P}(A)\mathbf{P}(B)|/\mathbf{P}(A) \to 0,$$

where the supremum is taken over all $A \in \mathfrak{M}^0_{-\infty}$, $\mathbf{P}(A) > 0$, and $B \in \mathfrak{M}^\infty_n$.
The aim of the present paper is to obtain estimates of the form

$$\Delta_n(a) \leqslant Cn^{-1/2}(1 + \|a\|^3) \lg^p n, \ p \geqslant 0.$$

THEOREM 1. *Suppose that the sequence $\{X_j\}_{j=1}^\infty$ satisfies the m–dependence condition, and that X_j's are not correlated, that is for any g, $h \in H$,*

$$\mathbf{E}\langle X_j, \ g\rangle\langle X_l, \ h\rangle = 0, \quad j \neq l. \tag{1}$$

Suppose also, that there exists a constant $M > 0$ such that

$$|X_j| \leqslant M \quad \text{a.s.} \tag{2}$$

Then there exists a constant $C > 0$, depending on the distribution of r.v. X_j, such that

$$\Delta_n(a) \leqslant Cm^3 n^{-1/2}(1 + \|a\|^3)\sqrt{\lg n}.$$

Let us somewhat weaken the condition (2) and introduce the following condition (L): there exists a constant $L > 0$ such that

$$\mathbf{E}|\langle X_j, h\rangle|^3 \leqslant L\|h\|\mathbf{E}\langle X_1, h\rangle^2$$

for any $h \in H$.

THEOREM 2. *Suppose that there exist $K > 0$ and $\beta > 0$ such that for all $m \geqslant 1$ the inequality*

$$\varphi(m) \leqslant K \exp\{-\beta m\} \tag{3}$$

*holds, where $\varphi(m)$ is the uniform mixing coefficient. Suppose also, that the conditions (1), (L) are satisfied, and that $\mathbf{E}\|X_j\|^3 < \infty$.
Then there exists a constant C, depending on the constants K, β and on the distribution of r.v. X_j, such that*

$$\Delta_n(a) \leqslant Cn^{-1/2}(1 + \|a\|^3) \lg^3 n.$$

The condition (L) is of purely technical character and is used for estimating the absolute value of the characteristic function of r.v. $\|S_n\|^2$ (see (Yurinskii, 1982, Lemma 2.2)). If this condition is excluded, we can obtain the following

THEOREM 3. *Suppose that the sequence* $\{X_j\}_{j=1}^{\infty}$ *satisfies* (1) *and* (3), *and that* $\mathbf{E}\|X_j\|^3 < \infty$. *Then there exists a constant* C, *depending on the constants* K *and* β *and on the distribution of r.v.* X_j, *such that*

$$\Delta_n(a) \leqslant Cn^{-1/2}(1 + \|a\|^3)\lg^6 n.$$

The non-correlatedness condition is connected with some technical difficulties that arise when one investigates the covariance operator of S_n. Using the results from (Bradley, 1981; Utev, 1989) we may exclude this also. In this case we impose the additional restriction on the mixing coefficient: $\varphi(1) < 1$.

THEOREM 4. *Suppose that* $\{X_j\}_{j=1}^{\infty}$ *satisfies the condition* (3), *and that* $\mathbf{E}\|X_j\|^3 < \infty$. *Suppose, moreover, that* $\varphi(1) < 1$. *Then there exists a constant* $C > 0$, *depending on the constants* K, β *and on the distribution of r.v.* X_j, *such that*

$$\Delta_n(a) \leqslant C(1 + \|a\|^3)n^{-1/2}\lg^6 n.$$

Remark. The conclusions of Theorems 2—4 remain valid, if we change in their assumptions the uniform mixing coefficient to absolute regularity coefficient.

2. THE METHOD OF PROOFS

The proofs of Theorems 1—4 are based on the investigation of the characteristic function (c.f.) of the r.v. $\|S_n + a\|^2 : f_n(t,a) = \mathbf{E}\exp\{it\|S_n + a\|^2\}$. To get an estimate for the difference between the c.f. $f_n(t,a)$ and the c.f. $f(t,a) = \mathbf{E}\exp\{it\|Y + a\|^2\}$ of the Gauss r.v. Y we use the fact of its stability as a solution of the differential equation

$$f'(t,a) = i\|a\|^2 f(t,a) - \left(\sum_{k=1}^{\infty} \frac{4a^{(k)2}t\lambda_k(1 - it\lambda_k)}{(1 - 2it\lambda_k)^2}\right) f(t,a) +$$

$$+ \left(\sum_{k=1}^{\infty} \frac{i\lambda_k}{1 - 2it\lambda_k}\right) f(t,a), \quad f(0,a) = 1,$$

where $\lambda_1 \geqslant \lambda_2 \geqslant \ldots$ are the eigen-values of the covariance operator of the r.v. Y, and $a^{(k)}$ is the projection of a to the eigen-vector e_k corresponding to λ_k. For any $X \in H$ we denote by $X^{(k)}$ its projection to e_k. We denote:

$$U_n^{(k)}(t,a) = \mathbf{E}(S_n^{(k)})^2 \exp\{it\|S_n + a\|^2\}, \tag{4}$$

$$V_n^{(k)}(t,a) = \mathbf{E}(S_n^{(k)}) \exp\{it\|S_n + a\|^2\}. \tag{5}$$

Using these notations, the derivative of the c.f. $f_n(t,a)$ can be represented in the form:

$$f_n'(t,a) = \sum_{k=1}^{\infty} iU_n^{(k)}(t,a) + 2i\sum_{k=1}^{\infty} a^{(k)}V_n^{(k)}(t,a) + i\|a\|^2 f_n(t,a). \tag{6}$$

It is evident that the right hand sides of the equalities (4), (5) can be rewritten in the following way:

$$U_n^{(k)}(t,a) = n^{-1/2}\sigma_n^{-1}\sum_{j=1}^{n} \mathbf{E}X_j^{(k)}S_n^{(k)} \exp\{it\|S_n + a\|^2\}, \tag{7}$$

$$V_n^{(k)}(t,a) = n^{-1/2}\sigma_n^{-1}\sum_{j=1}^{n} \mathbf{E}X_j^{(k)} \exp\{it\|S_n + a\|^2\}. \tag{8}$$

Using the "local sectioning techniques", proposed in (Stein, 1972) and developed in (Tikhomirov, 1980; 1986), from the equalities (7) and (8) we may deduce the following representation for the function $U_n^{(k)}(t,a)$ and $V_n^{(k)}(t,a)$:

$$U_n^{(k)}(t,a) = \lambda_k f_n(t,a) + 2it\lambda_k U_n^{(k)}(t,a) + 2ita^{(k)}\lambda_k V_n^{(k)}(t,a) + R_n^{(k)}(t,a), \tag{9}$$

$$V_n^{(k)}(t,a) = 2it\lambda_k a^{(k)} f_n(t,a) + 2it\lambda_k V_n^{(k)}(t,a) + Q_n^{(k)}(t,a). \tag{10}$$

From (9) and (10) it follows that

$$V_n^{(k)}(t,a) = \Big((2ita^{(k)}\lambda_k)/(1-2it\lambda_k)\Big) f_n(t,a) + \hat{R}_n^{(k)}(t,a), \tag{11}$$

$$U_n^{(k)}(t,a) = \Big(\lambda_k/(1-2it\lambda_k)\Big) f_n(t,a) -$$
$$-\Big(4t^2(a^{(k)})^2\lambda_k^2/(1-2it\lambda_k)^2\Big) f_n(t,a) + \hat{Q}_n^{(k)}(t,a). \tag{12}$$

Substituting the expressions obtained for $U_n^{(k)}(t,a)$ and $V_n^{(k)}(t,a)$ into the equality (6), we get the equation

$$f_n'(t,a) = i\|a\|^2 f_n(t,a) - \Big(\sum_{k=1}^{\infty} 4(a^{(k)})^2 t\lambda_k(1-it\lambda_k)/(1-2it\lambda_k)^2\Big) f_n(t,a) +$$

$$+\Big(\sum_{k=1}^{\infty}(i\lambda_k/(1-2it\lambda_k)\Big) f_n(t,a) + \tilde{R}_n(t,a). \tag{13}$$

Here the term $\tilde{R}_n(t,a)$ has an estimate

$$\tilde{R}_n(t,a) = O\Big((|t|n^{-1/2})^p\Big), \quad p \geqslant 2,$$

and this allows us to get the desired estimate for the difference between $f_n(t,a)$ and $f(t,a)$ in some segment $|t| \leqslant \gamma_1 n^{1/6}$. As for the domain

$$\gamma_1 n^{1/6} < |t| < \gamma_2 n^{1/2} \lg^{-p} n,$$

it is sufficient to have an appropriate estimate for $f_n(t,a)$ itself. To get this, some combination of symmetrization and local sectioning is used.
We give below a detailed proof of the Theorem 1 only.

3. THE PROOF OF THEOREM 1

Without loss of generality we can assume, that $\sigma_n = 1$. Note also that, under the assumptions of Theorem 1, the covariance operator for the sum S_n coincides with one for X_1.
We fix the following notations:

$$S_{j,0} = S_n,\ S_{j,\nu} = \sum_{l:|l-j|>\nu m,\ |l-l_0|>\nu m} n^{-1/2} X_l, \quad \nu = 1,\ldots,r,$$

$$\Delta_{j,\nu} = S_{j,\nu-1} - S_{j,\nu},\ \delta_{j,\mu,\nu} = S_{j,\mu-1} - S_{j,\nu}, \quad 1 \leqslant \mu \leqslant \nu \leqslant r$$

(the integers l_0 and r will be chosen later). Under the assumptions of Theorem 1, we have the following representation for $R_n^{(k)}(t,a)$:

$$R_n^{(k)}(t,a) = \sum_{\nu=1}^{5} \bar{\rho}_\nu^{(k)}(t,a) + \sum_{\nu=2}^{r+1} \tilde{\rho}_\nu^{(k)}(t,a) + \sum_{\nu=1}^{r} \hat{\rho}_\nu^{(k)}(t,a), \tag{14}$$

where

$$\bar{\rho}_1^{(k)}(t,a) = n^{-1/2} \sum_{j=1}^{n} \mathbf{E} X_j^{(k)} \Delta_{j,1}^{(k)} \exp\{it\|S_{j,2}+a\|^2\} - \lambda_k f_n(t,a),$$

$$\bar{\rho}_2^{(k)}(t,a) = n^{-1/2} \sum_{j=1}^{n} \mathbf{E} X_j^{(k)} \Delta_{j,1}^{(k)} \left(\exp\{2it\langle \delta_{j,1,2}, S_{j,3}+a\rangle\} - 1\right) \exp\{it\|S_{j,3}+a\|^2\},$$

$$\bar{\rho}_3^{(k)}(t,a) = n^{-1/2} \sum_{j=1}^{n} \mathbf{E} X_j^{(k)} \Delta_{j,1}^{(k)} \Big(\exp\{it(\|\delta_{j,1,3}\|^2 - \|\Delta_{j,3}\|^2)\} - 1\Big) \times$$
$$\times \exp\{2it\langle S_{j,3}+a, \delta_{j,1,2}\rangle + it\|S_{j,2}+a\|^2\},$$

$$\bar{\rho}_4^{(k)}(t,a) = n^{-1/2} \sum_{j=1}^{n} \mathbf{E} X_j^{(k)} \Delta_{j,1}^{(k)} \left(\exp\{2it\langle \delta_{j,1,2}, S_{j,3}+a\rangle\} - 1\right) \times$$
$$\times \left(\exp\{2it\langle \Delta_{j,3}, S_{j,3}+a\rangle\} - 1\right) \exp\{it\|S_{j,3}+a\|^2\},$$

$$\bar{\rho}_5^{(k)}(t,a) = n^{-1/2} \sum_{j=1}^{n} \mathbf{E} X_j^{(k)} \Delta_{j,1}^{(k)} \left(\exp\{2it\langle \delta_{j,1,2}, S_{j,3}+a\rangle\} - 1\right) \times$$

$$\times\left(\exp\left\{it\|\Delta_{j,3}\|^2\right\}-1\right)\exp\left\{2it\langle\Delta_{j,3},S_{j,3}+a\rangle+it\|S_{j,3}+a\|^2\right\},$$

$$\tilde{\rho}_2^{(k)}(t,a)=n^{-1/2}\sum_{j=1}^{n}\mathbf{E}X_j^{(k)}S_{j,1}^{(k)}\left(\exp\left\{2it\langle\Delta_{j,1},S_{j,r}+a\rangle\right\}-1\right)\times$$

$$\times\exp\left\{it\|S_{j,2}+a\|^2\right\}-2it\lambda_k U_n^{(k)}(t,a)-2it\lambda_k a_n^{(k)}V_n^{(k)}(t,a),$$

$$\tilde{\rho}_\nu^{(k)}(t,a)=n^{-1/2}\sum_{j=1}^{n}\mathbf{E}X_j^{(k)}S_{j,1}^{(k)}\prod_{\mu=1}^{\nu-1}\left(\exp\left\{2it\langle\Delta_{j,\mu},S_{j,r}+a\rangle\right\}-1\right)\times$$

$$\times\exp\left\{it\|S_{j,\nu}+a\|^2\right\},\quad \nu=3,\dots,r,$$

$$\tilde{\rho}_{r+1}^{(k)}(t,a)=n^{-1/2}\sum_{j=1}^{n}\mathbf{E}X_j^{(k)}S_{j,1}^{(k)}\prod_{\mu=1}^{r}\left(\exp\left\{2it\langle\Delta_{j,\mu},S_{j,r}+a\rangle\right\}-1\right)\times$$

$$\times\exp\left\{it\|S_{j,r}+a\|^2\right\},$$

$$\tilde{\rho}_\nu^{(k)}(t,a)=n^{-1/2}\sum_{j=1}^{n}\mathbf{E}X_j^{(k)}S_{j,1}^{(k)}\prod_{\mu=1}^{\nu-1}\left(\exp\left\{2it\langle\Delta_{j,\mu},S_{j,r}+a\rangle\right\}-1\right)\times$$

$$\times\Big(\exp\left\{it\left(\|\delta_{j,\nu,r}\|^2-\|\delta_{j,\nu+1,r}\|^2\right)\right\}-1\Big)\exp\left\{2it\langle\Delta_{j,\nu},S_{j,r}+a\rangle+\right.$$

$$\left.+it\|S_{j,\nu}+a\|^2\right\},\quad \nu=1,2,\dots,r.$$

Similarly, for $Q_n^{(k)}(t,a)$:

$$Q_n^{(k)}(t,a)=\sum_{\nu=2}^{r+1}\tilde{\varkappa}_\nu^{(k)}(t,a)+\sum_{\nu=1}^{r}\hat{\varkappa}_\nu^{(k)}(t,a), \tag{15}$$

where

$$\tilde{\varkappa}_2^{(k)}(t,a)=n^{-1/2}\sum_{j=1}^{n}\mathbf{E}X_j^{(k)}\left(\exp\left\{2it\langle\Delta_{j,1},S_{j,r}+a\rangle\right\}-1\right)\times$$

$$\times\exp\left\{it\|S_{j,2}+a\|^2\right\}-2it\lambda_k V_n^{(k)}(t,a)-2it\lambda_k a^{(k)}f_n(t,a),$$

$$\tilde{\varkappa}_\nu^{(k)}(t,a)=n^{-1/2}\sum_{j=1}^{n}\mathbf{E}X_j^{(k)}\prod_{\mu=1}^{\nu-1}\left(\exp\left\{2it\langle\Delta_{j,\mu},S_{j,r}+a\rangle\right\}-1\right)\times$$

$$\times\exp\left\{it\|S_{j,\nu}+a\|^2\right\},\quad \nu=3,\dots,r,$$

$$\tilde{\varkappa}_{r+1}^{(k)}(t,a)=n^{-1/2}\sum_{j=1}^{n}\mathbf{E}X_j^{(k)}\prod_{\mu=1}^{r}\left(\exp\left\{2it\langle\Delta_{j,\mu},S_{j,r}+a\rangle\right\}-1\right)\exp\left\{it\|S_{j,r}+a\|^2\right\},$$

$$\hat{\varkappa}_\nu^{(k)}(t,a)=n^{-1/2}\sum_{j=1}^{n}\mathbf{E}X_j^{(k)}\prod_{\mu=1}^{\nu-1}\left(\exp\left\{2it\langle\Delta_{j,\mu},S_{j,r}+a\rangle\right\}-1\right)\times$$

$$\times\left(\exp\left\{it(\|\delta_{j,\nu,r}\|^2-\|\delta_{j,\nu+1,r}\|^2)\right\}-1\right)\exp\left\{2it\langle\Delta_{j,\mu},S_{j,r}+a\rangle\right\}\times$$
$$\times\exp\left\{it\|S_{j,\nu}+a\|^2\right\},\quad \nu=1,\dots,r.$$

For the most of the summands on the right hand side of (14) and (15), the necessary estimates can be obtained, in fact, as in the one-dimensional case (see (Tikhomirov, 1980, Lemma 3.1)). The following inequalities hold:

$$|\hat{\rho}_\nu^{(k)}(t,a)|\leqslant C_1\lambda_k|t|m^2n^{-1/2}\left(C_2|t|mn^{-1/2}(1+\|a\|)\right)^{\nu-1},\quad \nu=2,\dots,r,\tag{16}$$

$$|\hat{\varkappa}_\nu^{(k)}(t,a)|\leqslant C_1\lambda_k^{1/2}|t|m^2n^{-1/2}\left(C_2|t|mn^{-1/2}(1+\|a\|)\right)^{\nu-1},\quad \nu=2,\dots,r,\tag{17}$$

$$|\bar{\rho}_{r+1}^{(k)}(t,a)|\leqslant C\lambda_k|t|m\left(C_2|t|mn^{-1/2}(1+\|a\|)\right)^{r-1}(1+\|a\|),\tag{18}$$

$$|\bar{\rho}_5^{(k)}(t,a)|\leqslant C\lambda_k t^2m^4n^{-3/2}(1+\|a\|),\tag{19}$$

$$|\bar{\rho}_4^{(k)}(t,a)|\leqslant C\lambda_k t^2m^3n^{-1}(1+\|a\|)^2.\tag{20}$$

For the function $\bar{\rho}_2^{(k)}(t,a)$, $\tilde{\rho}_\nu^{(k)}(t,a)$, $\tilde{\varkappa}_\nu^{(k)}(t,a)$ some further efforts are required, the corresponding argument is given below. Note that the sets of r.v.'s $X_j,\Delta_{j,1},\dots,$ $\Delta_{j,\nu-1}$ and of r.v.'s $S_{j,\nu},S_{j,r}$ are independent. Moreover, the variables

$$\xi_{l_0,\nu}=n^{-1/2}\sum_{l:|j-l|>\nu m,\ l<l_0-\nu m}X_l,$$

and

$$\eta_{l_0,\nu}=n^{-1/2}\sum_{l:|j-l|>\nu m,\ l\geqslant l_0+\nu m}X_l$$

are also independent. Hence, to obtain estimates for $\bar{\rho}_2^{(k)}(t,a)$, $\tilde{\rho}_\nu^{(k)}(t,a)$, $\tilde{\varkappa}_\nu^{(k)}(t,a)$ we can use the symmetrization argument (see, for example (Yurinskii, 1982, Lemma 2.1)) and obtain the following inequalities:

$$|\bar{\rho}_2^{(k)}(t,a)|\leqslant C\lambda_k|t|m^2n^{-1/2}(1+\|a\|)\mathbf{E}^{1/4}\exp\left\{2it\langle\tilde{\xi}_{l_0,3},\tilde{\eta}_{l_0,3}\rangle\right\}\tag{21}$$

(here and below $\tilde{\xi}_{l_0,\nu}(\tilde{\eta}_{l_0,\nu})$ denotes the symmetrization of $\xi_{l_0,\nu}(\eta_{l_0,\nu})$),

$$|\tilde{\rho}_\nu^{(k)}(t,a)|\leqslant C\lambda_k|t|^{\nu-1}m^{\nu-1}n^{-(\nu-2)/2}(1+\|a\|)^{\nu-1}\times$$
$$\times\mathbf{E}^{1/4}\exp\left\{2it\langle\tilde{\xi}_{l_0,\nu},\tilde{\eta}_{l_0,\nu}\rangle\right\},\quad \nu=3,4,\dots,r,\tag{22}$$

$$|\tilde{\varkappa}_\nu^{(k)}(t,a)|\leqslant C\lambda_k|t|^{\nu-1}m^{\nu-1}n^{-(\nu-2)/2}(1+\|a\|)^{\nu-1}\times$$
$$\times\mathbf{E}^{1/4}\exp\left\{2it\langle\tilde{\xi}_{l_0,\nu},\tilde{\eta}_{l_0,\nu}\rangle\right\},\quad \nu=3,\dots,r.\tag{23}$$

To obtain estimates for $\hat{\rho}_1^{(k)}(t,a)$ and $\hat{\varkappa}_1^{(k)}(t,a)$, we at first approximate the r.v. $S_{j,r}$, which is used in the definition of these functions, by the r.v. $S_{j,r+1}$, and secondly apply the same approach as was used for the above inequalities. Then we have

$$|\hat{\rho}_1^{(k)}(t,a)| \leqslant C\lambda_k\big(t^2m^2n^{-1}(1+\|a\|)^2+|t|mn^{-1/2}(1+\|a\|)\big)\times$$
$$\times\mathbf{E}^{1/4}\exp\left\{2it\langle\tilde{\xi}_{l_0,\nu},\tilde{\eta}_{l_0,\nu}\rangle\right\}, \tag{24}$$
$$|\hat{\varkappa}_1^{(k)}(t,a)| \leqslant C\lambda_k^{1/2}\big(t^2m^2n^{-1}(1+\|a\|)+|t|mn^{-1/2}(1+\|a\|)\big)\times$$
$$\times\mathbf{E}^{1/4}\exp\left\{2it\langle\tilde{\xi}_{l_0,\nu},\tilde{\eta}_{l_0,\nu}\rangle\right\}. \tag{25}$$

To conclude with the right hand sides of (14) and (15), we have now to obtain estimates for the functions $\bar{\rho}_1^{(k)}(t,a)$, $\tilde{\rho}_2^{(k)}(t,a)$, $\tilde{\varkappa}_2^{(k)}(t,a)$. Consider at first $\tilde{\rho}_2^{(k)}(t,a)$. Let us represent it in the form:

$$\tilde{\rho}_2^{(k)}(t,a)=\sum_{\nu=1}^{5} d_\nu^{(k)}(t,a), \tag{26}$$

where

$$d_1^{(k)}(t,a)=n^{-1/2}\sum_{j=1}^{n}\mathbf{E}X_j^{(k)}\delta_{j,1,2}^{(k)}\big(\exp\left\{2it\langle\Delta_{j,1},S_{j,r}+a\rangle\right\}-1\big)\times$$
$$\times\exp\left\{it\|S_{j,3}+a\|^2\right\},$$
$$d_2^{(k)}(t,a)=n^{-1/2}\sum_{j=1}^{n}\mathbf{E}X_j^{(k)}\delta_{j,1,2}^{(k)}\big(\exp\left\{2it\langle\Delta_{j,1},S_{j,r}+a\rangle\right\}-1\big)\times$$
$$\times\Big(\exp\left\{it(\|S_{j,2}+a\|^2-\|S_{j,3}+a\|^2)\right\}-1\Big)\exp\left\{it\|S_{j,3}+a\|^2\right\},$$
$$d_3^{(k)}(t,a)=n^{-1/2}\sum_{j=1}^{n}\mathbf{E}X_j^{(k)}\delta_{j,1,2}^{(k)}\big(\exp\left\{2it\langle\Delta_{j,1},S_{j,r}+a\rangle\right\}-1-$$
$$-2it\langle\Delta_{j,1},S_{j,r}+a\rangle\big)\exp\left\{it\|S_{j,2}+a\|^2\right\},$$
$$d_4^{(k)}(t,a)=2it\Big(n^{-1/2}\sum_{j=1}^{n}\mathbf{E}X_j^{(k)}\delta_{j,1,2}^{(k)}\langle\Delta_{j,1},S_{j,r}+a\rangle\times$$
$$\times\exp\left\{it\|S_{j,2}+a\|^2\right\}-\lambda_kU_n^{(k)}(t,a)\Big),$$
$$d_5^{(k)}(t,a)=2it\Big(n^{-1/2}\sum_{j=1}^{n}\mathbf{E}X_j^{(k)}\delta_{j,2}^{(k)}\langle\Delta_{j,1},a\rangle\times$$
$$\times\exp\left\{it\|S_{j,2}+a\|^2\right\}-\lambda_ka^{(k)}V_n^{(k)}(t,a)\Big).$$

Due to our choice of the basis, the coordinates of the vector X_j are non–correlated, and because of the condition (2), the coordinates of X_j are non–correlated with the coordinates of X_l, $l\neq j$. Moreover, as it was already noted, r.v.'s X_j, $\Delta_{j,1}$ and $S_{j,2}, S_{j,r}$ are independent. It follows, that

$$n^{-1/2}\sum_{j=1}^{n}\mathbf{E}X_j^{(k)}S_{j,2}^{(k)}\langle\Delta_{j,1},S_{j,r}\rangle\exp\left\{it\|S_{j,2}+a\|^2\right\}=$$

$$=\lambda_k n^{-1}\sum_{j=1}^{n}\mathbf{E}S_{j,2}^{(k)}S_{j,r}^{(k)}\exp\left\{it\|S_{j,2}+a\|^2\right\},\tag{27}$$

$$n^{-1/2}\sum_{j=1}^{n}\mathbf{E}X_j^{(k)}S_{j,2}^{(k)}\langle\Delta_{j,1},S_{j,r}+a\rangle\exp\left\{it\|S_{j,2}+a\|^2\right\}=$$

$$=\lambda_k a^{(k)}n^{-1}\sum_{j=1}^{n}\mathbf{E}S_{j,2}^{(k)}\exp\left\{it\|S_{j,2}+a\|^2\right\}.\tag{28}$$

Using these equalities for the function $d_\nu^{(k)}(t,a)$ with $\nu=4,5$, it is not difficult to get the following estimates

$$|d_4^{(k)}(t,a)|\leqslant C\lambda_k\Big[t^4m^3n^{-3/2}(1+\|a\|)^3+t^3m^2n^{-1}(1+\|a\|)^2\times$$

$$\times\mathbf{E}^{1/4}\exp\left\{2it\langle\tilde\xi_{l_0,3},\tilde\eta_{l_0,3}\rangle\right\}+t^2mn^{-1/2}(1+\|a\|)\mathbf{E}^{1/4}\exp\left\{2it\langle\tilde\xi_{l_0,3},\tilde\eta_{l_0,3}\rangle\right\}\Big],\tag{29}$$

$$|d_5^{(k)}(t,a)|\leqslant C\lambda_k a^{(k)}\Big[t^2m^2n^{-1}(1+\|a\|)^2+|t|mn^{-1/2}(1+\|a\|)\times$$

$$\times\mathbf{E}^{1/4}\exp\left\{2it\langle\tilde\xi_{l_0,3},\tilde\eta_{l_0,3}\rangle\right\}\Big].\tag{30}$$

For the functions $d_\nu^{(k)}(t,a)$, $\nu=1,2,3$, the following inequalities hold:

$$|d_1^{(k)}(t,a)|\leqslant C\lambda_k|t|m^2n^{-1/2}(1+\|a\|)\mathbf{E}^{1/4}\exp\left\{2it\langle\tilde\xi_{l_0,3},\tilde\eta_{l_0,3}\rangle\right\},\tag{31}$$

$$|d_2^{(k)}(t,a)|\leqslant C\lambda_k t^2m^3n^{-1}(1+\|a\|)^2,\tag{32}$$

$$|d_3^{(k)}(t,a)|\leqslant C\lambda_k t^2m^2n^{-1/2}(1+\|a\|)^2\mathbf{E}^{1/4}\exp\left\{2it\langle\tilde\xi_{l_0,3},\tilde\eta_{l_0,3}\rangle\right\}.\tag{33}$$

From the equality (26) and the inequalities (29)—(33) it follows that

$$|\tilde\rho_2^{(k)}(t,a)|\leqslant C\lambda_k\Big\{t^4m^3n^{-3/2}(1+\|a\|)^3+t^2m^3n^{-1}(1+\|a\|)^3+$$

$$+\Big[t^3m^2n^{-1}(1+\|a\|)^2+t^2m^2n^{-1/2}(1+\|a\|)^2+|t|m^2n^{-1/2}(1+\|a\|)\Big]\times$$

$$\times\mathbf{E}^{1/4}\exp\left\{2it\langle\tilde\xi_{l_0,3},\tilde\eta_{l_0,3}\rangle\right\}\Big\}.\tag{34}$$

The estimate for $\bar\rho_1^{(k)}(t,a)$ looks as follows:

$$|\bar\rho_1^{(k)}(t,a)|\leqslant C\lambda_k\Big[|t|m^2n^{-1}+t^2m^2n^{-1}(1+\|a\|)^2+$$

$$+|t|mn^{-1/2}(1+\|a\|)\mathbf{E}^{1/4}\exp\left\{2it\langle\tilde{\xi}_{l_0,3},\tilde{\eta}_{l_0,3}\rangle\right\}\Big]. \tag{35}$$

At last, for $\tilde{\varkappa}_2^{(k)}(t,a)$ one can obtain the estimate

$$|\tilde{\varkappa}_2^{(k)}(t,a)| \leqslant C\lambda_k\Big[t^2m^2n^{-1/2}(1+\|a\|)^2\mathbf{E}^{1/4}\exp\left\{2it\langle\tilde{\xi}_{l_0,3},\tilde{\eta}_{l_0,3}\rangle\right\}+$$

$$+t^2mn^{-1/2}(1+\|a\|)^2\mathbf{E}^{1/4}\exp\left\{2it\langle\tilde{\xi}_{l_0,r+1},\tilde{\eta}_{l_0,r,3}\rangle\right\}+t^2mn^{-1}(1+\|a\|)^2\Big]. \tag{36}$$

As we shall see below, to get the desired estimates it suffices to take $r = 4$. Combining the inequalities (16)—(36), we get the following estimates for $R_n^{(k)}(t,a)$ and $Q_n^{(k)}(t,a)$:

$$|R_n^{(k)}(t,a)| \leqslant C\lambda_k\sum_{\nu=2}^{4}|t|^\nu m^{\nu+1}n^{-\nu/2}(1+\|a\|)^{\nu-1}+$$

$$+t^4m^4n^{-3/2}(1+\|a\|)^5+t^2m^3n^{-1}(1+\|a\|)^3+$$

$$+\sum_{\nu=2}^{4}|t|^{\nu-1}m^{\nu-1}n^{-(\nu-1)/2}(1+\|a\|)^{\nu-1}\max_{2\leqslant\nu\leqslant5}\mathbf{E}^{1/4}\exp\left\{2it\langle\tilde{\xi}_{l_0,\nu},\tilde{\eta}_{l_0,\nu}\rangle\right\}, \tag{37}$$

$$|Q_n^{(k)}(t,a)| \leqslant C\lambda_k^{1/2}\sum_{\nu=2}^{4}|t|^\nu m^{\nu+1}n^{-\nu/2}(1+\|a\|)^{\nu-1}+$$

$$+t^2m^3n^{-1}(1+\|a\|)^2+t^4m^4n^{-3/2}(1+\|a\|)^4+$$

$$+\sum_{\nu=2}^{4}m^{\nu-1}|t|^{\nu-1}n^{-(\nu-1)/2}(1+\|a\|)^{\nu-1}\max_{2\leqslant\nu\leqslant5}\mathbf{E}^{1/4}\exp\left\{2it\langle\tilde{\xi}_{l_0,\nu},\tilde{\eta}_{l_0,\nu}\rangle\right\}. \tag{38}$$

These estimates, together with the equalities (9)—(13), give the following estimate

$$|\tilde{R}_n^{(k)}(t,a)| \leqslant C\Big[\sum_{\nu=2}^{4}|t|^\nu m^\nu n^{-\nu/2}(1+\|a\|)^\nu+$$

$$+\sum_{\nu=2}^{4}|t|^{\nu-1}m^{\nu-1}n^{-(\nu-1)/2}(1+\|a\|)^\nu\max_{2\leqslant\nu\leqslant5}\mathbf{E}^{1/4}\exp\left\{2it\langle\tilde{\xi}_{l_0,\nu},\tilde{\eta}_{l_0,\nu}\rangle\right\}+$$

$$+t^2m^4n^{-1}(1+\|a\|)^3+t^4m^4n^{-3/2}(1+\|a\|)^5\Big]. \tag{39}$$

Now, to transform the inequality (39) to the final form, we have to estimate the functions

$$g_\nu(t)=\mathbf{E}\exp\left\{2it\langle\tilde{\xi}_{l_0,\nu},\tilde{\eta}_{l_0,\nu}\rangle\right\},\quad \nu=1,2,\ldots,r. \tag{40}$$

From now on we shall suppose that $l_0 = [n/2]$. Denote $\mathbf{E}\exp\{2it\langle\tilde{\xi}_{l_0,\nu},y\rangle\}$ by $g_\nu(t,y)$. The r.v.'s $\tilde{\xi}_{l_0,\nu}$ and $\tilde{\eta}_{l_0,\nu}$ being independent, the following equality holds

$$g_\nu(t) = \mathbf{E} g_\nu(t, \tilde{\eta}_{l_0,\nu}), \quad \nu = 1, \dots, r. \tag{41}$$

Due to the results from (Heinrich, 1986) it follows that

$$|g_\nu(t, y)| = \exp\left\{-\delta^2 t^2 \langle By, y\rangle\right\}, \quad \nu = 1, 2, \dots, r, \tag{42}$$

for $|t| \leqslant \gamma m^{-2} n^{1/2} |y|^{-1}$, where B is the covariance operator for the r.v. X_1, δ is some absolute constant, $\gamma > 0$ is a constant depending on the distribution of X_1, only. From (41) and (42) it follows that, for $|t| \leqslant \gamma m^{-2} n^{1/2} \rho_n^{-1}$,

$$|g_\nu(t)| \leqslant \mathbf{E} \exp\left\{-\delta^2 t^2 \langle B\tilde{\xi}_{l_0,\nu}, \tilde{\eta}_{l_0,\nu}\rangle\right\} + \mathbf{P}\left\{|\tilde{\eta}_{l_0,\nu}| > \rho_n\right\}, \tag{43}$$

where $\rho_n = \left[C \lg^{1/2} n\right]$. Note that

$$\mathbf{E} \exp\left\{-\delta^2 t^2 \langle B\tilde{\eta}_{l_0,\nu}, \tilde{\eta}_{l_0,\nu}\rangle\right\} = \mathbf{E} \exp\left\{it\langle Y, \tilde{\eta}_{l_0,\nu}\rangle\right\},$$

where Y is the Gauss r.v. with the covariance operator B. Changing, in the above arguments, $\tilde{\xi}_{l_0,\nu}$ to $\tilde{\eta}_{l_0,\nu}$ and $\tilde{\eta}_{l_0,\nu}$ to Y, for $|t| \leqslant \gamma m^{-2} n^{1/2} \rho_n^{-1}$, we obtain

$$|g_\nu(t)| \leqslant \mathbf{E} \exp\left\{it\delta\langle Y_1, Y_2\rangle\right\} + \mathbf{P}\left\{|\tilde{\xi}_{l_0,\nu}| > \rho_n\right\} + \mathbf{P}\left\{|\tilde{\eta}_{l_0,\nu}| > \rho_n\right\}, \tag{44}$$

where Y_1, Y_2 are independent Gauss r.v. with the covariance operator B.
From the results of (Yurinskii, 1982) it follows that

$$\mathbf{E} \exp\left\{it\delta\langle Y_1, Y_2\rangle\right\} \leqslant \left(1 + (\delta_1 t)^2\right)^{-q/2}$$

provided $\lambda_q \geqslant \delta_1$. Finally, using exponential estimates for the probabilities $\mathbf{P}\{|\tilde{\xi}_{l_0,\nu}| > \rho_n\}$ and $\mathbf{P}\{|\tilde{\eta}_{l_0,\nu}| > \rho_n\}$, we obtain the inequalities

$$|g_\nu(t)| \leqslant (1 + (\delta_1 t)^2)^{-q/2} + C(q_1) n^{-q_1}, \ \nu = 1, 2, \dots, r, \tag{45}$$

where q_1 may be chosen sufficiently large.
From (13),(39),(40), (45) it follows, that for $|t| \leqslant \gamma n^{1/6} (1 + \|a\|)^{-1}$,

$$|f_n(t, a) - f(t, a)| \leqslant \int_0^{|t|} |\tilde{R}_n(u, a)| du \leqslant C_1 |t|^3 n^{-1} (1 + \|a\|)^2 +$$

$$+ C_2 m n^{-1/2} (1 + \|a\|)^2 \int_0^{|t|} \left(\frac{u du}{(1 + (\delta_1 u)^2)^{q/2}}\right) + C_3 |t| n^{-q_1}. \tag{46}$$

In order to estimate $|f_n(t, a)|$, we use the equality:

$$f_n(t, a) = \mathbf{E} \exp\left\{it\|S_{l_0,1} + a\|^2\right\} + \sum_{\nu=2}^{r} \mathbf{E} \prod_{\mu=1}^{\nu-1} \left(\exp\left\{2it\langle \Delta_\mu, S_{l_0,r} + a\rangle\right\} - 1\right) \times$$

$$\times \exp\left\{it\|S_{l_0,\nu}+a\|^2\right\}+\sum_{\nu=2}^{r}\mathbf{E}\prod_{\mu=1}^{\nu-1}\left(\exp\left\{2it\langle\Delta_\mu,S_{l_0,r}+a\rangle\right\}-1\right)\times$$

$$\times\Big(\exp\left\{it(\|\delta_{l_0,1,\nu}\|^2-\|\delta_{l_0,\nu+1,r}\|^2)\right\}-1\Big)\exp\left\{2it\langle\Delta_\nu,S_{l_0,r}+a\rangle+it\|S_{l_0,\nu}+a\|^2\right\}+$$

$$+\mathbf{E}\prod_{\mu=1}^{r}\left(\exp\left\{2it\langle\Delta_\mu,S_{l_0,r}+a\rangle\right\}-1\right)\exp\left\{it\|S_{l_0,\nu}+a\|^2\right\}. \tag{47}$$

The summands in the right hand side of (47) can be estimated without any difficulty, i.e., the corresponding argument being in essence the same as one used above when analyzing the representation (13). We formulate the final result: in the domain $|t|\leqslant \gamma m^{-2}n^{1/2}(1+\|a\|)^{-1}(\sqrt{\lg n})^{-1}$ the inequality

$$|f_n(t,a)|\leqslant\left(1+(\delta_1 t)^2\right)^{-q/8}+C_1n^{-q_1/4}+C_2|t|n^{-1/2}\rho_n(1+\|a\|)\times$$

$$\times\left(1+(\delta_1 t)^2\right)^{-q/8}+C_3|t|m^2n^{-1}\rho_n^2(1+\|a\|)+C_4\left(|t|mn^{-1/2}\rho_n(1+\|a\|)\right)^r \tag{48}$$

holds. Choosing $r\sim\gamma\lg n$, and taking γ sufficiently large, we can make the last summand in (49) decreasing in the domain $|t|\leqslant\gamma m^{-2}n^{1/2}\rho_n^{-1}(1+\|a\|)^{-1}$ more rapidly than any given power of n.
Now set $q=13$ and $q_1=8$. To estimate $\Delta_n(a)$, use the Esseen inequality. We obtain:

$$|\Delta_n(a)|\leqslant C_1m^2n^{-1/2}\rho_n(1+\|a\|)+C_2\int\limits_{|t|\leqslant T_1}\left(|f_n(t,a)-f(t,a)||t|^{-1}\right)dt+$$

$$+C_2\int\limits_{T_1\leqslant|t|\leqslant T_2}\left(|f_n(t,a)-f(t,a)||t|^{-1}\right)dt, \tag{49}$$

where $T_1=\gamma_1n^{1/6}(1+\|a\|)^{-1}$, $T_2=\gamma_2m^{-2}n^{1/2}\rho_n^{-1}(1+\|a\|)^{-1}$.
From the inequalities (46), (48), (49) it follows that

$$|\Delta_n(a)|\leqslant C_1m^2n^{-1/2}(1+\|a\|)(\lg n)^{1/2}+$$

$$+C_2mn^{-1/2}(1+\|a\|)^2+C_3n^{-1}+C_4m^2n^{-1/2}(1+\|a\|)^3(\lg n)^{1/2}$$

and this evidently gives the required estimate.

REFERENCES

Bradley, R.S. (1981). Central limit theorems under weak dependence. *J. Multivar. Analysis* **11**, 1–16.

Götze, F. (1979). Asymptotic expansions for bivariate von Mises functionals. *Z. Wahr. verw. Geb.* **50**, 333–355.

Heinrich, L. (1986). *Beitrage zur Summations theorie von Folgen und Feldern m–abhangiger Zufallsgrossen.* Facultät für Math., Bergakademie, Freiberg, Dissertation "B". 1–146.

Nagaev, S.V. (1985). On the rate of convergence to the normal law in Hilbert space. *Theory Probab. Appl.* **30**, 19–32.

Stein, Ch. (1972). A bound for the error in the normal approximation of a sum of dependent random variables. *In: Proc. Berkeley Symp. Math. Statist. and Probab.* **2**, 583–603.

Tikhomirov, A.N. (1980). On the rate of convergence in a central limit theorem for weakly dependent variables. *Theory Probab. Appl.* **25**, 800–818.

Tikhomirov, A.N. (1986). On the distribution of the maximal sum of weakly dependent variables. *Theory Probab. Appl.* **31**, 829–834.

Utev, S. (1989). Central limit theorem for dependent random variables. *Abstr. Fifth Intern. Vilnius Conf. Probab. Theory and Math. Statist.* **2**, 200–201.

Yurinskii, V.V. (1982). On the accuracy of the normal approximation of the sphere hitting probability. *Theory Probab. Appl.* **27**, 270–278.

Zalesskii, B.A. Sazonov, V.V. and Ulyanov, V.V. (1988). Normal approximation in Hilbert space. *Theory Probab. Appl.* **33**, 225–245.

Prob. Theory and Math. Stat., Vol. 2, pp. 495–504
B.Grigelionis *et al.* (Eds.)
1990 VSP/Mokslas

ON DISTRIBUTION OF ADDITIVE ARITHMETICAL FUNCTIONS ON THE SET OF SHIFTED PRIMES

N.M. TIMOFEEV
VGPI, pr. Stroitelei 11, 600024 Vladimir, USSR

ABSTRACT

We prove several results on the behaviour of additive functions on the sequence $\{p+a\}$, $a\neq 0$ In particular, we establish a complete analogue of the Ruzsa theorem. This result will be applicable to the study of additive function on shifted primes, both locally and globaly.

Let $g(n)$ be a real-valued additive arithmetical function. A function g is said to be additive if $g(a\cdot b)=g(a)+g(b)$ for every pair of positive integers a and b such that $(a,b)=1$.
In (Erdös, 1938) it is showed that if the two series

$$\sum_p \frac{1}{p}\min\left(1,g^2(p)\right),\quad \sum_{|g(p)|\leqslant 1}\frac{g(p)}{p} \tag{1}$$

converge then the additive function possesses a limit distribution, i.e., the frequencies

$$\frac{1}{x}\Big|\{n:n\leqslant x,\quad g(n)\leqslant u\}\Big| \tag{2}$$

converge weakly to a distribution function $F(u)$ as $x\to\infty$.

The distribution functions $F_x(u)$ converge weakly to the distribution function $F(u)$, if

$$\lim_{x\to\infty} F_x(u)=F(u),$$

for every point u at which $F(u)$ is continuous.
In (Erdös and Wintner, 1939) it was proved that if the frequencies (2) converge weakly to $F(u)$ then the series (1) converge.
The next result was established, independently, in (Elliott and Ryavec, 1971) and (Levin and Timofeev, 1971).
There exists a real function $A(x)$, defined for $x\geqslant 1$, for which the frequencies

$$\frac{1}{x}\Big|\{n:n\leqslant x,\quad g(n)-A(x)\leqslant u\}\Big|$$

converge weakly to a limit distribution as $x \to \infty$, iff there exist a constant s such that

$$\sum_p \frac{1}{p} \min\left(1, (g(p) - s\log p)^2\right) < \infty. \tag{3}$$

These two results as well as some other interesting facts can be deduced from

THEOREM (Ruzsa, 1980). *Let*

$$Q_h(x) = \sup_a \frac{1}{x}\left|\{n : n \leqslant x, \quad a \leqslant g(n) < a + h\}\right|.$$

For every additive function g we have

$$Q_1(x) \ll \frac{1}{\sqrt{W(x)}},$$

where

$$W(x) = \min\left(s^2 + \sum_{p \leqslant x} \frac{1}{p} \min\left(1, (g(p) - s\log p)^2\right)\right).$$

The involved constant is absolute.

We say that an additive function $g(n)$ has a limit distribution on the set of shifted primes, if the frequencies

$$\frac{1}{\pi(x)}\left|\{p : p \leqslant x, \quad g(p + a) \leqslant u\}\right|, \quad a \neq 0,$$

converge weakly as $x \to \infty$ to a distribution function $F(u)$.

In (Katai, 1968) it is shown that if the series (1) converge, then $g(p+1)$ has a limit distribution. The question of necessity of these conditions was posed by Erdös and Kubilius (Conjecture II) in (Katai, 1977, problem 3; Elliott, 1979; 1980). This is known to be true if one assumes that $g(n)$ is uniformly bounded on the prime numbers (Katai, 1969), or that $g(p) \geqslant 0$ for each prime p (Elliott, 1974a), or that $g(p+1)$ is bounded uniformly for all primes (Elliott, 1974b), or that for each $\varepsilon > 0$ (Timofeev, 1983)

$$\liminf_{x \to \infty} \sum_{p \leqslant x,\ g(p) < 0} \frac{1}{p} \min\left(1, \frac{|g(p)|}{\log^\varepsilon x}\right) = 0.$$

In this article we shall prove the following result, which is an analogue of the Ruzsa's theorem.

THEOREM 1. *Let $a \neq 0$ and*

$$R_h(x) = \sup_b \frac{1}{\pi(x)}\left|\{p : p \leqslant x, b \leqslant g(p + a) < b + h\}\right|.$$

For every additive function g we have

$$R_1(x) \ll \frac{\log^2(W(x)+2)}{\sqrt{W(x)}},$$

where

$$W(x) = \min_s \left(s^2 + \sum_{p \leqslant x} \frac{1}{p} \min\left(1, (g(p) - s\log p)^2\right)\right).$$

The involved constant is absolute.

Theorems 2—6 below can be deduced from Theorem 1, in the same way as in (Ruzsa, 1980) similar results were deduced in the case of natural numbers.

THEOREM 2. *In order that the frequencies*

$$\frac{1}{\pi(x)}\left|\{p : p \leqslant x, \quad g(p+a) \leqslant u\}\right|, \quad a \neq 0,$$

converge weakly to a limit distribution as $x \to \infty$, it is both necessary and sufficient that the series (1) *converge.*

Then the limit distribution is of a pure type, and will be continuous if and only if the series

$$\sum_{g(p)\neq 0} \frac{1}{p} \tag{4}$$

diverges.

THEOREM 3. *In order that there exists a real function $A(x)$, for which the frequencies*

$$\frac{1}{\pi(x)}\left|\{p : p \leqslant x, \quad g(p+a) - A(x) \leqslant u\}\right|, a \neq 0,$$

converge weakly to a limit distribution as $x \to \infty$, it is both necessary and sufficient that there is a constant $s = s_0$ for which the series (3) *converge.*

THEOREM 4. *$R_1(x) \to 0$ as $x \to \infty$, except there is a number $s = s_0$ such that the series* (3) *converge.*

THEOREM 5. *$R_h(x) \to 0$ as $x \to \infty$ and $h \to 0$, unless*

$$\sum_{g(p)\neq 0} \frac{1}{p} < \infty.$$

THEOREM 6. *Let $a \neq 0$ and*

$$E(x) = \sum_{p \leqslant x,\ g(p)\neq 0} \frac{1}{p}.$$

For every additive function g we have

$$\frac{1}{\pi(x)}\Big|\{p : p \leqslant x, \quad g(p+a) = b\}\Big| \ll \frac{\log^2 E(x)}{\sqrt{E(x)}}.$$

The involved constant is absolute.

Using Theorems 2 and 6, one can solve the problem 21 from (Elliott, 1979; 1980).

THEOREM 7. *The limit*

$$\lim_{x\to\infty} \frac{1}{\pi(x)}\Big|\{p : p \leqslant x, \quad g(p+a) = c\}\Big|, \; a \neq 0,$$

always exists. It is non–zero for at most countably many values c.

This limit is non–zero for a value of c if and only if the series (4) converge.
We shall deduce Theorem 1 from the following result.

THE MAIN THEOREM. *Let $f(n)$ be a multiplicative function, $|f(n)| \leqslant 1$, $a \neq 0$. Then*

$$\frac{1}{\pi(x)} \sum_{n\leqslant x,\; n\in r(y,z)} f(n+a) = \prod_{y<p\leqslant z}\left(1-\frac{1}{p}\right)\frac{1}{\pi(x)}\sum_{n\leqslant x} f(n) + R,$$

where

$$R = O\left(\frac{\log x}{\log z}\left(\frac{\log y}{\sqrt[21]{y}} + \frac{(\log\log x)^3}{\sqrt[3]{\log x}} + \exp\left(-\frac{\log x}{\log z}\right)\log y\right)\right),$$

$1 < y \leqslant \exp(\sqrt{\log x})$ and $r(y,z)$ denotes the set of all integers divisible by not primes p in the interval $y < p \leqslant z$.

The proof of this theorem will be published in the journal "Acta Arithmetica".
Now we consider the question when the frequencies

$$\frac{1}{\pi(x)}\Big|\{p : p \leqslant x, \quad g(p+1) - A(x) \leqslant uB(x)\}\Big|$$

converge weakly to a limit law assuming that the renormalising function $B(x)$ satisfies

$$\lim_{x\to\infty} \frac{B(x^y)}{B(x)} = 1$$

for each $y > 0$. We shall prove the Conjecture 15 from (Elliott, 1979; 1980).

THEOREM 8. *Let the measurable function $B(x)$ satisfy $B(x) \to \infty$ and, for each fixed positive number y, $B(x^y)/B(x) \to 1$ as $x \to \infty$.*
In order that, for some function $A(x)$ the frequencies

$$\frac{1}{\pi(x)}\Big|\{p : p \leqslant x, \quad g(p+1) - A(x) \leqslant uB(x)\}\Big|$$

possess a limit distribution as $x \to \infty$, it is both necessary and sufficient that there is a constant b such that the function $h(p) = g(p) - b\log p$ satisfies the following conditions:

there exists a non–decreasing function $M(u)$, $M(-\infty) = 0$, such that at every continuity point u of $M(u)$

$$\lim_{x\to\infty} M_x(u) = \sum_{p\leqslant x,\ h(p)\leqslant uB(x)} \frac{1}{p}\min\left(1, \frac{h^2(p)}{B^2(x)}\right) = M(u), \tag{5}$$

$$\lim_{x\to\infty} \frac{1}{\log x}\sum_{p\leqslant x} \frac{\log p}{p}\min\left(1, \frac{|h(p)|}{B(x)}\right) = 0. \tag{6}$$

When these conditions are satisfied, the function $A(x)$ may be chosen as

$$A(x) = b\log x + B(x)\sum_{p\leqslant x}\left(\frac{h(p)}{B(x)}\right)^{*}\frac{1}{p} + cB(x), \tag{7}$$

where $u^* = u$ if $|u| \leqslant 1$, $u^* = 1$ if $u > 1$ and $u^* = -1$ if $u < -1$. The limit law is proper if and only if $M(+\infty) > 0$.

DEDUCTION OF THEOREM 1 FROM THE MAIN THEOREM

For every d we have

$$\left|\{p : p \leqslant x,\quad |g(p+a) - d| \leqslant 1\}\right| \leqslant$$

$$\leqslant \left|\{n : n \leqslant x,\ n \in r(y,z),\ |g(n+a) - d| \leqslant 1\}\right| + z.$$

Applying the inequality

$$\int_{-1}^{1}(1-|t|)e^{ity}dt = \left(\frac{\sin(y/2)}{y/2}\right)^2 \geqslant \frac{1}{2}, \quad |y| \leqslant 1,$$

we get

$$\left|\{p : p \leqslant x,\quad |g(p+a) - d| \leqslant 1\}\right| \leqslant$$

$$\leqslant 2\sum_{n\leqslant x,\ n\in r(y,z)}\int_{-1}^{1} e^{it(g(n+a)-d)}(1-|t|)dt + z =$$

$$= 2\int_{-1}^{1} e^{-itd}(1-|t|)\sum_{n\leqslant x,\ n\in r(y,z)} e^{itg(n+a)}dt + z.$$

It follows from the main theorem that

$$\frac{1}{\pi(x)}\left|\{p :\ p \leqslant x,\quad |g(p+a) - d| \leqslant 1\}\right| \leqslant$$

$$\leqslant \frac{2}{\pi(x)} \prod_{y<p\leqslant z} \left(1-\frac{1}{p}\right) \int_{-1}^{1} e^{-itd}(1-|t|) \Big| \sum_{n\leqslant x} e^{itg(n)} dt + 2R \leqslant$$

$$\leqslant \frac{2}{\pi(x)} \prod_{y<p\leqslant z} \left(1-\frac{1}{p}\right) \sum_{n\leqslant x} \left(\left(\frac{g(n)-d}{2}\right)^{-1} \sin\frac{g(n)-d}{2} \right)^{2} + 2R$$

and, therefore, we have

$$R_1(x) \leqslant 4\frac{\log x}{\log z} \log y \sup_d \frac{1}{x} \sum_{n\leqslant x} \left(\frac{\sin\left(\frac{g(n)-d}{2}\right)}{\frac{g(n)-d}{2}} \right)^2 + 2R \leqslant$$

$$\leqslant 4\frac{\log x}{\log z} \log y \sup_d \sum_{-\infty}^{+\infty} \max_{k\leqslant y<k+1} \left(\frac{\sin(y/2)}{y/2}\right)^2 \frac{1}{x} \times$$

$$\times \Big| \{ n : n \leqslant x, \big|g(n) - (d+k)\big| \leqslant 1 \} \Big| + 2R.$$

Evidently $((\sin y)/y)^2 \leqslant \min(1, y^{-2})$, and hence

$$R_1(x) \leqslant 36 \frac{\log x}{\log z} \log y Q_1(x) + 2R.$$

Let

$$y = (W(x)+2)^{50}, \log z = \frac{\log x}{\log(W(x)+2)},$$

then

$$R_1(x) \ll \frac{1}{\sqrt{W(x)}} \log^2(W(x)+2),$$

by Ruzsa's Theorem.

This completes the proof of Theorem 1.

Proof of Theorem 8. Sufficiency. We shall deduce our theorem using Theorem 1 and the following lemma from (Timofeev, 1983).

LEMMA 1. *Let M be a family of multiplicative functions such that $|f(n)| \leqslant 1$,*

$$\sup_{f\in M} \sum_{p\leqslant x} |f(p)-1| \frac{\log p}{p} \leqslant \varepsilon(x) \log x, \tag{8}$$

where $\varepsilon(x) \to 0$ as $x \to \infty$. Then we have

$$\frac{1}{\pi(x)} \sum_{p\leqslant x} f(p+1) = \prod_{p\leqslant x} \left(1 + \sum_{r=1}^{\infty} \frac{f(p^r) - f(p^{r-1})}{\varphi(p^r)}\right) + o(1)$$

uniformly in $f \in M$.

Let $f(n) = \exp(ith(n)/B(x))$. Since,

$$\big| \exp(ita) - 1 \big| \leqslant \big(|t|+1\big) \min\big(1, |a|\big), \tag{9}$$

using (6), we get

$$\sum_{p\leqslant x}\left|\exp\left(it\frac{h(p)}{B(x)}\right)-1\right|\frac{\log p}{p}=o(\log x)$$

uniformly in t for $|t|\leqslant C$, where C is arbitrary. Applying Lemma 1, we obtain

$$\frac{1}{\pi(x)}\sum_{p\leqslant x}\exp\left(it\frac{h(p+1)}{B(x)}\right)=\exp\left(\sum_{p\leqslant x}\left(\exp\left(it\frac{h(p)}{B(x)}\right)-1\right)\frac{1}{p}\right)+o(1)$$

uniformly in $t:|t|\leqslant C$. Since $B(x)\to\infty$ as $x\to\infty$, using (7) we have

$$\frac{1}{\pi(x)}\sum_{p\leqslant x}\exp\left(it\frac{g(p+1)-A(x)}{B(x)}\right)=$$

$$=\frac{1}{\pi(x)}\sum_{p\leqslant x}\exp\left(it\frac{h(p+1)}{B(x)}-it\sum_{p\leqslant x}\left(\frac{h(p)}{B(x)}\right)^{*}\frac{1}{p}-itc\right)+o(1)=$$

$$=\exp\left(\sum_{p\leqslant x}\left(\exp\left(it\frac{h(p)}{B(x)}\right)-1-it\left(\frac{h(p)}{B(x)}\right)^{*}\right)\frac{1}{p}-itc\right)+o(1)$$

uniformly in t in the interval $|t|\leqslant C$. Applying (5), we get

$$\frac{1}{\pi(x)}\sum_{p\leqslant x}\exp\left(it\frac{g(p+1)-A(x)}{B(x)}\right)=\exp\left(\int_{-\infty}^{+\infty}\frac{e^{itu}-1-itu^{*}}{(u^{*})^{2}}dM_{x}(u)-itc\right)+$$

$$+o(1)=\exp\left(\int_{-\infty}^{+\infty}\frac{e^{itu}-1-itu^{*}}{(u^{*})^{2}}dM(u)-itc\right)+o(1)$$

uniformly in t with $|t|\leqslant C$, and the first part of Theorem 8 follows.

Proof of Theorem 8. Necessity. According to Lemma II.5 from (Elliott, 1979; 1980) there are continuous strictly increasing unbounded functions $y(x)$, $z(x)$, such that $y(x)<x<z(x)$,

$$\frac{\log y(x)}{\log x}\to 0,\ \frac{\log z}{\log x}\to\infty\quad(x\to\infty)$$

so that, as $x\to\infty, B(w)/B(x)\to 1$, uniformly in $y\leqslant w\leqslant z$. According to Theorem 1, there is a constant c such that exists $s(x)$ satisfying

$$\sum_{p\leqslant w}\frac{1}{p}\min\left(1,\left(\frac{g(p)-s(w)\log p}{B(x)}\right)^{2}\right)\leqslant c \tag{10}$$

for $w \in [y, z]$. Since

$$\min\left(1, (a+b)^2\right) \geqslant \frac{1}{2}\min(1, a^2) - \min(1, b^2),$$

for $u \in [1, 2]$ we have

$$c \geqslant \sum_{p \leqslant x} \frac{1}{p} \min\left(1, \left(\frac{g(p) - s(x)\log p}{B(x)}\right)^2\right) \geqslant$$

$$\geqslant \frac{1}{2} \sum_{p \leqslant x} \frac{1}{p} \min\left(1, \left(\frac{s(x) - s(x^u)}{B(x)} \log p\right)^2\right) -$$

$$- \sum_{p \leqslant x} \frac{1}{p} \min\left(1, \left(\frac{g(p) - s(x^u)\log p}{B(x)}\right)^2\right).$$

Hence

$$\sum_{p \leqslant x} \frac{1}{p} \min\left(1, \left(\frac{s(x) - s(x^u)}{B(x)} \log p\right)^2\right) \leqslant 4c. \tag{11}$$

Let $d = e^{6c}$. Let us show, that for $u \in [1, 2]$

$$\left|s(x) - s(x^u)\right| \leqslant d\frac{B(x)}{\log x} \quad (x \geqslant x_0). \tag{12}$$

Assume, on the contrary, that we can find an unbounded sequence x_l with corresponding numbers $u_l \in [1, 2]$ such that

$$\sum_{p \leqslant x} \frac{1}{p} \min\left(1, \left(\frac{s(x) - s(x^u)}{B(x)} \log p\right)^2\right) \geqslant \sum_{x^{1/d} \leqslant p \leqslant x} \frac{1}{p} \geqslant 5c.$$

This contradicts (11).

Let $t \geqslant x$ and $2^m \leqslant \log t / \log x$. We have

$$\left|s(t) - s(x)\right| \leqslant \sum_{i=1}^{m} \left|s(x^{2^i}) - s(x^{2^{i-1}})\right| + \left|s(x^{2^m}) - s(t)\right|.$$

Since $B(x^u)/B(x) \to 1$ uniformly in $u \in [1, 2]$, we obtain $B(x^u) \leqslant 3B(x)/2$ for $u \in [1, 2]$, and $B(x^{2^i}) \leqslant (3/2)^i B(x)$. From (12) it follows that

$$\left|s(t) - s(x)\right| \leqslant d_1 \sum_{i=1}^{\infty} \frac{B(x^{2^{i-1}})}{\log x^{2^{i-1}}} \leqslant d_1 \frac{B(x)}{\log x} \sum_{i=1}^{\infty} \left(\frac{3}{2}\right)^{i-1} \frac{1}{2^{i-1}}$$

for any $t \geqslant x$. Since $B(x)/\log x \to 0$ as $x \to \infty$, we have $s(x) \to b$ as $x \to \infty$ and $|b - s(x)| \leqslant d_2 B(x)/\log x$.

Hence, applying (10), we obtain

$$\sum_{p\leqslant z}\frac{1}{p}\min\left(1,\left(\frac{h(p)}{B(x)}\right)^2\right)\leqslant c_1,$$

where $h(p)=g(p)-b\log p$.
Since $\log z/\log x\to\infty$ as $x\to\infty$, there are functions $z_1(x), z_2(x)$ such that $x<z_1<z_2<z$

$$\log z_1/\log x\to\infty,\ \log z_2/\log z_1\to\infty$$

and

$$\sum_{z_1<p\leqslant z_2}\frac{1}{p}\min\left(1,\left(\frac{h(p)}{B(x)}\right)^2\right)\to 0\quad (x\to\infty). \tag{13}$$

Let $f(n,v,t)$ be the multiplicative function:

$$f(p^r,v,t)=\begin{cases}\exp\left(it\frac{h(p^r)}{B(x)}\right), & \text{if } v=z_2(x),\\ 1, & \text{othervise.}\end{cases}$$

According (9) and (13), $f(n,v,t)$ satisfies condition (8). Since $B(x)\to\infty$ as $x\to\infty$, using Lemma 1 we have

$$|\tau(t)|=\frac{1}{\pi(z_2)}\left|\sum_{p\leqslant z_2}\exp\left(it\frac{g(p+1)}{B(x)}\right)\right|+o(1)=$$

$$=\frac{1}{\pi(z_2)}\left|\sum_{p\leqslant z_2}\exp\left(it\frac{h(p+1)}{B(x)}\right)\right|+o(1)=$$

$$=\exp\left(-\sum_{p\leqslant z_2}\frac{1}{p}\left(1-\cos t\frac{h(p)}{B(x)}\right)\right)+o(1)$$

uniformly in $t\in[-t_0,t_0]$, $t_0\neq 0$, where $\tau(t)$ is the characteristic function of the limit distribution.
Thus,

$$\sum_{p\leqslant z_2}\frac{1}{p}\left(1-\cos t\frac{h(p)}{B(x)}\right)=\log\frac{1}{|\tau(t)|}+o(1)$$

uniformly in $t\in[-t_0,t_0]$.
Since $\log x/\log y\to\infty$ as $x\to\infty$, there are functions $y_1(x), y_2(x)$, such that $y<y_1<y_2<x$, $\log y_2/\log y_1\to\infty$, $\log x/\log y_2\to\infty$ and

$$\sum_{y_1<p\leqslant y_2}\frac{1}{p}\min\left(1,\left(\frac{h(p)}{B(x)}\right)^2\right)\to 0\quad (x\to\infty).$$

Hence, in the same way as before, we obtain

$$\sum_{p\leqslant y_2}\frac{1}{p}\left(1-\cos t\frac{h(p)}{B(x)}\right)=\log\frac{1}{|\tau(t)|}+o(1)$$

uniformly in $t \in [-t_0, t_0]$.
Thus we have proved that there exists a neighborhood of the point $t = 0$, where

$$\sum_{y_2 < p \leqslant x} \frac{1}{p}\left(1 - \cos t\frac{h(p)}{B(x)}\right) \to 0 \quad (x \to \infty)$$

uniformly in t. This implies (6) (e.g., (Levin and Timofeev, 1976, p.144)). Applying Lemma 1 we have

$$\tau(t) = \frac{1}{\pi(x)} \sum_{p \leqslant x} \exp\left(it\frac{g(p+1) - A(x)}{B(x)}\right) + o(1) =$$

$$= \exp\left(\sum_{p \leqslant x}\left(\exp\left(it\frac{h(p)}{B(x)}\right) - 1\right)\frac{1}{p} - it\big(A(x) - b\log x\big)\right) + o(1)$$

uniformly in t : $|t| \leqslant C$, where C arbitrary, and (5) follows (e.g., (Levin and Timofeev, 1976) or (Elliott, 1979; 1980)).
This completes the proof of Theorem 8.

REFERENCES

Elliott, P.D.T.A. (1974a). On the limiting distribution of $f(p+1)$ for non-negative additive functions. *Acta Arithm.* **25**, 259–264.

Elliott, P.D.T.A. (1974b). A conjecture of Katai. *Acta Arithm.* **26**, 11–20.

Elliott, P.D.T.A. (1979). *Probabilistic Number Theory.* I. Grundlehren Math. **239**, Springer, New—York Berlin.

Elliott, P.D.T.A. (1980). *Probabilistic Number Theory.* II. Grundlehren Math. **240**, Springer, New—York Berlin.

Elliott, P.D.T.A. and Ryavec, C. (1971). The distribution of the values of additive arithmetical functions. *Acta Math.* **126**, 143–164.

Erdös, P. (1938). On the density of some sequences of numbers. III. *J. London Math. Soc.* **13**, 119–127.

Erdös, P. and Wintner, A. (1939). Additive arithmetical functions and statistical independence. *Amer. J. Math.* **61**, 713–721.

Katai, J. (1968). On distribution of arithmetical functions on the set of prime plus one. *Comp. Math.* **19**, 278–289.

Katai, J. (1969). Some remarks on additive arithmetical functions. *Liet. Matem. Rink.* **9**, 515–518.

Katai, J. (1977). Research problems in number theory. *Bubl. Math.* **24**, 263–276.

Levin, B.V. and Timofeev, N.M. (1971). An analytic method in probabilistic number theory. *Uchen. Zap. Vladimirskogo Gos. Ped. Inst. Math.* **57**, 57–150.

Levin, B.V. and Timofeev, N.M. (1976). Some integral limit theorems for additive functions. *Liet. Matem. Rink.* **16**, 133–147.

Timofeev, N.M. (1983). Distribution of values of additive function on the sequence $\{p+1\}$. *Math. Notes* **33**, 933–942.

Ruzsa, I.Z. (1980). On the concentration of additive functions. *Acta Math. Sci. Hungar.* **36**, 215–232.

Prob. Theory and Math. Stat., Vol. 2, pp. 505–518
B.Grigelionis *et al.* (Eds.)
1990 VSP/Mokslas

A NEW LOOK ON SEMIMARTINGALE DISTRIBUTIONS WITH SEVERAL EXAMPLES

E.I. TROFIMOV

Institute of Mathematics and Mechanics,
Kazan University, 420008 Kazan, USSR

ABSTRACT

The construction of semimartingale distributions by means of the fractional Sobolev spaces is presented. This work concerns with general compactness criterias and simple examples of applications.

INTRODUCTION

In some recent works of the author it was shown, that every semimartingale $(X_t)_{[0,1]}$ is a Borelian random element in Banach spaces of Besov—Nikolskii $B^s_{p,q}[0,1]$ with appropriate p, q, s. New tightness criteria and functional limit theorems with respect to their normed topologies $\tau = \tau^s_{p,q}$ are obtained also. This paper is devoted to generalized description of weakly compact sets of processes, and to selected exposition of numerous applications. There we consider only the scalar processes.

First, for a better understanding, we explain some useful properties of τ–topologies, compare them with the Skorokhod topology **Sk**, and give examples of τ–continuous mappings (Section 1).

Secondly, we prove a number of inequalities for $\mathcal{L}^r(\mathbf{P}; B^s_{p,q})$ and $\mathcal{S}^r$–norms of special semimartingales, which shows that τ–construction becomes suitable for describing of semimartingale distributions (Section 2).

Next, we introduce a notion of stochastic duality and give new compactness criteria for semimartingales in such terms with sketches of proofs (Section 3). They seem to be original analogues of classical compactness theorems for vector measures.

The last Section 4 contains simple limit theorems for stochastic boundary value problems to partial differential operator, discrete space–time approximations of semimartingales and statistical processes. The references are restricted everywhere by several key works with further bibliography.

1. SUBSPACES OF BESOV—NIKOLSKII IN D[0,1]

1.1. *Basic notations*

First consider the standard Besov—Nikolskii and Sobolev—Slobodetskii spaces (for details see (Lions and Magenes, 1968; 1972; Besov et al., 1975; Triebel, 1986)). For any $a < b \in \mathbf{R}$, $1 \leqslant p,\ q < \infty$, and $s = 0$, let $B^s_{p,q} = B^s_{p,q}[a,b] = \mathcal{L}^p[a,b]$. For $0 < s < 1/p$, let $B^s_{p,q}$ be completion of $C^\infty[a,b]$ by the norm

$$\left\|x\middle|B^s_{p,q}\right\| \equiv \left\|x\middle|\mathcal{L}^p[a,b]\right\| + \left\|x\middle|b^s_{p,q}[a,b]\right\|$$

with further factorization, where

$$\left\|x\middle|b^s_{p,q}[a,b]\right\| \equiv \left(\int_{-1}^{1} |\theta|^{-1-qs}\, d\theta \left(\int_a^b dt \left|x(t+\theta) - x(t)\right|^p\right)^{q/p}\right)^{1/q};$$

and $x(u) - x(v) \equiv 0$, if $(u,v) \not\subset (a,b)$.
The spaces $H^s_2 = B^s_{2,2}$ are Hilbert spaces of Sobolev and Slobodetskii.
Let $D = D[0,1]$ be the class of all right continuous real functions with finite left hand limits, and let for the same p, q, s:

$$D\widetilde{B}^s_{p,q} \equiv \left\{x \in D[0,1] : \left\|x\middle|\widetilde{B}^s_{p,q}\right\| \equiv |x(0)| + \left\|\widetilde{x}\middle|b^s_{p,q}(0,2)\right\| < \infty\right\},$$

where $\widetilde{x} \equiv x$ over $[0,1]$, and $\widetilde{x} \equiv x(1)$ over $1 < t < 2$.
Let $\widetilde{\tau} = \widetilde{\tau}^s_{p,q}$ (respectively, $\widetilde{\lambda}$, **Sk**, **u**) be the topology of convergence in $D\widetilde{B}^s_{p,q}$ with respect to the norm (respectively, the measure $\overline{\lambda} \equiv \varepsilon_0 + \varepsilon_1 + \lambda$, where ε_a is the Dirac measure with $\operatorname{supp}(\varepsilon_a) = \{a\}$, λ is the Lebesgue measure on $(0,1)$; trace of the Skorokhod's topology over $D\widetilde{B}$; the uniform norm). Notation $D\widetilde{B}$ without a sign of topology means the space $(D\widetilde{B}, \widetilde{\tau})$. Finally we deal with classes of $\widetilde{\lambda}$-equivalence of functions as if they are true functions everywhere, it does not lead us to misunderstandings.

1.2. *On topologies*

Now some properties and examples showing what are the topologies τ , $\widetilde{\tau}$ with respect to **Sk** and **u** will be listed. Through simple but longer calculations (Trofimov, 1989) one may prove that:

a) every simple step function and every function with finite variation from D (denote these classes by E and V, respectively), and every function of the class $C^1[0,1]$ belong to $D\widetilde{B}$; each of the classes E, V is dense in $D\widetilde{B}$ and B;

b) there exist functions in B with infinite and oscillating discontinuites;

c) there exist continuous functions, which do not belong to $D\widetilde{B}$;

d) there exist τ–converging sequences from $D\widetilde{B}$, which do not converge in **u**–topology;

e) there exist **u**–converging sequences in $D\widetilde{B}$, which do not converge in $\widetilde{\tau}$–topology.

So, $C[0,1] \not\subset D\widetilde{B} \not\subset D$, the topologies $\widetilde{\tau}$ and **Sk** are uncomparable, and it is the same for $\widetilde{\tau}$ and **u**.

Remark. Roughly speaking, *b*–norms calculate a mean value of oscillations functions. That is nearly to crossing numbers and measure of occupation times for paths of processes. And one may expect, that $\widetilde{\tau}$–construction of functional distributions will become as natural one for semimartingales and Markov processes theory, indeed.

1.3. *Continuous mappings on* $D\widetilde{B}^{s}_{p,q}$

The following properties will be used below (for details and another properties of spaces B see (Lions and Magenes, 1968; 1972; Besov et al., 1975; Triebel, 1986)).

a) *Pointwise multiplication.* If $s > 0$, the mapping

$$(D\widetilde{B})^{\times 2} \ni (x,y) \rightsquigarrow \big(x(t)y(t)\big)_{[0,1]} \in D\widetilde{B}$$

is continuous, and the inequality of Peetré holds:

$$\big\|xy\big|\widetilde{B}\big\| \leqslant c\big\|x\big|\widetilde{B}\big\|\,\big\|y\big|\widetilde{B}\big\|.$$

In particular, the mapping

$$\mathbf{R}^{\times 2} \times D\widetilde{B} \ni (\alpha,\beta,x) \rightsquigarrow \big(I_{[\alpha,\beta)}x\big)_{[0,1]} \in D\widetilde{B}$$

is continuous. Every polinomial mapping $(D\widetilde{B})^{\times n} \rightsquigarrow D\widetilde{B}$ is continuous.

b) *Composition.* Denote by Lip the space of all bounded continuous functions on **R** with finite norm

$$\big\|f\big|\mathrm{Lip}\big\| \equiv \sup_{\alpha\in\mathbf{R}} \big|f(\alpha)\big| + \sup_{\alpha\neq\beta}\left|\frac{f(\alpha)-f(\beta)}{\alpha-\beta}\right|.$$

LEMMA. *For every $f \in$ Lip, $x \in D\widetilde{B}$, the composition $f \circ x$ belongs to $D\widetilde{B}$, and the mapping*

$$\mathrm{Lip} \times D\widetilde{B} \ni (f,x) \rightsquigarrow f\circ x \in D\widetilde{B}$$

is continuous.

Proof. Through immediate calculating we obtain the estimate

$$\big\|f\circ x\big|\widetilde{B}\big\| \leqslant \big\|f\big|\mathrm{Lip}\big\|\,\big\|x\big|\widetilde{B}\big\|.$$

c) *Variation.* Let

$$\mathrm{Var}(x) \equiv |x(0)| + \sup \sum_i |x(t_i) - x(t_{i+1})|,$$

where the supremum is calculated over all finite partitions $\{t_i\}_{i=0}^{I}$ of interval $[0,1]$, and $t_0 = 0$, $t_I = 1$. Then the function $(D\widetilde{B}, \varkappa) \ni x \rightsquigarrow \mathrm{Var}(x)$ is lower semi-continuous for $\varkappa = \widetilde{\tau}$, because it is such for $\varkappa = \widetilde{\lambda}$ (see (Meyer and Zheng, 1984)). The last property involves, that the class $G \subset D\widetilde{B}$ of all non-decreasing functions is $\widetilde{\lambda}$-closed.

d) *Embeddings.* For any $s' > s$ the embedding operator $B^{s'}_{p,q} \hookrightarrow B^{s}_{p,q}$ is continuous, and more that, compact. In contrast, the embeddings $D\widetilde{B}^{s'}_{p,q} \hookrightarrow D\widetilde{B}^{s}_{p,q}$ are continuous, only; the compactness property fails. Embeddings, which usually are called as trace, will be considered later in Section 4.

e) *Integration.* In contrast to **Sk**–construction, a most important class of τ– and $\widetilde{\tau}$–continuous mappings on scales $B^{\cdot}_{\cdot}$ and $D\widetilde{B}^{\cdot}$ presents a variety of integral (nonlinear and singular, in general) operators, in particular, pseudo–differential operators.
At last, using known embeddings between basic interpolation scales of B– and F–spaces (see (Triebel, 1986)) and another, more special spaces, one may transfer the results of $\widetilde{\tau}$–construction of distributions onto almost all functional spaces, which are of any interest in applications.

1.4. *Measurability*

Denote by $\sigma(E,\varkappa)$ the Borel σ–algebra over a given topological space $(E,\varkappa)$. If F is a subclass of D, denote by $\mathrm{C}(F)$ the σ–algebra over F, generated by all cylindrical sets of finite dimension. Let $\varkappa$ be any topology from the set $\widetilde{\lambda}$, $\widetilde{\tau}$, **Sk**, $\widetilde{\tau} \vee$ **Sk**, and let $\mathrm{M}^1_+(E,\varkappa)$ be the space of all Borelian probabilities over $(E,\varkappa)$.

LEMMA 1.4a (Trofimov, 1989). $\sigma(D\widetilde{B},\varkappa) = C(D\widetilde{B})$, $D\widetilde{B} \in (D,\widetilde{\lambda}) = \sigma(D,\mathbf{Sk})$.

COROLLARY 1.4b (cf. (Meyer and Zheng, 1984; Billingsley, 1975)). *For any $\bar{t} = (t_1,\ldots,t_k) \in [0,1]^k$ the projection $(D\widetilde{B},\varkappa) \ni x \rightsquigarrow (x(t_1),\ldots x(t_k)) \in \mathbf{R}^k$ is measurable.*

COROLLARY 1.4c (cf. (Meyer and Zheng, 1984; Billingsley, 1975)). *Every adapted random process $(X_t)_{[0,1]}$ with path in $(D\widetilde{B},\varkappa)$ induces a Borelian element in this space.*

COROLLARY 1.4d. *The classes of measures $\mathrm{M}^1_+(D\widetilde{B},\varkappa)$, where $\varkappa$ is equal to any from set $\widetilde{\lambda}$, $\widetilde{\tau}$, **Sk**, $\widetilde{\tau} \vee$ **Sk**, coincide to each other.*

COROLLARY 1.4e (cf. (Meyer and Zheng, 1984; Billingsley, 1975)). *The following three assertions are equivalent:*

(i) *two measures* $\mathbf{P}, \mathbf{Q} \in \mathbf{M}^1_+(D\widetilde{B})$ *coincide;*

(ii) *all of their finite dimensional distributions* (f.d.d.'s.) *coincide;*

(iii) *all of their f.d.d.'s. over any dense set* $U \in [0,1]$, $U \ni 1$, *coincide.*

Proof. By continuity from the right any k–dimensional distributions of $\mathbf{P}$ and $\mathbf{Q}$ are distinct in any point $\bar{t} \in [0,1]^k$, iff they are distinct over any closed interval $\mathcal{J} \subset [0,1]^k$ such, that $\bar{t} \in \mathcal{J}$, and $\widetilde{\lambda}(\mathcal{J}) > 0$.

2. INEQUALITIES. PATHS OF SEMIMARTINGALES

From now we restrict our consideration to the spaces $D\widetilde{H}^s_2$, $s < 1/2$. All main results below are valid for the spaces $D\widetilde{B}^s_{p,q}$ with $0 \leqslant s < 1/(p \vee (q \wedge 2))$, $1 \leqslant p$, $q < \infty$. In the present section adapted processes are studied only. By Corollary 1.4c such processes, say (X_t), with a.e. paths in $D\widetilde{H}$ may be considered as Borelian random element in $D\widetilde{H}$, which we denote by X too.

2.1. *Inequalities*

Let X be any adapted process on a given base $\beta = (\Omega, \mathbf{F}, (\mathbf{F}_t)_{[0,1]}, \mathbf{P})$ under the usual assumptions.

PROPOSITION (Trofimov, 1989). 1^0. *If* (X_t) *belongs to the class* $\mathcal{V}$ *of all adapted processes with finite variation, then (a.s.)*

$$\|X\cdot|\widetilde{H}^s\| \leqslant c_s \mathrm{Var}(X\cdot). \tag{1}$$

2^0. *If* X *belongs to the class* $\mathcal{M}^2_0$ *of all square integrable martingales with* $X_0 = 0$, *and* V *denotes either process* $[X,X]\cdot$, *or* $\langle X\rangle\cdot$ (the quadratic variation and its dual predictable projection, respectively), *then*

$$\mathbf{E}\Big(\|X\cdot|\widetilde{H}^s_2\|^2\Big) = \mathbf{E}\|V\cdot|\widetilde{B}^{2s}_{1,1}\|, \tag{2}$$

and for every $r \geqslant 2$

$$\mathbf{E}\Big(\|X\cdot|\widetilde{H}^s_2\|^r\Big) \leqslant c_s \mathbf{E}(|X_1|^r). \tag{3}$$

3^0. *If* X *belongs to the class* $\mathcal{S}^r$ *of special semimartingales with* $r \geqslant 2$ *and canonical decomposition* $X = A + M$, $A \in \mathcal{A}$ *is the subclass of predictable processes from* $\mathcal{V}$, $M \in \mathcal{M}^2_0$, *then*

$$\|X\cdot|\mathcal{L}^r(\mathbf{P};\ D\widetilde{H})\| \leqslant c_s \|X\cdot|\mathcal{S}^r\|, \tag{4}$$

where

$$\|X\cdot|\mathcal{S}^r\| \equiv \|\mathrm{Var}(A) + \langle M\rangle_1^{1/2}|\mathcal{L}^r(\mathbf{P})\|.$$

The inequalities (1)—(4) are obtained as the consequences of Jensen's inequality, Doob's inequality for square integrable martingales, and the Hirschmann's one:

$$\int_0^1 \left|x(t+\theta)-x(t)\right|^2 dt \leqslant c|\theta|\left(\mathrm{Var}(x)\right)^2.$$

So, every $\mathcal{S}^2$–semimartingale and process with finite variation are Borelian elements in $(D\widetilde{H}, \varkappa)$ for $\varkappa = \widetilde{\lambda}$, $\widetilde{\tau}$, $\mathbf{Sk}$, $\widetilde{\tau} \vee \mathbf{Sk}$.

2.2. *General semimartingales*

Since each semimartingale X of general kind belongs locally to the class $\mathcal{V} \oplus \mathcal{M}_0^2$, it induces, by Corollary 1.4c and Proposition 2.1, a Borelian element in $(D\widetilde{H}, \varkappa)$. Denote by $\mathcal{S}^0$ the class of all semimartingales over β, and by $\mathcal{S}^{0,(n)}$ the class $\mathcal{S}^0$ with respect to the base $\beta^n = (\Omega^n, \mathbf{F}^n, (\mathbf{F}_t^n)_{[0,1]}, \mathbf{P}^n)$ from a given family $\{\beta^n, n \in \mathcal{J}\}$, and by $\widetilde{\mathbf{\Pi}}_1^{(n)}$ any dense subclass of the class $\mathbf{\Pi}_1^{(n)}$ of all β^n–predictable processes with values in $[-1,1]$, endowed with simple convergence on $[0,1] \times \Omega'$ with certain $\Omega' \subset \Omega^n$, $\mathbf{P}^n(\Omega') = 1$.

Using Lenglart's inequality and domination relation between $\widetilde{H}$–norms of semimartingales and $\widetilde{B}$–norms of their predictable characteristics from Section 2.1 we obtain the following (see (Trofimov, 1989), for details)

PROPOSITION. 1^0. *Let* $X \in \mathcal{S}^2_{\mathrm{loc}}$, *and* $X = A + M$, $A \in \mathcal{A}$, $M \in \mathcal{M}^2_{0,\mathrm{loc}}$. *Then*

$$\mathbf{P}\left\{\left\|X\middle|\widetilde{H}\right\| \geqslant a\right\} < b/a + \mathbf{P}\left\{\mathrm{Var}(A) + \langle M\rangle_1^{1/2} > b\right\}, \quad a, b > 0.$$

2^0. *Let a set* $\left\{X^n \in \mathcal{S}^{0,(n)};\ n \in \mathcal{J}\right\}$ *be given. Then the set of r.v.'s* $\left\{\left\|X^n\middle|\widetilde{H}\right\|\right\}_{n\in\mathcal{J}}$ *is* $\mathcal{L}^0$*–bounded, iff the set of r.v.'s*

$$\mathcal{F}\left(\widetilde{\mathbf{\Pi}}_1^{(n)}, n \in \mathcal{J}\right) \equiv \left\{F^n \cdot X_1^n :\ F^n \in \widetilde{\mathbf{\Pi}}_1^{(n)},\ n \in \mathcal{J}\right\}$$

is $\mathcal{L}^0$*–bounded.*

A set of r.v.'s $\left\{\xi^n \in \mathcal{L}^0(\mathbf{P}^n)\right\}_{n\in\mathcal{J}}$ is $\mathcal{L}^0$–bounded, if for each $\varepsilon > 0$ there exist $m < \infty$ such, that $\sup\limits_n \mathbf{P}^n \left\{|\xi^n| > m\right\} < \varepsilon$. And as usually, $F \cdot X$ denotes the Ito's stochastic integral.

Remark. Let the bases β^n coincide for all $n \in \mathcal{J}$. Then $\mathcal{F}\left(\widetilde{\mathbf{\Pi}}_1^{(n)}, n \in \mathcal{J}\right)$ is $\mathcal{L}^0$–bounded, iff $\{X^n\}$ is bounded in $\mathcal{S}^0$ with respect to Emery's topology (Dellacherie and Meyer, 1980).

2.3. *Additive functionals*

If $X \in \mathcal{S}^0$ with $X_0 = 0$, then there exist a process $(L_t^a)_{[0,1]} \in \mathcal{V}_0^+$ with parameter $a \in \mathbf{R}$ such, that the random function $L^{\cdot}_{\cdot}$ is measurable, and (a.s.)

$$[X, X]_t = \int_{\mathbf{R}} L_t^a \, da \tag{6}$$

(see (Jacod, 1979; Metivier and Pellaumail, 1977)). If X belongs to the class $\mathcal{M}_0^{2,c}$ of continuous martingales from $\mathcal{M}_0^2$, then L may be interpreted either as a *density of occupation times*, or as a *local time process* of X.
For any bounded measurable function f look for the martingale $f(X) \cdot X$ and for the additive functional:

$$A_t^f = f(X)^2 \cdot \langle X \rangle_t = \int_{\mathbf{R}} L_t^a \left(f(a)\right)^2 da.$$

Using the identities (2), (6) we get a precise formulation of intuitive Remark 1.2:

LEMMA. *Let $X \in \mathcal{M}_0^2$. Then for each bounded measurable function f:*

$$\left\| f(X) \cdot X \cdot | \mathcal{L}^2 \left(\mathbf{P}; D\widetilde{H}_2^s\right) \right\|^2 = \left\| A^f_{\cdot} | \mathcal{L}^1 \left(\mathbf{P}; D\widetilde{B}_{1,1}^{2s}\right) \right\|.$$

3. STOCHASTIC DUALITIES. COMPACTNESS

We consider now vector measures in $\mathcal{L}^0(\mathbf{P})$ induced by semimartingales as in (Metivier and Pellaumail, 1977). Though, there is an essential difference between the theorem of Metivier and Pellaumail and our main Theorem 3.2a below: the last one assures weak compactness property of sets of semimartingales in distribution.

3.1. *Notations and definitions*

Let us introduce the canonical filtering base

$$\widehat{\beta} \equiv \left(\widehat{\Omega} \equiv D\widetilde{H},\ \widehat{\mathbf{F}} \equiv \mathrm{C}(D\widehat{H}),\ \left(\widehat{\mathbf{F}}_t \equiv \bigcap_{u>t} \mathrm{C}(D\widetilde{H}[0,u]) \right)_{0 \leqslant t \leqslant 1} \right)$$

and the canonical process $\widehat{X} = \widehat{X}(x,t) = x(t)$, $x \in D\widetilde{H}$.
Let $\mathbf{M}(D\widetilde{H})$ (respectively, $\mathbf{M}_+(D\widetilde{H})$) be the class of all signed (respectively, non-negative) Borelian bounded measures on $(D\widetilde{H}, \widetilde{\tau})$. For each $\mathbf{Q} \in \mathbf{M}(D\widetilde{H})$ denote through $(\widehat{\beta}, \mathbf{Q})$ the base $\widehat{\beta}$ with measure $\mathbf{Q}$ (where, as usually, each σ-algebra may be assumed to be completed with all $\mathbf{Q}$-evanescent sets from $\widehat{\Omega}$).

By theorems (9.19) (on restriction of filtration) and (10.37) (on images of bases for every semimartingale X on a given probabilistic base β) from (Jacod, 1979), the process $\widehat{X}$ is semimartingale with respect to the base $\left(\widehat{\beta}, \mathbf{L}(X; D\widetilde{H})\right)$. Seeking limit theorems for semimartingales one may deal with measures $\mathbf{Q}$ like $\mathbf{L}(X; D\widetilde{H})$ only.
So, let us call by *semimartingale measures* the zero–measure $\mathbf{O} \in \mathbf{M}(D\widetilde{H})$ and each measure $\mathbf{Q} \in \mathbf{M}(D\widetilde{H})$, for each $\widehat{X}$ is semimartingale with respect to bases $(\widehat{\beta}, \mathbf{Q}^+)$ and $(\widehat{\beta}, \mathbf{Q}^-)$, where $\mathbf{Q} = \mathbf{Q}^+ - \mathbf{Q}^-$, $\mathbf{Q}^+ \equiv \mathbf{Q}I_{\Omega^+}$, $\mathbf{Q}^- \equiv -\mathbf{Q}I_{\Omega^-}$, Ω^+, $\Omega^- \in \widehat{\mathbf{F}}$, is the Hahn—Jordan decomposition of $\mathbf{Q}$. Denote the set of all semimartingale measures through $\mathbf{SM}$.
Now, let $\widehat{\mathbf{\Pi}}_e$ (respectively, $\widehat{\mathbf{\Pi}}_{1,e}$) be the space of all $\widehat{\beta}$–predictable finite step processes (respectively, with values in $[-1,1]$) and for $F \in \widehat{\mathbf{\Pi}}_{1,e}$, let $F \cdot \widehat{X}_1$ be the finite Ito's integral sum. In virtue of local character of semimartingales and stochastic integrals (Meyer, 1976), the inducted distribution $\mathbf{Q}\left((F \cdot X_1)^{-1}\right)$, for any $\mathbf{Q} \in \mathbf{SM}$, belongs to $\mathbf{SM}$, also.

PROPOSITION. *Any measure $\mathbf{Q} \in \mathbf{M}(D\widetilde{H})$ belongs to the class $\mathbf{SM}$, iff the set of r.v.'s $\left\{F \cdot \widehat{X}_1^{\mathbf{Q}} : F \in \widehat{\mathbf{\Pi}}_{1,e}\right\}$ is $\mathcal{L}^0(\mathbf{Q})$–bounded.*

The proposition is an intrinsic consequence of the theorem on characterization of semimartingales (see (Dellacherie and Meyer, 1980), for example).

3.2. *Dualities and compactness*

Consider the class $\mathbf{SM}$ in duality with the linear class $\widehat{\mathbf{\Pi}}_e$ by means of the random bilinear functional:

$$\mathbf{SM} \times \widehat{\mathbf{\Pi}}_e \ni (\mathbf{Q}, F) \rightsquigarrow F \cdot \widehat{X}_1^{\mathbf{Q}} \in \mathcal{L}^0(\widehat{\Omega}, \mathbf{Q}).$$

Let $\mathbf{SM_U}$ (respectively, $\mathbf{SM}_\sigma$) be the class $\mathbf{SM}$ equipped with the topology of uniform (respectively, simple) convergence over the set $\widehat{\mathbf{\Pi}}_{1,e}$. On the other hand, $\mathbf{SM}$ may be equipped with trace of topology $w_{\widetilde{\tau}}$ of weak convergence in $\mathbf{M}(D\widetilde{H})$, induced by its canonical duality with the space $C(D\widetilde{H}, \widetilde{\tau})$ of all bounded continuous functions on $D\widetilde{H}$; denote this space by $\mathbf{SM_W}$. Any non–empty set E in $\mathbf{SM}_\sigma$ is relatively compact, iff it is bounded in $\mathbf{SM_U}$ (see (Metivier and Pellaumail, 1977)).

THEOREM 3.2a. *The set E is relatively compact in $\mathbf{SM_W}$, iff it is bounded in $\mathbf{SM_U}$.*

The assertion "if" is just reformulation of Theorem 3 (Trofimov, 1989). The direct Prokhorov's theorem, the property of compactness of embeddings $H^{s'} \subset H^s$, $s < s' < 1/2$, together with $\mathcal{L}^0$–boundedness assertion from Proposition 2.2 were used there. The assertion "only if" may be proved easily by contradiction.

COROLLARY 3.2b. *Any set* $E \in \mathbf{M}^1_+(D\widetilde{H})$ *is relatively compact in* $\mathbf{SM}^1_{+,\mathbf{W}} \equiv \equiv \mathbf{M}^1_+(D\widetilde{H}) \cap \mathbf{SM}_\mathbf{W}$, *iff it is bounded in* $\mathbf{SM}_\mathbf{U}$.

Remark. The theorem admits several extensions and improvements. For example, the theorem is true for quasi-normed spaces $B^s_{p,q}$, $-\infty < s < +\infty$, $0 < p,q < +\infty$ of Besov—Triebel (Triebel, 1986) instead of H^s_2. Denoting $DB^0_{0,0} \equiv (D, \widetilde{\lambda})$, we complete the picture, because boundedness assumption in $\mathbf{SM}_\mathbf{U}$ is exactly Stricker's condition (Stricker, 1985), which assures relative compactness of E in class $\mathbf{SM}$ with respect to the weak topology on the space $C(D, \widetilde{\lambda})$. Similarly, one could consider the stochastic duality with the random functional in $\mathcal{L}^0(\widehat{\Omega}, \mathbf{Q}; D\widetilde{H}) : (\mathbf{Q}, F) \rightsquigarrow F \cdot \widehat{X}$ — indefinited integral.

3.3. *Predictable criterias*

Let $X^n \in \mathcal{S}^{0,(n)}$ be given semimartingales on certain bases $\beta^n = (\Omega^n, \mathbf{F}^n, (\mathbf{F}^n_t)_{[0,1]}, \mathbf{P}^n)$, $n \in \mathbf{N}$ under usual assumptions; and let $X^n = V^n + M^n$, where $V^n \in \mathcal{V}^{(n)}$, $M^n \in \mathcal{M}^{2,(n)}_{0,\mathrm{loc}}$, $n \geqslant 1$, be some decompositions of them.

THEOREM 3.3a (Trofimov, 1989). *If the set of r.v.'s* $\{\mathrm{Var}(V^n) + \langle M^n \rangle_1\}_{n\in\mathbf{N}}$ *is* $\mathcal{L}^0$*-bounded, then it is bounded in* $\mathbf{SM}_\mathbf{U}$.

Now, let (B^n, C^n, ν^n) be triplet of predictable local characteristics of X^n, respectively, for certain odd, finite restriction function h. Introduce two assumptions, which seem to be a simplifications of the well-known **Sk**–tightness criteria of Jacod–Mémin–Metivier (Jacod et al., 1983):

$$\text{(A)} \quad \lim_{a\to\infty} \sup_n \mathbf{P}^n \left\{ |X^n_0| + \mathrm{Var}(B^n) + C^n_1 + \nu^n\big([0,1] \times (u^2 \wedge 1)\big) > a \right\} = 0.$$

$$\text{(B)} \quad \lim_{a\to\infty} \sup_n \mathbf{P}^n \left\{ \nu^n\big([0,1] \times \{|u| > a\}\big) > \varepsilon \right\} = 0 \quad \text{for each } \varepsilon > 0.$$

Now we shall show, that (A)+(B) implies $\widetilde{\tau}$-compactness of $\{\mathbf{L}(X^n)\}$. Let $X^{n,c}$ be the continuous martingale part of semimartingale X^n, and

$$U^n = \sum_{0<s\leqslant\cdot} (\Delta X^n_s - h(\Delta X^n_s)) \in \mathcal{V}^{d,(n)}_0, \quad Y^n \equiv X^n - X^n_0 - U^n \in \mathcal{S}^{2,(n)}_{0,\mathrm{loc}},$$

$$M^{d,n} = Y^n - B^n - X^{n,c} \in \mathcal{M}^{2,d,(n)}_{0,\mathrm{loc}}.$$

Using the previous theorem and linearity of $D\widetilde{H}$, it is easy to show, that:

(*) the assumption (A) on the set $\{\nu^n\}$ implies boundedness property of $\{\mathbf{L}(M^{d,n})\}_{n\in\mathbf{N}}$ in $\mathbf{SM}_\mathbf{U}$; and so, the assumption (A), in whole, implies boundedness property of $\{\mathbf{L}(X^n - U^n)\}_{n\in\mathbf{N}}$ in $\mathbf{SM}_\mathbf{U}$.

Using the Lenglart—Rebolledo inequality, as above in Proposition 2.2, one may show, that:

(**) assumptions (A) on measures ν^n with (B) imply $\mathcal{L}^0$–boundedness of the set $\{\mathrm{Var}(U^n)\}_{n\in\mathbf{N}}$, and so, by Theorem 3.3a imply boundedness of $\{\mathbf{L}(U^n)\}_{n\in\mathbf{N}}$ in the space $\mathbf{SM}_\mathbf{U}$.

Combining the conclusions (*) and (**), and Corollary 3.2b, we obtain.

THEOREM 3.3b. *The assumptions* (A) *and* (B) *imply the relative compactness of set* $\{\mathbf{L}(X^n)\}_{n\in\mathbb{N}}$ *in* $\mathbf{SM}^1_{+,\mathbf{W}}$.

The claim c) follows from lower semi-continuity of variation in 1.3, c).

COROLLARY 3.3c. *Let* $X^{(n)} \in \mathcal{V}^{(n),+}$, $n \geqslant 1$, *then the set* $\{\mathbf{L}(V^n)\}_{n\in\mathbb{N}}$ *is relatively compact in* $(\mathbf{M}^1_+(G), w_{\tilde{\tau}})$, *iff the set* $\{V^n_1\}_{n\in\mathbb{N}}$ *is* $\mathcal{L}^0$*-bounded.*

COROLLARY 3.3d. *Let* $X^n \in \mathcal{M}^{2,(n)}_{0,\mathrm{loc}}$, $|\Delta X^n| < c$, $n \geqslant 1$, *then the set* $\{\mathbf{L}(X^n)\}_{n\in\mathbb{N}}$ *is relatively compact in* $\mathbf{SM}^1_{+,\mathbf{W}}$, *iff the set* $\{\langle X^n\rangle_1\}_{n\in\mathbb{N}}$ *is* $\mathcal{L}^0$*-bounded.*

The claim 3.3d is essentialy specification of Corollaries 3.2b and 3.3c for martingales and their accompaining processes $\langle X^n\rangle$.

4. LIMIT THEOREMS IN EXAMPLES

We omit here all precise formulations on general functional limit theorems (FLTh's) for to be not out of frames of the work.

4.1. *Generalities*

Consider, at first, the method of proof of FLTh's by sheme: "tightness + weak finite dimensional convergence ⇒ weak functional convergence". Keeping in mind the Corollary 1.4e and combining the well-known theorems on finite dimensional convergence of processes, and **Sk**-limit theorems (see (Liptser and Shiryaev, 1986; Jacod and Shiryaev, 1987)) with further references) one may get from theorems of Section 3 a wide class of FLTh's on stable convergence to a process with conditionally independent increments with respect to topologies $\tilde{\tau}$ and $\tilde{\tau} \vee \mathbf{Sk}$. In particular, this way improved FLTh's for sums of triangular arrays of (dependent, in general) r.v.'s and vectors may bedrowed out (cf. ibid.). Using $\tilde{\tau}$-continuity of polynomial mappings one may obtain $\tilde{\tau}$-FLTh's for polynomial forms of vector processes, which are of impotance in numerous statistical applications.

On martingale method. It is well-known, that the most general approach to FLTh's gives so called martingale characterization method for Markov processes and semimartingales (see (Jacod and Shiryaev, 1987; Grigelionis et al., 1982; Ethier and Kurtz, 1986) for **Sk**-construction). Such kind theorems are obtained by the author for extended stable convergence (by Rényi and Aldous) built on topologies $\tilde{\lambda}$, $\tilde{\tau}$, **Sk**, $\tilde{\tau} \vee$ **Sk**, and some other. They are new even in the **Sk**-construction. Now, in contrast to the **Sk**-approach, the third important method of proofs for $\tilde{\tau}$-FLTh's consist in the use of the well developed *theory of summing of Banach space valued random variables*. That is the central LTh's, the laws of iterated logarithm, theorems on a rate of convergence (Araujo and Giné, 1980; Paulauskas

and Rachkauskas, 1987; Ledoux and Talagrand, 1988). In particular, a simple calculation shows, that the space H_2^s is a space of type and cotype 2.
So, we have got a lot of nice theorems on convergence and rates of convergence in case, when a sequence X^n of processes built up as partial sums of any anticipative processes. That is very typical in statistical applications (see Section 4.4).

4.2. *Coretraction limit theorem*

In a simple example we indicate now principial possibilities in boundary value problems with active stochastic processes at boundary manifolds, which offers the $\widetilde{\tau}$-approach. Avoiding technical complications we give an artificial setting of the problem here.
For integer $r \geqslant 0$ and any $\sigma \in (r+1/2,\ r+1)$ let $\widetilde{H}_2^{\sigma,2}$ be the factor-space of the standart space $H_2^s(\mathbf{R}^2)$ of Sobolev and Slobodetskii with respect to equivalence relation a.e. on $[0,1]^2$. Introduce product-spaces:

$$\Pi \equiv \prod_{j=0}^{r} \widetilde{H}_2^{\sigma-j-1/2}[0,1], \quad D\Pi \equiv \prod_{j=0}^{r-1} \widetilde{H}_2^{\sigma-j-1/2}[0,1] \times D\widetilde{H}_2^{\sigma-r-1/2}[0,1]$$

endowed with product-topologies (y.o).
It is known (Lions and Magenes, 1968; Besov et al., 1975; Triebel, 1986) that there exists a linear continuous and continuously inversible operator $\mathcal{R}$ (called as *retraction*, with inverse — as *coretraction*) acting from $\widetilde{H}_2^{\sigma,2}$ onto Π by the formula

$$(\mathcal{R}f)(u) = \left(f(u_1,0), \frac{\partial f}{\partial u_2}(u_1,0), \ldots, \frac{\partial^r f}{\partial u_2^r}(u_1,0) \right), \quad u_1 \in [0,1],$$

where u_i is the i-th increment of $u \in [0,1]^2$.
Consider any weakly convergent sequence of vector semimartingales

$$X^n = (X^{n,0}, \ldots, X^{n,r}) \to X^\infty \quad \text{(y.o.)}$$

on any given probabilistic bases with paths in $D\Pi$ (where $X^{n,r}$ may be a semimartingale of general kind, in accordance with the previous results). Consider random fields $Y^n \equiv \mathcal{R}^{-1}(X^n)$, $n = 1, 2, \ldots \infty$ with path in $\widetilde{H}_2^{\sigma,2}$.

PROPOSITION. $\mathbf{L}(Y^n) \to \mathbf{L}(Y^\infty)$ *in* $\mathbf{M}_+^1\left(\widetilde{H}_2^{\sigma,2}\right)$ *(y.o.)*

4.3. *Discrete space-time approximations*

Let a semimartingale $X \in \mathcal{S}^0$ on any base $\beta = (\Omega,\ \mathbf{F},\ (\mathbf{F}_t)_{[0,1]},\ \mathbf{P})$ be given. Suppose (for brevity, only), that X is continuous in probability. Consider any sequences of Markov shains (discrete times)

$$\left\{t_j^n : 0 \leqslant j \leqslant J^n < \infty\right\}, \quad 0 = t_0^n \leqslant t_1^n \leqslant \ldots \leqslant t_{J^n}^n = 1, \quad n \in \mathbf{N};$$

and consider adapted random interval partitions (discrete spaces)

$$A_i^{n,j} = \left(a_i^{n,j}, a_{i+1}^{n,j}\right], \quad i \in \mathbb{N}, \quad \bigcup_i A_i^{n,j} = \mathbf{R}$$

with random values $\alpha_i^{n,j} \in A_i^{n,j}$, where $a_i^{n,j}$, $\alpha_i^{n,j} \in \mathbf{F}_{t_j^n}$, $i, j, n \in \mathbb{N}$.
Suppose, that as $n \to \infty$

$$\delta^n \equiv \sup_{i,j}\left(a_{i+1}^{n,j} - a_i^{n,j}\right) \bigvee \sup_j \left(t_{j\,1}^n - t_j^n\right) \xrightarrow{\mathbf{P}} 0,$$

and for any r.v. ξ set $\xi_\omega^{(n,j)} \equiv \alpha_i^{n,j}$, iff $\xi_\omega \in A_i^{n,j}$.
Introduce discrete space–time approximations to X from the class $\mathcal{V}$:

$$X_t^n \equiv \sum_{j=0}^{J^n} \left(X_{t_j^n}\right)^{(n,j)} I_{[t_j^n, t_{j+1}^n)}(t), \quad 0 \leqslant t \leqslant 1.$$

PROPOSITION. $\mathbf{L}(X^n) \longrightarrow \mathbf{L}(X)$ *(y.o.) as* $n \to \infty$.

Proof. Since X is continuous in probability, $F \cdot X_1^n \xrightarrow{\mathbf{P}} F \cdot X_1$ for every $F \in \mathbf{\Pi}_{1,e}$ (the class of β–predictable step processes with values in $[-1, 1]$). In particular, $X_t^n \xrightarrow{\mathbf{P}} X_t$ for each t from certain set $U \in [0, 1]$ with $l(U) = 3$. So, it remains to prove compactness property for the set $\{\mathbf{L}(X^n)\}_{n \in \mathbb{N}}$.
The families $\{F \cdot X_1\colon\ F \in \mathbf{\Pi}_{1,e}\}$ and $\{F \cdot X_1^n\colon\ F \in \mathbf{\Pi}_{1,e}\}$ for each n are bounded. Hence $\{F \cdot X_1^n\colon\ F \in \mathbf{\Pi}_{1,e},\ n \in \mathbb{N}\}$ is $\mathcal{L}^0$–bounded, too, in virtue of the mentioned convergence in probability. By our $\widetilde{\tau}$–compactness criteria, the last boundedness property is enough.

4.4. *Statistical processes*

Consider an infinite sample of i.i.d. r.v.'s $\xi_1, \dots \xi_n, \dots$ with common distribution function F, supp$(F) \subset (0, 1)$. Let F^n be the standard empirical d.f. built upon the set $(\xi_i)_{i \leqslant n}$, and $s < 1/2$, as usually.

LEMMA 4.4a (Improved lemma of Glivenko and Cantelli). $F^n \to F$ *(a.s.) in the topology* $\mathbf{u} \vee \widetilde{\tau}$.

Proof. The family $\{F^n(\cdot\,, \omega)\}_{n \in \mathbb{N}}$ is compact in $(D\widetilde{H}, \widetilde{\tau})$ with probability 1 for $\mathrm{Var}(F^n) = 1$, and it remains to use the classical lemma.

THEOREM 4.4b (Improved Donsker's theorem). *The empirical process* $e_t^n \equiv$ $\equiv \sqrt{n}(F^n - F)$ *converges weakly to a centered Gaussian process with correlation function* $K(t, u) = F(t)(1 - F(u))$, $t \leqslant u$, *in topology* $w_{(\mathbf{u} \vee \widetilde{\tau})}$, *if* F *is continuous and in* $w_{(\widetilde{\tau} \vee \mathbf{Sk})}$, *if* F *is discontinuous.*

The assertion follows from the classical theorem (see (Billingsley, 1975)) and from the CLTh for the sums

$$e^n = n^{-1/2} \sum_{i=1}^{n} \eta_i,$$

where the summands $\eta_i \equiv (I_{(\xi_i,1]}(t) - F(t))$ are independent and uniformly bounded in $D\widetilde{H}$ (see (Araujo and Giné, 1980; Paulauskas and Rachkauskas, 1987; Ledoux and Talagrand, 1988) for the CLTh in spaces of type 2).
In this way one may obtain the $\widetilde{\tau}$- and $\mathbf{u} \vee \widetilde{\tau}$-compact LIL's for e^n (ibid). The $\widetilde{\tau}$-approach becomes as useful in study of quantile process.

Remark on modified empirical process. Let $(\xi_{1,n}, \ldots, \xi_{n,n})$, $n \geqslant 1$, be the order statistics, and let F^n be continuous broken-line increasing function with top-points $\widetilde{F}^n(0) = 0$, $\widetilde{F}^n(\xi_{i,n}) = i/(n+1)$, $i \leqslant n$, $\widetilde{F}^n(1) = 1$. And let $\widetilde{e}^n_\cdot \equiv \sqrt{n}(\widetilde{F}^n_\cdot - F_\cdot)$ be the modified empirical process (cf. (Billingsley, 1975, §13)). Then the improved lemma of Glivenko and Cantelli and the Donsker's theorem are true for $\widetilde{F}^n, \widetilde{e}^n$ instead of F^n, e^n, because $\|\widetilde{F}^n - F\|_u + \|F^n - F|\widetilde{H}\| \leqslant c/n$.

Modified empirical characteristic function. Let

$$\widetilde{\chi}^n_t = \frac{1}{\sqrt{2\pi}} \int_{\mathbf{R}} e^{itu} \widetilde{F}^n(du), \quad \widetilde{c}^n = \frac{1}{\sqrt{2\pi}} \int_{\mathbf{R}} e^{itu} \widetilde{e}^n(du), \ t \in \mathbf{R};$$

and let χ be the characteristic function of F. Since the mappings: $\widetilde{F}^n \to \widetilde{\chi}^n$, $\widetilde{e}^n \to \widetilde{c}^n$ from $D\widetilde{H}$ into $(H_2^s(\mathbf{R})_{\text{loc}}, \widetilde{\tau}_{\text{loc}})$ are continuous (a.s.), taking into account the remark and classical results (see (Csörgő, 1981), for example), we obtain: $\widetilde{\chi}^n \xrightarrow{\text{a.s.}} \chi(\widetilde{\tau}_{\text{loc}})$; $\widetilde{e}^n \to$ {the law of Gaussian Kiefer process} (in $w_{\widetilde{\eta}_{\text{loc}}}$); the $\widetilde{\tau}$-compact LIL for $\widetilde{e}^n$. Under some additional assumptions for d.f. F (see, ibid) we obtain: $\widetilde{\chi}^n \xrightarrow{\text{a.s.}} \chi((\mathbf{u} \vee \widetilde{\tau})_{\text{loc}})$, $\widetilde{e}^n \longrightarrow$ {the law of Kiefer process} (in $w_{(\mathbf{u} \vee \widetilde{\tau})_{\text{loc}}}$).
Integral type *test statistics* like Kramér—von Mises—Smirnov (Serfling, 1980) may be studied similarly, since they are $\widetilde{\tau}$-continuous functions over e^n.

REFERENCES

Araujo, A. and Giné, E. (1980). *The Central Limit Theorems for the Real and Banach Valued Random Variables.* Wiley, New York.

Besov, O.V., Il'in, V.P. and Nikolskii, S.M. (1975). *Integral Representations of Functions and Embedding Theorems.* Nauka, Moscow.

Bichteler, K. and Jacod, J. (1983). Random measures and stochastic integration. *Lect. Notes Contr. Inf. Sci.* **49**, 1–18.

Billingsley, P. (1975). *Convergence of Probability Measures.* Wiley, New York.

Csörgő, S. (1981). Limit behaviour of the empirical characteristic function. *Ann. Probab.* **9**, 138–144.

Dellacherie, C. and Meyer, P.A. (1980). *Probabilités et Potentiels. Vol. II,* Herman, Paris.

Ethier, S.N. and Kurtz, T.G. (1986). *Markov Processes. Characterization and Convergence.* Wiley, New York.

Grigelionis, B., Kubilius, K. and Mikulevičius, R. (1982). Martingale approach to functional limit theorems. *Usp. Mat. Nauk* **32**, 39–51.

Jacod, J. (1979). *Calcul Stochastique et Problemes de Martingales. Lect. Notes in Math.* **714**, Springer, Berlin–Heildelberg etc.

Jacod, J., Memin, J. and Metivier, M. (1983). On tightness and stopping times. *Stoch. Proc. Appl.* **14**, 109–146.

Jacod, J. and Shiryaev, A.N. (1987). *Limit Theorems for Stochastic Processes.* Springer, Berlin–Heidelberg etc.

Ledoux, M. and Talagrand, M. (1988). Characterization of the law of iterated logarithm in Banach spaces. *Ann. Probab.* **16**, 1242–1264.

Lions, J.L. and Magenes, E. (1968). *Problemes aux Limites non Homogénes et Applications. Vol. I,* Dunod, Paris.

Lions, J.L. and Magenes, E. (1972). *Problemes aux Limites non Homogénes et Applications. Vol. II,* Dunod, Paris.

Liptser, R.S. and Shiryaev, A.N. (1986). *Martingale Theory.* Nauka, Moskow.

Metivier, M. and Pellaumail, J. (1977). Measures stochastiques a valeurs dans des espaces L_0. *Z. Wahr. verw. Geb.* **40**, 101–114.

Meyer, P.A. (1976). Un cours sur les integrales stochastiques. *Lect. Notes in Math.* **511**, 245–400.

Meyer, P.A. and Zheng, W.A. (1984). Tightness criteria for laws of semimartingales. *Ann. Inst. H. Poincare* **20**, 353–372.

Mikulevichius, R. and Platen, E. (1988). Time discrete Taylor approximations for Ito processes with jumps component. *Math. Nachricht.* **138**, 93–104.

Paulauskas, V. and Rachkauskas, A. (1987). *Accuracy of Approximations in the Central Limit Theorems in Banach Spaces.* Mokslas, Vilnius.

Serfling, R.J. (1980). *Approximation Theorems of Mathematical Statistics.* Wiley, New York.

Stricker, C. (1985). Lois dé semimartingales et critéries de compacite. *Lect. Notes in Math.* **1123**, 209–217.

Triebel, H. (1986). *Theory of Function Spaces.* Birkhäuser, Berlin.

Trofimov, E.I. (1989). Semimartingales: L^2–construction of distributions. *Izvestia Vuzov. Matematika.* **2**, 68–78.

Yoeurp, C. (1978). Complements sur les temps locaux et les quasimartingales. *Asterisque* **52–53**, 197–218.

Prob. Theory and Math. Stat., Vol. 2, pp. 519–528
B.Grigelionis *et al.* (Eds.)
1990 VSP/Mokslas

CENTRAL LIMIT THEOREM FOR DEPENDENT RANDOM VARIABLES

S.A. UTEV
Institute of Mathematics, 630090 Novosibirsk, USSR

ABSTRACT

This paper is devoted to the central limit theorem (CLT) for sums of dependent random variables. We treat mainly the case of non–stationary sequences or triangular arrays of r.v.'s.

1. INTRODUCTION

The main difference of the results obtained in this paper from the known ones is that we do not impose conditions on the behavior of the variances of the sums under consideration. This fact allows us to study triangular arrays of strongly dependent r.v.'s such as m_n–dependent r.v.'s as $m_n \to \infty$.
Let $X = \{X_{1,n}, \dots, X_{k_n,n},\ n \geqslant 1\}$ be a triangular array of r.v.'s such that

$$\mathbf{E}X_{i,n} = 0, \quad \mathbf{E}X_{i,n}^2 < \infty$$

for all $n \geqslant 1, i = 1, \dots, k_n$. By $\sigma(X_i, i \in U)$ we denote the σ–algebra generated by the r.v.'s with indices in the set U. Let

$$S_n = \sum_{i=1}^{k_n} X_{i,n}, \quad \sigma_n^2 = \mathbf{E}S_n^2.$$

Throughout this paper we define

$$a \wedge b = \min(a, b), \quad X_{i,n} = 0,\ i > k_n, \quad \sum_{i \geqslant a}^{b} g_i = 0$$

for all $b < a$ and all g_i; $I(A)$ is the indicator function of an event A, $[x]$ is the integer part of x.
Section 2 of this work is a short survey on the CLT for the so–called φ–mixing triangular arrays and φ–mixing sequences of r.v.'s. It is followed in Section 3 by a discussion of applications to m_n–dependent r.v.'s and to functionals defined on Markov chains. In Section 4 we suggest an approach to the proof of the CLT for sums of dependent r.v.'s specified to the example of the so–called p–mixing

triangular arrays. Finally, Section 5 is a collection of inequalities for moments of the sums of dependent r.v.'s.

2. φ–MIXING TRIANGULAR ARRAYS

By

$$\varphi(\mathcal{F},\mathcal{G}) = \sup\left\{|\mathbf{P}(B|A) - \mathbf{P}(B)|,\ A \in \mathcal{F},\ B \in \mathcal{G},\ \mathbf{P}(A) > 0\right\}$$

we denote the φ–mixing coefficient. Put

$$\varphi_n(X,k) \equiv \varphi_n(k) = \sup\left\{\varphi\Big(\sigma(X_{i,n},\ i \leqslant s), \sigma(X_{i,n}, i \geqslant s+k)\Big),\quad s \geqslant 1\right\}$$

$$\varphi(k) = \sup_n \varphi_n(k), \quad k = 1,2,\dots\ .$$

The φ–mixing condition means that

$$\varphi(k) \to 0 \quad \text{as} \quad k \to \infty. \tag{2.1}$$

THEOREM 2.1. *Let the condition* (2.1) *be satisfied and assume that*

$$\sigma_n^{-2}\sum_{i=1}^{k_n}\mathbf{E}\left\{X_{i,n}^2 I(|X_{i,n}| > \varepsilon\sigma_n)\right\} \to 0$$

as $n \to \infty$, for every $\varepsilon > 0$. Then the CLT *holds, i.e.,*

$$S_n/\sigma_n \xrightarrow{\omega} \eta,$$

where η has the standard normal distribution.

Consider now the so–called φ–mixing sequences of r.v.'s.
Let $X = \{X_n, n \geqslant 1\}$ be a sequence of r.v.'s. We may treat it as a triangular array with

$$X_{i,n} = X_i,\ i = 1,\dots,n,\ k_n = n,\ n = 1,2,\dots\ .$$

The φ–mixing sequences of r.v.'s are the ones which satisfy condition (2.1).

COROLLARY 2.2. *Let* (2.1) *be satisfied and*

$$\sigma_n^{-2}\sum_{i=1}^{n}\mathbf{E}\left\{X_i^2 I(|X_i| > \varepsilon\sigma_n)\right\} \to 0$$

as $n \to \infty$, for every $\varepsilon > 0$. Then the CLT *holds.*

The φ–mixing stationary sequences are investigated more completely as well as φ–mixing sequences close to stationary ones, if the following supplementary condition is satisfied $\sigma_n^2 = nh(n)$ (see (Ibragimov, 1962; Peligrad, 1985)).

3. m_n–DEPENDENT R.V.'S AND FUNCTIONALS DEFINED ON MARKOV CHAINS

A) *m_n–dependent r.v.'s.* The condition of m_n–dependence means that the σ–algebras $\sigma(X_{i,n},\ i \leqslant s)$ and $\sigma(X_{i,n},\ i \geqslant s + m_n + 1)$ are independent for every s'. In general, this condition does not imply (2.1). But, if we define

$$Y_{k,n} = X_{(k-1)(m_n+1)+1} + \ldots + X_{k(m_n+1)},$$

$$r_n = \left[n/(m_n+1)\right] + 1, \quad k = 1, \ldots, r_n,$$

then $Y = \{Y_{1,n}, \ldots, Y_{r_n,n},\ n \geqslant 1\}$ will be a φ–mixing triangular array of r.v.'s with zero means and finite second moments. Thus from Theorem 2.1 and Lemma 5.3 (see Section 5) we obtain the following result.

COROLLARY 3.1. *Let*

$$(m_n+1)\sigma_n^{-2} \sum_{i=1}^{k_n} \mathbf{E}\Big\{X_{i,n}^2 I(|X_{i,n}| > \varepsilon\sigma_n(m_n+1)^{-1})\Big\} \to 0$$

as $n \to \infty$ for every $\varepsilon > 0$. Then the CLT *holds.*

The result above extends the results from (Orey, 1958; Berk, 1973).

B) *Functionals defined on Markov chains.* Let $X_{i,n} = f_{i,n}(Z_{i,n})$, where

$$\{Z^n,\ n \geqslant 1\} \equiv \{Z_{1,n}, \ldots, Z_{k_n,n};\ Q_n,\ n \geqslant 1\}$$

is a sequence of Markov chains with the state space Q_n and let $f_{1,n}(q_1,\omega), \ldots, f_{k_n,n}(q_{k_n},\omega)$ be independent random processes which are also independent of the chain $Z^{(n)}$ for each n.

Suppose that

$$\varphi_n(Z^{(n)}, k) \leqslant g(\alpha_n k), \tag{3.1}$$

where $g(x)$ is a non–negative function, $g(x) \to 0$ as $x \to \infty$, $\alpha_n > 0$ and $\varphi_n(Z^{(n)}, k)$ is the φ–mixing coefficient constructed according to the chain $Z^{(n)}$.

As above condition (3.1) does not in general imply (2.1) and we use it in an analogous way.

COROLLARY 3.2. *Let* (3.1) *be satisfied and*

$$\alpha_n^{-1}\sigma_n^{-2} \sum_{i=1}^{k_n} \mathbf{E}\Big\{X_{i,n}^2 I(|X_{i,n}| > \varepsilon\sigma_n\alpha_n)\Big\} \to 0$$

as $n \to \infty$, for every $\varepsilon > 0$. Then the CLT *holds.*

This result extends the results from (Dobrushin, 1956). Proofs of the results presented in Sections 2 and 3 can be found in (Utev, 1985; 1989).

4. ρ–MIXING TRIANGULAR ARRAYS

Let

$$\rho(\mathcal{F},\mathcal{G}) = \sup\left\{|\mathrm{corr}(X,Y)| : X \in L_2(\mathcal{F}), Y \in L_2(\mathcal{G})\right\},$$

$$\rho(k) = \sup\left\{\rho(\sigma(X_{i,n}, i \leqslant s), \sigma(X_{i,n}, i \geqslant s+k)); s \geqslant 1, n \geqslant 1\right\}.$$

The ρ–mixing condition means that

$$\rho(k) \to 0 \quad \text{as} \quad k \to \infty. \tag{4.1}$$

THEOREM 4.1. *Let*

$$\sum_{k=1}^{\infty} \rho(2^k) < \infty, \tag{4.2}$$

$$\liminf_{n\to\infty} \sum_{i=1}^{\sigma_n^2/k_n} \mathbf{E}X_{i,n}^2 > 0, \tag{4.3}$$

$$\sigma_n^{-2} \sum_{i=1}^{k_n} \mathbf{E}\left\{X_{i,n}^2 I(|X_{i,n}| > \varepsilon\sigma_n)\right\} \to 0 \tag{4.4}$$

as $n \to \infty$, for every $\varepsilon > 0$. Then the CLT *holds.*

There is a well–known example of strictly stationary ρ–mixing sequence of r.v.'s satisfying conditions (4.3), (4.4) and $\rho(k) = O(1/\ln k)$ (i.e., (4.1) is valid but (4.2) is violated) for which the CLT does not hold.

Let

$$B_n^2 = \mathbf{E}X_{1,n}^2 + \ldots + \mathbf{E}X_{k_n,n}^2.$$

COROLLARY 4.2. *Suppose that*

$$\sum_{k=1}^{\infty} \rho(2^k) < \infty,$$

$$\limsup_{n\to\infty} \max_{i\leqslant k_n-m} \mathrm{corr}(X_{i,n}, X_{i+m,n}) = 0, \quad \forall m = 1,2,\ldots,$$

$$B_n^{-2} \sum_{i=1}^{k_n} \mathbf{E}\{X_{i,n}^2 I(|X_{i,n}| > \varepsilon B_n)\} \to 0$$

as $n \to \infty$, for every $\varepsilon > 0$. Then

$$\lim_{n\to\infty} \sigma_n^2/B_n^2 = 1 \tag{4.5}$$

and the CLT *holds.*

The last corollary is an answer to the question of R. Bradley (raised in a private conversation). Such problems arise in the context of kernel type estimators for

probability density in the case of dependent observations (see (Bradley, 1983)). The assertion (4.5) is the corollary of Theorem 5.4 (see Section 5).
See also (Ibragimov, 1975; Peligrad, 1986) for related questions.

Proof of Theorem 4.1. The first step is the standard procedure of truncation. By (4.4) we find a sequence of positive numbers $\{\varepsilon_n, n \geqslant 1\}$ such that $\varepsilon_n \to 0$,

$$\sigma_n^{-2} \sum_{i=1}^{k_n} \mathbf{E}\Big\{X_{i,n}^2 I(|X_{i,n}| > \varepsilon_n \sigma_n)\Big\} \to 0$$

as $n \to \infty$. By Theorem 5.4 (see Section 5) we have

$$\sigma_n^{-2}\mathbf{E}\left(\sum_{i=1}^{k_n}\left(X_{i,n}I(|X_{i,n}| > \varepsilon_n\sigma_n) - \mathbf{E}\{X_{i,n}I(|X_{i,n}| > \varepsilon_n\sigma_n)\}\right)\right)^2 \leqslant$$

$$\leqslant c\sigma_n^{-2} \sum_{i=1}^{k_n} \mathbf{E}\Big\{X_{i,n}^2 I(|X_{i,n}| > \varepsilon_n \sigma_n)\Big\} \to 0$$

as $n \to \infty$. Therefore, without loss of generality, we may assume instead of (4.3), (4.4) that

$$\sigma_n \equiv 1, \quad \sum_{i=1}^{k_n} \mathbf{E}|X_{i,n}|^3 \to 0 \quad \text{as } n \to \infty, \tag{4.6}$$

$$\limsup_{n\to\infty} \sum_{i=1}^{k_n} \mathbf{E}X_{i,n}^2 < \infty.$$

The next step is a modification of the method of composition for characteristic function. Let

$$d_t(X, Y) = \Big|\mathbf{E}e^{itX} - \mathbf{E}e^{itY}\Big|.$$

Then it is enough to prove that for all t,

$$d_t\left(\sum_{i=1}^{k_n} X_{i,n}, \eta\right) \to 0$$

as $n \to \infty$, where η has the standard normal distribution. Now we obtain a recurrent inequality for the considered distance.
We describe first some simple properties of the introduced distance. Let X, X_1, X_2, Y_1, Y_2 be random variables with zero means and finite second moments. We assume that the r.v.'s Y_1, Y_2 are independent and that the distribution of X_j coincides with that of Y_j, $j = 1, 2$. Put

$$A_t(X) = d_t(X, \eta\mathbf{E}X^2).$$

LEMMA 4.3. *The following inequalities are valid:*

$$A_t(X) \leqslant \frac{2}{3}|t|^3\mathbf{E}|X|^3,$$

$$A_t(Y_1+Y_2) \leqslant A_t(Y_1)+A_t(Y_2),$$

$$d_t(X_1+X_2,X_1) \leqslant \frac{t^2}{2}\Big(\mathbf{E}X_2^2+(\mathbf{E}X_1^2\mathbf{E}X_2^2)^{1/2}\Big),$$

$$d_t(\eta a,\eta b) \leqslant \frac{t^2}{2}|a^2-b^2|,$$

$$d_t(X_1+X_2,Y_1+Y_2) \leqslant 8t^2\rho(\sigma(X_1),\sigma(X_2))(\mathbf{E}X_1^2\mathbf{E}X_2^2)^{1/2}.$$

Proof of Lemma 4.3. Let us deduce, e.g., the last inequality (the others are simpler). We have

$$d_t(X_1+X_2,Y_1+Y_2) = \big|\mathbf{E}(e^{itX_1}-1)(e^{itX_2}-1)-$$

$$-\mathbf{E}(e^{itX_1}-1)\mathbf{E}(e^{itX_2}-1)\big| \leqslant$$

$$\leqslant 8\rho\Big(\sigma(X_1),\sigma(X_2)\Big)\Big(\mathbf{E}|e^{itX_1}-1|^2\mathbf{E}|e^{itX_2}-1|^2\Big)^{1/2} \leqslant$$

$$\leqslant 8t^2\rho(\sigma(X_1),\sigma(X_2))(\mathbf{E}X_1^2\mathbf{E}X_2^2)^{1/2}.$$

Thus lemma is proved.
It is convenient to write the main inequality in the form of a separate lemma.

LEMMA 4.4. *The following inequality is valid*

$$A_t\left(\sum_{i=a+1}^{a+v} X_{i,n}\right) \leqslant c_1|t|^3h^6\sum_{i=a+1}^{a+v}\mathbf{E}|X_{i,n}|^3+$$

$$+c_2c(\rho)\left(h^{-1/9}+\sum_{k:2^k\geqslant h}\rho(2^k)\right)\sum_{i=a+1}^{a+v}\mathbf{E}X_{i,n}^2,$$

where

$$c(\rho)=\exp\left(c_0\left(1+\sum_{k=1}^{\infty}\rho(2^k)\right)\right),$$

c_i are absolute constants, h is an arbitrary natural number.

Before proving this lemma we finish the proof of Theorem 4.1. By Lemma 4.4 and (4.6) we have

$$d_t\left(\sum_{i=1}^{k_n}X_{i,n},\eta\right) = A_t\left(\sum_{i=1}^{k_n}X_{i,n}\right) \leqslant c_3(\rho,t)\left(h^6\sum_{i=1}^{k_n}\mathbf{E}|X_{i,n}|^3+\delta(h)\right),$$

where

$$\delta(h)=h^{-1/9}+\sum_{k:2^k\geqslant h}\rho(2^k)\to 0$$

as $h\to\infty$, by condition (4.2).

Consequently

$$\inf_{h\geqslant 1}\left(h^6\sum_{i=1}^{k_n}\mathbf{E}|X_{i,n}|^3+\delta(h)\right)\to 0 \text{ as } n\to\infty,$$

since

$$\sum_{i=1}^{k_n}\mathbf{E}|X_{i,n}|^3\to 0 \text{ as } n\to\infty.$$

Now we shall prove Lemma 4.4.

Proof of Lemma 4.4. Let $v\leqslant 4^9h^3$. By Lemma 4.3, we have

$$A_t\left(\sum_{i=a+1}^{a+v}X_{i,n}\right)\leqslant\frac{2}{3}|t|^3\mathbf{E}\left|\sum_{i=a+1}^{a+v}X_{i,n}\right|^3\leqslant c_5h^6\sum_{i=a+1}^{a+v}\mathbf{E}|X_{i,n}|^3$$

since $|x|^3$ is a convex function.
Let now $v>4^9h^3$. Without loss of generality assume that $a=0$, and define

$$m=[v^{1/3}], B=\{u:2^{-1}v(1-v^{-1/3})\leqslant um\leqslant 2^{-1}v\},$$

$$A=\left\{u:0\leqslant u\leqslant v,\sum_{i=um+1}^{(u+1)m}\mathbf{E}X_{i,n}^2\leqslant(m/v)^{1/3}\sum_{i=1}^{v}\mathbf{E}X_{i,n}^2\right\}.$$

By the construction, there is point $u\in A\cap B$ (see (Utev, 1989)).
We start from the following simple identity:

$$Q\equiv\sum_{i=1}^{v}X_{i,n}=\sum_{i=1}^{um}X_{i,n}+\sum_{i=um+1}^{(u+1)m}X_{i,n}+\sum_{i=(u+1)m+1}^{v}X_{i,n}\equiv$$

$$\equiv Q_1+Q_2+Q_3. \tag{4.7}$$

By Lemma 4.3

$$d_t(Q,Q_1+Q_3)=d_t(Q,Q-Q_2)\leqslant\frac{t^2}{2}\left(\mathbf{E}Q_2^2+(\mathbf{E}Q_2^2\mathbf{E}Q^2)^{1/2}\right).$$

By Theorem 5.4 (see Section 5)

$$\mathbf{E}\left(\sum_{i=r+1}^{r+u}X_{i,n}\right)^2\leqslant c(\rho)\sum_{i=r+1}^{r+u}\mathbf{E}X_{i,n}^2. \tag{4.8}$$

Thus

$$d_t(Q,Q_1+Q_3)\leqslant c_6c(\rho)t^2v^{-1/9}\sum_{i=1}^{v}\mathbf{E}X_{i,n}^2. \tag{4.9}$$

Given the r.v.'s Q_1,Q_3 we define two independent r.v.'s g_1,g_3 such that the distribution of g_i coincides with that of Q_i, $i=1,3$. Lemma 4.3 and inequality (4.8) yield

$$d_t(Q_1+Q_3,g_1+g_3)\leqslant 8\rho(m)t^2(\mathbf{E}Q_1^2\mathbf{E}Q_3^2)^{1/2}\leqslant$$

$$\leqslant c_7 t^2 c(\rho)\rho(m)\sum_{i=1}^{v}\mathbf{E}X_{i,n}^2, \tag{4.10}$$

$$A_t(g_1+g_3)\leqslant A_t(g_1)+A_t(g_3). \tag{4.11}$$

Finally, by the same lemma and the construction,

$$d_t\big(\eta,\eta\mathbf{E}(g_1+g_3)^2\big)\leqslant\frac{t^2}{2}\Big|1-\mathbf{E}(g_1+g_3)^2\Big|\leqslant$$

$$\leqslant c_8 t^2 c(\rho) v^{-1/9}\sum_{i=1}^{v}\mathbf{E}X_{i,n}^2. \tag{4.12}$$

Combining estimates (4.9)—(4.12), we get the following recurrent inequality:

$$A_t\left(\sum_{i=1}^{v}X_{i,n}\right)\leqslant A_t\left(\sum_{i=1}^{um}X_{i,n}\right)+A_t\left(\sum_{i=(u+1)m+1}^{v}X_{i,n}\right)+$$

$$+c_9 t^2 c(\rho)\Big(v^{-1/9}+\rho([v^{1/3}])\Big)\sum_{i=1}^{v}\mathbf{E}X_{i,n}^2$$

for $v>4^9h^3$.

The last step is an induction reasoning which is omitted. Lemma 4.4 and hence Theorem 4.1 are proved.

5. BOUNDS FOR MOMENTS

Let $\{X_n : n\geqslant 1\}$ be a sequence of r.v.'s and $\varphi(k)$, $\rho(k)$ be the φ–mixing, ρ–mixing coefficients constructed according to these r.v.'s.

Let

$$S_n = X_1+\ldots+X_n,\quad n=1,2,\ldots\ .$$

The following Theorem 5.1 and Lemmas 5.2, 5.3 are useful tools in the proofs of the CLT under the φ–mixing condition.

THEOREM 5.1. *Let $\mathbf{E}X_i=0$, $i=1,\ldots,n$, and $\varphi(p)\leqslant 1/5$ for some natural p. Then for any $t\geqslant 2$ the inequality*

$$\mathbf{E}\max_{1\leqslant k\leqslant n}|S_k|^t\leqslant (c_{10}t)^t\Big(p^t\mathbf{E}\max_{1\leqslant i\leqslant n}|X_i|^t+(\mathbf{E}S_n^2)^{t/2}\Big).$$

holds

LEMMA 5.2. *Let $\varphi(p)<1/2$ for some natural p. Then*

$$\mathbf{E}\left(\sum_{i=1}^{n}|X_i|\right)^2\leqslant c(\varphi,p)\left(\sum_{i=1}^{n}\mathbf{E}X_i^2+\left(\sum_{i=1}^{n}\mathbf{E}|X_i|\right)^2\right).$$

LEMMA 5.3. *For any* $\varepsilon \geqslant 0$

$$\mathbf{E}\{S_n^2 I(|S_n| > \varepsilon n)\} \leqslant 2n \sum_{i=1}^{n} \mathbf{E}\{X_i^2 I(|X_i| > \varepsilon/2)\}.$$

The following Theorems 5.4, 5.5 are useful in proving the CLT under the ρ–mixing condition. Let

$$\sum_{k=1}^{\infty} \rho(2^k) < \infty, \ \mathbf{E}X_i = 0, \ i = 1, 2, \dots .$$

THEOREM 5.4. *The following inequalities are valid*

$$\mathbf{E}\left|\sum_{i=1}^{n} X_i\right|^2 \leqslant c(\rho) \sum_{i=1}^{n} \mathbf{E}X_i^2,$$

$$\left|\mathbf{E}\left(\sum_{i=1}^{n} X_i \sum_{j=i+h}^{n} X_j\right)\right| \leqslant c_1 c(\rho) \left(\sum_{k:2^k \geqslant h} \rho(2^k)\right) \sum_{i=1}^{n} \mathbf{E}X_i^2,$$

where

$$c(\rho) = \exp\left(c_0\left(1 + \sum_{k=1}^{\infty} \rho(2^k)\right)\right),$$

c_i *are absolute constants,* h *is an arbitrary natural number.*

THEOREM 5.5. *Let* $\rho(1) < 1$. *Then*

$$\mathbf{E}\left|\sum_{i=1}^{n} X_i\right|^2 \geqslant c \sum_{i=1}^{n} \mathbf{E}X_i^2,$$

where $c > 0$ *depends only on the sequence* $\{\rho(k), k \geqslant 1\}$.

The last result extends the results of (Bradley, 1988).
Proof of the above inequalities, except Theorem 5.5, are presented in (Utev, 1985; 1989).

REFERENCES

Berk, K.N. (1973). A central limit theorem for m–dependent random variables with unbounded m. *Ann. Probab.* 1, 352–354.

Bradley, R.C. (1983). Asymptotic normality of some kernel–type estimators of probability density. *Statist. Probab. Letters* 1, 295–300.

Bradley, R.C. (1988). A central limit theorem for stationary ρ–mixing sequences with infinite variance. *Ann. Probab.* 16, 313–332.

Dobrushin, R.L. (1956). Central limit theorem for non–stationary chains. I, II. *Theory Probab. Appl.* 1, 65–80, 329–383.

Ibragimov, I.A. (1962). Some limit theorem for stationary processes. *Theory Probab. Appl.* 7, 349–382.

Ibragimov, I.A. (1975). A note on the central limit theorem for dependent random variables. *Theory Probab. Appl.* **20**, 135–141.

Orey, S. (1958). A central limit theorem for m–dependent random variables. *Duke Math. J.* **25**, 543–546.

Peligrad, M. (1985). An invariance principle for φ–mixing sequences. *Ann. Probab.* **13**, 1304–1313.

Peligrad, M. (1986). Recent advances in the central limit theorem and its weak invariance principle for mixing sequences of random variables. In: *Dependence in Probability and Statistics*, Birkhäuser, Basel—Stuttgart, pp. 193–223.

Utev, S.A. (1985). Inequalities for sums of weakly dependent random variables and estimates of the convergence rate in the invariance principle. In: *Limit Theorems for Sums of Random Variable*, Optimization Software, New York, pp. 73–114.

Utev, S.A. (1989). Sums of random variables with φ–mixing condition. In: *Asymptotical Analysis of Stochastic Processes Distributions*, Novosibirsk.

Prob. Theory and Math. Stat., Vol. 2, pp. 529–534
B.Grigelionis *et al.* (Eds.)
1990 VSP/Mokslas

RÉNYI DISTANCES OF SOME DIFFUSION PROCESSES

I. VAJDA
Institute of Information Theory and Automatization,
Pod vodárenskou věží 4, 182 08 Praha 8, Czechoslovakia.

ABSTRACT

The evaluation of the Rényi distances $R(a)$ of any order $a > 0$ for the distributions of the diffusion processes are obtained.

1. INTRODUCTION

We consider solutions $\mathbf{X}_j = (X_j(t)|0 \leqslant t \leqslant T)$, $T > 0$, of the linear stochastic differential equations

$$dX_j(t) = -\alpha_j(t)X_j(t)dt + \beta(t)dW(t), \quad j = 1, 2,$$

where $(W(t)|0 \leqslant t \leqslant T)$ is the standard Wiener process and $\alpha_j(t)$ and $\beta(t) > 0$ are continuous, with $X_j(0) \equiv \mathcal{N}(\mu_j, \sigma^2)$. We shall evaluate the Hellinger integrals $H(a)$ and the Rényi distances $R(a)$ of any order $a > 0$ for the distributions $P_{1,T}, P_{2,T}$ of the processes $\mathbf{X}_1, \mathbf{X}_2$.
In (Liese and Vajda, 1987) the Hellinger integral of order $a > 0$ for distributions P_1, P_2 with densities p_1, p_2 was assumed to be ∞ if $a > 1$ and, at the same time, P_1 is not absolutely continuous with respect to P_2. Otherwise it was defined by

$$H(a) = \int \chi_{\{p_1 p_2 > 0\}} p_1^a p_2^{1-a} d\mu,$$

where μ is the Radon—Nikodym dominating measure. The Rényi distance $R(a)$ has been defined on p. 37 of the same book, $a > 0$, $a \neq 1$; it is equal to

$$R(a) = -\frac{\ln H(a)}{a(1-a)}.$$

According to (2.14) from (Liese and Vajda, 1987, p. 38), the limit

$$R(1) = \lim_{a \uparrow 1} R(a)$$

always exists and is equal to the Kullback—Leibler's $I(P_1||P_2)$.

2. THE METHOD

The Rényi distances of diffusion processes have been recently evaluated in several papers (see (Vajda, 1990; Michalek, 1990) and some former papers cited there). However, the complete spectrum of the Rényi distances was so far obtained only for the case where $\mathbf{X}_2$ is a deterministic shift of $\mathbf{X}_1$, i.e., where $X_2(t) = X_1(t) + m(t)$ and m depends on t only. Our result is obtained by applying the same elementary method as in (Vajda, 1990) to the present situation where not only the mean values but also the covariance structure of the processes $\mathbf{X}_1, \mathbf{X}_2$ is to some extent different. The general case seems to be too complicated for this method. The method in the frame of the present paper reduces to the following three facts.

LEMMA 1. *Let $P_{j,n}$ for $j = 1,2$ be the distributions of the random vectors $(X_j(t_0), \ldots, X_j(t_n))$ with $t_i = iT/n$ and $H_n(a)$ and $R_n(a)$ be the corresponding Hellinger integrals and Rényi distances. If there exists*

$$\lim_{n\to\infty} H_n(a) \quad \text{or} \quad \lim_{n\to\infty} R_n(a)$$

then this limit equals $H(a)$ or $R(a)$, respectively.

Proof. The sequence $H_n(a)$ contains a monotone subsequence yielded by dyadically rational times from $[0,T]$. Due to the stochastic continuity of the processes $\mathbf{X}_1, \mathbf{X}_2$, the limit of this subsequence equals $H(a)$ (cf. the monotonicity theorem in (Liese and Vajda, 1987, p.17)). Thus $H(a)$ is the only possible limit of $H_n(a)$. Analogous property for the Rényi distances is an obvious consequence.

LEMMA 2. *Let $\psi_j(y)$ be the normal densities with real means μ_j and positive variances σ_j^2. Then their Hellinger integral*

$$\int_{\mathbf{R}} \psi_1^a(y)\, \psi_2^{1-a}(y) dy$$

equals to ∞ or

$$\left(\frac{\sigma_2^{2a}\sigma_1^{2(1-a)}}{a\sigma_2^2 + (1-a)\sigma_1^2}\right) \exp\left\{-\frac{1}{2}\frac{a(1-a)(\mu_2-\mu_1)^2}{a\sigma_2^2 + (1-a)\sigma_1^2}\right\},$$

depending on whether $a\sigma_2^2 + (1-a)\sigma_1^2 \leqslant 0$ or not.

Proof. See (2.24) in (Liese and Vajda, 1987, pp. 44–45).

LEMMA 3. *If in the previous lemma $\sigma_1 = \sigma_2 = \sigma$ and $\gamma \geqslant 0$, then*

$$\int_{\mathbf{R}} \psi_1^a(y)\psi_2^{1-a}(y) \exp\left\{-\frac{1}{2}\frac{\gamma y^2}{\sigma^2}\right\} dy$$

is either ∞ *or*

$$\frac{1}{\sqrt{1+\gamma}}\exp\left\{-\frac{1}{2\sigma^2}\left[\frac{\gamma}{1+\gamma}(a\mu_1^2+(1-a)\mu_2^2)+\frac{a(1-a)(\mu_1-\mu_2)^2}{1+\gamma}\right]\right\},$$

depending on whether $1+\gamma \leqslant 0$ *or not.*

Proof. The integrand here is a constant time the exponential function of the argument

$$-\frac{1}{2\sigma^2}\{a(y-\mu_1)^2+(1-a)(y-\mu_2)^2+\gamma y^2\}.$$

This is easily rewritten as

$$-\frac{1}{2\sigma^2}\left\{(1+\gamma)\left(y-\frac{a\mu_1+(1-a)\mu_2}{1+\gamma}\right)^2+[\cdot/.]\right\}$$

where $[\cdot/.]$ denotes the corresponding term figuring in Lemma 3. The rest is clear since the exponential function is normed by $(2\pi\sigma^2)^{-1/2}$.

3. THE RESULTS

THEOREM 1. *The Rényi distance* $R(a)$ *of the distributions of the processes* $\mathbf{X}_1, \mathbf{X}_2$ *is given for all* $a>0$ *by*

$$R(a)=\frac{(\mu_1-\mu_2)^2}{2\sigma^2}+\frac{1}{2}\int_0^T\left(\frac{\alpha_1(t)-\alpha_2(t)}{\beta(t)}\right)^2 dt. \tag{1}$$

Proof. It is well-known (cf. e.g., (Gikhman and Skorokhod, 1968)) that $\mathbf{X}_j$ are zero-mean Gauss—Markov processes with covariances for all $0\leqslant s\leqslant t\leqslant T$ defined by

$$R_j(s,t)=V_j(s)\exp\left\{-\int_s^t \alpha_j(u)du\right\}, \tag{2}$$

where the variances $V_j(t)=\mathbf{E}X_j^2(t)=R_j(t,t)$ are equal to

$$\exp\left\{-2\int_0^t \alpha_j(u)\,du\right\}\left\{\sigma^2+\int_0^t\left[2\int_0^s \alpha_j(u)\,du\right]\beta^2(s)\,ds\right\}. \tag{3}$$

It is also well-known that the transition probability densities $p_j(s,x;t,y)$ of the processes $\mathbf{X}_j$ are given for every $0\leqslant s<t\leqslant T$ and every $x,y\in\mathbf{R}$ by

$$\frac{1}{\sigma_j(s,t)}\varphi\left(\frac{y-\mu_j(s,t)x}{\sigma_j(s,t)}\right), \tag{4}$$

where φ is the standard normal density and

$$\mu_j(s,t) = \exp\left\{-\int\limits_s^t \alpha_j(u)\,du\right\},$$

$$\sigma_j^2(s,t) = V_j(t) - V_j(s)\mu_j^2(s,t). \tag{5}$$

By (3) and (5) we have for every $0 \leqslant s < T$ and every j

$$\lim_{t\downarrow s}\frac{1-\mu_j(s,t)}{t-s} = \alpha_j(s), \quad \lim_{t\downarrow s}\frac{\sigma_j^2(s,t)}{t-s} = \beta^2(s), \tag{6}$$

where the convergence is uniform with respect to s. By (4) the densities $p_j(x_0,\ldots,x_n)$ of the distributions $P_{j,n}$ considered in Lemma 1 are equal to

$$\frac{1}{\sigma}\varphi\left(\frac{x_0-\mu_j}{\sigma}\right\}\prod_{i=1}^n\frac{1}{\sigma_j(i,n)}\varphi\left(\frac{x_i-\mu_j(i,n)x_{i-1}}{\sigma_j(i,n)}\right), \tag{7}$$

where, for $t_i = iT/n$,

$$\mu_j(i,n) = \mu_j(t_{i-1},t_i), \quad \sigma_j(i,n) = \sigma_j(t_{i-1},t_i), \quad i=1,\ldots,n.$$

By (6) it follows from this that if $n\to\infty$ then, uniformly for all $i=1,\ldots,n$,

$$\mu_j(i,n) = 1-\frac{\alpha_j(t_{i-1})}{n}+o\left(\frac{1}{n}\right), \quad \sigma_j^2(i,n) = \frac{\beta^2(t_{i-1})}{n}+o\left(\frac{1}{n}\right). \tag{8}$$

Therefore, in the notation of Lemma 1,

$$H_n(a) = \int\limits_{\mathbf{R}^{n+1}} h_{x_0}\prod_{i=1}^n h_{x_{i-1},x_i}(i,n)\,dx_0\ldots dx_n, \tag{9}$$

where $h_{x_0} = \psi_1^a(x_0)\psi_2^{1-a}(x_0)$ with ψ_1,ψ_2 as in Lemma 3, and

$$h_{z,y}(i,n) = \psi_{1,i,n}^a(y)\psi_{2,i,n}^{1-a}(y),$$

where $\psi_{j,i,n}(y)$ stands for the normal density with mean $\mu_j(i,n)z$ and variance $\sigma_j^2(i,n)$.

Now we shall carry out step–by–step the integration in (9). It follows from Lemma 2 and from (8) that

$$\int\limits_{\mathbf{R}} h_{x_{n-1},x_n}(n,n)dx_n =$$

$$= \left(1+o\left(\frac{1}{n}\right)\right)\exp\left\{-\frac{a(1-a)}{2n}\frac{\left(\alpha^2(t_{n-1})+o_{n-1}(1/n)\right)x_{n-1}^2}{\beta^2(t_{n-1})+o_{n-1}(1)}\right\},$$

where

$$\alpha(t) = \alpha_1(t)-\alpha_2(t) \tag{10}$$

and where the variables $o(1/n), o_{n-1}(1/n)$ and $o_{n-1}(1)$ do not depend on x_{n-1}. Putting in Lemma 3

$$\sigma^2 = \frac{\beta^2(t_{n-2})}{n}+o\left(\frac{1}{n}\right), \quad \gamma = \frac{\sigma^2 a(1-a)}{n}\frac{\left(\alpha^2(t_{n-1})+o_{n-1}(1/n)\right)}{\beta^2(t_{n-1})+o_{n-1}(1)}$$

and using (8) and the last result we obtain

$$\int_{\mathbf{R}^2} h_{x_{n-2},x_{n-1}}(n-1,n)h_{x_{n-1},x_n}(n,n)dx_n dx_{n-1} =$$

$$= \left(1+o\left(\frac{1}{n}\right)\right)\exp\left\{-\frac{a(1-a)}{2n}\left[\frac{\alpha^2(t_{n-1})+o_{n-1}(1/n)}{\beta^2(t_{n-1})+o_{n-1}(1)}+\right.\right.$$

$$\left.\left.+\frac{\left(\alpha^2(t_{n-2})+o_{n-2}(1/n)\right)x_{n-2}^2}{\beta^2(t_{n-2})+o_{n-2}(1)}\right]\right\},$$

where the values $o(1/n)$ and $o_{n-1}(1/n), o_{n-1}(1)$ differ from those above but possess similar properties.

We see that the last Hellinger integral differs from the previous one by the factor

$$\exp\left\{-\frac{a(1-a)}{2n}\frac{\left(\alpha^2(t_{n-1})+o_{n-1}(1/n)\right)}{\beta^2(t_{n-1})+o_{n-1}(1)}\right\}$$

not depending on any argument from the integration domain in (9), except that t_{n-1}, x_{n-1} are to be replaced in the second factor by t_{n-2}, x_{n-2}. Therefore we see that, repeating the same step $n-2$ times, we obtain at last from (9)

$$H_n(a) = \left(1+o\left(\frac{1}{n}\right)\right)\exp\left\{-\frac{a(1-a)}{2n}\sum_{i=1}^{n}\frac{\alpha^2(t_i)+o_i(1/n)}{\beta^2(t_i)+o_i(1)}\right\}I_n(a),$$

where

$$I_n(a) = \int_{\mathbf{R}} h_{x_0}\exp\left\{-\frac{a(1-a)}{2n}\frac{\left(\alpha^2(t_{n-1})+o_0(1/n)\right)x_0^2}{\beta^2(t_{n-1})+o_0(1)}\right\}dx_0 =$$

$$= \int_{\mathbf{R}} h_{x_0}dx_0 + o(1)$$

so that, by Lemma 2,

$$I_n(a) = \exp\left\{-\frac{a(1-a)}{2}\frac{(\mu_1-\mu_2)^2}{\sigma^2}\right\}+o(1).$$

Therefore we may summarize the above considerations as follows

$$R_n(a) = \frac{(\mu_1-\mu_2)^2}{\sigma^2}+\frac{1}{2n}\sum_{i=1}^{n}\frac{\alpha^2(t_i)+o_i(1/n)}{\beta^2(t_i)+o_i(1)}+o(1).$$

Since all variables $o_i(1/n), o_i(1)$ tend to zero uniformly in i, the second right-hand term tends to the integral considered in the Theorem 1 (cf. (10)). The Theorem 1 thus holds by Lemma 1.

Denote

$$\rho_T = \int_0^T \left(\frac{\alpha_1(t) - \alpha_2(t)}{\beta(t)} \right)^2 dt.$$

The condition $\rho_T \to \infty$ as $T \to \infty$ is obviously necessary and sufficient for the asymptotic discernibility of the processes $\mathbf{X}_1, \mathbf{X}_2$ with zero error. To be more precise, consider the set Φ of all tests φ defined on the sample space of the processes $\mathbf{X}_1, \mathbf{X}_2$ and taking on values in $[0, 1]$. We are interested in the errors of testing the simple hypothesis that the observed process is $\mathbf{X}_1$ against the simple alternative that it is $\mathbf{X}_2$. Namely let

$$\beta_T = \inf \left\{ \int (1 - \varphi) dP_{2,T} \mid \varphi \in \Phi, \int \varphi dP_{1,T} \leqslant \alpha \right\}$$

be the Neyman—Pearson error and

$$\varepsilon_T = \inf \left\{ \alpha \int \varphi dP_{1,T} + (1 - \alpha) \int (1 - \varphi) dP_{2,T} \mid \varphi \in \Phi \right\}$$

be the Bayes error.

THEOREM 2. *If ρ_T tends to infinity as $T \to \infty$ then*

$$\lim_{T \to \infty} \beta_T^{1/\rho_T} = \exp\left\{ -\frac{1}{2} \right\}$$

and

$$\lim_{T \to \infty} \varepsilon_T^{1/\rho_T} = \exp\left\{ -\frac{1}{8} \right\}.$$

Proof. It follows from Theorem 1 that all the assumptions of the generalized theorems of Chernoff and Stein proved in Section 12 of (Vajda, 1989) hold. Thus the present asymptotic relations follow directly from there.

REFERENCES

Gikhman, I.I. and Skorokhod, A.V. (1968). *Stochastic Differential Equations.* Naukova Dumka, Kiev.

Liese, F. and Vajda, I. (1987). *Convex Statistical Distances.* Teubner, Leipzig.

Michalek, J. (1990). *I*-divergence of some diffusion processes. *Probl. Control and Inform. Theory.* (to appear).

Vajda, I. (1989). *Theory of Statistical Inference and Information.* Kluwer, Dordrecht.

Vajda, I. (1990). Distances and discrimination rates of stochastic processes. *Stoch. Processes Appl.* (to appear).

Prob. Theory and Math. Stat., Vol. 2, pp. 535–541
B.Grigelionis *et al.* (Eds.)
1990 VSP/Mokslas

ESTIMATES FOR THE SMOOTH FUNCTIONAL IN THE CLASSICAL CASE

G. VALIUKEVIČIUS
Vilnius, Lithuania

ABSTRACT

The distance between certain semimartingale and the solution of stochastic differential equation with smooth coefficients is estimated. The distance is calculated as difference of corresponding averages of smooth functionals over the space of Skorokhod trajectories.

For the given process X with values in G written by means of the stochastic integrals

$$X_{t_1} = X_{t_0} + \int_{t_0}^{t_1} a_t dt + \int_{t_0}^{t_1} b_t dW_t + \int_{t_0}^{t_1}\int h(q - \mathbf{1}_O(h)n_t^c)(dt, dh),$$

we calculate the distance to some "nice" process $Y_t \in G$

$$Y_{t_1} = Y_{t_0} + \int_{t_0}^{t_1} \bar{a}(Y_t)dt + \int_{t_0}^{t_1} \bar{b}(Y_t)dW_t + \int_{t_0}^{t_1}\int h(\bar{q}(Y_{t-}) - \mathbf{1}_O(h)\bar{n}_t^c(Y_{t-}))(dt, dh).$$

The distance is calculated as the average of smooth functionals F over the space of Skorokhod trajectories $D_{[0,T]}(G)$. The space G will be taken $\mathbf{R}$ only for the sake of simplicity. We demand the functional F to be continuous over the Skorokhod topology and to have bounded Frechet derivatives over the uniform convergence norm $\|x\| = \sup\limits_{0 \leqslant t \leqslant T} |x_t|$.

The constants for the bounds of derivatives will appear in the estimate of distance

$$|F(x)| \leqslant C_0, \quad \|F^{(k)}(x)\| = \sup_{\|h\| \leqslant 1} |F^{(k)}(x)(h)^k| \leqslant C_k, \quad 1 \leqslant k \leqslant 3 + n.$$

The nice process must be defined by the fields $\bar{a}, \bar{b}, \bar{q}$ over our space G. We can begin its trajectories from deliberately taken initial point and in this way we get the flow of the diffeomorphisms

$$\Phi_{t_2}^{t_1} : G \to G, \quad Y_t = \Phi_t^{t_0}(Y_{t_0}).$$

We demand the smoothness for the averaged test function. This function also may be random:

$$f_t(p) = \mathbf{E}\Big\{f(\Phi_T^t(p))\Big|F_t\Big\}, \quad p \in G.$$

With the bounds for the derivatives of diffeomorphisms

$$|\partial\Phi_{t_2}^{t_1}| \leqslant d_1, \ |\partial^2\Phi_{t_2}^{t_1}| \leqslant d_2, \ldots, |\partial^n\Phi_{t_2}^{t_1}| \leqslant d_n, \quad t_0 \leqslant t_1 \leqslant t_2 \leqslant T,$$

we shall have new bounds for the averaged test functions

$$\Big|\,\triangle\,\partial f_t(p)\Big| = \Big|\mathbf{E}\Big\{\,\triangle\,\Phi_T^t(p) f(\Phi_T^t(p))\Big|F_t\Big\}\Big| \leqslant \Delta d_1 \sup_{p\in G}|\partial f(p)|,$$

in the same way

$$\Big|\,\Delta^2\,\partial^2 f_t(p)\Big| \leqslant \Delta^2\Big(d_1^2 \sup_{p\in G}|\partial^2 f(p)| + d_2 \sup_{p\in G}|\partial f(p)|\Big),$$

and so on.
Let us denote new constants

$$\begin{aligned}
\bar{C}_0 &= C_0,\\
\bar{C}_1 &= d_1 C_1,\\
\bar{C}_2 &= d_2 C_1 + d_1^2 C_2,\\
\bar{C}_3 &= d_3 C_1 + 3d_2 d_1 C_2 + d_1^3 C_3,\\
\bar{C}_4 &= d_4 C_1 + 4d_3 d_1 C_2 + 6d_2 d_1^2 C_4 + d_1^4 C_4 + 3d_2^2 C_2,\\
\bar{C}_5 &= d_5 C_1 + 5d_4 d_1 C_2 + 10d_3 d_2 C_2 + 10d_3 d_1^2 C_3 + 10d_2 d_1^3 C_4 + d_1^5 C_5 +\\
&\quad + 15d_2^2 d_1 C_3,\\
\bar{C}_6 &= d_6 C_1 + 6d_5 d_1 C_2 + 15d_4 d_2 C_2 + 20d_3^2 C_2 + 15d_4 d_1^2 C_3 + 60d_3 d_2 d_1 C_3 +\\
&\quad + 15d_2^3 C_3 + 20d_3 d_1^3 C_4 + 45d_2^2 d_1^2 C_4 + 15d_2 d_1^4 C_5 + d_1^6 C_6.
\end{aligned}$$

We also demand some regularity properties of the average test function. Markov process Y_t provides such properties.
The estimate with the variational distance between the characteristics is

$$\Big|\mathbf{E}\{F(X) - F(Y)\}\Big| \leqslant$$

$$\leqslant \bar{C}_1\mathbf{E}\Big\{\int_0^T |a_t - \bar{a}(X_t)|dt + \sum_{0\leqslant t\leqslant T}\Big|\int_{h\in O} h(n_t^{\triangle} - \bar{n}_t^{\triangle}(X_{t-}))(dh)\Big|\Big\} +$$

$$+\frac{\bar{C}_2}{2!}\mathbf{E}\Big\{\int_0^T (b_t^2 - \bar{b}^2(X_t))dt + \int_{h\in O} h^2(n_t^c - \bar{n}_t^c(X_{t-}))(dh) +$$

$$+ \sum_{0\leqslant t\leqslant T}\Big|\int_{h\in O} h^2(n_t^{\triangle} - \bar{n}_t^{\triangle}(X_{t-}))(dh)\Big|\Big\} +$$

$$+\frac{\bar{C}_3}{3!}\mathbf{E}\Big\{\int_0^T\Big(\Big|\int_{h\in O} h^3(n_t^c-\bar{n}_t^c(X_{t-}))(dh)\Big|+$$

$$+\sum_{0\leqslant t\leqslant T}\Big|\int_{h\in O} h^3(n_t^\Delta-\bar{n}_t^\Delta(X_{t-}))(dh)\Big|\Big)dt\Big\}+$$

$$+\frac{\bar{C}_4}{4!}\mathbf{E}\Big\{\int_0^T\Big(\Big|\int_{h\in O} h^4(n_t^c-\bar{n}_t^c(X_{t-}))(dh)\Big|+$$

$$+\sum_{0\leqslant t\leqslant T}\Big|\int_{h\in O} h^4(n_t^\Delta-\bar{n}_t^\Delta(X_{t-}))(dh)\Big|\Big)dt\Big\}+$$

$$+2C_0\mathbf{E}\Big\{\int_0^T\Big(\Big|\int_{h\notin O}(n_t^c-\bar{n}_t^c(X_{t-}))(dh)\Big|+\sum_{0\leqslant t\leqslant T}\Big|\int_{h\notin O}(n_t^\Delta-\bar{n}_t^\Delta(X_{t-}))(dh)\Big|\Big)dt\Big\}.$$

The application of partial integration formula gives more sensible estimate. We denote

$$A_S^{(1)}=\int_0^S(a_t-\bar{a}(X_t))dt+\sum_{0\leqslant t\leqslant S}\int_{h\in O} h\Big(n_t^\Delta-\bar{n}_t^\Delta(X_{t-})\Big)(dh),$$

$$A_S^{(2)}=\int_0^S\Big(b_t^2-\bar{b}^2(X_t)\Big)dt+\int_0^S\int_{h\in O} h^2\Big(n_t^c-\bar{n}_t^c(X_{t-})\Big)(dt,dh)+$$

$$+\sum_{0\leqslant t\leqslant S}\int_{h\in O} h^2\Big(n_t^\Delta-\bar{n}_t^\Delta(X_{t-})\Big)(dh),$$

$$A_S^{(3)}=\int_0^S\int_{h\in O} h^3\Big(n_t^c-\bar{n}_t^c(X_{t-})\Big)(dt,dh)+\sum_{0\leqslant t\leqslant S}\int_{h\in O} h^3\Big(n_t^\Delta-\bar{n}_t^\Delta(X_{t-})\Big)(dh).$$

Then the estimate can be written for $\frac{1}{p}+\frac{1}{q}=1,\ \ 1\leqslant p\leqslant\infty$

$$\Big|\mathbf{E}\{F(X)-F(Y)\}\Big|\leqslant(C_1+C_1^*)\mathbf{E}\Big\{\sup_{0\leqslant t\leqslant T}|A_t^{(1)}|^p\Big\}^{1/p}+$$

$$+\frac{1}{2!}(C_2+C_2^*)\mathbf{E}\Big\{\sup_{0\leqslant t\leqslant T}|A_t^{(2)}|^p\Big\}^{1/p}+\frac{1}{3!}(C_3+C_3^*)\mathbf{E}\Big\{\sup_{0\leqslant t\leqslant T}|A_t^{(3)}|^p\Big\}^{1/p}+$$

$$+\frac{\bar{C}_4}{4!}\mathbf{E}\Big\{\int_0^T\Big(\Big|\int_{h\in O} h^4\Big(n_t^c-\bar{n}_t^c(X_{t-})\Big)(dh)\Big|+$$

$$+\sum_{0\leqslant t\leqslant T}\Big|\int_{h\in O} h^4\Big(n_t^{\Delta}-\bar n_t^{\Delta}(X_{t-})\Big)(dh)\Big|\Big)dt\Big\}+$$

$$+2C_0\mathbf{E}\Big\{\int_0^T\Big(\Big|\int_{h\notin O}\Big(n_t^c-\bar n_t^c(X_{t-})\Big)(dh)\Big|+\sum_{0\leqslant t\leqslant T}\Big|\int_{h\notin O}\Big(n_t-\bar n_t(X_{t-})\Big)(dh)\Big|\Big)dt\Big\},$$

with new constants

$$C_i^*=\mathbf{E}\Big\{\Big(\int_0^T\Big|a-\bar a(X_t)\Big|dt+\sum_{0\leqslant t\leqslant T}\Big|\int_{h\in O} h\Big(n_t^{\Delta}-\bar n_t^{\Delta}(X_{t-})\Big)(dh)\Big|\Big)^q\Big\}^{1/q}+$$

$$+\frac{\bar C_{i+2}}{2!}\mathbf{E}\Big\{\Big(\int_0^T\Big|b_t^2-\bar b^2(X_t)\Big|dt+\Big|\int_{h\in O} h^2\Big(n_t^c-\bar n_t^c(X_{t-})\Big)(dh)\Big|+$$

$$+\sum_{0\leqslant t\leqslant T}\Big|\int_{h\in O} h^2\Big(n_t^{\Delta}-\bar n_t^{\Delta}(X_{t-})\Big)(dh)\Big|\Big)^q\Big\}^{1/q}+$$

$$+\frac{\bar C_{i+3}}{3!}\mathbf{E}\Big\{\Big(\int_0^T\Big(\Big|\int_{h\in O} h^3\Big(n_t^c-\bar n_t^c(X_{t-})\Big)(dh)\Big|+$$

$$+\sum_{0\leqslant t\leqslant T}\Big|\int_{h\in O} h^3\Big(n_t^{\Delta}-\bar n_t^{\Delta}(X_{t-})\Big)(dh)\Big|\Big)dt\Big)^q\Big\}^{1/q}+$$

$$+2\bar C_i\mathbf{E}\Big\{\Big(\int_0^T\Big(\Big|\int_{h\notin O}\Big(n_t^c-\bar n_t^c(X_{t-})\Big)(dh)\Big|+$$

$$+\sum_{0\leqslant t\leqslant T}\Big|\int_{h\notin O}\Big(n_t^{\Delta}-\bar n_t^{\Delta}(X_{t-})\Big)(dh)\Big|\Big)dt\Big)^q\Big\}^{1/q}.$$

To prove this we approximate every Skorokhod trajectory $x\in D(G)\uparrow[0,T]$ by step trajectories $x^{(n)}$:

$$x_t^{(n)}:=x_{t_k^{(n)}},\quad t_k^{(n)}\leqslant t<t_{k+1}^{(n)}.$$

We will have the convergence in the Skorokhod topology, therefore every continuous functional F is replaced by the cylindrical one with a convergence over each trajectory

$$F^{(n)}(x_1,\dots,x_n)=F(x^{(n)})\to F(x)\in\mathbf{R},\quad x\in D.$$

The partial derivatives of such cylindrical functional is calculated by constructing the needed trajectory of the difference $h(\Delta)\in D$

$$\Delta\partial_k F^{(n)}(x_{t_1},\dots,x_{t_n})=h(\Delta)\partial F(x^{(n)}),$$

$$h(\Delta)_t = \Delta, \quad t_k^{(n)} \leqslant t < t_{k+1}^{(n)} = 0, \quad \text{elsewhere}.$$

We shall explain the given formula only for the classical case, when our processes are defined by ordinary differential equations

$$X_{t_1} = X_{t_0} + \int_{t_0}^{t_1} a_t \, dt, \quad Y_{t_1} = Y_{t_0} + \int_{t_0}^{t_1} \bar{a}(Y_t) \, dt.$$

Exploring the changing of the initial trajectories we get the formula for the test function

$$f(X_T) - f(Y_T) = f(X_T) - f(\Phi_T^{t_0}(X_0)) = \int_{t_0}^{T} \Big(a_t - \bar{a}(X_t)\Big) D\Phi_T^t(X_t) Df(\Phi_T^t(X_t)) \, dt.$$

The same formula is also valid for the cylindrical functional

$$\Delta F^n = F^{(n)}(X_{t_1}, \dots, X_{t_n}) - F^{(n)}(Y_{t_1}, \dots, Y_{t_n}) =$$

$$= \int_{t_0}^{T} \Big(a_t - \bar{a}(X_t)\Big) \sum_{k: t \leqslant t_k} D\Phi_{t_k}^t(X_t) \partial_k F^{(n)}\Big(X_{t_1}, X_{t_2}, \dots, \Phi_{t_{n-1}}^t(X_t), \Phi_{t_n}^t(X_t)\Big) \, dt.$$

Then

$$|\Delta F^n| \leqslant \int_{t_0}^{T} \Big|a_t - \bar{a}(X_t)\Big| d_1 \sum_{k: t \leqslant t_k} \Big|\partial_k F^{(n)}\Big(X_{t_1}, X_{t_2}, \dots, \Phi_{t_{n-1}}^t(X_t), \Phi_{t_n}^t(X_t)\Big)\Big| \, dt \leqslant$$

$$\leqslant d_1 C_1 \int_{t_0}^{T} |a_t - \bar{a}(X_t)| \, dt.$$

Now we can work with the sum $\sum \partial_k F^{(n)}$ by exchanging the sign of argument $(a_t - \bar{a}(X_t))$.
Therefore

$$\Big|\mathbf{E}\Big\{F^{(n)}(X_{t_1}, \dots, X_{t_n}) - F^{(n)}(Y_{t_1}, \dots Y_{t_n})\Big\}\Big| \leqslant \bar{C}_1 \mathbf{E}\Big\{\int_0^T \Big|a_t - \bar{a}(X_t)\Big| dt\Big\}.$$

By the Lebesgue theorem of dominated convergence we have the same estimate for the integrals of initial functional.
Partial integration gives us estimates with integrated difference of characteristics

$$F^{(n)}(X_{t_1}, \dots, X_{t_n}) - F^{(n)}(Y_{t_1}, \dots, Y_{t_n}) =$$

$$= \int_{t_0}^{T} \Big(a_t - \bar{a}(X_t)\Big) \sum_{k: t \leqslant t_k} D\Phi_{t_k}^t(X_t) \partial_k F^{(n)}\Big(X_{t_1}, X_{t_2}, \dots, \Phi_{t_{n-1}}^t(X_t), \Phi_{t_n}^t(X_t)\Big) \, dt =$$

$$= \sum_{1\leqslant k\leqslant n} \int\limits_0^{t_k} \Big(a_t - \bar{a}(X_t)\Big) D\Phi^t_{t_k}(X_t)\times$$

$$\times\partial_k F^{(n)}\Big(X_{t_1},\ldots,\Phi^t_{t_k}(X_t),\ldots,\Phi^t_{t_{n-1}}(X_t),\ \Phi^t_{t_n}(X_t)\Big)\,dt =$$

$$= \sum_{1\leqslant k\leqslant n} A_{t_k}\partial_k F^{(n)}\Big(X_{t_1},\ldots,X_{t_k},\Phi^{t_k}_{t_{k+1}}(X_{t_k}),\ldots,\Phi^{t_k}_{t_n}(X_{t_k})\Big)+$$

$$+\int\limits_0^{t_k} A_t\Big(\frac{d}{dt}D\Phi^t_{t_k}(X_t)\Big)\partial_k F^{(n)}\Big(X_{t_1},\ldots,\Phi^t_{t_k}(X_t),\ldots,\Phi^t_{t_{n-1}}(X_t),\Phi^t_{t_n}(X_t)\Big)\,dt,$$

where $A_s = \int\limits_0^s \Big(a_t - \bar{a}(X_t)\Big)\,dt$ and

$$\frac{d}{dt}D\Phi^t_{t_k}(X_t) = \Big(a_t - \bar{a}(X_t)\Big)\sum_{l:t\leqslant t_l} D\Phi^t_{t_k}(X_t)D\Phi^t_{t_l}(X_t)\times$$

$$\times\partial_l\partial_k F^{(n)}\Big(X_{t_1},\ldots,\Phi^t_{t_k}(X_t),\ldots,\Phi^t_{t_n}(X_t)\Big)+$$

$$+D^2\Phi^t_{t_k}(X_t)\partial_k F^{(n)}\Big(X_{t_1},\ldots,\Phi^t_{t_k}(X_t),\ldots,\Phi^t_{t_n}(X_t)\Big).$$

The first sum will be changed by

$$\sum_{1\leqslant k\leqslant n} A_{t_k}\Big(\partial_k F^{(n)}\Big(X_{t_1},\ldots,\Phi^{t_k}_{t_{k+1}}(X_{t_k}),\ldots,\Phi^{t_k}_{t_n}(X_{t_k})\Big)\Big) =$$

$$= \sum_{1\leqslant k\leqslant n} A_{t_k}\Big(\partial_k F^{(n)}(X_{t_1},\ldots,X_{t_n}) + \text{difference}\Big)$$

where the difference is equal to

$$\partial_k F^{(n)}\Big(X_{t_1},\ldots,X_{t_k},\Phi^{t_k}_{t_{k+1}}(X_{t_k}),\ldots,\Phi^{t_k}_{t_n}(X_{t_k})\Big) - \partial_k F^{(n)}\Big(X_{t_1},\ldots,X_{t_n}\Big) =$$

$$= \int\limits_{t_k}^{T}\Big(a_t - \bar{a}(X_t)\Big)\sum_{l:t\leqslant t_l} D\Phi^t_{t_l}(X_t)\times$$

$$\times\partial_l\partial_k F^{(n)}\Big(X_{t_1},\ldots,X_{t_k},\ldots,\Phi^t_{t_{n-1}}(X_t),\Phi^t_{t_n}(X_t)\Big)\,dt.$$

There are the same derivatives of the pushed forward cylindrical functional $F^{(n)}$, only with the assumption $\Phi^t_{t_k} = 1,\ t \geqslant t_k$.
Now we apply the Hölder inequality

$$\mathbf{E}\Big\{|AF|\Big\} \leqslant \mathbf{E}\Big\{|A|^p\Big\}^{1/p}\mathbf{E}\Big\{|F|^q\Big\}^{1/q},\ \ \frac{1}{p}+\frac{1}{q}=1,\ \ 1\leqslant p\leqslant\infty.$$

Therefore we obtain an estimate

$$\left|\mathbf{E}\left\{F^{(n)}(X_{t_1},\dots,X_{t_n})-F^{(n)}(Y_{t_1},\dots,Y_{t_n})\right\}\right|\leqslant$$

$$\leqslant \mathbf{E}\left\{\sup_{0\leqslant t\leqslant T}|A_t|^p\right\}^{1/p}\left\{\mathbf{E}\Big(\sum_{1\leqslant k\leqslant n}\left|\partial_k F^{(n)}(X_{t_1},\dots,X_{t_n})\right|\Big)^q\right\}^{1/q}+$$

$$+\Big\{\mathbf{E}\Big(\sum_{1\leqslant k\leqslant n}\Big[\int_{t_k}^{T}|a_t-\bar a(X_t)|\sum_{l:t\leqslant t_l}\left|D\Phi^t_{t_l}(X_t)\partial_l\partial_k F^{(n)}(X_{t_1},\dots,\Phi^t_{t_n}(X_t))\right|dt+$$

$$+\int_0^{t_k}|a_t-\bar a(X_t)|\sum_{l:t\leqslant t_l}\Big|D\Phi^t_{t_k}(X_t)\partial_l\partial_k F^{(n)}(X_{t_1},\dots,\Phi^t_{t_k}(X_t),\dots,\Phi^t_{t_n}(X_t))+$$

$$+D^2\Phi^t_{t_k}(X_t)\partial_k F^{(n)}(X_{t_1},\dots,\Phi^t_{t_k}(X_t),\dots,\Phi^t_{t_n}(X_t))\Big|\,dt\Big]\Big)^q\Big\}^{1/q}\leqslant$$

$$\leqslant \mathbf{E}\left\{\sup_{0\leqslant t\leqslant T}|A_t|^p\right\}^{1/p}\left(C_1+\left(d_1^2C_2+d_2C_1\right)\mathbf{E}\left\{\Big(\int_0^T|a_t-\bar a(X_t)|\,dt\Big)^q\right\}^{1/q}\right).$$

This estimate will be valid also for a initial functional F.

In previous work (Valiukevičius, 1989) we intended to give calculus for the general case announced at the beginning of this work. We must agree that such calculations are difficult not only to write but also to read. The method for calculus of averaged test functions was initiated in earlier work (Valiukevičius, 1983).

REFERENCES

Valiukevičius, G. (1983). *Calculation of Averaged Test Functions. Preprint*, Vilnius.

Valiukevičius, G. (1989). Estimates for smooth functionals. In: *Conspects on Analysis*, pp. 22–30. Vilnius.

Prob. Theory and Math. Stat., Vol. 2, pp. 542–551
B.Grigelionis *et al.* (Eds.)
1990 VSP/Mokslas

ON LARGE DEVIATIONS IN AVERAGING PRINCIPLE FOR STOCHASTIC DIFFERENTIAL EQUATIONS WITH PERIODIC COEFFICIENTS.1

A.Yu. VERETENNIKOV
Institute for Control Sciences, Profsoyuznaya 65, 117806 Moscow, USSR

ABSTRACT

Large deviations in averaging principle for stochastic differential equations in Euclidean space are studied. The considered phase space is not compact — this is the main difference from the usual theory since Frobenius type theorem is not valid.

The problem of large deviations in the form of Wentzell and Freidlin for stochastic differential equations systems averaging principle is studied. We consider the case when phase space of "fast" process is not compact. The periodicity of coefficients does not complicates the situation but widens the area of possible applications. In this paper the main auxiliary result of Freidlin's large deviations theory — "condition F" (equality (1.9) in Theorem 1.1) — is established. This result was not known for noncompact spaces. The proof of the main theorem (Theorem 1.2) follows from "condition F" by standard way (see (Freidlin, 1978; Wentzell and Freidlin, 1984)).

1. MAIN ASSUMPTIONS AND RESULTS

The following stochastic differential equations systems are considered:

$$dX_t^\varepsilon = b(t, X_t^\varepsilon, Y_t^\varepsilon)dt, \quad X_0^\varepsilon = x,$$

$$dY_t^\varepsilon = \varepsilon^{-1}\big(C(t/\varepsilon)B(t/\varepsilon, X_t^\varepsilon, Y_t^\varepsilon) + g(t/\varepsilon, Y_t^\varepsilon)\big)dt + \varepsilon^{-1}C(t/\varepsilon, Y_t^\varepsilon)dw_t, \tag{1.1}$$

$$X_0^\varepsilon = x, \quad Y_0^\varepsilon = y,$$

where X_t^ε, x are points in E^d, Y_t^ε, y are points in $E^l (d, l \geqslant 1)$, coefficients b, B, g, C are bounded Borel functions, $b(t,x,y)$ is a d-dimensional vector, $B(t,x,y)$ and $g(t,y)$ are l-dimensional vectors, $C(t,y)$ is a matrix $l \times l$, w_t is an l-dimensional Wiener process, ε is a small parameter. We suppose that for any $r > 0$

$$\inf_{|y|\leqslant r} \inf_{t\geqslant 0} \inf_{\lambda \in E^l} \lambda CC^*(t,y)\lambda > 0. \tag{1.2}$$

We suppose also that for each $R > 0$ there exists such a measure μ ($\mu \geqslant 0$) on $U_R = \{y : y \in E^l, \ |y| \leqslant R\}$ that for any $t \geqslant 0, \ y \in U_R, \ z$

$$0 < N^{-1} \leqslant \frac{\mathbf{P}_{t,y}(y_{t+1}^z \in dy)}{\mu(dy)} \leqslant N < \infty \tag{1.3}$$

where y_t^z is a solution of stochastic differential equation

$$dy_t^z = (C(t, y_t^z)B(t, z, y_t^z) + g(t, y_t^z))dt + C(t, y_t^z)dw_t, \ y_0^z = y. \tag{1.4}$$

If coefficients are smooth enough and the matrix CC^* is non–degenerate then the condition (1.3) holds with μ—Lebesgue measure as is known. Conversely if coefficients are not smooth then (1.3) may fails (Safonov, 1981). If functions C, B and g do not depend on t and the matrix CC^* is non–degenerate and uniform continuous then (1.3) holds; this follows from existence of the positive density of y_t^z (Stroock and Varadhan, 1979) and Hölder continuity of the function $\mathbf{P}_{t,y}(y_{t+1}^z \in \Gamma)$, $\Gamma \in \mathcal{B}(E^l)$, uniform in z (Krylov and Safonov, 1981). We suppose that the weak uniqueness of the solution of stochastic Ito equation

$$dy_t = g(t, y_t)dt + C(t, y_t)dw_t, \ y_0 = y \tag{1.5}$$

holds. (One of the possible sufficient conditions for such uniqueness is, for example, condition (1.2) and uniform continuity of $CC^*(t, \cdot)$ (Stroock and Varadhan, 1979; Veretennikov, 1982).) The function $b(t, x, y)$ satisfies Lipschitz condition:

$$\left|b(t, x, y) - b(t, x', y)\right| \leqslant N|x - x'| \tag{1.6}$$

and continuous in t uniformly with respect to x, y.
If (1.2) holds then there exists (weak) solution of (1.5) (Krylov, 1980). The existence of solution of equation (1.1) with $B \equiv 0$ follows from (1.6). The existence of the solution of equation (1.1) with any bounded B follows from Girsanov's transformation. The weak uniqueness is checked with the help of Girsanov's transformation as well. For every $\varepsilon > 0$ the solution of (1.1) exists on some probability space which may depend on ε but we omit the index ε in all symbols $\mathbf{P}, \mathbf{E}, w$ et al. Coefficients B, g and C are supposed to be periodic in t with the period 1.
At last, we formulate conditions which characterize stability properties of the "fast" process Y_t^ε (or y_t^z). Let us denote

$$\tau_R = \inf(k = 0, 1, 2, \ldots : \ |y_k^z| \leqslant R).$$

We suppose that for every $\alpha > 0$ there exist such $\varepsilon > 0$ and $R_\alpha > 0$ that for any $R \geqslant R_\alpha, \ z, \ y$

$$E_{0,y} \exp(\alpha \tau_R) \leqslant \exp\left(\varepsilon(|y| - R)_+\right), \tag{1.7}$$

and there exists $C > 0$ such that for any $\varepsilon > 0, \ R \geqslant R_\alpha, \ y, \ z, \ t$

$$E_{0,y} \exp\left(\varepsilon(|y_t^z| - R)_+\right) \leqslant C \exp\left(\varepsilon(|y| - R)_+\right). \tag{1.8}$$

The sufficient condition for (1.7) and (1.8) to hold is the following equality:

$$\lim_{y\to\infty} \sup_{x,t} \frac{(g(t,y), y/|y|)}{\|CC^*(t,y)\|} = -\infty$$

(cf. (Veretennikov, 1988a)).
Let $\alpha \in E^l$, $x, z \in E^d$, $y \in E^l$, $t' \geqslant 0$.

THEOREM 1.1. *Let conditions (1.2), (1.3), (1.6)—(1.8) hold. Then there exists a limit*

$$\mu(t', x, z, \alpha) = \lim_{t\to\infty} t^{-1} \ln \mathbf{E}_{0,y} \exp\left(\int_0^t \alpha b(t', x, y_s^z) ds\right), \tag{1.9}$$

where the function $\mu(t', x, z, \alpha)$ is continuous in (t', x, z, α) and convex in α. Moreover, for any $\alpha, C, \delta, \nu > 0$ there exist $t_\delta \geqslant 0$, $R > 0$, $C(R) > 0$ such that for any x, y, z, t' and $t \geqslant t_\delta$ the following estimates hold:

$$-\delta - C(R)t^{-1} + t^{-1} \ln \max\{\exp(-\|\alpha b\|_B t), 1 - \exp(-Ct + \nu(|y| - R + 1)_+\} \leqslant$$

$$\leqslant t^{-1} \ln \mathbf{E}_{0,y} \exp\left(\int_0^t \alpha b(t', x, y_s^z) ds\right) - \mu(t', x, z, \alpha) \leqslant \tag{1.10}$$

$$\leqslant \delta + C(R)t^{-1} + t^{-1} \ln \min\{\exp(\|\alpha b\|_B t), 1 + \exp(-Ct + \nu(|y| - R + 1)_+)\}.$$

Let $L(t', x, \beta) = \sup(\alpha\beta - \mu(t', x, x, \alpha))$. This function is lower semi-continuous in (x, β). Let

$$S_{0,T}(\varphi) = \int_0^T L(s, \varphi_s, \dot{\varphi}_s) ds \tag{1.11}$$

if φ is absolutely continuous, and $S_{0,T} = +\infty$ for the remaining $\varphi \in C[0, T; E^d]$.

THEOREM 1.2. *Let conditions (1.2), (1.3), (1.6)—(1.8) be satisfied. Then $\varepsilon^{-1} S_{0,T}(\varphi)$ is the action functional for the family of processes $(X_t^\varepsilon, 0 \leqslant t \leqslant T)$ in the space $C[0, T; E^d]$ as $\varepsilon \to 0$, i.e., for any $x \in E^d, s \geqslant 0$ the set*

$$\Phi_x(s) = \left(\varphi : \varphi \in C[0, T; E^d],\ \varphi_0 = x,\ S_{0,T}(\varphi) \leqslant s\right)$$

is compact in $C[0, T; E^d]$ and for any $r, s, \delta, \gamma > 0$ there exists ε_0 such that for any $\varepsilon \in (0, \varepsilon_0)$ the following inequalities hold:

$$\inf_{|y|\leqslant r} \mathbf{P}^\varepsilon(\rho_{0,T}(X^\varepsilon, \varphi) < \delta) \geqslant \exp(-\varepsilon^{-1}(S_{0,T}(\varphi) + \gamma)), \tag{1.12}$$

$$\sup_{|y|\leqslant r} \mathbf{P}^\varepsilon(\rho_{0,T}(X^\varepsilon, \Phi_x(s)) \geqslant \delta) \leqslant \exp(-\varepsilon^{-1}(s - \gamma)), \tag{1.13}$$

where ρ is the Euclidean distance in $C[0,T;E^d]$, $\varphi \in C[0,T;E^d]$, $\varphi_0 = x$.

2. AUXILIARY RESULTS

Fix $a > 0$, $\alpha \in E^d$, $|\alpha| < a$, and let

$$\tau_0 = 0,\ \tau = \inf\left(k = 0,1,2,\ldots : |y_k^z| \leqslant r - 1\right),$$

$$\gamma_n^0 = \tau_n + 1,\ \gamma_n^{i+1} = \vartheta_{\gamma_n^i}\tau + 1,\ \tau_{n+1} = \inf\left(\gamma_n^i : |y_{\gamma_n^i}^z| \leqslant R,\ i = 0,1,\ldots\right),\quad n \geqslant 0$$

(ϑ is a shift operator).

LEMMA 2.1. *Let conditions (1.2), (1.3), (1.6)—(1.8) be satisfied. Then for any $a > 0$, there exists $R_a > 0$ such that for any $|\alpha| < a$, $R \geqslant R_a$ and for any t', x, z the following limit exists:*

$$\mu_R(t', x, z, \alpha) = \lim_{n\to\infty} n^{-1} \ln \mathbf{E}_{0,y} \exp\left(\int_0^{\tau_n} \alpha b(t', x, y_s^z) ds\right), \tag{2.1}$$

the function $\mu_R(t', z, \alpha)$ is differentiable in α as $|\alpha| < a$, $R \geqslant R_a$, and convex in α ($\alpha \in E^d$). Moreover, for any $a > 0, x, z, t', |\alpha| < a$, $R \geqslant R_a$ there exists $C(R)$ such that for any $|y| \leqslant R$, n

$$\exp\left(n\mu_R(t', x, z, \alpha) - C(R)\right) \leqslant \mathbf{E}_{0,y} \exp\left(\int_0^{\tau_n} \alpha b(t', x, y_s^z) ds\right) \leqslant$$

$$\leqslant \exp\left(n\mu_R(t', x, z, \alpha) + C(R)\right). \tag{2.2}$$

Let us denote $n(t) = \sup(k = 0,1,2,\ldots : \tau_k \leqslant t)$.

LEMMA 2.2. *Let conditions (1.2), (1.3), (1.6)—(1.8) be satisfied. Then for any $R \geqslant R_a$*

$$\mathbf{P}\left(\frac{n(t)}{t} \to \nu_R,\ t \to \infty\right) = 1,$$

where ν_R^{-1} is the expectation of $(\tau_{n+1} - \tau_n)$ for stationary distributed (X_{τ_n}). Moreover, for any $\delta, C, \nu > 0$ there exists $R_0, t(\delta)$ such that for any $R \geqslant R_0$, $t \geqslant t(\delta)$, y, x, z

$$\lim_{R\to\infty} \nu_R = 1, \tag{2.3}$$

and for any y

$$\lim_{R\to\infty} \sup_n \left|\frac{1}{n}\mathbf{E}_{0,y}\tau_n - 1\right| = 0. \tag{2.4}$$

LEMMA 2.3. *Let conditions (1.7), (1.8) be satisfied. Then for any $c, \nu > 0$ and $C > 1$, there exists $R_0 > 0$ such that for any $R \geqslant R_0$, x, y, z*

$$\mathbf{E}_{0,y} \exp\left(c(\tau_1 - 1)\right) \leqslant C \exp\left(\nu(|y| - R + 1)_+\right). \tag{2.5}$$

Denote $B(U_R)$ the space of all bounded Borel functions on U_R, where $U_R = = (y \in E^l : |y| \leqslant R)$, and $B^+(U_R)$ the cone of all non-negative functions on U_R. Fix $R > 0$ and consider the semi-group of operators $(T_n^\alpha, n = 0, 1, 2, \dots)$ on $B(U_R)$:

$$T_n^\alpha f(y) = \mathbf{E}_{0,y} f(y_{\tau_n}^z) \exp\left(\int_0^{\tau_n} \alpha b(t', x, y_s^z) ds \right).$$

Conditions (1.2), (1.3), (1.6)—(1.8) are supposed to be fulfilled. As b is bounded, then by Lemma 2.3 for any $|\alpha| < a$, t', x, z

$$T_n^\alpha B(U_R) \subset B(U_R), \quad n \geqslant 0,$$

if R is large enough, and so all T_n^α are positive operators on $B(U_R)$ as $R \geqslant R_a$ (Krasnosel'skii et al., 1989).
Positive operators T^α is called u-bounded (see (Krasnosel'skii et al., 1989)) with $u \in B^+(U_R)$ if for any $f \in B^+(U_R)$, $f \neq 0$, there exists $c(f) > 0$ such that $c(f)^{-1} u \leqslant T^\alpha f \leqslant c(f) u$.

LEMMA 2.4. *Let conditions* (1.2), (1.3), (1.6)—(1.8) *be satisfied. Then for any* $|\alpha| < a$, $R \geqslant R_a$ *operator* T^α *is* 1*-bounded.*

LEMMA 2.5. *Let conditions of Theorem* 1.1 *be satisfied. Then for any* $R \geqslant R_a$ *the spectral radius* $r(T^\alpha)$ *is a simple eigen-value of* T^α *and its eigen-function* $e_R \in B^+(U_R)$. *The function* $r(T^\alpha)$ *is differentiable in* α ($|\alpha| < a$) *and convex in* α ($|\alpha| \leqslant a$).

3. PROOFS OF AUXILIARY RESULTS

Proof of Lemma 2.4. We use the estimate of Lemma 2.3. By virtue of (2.5), for every $c > 0$ and $y \in E^l$ we have

$$\mathbf{E}_{0,y} \exp(c\tau_1) \leqslant C \exp(c) \exp(\nu(|y| - R + 1)_+)$$

for some $C > 1$, $N > 0$ and $R \geqslant R_a$ ($|\alpha| < a$). So, if $y \in U_R$, we estimate

$$T^\alpha f(y) = \mathbf{E}_{0,y} f(y_{\tau_1}^z) \exp\left(\int_0^{\tau_1} \alpha b(t', x, y_s^z) ds \right) \leqslant \|f\|_{B(U_R)} \mathbf{E}_{0,y} \exp(\|\alpha b\|_B \tau_1) \leqslant$$

$$\leqslant C \exp\left(a \|b\|_B + \nu\right) \|f\|_{B(U_R)}.$$

Let us now establish the lower estimate. Fix some $y_0 \in U_R$. If $f \geqslant 0$ and $f \neq 0$ then there exist $\delta > 0$ and Lebesgue set $U \subset U_R$ such that $f|_{U_R} \geqslant \delta$ and, by virtue of (1.3),

$$q_R = \inf_{y \in U_R} \inf_z \mathbf{P}_{0,y}(y_1^z \in U) > 0.$$

Then

$$T^{\alpha}f(y) \geqslant \mathbf{E}_{0,y}f(y_1^z)\exp\Big(\int_0^1 \alpha b(t',x,y_s^z)dsI(\tau=1)\Big) \geqslant$$

$$\geqslant \exp\big(-a\|b\|_B\big)\delta\mathbf{E}_{0,y}I(y_1^z \in U) \geqslant \exp(-a\|b\|_B)\delta q_R > 0.$$

Lemma 2.4 is proved.

Proof of Lemma 2.3. We have

$$\tau_1 = \inf(\gamma_0^i : |y_{\gamma_0^i}^z| \leqslant R,\ i = 0,1,2,\ldots),$$

and (1.7), (1.8) imply $\mathbf{P}(\tau_1 < \infty) = 1$. Let us omit index 0 in γ_0^i. We have

$$\tau_1 = \sum_{i=1}^{\infty}(\gamma^i - \gamma^{i-1})I(\gamma^i \leqslant \tau_1).$$

By virtue of (1.8) we obtain

$$p_R = \sup_{|y|\leqslant R-1}\mathbf{P}_{0,y}(|y_1^z| > R) \leqslant \sup_{|y|\leqslant R-1}\exp(-\varepsilon R)\mathbf{E}_{0,y}\exp(\varepsilon|y_1^z|) \leqslant$$

$$\leqslant \sup_{|y|\leqslant R-1}\exp(-\varepsilon)\mathbf{E}_{0,y}\exp(\varepsilon(|y_1^z| - R + 1)_+) \leqslant$$

$$\leqslant \exp(-\varepsilon)\sup_{|y|\leqslant R-1}C\exp(\varepsilon(|y| - R + 1)_+) = C\exp(-\varepsilon),$$

where ε may be chosen arbitrary large by the choice of R, so the value p_R may be chosen arbitrary small. Further we have

$$\mathbf{E}_{0,y}\exp(c(\tau_1 - 1)) = \mathbf{E}_{0,y}\exp\Big(c\Big(\sum_{i=0}^{\infty}(\gamma^{i+1} - \gamma^i)I(\gamma^i < \tau_1) - 1\Big)\Big) =$$

$$= \mathbf{E}_{0,y}\exp\Big(c\Big((\gamma^1 - 1) + \sum_{i=1}^{\infty}(\gamma^{i+1} - \gamma^i)I(\gamma^i < \tau_1)\Big)\Big) =$$

$$= \mathbf{E}_{0,y}\exp\Big(c\Big(\vartheta_1\tau + \sum_{i=1}^{\infty}(\gamma^{i+1} - \gamma^i)\prod_{j=1}^{i}I(|y_{\gamma^j}^z| > R)\Big)\Big).$$

Let $N \geqslant 1$. Then

$$\exp\Big(c\sum_{i=1}^{N}(\gamma^{i+1} - \gamma^i)\prod_{j=1}^{i}I(|y_{\gamma^j}^z| > R)\Big) \leqslant$$

$$\leqslant \sum_{i=1}^{N}\exp\Big(c\sum_{j=1}^{i}(\gamma^{j+1} - \gamma^j)\Big)\prod_{k=1}^{i}I(|y_{\gamma^k}^z| > R) + 1,$$

and so,

$$\exp\left(c\sum_{i=1}^{\infty}(\gamma^{i+1}-\gamma^{i})\prod_{j=1}^{i}I(|y^{z}_{\gamma^{j}}|>R)\right)\leqslant$$

$$\leqslant 1+\sum_{i=1}^{\infty}\exp\left(c\sum_{j=1}^{i}(\gamma^{j+1}-\gamma^{j})\right)\prod_{k=1}^{i}I(|y^{z}_{\gamma^{k}}|>R).$$

Thus,

$$\mathbf{E}_{0,y}\exp(c(\tau_1-1))\leqslant$$

$$\leqslant 1+\sum_{i=1}^{\infty}\mathbf{E}_{0,y}\exp\left(c\vartheta_1\tau+c\sum_{j=1}^{i}(\gamma^{j+1}-\gamma^{j})\right)\prod_{k=1}^{i}I(|y^{z}_{\gamma^{k}}|>R).$$

By Hölder inequality we have

$$\mathbf{E}_{0,y}\exp\left(c\vartheta_1\tau+c\sum_{j=1}^{i}(\gamma^{j+1}-\gamma^{j})\right)\prod_{k=1}^{i}I(|y^{z}_{\gamma^{k}}|>R)\leqslant$$

$$\leqslant\left(\mathbf{E}_{0,y}\exp\left(2c\vartheta_1\tau+2c\sum_{j=1}^{i}(\gamma^{j+1}-\gamma^{j})\right)\right)^{1/2}\left(\mathbf{E}_{0,y}\prod_{k=1}^{i}I(|y^{z}_{\gamma^{k}}|>R)\right)^{1/2}.$$

By strong Markov property,

$$\mathbf{E}_{0,y}\prod_{k=1}^{i}I(|y^{z}_{\gamma^{k}}|>R)\leqslant p_R^{i}.$$

So we have

$$\mathbf{E}_{0,y}\exp\left(c\vartheta_1\tau+c\sum_{j=1}^{i}(\gamma^{j+1}-\gamma^{j})\right)\prod_{k=1}^{i}I(|y^{z}_{\gamma^{k}}|>R)\leqslant$$

$$\leqslant p_R^{1/2}\left(\mathbf{E}_{0,y}\exp\left(2c\vartheta_1\tau+2c\sum_{j=1}^{i}(\gamma^{j+1}-\gamma^{j})\right)\right)^{1/2}.$$

As γ^j-1 is a stopping time, then by virtue of (1.7), (1.8), strong Markov property and induction we obtain

$$\mathbf{E}_{0,y}\exp\left(2c\vartheta_1+2c\sum_{j=1}^{i}(\gamma^{j+1}-\gamma^{j})\right)\leqslant C\exp(2ic)\exp\left((|y|-R+1)_+\right)$$

where $C>1$ and $C-1$ may be chosen arbitrary small as $R\to\infty$. Thus,

$$\mathbf{E}_{0,y}\exp\left(c(\tau_1-1)\right)\leqslant C_0=1+\sum_{i=1}^{\infty}p_R^{i/2}C^{1/2}\exp(ic)\exp(\nu(|y|-R+1)_+)$$

and the value $C_0 - 1$ here may be chosen arbitrary small as $R \to \infty$. Lemma 2.3 is proved.

The proof of Lemma 2.5 follows from Lemma 2.4 and Theorems 11.1 and 9.7 from (Krasnosel'skii et al., 1989). Lemma 2.2 is proved with the help of similar ideas. The proof of Lemma 2.1 follows from Lemmas 2.5 and 2.2 and the definitions of eigen–values and eigen–functions (see (Wentzell and Freidlin, 1984; Chapter 7)).

4. PROOF OF THEOREM 1.1

By virtue of Lemmas 2.1 and 2.2, we obtain the existence of the limit (1.9) and convexity in α, with the help of the equality

$$t^{-1} \ln \mathbf{E}_{0,y} \exp\left(\int_0^t \alpha b(t', x, y_s^z) ds \right) =$$

$$= \nu_R \frac{1}{\nu_R t} \ln \left\{ \mathbf{E}_{0,y} \exp\left(\int_0^t \alpha b(t', x, y_s^z) ds I\left(|t - \tau_{[\nu_R t]}| > \delta t \right) \right) + \right.$$

$$\left. + \mathbf{E}_{0,y} \exp\left(\left(\int_0^{\tau_{[\nu_R t]}} + \int_{\tau_{[\nu_R t]}}^t \right) \alpha b(t', x, y_s^z) ds I\left(|t - \tau_{[\nu_R t]}| \leqslant \delta t \right) \right) \right\}$$

for any $\delta > 0$ and R sufficiently large.

To prove the continuity of $\mu(t', x, z, \alpha)$ (locally uniform in t'), let us use the representation

$$\mu(t', x, z, \alpha) = \lim_{t \to \infty} t^{-1} \ln \mathbf{E}_{0,y} \exp\left\{ \int_0^t \left(\alpha b(t', x, y_s) - \frac{1}{2} |B(s, z, y_s)|^2 \right) ds + \right.$$

$$\left. + \int_0^t B(s, z, y_s) dw_s \right\},$$

where $y_s = y_s^0$. We obtain

$$\left| \mu(t', x, z, \alpha) - \mu(t', x', z', \alpha') \right| = \lim_{t \to \infty} t^{-1} \left| \ln \mathbf{E}_{0,y} \exp\left(\int_0^t \left(\left(\alpha b(t', x, y_s) - \right. \right. \right. \right.$$

$$\left. - \alpha' b(t', x', y_s) \right) - \frac{1}{2} \left(|B(s, z, y_s)|^2 - |B(s, z', y_s)|^2 \right) \Big) ds +$$

$$+ \int_0^t \Big(B(s, z, y_s) - B(s, z', y_s) \Big) dw_s \Big|.$$

Now by Lipschitz condition, as $|\alpha|, |\alpha'| < a$, we have

$$\left|\int_0^t \left(\left(\alpha b(t',x,y_s) - \alpha' b(t',x',y_s)\right) - \frac{1}{2}\left(|B(s,z,y_s)|^2 - |B(s,z',y_s)|^2\right)\right) ds\right| \leqslant$$

$$\leqslant ta\|b\|_B\,|x-x'| + t\|b\|_B\,|\alpha-\alpha'| + Ct\|b\|_B\,|z-z'|.$$

Now,

$$\mathbf{E}_{0,y}\exp\left(\int_0^t \left(B(s,z,y_s) - B(s,z',y_s)\right) dw_s\right) =$$

$$= \mathbf{E}_{0,y}\exp\left(\int_0^t \left(B(s,z,y_s) - B(s,z',y_s)\right) dw_s - \right.$$

$$-\frac{1}{2}\int_0^t \left|B(s,z,y_s) - B(s,z',y_s)\right|^2 ds +$$

$$\left. +\frac{1}{2}\int_0^t \left|B(s,z,y_s) - B(s,z',y_s)\right|^2 ds\right) \leqslant \exp\left(Ct|z-z'|^2\right)$$

and

$$\mathbf{E}_{0,y}\exp\left(\int_0^t \left(B(s,z,y_s) - B(s,z',y_s)\right) dw_s\right) \geqslant \exp\left(-Ct|z-z'|^2\right).$$

So we obtain continuity in (x, z, α). Theorem 1.1 is proved.
The proof of Theorem 1.2 follows from the considerations in the proof of Theorem 7.9.2 from (Wentzell and Freidlin, 1984) using Theorem 1.1 and condition (1.8) instead of "condition *F*".

REFERENCES

Freidlin, M.I. (1978). Averaging principle and large deviations. *Russian Math. Surveys* **33** **(5)**, 117–176.

Gärtner, J. (1977). On large deviations from the unvariant measure. *Theory Probab. Appl.* **22**, 24–39.

Krasnosel'skii, M.A., Lifshitz, E.A. and Sobolev, A.V. (1989). *Positive Linear Systems*. Helderman, Berlin.

Krylov, N.V. (1980). *Controlled Diffusion Processes*. Springer, Berlin—New York.

Krylov, N.V. and Safonov, M.V. (1981). A certain property of solutions of parabolic equations with measurable coefficients. *Math. USSR Izv.* **16** **(1)**, 151–164.

Safonov, M.V. (1981). Example of diffusion process with singular distribution in the fixed time. In: *Abstr. Third Intern. Vilnius Conf. Theory. Probab. Math. Statist.*, Inst. Math. Cybern., Vilnius, pp. 133–134.

Stroock, D.W. and Varadhan, S.R.S. (1979). ***Multidimensional Diffusion Processes.*** Springer, Berlin—New York.

Varadhan, S.R.S. (1984). ***Large Deviations and Applications.*** SIAM, Philadelphia.

Veretennikov, A.Yu. (1982). Parabolic equations and Ito stochastic equations with coefficients discontinuous in the time variable. ***Math. Notes*** **31**, 278–283.

Veretennikov, A.Yu. (1988a). On mixing rate and averaging principle for hypoelliptic stochastic differential equations. ***Math. USSR Izv.*** **52(5)**.

Veretennikov, A.Yu. (1988b). Bounds for the mixing rate in the theory of stochastic equations. ***Theory Probab. Appl.*** **32**, 273–281.

Wentzell, A.D. and Freidlin, M.I. (1984). ***Random Perturbations of Dynamical Systems.*** Springer, Berlin—New York.

Prob. Theory and Math. Stat., Vol. 2, pp. 552–562
B.Grigelionis *et al.* (Eds.)
1990 VSP/Mokslas

ON LOGARITHMIC REFINING TERMS IN LIMIT THEOREMS TAKING INTO ACCOUNT LARGE DEVIATIONS OF SUMS OF INDEPENDENT RANDOM VARIABLES

V.V. VINOGRADOV
Department of Probability, Faculty of Mathematics and Mechanics,
Moscow State University, Lenin Hills, 119899 Moscow, USSR

ABSTRACT

Consider $\{X_n, n \geqslant 1\}$ – i.i.d.r.v. with power tails of index $\alpha \geqslant 3$ being an integer. Asymptotic expansions for distribution function of properly centered and normalized sum $X_1 + \ldots + X_n$ in different ranges of deviations are constructed. All the expansions contain terms having logarithms as their factors. A new concept of pseudomoments is proposed. The interplays between theorems both on large and normal deviations, and theorems acting on the whole x–axis are demonstrated.

INTRODUCTION

Let $\{X_n, n \geqslant 1\}$ be i.i.d.r.v. with a distribution function $F(x)$ having the following asymptotics

$$1 - F(x) = c_\alpha x^{-\alpha} + o(x^{-\alpha}); \tag{0.1}$$

$$F(-x) = d_\alpha x^{-\alpha} + o(x^{-\alpha}) \tag{0.1'}$$

as $x \to \infty$, where $\alpha > 0, c_\alpha + d_\alpha > 0$. Below we shall refer to such tails of F as power tails.
The case of power tails is of much interest by the reason that the construction of asymptotic expansions both in limit theorems on weak convergence (i.e., theorems on normal deviations) for distribution functions of properly normalized sums $S_n = X_1 + \ldots + X_n$ (see, for example (Cramér, 1963)) in limit theorems on large deviations for S_n (see (Vinogradov, 1985; 1988)) under mild supplementary constraints on their behaviour (as compared with (0.1) and (0.1′)) is rather simple. Moreover, just in this case the connection of theorems on normal deviations with nonuniform estimates of the remainder term (i.e., NUE–theorems) with those on large deviations is clearly seen. Let us elucidate this connection on a simple example.

PROPOSITION 0.1. *Let conditions* (0.1) *and* (0.1′) *be satisfied with* $\alpha \in (0,1) \cup (1,2)$; $MX_i = 0$ *in the case* $\alpha \in (1,2)$; $S_n^* = S_n/n^{1/\alpha}$.

Then for any $n \geqslant 1$ integer and for any x real the following assertion is true:

$$|F_{S_n^*}(x) - G_\alpha(x)| \leqslant \delta(n)/(1+|x|)^\alpha, \tag{0.2}$$

where $\delta(n) \to 0$ as $n \to \infty$; $G_\alpha(\cdot)$ is the stable distribution function with index α (the limit for the random sequence S_n^*); we denote the distribution function of any random variable Z by F_z throughout the work.

Proof of Proposition 0.1. Clearly, for x not tending to infinity it coincides with the well-known theorem on weak convergence to a stable limit distribution. On the other hand, for $|x| \to \infty$ we set $y = xn^{1/\alpha}$ turning estimate (0.2) into the following one:

$$|F_{S_n}(y) - G_\alpha(y/n^{1/\alpha})| \leqslant \delta(n) n y^{-\alpha} \tag{0.2'}$$

as $n \to \infty$, $|y|/n^{1/\alpha} \to \infty$. Estimate (0.2′) is valid in view of the theorems on large deviation probabilities (LDP) for the sequence S_n (see, for example (Heyde, 1968)) and the tail behaviour of the function $G_\alpha(\cdot)$ (see (Gnedenko and Kolmogorov, 1954, Section 35, Theorem 5)).

The principal approach of this work consists in simultaneous considering the NUE–, LD–theorems, and theorems with uniform estimates of the remainder term (i.e., UE–theorems). It is interesting to note that theorems of the first class mentioned above can yield integral theorems acting on the whole x–axis. The first theorem acting on the whole x–axis has been proved by (Linnik, 1961) (see also (Ibragimov and Linnik, 1971, Chapter 14)).

The following important result describing the asymptotics of $P\{S_n > y\}$ in the full range of large deviations is conceptually close both to theorems acting on the whole x–axis and to NUE–theorems.

PROPOSITION 0.2 (see (A.Nagaev, 1970; S.Nagaev, 1979, Theorem 1.9)). *Let conditions* (0.1) *and* (0.1′) *be satisfied with* $\alpha > 2$; $MX_i = 0$, $DX_i = 1$. *Then the following assertion is true:*

$$P\{S_n > y\} = (1 - \Phi(y/n^{1/2}))(1 + r_1(n,y)) + nP\{X_i > y\}(1 + r_2(n,y)), \tag{0.3}$$

where $r_1(n,y) = o(1)$, $r_2(n,y) = o(1)$ *as* $n \to \infty$, $y/n^{1/2} \to \infty$, *and* $\Phi(\cdot)$ *is the Laplace function.*

We shall need the following convergence rate estimates both in (0.3) and in its left–hand analogue for $y \geqslant (Kn \ln n)^{1/2}$ in the proof of our main results. Set $\Delta_+(n,y) = P\{S_n > y\} - nc_\alpha y^{-\alpha}$ and $\Delta_-(n,y) = P\{S_n < -y\} - nd_\alpha y^{-\alpha}$, where $n \geqslant 1$ is any integer and $y > 0$ is any real number.

PROPOSITION 0.3 (see (Vinogradov, 1988, Theorem 1)). *Under the same assumptions there exist positive constants* K_1, K_2 *and* K_3 *such that the following assertions are true:*

$$\begin{aligned}|\Delta_+(n,y)| \leqslant K_2 n y^{-\alpha}\{(y/n^{1/2})^{-\alpha/(1+\alpha)} +\\ + \exp(-K_3(y^2/n)^{1/(1+\alpha)})\} + n \sup_{x \geqslant y/3} |1 - F(x) - c_\alpha x^{-\alpha}|;\end{aligned} \tag{0.4}$$

$$|\Delta_-(n,y)| \leqslant K_2 n y^{-\alpha}\{(y/n^{1/2})^{-\alpha/(1+\alpha)} + \exp(-K_3(y^2/n)^{1/(1+\alpha)})\} + n \sup_{x \leqslant -y/3} |F(x) - d_\alpha |x|^{-\alpha}| \tag{0.4'}$$

for any $n \geqslant 1$ integer and $y \geqslant (K_1 n \ln n)^{1/2}$.

Let us note that the third terms in the right–hand sides of (0.4) and (0.4′) are "unremovable errors" generated by the lack of perfect information on the tail behaviour of $F(x)$.
It seems reasonable to assume that if more precise information on the tail behaviour of function F is available (as compared with (0.1) and (0.1′)) these terms can be written down more precisely, i.e., the further refinements for $P\{S_n > y\}$ can be deduced from them. For the special case $\alpha > 0$, α is not an integer, this has been carried out in (Vinogradov, 1988) under the following supplementary condition:

$$1 - F(x) = \sum_{i=1}^{l} c_{\alpha_i} x^{-\alpha_i} + o(x^{-r}) \tag{0.5}$$

as $x \to \infty$, where $\alpha = \alpha_1 < \alpha_2 < \ldots < \alpha_l \leqslant r$.
One of the main points in the proof of the above mentioned result for $\alpha > 2$, α is not an integer (see (Vinogradov, 1988, Theorem 3)) consists in application of NUE–theorems with asymptotically precise estimate of the remainder term (in the sense of dependence on n and y) arising from Theorem 4 in (Bikjalis, 1966) and from Theorem 1 in (Osipov, 1972). Hence, a problem to establish their analogues for an integer $\alpha = \alpha_1 \geqslant 3$ comes up naturally. This problem is actually solved in Theorems 1.1 and 2.1 of Sections 1 and 2, respectively. However, the principal aim of this work is not only to supplement the investigations for non–integer values of the parameter α with those for integer values, but to find terms of new kind of asymptotic expansions for $P\{S_n > y\}$. Our Theorems 1.1, 1.2 and 2.1 are of this type, since terms with logarithms as their factors arise in the asymptotic expansions of these theorems.
There are numerous papers considering NUE– and UE–theorems for an integer $\alpha \geqslant 3$ or some variant thereof (including (Bikjalis, 1966; Ibragimov, 1967; Osipov, 1972 and Michel, 1976)). Theorem 1 in (Ibragimov, 1967) is the most important known result for the current investigation, since Theorem 1.1 in the next section being its nonuniform analogue.
The final Section 2 deals with construction of asymptotic expansions for LDP of S_n under condition (0.5) with an integer $\alpha = \alpha_1 \geqslant 3$. As well as in the proof of NUE–theorem 1.1, we have to apply UE–theorem 1.2 to prove LD–theorem 2.1.
It should be emphasized that in the asymptotic expansions of all the theorems proved in the present work terms with logarithms as their factors arise. The comparison of Theorems 1.1 and 2.1 with Proposition 0.2 provides information about the remainder terms $r_1(\cdot,\cdot)$ and $r_2(\cdot,\cdot)$ of (0.3): under mild supplementary constraints (compare (1.1) and (1.1′) with (0.1) and (0.1′)) the first one is completely clarified by Theorem 1.1, while the second one is partly clarified by Theorem 2.1. Clearly, the asymptotic expansion of Theorem 1.1 can be effectively applied for computations of $P\{S_n > y\}$ in the range of normal and moderate deviations (i.e., for $y \leqslant (Kn \ln n)^{1/2}$), while that of Theorem 2.1 can be used for $y \geqslant K n^{1/2+\theta}$.

1. ASYMPTOTIC EXPANSIONS IN THE CLT TAKING INTO ACCOUNT LARGE DEVIATIONS

First we introduce some auxiliary notation. Let us denote any sequence of non–negative numbers tending to zero as $n \to \infty$ by $\{\delta(n)\}$. By letter K we denote various positive constants depending on the distribution function $F(\cdot)$, but not on the parameters n and y (different positive constants entering in the same formula will be denoted by $K, K_1, K_2, \ldots$). Set $a \vee b = \max(a,b)$; $a \wedge b = \min(a,b)$, $F_n(x) = = \mathrm{P}\{S_n/n^{1/2} \leqslant x\}$, and $f(t)$ stands for the characteristic function of r.v. X_1.
Throughout this section we set mild supplementary constraints on the tail behaviour of the common distribution function $F(x)$ of independent random sequence $\{X_n, n \geqslant 1\}$ as compared with (0.1) and (0.1′). Namely, let us assume that

$$1 - F(x) = c_\alpha x^{-\alpha} + O(x^{-\alpha-\kappa}); \tag{1.1}$$

$$F(-x) = d_\alpha x^{-\alpha} + O(x^{-\alpha-\kappa}) \tag{1.1'}$$

as $x \to \infty$, where $\alpha \geqslant 3$ is any integer, $c_\alpha + d_\alpha > 0$, and $\kappa \in (0,1)$ is any real number. Set

$$b_\alpha = \int\limits_{-\infty}^{-1} x^\alpha d(F(x) - d_\alpha |x|^{-\alpha}) + \int\limits_{-1}^{1} x^\alpha \, dF(x) + \int\limits_{1}^{\infty} x^\alpha \, d(F(x) + c_\alpha x^{-\alpha}), \tag{1.2}$$

where the first and the third integrals are understood as improper Stieltjes integrals. Below we shall refer to b_α as the αth pseudomoment. Obviously, b_α is finite under conditions (1.1) and (1.1′), since all the integrals on the right–hand side of (1.2) being convergent.
Let us introduce three auxiliary sequences $\{E_+(k), k \geqslant 1\}, \{E_-(k), k \geqslant 1\}$, and $\{J_k, k \geqslant 1\}$:

$$E_+(k) = \int\limits_0^1 \left(e^{iz} - \sum_0^k (iz)^m/m! \right) d(-z^{-k}) + \\ + \int\limits_1^\infty \left(e^{iz} - \sum_0^{k-1} (iz)^m/m! \right) d(-z^{-k});$$

$$E_-(k) = \int\limits_0^1 \left(e^{-iz} - \sum_0^k (-iz)^m/m! \right) d(-z^{-k}) + \\ + \int\limits_1^\infty \left(e^{-iz} - \sum_0^{k-1} (-iz)^m/m! \right) d(-z^{-k});$$

$$J_k = \begin{cases} \int\limits_0^\infty \left(\cos z - \sum\limits_{m=0}^{(k-1)/2} (-1)^m z^{2m}/(2m)! \right) d(-z^{-k}), & \text{for odd } k; \\ \int\limits_0^\infty \left(\sin z - \sum\limits_{m=0}^{(k-2)/2} (-1)^{m+1} z^{2m+1}/(2m+1)! \right) d(-z^{-k}), & \text{for even } k. \end{cases}$$

After simple calculations we obtain the following relationships between these sequences:

$$J_k = (E_+(k) + E_-(k))/2 \tag{1.3}$$

for odd k, and

$$J_k = (E_+(k) - E_-(k))/2i \tag{1.3'}$$

for even k.
Obviously, $J_1 = -\pi/2$ and $J_{k+1} = (-1)^{k+1} J_k/k$. From this version of the definition of $\{J_k\}$ by recursion we readily infer that for any integer $k \geqslant 1$

$$J_k = (-1)^{k(k+1)/2} \pi/(2(k-1)!). \tag{1.4}$$

Let us also introduce three families of functions $\{A_k(x)\}, \{B_k(x)\}, \{D_k(x)\}$ (where $k = 1, 2, \dots; \ -\infty < x < \infty$) as follows:

$$A_k(x) = \frac{1}{2\pi} \int\limits_{-\infty}^{\infty} e^{-t^2/2 - itx} E_{\operatorname{sgn} t}(k) |t|^k \, dt;$$

$$B_k(x) = \frac{1}{2\pi} \int\limits_{-\infty}^{\infty} e^{-t^2/2 - itx} E_{-\operatorname{sgn} t}(k) |t|^k \, dt;$$

$$D_k(x) = \frac{1}{2\pi} \int\limits_{-\infty}^{\infty} e^{-t^2/2 - itx} (it)^k \ln \frac{1}{|t|} \, dt.$$

Functions $A_k(\cdot), B_k(\cdot), D_k(\cdot)$ are Fourier's transforms of absolutely integrable functions, and therefore are continuous, bounded and tend to zero as $|x| \to \infty$. Naturally, their properties listed above are also valid for their linear combination

$$h_\alpha(x) = c_\alpha A_\alpha(x) + d_\alpha B_\alpha(x) + e_\alpha D_\alpha(x)/(\alpha - 1)!, \tag{1.5}$$

where $e_\alpha = c_\alpha + (-1)^\alpha d_\alpha$. We set $H_\alpha(x) = \int_{-\infty}^{x} h_\alpha(v)\, dv$; this integral is finite for any value of x in view of the properties of h_α. Moreover, $H_\alpha(-\infty) = H_\alpha(\infty) = 0$.

Let

$$\Delta_{n,\alpha}(x) = F_n(x) - \Phi(x) - \sum_{\nu=1}^{\alpha-2} Q_\nu(x)/n^{\nu/2} +$$
$$+ \frac{e_\alpha \ln n}{2(\alpha-1)!n^{(\alpha-2)/2}}\Phi^{(\alpha)}(x) - \frac{H_\alpha(x)}{n^{(\alpha-2)/2}},$$

where the functions $Q_\nu(\cdot)$ are expressed in terms of the moments of r.v. X_1 by formulas which can be found in (Petrov, 1975, Chapter 5); we should use the α-th pseudomoment b_α instead of the α-th moment when constructing function $Q_{\alpha-2}(\cdot)$.
Let us present now the main result of this section.

THEOREM 1.1. *Let conditions (1.1) and (1.1′) be met with an integer $\alpha \geqslant 3$; $MX_i = 0$, $DX_i = 1$. Let us assume, in addition, that r.v. X_1 is not a lattice variable if $\alpha = 3$ or condition (C) is fulfilled if $\alpha \geqslant 4$. Then*

a) *for any $n \geqslant 1$ integer and for any x real the following assertion is true:*

$$|\Delta_{n,\alpha}(x)| \leqslant \delta(n)n^{-(\alpha-2)/2}(1+|x|)^{-\alpha}; \tag{1.6}$$

b) *function $H_\alpha(\cdot)$ has the following asymptotics at the tails:*

$$H_\alpha(\infty) - H_\alpha(x) = -H_\alpha(x) = c_\alpha x^{-\alpha} + O(x^{-\alpha-1}\ln x);$$
$$H_\alpha(-x) = d_\alpha x^{-\alpha} + O(x^{-\alpha-1}\ln x)$$

as $x \to \infty$.

To prove Theorem 1.1 we shall need the following UE–theorem describing the asymptotic behaviour of function $|\Delta_{n,\alpha}(x)|$. Besides, this result is of a particular value.

THEOREM 1.2. *Under the same assumptions for any $n \geqslant 1$ integer and for any real x the following assertion is true:*

$$|\Delta_{n,\alpha}(x)| \leqslant Kn^{-(\alpha+\kappa-2)/2}. \tag{1.7}$$

Note the better rate on n in estimate (1.7) as compared with (1.6). That makes it possible to establish (1.6) not only in the range of normal deviations, but also for $|x| \to \infty, x = o(n^{\kappa/2\alpha})$.
Proof of Theorem 1.2 is straightforward and relies on the following representation of $f(t)$ near zero:

$$f(t) = 1 + \sum_{k=2}^{\alpha} b_k(it)^k/k! + \frac{e_\alpha}{(\alpha-1)!}(it)^\alpha \ln\frac{1}{|t|} +$$
$$+|t|^\alpha(c_\alpha E_{\operatorname{sgn} t}(\alpha) + d_\alpha E_{-\operatorname{sgn} t}(\alpha)) + O(|t|^{\alpha+\kappa})$$

as $|t| \to 0$.

Proof of Theorem 1.1 consists of two steps. b) Let us split the proof of this statement into three parts. Namely, the tail behaviour of functions $D_\alpha(\cdot), A_\alpha(\cdot), B_\alpha(\cdot)$ as $|x| \to \infty$ are discovered in b1)—b3) respectively; their combination with (1.5) yields the assertion of b).
b1) In order to express $D_\alpha(\cdot)$ in a more convenient form we make the change of variables $v = -x$ in the integral from $-\infty$ to 0.
We get:

$$D_\alpha(x) = -\frac{i^\alpha}{2\pi} \int_0^\infty e^{-t^2/2} t^\alpha \ln t (e^{-itx} + (-1)^\alpha e^{itx})\, dt. \tag{1.8}$$

Set

$$I_\alpha(x) = \int_0^\infty e^{-t^2/2+itx} t^\alpha \ln t \, dt.$$

Using this notation we readily deduce from (1.1) and the properties of function e^{itx} that

$$\begin{aligned} D_\alpha(x) &= i^{\alpha+1} \mathrm{Im} I_\alpha(x)/\pi \quad \text{for} \quad \text{odd} \quad \alpha; \\ D_\alpha(x) &= -i^\alpha \mathrm{Re} I_\alpha(x)/\pi \quad \text{for} \quad \text{even} \quad \alpha. \end{aligned} \tag{1.9}$$

Therefore, we should investigate the asymptotics of $I_\alpha(\cdot)$. We apply (Fedoruk, 1987, Chapter 3, Lemma 1.4) for odd α and obtain

$$\mathrm{Im} I_\alpha(x) = \pm i^{\alpha+1} \Gamma(\alpha+1) \pi x^{-\alpha-1}/2 + O(|x|^{-\alpha-2} \ln |x|) \tag{1.10}$$

as $x \to \pm\infty$, while for even α

$$\mathrm{Re} I_\alpha(x) = i^{\alpha+2} \Gamma(\alpha+1) \pi |x|^{-\alpha-1}/2 + O(|x|^{-\alpha-2} \ln |x|) \tag{1.10$'$}$$

as $x \to \pm\infty$.
Combining (1.9), (1.10), (1.10′) we infer that

$$D_\alpha(x) = \pm \alpha! x^{-\alpha-1}/2 + O(|x|^{-\alpha-2} \ln |x|) \tag{1.11}$$

for odd α;

$$D_\alpha(x) = \alpha! |x|^{-\alpha-1}/2 + O(|x|^{-\alpha-2} \ln |x|)$$

for even α as $x \to \pm\infty$.
b2) To clarify the tail behaviour of $A_\alpha(x)$ we make the same change of variables $v = -x$ in the integral from $-\infty$ to 0, as in b1). We get:

$$A_\alpha(x) = \frac{1}{2\pi} \int_0^\infty e^{-t^2/2} t^\alpha (E_+(\alpha) e^{-itx} + E_-(\alpha) e^{itx})\, dt.$$

Set

$$I_+(\alpha;x) = \int_0^\infty e^{-t^2/2+itx} t^\alpha \, dt; \tag{1.12}$$

$$I_-(\alpha;x) = \int_0^\infty e^{-t^2/2-itx} t^\alpha \, dt. \tag{1.12'}$$

Obviously,

$$\mathrm{Re} I_+(\alpha;x) = \mathrm{Re} I_-(\alpha;x); \tag{1.13}$$

$$\mathrm{Im} I_+(\alpha;x) = -\mathrm{Im} I_-(\alpha;x). \tag{1.13'}$$

In this notation, $A_\alpha(\cdot)$ may be represented as

$$A_\alpha(x) = \Big(E_+(\alpha)I_-(\alpha;x) + E_-(\alpha)I_+(\alpha;x)\Big)/(2\pi). \tag{1.14}$$

Integrating by parts we readily obtain

$$I_+(\alpha;x) = \alpha!(-ix)^{-\alpha-1} + O(|x|^{-\alpha-2})$$

as $x \to \pm\infty$. Hence,

$$I_+(\alpha;x) = \mathrm{Re} I_+(\alpha;x) + O(|x|^{-\alpha-2})$$

for odd α;

$$I_+(\alpha;x) = i\mathrm{Im} I_+(\alpha;x) + O(|x|^{-\alpha-2}) \tag{1.15}$$

for even α as $x \to \pm\infty$, as well as

$$\mathrm{Re} I_+(\alpha;x) = \alpha!(-ix)^{-\alpha-1} + O(|x|^{-\alpha-2}) \tag{1.16}$$

for odd α;

$$i\mathrm{Im} I_+(\alpha;x) = \alpha!(-ix)^{-\alpha-1} + O(|x|^{-\alpha-2}) \tag{1.16'}$$

for even α.

Combining (1.13), (1.13′), (1.14), (1.15), (1.16), (1.16′) we infer that

$$A_\alpha(x) = \begin{cases} \alpha!(E_+(\alpha) + E_-(\alpha))(-ix)^{-\alpha-1}/(2\pi) + O(|x|^{-\alpha-2}) & \text{for odd } \alpha; \\ \alpha!(E_-(\alpha) - E_+(\alpha))(-ix)^{-\alpha-1}/(2\pi) + O(|x|^{-\alpha-2}) & \text{for even } \alpha \end{cases}$$

as $x \to \pm\infty$.

Let us combine this formula with relations (1.3), (1.3′), (1.4). We obtain

$$A_\alpha(x) = \alpha x^{-\alpha-1}/2 + O(|x|^{-\alpha-2}) \tag{1.17}$$

as $x \to \pm\infty$.

b3) The investigation of the tail behaviour of $B_\alpha(\cdot)$ is quite similar to that of $A_\alpha(\cdot)$ and therefore is omitted. Finally we obtain

$$B_\alpha(x) = \alpha(-x)^{-\alpha-1}/2 + O(|x|^{-\alpha-2}) \tag{1.17'}$$

as $x \to \pm\infty$.

The rest of the proof of b) is trivial: one has to combine (1.11), (1.17), (1.17′) and then apply (1.5). Hence,

$$-H_\alpha(x) = \int_x^\infty h_\alpha(v)\,dv = c_\alpha x^{-\alpha} + O(x^{-\alpha-1}\ln x); \tag{1.18}$$

$$H_\alpha(-x) = \int_{-\infty}^{-x} h_\alpha(v)\,dv = d_\alpha x^{-\alpha} + O(x^{-\alpha-1}\ln x) \tag{1.18'}$$

as $x \to \infty$. That concludes the proof of b).

a) Let us keep in mind that uniform estimate (1.7) yields nonuniform estimate (1.6) both for the full range of normal deviations and for $|x| \to \infty$, $x = o(n^{\kappa/2\alpha})$. Therefore, only the case $|x| \geqslant Kn^{\kappa/2\alpha}$ remains to be considered. Combining (0.4), (0.4′), (1.18) and (1.18′) we infer that

$$\left|\mathbf{P}\{S_n < -y\} - \frac{H_\alpha(-y/n^{1/2})}{n^{(\alpha-2)/2}}\right| \vee \left|\mathbf{P}\{S_n > y\} + \frac{H_\alpha(y/n^{1/2})}{n^{(\alpha-2)/2}}\right| \leqslant$$

$$\leqslant K_2 n y^{-\alpha}\{(y/n^{1/2})^{-\alpha/(1+\alpha)} + \exp(-K_3(y^2/n)^{1/(1+\alpha)})\} + \tag{1.19}$$

$$+K_4 n y^{-\alpha-\kappa} + K_5 n y^{-\alpha}(y/n^{1/2})^{-1}\ln(y/n^{1/2}).$$

Now we apply the obvious identities $\mathbf{P}\{S_n \leqslant -y\} = F_n(-x)$ and $\mathbf{P}\{S_n > y\} = 1 - F_n(x)$ to the left-hand side of this inequality. Taking into account the exponential descent of the Laplace function $\Phi(\cdot)$ at the tails, we may rewrite (1.19) as follows:

$$|F_n(x) - \Phi(x) - H_\alpha(x)n^{-(\alpha-2)/2}| \leqslant$$

$$\leqslant K_2 n^{-(\alpha-2)/2}|x|^{-\alpha}\Big\{|x|^{-\alpha/(1+\alpha)} + \exp\big(-K_3|x|^{2/(1+\alpha)}\big) + |x|^{-\kappa} + |x|^{-1}\ln|x|\Big\}$$

as $n \to \infty$ for $|x| \geqslant (K\ln n)^{1/2}$. It remains to combine the latter inequality with the exponential estimates for the tails of functions $Q_\nu(\cdot)$ and $\Phi^{(\alpha)}(\cdot)$.

2. ASYMPTOTIC EXPANSIONS OF LDP FOR S_n

In the previous section we constructed asymptotic expansions for the distribution function of the normalized sum $S_n/n^{1/2}$ which hold under conditions (1.1) and

(1.1′) and provide asymptotics of LDP for S_n up to equivalence. On the other hand, we can not guarantee their accuracy better than up to $Kny^{-\alpha-\kappa}$ (that is not so much different from $o(ny^{-\alpha})$) at least because that is the case even for $n = 1$. Therefore, in order to get a more detailed information on the asymptotic behaviour of the right–hand deviations for S_n it seems reasonable to set supplementary constraints on the asymptotics of $1 - F(x)$ as $x \to \infty$.

To this aim, both for non–integer values of parameter α and for integer ones it should be required that the right–hand tail of the distribution function $F(x)$ can be expanded via negative powers of x as $x \to \infty$. Set $\alpha = \alpha_1$.

THEOREM 2.1. *Let conditions* (0.5) *and* (1.1) *be met with an integer* $\alpha_1 \geqslant 3$; $MX_i = 0$, $DX_i = 1$. *Let us assume, in addition, that r.v.* X_1 *is not a lattice variable if* $\alpha_1 = 3$ *or condition* (C) *be fulfilled if* $\alpha \geqslant 4$. *Let* $\Theta > 0$ *be any real number.*

Then for any integer $n \geqslant 1$ and for any real $y > 0$ the following assertions are true:

$$P\{S_n > y\} = n\sum_{i=1}^{l} c_{\alpha_i} y^{-\alpha_i}\left(1 + \sum_{m=2}^{m(\alpha_1,\Theta)} (-1)^m C^m_{-\alpha_i} y^{-m}(n-1)^{m/2} L(m, n-1)\right) +$$

$$+\bar{r}_1(n,y) + \bar{r}_2(n,y), \tag{2.1}$$

where

$$L(m, 0) = 1 \quad \text{for any integer} \quad m \geqslant 2;$$

$$L(2, n-1) = 1 \quad \text{for any integer} \quad n \geqslant 1;$$

$$L(m, n-1) = \int_{-\infty}^{\infty} v^m \, d\left(\Phi(v) + \sum_{\nu=1}^{m-2} Q_\nu(v)/(n-1)^{\nu/2}\right)$$

if $n \geqslant 2$ for any $1 < m \leqslant \alpha_1 - 1$;

$$L(m, n-1) = \int_{-\infty}^{\infty} v^m \, d\left(\Phi(v) + \sum_{\nu=1}^{\alpha_1-2} Q_\nu(v)/(n-1)^{\nu/2} - \frac{e_\alpha \ln n}{2(\alpha-1)! n^{(\alpha-2)/2}} \Phi^{(\alpha)}(v)\right)$$

if $n \geqslant 2$ for any $m \geqslant \alpha_1$; $m(\alpha_1, \Theta) = \alpha_1 + 1 + [(\alpha_1 \pm 2)/2\Theta]$; $C^m_{-\alpha_i}$ stands for the coefficient at t^m in the Taylor expansion of $(1+t)^{-\alpha_i}$ near zero;

$$|\bar{r}_1(n,y)| \leqslant n \sup_{x \geqslant y/3} |1 - F(x) - \sum_{i=1}^{l} c_{\alpha_i} x^{-\alpha_i}|;$$

$$|\bar{r}_2(n,y)| \leqslant K_1(ny^{-\alpha_1})^2$$

for any integer $n \geqslant 1$ and $y \geqslant K(\Theta)n^{1/2+\Theta}$, where $K(\cdot)$ is some function from $\mathbf{R}^1_+$ to $\mathbf{R}^1_+$.

Proof of Theorem 2.1 relies on the same techniques as in (Vinogradov, 1985) and Section 6.3 of (Wentzell, 1986), and therefore is omitted.

Acknowledgement

The author thanks I.A. Ibragimov and A.D. Wentzell for helpful discussions.

REFERENCES

Bikjalis, A. (1966). Estimates of the remainder term in the central limit theorem. *Liet. Matem. Rink.* **6**, 323–346.

Cramér, H. (1963). On asymptotic expansions for sums of independent random variables with a limiting stable distribution. *Sankhya* **A25**, 13–24.

Fedoruk, M.V. (1987). *Asymptotics: Integrals and Series.* Nauka, Moscow.

Gnedenko, R.V. and Kolmogorov, A.N. (1954). *Limiting Distributions for Sums of Independent Random Variables.* Addison–Wesley, Cambridge.

Heyde, C.C. (1968). On large deviation probabilities in the case of attraction to a non–normal stable law. *Sankhya* **A30**, 253–258.

Ibragimov, I.A. (1967). On Chebyshev–Cramér asymptotic expansions. *Theory Probab. Appl.* **12**, 455–469.

Ibragimov, I.A. and Linnik, Yu.V. (1971). *Independent and Stationary Sequences of Random Variables.* Wolters–Noordhoff, Groningen.

Linnik, Yu. V. (1961). On the probability of large deviations for the sums of independent variables. In: *Proc. Fourth Berkeley Symp. on Math. Statist. and Probab.*, 1, 2, Univ. California Press, Berkeley–L.A., pp. 289–306.

Michel, R. (1976). On the accuracy of nonuniform Gaussian approximation to the distribution functions of sums of independent and identically distributed random variables. *Z. Wahrsch. verw. Geb.* **35**, 337–347.

Nagaev, A.V. (1970). Large deviation probabilities for sums of independent random variables *D. Sci. Thesis*, Tashkent.

Nagaev, S.V. (1979). Large deviations of sums of independent random variables. *Ann. Probab.* **7**, 745–789.

Osipov, L.V. (1972). On asymptotic expansions for the distribution function of sum of independent random variables with nonuniform estimates of the remainder term. *Bull. Leningrad Univ.* **1**, 51–59.

Petrov, V.V. (1975). *Sums of Independent Random Variables.* Springer, Berlin.

Vinogradov, V.V. (1985). Asymptotic expansions of the probabilities of large deviations for sums of independent random variables under violation of Cramér's condition. *Theory Probab. and Math. Statist.* **31**, 21–27.

Vinogradov, V.V. (1988). On asymptotic expansions in limit theorems for sums of independent random variables in the case of power tails. *Depon. VINITI* No 8859–B88.

Wentzell, A.D. (1990). *Limit Theorems on Large Deviations for Markov Stochastic Processes.* Kluwer, Dordrecht.

Prob. Theory and Math. Stat., Vol. 2, pp. 563–568
B.Grigelionis *et al.* (Eds.)
1990 VSP/Mokslas

ON STOCHASTIC CONVOLUTIONS MODELS

V.E. VOL'KOVICH
Chilanzar, kv. 23, building 1A, apt. 5, 700139 Tashkent, USSR.

ABSTRACT

We study models of stochastic convolutions which generalized and unify many convolutions known before.

The theory of stochastic convolutions arise from the study of various operations on probability measures which appear in many branches of mathematics. An example of such operations is provided by hypergroups. In the field of hypergroups analogs of its convolution of measures were studied on such structures as homogeneous space, generalized translation space, Sturm–Liouville systems, spherical systems and others. Typical examples of hypergroups are provided by the co-set space $G//H$ of a locally compact group G by a compact subgroup H of G. Thus, some results from the probability theory on hypergroups find application to the problem of extension in orthogonal polynomials (Lasser, 1983a; 1983b; 1984). Many facts of the probability theory on hypergroups have been accounted in (Heyer, 1983; 1984).
Other examples of stochastic convolutions are provided by the Urbanik convolutions. These convolutions have been introduced in (Urbanik, 1964) by a set of algebraic and topological axioms. The properties of the Urbanik convolutions have been studied by many authors. Examples of those regular convolutions which are not hypergroups' convolutions can be found in (Kucharczak and Urbanik, 1987; Vol'kovich, 1984).
Yet other convolutions of measures are known (see, for example (Askey and Gasper, 1977; Levitan, 1960)). Although these operations have been introduced by various tools, their properties are similar to the classical convolutions properties.
In this paper we shall study generalized models of convolutions which make it possible to consider the above mentioned convolutions from one point of view.
Let X be a locally compact of a complete metric separable space. By $W(X)$ we denote the Banach space of all bounded Radon measures on X and by $B(X)$ its subset of all probability measures on X furnished with the weak topology τ. The Dirac measure at a point $x \in X$ will be denoted by δ_x. Also we use the following notation. $L(X)$ is the Banach space of all Borel complex bounded function on X, $\|f\|$ =supvrai (f), $C(X) \subset L(X)$ is the subspace of all continuous functions in $L(X)$. If X is a locally compact space, then $\tilde{C}(X)$ is the space of all continuous functions. Its topology is the topology of uniform convergence on compact subsets of X.

A binary operation $\circ$ on $B(X)$ is called a stochastic convolution (s.c.) if it has the following properties:

a) $(B(X), \circ)$ is a commutative topological semigroup with unit δ_{x_0}, where $x_0 \in X$.

b) There exists a continuous map $x \to x^*$ from X into X such that for all $x \in X$, $x = x^{**}$ and $\mu_1^* \circ \mu_2^* = (\mu_1 \circ \mu_2)^*$, where $\mu^*(A) = \mu(A^*)$.

c) For all $a, b \geqslant 0$, $a + b = 1$, and for $\mu_1, \mu_2, \mu_3 \in B(X)$

$$(a\mu_1 + b\mu_2) \circ \mu_3 = a(\mu_1 \circ \mu_3) + b(\mu_2 \circ \mu_3).$$

The term "stochastic" is used because typically the measure $\mu = \delta_{x_1} \circ \delta_{x_2}$ is not a Dirac measure.
We shall say, that $\circ$ is weakly regular, if there exist a locally compact separable space M and a function $\omega \in L(X \times M)$ such that the map

$$\Phi(\mu, m) = \int_M \omega(x, m)\mu(dx), \quad \mu \in B(X),$$

of $B(X)$ into $L(M)$ has the following properties:

a) $\mu_3 = \mu_2 \circ \mu_1$ if and only if

$$\Phi(\mu_3, m) = \Phi(\mu_2, m)\Phi(\mu_1, m)$$

for all $m \in M$; $\mu_1, \mu_2, \mu_3 \in B(X)$;

b) If $\eta = \mu^*$, then for all $m \in M$, $\mu \in B(X)$

$$\Phi(\eta, m) = \overline{\Phi(\mu, m)}.$$

Then $\Phi(\mu, m)$ be an algebraic homomorphism of the ring $R = (W(X), \circ, *, +)$ into $L(M)$. We have the ring because it can be extended to a s.c. in the set $W(X)$. The inequality $|\omega(x, m)| \leqslant 1$ for all $x \in X$ and $m \in M$ is true.
The integer transformation $\Phi(\mu, m)$ can be used to study the s.c. Usually it is supposed that ω is a continuous kernel. The idea of using Borel kernels to define generalized convolutions is due to Urbanik (Urbanik, 1986). The most natural topology on R is the weak Φ-topology, τ_ω defined by the functions $\Phi(\mu, m)$. This topology is the weakest one in which the functions $\Phi(\mu, m)$ are continuous at every $\mu \in W(X)$. A topology τ on R is regular if the following property is fulfilled: τ is defined by a set of seminorms $\|\cdot\|_i$ on R with

a) $\|\mu_1 \circ \mu_2\|_i \leqslant \|\mu_1\|_i \|\mu_2\|_i$ for all $i \in Z$, $\mu_1, \mu_2 \in W(X)$.

b) If τ^* is the topology on $B(X)$ which is defined by τ, then τ^* is the standard weak topology on $B(X)$.
Also we shall call a s.c. $\circ$ regular if τ_ω is a regular topology.

THEOREM 1. *A s.c. $\circ$ is regular, if and only if there exists a regular topology τ on R such that*

$$\sup_i \|\mu \circ \mu^*\|_i = \sup_i \|\mu\|_i \sup_i \|\mu^*\|_i$$

for every i and $\mu \in R$.

An analog of the Gelfand—Naimark theorem for seminormed rings (Sya Do Shin, 1959) has been used to prove this theorem.
We shall introduce the dual object for a regular s.c. $\circ$. We shall say that $\circ$ is binary s.c. if $\circ$ is regular s.c. and for every $m_1, m_2 \in M^0$ and x there exists $\sigma(m_1, m_2, \cdot) \in B(M^0)$ such that

$$\omega(x, m_1)\omega(x, m_2) = \int\limits_{M^0} \omega(x, m)\sigma(m_1, m_2, dm). \tag{1}$$

In this case a s.c. $\circ'$ on $B(M^0)$ can be defined by (1). In general $X \subset M^{0'}$. A s.c. $\circ$ is called strong if $X = M^{0'}$. In general a s.c. is not a binary s.c. For binary s.c. one can construct the theory of positive and negative definite functions. In the theory of hypergroup the object analogous to the binary s.c. has been called the dual hypergroup with respect to point-wise multiplication (see (Lasser, 1983a,b; 1984; Heyer, 1984)). The proposed approach to the definition of a binary object is similar that of Vershik and Kerov (Vershik, 1972; Kerov, 1974). For example, a continious function f on X is called $\circ$ positive definite ($\circ$ p.d.) if for each set of complex numbers $\{c_i\}_{i=1}^n$ and points $\{x_i\}_{i=1}^n$ in X one has

$$\sum_{i,j=1}^{n} T^{x_i} f(x_j^*)c_i\bar{c}_j \geqslant 0,$$

where for $f \in L(X)$ and $x_1, x_2 \in X$

$$T^{x_1} f(x_2) = \int\limits_{X} f(x)(\delta_{x_1} \circ \delta_{x_2})(dx).$$

THEOREM 2. *Let $\circ$ be a binary strong s.c. Then f is $\circ'$ p.d. if and only if $f(m) = \Phi(\mu, m)$ for an $m \in M^0$ and some positive measure μ.*

In (Vol'kovich, 1985) we introduced a model of symmetric stochastic operations. In our present terms these operations are regular s.c. for which identical involution map and the following axiom are fulfilled:

A. If $\hat{B}$ is a relatively compact subset, then the set $F(\hat{B})$ of all divisors in the sence of $\circ$ in $B(X)$ is a relatively compact set too.
If a s.c. is a binary s.c., then a sufficient condition can be given for A to be true.

THEOREM 3. *Let $\circ$ be a binary strong s.c. and for all $t \in \mathbf{R}_+$ there exist measures μ_t in $B(M^0)$ such that*

$$\Phi'(\mu_t, x) = \int\limits_{M^0} \omega(x, m)\mu_t(dm) = \exp(-t\varphi(x))$$

for every $x \in X$. Here $\varphi(x)$ is a positive continuous function and for every $c \geqslant 0$ sets

$$A_c = \{x \in X : \varphi(x) \leqslant c\}$$

are compact.

For example, the binary strong polynomial hypergroups convolution and regular Urbanik's convolutions satisfy the conditions of this theorem. Below we shall assume that $\circ$ is a binary s.c. Infinitely divisible (i.d.) measure γ is called Gaussian if $\gamma = \mu \circ e(F)$, where μ is i.d., implies $F = \delta_{x_0}$. The probability measure $e(F)$ is defined by

$$\Phi(e(F), m) = \exp\left(\int\limits_X (\omega(x,m) - 1)F(dx)\right),$$

where F is a non–negative σ–finite measure on X. Of course, we suppose also that F satisfies the following conditions:

a) $\int\limits_{X \setminus x_0} (1 - \omega(x,m))F(dx) < \infty$ for all $m \in M$;

b) $F(X\setminus V) < \infty$ for any neighbourhood V of x_0 in X. We shall call F a spectral measure. It is an analog of the Poisson measure if F is finite. Here $\Phi(\gamma, m) \neq 0$ for all $m \in M$ and the function $f(m) = \log \Phi(\gamma, m)$ is a solution of the following equation

$$\int\limits_{M^0} f(m)(\delta_{m_1} \circ' \delta_{m_2})(dm) = f(m_1) + f(m_2).$$

THEOREM 4. *Let μ be an i.d. measure. Then there exist an idempotent measure η, a Gaussian measure γ and a spectral measure F such that measure $\mu = \eta \circ \gamma \circ e(F)$. The measures η, γ, F are determined uniquely by the measure μ.*

This theorem is an analog of the Lévy—Khinchine formulae for binary strong s.c. An analog of such formulae for binary polynomial hypergroups can be found in Theorem 4 without property F of hypergroup.
If $\circ$ is the regular Urbanik's convolution, then the measure γ is nontrivial Gaussian if and only if γ is the characteristic measure of $\circ$. The regular Urbanik's convolutions are binary strong convolutions and $M = X = \mathbf{R}_+$ we have

$$\Phi(\gamma, t) = \exp(-ct^\alpha),$$

here $c > 0$, and α is the characterization exponent of s.c. $\circ$. In this case the definition of i.d. distribution can be generalized. We shall call μ a γ i.d. measure if for every n there exist $a_n \geqslant 0$, $\mu_n \in B(\mathbf{R}_+)$ such that

$$e^{-a_n t^\alpha} \Phi(\mu, t) = \Phi^n(\mu_n, t)$$

for all $t \in \mathbf{R}_+$. If $a_n = O(1)$, as $n \to \infty$ then μ is an ordinary i.d. measure. Suppose

$$a_n = O(1) n^{1-\frac{1}{m}} \tag{2}$$

for some $m \geqslant 1$. In this case an analog of Lévy–Khinchin formulae can be found. Let

$$\omega_k(t) = \omega(t) - \sum_{j=0}^{k} a_j t^{\alpha_j}$$

for some $k \geqslant 0$ and $t \in \mathbf{R}_+$. The set $(a_0, \ldots, a_k)$ is defined by the property $\omega_k = o(t^{\beta+\varepsilon})$ as $t \to 0$ for every $\varepsilon > 0$, where β is maximal. In this case for all $t \geqslant 0$, we have

$$\log \Phi(\mu, t) = \sum_{j=1}^{m-1} b_j t^{\alpha_j} + \int_0^1 \omega_{m-1}(ut) G_1(du) + \int_1^\infty \omega_0(ut) G_2(du),$$

where G_2 is a finite measure on $[1, \infty)$, G_1 is a σ–finite measure on (0,1) and $G_1([c,1]) < \infty$, $0 < c < 1$,

$$\left| \int_0^1 \omega_{m-1}(ut) G_1(du) \right| < \infty$$

for all $t \in \mathbf{R}_+$.

We shall call μ a γ–stable law if for every $a, b \geqslant 0$ there exists $d > 0$ such that

$$\Phi(at, \mu)\Phi(bt, \mu) = \Phi(ct, \mu) e^{-dt^\alpha}, \; t \in \mathbf{R}_+.$$

Such measures satisfy (2). Moreover,

$$\Phi(\mu, t) = \exp(-c_1 t^\alpha - c_2 t^\beta), \; t \in \mathbf{R}_+,$$

where $c_1, c_2 \geqslant 0$, $\alpha \leqslant \beta$ and $c_1 > 0$ if $\alpha < \beta$. For ordinarily used convolutions these distributions have arised in problems of characterization of probability laws by stochastic properties of linear statistics (Kagan et al., 1972, Chapter 2). In this case

$$\Phi(|\mu|^2, t) = \exp(-c_1 t^2 - c_2 t^\beta), \; t \in \mathbf{R}_+,$$

where $c_2 > 0$, $\beta \neq 2k$, $k \in \mathbf{Z}$, $[\beta/2]$ is even number. If $\beta > 4$, then $c_1 > 0$.

REFERENCES

Askey, R. and Gasper, G. (1977). Convolution structures for Laguerre polynomials. *J. D'Analyse Math.* **37**, 48–68.

Heyer, H. (1983). Convolution semigroups of probability measures on Gelfand pairs. *Expos. Math.* **1**, 3–45.

Heyer, H. (1984). Probability theory on hypergroups: a servey. *Lect. Notes. in Math.* **1064**, 481–550.

Kagan, A.M., Linnik, Yu.V. and Rao, C.R. (1972). *Characterization Problems of Mathematical Statistics.* Nauka, Moscow.

Kerov, S.V. (1974). Duality of finite dimensional C–algebras. *Vestnik Leningrad State Univ.* **7**, 23–29.

Kucharczak, J. and Urbanik, K. (1987). Transformations preserving weak stability. *Bull. Pol. Acad. Math.*, 475–486.

Lasser, R. (1983a). Orthogonal polynomials and hypergroups. *Rend. Math.* **3**, 185–209.

Lasser, R. (1983b). Bochner theorems for hypergroups and their applications to orthogonal polynomial expensions. *J. Appl. Theor.* **37**, 311–325.

Lasser, R. (1984). On the Lévy–Khinchin formula for commutative hypergroups. *Lect. Notes Math.* **1064**, 298–308.

Levitan, B. (1960). Onnone class of solutions for Kolmogorov–Smolukhovsky equation. *Vestnik Leningrad State Univ.*, **7**, 81–115.

Sya Do Shin (1959). On semi–normalized rings with involution. *Izvestiya Acad. Sci. USSR.* **23**, 509–528.

Urbanik, K. (1964). Generalized convolutions. *Studia Math.* **23**, 217–245.

Urbanik, K. (1986). Generalized convolutions. IV. *Studia Math.* **83**, 57–95.

Vershik, A.M. (1972). Geometric theory for state, von Neimann boundary duality of C–algebras. *Notes of Sci. Seminars of Leningrad Branch Math. Inst.* **29**, 147–154.

Vol'kovich, V.E. (1984). Multidimensional$B$–stable distribution and realizations of generalized convolutions. Problems of stability of stochastic models. In: *Proc. VNIISI Seminar*, VNIISI, Moscow, pp. 40–54.

Vol'kovich, V.E. (1985). On infinitely decomposable measures in algebras with stochastic convolutions. Problems of stability of stochastic models. In: *Proc. VNIISI Seminar*, VNIISI, Moscow, pp. 15–24.

Prob. Theory and Math. Stat., Vol. 2, pp. 569–578
B.Grigelionis *et al.* (Eds.)
1990 VSP/Mokslas

SOME PROPERTIES OF NON–HOMOGENEOUS TRANSIENT RENEWAL PROCESSES

W. WAJDA
Adam Mickiewicz University, Institute of Mathematics,
Matejki 48/49, 60–769 Poznań, Poland

ABSTRACT

In this paper a non–homogeneous transient renewal process is defined. This process is generated by a sequence of random variables with distributions belonging to the subexponential class $\mathcal{F}$ introduced in (Chistyakov, 1964). For these processes the asymptotic behaviour of the distributions of forward delay, backward delay and the moments of the renewal process are studied.

1. INTRODUCTION

Define a sequence $(T_n)_{n\in\mathbb{N}}$ of independent and non–negative random variables with distributions $(F^n)_{n\in\mathbb{N}}$ which can vary with n. For any $n \in \mathbb{N}$, let F^n be the defective distribution on $(0,\infty)$ such that $F^n(0+) = 0$, $F^n(t) < F^n(\infty) = \alpha < 1$ for all $t \geqslant 0$. Consider the sequence of random variables $(S_n)_{n\in\mathbb{N}}$ defined as follows

$$S_0 = 0, \qquad S_n = \sum_{k=1}^{n} T_k \quad \text{for} \quad n \geqslant 1,$$

and denote the distribution of S_n by $F^{1,n}$, so

$$F^{1,1} = F^1, \; F^{1,n}(t) = \int_0^t F^n(t-u)F^{1,n-1}(du) \quad \text{for} \quad n > 1.$$

A random process of the following form $N(t) = \sup\{n\colon S_n \leqslant t\}$ is called the non–homogeneous renewal process. If there exists the smallest index J such that $T_J = \infty$ then such process is called non–homogeneous transient renewal process. The renewal function associated with (F^n) is

$$H(t) = \sum_{n=1}^{\infty} F^{1,n}(t).$$

Assume the existence of a real $\beta > 0$ such that for some divergent series $(t_n)_{n\in\mathbb{N}}$ of non–negative reals $1 - F^n(t_n) \geqslant \beta$ for all $n \in \mathbb{N}$. Then $H(t) < \infty$ as it is shown in (Wajda, 1986).

Also define the forward delay $\xi_t = S_{N(t)+1} - t$ and the backward delay $\eta_t = t - S_{N(t)}$. The properties of $N(t)$, ξ_t, η_t will be studied under the condition that the proper distributions $\alpha^{-1} F^n$ belong to the subexponential class $\mathcal{F}$. The class $\mathcal{F}$ introduced in (Chistyakov, 1964) has some applications in the theory of branching processes, queues, random walks and risks. Let G be a proper distribution on $(0,\infty)$ such that $G(x) < 1$ for $x \geqslant 0$; then G belongs to $\mathcal{F}$ if for any $n \in \mathbb{N}$

$$\lim_{x\to\infty} \frac{1-G^{n*}(x)}{1-G(x)} = n. \tag{1.1}$$

Chistyakov proved that the equality (1.1) is true if and only if

$$\lim_{x\to\infty} \frac{1-G^{2*}(x)}{1-G(x)} = 2.$$

He also showed that, if $G \in \mathcal{F}$ then for any $a \in \mathbb{R}$

$$\lim_{x\to\infty} \frac{1-G(x+a)}{1-G(x)} = 1. \tag{1.2}$$

Teugels in (Teugels, 1975) proved the following fact:

LEMMA 1.1. *If $G \in \mathcal{F}$ and $\lim_{x\to\infty} \frac{1-F(x)}{1-G(x)} = c > 0$, then $F \in \mathcal{F}$.*

In this paper we shall assume that for all $m, n \in \mathbb{N}$

$$\lim_{t\to\infty} \frac{1-\alpha^{-1}F^m(t)}{1-\alpha^{-1}F^n(t)} = 1. \tag{1.3}$$

Examples of improper distributions F^n satisfying (1.3) whose rescaled versions belong to $\mathcal{F}$ are

1°. $\alpha - F^n(t) \sim \dfrac{c}{t^{1+b} + nt}$, $\quad b, c > 0$,

2°. $\alpha - F^n(t) \sim \left(t^c + nt^b\right)\exp\left(-t^\delta\right)$, $0 < \delta < 1$, $0 < b < c$,

3°. $\alpha - F^n(t) \sim x^{-b} L_n(x)$, $b > 0$, L_n are slowly varying and $L_n(x)/L_m(x) \to 1$.

Now, we present some additional facts concerned with the class $\mathcal{F}$.

LEMMA 1.2. *If $G^n \in \mathcal{F}$ for any $n \in \mathbb{N}$, then for any $k \in \mathbb{N}$*

$$\lim_{t\to\infty} \int_0^t \frac{1-G^{k+1}(t-n)}{1-G^{k+1}(t)} G^{1,k}(dn) = 1. \tag{1.4}$$

Proof. First we find a lower bound. For all $t \geqslant 0$

$$J(t) \equiv \int_0^t \frac{1-G^{k+1}(t-n)}{1-G^{k+1}(t)} G^{1,k}(dn) \geqslant G^{1,k}(t).$$

On the other hand for any $x \in (0,t)$ we have

$$J(t) = \int_0^x \frac{1 - G^{k+1}(t-n)}{1 - G^{k+1}(t)} G^{1,k}(dn) + \int_x^t \frac{1 - G^{k+1}(t-n)}{1 - G^{k+1}(t)} G^{1,k}(dn).$$

Since the function $(1 - G^{k+1}(t-n))/(1 - G^{k+1}(t))$ has a finite value at any t and there exists a finite limit of this function, there exists a real K such that

$$\frac{1 - G^{k+1}(t-n)}{1 - G^{k+1}(t)} < K \quad \text{for all} \quad t \geqslant 0.$$

Now we can write

$$J(t) \leqslant \frac{1 - G^{k+1}(t-x)}{1 - G^{k+1}(t)} G^{1,k}(x) + K\left(G^{1,k}(t) - G^{1,k}(x)\right).$$

As $t \to 0$, we have

$$1 \leqslant \varliminf_{t\to\infty} J(t) \leqslant \varlimsup_{t\to\infty} J(t) \leqslant G^{1,k}(x) + K\left(1 - G^{1,k}(x)\right)$$

and, finally, letting $x \to \infty$ we get (1.4).

COROLLARY 1.1. *If $G^n \in \mathcal{F}$ for any $n \in \mathbb{N}$, then for any $k \in \mathbb{N}$, and $\varepsilon > 0$ there exists $\delta(\varepsilon, k)$ such that*

$$\lim_{t\to\infty} \int_{\delta(\varepsilon,k)}^t \frac{1 - G^{k+1}(t-n)}{1 - G^{k+1}(t)} G^{1,k}(dn) < \varepsilon. \tag{1.5}$$

Proof. Choose $\delta = \delta(\varepsilon, k)$ such that $G^{1,k}(\delta) > 1 - \varepsilon$. Then

$$J(\delta) \geqslant G^{1,k}(\delta) > 1 - \varepsilon$$

and (1.5) follows.

LEMMA 1.3. *If $G^n \in \mathcal{F}$ for any $n \in \mathbb{N}$ and for all $n, m \in \mathbb{N}$*

$$\lim_{x\to\infty} \frac{1 - G^n(x)}{1 - G^m(x)} = 1$$

then

$$\lim_{x\to\infty} \frac{1 - G^{1,n}(x)}{1 - G^m(x)} = n. \tag{1.6}$$

Proof. For $n = 1$ the equality (1.6) is obviously true. Suppose that (1.6) is true for some $n \in \mathbb{N}$. Then by Lemma 1.2 we have

$$\lim_{x\to\infty} \frac{1 - G^{1,n+1}(x)}{1 - G^m(x)} = \lim_{x\to\infty} \frac{1 - G^{1,n}(x) + G^{1,n}(x) - G^{1,n+1}(x)}{1 - G^m(x)} =$$

$$= n + \lim_{x\to\infty} \frac{1-G^{n+1}(x)}{1-G^{m}(x)} \int_0^x \frac{1-G^{n+1}(x-u)}{1-G^{n+1}(x)} G^{1,n}(du) = n+1.$$

LEMMA 1.4. *If $\alpha^{-1}F^n \in \mathcal{F}$ for each $n \in \mathbb{N}$ and assumption (1.3) holds, then $(H(\infty))^{-1}\, H \in \mathcal{F}$.*

Proof. First we notice that by Lemma 1.3 for $k \in \mathbb{N}$,

$$\lim_{x\to\infty} \frac{\alpha^k - F^{1,k}(x)}{\alpha - F^1(x)} = \lim_{x\to\infty} \alpha^{k-1} \frac{1-\alpha^{-k}F^{1,k}(x)}{1-\alpha^{-1}F^1(x)} = k\alpha^{k-1}. \tag{1.7}$$

It follows that

$$\frac{1-H(x)/H(\infty)}{1-\alpha^{-1}F^1(x)} = \frac{\alpha}{H(\infty)} \sum_{k=1}^{\infty} \frac{\alpha^k - F^{1,k}(x)}{\alpha - F^1(x)},$$

$$\lim_{x\to\infty} \frac{1-H(x)/H(\infty)}{1-\alpha^{-1}F^1(x)} = \lim_{x\to\infty} \frac{\alpha}{H(\infty)} \sum_{k=1}^{\infty} \frac{\alpha^k - F^{1,k}(x)}{\alpha - F^1(x)} =$$

$$= \frac{\alpha}{H(\infty)} \sum_{k=1}^{\infty} k\alpha^{k-1} = \frac{1}{1-\alpha}$$

and, by Lemma 1.1, $(H(\infty))^{-1}H \in \mathcal{F}$.

2. SOME PROPERTIES OF TRANSIENT PROCESSES

We restrict our attention to the processes in which all lifetimes which began before time t are finite. Such processes are called to be "alive" at time t.
Let $A(t) = 1$ if and only if $S_{N(t)+1} < \infty$. Consider $q(t) \equiv \mathbf{P}\{A(t) = 1\}$:

$$q(t) = \mathbf{P}\{S_{N(t)+1} < \infty\} = \alpha - F^1(t) +$$

$$+ \sum_{n=1}^{\infty} \int_0^t (\alpha - F^{n+1}(t-x)) F^{1,n}(dx) =$$

$$= \alpha + \alpha H(t) - H(t) = (1-\alpha)\big(H(\infty) - H(t)\big).$$

Because $q(t)$ depends on the tail of H, the conditional properties of ξ_t, η_t and $N(t)$ will depend upon various ratios involving the tails of (F^n) and H.
We shall see that conditioning $\alpha^{-1}\ F^n \in \mathcal{F}$ on $\{A(t) = 1\}$ does not appreciably affect the moments of $N(t)$. The conditional and unconditional moments have the same finite limits. Also, the conditional asymptotic distributions of the forward and backward delays are degenerated.

LEMMA 2.1. *If $\alpha^{-1}F^n \in \mathcal{F}$ for any $n \in \mathbb{N}$ and assumption (1.3) holds, then*

$$\lim_{t\to\infty} \mathbf{P}\{T_{N(t)+1} > t \mid A(t) = 1\} = 1.$$

Proof. By (1.3) and (1.7)

$$\mathbf{P}\left\{T_{N(t)+1} > t \mid A(t) = 1\right\} =$$

$$= \frac{1}{q(t)}\left(\alpha - F^1(t) + \sum_{k=1}^{\infty}\int_0^t \left(\alpha - F^{k+1}(t)\right) F^{1,k}(dx)\right) =$$

$$= \frac{\alpha - F^1(t)}{q(t)} + \frac{\sum_{k=1}^{\infty}\left(\alpha - F^{k+1}(t)\right) F^{1,k}(t)}{q(t)} =$$

$$= \frac{\alpha - F^1(t)}{q(t)} + \frac{\sum_{k=1}^{\infty}\left(\alpha - F^{k+1}(t)\right) F^{1,k}(t)}{(1-\alpha)(\alpha - F^1(t))}\frac{\alpha - F^1(t)}{H(\infty) - H(t)} \to$$

$$\to 1 - \alpha + \frac{\alpha}{(1-\alpha)^2}(1-\alpha)^2 = 1.$$

Therefore, if the process is alive at some distant time t, this seems to be because one of the lifetimes began prior to the instant t is itself longer than t. Evidently we have not observed a large number of moderate lifetimes; processes built up of many such "reasonable" T_n would have died before reaching time t.
In view of this result about $T_{N(t)+1} = \eta_t + \xi_t$, it is not surprising that the forward and backward delays have degenerated limiting distributions.

LEMMA 2.2. *Suppose that $\alpha^{-1}F^n \in \mathcal{F}$ for any $n \in \mathbb{N}$ and assumption* (1.3) *holds. Then for all fixed x*

$$\lim_{t\to\infty} \mathbf{P}\{\xi_t \leqslant x | A(t) = 1\} = 0,$$
$$\lim_{t\to\infty} \mathbf{P}\{\eta_t \leqslant x | A(t) = 1\} = 0.$$

Proof. Take any $x \geqslant 0$ and observe that

$$\mathbf{P}\{\xi_t \leqslant x\} = F^1(t+x) - F^1(t) +$$

$$+ \sum_{n=1}^{\infty}\int_0^t \left(F^{n+1}(t+x-u) - F^{n+1}(t-u)\right) F^{1,n}(du) =$$

$$= F^1(t+x) - F^1(t) + \sum_{n=1}^{\infty}\left(\int_0^{t+x} F^{n+1}(t+x-u)F^{1,n}(du) -\right.$$

$$\left. - \int_t^{t+x} F^{n+1}(t+x-u)F^{1,n}(du) - \int_0^t F^{n+1}(t-u)F^{1,n}(du)\right) =$$

$$= H(t+x) - H(t) - \sum_{n=1}^{\infty}\int_t^{t+x} F^{n+1}(t+x-u)F^{1,n}(du).$$

Hence,

$$\mathbf{P}\{\xi_t \leqslant x \mid A(t) = 1\} =$$

$$= \frac{H(t+x) - H(t) - \sum\limits_{n=1}^{\infty} \int\limits_{t}^{t+x} F^{n+1}(t+x-u) F^{1,n}(du)}{(1-\alpha)(H(\infty) - H(t))} \leqslant$$

$$\leqslant \frac{H(t+x) - H(t)}{(1-\alpha)(H(\infty) - H(t))} \to 0 \quad \text{as} \quad t \to \infty,$$

by Lemma 1.4 and equality (1.2).
Similarly, for a fixed $x \in (0,t)$ we get

$$\mathbf{P}\{\eta_t \leqslant x \mid A(t) = 1\} =$$

$$= \frac{\alpha(H(t) - H(t-x)) - \sum\limits_{n=1}^{\infty} \int\limits_{t-x}^{t} F^{n+1}(t-u) F^{1,n}(du)}{(1-\alpha)(H(\infty) - H(t))} \leqslant$$

$$\leqslant \frac{\alpha}{1-\alpha} \frac{H(t) - H(t-x)}{H(\infty) - H(t)} \to 0 \text{ as } t \to \infty.$$

The above lemmas warn us that even transient processes alive at time t do not behave like proper processes.

THEOREM 2.1. *Suppose that $\alpha^{-1}F^n \in \mathcal{F}$ for any $n \in \mathbb{N}$ and assumption (1.3) holds. Then*

$$\lim_{t\to\infty} \mathbf{P}\{\eta_t \leqslant t - c \mid A(t) = 1 \wedge N(t) \geqslant 1\} = \frac{H(\infty) - H(c)}{H(\infty)}.$$

Proof. Let us take any $c \leqslant t$. Then

$$B(t,c) \equiv \mathbf{P}\{\eta_t \leqslant t - c \mid A(t) = 1 \wedge N(t) \geqslant 1\} =$$

$$= \sum_{n=1}^{\infty} \int\limits_{c}^{t} \left(\alpha - F^{n+1}(t-u)\right) F^{1,n}(du) \Big/ \sum_{n=1}^{\infty} \int\limits_{0}^{t} \left(\alpha - F^{n+1}(t-u)\right) F^{1,n}(du).$$

Therefore for $x \in (c,t)$ we have

$$B(t,c) = \frac{\sum\limits_{n=1}^{\infty} \Big(\int\limits_{c}^{x} (\alpha - F^{n+1}(t-u)) F^{1,n}(du) + \int\limits_{x}^{t} (\alpha - F^{n+1}(t-u)) F^{1,n}(du) \Big)}{F^1(t) - (1-\alpha)H(t)}$$

and

$$B(t,c) \leqslant \left(F^1(t) - (1-\alpha)H(t)\right)^{-1} \sum_{n=1}^{\infty} \Big(\left(\alpha - F^{n+1}(t-x)\right) \left(F^{1,n}(t) - F^{1,n}(c)\right) +$$

$$+\alpha^{n+1}\int_x^t \left(1-\frac{F^{n+1}(t-u)}{\alpha}\right)\frac{F^{1,n}(du)}{\alpha^n}\bigg).$$

Corollary 1.1 ensures that we can choose x to make the last integral less than an arbitrarily small ε. Hence

$$B(t,c) \leqslant \sum_{n=1}^{\infty} \frac{\alpha - F^{n+1}(t-x)}{H(\infty)-H(t)} \frac{H(\infty)-H(t)}{F^1(t)-(1-\alpha)H(t)} \times$$

$$\times \left(F^{1,n}(t) - F^{1,n}(c)\right) + \sum_{n=1}^{\infty} \varepsilon \alpha^{n+1}$$

and

$$\overline{\lim_{t\to\infty}} B(t,c) \leqslant \sum_{n=1}^{\infty} (1-\alpha)^2 \frac{1}{\alpha(1-\alpha)} \left(F^{1,n}(t) - F^{1,n}(c)\right) =$$

$$= \frac{1-\alpha}{\alpha}\left(H(\infty)-H(c)\right).$$

On the other hand,

$$B(t,c) \geqslant \left(F^1(t) - (1-\alpha)H(t)\right)^{-1} \sum_{n=1}^{\infty} \left(\alpha - F^{n+1}(t-c)\right)\left(F^{1,n}(t) - F^{1,n}(c)\right)$$

and

$$\underline{\lim_{t\to\infty}} B(t,c) \geqslant \frac{1-\alpha}{\alpha}\left(H(\infty)-H(c)\right).$$

Thus the proof is completed.

THEOREM 2.2. *Suppose that $\alpha^{-1}F^n \in \mathcal{F}$ for any $n \in \mathbf{N}$ and assumption (1.3) holds. Then for any non-negative function $u(t)$ and $i,j \in \mathbf{N}$*

$$\lim_{t\to\infty} \mathbf{P}\{\xi_t \leqslant u(t) \mid A(t)=1\} =$$

$$= 1 - \lim_{t\to\infty} \frac{F^i(\infty) - F^i(t+u(t))}{F^j(\infty) - F^j(t)}.$$

Proof. It is clear that

$$\mathbf{P}\{\xi_t \leqslant u(t) \mid A(t)=1\} =$$

$$= \frac{H(t+u(t)) - H(t) - \sum_{n=1}^{\infty} \int_t^{t+u(t)} F^{n+1}(t+u(t)-v)\, F^{1,n}(dv)}{(1-\alpha)(H(\infty)-H(t))} =$$

$$= \frac{H(t+u(t))-H(t)}{H(\infty)-H(t)} + \sum_{n=1}^{\infty} \int_{t}^{t+u(t)} \frac{(\alpha - F^{n+1}(t+u(t)-v)\, F^{1,n}(dv)}{(1-\alpha)(H(\infty)-H(t))}.$$

The second summand can be transformed as follows

$$J(t) = \frac{1}{1-\alpha} \sum_{n=1}^{\infty} \int_{t}^{t+u(t)} \frac{\alpha - F^{n+1}(t+u(t)-v)}{H(\infty)-H(t+u(t)-v)} \times$$

$$\times \frac{H(\infty)-H(t+u(t)-v)}{H(\infty)-H(t+u(t))} F^{1,n}(dv).$$

The function $(\alpha - F^{n+1}(v))/(H(\infty)-H(v))$ is finite for all v and has a limit as $v \to \infty$. Hence, by (1.3), there exists K such that

$$(\alpha - F^{n+1}(v))/(H(\infty)-H(v)) < K$$

for all v and n.
Thus

$$J(t) \leqslant \frac{K}{1-\alpha} \int_{t}^{t+u(t)} \frac{H(\infty)-H(t+u(t)-v)}{H(\infty)-H(t+u(t))} H(dv)$$

which — by Corollary 1.1 — converges to 0 as $t \to \infty$.
Therefore,

$$\lim_{t\to\infty} \mathbf{P}\{\xi_t \leqslant u(t) \mid A(t)=1\} = \lim_{t\to\infty} \frac{H(t+u(t))-H(t)}{H(\infty)-H(t)} =$$

$$= 1 - \lim_{t\to\infty} \frac{H(\infty)-H(t+u(t))}{H(\infty)-H(t)} = 1 - \lim_{t\to\infty} \frac{F^{i}(\infty)-F^{i}(t+u(t))}{F^{j}(\infty)-F^{j}(t)}$$

for any i, j.
Now we turn to the moments of $N(t)$.

Lemma 2.3. *Suppose that $\alpha^{-1}F^{n} \in \mathcal{F}$ for any $n \in \mathbf{N}$ and assumption (1.3) holds. Then*

$$\lim_{t\to\infty} \mathbf{P}\{N(t)=k \mid A(t)=1\} = \lim_{t\to\infty} \mathbf{P}\{N(t)=k\} = \alpha^{k}(1-\alpha).$$

Proof. It is clear that

$$P\{N(t)=k|A(t)=1\} = (q(t))^{-1} \int_{0}^{t} (\alpha - F^{k+1}(t-x))F^{1,k}(dx) =$$

$$= (q(t))^{-1}(\alpha F^{1,k}(t) - F^{1,k+1}(t)) =$$

$$= \left(\frac{\alpha^{k+1} - F^{1,k+1}(t)}{\alpha - F^1(t)} - \frac{\alpha(\alpha^k - F^{1,k}(t))}{\alpha - F^1(t)} \right) \frac{\alpha - F^1(t)}{(1-\alpha)(H(\infty) - H(t))} \to$$

$$\to (\alpha^k(k+1) - \alpha^k k)(1-\alpha) = \alpha^k(1-\alpha) \text{ as } \quad t \to \infty.$$

On the other hand,

$$\mathbf{P}\{N(t) = k\} = F^{1,k}(t) - F^{1,k+1}(t) \to \alpha^k(1-\alpha) \text{ as } t \to \infty.$$

Remark. If $N(t)$ is the renewal process for proper distributions then $\lim_{t\to\infty} \mathbf{P}\{N(t) = k\} = 0$ for all fixed k.

THEOREM 2.3. *Let $k \in \mathbf{N}$.*

(i) *If $\alpha^{-1}F^n \in \mathcal{F}$ for any $n \in \mathbf{N}$ and* (1.3) *holds, then*

$$\lim_{t\to\infty} \mathbf{E}(N(t)^k | A(t) = 1) = \lim_{t\to\infty} \mathbf{E}(N(t)^k) =$$

$$= \sum_{i=1}^{k} \frac{(-1)^i \alpha^i}{(1-\alpha)^i} \sum_{j=1}^{i} (-1)^j \binom{i}{j} j^k. \qquad (2.1)'$$

(ii) *If equality* (2.1) *is true for any $k \in \mathbf{N}$, then $(H(\infty))^{-1}H \in \mathcal{F}$.*

Proof. Suppose that $\alpha^{-1}F^n \in \mathcal{F}$ and (1.3) holds for any n.
Then by Lemma 2.3

$$\lim_{t\to\infty} \mathbf{E}\left(N(t)^k \mid A(t) = 1\right) = \sum_{n=1}^{\infty} n^k \alpha^n (1-\alpha) = \lim_{t\to\infty} \mathbf{E}\left(N(t)^k\right)$$

and finally

$$\lim_{t\to\infty} \mathbf{E}\left(N(t)^k \mid A(t) = 1\right) = \lim_{t\to\infty} \mathbf{E}\left(N(t)^k\right) =$$

$$= \sum_{i=1}^{k} \frac{(-1)^i \alpha^i}{(1-\alpha)^i} \sum_{j=1}^{i} (-1)^j \binom{i}{j} j^k.$$

On the other hand, let $k = 1$

$$\mathbf{E}\,(N(t) \mid A(t) = 1) =$$

$$= \mathbf{E} \sum_{n=1}^{\infty} I\{S_n \leqslant t \wedge A(t) = 1\}(q(t))^{-1} =$$

$$= \sum_{n=1}^{\infty} \int_0^t \frac{q(t-u)}{q(t)} F^{1,n}(du) = \int_0^t \frac{H(\infty) - H(t-u)}{H(\infty) - H(t)} H(du) =$$

$$= \frac{(H(\infty))^2 - H^{2*}(t)}{H(\infty) - H(t)} - \frac{(H(\infty))^2 - H(\infty)H(t)}{H(\infty) - H(t)} =$$

$$= \frac{(H(\infty))^2 - H^{2*}(t)}{H(\infty) - H(t)} - \frac{\alpha}{1-\alpha}.$$

By assumption $\lim_{t\to\infty} \mathbf{E}(N(t) \mid A(t) = 1) = \alpha/(1-\alpha)$, so

$$\lim_{t\to\infty} \frac{(H(\infty))^2 - H^{2*}(t)}{H(\infty) - H(t)} = 2\alpha/(1-\alpha).$$

This implies $(H(\infty))^{-1} H \in \mathcal{F}$.
Similar results for homogeneous transient renewal processes are obtained in (Murphree, 1987).

REFERENCES

Chistyakov, V.P. (1964). A theorem on sums of independent positive random variables. *Theory Probab. Appl.* **9**, 640–648.

Murphree, E.S. (1987). Transient renewal processes in the subexponential case. *J. Appl. Probab.* **24**, 88–96.

Teugels, J. (1975). On the class of subexponential distributions. *Ann. Probab.* **3**, 1000–1011.

Wajda, W. (1986). The renewal functions for the non-homogeneous renewal process. *Fasciculi Mathematici* **16**, 95–100.

Prob. Theory and Math. Stat., Vol. 2, pp. 579–584
B.Grigelionis *et al.* (Eds.)
1990 VSP/Mokslas

SOME LARGE DEVIATION RESULTS FOR U–STATISTICS

W. WOLF
Technische Universität, Dresden, DDR.

ABSTRACT

Using uniform and non–uniform estimates of the difference between the distribution functions of a normalized U–statistics and a given stable distribution function with characteristic exponent α, $1 < \alpha < 2$, we determine the range of large deviations.

I. Let $X_1, \ldots, X_n$ be i.i.d. r.v.'s with the common distribution function (d.f.) F. If F belongs to the normal domain of attraction of a stable law G_α, we write $F \in D(G_\alpha)$ or $X \in D(G_\alpha)$. Let $F_n(x) = \mathbf{P}(X_1 + \ldots + X_n - A_n < xB_n)$ denote the standardized sum with certain centering and norming constants the distribution function of A_n and $B_n > 0$, respectively. Finally, let $G_{\alpha\beta}(x) = G_\alpha(x)$ denote the stable d.f. with the characteristic exponent $1 < \alpha \leqslant 2$, the scale parameter $\lambda \geqslant 0$, the skewness parameter $\beta \in [-1, 1]$ and the characteristic function (c.f.)

$$g_\alpha(t; \lambda, \beta) = \begin{cases} \exp\Big(- \lambda |t|^\alpha \Big(1 - \beta \operatorname{sgn} t \ \tan \alpha\pi/2\Big)\Big), & 1 < \alpha < 2, \\ \exp(-\lambda\, t^2), & \alpha = 2. \end{cases}$$

Let $r \geqslant 0$ be a real number and $k \geqslant 0$ be an integer. For $F \in D(G_{\alpha\beta})$ we introduce the following pseudomoments (p.m.'s):
$\nu_r = \nu_r(F - G_{\alpha\beta}) = \int |x|^r \Big| d\big(F(x) - G_{\alpha\beta}(x)\big)\Big|$ — absolute p.m. of r–th order,
$\mu_k = \mu_k(F - G_{\alpha\beta}) = \int x^r d\big(F(x) - G_{\alpha\beta}(x)\big)$ — p.m. of k–th order,
$\kappa_r = \int x^{r-1} \big|F(x) - G_{\alpha\beta}(x)\big| dx$, $r \geqslant 1$, — absolute difference moment of r–th order.
P.m.'s were introduced by Bergström (Bergström, 1952) and Zolotarev (Zolotarev 1962; 1965) and studied by several authors. For a complete investigation see (Christoph, 1989).
Let $H(x_1, \ldots, x_m)$ be a symmetric function of m arguments. For $n \geqslant 2$ define a U–statistic by

$$U_n = \binom{n}{m}^{-1} \sum_{1 \leqslant i_1 < \ldots < i_m \leqslant n} H(X_{i_1}, \ldots, X_{i_m}), \quad 1 < m \leqslant n.$$

If $\mathbf{E}|H(X_1,\dots,X_m)| < \infty$ then, without loss of generality, put $\mathbf{E}U_n = \mathbf{E}H(X_1,\dots,X_m) = 0$ and we can decompose U_n as follows (Hoeffding, 1948)

$$U_n = S_n + R_n, \text{ where } S_n = \frac{m}{n}\sum_{i=1}^{n} Q_1(X_i), \quad R_n = \sum_{l=2}^{m}\binom{m}{l}U_n^{(l)},$$

$$U_n^{(l)} = \binom{n}{l}^{-1} \sum_{1\leqslant i_1<\dots<i_l\leqslant n} H_l(X_{i_1},\dots,X_{i_l})$$

with

$$H_l(X_{i_1},\dots,X_{i_l}) = \sum_{j=1}^{l}(-1)^{l-j} \sum_{1\leqslant i_1\leqslant k_1<\dots<k_j\leqslant i_l<n} Q_j(X_{k_1},\dots,X_{k_j}),$$

$$Q_j(X_1,\dots,X_j) = \mathbf{E}\big(H(X_1,\dots,X_m)/X_1,\dots,X_j\big).$$

There is an important difference between p.m.'s and moments.

i) If we study the closeness of $F_n(x)$ or $\mathbf{P}(B_nU_n < x)$ to the normal d.f. $G_2(x) = \Phi(x)$ under moment conditions then we have the following Berry—Esseen estimates (Berry, 1941; Callaert and Janssen, 1978; Esseen, 1945)

$$\big|F_n(x) - \Phi(x)\big| \leqslant C_1\frac{\mathbf{E}|X_1|^3}{\sqrt{n(DX_1)^3}}, \tag{1}$$

$$\big|\mathbf{P}(B_nU_n < x) - \Phi(x)\big| \leqslant C_2\frac{\mathbf{E}\big|Q_1(X_1)\big|^3}{\sqrt{n\big(DQ_1(X_1)\big)^3}}. \tag{2}$$

These inequalities do not take into account how near is F or $\mathbf{P}\big(Q_1(X_1) < x\big)$ to the d.f. $\Phi(x)$. Throughout the paper, let $C_i, \bar{C}_i$, $i = 1,2,\dots$, denote absolute constants, which do not depend on n and x.

ii) It is well-known that if $X \in D(\Phi)$, then $\mathbf{E}|X_1+\dots+X_n|^k < \infty$ is equivalent to $\mathbf{E}|X_1|^k < \infty$. But, in the case $X \in D(G_\alpha)$, $\alpha < 2$, if the relations $\nu_r < \infty$ and $\nu_s = \infty$ hold for some $s > r > \alpha$, then for $X_1 + X_2$ both cases may occur: $\nu_r(F_2 - G_{\alpha\beta}) = \infty$ as well $\nu_s(F_2 - G_{\alpha\beta}) < \infty$ (see (Christoph, 1989)).

II. Let $X \in D(\Phi)$ and suppose that $\mathbf{E}|X|^3 < \infty$, $\mathbf{E}X = 0$, then the relation (1) and, moreover, the following non-uniform estimate (Nagaev, 1965)

$$\big|F_n(x) - \Phi(x)\big| \leqslant \bar{C}_1\big(\sqrt{n}(1+|x|^3)\big)^{-1} \tag{1'}$$

hold.

If $X \in D(G_\alpha)$, $0 < \alpha < 2$, then under some p.m.-conditions similar estimates hold

$$\big|F_n(x) - G_\alpha(x)\big| \leqslant C_3 n^{-(r-\alpha)/\alpha}, \tag{3}$$

$$\big|F_n(x) - G_\alpha(x)\big| \leqslant \bar{C}_3\frac{n^{-(r-\alpha)/\alpha}}{(1+|x|^r)}, \tag{3'}$$

where $\alpha < r \leqslant 1 + [\alpha]$, $B_n = n^{-1/\alpha}$ (see (Paulauskas, 1974 a,b)).
From the relations (1) or (1′) we have, by simple computations, that for arbitrary ε, $0 < \varepsilon < 1$,

$$(1 - F_n(x))/(1 - \Phi(x)) \to 1$$

in the range

$$1 < x < (1 - \varepsilon)\sqrt{\ln\, n}. \tag{4}$$

In contradiction to the former case, when $\alpha < 2$, from the estimates (3) and (3′) we derive that large deviations hold in different ranges (see below).

i) From (3) it follows that $(1 - F_n(x))/(1 - G_\alpha(x)) \to 1$ in the range

$$x = o\left(n^{r-\alpha/2\alpha}\right), \quad x > 1, \quad n \to \infty. \tag{5}$$

ii) From (3′) it follows that $(1-F_n(x))/(1-G_\alpha(x)) \to 1$ for all $x = x(n)$, where $x(n)$ denotes any monotone sequence tending to infinity. Hence, non-uniform Berry—Esseen-type estimates in the case $0 < \alpha < 2$ play an important role in connection with large deviations.
For other differences between the normal limit law and the remaining stable limit laws we refer to (Christoph, 1987; 1989).
Using for U-statistics the Berry—Esseen-bounds (2) or the inequality

$$\left|\mathbf{P}(B_n U_n < x) - \Phi(x)\right| \leqslant \bar{C}_2\left(1 + |x|^3\right)^{-1}, \tag{6}$$

we obtain (Callaert and Jansen, 1978; Zhao Lincheng and Chen Xiru, 1983) that large deviations for them hold in the range (4).
The result about large deviations when the distribution function of a normalized U-statistic is approximated by a stable d.f. G_α, $1 < \alpha < 2$, is given in the next section.

III. In the case when the d.f. $\mathbf{P}(Q_1(X_1) < x)$ belongs to the normal domain of attraction of a stable law G_α, $1 < \alpha < 2$, we know that the constants B_n in the normalized sum $B_n S_n$ are of the form $B_n = n^{1-(1/\alpha)}/m$. There exist the following Berry—Esseen-type results:

THEOREM 1. *Suppose that for given* $\alpha \in (1, 2]$
A) $Q_1(X_1) \in D(G_\alpha)$,
B) $M_\delta = \mathbf{E}\left|H(X_1, \dots, X_m)\right|^\delta < \infty$ *for some* $\delta \in \left(\frac{2\alpha}{1+\alpha}, \alpha\right]$.
Assume that

(i) $1 < \alpha < 2$ *and* $\kappa_r < \infty$ *for some* $r \in (\alpha, 2]$,

(ii) $1 < \alpha \leqslant 2$, $\mu_2 = 0$ *and* $\kappa_r < \infty$ *for some* $r \in (2, 1 + \alpha]$.
Then in both cases (i) *and* (ii), *we have*

$$\sup_x \left|\mathbf{P}(B_n U_n < x) - G_\alpha(x)\right| \leqslant C_4 n^{-p/\alpha} \tag{7}$$

for $n \geqslant m$, *where* $p = \min\big(r - \alpha,\ 1/3(\delta(1+\alpha) - 2\alpha)\big)$. *Furthermore, in the cases*

(iii) $1 < \alpha < 2$ *and* $\nu_r < \infty$ *for some* $r \in (\alpha, 2]$,

(iv) $1 < \alpha \leqslant 2$, $\mu_2 = 0$ *and* $\nu_r < \infty$ *for some* $r \in (2, 1+\alpha]$
we have for all $n \geqslant m$ *and* $x \in \mathbf{R}^1$

$$\big|\mathbf{P}(B_n U_n < x) - G_\alpha(x)\big| \leqslant \bar{C}_4 n^{-p/\alpha}\big(1 + |x|\big)^{-1} \tag{8}$$

and

$$\big|\mathbf{P}(B_n U_n < x) - G_\alpha(x)\big| \leqslant C_5 n^{-p/\alpha}\big(1 + |x|\big)^{-\delta}, \tag{9}$$

respectively.

Because of the simplicity of the method of the proof of Theorem 1, the obtained rates are far away from the optimal ones in (3), (3′) and (1), (1′), (2), (6), respectively, for $1 < \alpha < 2$ and $\alpha = 2$. We emphasize the fact that the bounds (7), (9) hold under much weaker conditions that the bounds (2) and (6). Using the estimates in Theorem 1 together with those on the tail behaviour of $1 - G_\alpha(x)$ as $x \to \infty$ we obtain

THEOREM 2. *Let the conditions of Theorem 1 be fulfilled for a given* α, $1 < \alpha < 2$. *Then as* $n \to \infty$

1) *from the uniform bound* (7) *we have that*

$$\frac{\mathbf{P}(B_n U_n \geqslant x)}{1 - G_\alpha(x)} \to 1 \quad \textit{in the range} \quad x = o\big(n^{p/2\alpha}\big),$$

2) *from the non-uniform bound* (8) *we have that*

$$\frac{\mathbf{P}(B_n U_n \geqslant x)}{1 - G_\alpha(x)} \to 1 \quad \textit{in the range} \quad x = o\big(n^{p/\alpha(\alpha-1)}\big),$$

3) *from the non-uniform bound* (9) *we have that*

$$\frac{\mathbf{P}(B_n U_n \geqslant x)}{1 - G_\alpha(x)} \to 1 \quad \textit{in the range} \quad x = o\big(n^{p/\alpha(\alpha-\delta)}\big)$$

if $\delta < \alpha$, *and for all real* x *if* $\delta = \alpha$.

IV. *Example.* Let $X_1, \ldots, X_n$ be i.i.d. r.v.'s with the common d.f.

$$\mathbf{P}(X_1 < x) = \begin{cases} y|x|^{-3/2}, & x \leqslant -b^{-2/3}, \\ 0.5, & -b^{-2/3} < x \leqslant b^{-2/3}, \\ 1 - yx^{-3/2}, & x > b^{-2/3}, \end{cases}$$

where

$$y = \frac{1}{2}\big(7^{-3/2}\pi^{-1/2}\sin\frac{3}{4}\pi\big)^{7/15}, \quad b = (2y)^{-1}.$$

Consider the kernel function

$$H(x_1, x_2) = |x_1|^{8/7} x_2 + |x_2|^{8/7} x_1.$$

We have

$$\mathbf{E}X_1 = 0, \mathbf{E}\big(|X_1|^{8/7}\big) = 21/5(2y)^{16/21} = c.$$

Therefore $Q_1(X_1) = \mathbf{E}\big(H(X_1, X_2)/X_1\big) = cX_1$. Obviously, $Q_1(X_1) \in D(G_{3/2,\,0})$. In (Danz and Wolf, 1989) it is shown, that $\nu_2 < \infty$, $\kappa_2 < \infty$ and

$$\mathbf{E}\big|H(X_1, X_2)\big|^{5/4} < \infty.$$

We have $\alpha = 3/2,\ r = 2, \delta = 5/4,\ p = 1/24$,

$$B_n = \frac{1}{2} n^{1-2/3},\ S_n = \frac{2}{n} \sum_{i=1}^{n} Q_1(X_i),\ B_n S_n = n^{-2/3} \mathbf{E}\big(|X_1^{8/7}|\big) \sum_{i=1}^{n} X_i$$

and

$$B_n U_n = \frac{n^{-2/3}}{(n-1)} \sum_{1 \leqslant i_1 < i_2 \leqslant n} H(X_{i_1}, X_{i_2}).$$

From (3) we obtain that $\mathbf{P}(B_n S_n \geqslant x)/(1 - G_{3/2}(x)) \to 1$ in the range (5): $x = o(n^{1/6})$. But the non-uniform estimate (3′) gives that $\mathbf{P}(B_n S_n \geqslant x)/\big(1 - G_{3/2}(x)\big) \to 1$ for all x. For the normalized U-statistic $B_n U_n$ we get large deviations in much smaller ranges. Applying the first and second assertion of Theorem 2 we derive large deviations in $x = o(n^{1/72})$ and $x = o(n^{1/18})$, respectively.

REFERENCES

Bergström, H. (1952). On distribution functions with a limiting stable distribution function. *Arkiv Mathematik* **25**, 463–476.

Berry, A.C. (1941). The accuracy of the Gaussian approximation to the sum of independent variates. *Trans. Amer. Math. Soc.* **49**, 122–136.

Callaert, H. and Janssen, P. (1978). The Berry–Esseen theorem for U–statistics. *Ann. Stat.* **6**, 417–421.

Christoph, G. (1985). Zur Berechnung der Pseudomomente von Zufallsgrössen. *Wiss. Zeitschrift TU Dresden* **34**, 78.

Christoph, G. (1987). *Beiträge zur Summationstheorie von Folgen unabhängiger Zufallsgrössen im Falle eines stabilen Grenzgesetzes.* Disertation B. TU Dresden.

Christoph, G. (1989). On some differences in limit theorems with a normal or a non–normal stable limit law (to appear).

Danz, S. and Wolf, W. (1989). On the rate of convergence of standardized U–statistics to stable limit laws. *Wiss. Zeitschrift TU Dresden* **38**, 143–145.

Esseen C.G. (1945). Fourier analysis of distribution functions. A mathematical study of the Laplace—Gaussian law. *Acta Math.* **77**, 1–125.

Heinrich, L. and Wolf, W. (1989). In: *Proceedings of the Fourth Prague Symposium on Asymptotic Statistics,* Charles University, Prague, pp. 249–257.

Hoeffding, W. (1948). A class of statistics with asymptotically normal distribution. *Ann. Math. Stat.* **19**, 293–325.

Nagaev, S.V. (1965). Some limit theorems on large deviations. *Theory Probab. Appl.* **10**, 231–254.

Paulauskas, V.I. (1974a). Estimates of the remainder term in a limit theorem with stable limit law. *Liet. Matem. Rink.* **14**, 165–188.

Paulauskas, V.I. (1974b). Uniform and non-uniform estimates of the remainder term in a limit theorem with stable limit law. *Liet. Matem. Rink.* **14**, 171–185.

Zhao Lincheng and Chen Xiru. (1983). Non-uniform convergence rates for distributions of *U*-statistics. *Scientia Sinica (ser. A)* **26**, 795–810.

Zolotarev, V.M. (1962). An analogue of the Edgeworth Cramér asymptotic expansion in the case of nearness to stable distribution laws. In: *Abstracts of the VI. All-Union-Conference on Probab. Theory and Math. Stat.* Vilnius, pp. 49–50.

Zolotarev, V.M. (1965). On the nearness of distributions of two independent random variables. *Theory Prob. Appl.* **10**, 519–526.

Prob. Theory and Math. Stat., Vol. 2, pp. 585–589
B.Grigelionis *et al.* (Eds.)
1990 VSP/Mokslas

OSIPOV'S THEOREM REVISITED

V.V. YURINSKY
Institute of Mathematics, Siberian Division of the Acad. Sci. USSR
Universitetsky prospect 4, 630090 Novosibirsk, USSR

ABSTRACT

Let $X_1, X_2, \ldots$ be i.i.d. Hilbert space valued random variables such that $\mathbf{E}X_1 = 0$, $\mathbf{E}\exp\{A \mid X_1 \mid\} < \infty$ for some $A > 0$. Denote $p_n(r) = \mathbf{P}\Big(\mid n^{-1/2}(X_1 + \ldots + X_n) \mid \geqslant r\, v_1\Big)$, $p_\infty(r) = \mathbf{P}(\mid Y \mid > r\, v_1)$, where v_1 is the largest eigenvalue of the covariance operator V of X_1 and Y is $(0, V)$ Gaussian. It is proved that for some positive constants c_i, if $c_0 \leqslant r \leqslant c_1 n^{1/2}$ then $p_n(r) = (1 + g)\, I(r\, n^{-1/2}) p_\infty(r)$, where $\mid g \mid \leqslant c_2 r\, n^{-1/2}$ and I is some average of a Hilbert space version of the Cramér series.

1. Let $X, X_1, X_2, \ldots$ be i.i.d. random vectors (RV) taking values in a real separable Hilbert space H with inner product $(\cdot,\cdot)$ and norm $\mid \cdot \mid$. Assume that

$$\exists A_1 > 0 : \mathbf{E}\exp(A_1 \mid X \mid) = A_2 < \infty, \quad \mathbf{E}X = 0. \tag{1}$$

The covariance operator of X is defined by

$$(Vf, g) = \mathbf{E}(X, f)(X, g). \tag{2}$$

It is well-known that the sum

$$Z_n = n^{-1/2}(X_1 + \ldots + X_n)$$

converges in distribution to a Gaussian H-valued RV Y with the mean and covariances of X:

$$\mathbf{E}Y = 0, \quad \mathbf{E}(Y, f)(Y, g) = (Vf, g). \tag{3}$$

The purpose of this communication is to compare the probabilities

$$p_n(r) = \mathbf{P}(\mid Z_n \mid \geqslant r\, v_1), \quad p_\infty(r) = \mathbf{P}(\mid Y \mid \geqslant r\, v_1) \tag{4}$$

for large values of r and n; here as elsewhere

$$v_1^2 = \ldots = v_d^2 > v_{d+1}^2 \geqslant \ldots \tag{5}$$

are the eigenvalues of V and d is the multiplicity of the largest eigenvalue. Some notation is necessary to state the main result. To simplify it is assumed in the sequel that

$$v_1^2 = 1. \tag{6}$$

The d–dimensional invariant subspace of V corresponding to the largest eigenvalue is

$$E^0 = \{x : V\,x = x\} \tag{7}$$

while $S = \{e \in E^0 : |\,e\,| = 1\}$, $|\,S\,|$ and $ds(e)$ are, respectively, its unit sphere, the surface area and the surface element of the unit sphere.
Large deviations of Z_n are described in terms of the "Cramér's series" $l(t,e)$. Its definition is as follows.
Let a be a real "angle" and e, f such unit vectors that $e \in E^0$ while f is orthogonal to E^0. The operator $T = T_{e,f,a}$ is the 2–dimensional rotation in H which transforms e, f into

$$\widehat{e} = (\cos a)\,e + (\sin a)\,f, \quad \widehat{f} = -(\sin a)\,e + (\cos a)\,f \tag{8}$$

leaving unchanged the orthocomplement of the linear hull of e, f.
The "Cramér's series" of X is the functional

$$l(t,e) = L^+(t,e) + t^2/2, \quad t \in \mathbf{R}_+, \tag{9}$$

where

$$L^+(t,e) = \sup_a \sup_{f \perp E^0} L(t,e,f,a),$$

$$L(t,e,f,a) = \inf_{h \in E^0} \Big(-(te, Th) + \ln \mathbf{E}\, \exp((X, Th))\Big).$$

The relation between the asymptotics of probabilities (4) is described by

THEOREM 1. *There exist positive constants c_i such that the restriction $c_0 \leqslant r \leqslant c_1 n^{1/2}$ implies the equality*

$$p_n(r) = (1+g)\,I\,(r\,n^{-1/2})p_\infty(r), \tag{10}$$

where $g = g(r,n)$ admits the estimate

$$|\,g\,| \leqslant c_2 r\,n^{-1/2} \tag{11}$$

and I is the average of the "Cramér's series" (9) over the unit sphere of E^0:

$$I(t) = |\,S\,|^{-1} \int_S \exp\Big(n\,l\,(t,e)\Big)\,ds(e) \tag{12}$$

(see (15) for $d = 1$).

The asymptotics (10) was initially derived by L.V. Osipov in a somewhat cruder form (Osipov, 1977; 1978). No relative error estimate was given and the conditions were slightly more restrictive than (1). For the deviations range in which the influence of the non–Gaussian factor I of (10) is negligible p_∞ was shown to approximate p_n under even milder restrictions that (1) in (Bentkus, 1986; Račkauskas, 1988); these papers contain an error estimate as well and consider the problem of large deviations in a more general setting. Sharp error estimates for the "quasi–Gaussian" range of large deviations are also due to B.A. Zalessky. A preliminary version of Theorem 1 was published in preprint form (Yurinsky, 1988) with RHS of (11) containing additional summand $r^{-2}\ln r$.

2. The proof of Theorem 1 is a version of Osipov's original reasoning. The progress is due primarily to the use of some tools designed by F. Götze to evaluate the convergence rate in the infinite–dimensional central limit theorem. Although fairly straightforward it is too long to be reproduced here in detail. It seems however worthwhile to give a summary of its main steps.

A) In the computation of the probability p_n of (4) it is convenient to focus on exits of Z_n into the exterior of the ball $\{|x|<r\}$ near a given point $r\,e,\ e\in S$ of its boundary.

In what follows use is made of the following orthogonal projection operators:

$$\underline{P}: H \to E^0;\quad \underline{P}_e : x \mapsto (x,e)e,\quad |e|=1;$$

$$\underline{Q} = \underline{Q}_e \equiv \underline{P} - \underline{P}_e,\quad e\in S;\quad \underline{R} = I - \underline{P}: H \to H\ominus E^0. \tag{13}$$

Let $q: \mathbf{R}\to\mathbf{R}_+$ be an even smooth function. On $\mathbf{R}_+$ q decreases and $q(u)=0$ for $u>1$;

$$\bar q = \int_{\mathbf{R}^{d-1}\cap\{|z|\leqslant 1\}} q\Big(|z|^2\Big)dz > 0.$$

The function $W:\mathbf{R}\to[0,1]$ is also smooth; $W(u)=0$ for $u<-1$, $W(u)=1$ for $u>1$ and $W(-u)+W(u)=1$ on $\mathbf{R}$.

In the case of multiple largest eigenvalue of V $(d\neq 1)$ the pair of functions q, W is used to introduce the "*focussing decomposition*"

$$p_n(r) = |S|^{-1}\int_S \mathbf{E}q^0(e,Z_n)ds(e)\cdot(1+g') \tag{14}$$

where $|g'|\leqslant ct,\ t=r\,n^{-1/2}$,

$$q^0(e,x) = I_{B_r^c}(x)q(|\,r\underline{Q}_e x\,|^2)\hat q(|\,r\,\underline{P}x\,|^2)W((x,e))W_v(|\,\underline{R}x\,|).$$

In the RHS of (14) the factor $I_{\{.\}}$ is the indicator function of the set $B_r^c = \{|x|\geqslant r\}$,

$$\hat q(u) = \Big(|S|^{-1}\int_S W((y,e))q(|\,\underline{Q}_e y\,|^2)d\,s\,(e)\Big)^{-1}$$

with the vector $y\in E^0$ chosen arbitrarily to satisfy $|y|^2=u$, and W_v is one more smooth cutoff function introduced to cope with large values of deviations off

the "main" subspace E^0. This last factor is responsible for the appearance of the error term g' in (14).
The function $\hat{q}$ in (14) is smooth. Its asymptotics for large values of the argument is readily derived via Laplace's method. The function $\hat{q}$ grows like $u^{(d-1)/2}$ for $u \to \infty$ and its k-th derivative has the order $u^{(d-1-2k)/2}$.
The decomposition (14) is obvious in the case $d = 1$ if understood as follows: $S = \{e_1, -e_1\}$ for $d = 1$ and

$$|S|^{-1} \int_S f(e)ds(e) = \left(f(+e_1) + f(-e_1)\right)/2. \tag{15}$$

Under this agreement one may still use (14) where $\underline{Q}_e \equiv 0$.
B) When calculating the integrand of (14) it is convenient to pass to a new set of i.i.d. RV's $\hat{X}, \hat{X}_1, \ldots$ whose distribution is obtained from that of X through *Cramér's transform*:

$$\mathbf{P}(\hat{X} \in dx) = m^0(h)^{-1} \exp((h, x))\mathbf{P}(X \in dx), \tag{16}$$

where $m^0(h) = \mathbf{E}\exp((h, X))$.
This leads to the representation

$$\mathbf{E}q^0(e, Z_n) = \exp\Big(n\big(-(\mathbf{E}\hat{X}, h) + \ln m^0(h)\big)\Big)\times$$
$$\times \mathbf{E}\exp\big(-(n^{1/2}h, Z_n^0)\big)q^0(e, n^{1/2}\mathbf{E}\hat{X} + Z_n^0), \tag{17}$$

where

$$Z_n^0 = n^{-1/2}(X_1^0 + \ldots + X_n^0), \quad X_i^0 = \hat{X}_i - \mathbf{E}\hat{X}.$$

To make the most of the Gaussian approximation of the central limit theorem for Z_n^0 the parameter h in (16) should be chosen carefully.
Following the steps of Osipov's original work one may prove that there exists a choice of h which places the barycentre $n^{1/2}\mathbf{E}\hat{X}$ quite close to $r\,e$. Namely, for each pair $e \in S$, $t = r\,n^{-1/2}$ there exist an "angle" $a^+ \in \mathbf{R}$, a unit vector f^+ orthogonal to E^0 and a vector $h^+ \in E^0$ such that for the vector $\hat{h} = T^+h$ obtained using the 2-dimensional rotation of **1** (cf.(8)) the following conditions are fulfilled:

i) $n\big(-(\hat{h}, \mathbf{E}\hat{X}) + \ln m^0(h)\big) = -r^2/2 + n\,l(t, e)$;

ii) $n^{1/2}\mathbf{E}\hat{X} = r\,\hat{e}$.

For the optimal $a^+, f^+, \hat{h}$ one also obtains the estimates

$$(a^+)^2 + |\hat{h}| + |\underline{Q}_e\hat{h}|^{1/2} \leqslant ct, \quad (\hat{e}, \hat{h}) = t(1 + g'), \quad |g'| \leqslant ct.$$

For the optimal $\hat{h}$ there is also the relation $\hat{h} - \underline{Q}_e\hat{h} = \text{const}\cdot\hat{e}$, so that in (17)

$$(n^{1/2}\hat{h}, Z_n^0) = r(1 + g'')(\hat{e}, Z_n^0) + g'''$$

with $|g''| + |g'''| \leqslant ct$ and g'' non-random.

The task of evaluating the integrand in (14) is thus reduced to the evaluation of

$$\mathbf{E}\exp\big(-(1+g'')r(\hat{e},Z_n^0)\big)q^0(e,r\,\hat{e}+Z_n^0). \tag{18}$$

C) The obvious thing to do is now to replace Z_n^0 in (18) by a suitable Gaussian RV. The Cramér transform influences somewhat the covariance operator of X^0. Nevertheless this influence can be checked to be negligible for the purposes of this calculation, so the suitable choice of Gaussian approximation for Z_n^0 is again the Y of (3).

Although quite natural, the use of the central limit theorem in the context of this work displays some special features owing to the fact that one needs not the usual convergence rate but a convergence rate for the mean of a functional containing both an exponentially growing factor and a discontinuity. To evaluate the Gaussian approximation error here one may use F.Götze technique.

D) The last step in the proof of Theorem 1 is the calculation of the Gaussian average

$$M(g'',a+,r)=\mathbf{E}\exp\big(-(1+g'')r(\hat{e},Y)\big)q^0(e,r\,\hat{e}+Y).$$

If the position of $\hat{e}$ were not "awry" with respect to that of e the task would have been reduced easily to the computation of large deviations asymptotics for Y. The wrong geometry demands some extra work. Its content is the study of stability of the well-known asymptotic representations (Zolotarev, 1961; Hoeffding, 1964). Technically this part is close to (Zolotarev, 1961).

REFERENCES

Bentkus, V.J. (1986). On large deviations in Banach spaces. *Theory Probab. Appl.* **31**.

Hoeffding, W. (1964). Of a theorem of V.M.Zolotarev. *Theory Probab.Appl.* **9**, 96–99.

Osipov, L.V. (1977). On large deviations for sums of independent random vectors. In: *Abstr. Second Vilnius Conf. on Probab. and Statist.* **2**, Inst. Math. and Cybern., Vilnius, pp. 95–96.

Osipov, L.V. (1978). On large deviations probabilities for sums of independent random vectors. *Theory Probab. Appl.* **23**, 490–506.

Račkauskas, A.J. (1988). Probabilities of large deviations in the Linnik areas in Hilbert space. *Liet. Mat. Rink.* **28**, 520–533.

Yurinsky, V.V. (1988). On the asymptotics of large deviations in a Hilbert space. *Preprint* **23**, Inst. Math. Acad. Sci. USSR Siberian Division, Novosibirsk.

Zolotarev, V.M. (1964). Of one probabilistic problem. *Theory Probab. Appl.* **6**, 219–222.

Prob. Theory and Math. Stat., Vol. 2, pp. 590–601
B.Grigelionis *et al.* (Eds.)
1990 VSP/Mokslas

GIBBS MEASURES FOR THE CURIE—WEISS RANDOM FIELD ISING MODEL

V.A. ZAGREBNOV[1], J.M.G. AMARO de MATOS[2*] and A.E. PATRICK[1]
[1]Laboratory of Theoretical Physics, Joint Institute for Nuclear Research, 141980 Dubna, USSR,
[2]Sonderforschungsbereich 123, Universität Heidelberg, Im Neuenheimer Feld 294, Heildelberg, D–6900, Federal Republic of Germany.**

ABSTRACT

The set of all infinite–volume equilibrium (Gibbs) states is constructed for the Curie—Weiss version of the random–field Ising model. It turns out that these states are random: they are governed by parameters with non–degenerate distributions. The corresponding observables (like magnetization) are also random, but with simple discrete distribution. This generalized self–averaging property will be called conditional (or partial) self–averaging.

1. INTRODUCTION

The purpose of this contribution is to construct and describe the set of infinite–volume equilibrium states (limiting Gibbs measures) for the Curie—Weiss version of the random field Ising model (RFIM).

The Curie—Weiss RFIM has been studied from different points of view (Angelescu and Zagrebnov, 1985; 1986; Salinas and Wreszinski, 1985) including the problem of fluctuations (Amaro de Matos and Perez, 1989), but as far as we know, no complete constructive description of all its equilibrium states exists. In order to fill this gap, we propose in Section 2 a general line of reasoning to define and describe the limiting Gibbs measures for the Curie—Weiss RFIM. According to our arguments (see Sections 2 and 3) the limiting state will in general depend on random parameters defined in a probability space $(X, \mathcal{A}, \rho)$, to be specified later. Here it has to be stressed that the proof of our main theorem will be based on a careful analysis (Amaro de Matos and Perez, 1989) of the fluctuations of the free energy.

For the explicit case of the Curie—Weiss RFIM it turns out that the limiting Gibbs states are random measures with non–degenerate distributions.

Let us recall what the general understanding of self–averaging is related to. In a finite–volume the free energy density is a *random* variable *depending* on the external field configurations. In the infinite–volume limit, it converges to a *non–random*

---

* Supported by CNPq,
**Present address: INSEAD, Boul. de Constance, Fontainebleau, 77305, France.

value defined only by the probability distribution of this field. In other words, one could say that this limit is a *random variable* with *degenerate* distribution, i.e., with a probability density with support in one point (an atomic distribution with one atom).

Definition 1.1. Let $\{\phi_n\}_{n\geqslant 1}$ be a sequence of random variables defined on a probability space $(\mathcal{R}, \mathcal{F}, \lambda)$ and ϕ be a random variable defined on a probability space $(\mathcal{R}', \mathcal{F}', \lambda')$. We say that this sequence is *partially self-averaging* if:
(i) $\phi_n \xrightarrow{d} \phi$ in distribution as $n \to \infty$;
(ii) there exists a partition $\mathcal{D}_\phi = \{D_1, D_2, \ldots\}$ of $\mathcal{R}'$, generated by the random variable ϕ, which is finite, i.e.,

$$\phi(\omega' \in \mathcal{R}') = \sum_{i=1}^{k} y_i I_{D_i}(\omega'),$$

where $I_D(\cdot)$ is the indicator of D and $\bigcup_{i=1}^{k} D_i = \mathcal{R}'$.
If $k = 1$ and $\mathcal{R} = \mathcal{R}'$, then one gets the standard self-averaging. In this case the d-convergence in (i) is equivalent to the convergence in probability (Shiryaev, 1984) and, moreover, one can often prove the convergence with probability 1 (see e.g., (Pastur, 1980)).

Remark 1.1. The above Definition 1.1 is equivalent to the following construction. There exists a sequence of finite partitions $\{\mathcal{D}_n = \{D_i^{(n)}\}_{i=1}^{k}\}_{n\geqslant 1}$ of the space $\mathcal{R}$ such that for the conditional expectations $\{\mathbf{M}(\phi_n \mid \mathcal{D}_n\}_{n\geqslant 1}$ we have convergence in probability

$$\lim_{n\to\infty} (\phi_n - \mathbf{M}(\phi_n \mid \mathcal{D}_n)) \stackrel{\lambda}{=} 0$$

and $\lim_{n\to\infty} \lambda(D_i^{(n)}) = \lambda'(D_i)$, for each $i = 1, 2, \ldots, k$. Therefore we call this property of the sequence $\{\phi_n\}_{n\geqslant 1}$, *the conditional self-averaging*.
In Section 4 a general quasi-average approach is developed to single out the pure states and their convex linear combinations. Section 5 is devoted to examples and concluding remarks. We show there that for some range of parameters the magnetization of RFIM in the field $\{h_i\}_{i\in\mathbf{Z}}$ is *conditionally self-averaging* (but not self-averaging) in the thermodynamic limit $\Lambda \uparrow \mathbf{Z}$ with the partitions $\mathcal{D}_n = \{D_+^{(n)}, D_-^{(n)}\}$, where

$$D_\pm^{(n)} = \left\{(h_i)_{i\in\Lambda} \;:\; \operatorname{sign}\sum_{i\in\Lambda} h_i = \pm 1\right\}, \; n = |\Lambda|.$$

2. GENERAL FORMALISM AND THE SET-UP OF THE PROBLEM

Let $\Lambda \subset \mathbf{Z}$ be a finite subset of the lattice $\mathbf{Z}$ with cardinality $|\Lambda| = n$. Then the free Ising spin system in the external field $\mathbf{h} \in \mathbf{R}^{\mathbf{Z}}$ is defined by the Hamiltonian

$$H_\Lambda^{(0)}\left(\mathbf{s}^\Lambda \mid \mathbf{h}\right) = -\sum_{j\in\Lambda} h_j s_j, \quad s_j = \pm 1. \tag{1}$$

Hence, according to (1), for each fixed configuration $\mathbf{h}$, the finite–volume free Gibbs measure is defined by

$$P_\Lambda^{(0)}\left(\mathbf{s}^\Lambda \mid \mathbf{h}\right) = \prod_{j\in\Lambda} \frac{e^{\beta s_j h_j}}{2\cosh \beta h_j}.$$

Then, by the Kolmogorov existence theorem (Shiryaev, 1984), for any fixed $\mathbf{h}$ we can construct a unique probability measure $P^{(0)}(\cdot)$ on the measurable space (Ω, Σ) (here $\Omega = \{-1,1\}^{\mathbf{Z}}$ is the space of spin configurations and Σ is the σ–algebra generated by the cylinder sets $C_V = \{\mathbf{s} \in \Omega\colon \{s_i = \bar{s}_i\}_{i\in V}\}$, card $V(\subset \mathbf{Z}) < \infty$) such that

$$P^{(0)}(C_V \in \Sigma \mid \mathbf{h}) = \prod_{j\in V} \frac{e^{\beta \bar{s}_j h_j}}{2\cosh \beta h_j}. \tag{2}$$

This random infinite product measure (indexed by the configurations of the external field $\mathbf{h}$) is the limiting Gibbs measure for the free Hamiltonian (1).
The Curie—Weiss RFIM corresponds to the perturbation of (1) by the Curie—Weiss ferromagnetic interaction, which gives the new Hamiltonian:

$$H_\Lambda(\mathbf{s}^\Lambda \mid \mathbf{h}) = -\frac{1}{2|\Lambda|}\sum_{i,j\in\Lambda} s_i s_j - \sum_{j\in\Lambda} h_j s_j. \tag{3}$$

Then the finite–volume Gibbs measure can be expressed as

$$P_\Lambda(\mathbf{s}^\Lambda \mid \mathbf{h}) = \int_{\mathbf{R}} \mu_\Lambda(dy \mid \mathbf{h}) P_\Lambda^{(0)}\left(\mathbf{s}^\Lambda \mid \mathbf{h} + y\right), \tag{4}$$

where $(\mathbf{h} + y)_j = h_j + y$ and $\mu_\Lambda(dy \mid \mathbf{h})$ is a probability measure on $\mathbf{R}$ with the density

$$\frac{\mu_\Lambda(dy \mid \mathbf{h})}{dy} = \exp\left(-\beta n G_\Lambda(y \mid \mathbf{h})\right)\left(\int_{\mathbf{R}} dy \exp\left(-\beta n G_\Lambda(y \mid \mathbf{h})\right)\right)^{-1}. \tag{5}$$

Here the function $G_\Lambda(y \mid \mathbf{h})$ is given by

$$G_\Lambda(y \mid \mathbf{h}) = \frac{y^2}{2} - \frac{1}{\beta n}\sum_{j\in\Lambda} \ln\cosh\beta(y + h_j). \tag{6}$$

Now, for any configuration $\mathbf{h}$, the given finite–volume Gibbs measure (4) can be extended to cylinder sets $C \subset \Omega$ by $\bar{P}_\Lambda$. Note that Ω is compact in the product topology and simultaneously a compact metric space. This metric can be defined by the formula

$$\rho(\mathbf{s}, \mathbf{s}') = \sum_{j\in\mathbf{Z}} |s_j - s_j'|\, 2^{-\|j\|},$$

where $||\cdot||$ denotes the Euclidean norm in $\mathbf{Z}$. Let $\mathcal{M}(\Omega)$ denote the set of probability measures on $\Sigma \subset \Omega$, and $P_n \Rightarrow P$ denote the weak convergence of probability measures. By the Stone—Weierstrass theorem, one has

PROPOSITION 2.1 (Shiryaev, 1984). *The set $\mathcal{M}(\Omega)$ is compact with respect to the topology of weak convergence and $P_n \Rightarrow P$ in $\mathcal{M}(\Omega)$ iff $P_n(C) \to P(C)$ for all cylinder sets $C \subset \Omega$.*

Then, by Proposition 2.1, there are subsequences $\{\bar{P}_{\Lambda_\alpha}\}_{\Lambda_\alpha \subset \mathbf{Z}}$ such that $\bar{P}_{\Lambda_\alpha} \Rightarrow$ $\Rightarrow P_\alpha$. From the explicit formulas (4) and (5) it follows that P_α are (linear convex) combinations of the shifted infinite product measures $P_y^{(0)}(\mathbf{s} \mid \mathbf{h}) \equiv P^{(0)}(\mathbf{s} \mid \mathbf{h}+y)$, i.e.,

$$P_\alpha(A \mid \mathbf{h}) = \int_{\mathbf{R}} \mu_\alpha(dy \mid \mathbf{h}) P_y^{(0)}(A \mid \mathbf{h}), \tag{7}$$

where $\mu_\alpha = w - \lim_{\Lambda_\alpha \uparrow \mathbf{Z}} \mu_{\Lambda_\alpha}$ is a probability measure on $(\mathbf{R}, \mathcal{B}(\mathbf{R}))$, and $A \in \Sigma$.

THEOREM 2.1. *Let the probability measure $d\nu$ on $(\mathbf{R}, \mathcal{B}(\mathbf{R}))$ be such that $\int_{\mathbf{R}} |x| d\nu(x) < \infty$ and λ is the product–measure*

$$d\lambda(\mathbf{h}) = \prod_{i \in \mathbf{Z}} d\nu(\mathbf{h}_i)$$

on $(\mathcal{R}, \mathcal{F})$. Then for λ–almost all $\mathbf{h} \in \mathcal{R}$ there are subsequences $\{\Lambda_\alpha(\mathbf{h}) \uparrow \mathbf{Z}\}_\alpha$, such that

$$\mu_{\Lambda_\alpha(\mathbf{h})}(dy \mid \mathbf{h}) \Rightarrow \mu_\alpha(dy)$$

and

$$\operatorname{supp} \mu_\alpha \subset M(\nu, \beta) = \{y_0 \in \mathbf{R}: \min G(y) = G(y_0)\}, \ \lambda - \text{a.s.},$$

where

$$G(y) = \frac{y^2}{2} - \frac{1}{\beta} \int_{\mathbf{R}} d\nu(x) \ln \cosh \beta(y + x). \tag{8}$$

Proof. Note that the sequence $\{h_j\}_{j \in \mathbf{Z}}$ is stationary and ergodic. Then, by the individual (Birkhoff—Khinchine) ergodic theorem (see (Shiryaev, 1984)), the function $G_\Lambda(y \mid \mathbf{h})$ in (6) converges λ–a.s. to $G(y)$ in (8), for every fixed $y \in \mathbf{R}$. Because a countable union of null sets is a null set, the same is true for a countable dense set $N \subset \mathbf{R}$. Then, using the fact that

$$\left|\partial_y \left[G_\Lambda(y \mid \mathbf{h}) - y^2/2\right]\right| \leqslant 1,$$

one obtains the λ–a.s. convergence $G_\Lambda(y \mid \mathbf{h}) \to G(y)$, uniformly on compacts for all $y \in \mathbf{R}$. Hence, for any $\varepsilon > 0$ there exists a compact $K \subset \mathbf{R}$ such that $\mu_\Lambda(\mathbf{R} \setminus K \mid \mathbf{h}) < \varepsilon$ for λ–almost all configurations $\mathbf{h} \in \mathcal{R}$ and all sufficiently large $\Lambda \subset \mathbf{Z}$. Now the first assertion of the theorem is a consequence of Prokhorov's compactness theorem (Shiryaev, 1984) for each $\mathbf{h}$ from the above mentioned set of almost all configurations. Then combining the uniform convergence $G_\Lambda(y \mid \mathbf{h}) \to$ $\to G(y)$ with Laplace's asymptotic method for (5) one gets that

$$\lim_{\Lambda \uparrow \mathbf{Z}} \int_{\mathbf{R}} \mu_\Lambda(dy \mid \mathbf{h}) g(y) \longrightarrow 0 \quad \lambda - \text{a.s.},$$

for every continuous function $g(y)$ with compact support such that supp $g \cap$ $\cap M(\nu, \beta) = \{\varnothing\}$. Thus, every limit point $\mu_\alpha(dy)$ of $\{\mu_\Lambda(dy \mid \mathbf{h})\}_{\Lambda \subset \mathbf{Z}}$ has support in $M(\nu, \beta)$.

COROLLARY 2.1. *The limiting Gibbs measures defined as the weak limit points* $\{P_\alpha\}$ *of the sequence* $\{\bar{P}_\Lambda\}_{\Lambda \subset \mathbf{Z}}$ *have the form* (7), *where* $\{\mu_\alpha(dy \mid \mathbf{h})\}_\alpha$ *are defined by Theorem* 2.1.

On the other hand, exploiting the explicit formula (4) which connects the finite-volume Gibbs measure with the product measure $P^{(0)}(\pi_\Lambda \mathbf{s} \mid \mathbf{h} + y)$ (here $\pi_\Lambda : \Omega \to$ $\to \{-1, 1\}^\Lambda$) and using Kolmogorov existence theorem, we can construct the *limiting Gibbs measures* for (3) if the sequence $\{\mu_\Lambda(dy \mid \mathbf{h})\}_{\Lambda \subset \mathbf{Z}}$ converges (in some sense) to a probability measure which, in general, is random: $\mu_\Lambda \to \mu_\omega$, where μ_ω is defined on the probability space $(X, \mathcal{A}, \rho)$ (see Section 3). To this end, one has to verify only the consistency of the family

$$\{P_\Gamma^\omega(\cdot \mid \mathbf{h}) = \int_{\mathbf{R}} \mu_\omega(dy \mid \mathbf{h}) P_y^{(0)}(\pi_\Gamma \cdot \mid \mathbf{h})\}_{\Gamma \subset \mathbf{Z}}.$$

Therefore, in general case one gets

$$P_\omega(A \mid \mathbf{h}) = \int_{M(\nu,\beta)} \mu_\omega(dy \mid \mathbf{h}) P_y^{(0)}(A \mid \mathbf{h}), \tag{9}$$

where $A \in \Sigma$. For the Curie—Weiss RFIM, the measure μ_ω can be described in the explicit way (see Sections 3—5).

Let $\mathcal{M}_\mathbf{h}(\Omega)$ denote the set of all limiting Gibbs measures (states) for the Curie—Weiss RFIM. Then, from the explicit formulas (7) and (9) we conclude that $\mathcal{M}_\mathbf{h}(\Omega)$ is the closed convex hull (mixture) of its *extremal points* $\mathcal{E}_\mathbf{h}(\Omega) = \{P_y^{(0)}\}_{y \in M(\nu,\beta)}$. Each product measure $P_y^{(0)}$ is clearly *m-fold mixing*:

$$\int_\Omega dP_y^{(0)}(\mathbf{s} \mid \mathbf{h}) g_1(\mathbf{s}) \ldots g_m(\mathbf{s}) = \prod_{k=1}^{m} \int_\Omega dP_y^{(0)}(\mathbf{s} \mid \mathbf{h}) g_k(\mathbf{s}) \tag{10}$$

for any set $\{g_k\}_{k=1}^m \in \mathcal{C}(\Omega)$ of continuous functions with disjoint supports. This property reflects the *purity* of the limiting Gibbs states for the Curie—Weiss RFIM (Angelescu and Zagrebnov, 1985; Brankov et al., 1986). Hence, in agreement with the common knowledge (Ruelle, 1969), each pure measure (phase) from the set $\mathcal{M}_\mathbf{h}(\Omega)$ is one of its extremal points because the decomposition $P_{y_0}^{(0)} = \lambda P_1 + (1 -$ $-\lambda) P_2$ with $\lambda \in (0, 1)$ and $P_1, P_2 \in \mathcal{M}_\mathbf{h}(\Omega)$ implies $P_1 = P_2 = P_{y_0}^{(0)}$.

3. THE LIMITING GIBBS STATES

According to the last section, the problem to prove the existence of the limiting measure is clearly related to the asymptotic control of integrals of the type

$$\int_{\mathbf{R}} t(y)\mu_{\Lambda}(dy \mid \mathbf{h}) \tag{11}$$

as $n \to \infty$, with $\mu_{\Lambda}(dy \mid \mathbf{h})$ given by (5) and $t \in C(\mathbf{R})$.
Then we could expect to evaluate (11) by means of Laplace's asymptotic method, but this is not straightforward, since the measures μ_{Λ} are random (they depend on $\mathbf{h}$). Nevertheless, in a recent paper (Amaro de Matos and Perez, 1989) it has been shown that the method is still applicable together with the control of the rate of the convergence $G_{\Lambda}(y \mid \mathbf{h}) \to G(y)$. So, the limiting measure

$$\lim_{\Lambda \uparrow \mathbf{Z}} \mu_{\Lambda}(dy \mid \mathbf{h})$$

may clearly be a random measure.
Then, the important point must be the study of the fluctuations of $G_{\Lambda}(y \mid \mathbf{h})$ around $G(y)$. To describe them, we first introduce the *type* of a minimum of G.

Definition 3.1. Suppose that the number of minima of $G(y)$ is finite (say, m). Then, each one of them is labelled as y_0^i, $i = 1, 2, \ldots, m$. We say that $k_i \in \mathbf{Z}$ is *the type of the minimum* y_0^i if the following expansion holds:

$$G(y) = G(y_0^i) + \frac{1}{(2k_i)!}\left(y - y_0^i\right)^{2k_i} \partial_y^{2k_i} G(y_0^i) + o\left[\left(y - y_0^i\right)^{2k_i}\right]$$

as $y \to y_0^i$, with $\partial_y^{2k_i} G(y_0^i) > 0$.
On the other hand, the convergence of $G_{\Lambda}(y \mid \mathbf{h})$ and all its derivatives is uniform on compacts λ–a.s. (Angelescu and Zagrebnov, 1985; Amaro de Matos and Perez, 1989). As a consequence, for each global minimum y_0^i it is possible to construct a compact set containing y_0^i and a sequence of minima of $G_{\Lambda}(y \mid \mathbf{h})$ converging to y_0^i. Let us denote these sequences by $\{y_n^i\}$, $i = 1, 2, \ldots, m$. Before studying the fluctuations of the free energy $G_{\Lambda}(y \mid \mathbf{h})$, we need the following

PROPOSITION 3.1 (Amaro de Matos and Perez, 1989). *If y_0^i is a minimum of the type k_i, then*

$$n^{[2(2k_i-1)]^{-1}}\left(y_n^i - y_0^i\right) = v_{k_i}(\mathbf{h}) + o(1) \tag{12}$$

in distribution, as $n \to \infty$, where $v_{k_i}(\mathbf{h})$ is a random variable with the probability measure

$$\mu_i(dx) = c_i x^{(2k_i-2)} \exp\left[-x^{2(2k_i-1)}\right] dx.$$

We shall denote this by

$$v_{k_i}(\mathbf{h}) \sim c_i x^{(2k_i-2)} \exp\left[-x^{2(2k_i-1)}\right] dx. \tag{13}$$

Now we can establish

LEMMA 3.1. *If y_0^i is a minimum of the type k_i, then*

$$G_\Lambda(y_n^i \mid \mathbf{h}) - G(y_0^i) = \frac{u_0^i}{\sqrt{n}} + o\left(\frac{1}{\sqrt{n}}\right)$$

in distribution, as $n \to \infty$, with

$$u_0^i \sim \mathcal{N}\Big\{0, \beta^{-2} \int d\nu(h)\Big[\ln\cosh\big[\beta(y_0^i + h)\big]\Big]^2 -$$

$$-\Big[\beta^{-1} \int d\nu(h)\ \ln\cosh\big[\beta(y_0^i + h)\big]\Big]^2\Big\}, \tag{14}$$

where $\mathcal{N}(a, b)$ denotes a Gaussian measure with mean a and variance b.

Remark 3.1. In general $\{u_0^i\}_{i=1}^m$ is a set of *dependent* Gaussian random variables with mean zero and variances $\{D_i\}_{i=1}^m$ defined by (14). We shall illustrate this dependence with an explicit example in Section 5.
We now return to the control of the integral in (11). For this we need

PROPOSITION 3.2 (Amaro de Matos and Perez, 1989). *For each $i = 1, 2, \ldots, m$, there exists a sufficiently small $\bar{\theta}_i > 0$ such that, as $n \to \infty$,*

$$n\Big[G_\Lambda\Big(\frac{s}{n^{\rho_i}} + y_n^i \mid \mathbf{h}\Big) - G_\Lambda(y_n^i \mid \mathbf{h})\Big] = \lambda_i(\mathbf{h})\frac{s^2}{2} + O\Big(\frac{|s|^3}{n^{3\rho_i}}\Big) + P_1(s) \tag{15}$$

for $|s| < n^{\rho_i}\bar{\theta}_i$, and

$$n\Big[G_\Lambda\Big(\frac{s}{n^{\rho_i}} + y_n^i \mid \mathbf{h}\Big) - G_\Lambda(y_n^i \mid \mathbf{h})\Big] \geqslant \frac{1}{2}\lambda_i(\mathbf{h})\frac{s^2}{2} + P_2(s) \tag{16}$$

for $|s| < n^{\rho_i}\bar{\theta}_i$, where P_1 and P_2 are polynomials of degree 2 and 3, respectively, $\rho_i = k_i/2(2k_i - 1)$ and $\lambda_i(\mathbf{h}) \geqslant 0$ is the random variable given by

$$\lambda_i(\mathbf{h}) = \partial_y^{2k_i} G(y_0^i)[v_{k_i}(\mathbf{h})]^{2k_i - 2}/(2k_i - 2).$$

By a suitable change of variables, (16) will allow the use of the dominant convergence theorem and (15) will control the integrand in (11). The result is stated as

THEOREM 3.1. *Let $\{\mu_\Lambda(dy \mid \mathbf{h})\}_{\Lambda \subset \mathbf{Z}}$ be the random probability measures defined on $(\mathbf{R}, \mathcal{B}(\mathbf{R}))$ by (5). Then for each $t \in C(\mathbf{R})$ one has the limit*

$$\bar{t} = \lim_{\Lambda \uparrow \mathbf{Z}} \int_{\mathbf{R}} t(y)\mu_\Lambda(dy \mid \mathbf{h}) = \sum_{j=1}^m b_j t\big(y_0^i\big) \tag{17}$$

in the sense of the convergence in distribution, with $\sum_{i=1}^m b_i = 1$ and $\{y_0^i\}_{i=1}^m = M(\nu, \beta)$. The random variables $b_i(\omega) = \{1$ or 0 with probabilities p_i and $1 -$

$-p_i$, respectively$\}$ are defined on the probability space $(X, \mathcal{A}, \rho)$, where X can be identified with $M(\nu, \beta)$. The probabilities p_i are given by

$$p_i = \mathbf{Pr}\Big\{u_0^i > u_0^j, \text{ for all } j \neq i\Big\}, \ i = 1, 2, \ldots, m, \tag{18}$$

where u_0^i are defined in Lemma 3.1.

COROLLARY 3.1. *The limiting Gibbs state $P_\omega(\cdot \mid \mathbf{h})$ (9) exists for almost all $\mathbf{h}$ and is given by*

$$P_\omega(A \mid \mathbf{h}) = \sum_{i=1}^{m} b_i(\omega) P^{(0)}_{y_0^i}(A \mid \mathbf{h}), \tag{19}$$

where $\omega \in X$ and $A \in \Sigma$.

Comparing (19) and (9) with (5), we see that in accordance with Definition 1.1 for any Borel set $B \subset \mathbf{R}$ the sequence $\mu_\Lambda(B \mid \mathbf{h})$ is partially self–averaging as $\Lambda \uparrow \mathbf{Z}$ if $m > 1$. The limiting state is a convex random linear combination of the product measures $P^{(0)}_{y_0^i}(\cdot \mid \mathbf{h})$ given in (7). These components of $P_\omega(\cdot \mid \mathbf{h})$ are the so–called "pure states" (see Section 2).
Before finishing this section let us stress some points. If the set $M(\nu, \beta)$ degenerates to one single point, then one gets for $\{\mu_\Lambda(B \mid \mathbf{h})\}_\Lambda$ a standard self–averaging. The corresponding limiting Gibbs state $P_\omega(\cdot \mid \mathbf{h}) = P^{(0)}_{y_0}(\cdot \mid \mathbf{h})$, given by (19) above, is unique and pure. For almost every $\mathbf{h} \in (\mathbf{R}^{\mathbf{Z}}, \mathcal{F}, \lambda)$, it can be constructed as a weak limit of the sequence of finite volume Gibbs measure on (Ω, Σ) defined by (4). On the contrary, if $\text{card}M(\nu, \beta) > 1$, this sequence has no limit except by subsequences (see Theorem 2.1 and Corollary 2.1). However, considering it as a sequence of *random* measures, by Theorem 2.1 and equation (4), we get that the projections

$$\{P_\Lambda(\pi_\Gamma \cdot \mid \mathbf{h})\}_{\Gamma \subset \Lambda, \Lambda \subset \mathbf{Z}}$$

converge in distribution on cylinder sets $C \subset \Omega$ to the consistent family of random marginals $\{P_\Gamma^\omega\}_{\Gamma \subset \mathbf{Z}}$. Besides the trivial dependence on $\mathbf{h}$ via the product measure (4), they depend also on $\omega \in X$. This is a key of the idea how to reconstruct the limiting Gibbs measures with the help of the Kolmogorov existence theorem. If we apply this more general construction to the case $\text{card}M(\nu, \beta) = 1$, we regain the result described before.

4. THE QUASI–AVERAGE PROCEDURE

It is now clear that the magnetization $\left\langle \frac{1}{n} \sum_{i=1}^{n} s_i \right\rangle_\Lambda$ (the Gibbs expectation of $\frac{1}{n} \sum_{i=1}^{n} s_i$) has no pointwise convergence as $n \to \infty$, "jumping" from one minimum of $G(y)$ to another and being in this limit a random variable with a non–trivial distribution.

The addition in the Hamiltonian (3) of the term

$$\Delta = \frac{\tau}{n}\left[\sum_{i=1}^{n}\left(s_i - y_0^i\right)\right]^2,$$

where $\tau > 0$ and y_0^i is one of these minima, will clearly disturb this situation, leading to one single value for the magnetization in the thermodynamic limit, namely

$$\left\langle \frac{1}{n}\sum_{i=1}^{n} s_i \right\rangle \to y_0^i.$$

Then we can define the perturbed hamiltonian $\bar{H}_\Lambda$ as

$$\bar{H}_\Lambda = H_\Lambda + \sum_{j=1}^{m}\frac{\tau_j}{n^\rho}\left[\sum_{i=1}^{n}\left(s_i - y_0^i\right)\right]^2,$$

with $\rho > 1$, and check whether variations of τ_j, $j = 1, \ldots, m$, would allow the resulting limiting measure to scan over all the minima y_0^i of the original Hamiltonian H_Λ in an arbitrary way (Brankov et al., 1986). In the other words, we expect that this procedure allow us to renormalize b_j or p_i in (17), at least for some value of ρ. We shall refer the reader to our paper (Amaro de Matos et al., 1989) for the details of this analysis.

Let us now describe the situation one finds for different values of ρ:

a) $\rho > 2$.

In this case the values of b_j in (17) and their probabilities are left unchanged.

b) $2 \geqslant \rho \geqslant 3/2$.

For these values of ρ we are able to change the weights b_j in Theorem 3.1 according to our choice of the $\{\tau_j\}$.

c) $3/2 > \rho > 1$.

In this case we may choose which y_0^i (at most two) will dominate. The weights b_j corresponding to all the others will be necessarily zero.

5. CONCLUSIONS

We illustrate here our results for the simple case of dichotomic fields, with probability density

$$\frac{d\nu(h)}{dh} = \frac{1}{2}\left[\delta(h - H) + \delta(h + H)\right].$$

It is clearly that $G_\Lambda(y \mid \mathbf{h})$ in (6) may be rewritten in this case as

$$G_\Lambda(y \mid \mathbf{h}) = G(y) - \frac{1}{2\beta H}\left(\frac{1}{n}\sum_{i=1}^{n} h_i\right) f(y), \tag{20}$$

where $G(y)$ is the even function given by (8), and $f(y)$ is the odd function

$$f(y) = \ln\left\{\frac{\cosh[\beta(y + H)]}{\cosh[\beta(y - H)]}\right\}.$$

This model has been carefully studied in (Salinas and Wreszinski, 1985), and it has been shown that there is $H_c > 0$ such that for $H > H_c$ the function $G(y)$ will have only one global quadratic minimum ($m = 1, k = 1$) at $y = 0$. If the structure of the minima being so, we say that the system is in the paramagnetic phase.
For $0 \leqslant H \leqslant H_c$ there is an increasing continuous function $\beta_c(H)$ such that $\beta_c(0) = 1$ and $\beta_c(H_c) = \infty$; for $\beta < \beta_c(H)$ the system is in the paramagnetic phase, while for $\beta > \beta_c(H)$ function $G(y)$ has two different symmetric global minima of the quadratic type ($m = 2,\ k_1 = k_2 = 1$).
At $\beta = \beta_c(H)$ the situation is the following. There is $H_t > 0,\ H_t < H_c$, such that if $H < H_t$, function $G(y)$ has one global quadratic minimum ($m = 1,\ k = 2$) at $y = 0$. If $H = H_t$, again there will be only one global minimum at $y = 0$ but of the type $k = 3$. This is a so-called tricritical point. Finally, if $H_c > H > H_t$, then $G(y)$ will have three quadratic global minima ($m = 3,\ k_1 = k_2 = k_3 = 1$), one at $y = 0$ and the other two symmetric.
We now turn to the evaluation of the probabilities p_i in Theorem 3.1 for this model. In all the cases where $G(y)$ has only one global minimum ($m = 1$) $b_1 = 1$ with probability 1, which follows trivially from the normalization condition $\sum_{i=1}^{m} b_i = 1$.
If $m = 2$, a completely different situation arises. Let

$$c_n = \text{sign}\frac{1}{\sqrt{n}}\left(\sum_{i=1}^{n} h_i\right)$$

and

$$g_n^i = \exp\{-\beta n G_\Lambda(y_n^i|\mathbf{h})\}\left(\sum_{k=1,2}\exp\{-\beta n G_\Lambda(y_n^k|\mathbf{h})\}\right)^{-1}.$$

Then by (20) we have

$$\{\mathbf{h}\ : G_\Lambda(y_n^1 \mid \mathbf{h}) > G_\Lambda(y_n^2 \mid \mathbf{h})\} = \{\mathbf{h}\ : c_n = 1\}.$$

Therefore, by the CLT for $\frac{1}{\sqrt{n}}\sum_{i=1}^{n} h_i$, one gets that $p_1 = p_2 = 1/2$. According to the line of reasoning of Theorem 3.1 (see (Amaro de Matos et al., 1989)) we get that

$$m_n(\mathbf{h}) = \sum_{j=1,2}\frac{1}{n}\sum_{i=1}^{n}\tanh\beta(Jy_n^i + h_i)g_n^i + o(1)$$

as $n \to \infty$.
By this theorem $g_n^1 - I_{\{u_0^1 > u_0^2\}} \xrightarrow{\lambda} 0$. Consequently, $g_1^n - I_{\{c_n = 1\}} \xrightarrow{\lambda} 0$. Therefore, by Definition 1.1 (see also Remark 1.1)

$$\lim_{n\to\infty}[m_n(\mathbf{h}) - M(m_n(\mathbf{h}) \mid c_n)] \xrightarrow{\lambda} 0,$$

i.e., $D_\pm^{(n)}$ are generated by the random variable c_n (see Section 1), and we get a conditional (or partial) self-averaging of magnetization.
For $m = 3$ the same reasoning applies, but now there are three minima at $y_0^1 = y^* > 0$, $y_0^2 = 0$ and $y_0^3 = -y^*$. The function $f(y)$ (see (20)) is increasing.

Hence, $y = y_n^2$ can not be unique point of the absolute minimum and $y = y_n^{1,3}$ will be absolute minimum point for $c_n = \pm 1$, correspondingly. Therefore, in this case $p_1 = p_3 = 1/2$ and $p_2 = 0$. Similarly to the above line of reasoning, one gets:

$$\lim_{n\to\infty} [m_n(\mathbf{h}) - M(m_n(\mathbf{h}) \mid c_n)] \xrightarrow{\lambda} 0.$$

In the situation when $m = 1$, according to Definition 1.1, we say that the magnetization is self-averaging, since $\left\langle \frac{1}{n} \sum_{i=1}^{n} s_i \right\rangle_\Lambda$ converges to a constant λ–a.s. For finite but large enough n, the probability density of $\left\langle \frac{1}{n} \sum_{i=1}^{n} s_i \right\rangle_\Lambda$ is concentrated in a peak around this constant and converges to a degenerate distribution as described in Theorem 3.1.

Still according to this theorem, if $m > 1$ then

$$\lim_{n\to\infty} \left\langle \frac{1}{n} \sum_{i=1}^{n} s_i \right\rangle = \sum_{j=1}^{m} b_j(\omega) y_0^j$$

in distribution. However, b_j are random variables which can be defined on a completely different probability space than the one on which the random field $\mathbf{h}$ is defined, as it was described at the end of Section 2, with probabilities that we have just calculated in this simple example. Again, for finite but sufficiently large n, the density of the probability distribution of $\left\langle \frac{1}{n} \sum_{i=1}^{n} s_i \right\rangle\Big|_\Lambda$ will be concentrated in peaks around the y_0^i with relative weights given by the ratio of the probabilities p_j. Therefore, the limiting magnetization is now a random variable with non–degenerate distribution, i.e., it manifests the partial (or conditional) self–averaging property. An application of the quasi–average procedure allows one to select any pure phase and restore the standard self–averaging property.

Acknowledgments

The authors acknowledge the hospitality of both institutions where this work has been done: J.A.M. thanks the Joint Institute for Nuclear Research (Dubna), and V.A.Z. the S.F.B. 123 of the Heidelberg University. They also would like to thank Dr. R. Kühn for helpful discussions and Dr. D. Surgailis for valuable remarks.

REFERENCES

Amaro de Matos, J., Patrick, A.E. and Zagrebnov, V.A. (1989). SFB 123, *Preprint* 539, University of Heidelberg *(subm. to J. Stat. Phys.)*.

Amaro de Matos, J. and Perez, J.F. (1989). SFB 123, *Preprint* 538, University of Heidelberg *(subm. to Commun. Math. Phys.)*.

Angelescu, N. and Zagrebnov, V.A. (1985). A generalized quasi–average approach to the description of the limit states of the n–vector Curie—Weiss ferromagnet. *J. Stat. Phys.* **41**, 323.

Angelescu, N. and Zagrebnov, V.A. (1986). Quasi–average approach to the description of the limit states of the n–vector Curie—Weiss ferromagnet. In: *Probab. Theory and Math. Stat., Proc. Fourth Vilnius Conf. on Probab. Theory and Math. Stat.* **1**, pp. 69–84, VNU Science Press, Utrecht.

Brankov, J.G., Zagrebnov, V.A. and Tonchev, N.S. (1986). Limit Gibbs states for Curie—Weiss—Ising model. *Theor. and Math. Phys.* **66**, 72.

Pastur, L.A. (1980). Spectral properties of disordered system in one–body approximation. *Commun. Math. Phys.* **75**, 179.

Ruelle, D. (1969). *Statistical Mechanics.* W.A. Benjamin, Reading, Massachusetts.

Salinas, S.R. and Wreszinski, W.F. (1985). On the mean–field Ising model in a random external field. *J. Stat. Phys.* **41**, 299.

Shiryaev, A.N. (1984). *Probability.* Springer, New York.

Prob. Theory and Math. Stat., Vol. 2, pp. 602–608
B.Grigelionis *et al.* (Eds.)
1990 VSP/Mokslas

ON THE APPROXIMATION OF CONVOLUTIONS BY INFINITELY DIVISIBLE DISTRIBUTIONS

A.Yu. ZAITSEV
Leningrad Branch of Steklov Mathematical Institute, Fontanka 27,
191011 Leningrad, USSR

ABSTRACT

This paper deals with the problem of approximation of convolutions of probability distributions by infinitely divisible laws. We give an example of probability distribution G such that the distance in variation between n–fold convolution G^n and an arbitrary infinitely divisible distribution is larger than 1/14 for any n. We also consider some new non–uniform estimates in multidimensional limit theorems.

Let $\mathfrak{F}_k$ be the set of all probability distributions on the σ–field $\mathfrak{B}_k$ of Borel subsets of the Euclidean space $\mathbf{R}^k$, $\mathfrak{F}_k^s \subset \mathfrak{F}_k$ be the set of symmetric distributions in $\mathfrak{F}_k$, $\mathfrak{D}_k$ be the set of infinitely divisible distributions. For $0 < \alpha < 2$ we denote

$$\mathfrak{F}_k^{(\alpha)} = \{F \in \mathfrak{F}_k^s \ : \ \hat{F}(t) \geqslant -1 + \alpha \text{ for all } t \in \mathbf{R}^k\},$$

where $\hat{F}(t)$, $t \in \mathbf{R}^k$, is the characteristic function of a distribution $F \in \mathfrak{F}_k$. The notation $x \in \mathbf{R}^k$ will indicate that $x = (x_1, \dots, x_k)$ with $x_j \in \mathbf{R}^1$, $j = 1, \dots, k$. For $x, y \in \mathbf{R}^k$ we write $x < y$, if $x_j < y_j$ for all $j = 1, \dots, k$, and $(x, y) = x_1 y_1 + \dots + x_k y_k$, $||x|| = (x, x)^{1/2}$. For $F \in \mathfrak{F}_k$, $\lambda \geqslant 0$ we denote $F(x) = F\{\{u \in \mathbf{R}^k : u < x\}\}$, $x \in \mathbf{R}^k$, the corresponding distribution function;

$$e(\lambda F) = e^{-\lambda} \sum_{m=0}^{\infty} \frac{F^m}{m!} \in \mathfrak{D}_k$$

(products and powers of measures are understood in the convolution sense). The notation $c_1(\cdot), c_2(\cdot), \dots$ will be used for different positive quantities depending only on the indicated argument. In particular, $c_l(k)$ depend only on the dimension k and c_m are absolute constants.

We shall consider the following characteristics of proximity of distributions $F, G \in \mathfrak{F}_k$: the uniform distance

$$\rho(F, G) = \sup_{x \in \mathbf{R}^k} |F(x) - G(x)|$$

and the distance in variation

$$\sigma(F,G) = \sup_{X\in\mathfrak{B}_k} |F\{X\} - G\{X\}|.$$

In the sequel, $\mathbf{Z}$ is the set of all integers, $\mathbf{N}$ is the set of all positive integers, $\mathcal{L}(\xi)$ is the distribution of a random vector ξ, E_a is the distribution concentrated at a point $a \in \mathbf{R}^k$, $E = E_0$,

$$Q(F,y) = \sup_x F\{[x, x+y]\}, \; y \geqslant 0,$$

is the concentration function of a distribution $F \in \mathfrak{F}_1$, $\lambda\{X\}$ is the Lebesgue measure of a set $X \in \mathfrak{B}_1$.

In this paper we consider the problem of approximation of the sequence of n-fold convolutions $F, F^2, \dots, F^n, \dots$ of an arbitrary distribution $F \in \mathfrak{F}_k$ by infinitely divisible distributions. There is known the exhaustive solution of this problem for the uniform distance in the one-dimensional case. Prokhorov in (Prokhorov, 1955) showed that for any $F \in \mathfrak{F}_1$

$$\rho(F^n, \mathfrak{D}_1) = \inf_{D\in\mathfrak{D}_1} \rho(F^n, D) \to 0$$

as $n \to \infty$. Moreover,

$$\psi(n) = \sup_{F\in\mathfrak{F}_1} \rho(F^n, \mathfrak{D}_1) \to 0$$

as $n \to \infty$ (the first uniform limit theorem (Kolmogorov, 1956)) and $c_1 n^{-2/3} \leqslant \psi(n) \leqslant c_2 n^{-2/3}$ for all $n \in \mathbf{N}$ ((Arak, 1981; 1982) see, also (Arak and Zaitsev, 1986)). It would be natural to consider analogous problems for the distance in variation. In (Prokhorov, 1955) there is formulated the statement that $\sigma(F^n, \mathfrak{D}_1) \to 0$ as $n \to \infty$, when a distribution $F \in \mathfrak{F}_1$ is either discrete or having the non-degenerate absolutely continuous part. Prokhorov (Prokhorov, 1960) and Kolmogorov (Kolmogorov, 1963) asked whether the function

$$\psi^*(n) = \sup_{F\in\mathfrak{F}_1} \sigma(F^n, \mathfrak{D}_1),$$

were convergent to zero as $n \to \infty$. For the negative answer to this question it would be sufficient to construct a sequence $\{F_n\}_{n=1}^{\infty} \subset \mathfrak{F}_1$ for which $\sigma(F_n^n, \mathfrak{D}_1)$ does not converge to zero. It happens that there exists a distribution $G \in \mathfrak{F}_1$ such that $\sigma(G^n, \mathfrak{D}_1) \geqslant 1/14$ for any $n \in \mathbf{N}$ (see Theorem 1). For $r \in \mathbf{N}$, $x \in \mathbf{R}^1$ let

$$f_r(x) = \sum_{l=1}^{100r} 2^{-4000r^3 - 40rl} \chi\left(\left[\frac{l-1}{100r}, \frac{l}{100r}\right), x\right), \; f(x) = \sum_{r=1}^{\infty} f_r(x),$$

where $\chi(X, \cdot)$ is the indicator function of a set X.

THEOREM 1. *Let ξ be a random variable with uniform distribution on the interval $[0,1]$, $\eta = f(\xi)$, $G = \mathcal{L}(\eta)$. Then $\sigma(G^n, \mathfrak{D}_1) \geqslant 1/14$ for any $n \in \mathbf{N}$. So,*

$$\psi^*(n) \geqslant \psi^* = \sup_{F\in\mathfrak{F}_1} \inf_{n\in\mathbf{N}} \sigma(F^n, \mathfrak{D}_1) \geqslant 1/14.$$

It is evident that the distribution G is singular. In connection with already mentioned result from (Prokhorov, 1955) the reader can ask whether the convergence $\sigma(F^n, \mathfrak{D}_1) \to 0$ is uniform over the class $\mathfrak{F}_c \subset \mathfrak{F}_1$ of absolutely continuous distributions or over the class $\mathfrak{F}_d \subset \mathfrak{F}_1$ of discrete distributions. In order to decline this hypothesis too it is enough to modify slightly the definition of the random variable η from Theorem 1.

THEOREM 2. *Let ξ be a random variable with uniform distribution on the interval $[0,1]$, $n \in \mathbb{N}$, $\eta_{n1} = \sum_{r=1}^{n} f_r(\xi)$, η_{n2} be a random variable which is independent of ξ and uniformly distributed on the interval $[-2^{-\alpha_n}n^{-1},\ 2^{-\alpha_n}n^{-1}]$, where $\alpha_n = 4000n^3 + 12000n^2$. Let*

$$G_{n1} = \mathcal{L}(\eta_{n1}),\ G_{n2} = \mathcal{L}(\eta_{n1} + \eta_{n2}).$$

Then

$$\min\left\{\sigma(G_{n1}^n, \mathfrak{D}_1),\ \sigma(G_{n2}^n, \mathfrak{D}_1)\right\} \geqslant 1/14.$$

It is clear that the distributions G_{n1} are discrete and the distributions G_{n2} are absolutely continuous. Denote

$$\psi_c(n) = \sup_{F \in \mathfrak{F}_c} \sigma(F^n, \mathfrak{D}_1),\ \psi_d(n) = \sup_{F \in \mathfrak{F}_d} \sigma(F^n, \mathfrak{D}_1).$$

Theorem 2 shows that $\min\{\psi_d(n),\ \psi_c(n)\} \geqslant 1/14$ for any $n \in \mathbb{N}$.
The proof of Theorems 1 and 2 will be published later. It is based on the following elementary lemma. Denote

$$\Lambda = \{X \subset \mathbf{R}^1 : X = a\mathbf{Z} + K \text{ where } a \in \mathbf{R}^1,\ K \text{ is an arbitrary compact set}\}.$$

LEMMA 1. *Let $H \in \mathfrak{F}_1$, $F = H^2$, $X \in \Lambda$ and $F\{X\} \geqslant 1 - \varepsilon$ where $0 < \varepsilon < 1/4$. Then there exists $x \in (X + X - X)/2$ such that*

$$H\{X - X + x\} \geqslant 1 - \varepsilon\left(1 - \varepsilon + \sqrt{1 - 4\varepsilon + \varepsilon^2}\right)^{-1}.$$

COROLLARY 1. *Suppose that the conditions of Lemma 1 are satisfied and $0 < \varepsilon < 1/12$. Then*

$$H\{B(X)\} \geqslant 1 - \frac{2}{3}\varepsilon,$$

where $B(X) = (X + X + X + X - X - X - X)/2$.

Now we present the main idea of the proof of Theorem 1. Let the conditions of this theorem be satisfied. Fix $n \in \mathbb{N}$ and suppose that there exists a distribution $D_0 \in \mathfrak{D}_1$ such that $\sigma(G^n, D_0) < 1/14$. Then we construct some sets $X_0, X_1, \ldots, X_{100n}$ having the following properties:

$$(G^n)\{X_j\} \geqslant 1 - \frac{1}{100}, \quad \bigcap_{j=0}^{100n} B^m(X_j) = \varnothing, \quad j = 0, 1, \ldots, 100n, \tag{1}$$

where $m = m(n) = [3\ln 17n] + 1$ ($[b]$ denotes the integer part of a number $b \in \mathbf{R}^1$). It is clear that

$$D_0\{X_j\} \geqslant 1 - \frac{1}{100} - \frac{1}{14} > 1 - \frac{1}{12}, \quad j = 0, \ldots, 100n. \tag{2}$$

Since the distribution D_0 is infinitely divisible, it can be represented in the form $D_0 = D_1^2 = \ldots = D_l^{2^l} = \ldots$, where $D_l \in \mathfrak{D}_1$. Using (2) and Corollary 1 we obtain by induction that

$$D_m\{B^m(X_j)\} \geqslant 1 - \frac{1}{12}\left(\frac{2}{3}\right)^m \geqslant 1 - \frac{1}{200n}, \quad j = 0, 1, \ldots, 100n. \tag{3}$$

From (3) it follows that

$$\bigcap_{j=0}^{100n} B^m(X_j) \neq \varnothing. \tag{4}$$

Comparing (1) with (4) we obtain a contradiction. Hence $\sigma(G^n, \mathfrak{D}_1) \geqslant 1/14$.
The constant 1/14 is not the smallest for which one can prove analogous results. It can be shown that for any $\delta > 0$ there exists a distribution $G_\delta \in \mathfrak{F}_1$ such that $\sigma(G_\delta^n, \mathfrak{D}_1) \geqslant 1/4 - \delta$ for all $n \in \mathbb{N}$. The distribution can be taken in the form $G_\delta = \mathcal{L}(\eta_\delta)$ where

$$\eta_\delta = \sum_{r=1}^{\infty} \sum_{l=1}^{c_3(\delta)r} 2^{-c_3(\delta)c_4(\delta)r^3 - c_4(\delta)rl} \chi\left(\left[\frac{l-1}{c_3(\delta)r}, \frac{l}{c_3(\delta)r}\right), \xi\right),$$

ξ is uniformly distributed on [0, 1] and $c_3(\delta)$, $c_4(\delta) \in \mathbb{N}$ are sufficiently large for small δ. On the other hand, $\sigma(F^n, \mathfrak{D}_1) \leqslant 1 - e^{-1}$ for any $F \in \mathfrak{F}_1$, $n \in \mathbb{N}$, because $\sigma(F^n, e(F^n)) \leqslant 1 - e^{-1}$. Thus, denoting

$$\psi^{**} = \sup_{F \in \mathfrak{F}_1} \sup_{n \in \mathbb{N}} \sigma(F^n, \mathfrak{D}_1),$$

we have

$$1/4 \leqslant \psi^* \leqslant \psi^{**} \leqslant 1 - e^{-1}.$$

It is interesting to find the exact values of ψ^* and ψ^{**}.
The second part of this paper deals with the approximation of F^n by accompanying laws $e(nF)$ in the case when $F \in \mathfrak{F}_k^{(\alpha)}$, $0 < \alpha < 2$. The main result is contained in the following theorem.

Theorem 3. *Let $0 < \alpha < 2$, $F \in \mathfrak{F}_k^{(\alpha)}$, $n \in \mathbb{N}$. Then*

$$\rho(F^n, e(nF)) \leqslant c_5(k)(n^{-1} + \exp(-n\alpha + c_6 k \ln^3 n)). \tag{5}$$

Now we present some historical remarks. The first estimates in multidimensional version of the first uniform limit theorem of Kolmogorov was obtained in (Sazonov,

1972) and (Presman, 1973). The best known result is due to (Presman, 1973). He proved that

$$\inf_{a\in\mathbf{R}^k} \rho(F^n, E_{na}e(nFE_{-a})) \leqslant c_7(k)n^{-1/3} \tag{6}$$

for any $F \in \mathfrak{F}_k$. It is known that the inequality is optimal with respect to order and cannot be essentially strengthened in general case. Sharpening this inequality becomes possible if we suppose, for example, that $F \in \mathfrak{F}_k^s$.
In one-dimensional case (Ibragimov and Presman, 1973) showed that

$$\rho(F^n, e(nF)) \leqslant c_8 n^{-1/2}(\ln n + 1)^{1/2} \tag{7}$$

for any $F \in \mathfrak{F}_1^s$. T.V.Arak in (Arak, 1980) proved that

$$\sup_{F\in\mathfrak{F}_1^{(1)}} \rho(F^n, e(nF)) \leqslant c_9 n^{-1}. \tag{8}$$

Using the inequality (8), the author (Zaitsev, 1981) obtained the inequality

$$\sup_{F\in\mathfrak{F}_k^s} \rho(F^n, e(nF)) \leqslant c_{10}(k)n^{-1/2} \tag{9}$$

for $k = 1$. The class $\mathfrak{F}_1^{(\alpha)}$, $0 < \alpha < 2$, was considered in (Zaitsev, 1983). There was announced the inequality

$$\sup_{F\in\mathfrak{F}_k^{(\alpha)}} \rho(F^n, F^{n+1}) \leqslant c_{11}(k)(n^{-1} + \exp(-n\alpha + c_{12}k\ln^3 n)) \tag{10}$$

for $k = 1$ (the detailed proof is contained in (Arak and Zaitsev, 1986)). Recently, in (Čekanavičius, 1987; 1989) were obtained some generalizations of the inequality (8). In particular, he proved that

$$\sup_{F\in\mathfrak{F}_1^{(\alpha)}} \rho(F^n, e(nF)) \leqslant c_{13}\alpha^{-9/2}(|\ln\alpha| + 1)n^{-1},\ 0 < \alpha < 2. \tag{11}$$

The author in (Zaitsev, 1988) proved the inequalities (9) and (10) for arbitrary $k \in \mathbf{N}$.
It is known that the inequality (9) is optimal with respect to order. Theorem 3 shows that the separation from -1 of the characteristic function of a distribution $F \in \mathfrak{F}_k^s$ leads to more quick decrease of $\rho(F^n, e(nF))$ than the inequality (9) is able to provide. The inequality (5) can be considered as a multidimensional generalization of the inequalities (8) and (11). Note that the one-dimensional variant of the inequality (5) is essentially more exact than (11) with respect to the character of dependence of the right-hand side on the parameter α.
The proof of Theorem 3 is published in (Zaitsev, 1989b). It is based on a new method which was already used in (Zaitsev, 1988; 1989a). This method can be considered as an improved version of one-dimensional "triangle function method" introduced and for the first time applied in (Arak, 1980) for the proof of inequality (8). It happens that this method allows to obtain a kind of non-uniform bounds for the closeness of compared distributions on convex polytopes.

Let

$$\mathfrak{X}_m = \{X \subset \mathbf{R}^k : X = \{x \in \mathbf{R}^k : a_j \leqslant (x, t_j) \leqslant b_j,\ t_j \in \mathbf{R}^k,$$
$$-\infty \leqslant a_j \leqslant b_j \leqslant \infty,\ j = 1, \ldots, m\}\},\ m = 1, 2, \ldots .$$

For $G = \mathcal{L}(\xi) \in \mathfrak{F}_k$, $X \in \mathfrak{X}_m$ let

$$q(G, X) = \inf_{t \in \mathbf{R}^k,\ \|t\|=1} Q\Big(\mathcal{L}((\xi, t)),\ \lambda\Big\{\{(t, x) : x \in X\}\Big\}\Big).$$

Theorem 4. *Let $F \in \mathfrak{F}_k^{(\alpha)}$, $0 < \alpha < 2$; $m, n \in \mathbf{N}$; $X \in \mathfrak{X}_m$, $q = q(e(nF), X)$. Then*

$$|(F^n)\{X\} - (e(nF))\{X\}| \leqslant$$
$$\leqslant c_{14}(m)\Big(n^{-1} q^{1/5}(|\ln q| + 1)^{17(m+1)/5} + \exp(-n\alpha + c_{15} m \ln^3 n)\Big). \qquad (12)$$

The right–hand side of the inequality (12) depends on q and can be small in comparison with that of (5) when the parameter $q = q(e(nF), X)$ is small. We can formulate analogous refinements for other results obtained with the help of the same method.

Theorem 5. *Let the conditions of Theorem 4 be satisfied. Then*

$$|(F^n)\{X\} - (F^{n+1})\{X\}| \leqslant$$
$$\leqslant c_{16}(m)\Big(n^{-1} q^{1/3}(|\ln q| + 1)^{3m+1} + \exp(-n\alpha + c_{17} m \ln^3 n)\Big). \qquad (13)$$

Theorem 6. *Let the distributions $G_i \in \mathfrak{F}_k$ be represented in the form $G_i = (1 - p_i)E + p_i V_i$ where $0 \leqslant p_i \leqslant 1$, $V_i \in \mathfrak{F}_k$, $i = 1, \ldots, n$. Let $G = \prod_{i=1}^{n} G_i$, $D = \prod_{i=1}^{n} e(G_i)$, D_0 be the distribution with characteristic function*

$$\hat{D}_0(t) = \prod_{i=1}^{n} \exp\Big(- p_i(1 - p_i)(1 - \mathrm{Re}\hat{V}_i(t))\Big).$$

Let $m \in \mathbf{N}$, $X \in \mathfrak{X}_m$, $q = q(D_0, X)$. Then

$$|G\{X\} - D\{X\}| \leqslant c_{18}(m) q^{1/3}(|\ln q| + 1)^{3m+1} \max_{1 \leqslant i \leqslant n} p_i. \qquad (14)$$

Theorems 4, 5, 6 are more general than corresponding non–uniform estimates obtained in (Zaitsev, 1988; 1989a; 1989b). Note that the dependence of the right–hand sides of the inequalities (12)—(14) on q is, apparently, not the best possible.

REFERENCES

Arak, T.V. (1980). On the approximation of n-fold convolutions of distributions having non-negative characteristic functions with accompanying laws. *Theory Probab. Appl.* **25**, 225–246.

Arak, T.V. (1981). On the rate of convergence in Kolmogorov's uniform limit theorem. I; II. *Theory Probab. Appl.* **26**, 225–245; 449–463.

Arak, T.V. (1982). Improvement of the lower bound for the rate of convergence in Kolmogorov's uniform limit theorem. *Theory Probab. Appl.* **27**, 767–772.

Arak, T.V. and Zaitsev, A.Yu. (1986). Uniform limit theorem for sums of independent random variables. *Proc. Steklov Inst. Math.* **174**, 1–216.

Čekanavičius, V. (1987). Approximation by generalized measures of Poisson type. *Thesis.* Vilnius, 1–157.

Čekanavičius, V. (1989). Approximations with accompanying distributions and asymptotic expansions. I; II. *Liet. Matem. Rink.* **29**, 171–178; 402–414.

Ibragimov, I.A. and Presman, E.L. (1973). On the rate of approach of the distributions of sums of independent random variables to accompanying distributions. *Theory Probab. Appl.* **18**, 753–766.

Kolmogorov, A.N. (1956). Two uniform limit theorems for sums of independent random variables. *Theory Probab. Appl.* **1**, 426–436.

Kolmogorov, A.N. (1963). Approximation to distributions of sums of independent terms by means of infinitely divisible distributions. *Proc. Moscow Math. Soc.* **12**, 437–451.

Presman, E.L. (1973). On a multidimensional variant of Kolmogorov's uniform limit theorem. *Theory Probab. Appl.* **18**, 396–402.

Prokhorov, Yu.V. (1955). On sums of identically distributed random variables. *Soviet Math. Dokl.* **105**, 645–647.

Prokhorov, Yu.V. (1960). On Kolmogorov's uniform limit theorem. *Theory Probab. Appl.* **5**, 103–113.

Sazonov, V.V. (1972). On the rate of approach of the distributions of sums of independent random variables to the class of infinitely divisible distributions in multidimensional case. *Proc. Tbilisi Univ.* **A4 (146)**, 29–37.

Zaitsev, A.Yu. (1981). Some properties of n-fold convolutions of distributions. *Theory Probab. Appl.* **26**, 152–156.

Zaitsev, A.Yu. (1983). Estimates of the closeness of successive convolutions of symmetric distributions. *Theory Probab. Appl.* **28**, 184–185.

Zaitsev, A.Yu. (1988). Estimates for the closeness of successive convolutions of multidimensional symmetric distributions. *Probab. Theory Rel. Fields* **79**, 175–200.

Zaitsev, A.Yu. (1989a). Multivariate version of the second Kolmogorov's uniform limit theorem. *Theory Probab. Appl.* **34**, 128–151.

Zaitsev, A.Yu. (1989b). On the approximation of convolutions of multidimensional symmetric distributions by accompanying laws. *Notes Sci. Seminars Leningrad Branch Steklov Math. Inst.* **177**, 55–72.

Prob. Theory and Math. Stat., Vol. 2, pp. 609–616
B.Grigelionis *et al.* (Eds.)
1990 VSP/Mokslas

BANACH SPACE OF MEASURES

Z.S. ZERAKIDZE
Tbilisi State University, Tbilisi, USSR

ABSTRACT

This paper is devoted to the investigation of the structure of families of probability measures, which was started in (Ibramhalilov and Skorokhod, 1980; Zerakidze, 1984, 1986a, 1986b).

1. WEAK AND STRONG DIVISIBILITY OF MEASURE FAMILIES

Let $(X, \mathcal{B})$ be a measurable space with a family $\{\mu_\alpha\}_{\alpha\in\mathcal{A}}$ of probability measures on it.

Let us give some definitions.

Definition 1. The family $\{\mu_\alpha\}_{\alpha\in\mathcal{A}}$ of probability measures is ***mutually singular*** if $(\forall i)(\forall j)$

$$(i \in \mathcal{A} \ \&\ j \in \mathcal{A}, \quad i \neq j \Rightarrow \text{ the measures } \mu_i \text{ and } \mu_j \text{ are orthogonal}).$$

Definition 2. The family $\{\mu_\alpha\}_{\alpha\in\mathcal{A}}$ of probability measures is called ***weakly divisible*** if there exists a family of $\mathcal{B}$–measurable sets $\{X_\alpha\}_{\alpha\in\mathcal{A}}$ such that $(\forall i)(\forall j)$

$$(i \in \mathcal{A} \ \&\ j \in \mathcal{A} \Rightarrow \mu_i(X_j) = \delta(i,j)),$$

where $\delta(i,j)$ denotes Kronecker's function on the Cartesian square $\mathcal{A}^2$ of the set $\mathcal{A}$.

Definition 3. The family $\{\mu_\alpha\}_{\alpha\in\mathcal{A}}$ of probability measures is ***strongly divisible*** if there exists a family of mutually disjoint $\mathcal{B}$–measurable sets $\{X_\alpha\}_{\alpha\in\mathcal{A}}$ such that the following relation $(\forall j)$

$$(i \in \mathcal{A} \Rightarrow \mu_i(X_i) = 1)$$

holds.

Definition 4. A family of σ–finite measures $\{\mu_\alpha\}_{\alpha\in\mathcal{A}}$ on $(X, \mathcal{B})$ is called ***weakly divisible*** if there exists a family of $\mathcal{B}$–measurable sets $\{X_\alpha\}_{\alpha\in\mathcal{A}}$ such that

$$\mu_\beta(X - X_\beta) = 0, \quad \mu_\beta(X_\alpha) = 0, \quad \forall \beta \neq \alpha.$$

Definition 5. A family of σ–finite measures $\{\mu_\alpha\}_{\alpha\in\mathcal{A}}$ on (X, B) is called ***strongly divisible*** if there exists a family of mutually disjoint $\mathcal{B}$–measurable sets $\{X_\alpha\}_{\alpha\in\mathcal{A}}$ such that

$$\mu_\beta(X - X_\beta) = 0, \quad \mu_\beta(X_\alpha) = 0, \quad \forall \beta \neq \alpha.$$

The parametric set $\mathcal{A}$ is assumed to have a σ–algebra $\mathcal{B}_1$ which contains all finite subsets of $\mathcal{A}$.

Definition 6. We say that the family of probability measures $\{\mu_\alpha\}_{\alpha\in\mathcal{A}}$ ***admits a consistent parameter estimator*** if there exists a measurable mapping g of $(X, \mathcal{B})$ into $(\mathcal{A}, \mathcal{B}_1)$ such that $(\forall i)$

$$(i \in \mathcal{A} \Rightarrow \mu_i(\{x : x \in X \ \& \ g(x) = i\}) = 1).$$

The validity of the following assertions can be easily verified:

1) If the family $\{\mu_\alpha\}_{\alpha\in\mathcal{A}}$ of probability measures is weakly divisible then it is mutually singular.

2) If the family $\{\mu_\alpha\}_{\alpha\in\mathcal{A}}$ of probability measures is strongly divisible then it is also weakly divisible.

3) If for the family $\{\mu_\alpha\}_{\alpha\in\mathcal{A}}$ of probability measures a consistent estimator exists then this family is strongly divisible.
It is well–known that a strongly divisible family $\{\mu_\alpha\}_{\alpha\in\mathcal{A}}$ of probability measures exists which had no consistent estimator.

Example 1. Let $\mathbf{Z}$ be the set of all integer numbers and $(\forall i)$

$$(i \in \mathbf{Z} \Rightarrow X_i = [0, 1]).$$

Put

$$X = \prod_{i\in\mathbf{Z}} X_i$$

and $(\forall k)$

$$(k \in \mathbf{Z} \Rightarrow \mathcal{A}_k = \prod_{i\geqslant k} X_i).$$

Let

$$\mathcal{A} = \bigcup_{k\in\mathbf{Z}} \mathcal{A}_k.$$

For any index $i \in \mathcal{A}_k$ the representation $i = (i_n, i_{n+1}, \ldots)$
holds, where $(\forall k)$

$$(k \geqslant n \Rightarrow i_k \in [0, 1]).$$

Let for any natural $k \geqslant n$ the symbol δ_k denotes the probability measure concentrated at i_k. Let for any $k < n$ measure μ_k be the Lebesgue measure on $[0, 1]$

$$\mu_i = \Big(\prod_{k<n} \mu_k\Big)\Big(\prod_{k\geqslant n} \delta_k\Big).$$

The family $\{\mu_i\}_{i\in\mathcal{A}}$ is singular but not strongly divisible.

Example 2. Let X be the square $[0,1] \times [0,1]$, $\mathcal{B}$ be the σ–algebra of Borel sets in this square, $\mathcal{A} = [0,1] \cup [2,3]$. Assume that if $\alpha \in [0,1]$ then μ_α is the Lebesgue measure on the interval $X_\alpha = \{0 \leqslant x \leqslant 1;\ y = \alpha\}$ and if $\alpha \in [2,3]$, then μ_α is the Lebesgue measure on the interval $X_\alpha = \{x = \alpha - 2; 0 \leqslant y \leqslant 1\}$. Since $X_\alpha \cap X_{\alpha'}$ is either empty or consists of one point we have

$$\mu_{\alpha'}(X_\alpha) = \begin{cases} 1, & \text{if } \alpha' = \alpha, \\ 0, & \text{if } \alpha' \neq \alpha. \end{cases}$$

Thus the family $\{\mu_\alpha\}_{\alpha \in [0,1] \cup [2,3]}$ of probability measures is weakly divisible.

Example 3. Let X be the square $[0,1] \times [0,1]$, $\mathcal{B}$ be the σ–algebra of Borel sets in this square, $\mathcal{A} = [0,1]$ and for $\alpha \in [0,1]$ measure μ_α be the Lebesgue measure on the interval $X_\alpha = \{0 \leqslant x \leqslant 1;\ y = \alpha\}$. Since $X_\alpha \cap X_{\alpha'} = \varnothing$, $\alpha \neq \alpha'$,

$$\mu_{\alpha'}(X_\alpha) = \begin{cases} 1, & \text{if } \alpha = \alpha', \\ 0, & \text{if } \alpha \neq \alpha'. \end{cases}$$

This family $\{\mu_\alpha\}_{\alpha \in [0,1]}$ of probability measures is strongly divisible.

THEOREM 1. *If a family* $\{\mu_k\}_{k \in \mathbb{N}}$, $\mathbb{N} = \{1, 2, \ldots\}$ *of finite measures defined on* $(X, \mathcal{B})$ *is mutually singular and*

$$\sum_{k=1}^{\infty} \mu_k < \infty.$$

Then μ_i *is singular*

$$\sum_{k \neq i} \mu_k.$$

Proof. Since $\{\mu_k\}_{k \in \mathbb{N}}$ is a sequence of finite mutually singular measures, there exist $\mathcal{B}$–measurable sets $\tilde{C}_{ik}$ such that $\mu_k(\tilde{C}_{ik}) = 0$ and $\mu_i(X - \tilde{C}_{ik}) = 0$, $k \neq i$. Define $\tilde{C}_i = \bigcup_{k \neq i} (X - \tilde{C}_{ik})$. Evidently, $\mu_i(\tilde{C}_i) = 0$, $\mu_k(X - \tilde{C}_i) = 0$ for all $k \neq i$, i.e., there exist $\mathcal{B}$–measurable sets $\tilde{C}_i$ and $X - \tilde{C}_i$ such that $\mu_i(\tilde{C}_i) = 0$ and $\sum_{k \neq i} \mu_k(X - \tilde{C}_i) = 0$. This means that μ_i and $\sum_{k \neq i} \mu_k$ are singular, and the theorem is proved.

THEOREM 2. *Let* $\{\mu_k\}_{k \in \mathbb{N}}$, $\mathbb{N} = \{1, 2, \ldots\}$ *be a family of finite mutually singular measures defined on* $(X, \mathcal{B})$ *and* $\sum_{k=1}^{\infty} \mu_k < \infty$. *Then the family* $\{\mu_k\}_{k \in \mathbb{N}}$ *of finite measures is strongly divisible.*

Proof. Applying Theorem 1 we can choose some $\mathcal{B}$–measurable sets $\tilde{C}_i$ such that $\sum_{k \neq i} \mu_k(\tilde{C}_i) = 0$ and $\mu_i(X - \tilde{C}_i) = 0$. Then we take the set

$$C_i = \tilde{C}_i \bigcap \Big[\bigcap_{k \neq i} (X - \tilde{C}_k) \Big].$$

Obviously $C_i \cap C_j = \varnothing$ if $i \neq j$, and $\mu_k(C_i) = 0$ if $k \neq i$,

$$X - C_i = \left(X - \tilde{C}_i\right) \cup \left(\bigcup_{k \neq i} \tilde{C}_k\right),$$

therefore

$$\mu_i(X - C_i) \leqslant \mu_i(X - \tilde{C}_i) + \sum_{k \neq i} \mu_i(\tilde{C}_k) = 0,$$

i.e., the family of measures $\{\mu_k\}_{k \in \mathbb{N}}$ is strongly divisible. The theorem is proved.

THEOREM 3. *Let* $\{\mu_k\}_{k \in \mathbb{N}}$, $\mathbb{N} = \{1, 2, \ldots\}$ *be a family of mutually singular probability measures defined on* $(X, \mathcal{B})$. *Then this family is strongly divisible.*

Proof. Define $\{\tilde{\mu}_k\} = \mu_k / 2^k$. Since the family $\{\mu_k\}_{k \in \mathbb{N}}$ of probability measures is mutually orthogonal and the finite measure $\tilde{\mu}_k$ is equivalent to the probability measure μ_k for any $k \in \mathbb{N}$, the family $\{\tilde{\mu}_k\}_{k \in \mathbb{N}}$ of finite measures is mutually singular and by Theorem 2 it is strongly divisible. Hence there exist mutually disjoint $\mathcal{B}$-measurable sets C_k such that $\tilde{\mu}_k(X - C_k) = 0$, $\tilde{\mu}_k(C_i) = 0$, $\forall i$, $k \in \mathbb{N}$, $i \neq k$. But since $\tilde{\mu}_k$ is equivalent to μ_k, we have

$$\mu_k(X - C_k) = 0, \quad \forall k \in \mathbb{N}, \quad \mu_k(C_i) = 0, \quad k \neq i.$$

This means that the family $\{\mu_k\}_{k \in \mathbb{N}}$ of probability measures is strongly divisible. The theorem is proved.

LEMMA 1. *Any non-degenerated* σ*-finite measure* μ *is equivalent to some probability measure.*

Proof. Since the measure μ is σ-finite there exists a countable family $(X_k)_{k \in \mathbb{N}}$ of μ-measurable sets forming a partition of the space X and $(\forall k)$

$$(k \in \mathbb{N} \Rightarrow \mu(X_k) < \infty).$$

Choose a sequence of non-negative numbers $\tilde{\rho}_i$ such that $\sum_{i=1}^{\infty} \rho_i \mu(X_i) = 1$ and put

$$\tilde{\mu}(B) = \sum_{k=1}^{\infty} \rho_i \mu(X_i \cap B).$$

Obviously, $\tilde{\mu}$ is a probability measure and it is equivalent to the measure μ. The lemma is proved.

THEOREM 4. *Let* $\{\mu_k\}_{k \in \mathbb{N}}$, $\mathbb{N} = \{1, 2, \ldots\}$ *be a family non-degenerated* σ*-finite mutually singular measures on* $(X, \mathcal{B})$. *Then this family is strongly divisible.*

Proof. By Lemma 1 the non-degenerate σ-finite measure μ_k is valent to some probability measure $\tilde{\mu}_k$ for any $k \in \mathbb{N}$. Since the family $\{\mu_k\}_{k \in \mathbb{N}}$ of σ-finite measures is mutually singulars, the family $\{\tilde{\mu}_k\}_{k\in\mathbb{N}}$ of probability measures is also mutually singular. By Theorem 3 the family $\{\tilde{\mu}_k\}_{k\in\mathbb{N}}$ is strongly divisible, i.e., there exist–measurable mutually disjoint sets C_k such that $\tilde{\mu}_k(X - C_k) = 0, \forall k \in \in \mathbb{N}$, $\tilde{\mu}_k(C_i) = 0$, $k \neq i$. But since $\tilde{\mu}_k$ is equivalent to $\tilde{\mu}_k$ for any $k \in \mathbb{N}$, we have

$$\mu_k(X - C_k) = 0, \quad \forall k \in \mathbb{N}, \quad \mu_k(C_i) = 0, \quad k \neq i,$$

and this means that the family $\{\mu_k\}_{k\in\mathbb{N}}$ is strongly divisible. The theorem is proved.

2. BANACH SPACE OF MEASURE STRUCTURE

Let M^σ be the linear space of all σ–finite signed measures on $\mathcal{B}$.

Definition 7. A linear sub–space of measures $M_{\mathcal{B}} \subset M^\sigma$ is called a ***Banach space of measures*** if:

1) on $M_{\mathcal{B}}$ a norm $\|\mu\|$, $\mu \in M_{\mathcal{B}}$ can be introduced such that $M_{\mathcal{B}}$ will be a Banach space and for all mutually singular measures μ and ν for any real number $\lambda \in \mathbb{R}$ the following inequality

$$\|\mu + \lambda\nu\| \geqslant \|\mu\|$$

holds;

2) if $\nu \in M_{\mathcal{B}}$ and $|f| \leqslant 1$ then $\nu_f(A) = \int\limits_A f(x)\nu(dx) \in M_{\mathcal{B}}$ and $\|\nu_f\| \leqslant \|\nu\|$;

3) if $\nu_n \in M_{\mathcal{B}}$ are such that $\nu_n(X) < \infty$, $\nu_n \downarrow 0$ $\forall n \in \mathbb{N}$, in variation and $\nu_n(X) < \infty$, $\forall n \in \mathbb{N}$ then

$$l^*(\nu_n) \to 0, \quad \forall l^* \in M_{\mathcal{B}}^*,$$

where $M_{\mathcal{B}}^*$ is the conjugate space of $M_{\mathcal{B}}$.

Let μ be some measure. Let $M_{\mathcal{B}}(\mu)$ denote the Banach space of measures ν of the form $\nu(B) = \int\limits_B f(x)\mu(dx)$, where $\int |f(x)|\mu(dx) < \infty$, with the norm

$$\|\nu\| = \int |f(x)|\mu(dx).$$

Definition 8. Let $\{\mu_\alpha\}_{\alpha\in\mathcal{A}}$ be a family of measures defined on $\mathcal{B}$. The family $\{\mu_\alpha\}_{\alpha\in\mathcal{A}}$ is called a ***frame of the Banach space*** $M_{\mathcal{B}}$ if

$$M_{\mathcal{B}} = \bigoplus_{\alpha\in\mathcal{A}} M_{\mathcal{B}}(\mu_\alpha),$$

i.e., $M_{\mathcal{B}}$ is the direct sum of the Banach spaces $M_{\mathcal{B}}(\mu_\alpha)$.

THEOREM 5. *Let $M_{\mathcal{B}}$ be a Banach space of measures. Then there exists a family of mutually orthogonal non-negative measures $\{\eta_\alpha\}_{\alpha\in\mathcal{A}}$ such that*

$$M_{\mathcal{B}} = \bigoplus_{\alpha\in\mathcal{A}} M_{\mathcal{B}}(\eta_\alpha),$$

i.e., the family $\{\eta_\alpha\}_{\alpha\in\mathcal{A}}$ is a "frame" of the Banach space $M_{\mathcal{B}}$.

Proof. Let K_+ be the cone of probability measures from $M_{\mathcal{B}}$. For every measure $\psi \in M_{\mathcal{B}}$ there exists a measure $\eta \in K_+$ such that $\psi(B) = \int_B g(x)\eta(dx)$. We find a family $\{\eta^*_\alpha\}_{\alpha\in\mathcal{A}}$ to which K_+ is subordinate, i.e., $\forall\nu \in K_+$ there exist a countable subset $A_1 \subset \mathcal{A}$ and real $\mathcal{B}$–measurable functions $g_\alpha(x)$, $\alpha \in A_1$ such that

$$\nu(B) = \sum_{\alpha\in A_1} \int_B g_\alpha(x)\eta^*_\alpha(dx).$$

Let $\mathcal{P} = \{\mu_\alpha\}_{\alpha\in\mathcal{A}}$ be a family of probability measures belonging to the cone K_+ where $\mathcal{A}$ is an interval of ordinal numbers. Let us complete order $\mathcal{A}$. We construct mutually orthogonal measures η^*_α by means of transfinite induction in the following manner. Put $\eta^*_1 = \mu_{\alpha_1}$. Let η^*_γ be defined for $\gamma < \alpha$ (they are mutually orthogonal). Then no more than a countable set Γ_α of such $\gamma < \alpha$ that η^*_γ are not orthogonal to μ_α can be found. Put

$$\hat\eta_\alpha(A) = \mu_\alpha(A) - \sum_{\gamma\in\Gamma_\alpha} \int_A \frac{d\mu_\alpha}{d\eta^*_\gamma}(x)\eta^*_\gamma(dx), \tag{1}$$

and

$$\eta^*_\alpha = \begin{cases} 0, & \text{if } \hat\eta_\alpha(X) = 0 \\ \hat\eta_\alpha/\hat\eta_\alpha(X), & \text{if } \hat\eta_\alpha(X) \neq 0. \end{cases}$$

The sum in (1) is finite since the measures η^*_γ are mutually orthogonal. The equality

$$\mu_\alpha(A) = \sum_{\gamma\in\Gamma_\alpha} \int_A \frac{d\mu_\alpha}{d\eta^*_\gamma}(x)\eta^*_\gamma(dx) + \hat\eta_\alpha(A) \tag{2}$$

is true and η^*_α is orthogonal to η^*_γ for $\gamma < \alpha$ by the construction. Thus we have that any measure $\mu_\alpha \in K_+$ is subordinate to the family of mutually orthogonal probability measures $\{\eta^*_\alpha,\ \alpha \in \mathcal{A}\}$. It is evident from formula (2) that $\eta^*_\alpha < \mu_\alpha$ $\forall\alpha \in \mathcal{A}$ and η^*_α is absolutely continuous w.r.t. μ_α, hence

$$d\eta^*_\alpha/d\mu_\alpha(x) \leqslant 1, \quad \forall\alpha \in \mathcal{A}.$$

This implies that $\eta^*_\alpha \in K_+$, $\forall\alpha \in \mathcal{A}$.

2. Let H^*_α be the set of measures from $M_{\mathcal{B}}$ absolutely continuous w.r.t. η^*_α. Consider the measure $\nu_C(B) = \int_B I_C(x)\eta^*_\alpha(dx)$ and put $\xi(C) = l(\nu_C)$, where l is a positive continuous linear functional on $M_{\mathcal{B}}$ such that $\|l\| = 1$ and $l(\nu_C) = \|\nu_C\|$

for any $\nu_C \neq 0$. Measure ν_C is additive and if $C_n \downarrow \varnothing$ we have $\nu_{C_n} \downarrow 0$ and $l(\nu_{C_n}) \to 0$. Therefore $\xi(C)$ is a measure. If $\xi(C) = 0$, then $\nu_C = 0$, therefore $\eta^*_\alpha(C) = 0$. So that $\eta^*_\alpha \sim \xi$. Hence

$$l(\nu_C) = \int_C g_\alpha(x)\eta^*_\alpha(dx) = \|\nu_C\|,$$

where $g_\alpha(x) > 0$ almost everywhere w.r.t. the measure η^*_α.

3. Let $\nu^\alpha_f(B) = \int_B f(x)\eta^*_\alpha(dx)$. By F_0 we denote the space of finite-valued measurable functions. If $f(x) = a_k$ for $x \in B_k$, $k = 1, \ldots, m$, $B_i \cap B_j = \varnothing$, $i \neq j$, $\bigcup_{j=1}^n B_j = X$, then

$$l(\nu^\alpha_f(B)) = l\left(\sum_{k=1}^n a_k \int_{B_k} \eta^*_\alpha I_{B_k}\right) = \sum_{k=1}^n a_k l\left(\nu_{B_k}\right) = \sum_{k=1}^n a_k \|\nu_{B_k}\| =$$

$$= \sum_{k=1}^n a_k \int_{B_k} g_\alpha(x)\eta^*_\alpha(dx) = \int f(x) g_\alpha(x)\eta^*_\alpha(dx).$$

But on the other hand, by definition $l\left(\nu^\alpha_f\right) = \|\nu^\alpha_f\|$ and, hence

$$\|\nu^\alpha_f\| = \int \left|f(x)g_\alpha(x)\right|\eta^*_\alpha(dx).$$

4. Let $f_n \geqslant 0$, $\nu^\alpha_f \in H^*_{B_\alpha}$, $f_n \uparrow f$, $f_n \in F_0$. We have $\|\nu_{f_n}\| \leqslant \|\nu_f\|$ and $\nu_{f-f_n} \downarrow 0$ in variation. Thus

$$\lim_{n\to\infty} l^*(\nu_{f-f_n}) = 0, \quad \forall l \in M^*_B,$$

and hence

$$\lim_{n\to\infty} l^*(\nu_{f_n}) = l^*(\nu_f).$$

In particular (3) also holds for the functional l, i.e., $\lim_{n\to\infty} l(\nu_{f_n}) = l(\nu_f)$. But on the other hand

$$\|\nu_t\| = l(\nu_f) = \lim_{n\to\infty} l(\nu_{f_n}) = \int \left|f(x)g_\alpha(x)\right|\eta^*_\alpha(dx). \tag{4}$$

This formula can be extended to real valued f in an obvious way. We have established that $H^*_\alpha \subset M_B(\eta_\alpha)$ and

$$H^*_\alpha = \left\{\nu_f \colon \int |f \cdot g_\alpha| \eta_\alpha < \infty\right\}.$$

Put

$$\eta_\alpha(C) = \int_C \frac{1}{g_\alpha(x)} \eta^*_\alpha(dx). \tag{5}$$

The measure η_α defined by formula (5) is σ–finite, since $g_\alpha(x) > 0$ almost everywhere w.r.t. to the measure η^*_α. It can be easily shown that $H^*_\alpha = M_B(\eta_\alpha)$. This completes the proof of the theorem.
Applying Theorem 5 one can easily prove the following theorem.

THEOREM 6. *Let M_B be a Banach space of measures. Then there exists a family of mutually orthogonal non–negative probability measures $\left\{\eta^{**}_\alpha\right\}_{\alpha\in A}$ such that $M_B = \oplus M_B\left(\eta^{**}_\alpha\right)$, i.e., the family $\left\{\eta^{**}_\alpha\right\}_{\alpha\in A}$ is a frame of M_B.*

REFERENCES

Ibramhalilov, I.Sh. and Skorokhod, A.V. (1980). *Consistent Estimators of the Parameters of Random Processes.* Naukova Dumka, Kiev.

Zerakidze, Z.S. (1984). Construction of statistical structures. *Commun. Acad. Sci. of Georgian SSR* **113**, 37–39.

Zerakidze, Z.S. (1986a). Construction of statistical structures. *Theory Probab. Appl.* **30.**

Zerakidze, Z.S. (1986b). Hilbert space of measures. *Ukr. Mat. J.* **38**, 148–154.